5 Unit Edition

EarthComm®

EARTH SYSTEM SCIENCE IN THE COMMUNITY

Michael J. Smith Ph.D.
American Geological Institute

John B. Southard Ph.D.
Massachusetts Institute of Technology

Ruta Demery
Editor

DEVELOPED BY THE AMERICAN GEOLOGICAL INSTITUTE
IN ASSOCIATION WITH IT'S ABOUT TIME PUBLISHING

It's About Time, Inc.
84 Business Park Drive, Armonk, NY 10504
Phone (914) 273-2233 Fax (914) 273-2227
Toll Free (888) 698-TIME
www.Its-About-Time.com

It's About Time Founder
Laurie Kreindler

Project Manager
Ruta Demery

Design
John Nordland

Studio Manager
Jon Voss

Creative Artwork
Tomas Bunk

Safety Reviewer
Dr. Ed Robeck

Production
Burmar Technical Corporation

Technical Art
Stuart Armstrong

Senior Photo Consultant
Bruce F. Molnia

Photo Research
Eric Shih

Associate Editor
Al Mari

Printed and bound in the United States of America

ISBN #1-58591-325-1

1 2 3 4 5 VH 07 06 05 04

This project was supported, in part, by the
National Science Foundation (grant no. ESI-9452789)

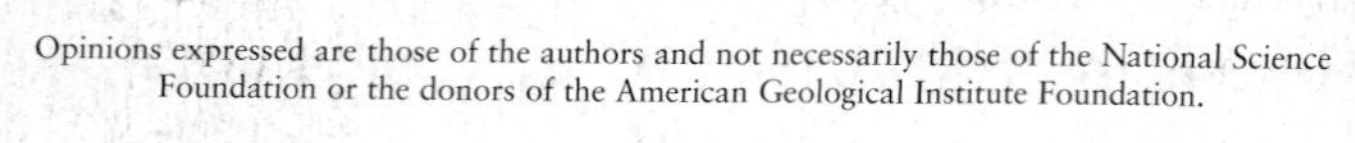

Opinions expressed are those of the authors and not necessarily those of the National Science Foundation or the donors of the American Geological Institute Foundation.

Unit 1 Earth's Dynamic Geosphere

Table of Contents

Unit II Understanding Your Environment

Unit III Earth's Fluid Spheres

Table of Contents

1 Oceans... and Your Community

2 Severe Weather... and Your Community

3 The Cryosphere... and Your Community

Energy Resources... and Your Community

Mineral Resources... and Your Community

Water Resources... and Your Community

Unit IV Earth's Natural Resources

Unit V Earth System Evolution

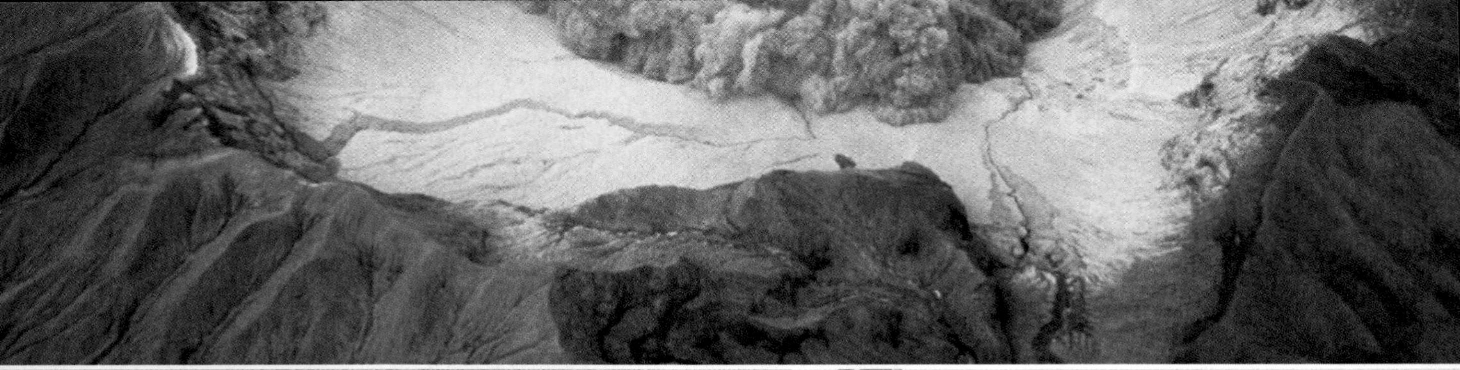
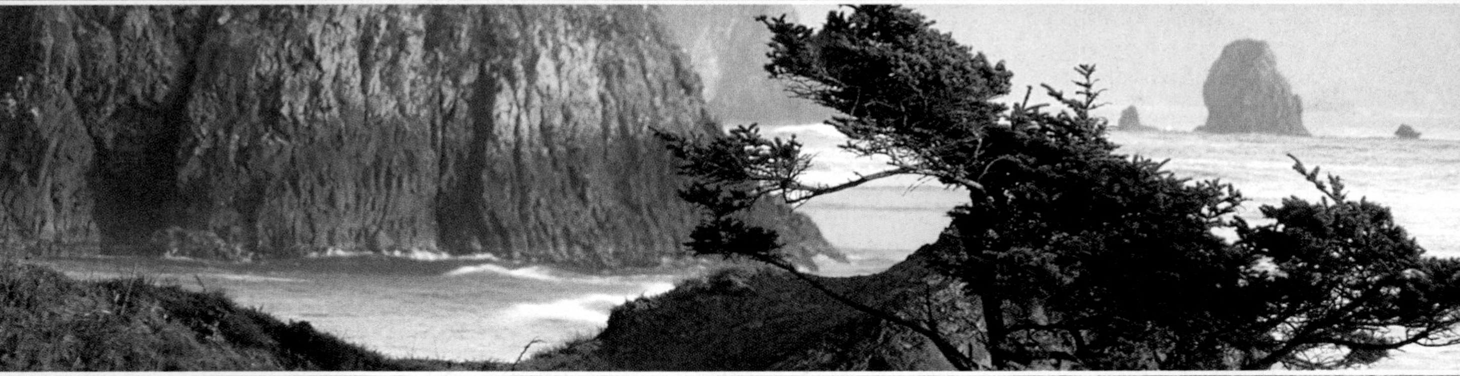
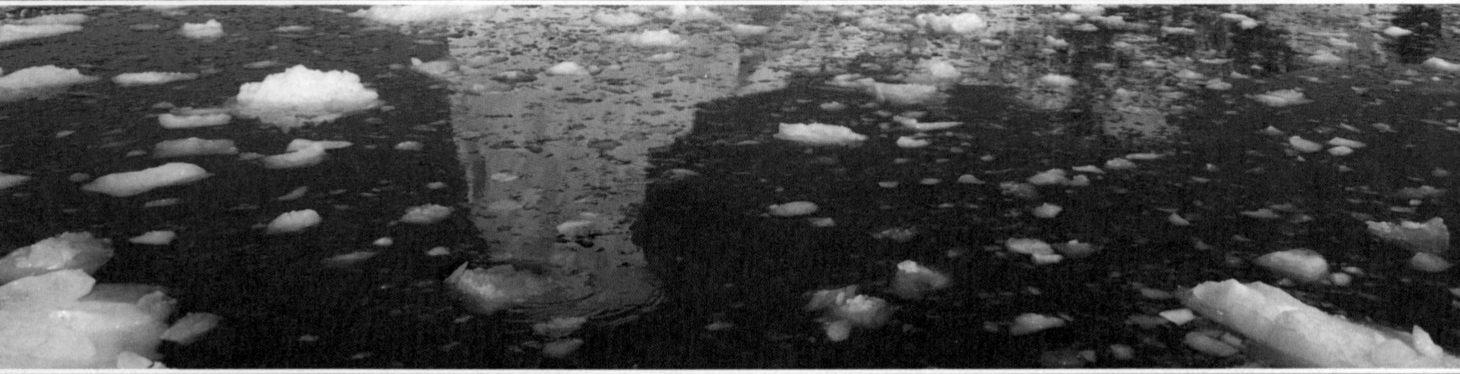
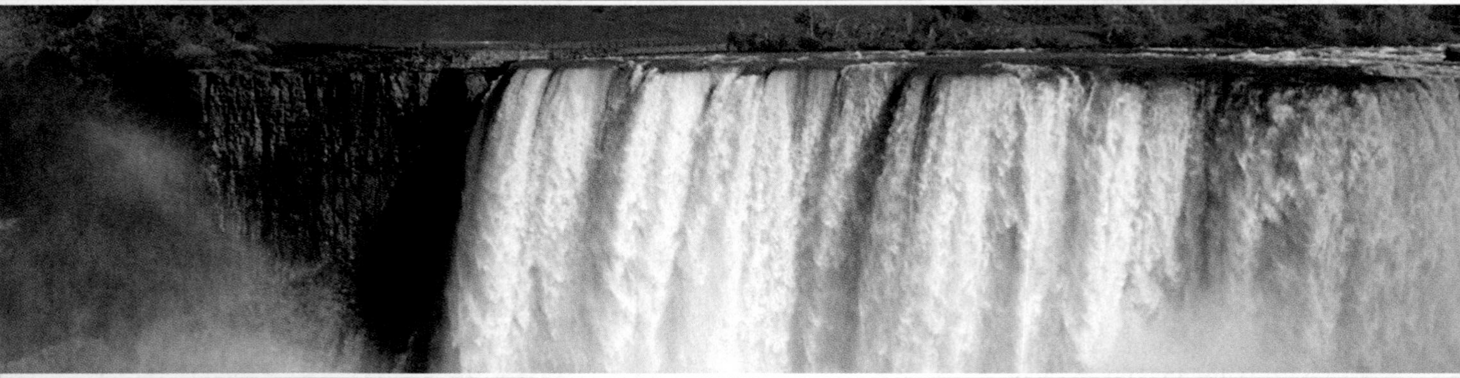
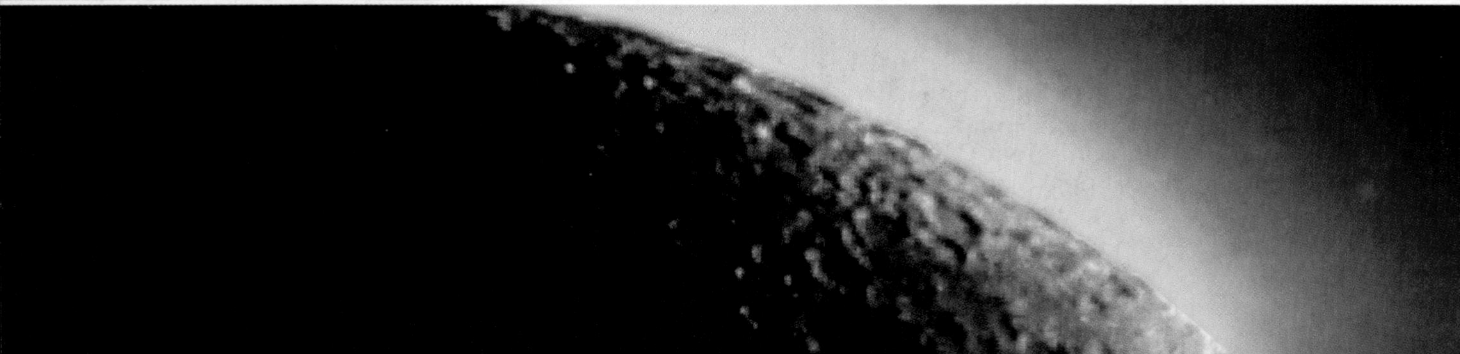

Principal Investigator

Michael Smith is Director of Education at the American Geological Institute in Alexandria, Virginia. Dr. Smith worked as an exploration geologist and hydrogeologist. He began his Earth Science teaching career with Shady Side Academy in Pittsburgh, PA in 1988 and most recently taught Earth Science at the Charter School of Wilmington, DE. He earned a doctorate from the University of Pittsburgh's Cognitive Studies in Education Program and joined the faculty of the University of Delaware School of Education in 1995. Dr. Smith received the Outstanding Earth Science Teacher Award for Pennsylvania from the National Association of Geoscience Teachers in 1991, served as Secretary of the National Earth Science Teachers Association, and is a reviewer for Science Education and The Journal of Research in Science Teaching. He worked on the Delaware Teacher Standards, Delaware Science Assessment, National Board of Teacher Certification, and AAAS Project 2061 Curriculum Evaluation programs.

Senior Writer

Dr. Southard received his undergraduate degree from the Massachusetts Institute of Technology in 1960 and his doctorate in geology from Harvard University in 1966. After a National Science Foundation postdoctoral fellowship at the California Institute of Technology, he joined the faculty at the Massachusetts Institute of Technology, where he is currently Professor of Geology Emeritus. He was awarded the MIT School of Science teaching prize in 1989 and was one of the first cohorts of first MacVicar Fellows at MIT, in recognition of excellence in undergraduate teaching. He has taught numerous undergraduate courses in introductory geology, sedimentary geology, field geology, and environmental Earth Science both at MIT and in Harvard's adult education program. He was editor of the Journal of Sedimentary Petrology from 1992 to 1996, and he continues to do technical editing of scientific books and papers for SEPM, a professional society for sedimentary geology. Dr. Southard received the 2001 Neil Miner Award from the National Association of Geoscience Teachers.

Safety Reviewer Dr. Edward Robeck, Salisbury University, MD

PRIMARY AND CONTRIBUTING AUTHORS

Earth's Dynamic Geosphere

Daniel J. Bisaccio
Souhegan High School
Amherst, NH

Steve Carlson
Middle School, OR

Warren Fish
Paul Revere School
Los Angeles, CA

Miriam Fuhrman
Carlsbad, CA

Steve Mattox
Grand Valley State University

Keith McKain
Milford Senior High School
Milford, DE

Mary McMillan
Niwot High School
Niwot, CO

Bill Romey
Orleans, MA

Michael Smith
American Geological Institute

Tom Vandewater
Colton, NY

Understanding Your Environment

Geoffrey A. Briggs
Batavia Senior High School
Batavia, NY

Cathey Donald
Auburn High School
Auburn, AL

Richard Duschl
Kings College
London, UK

Fran Hess
Cooperstown High School
Cooperstown, NY

Laurie Martin-Vermilyea
American Geological Institute

Molly Miller
Vanderbilt University

Mary-Russell Roberson
Durham, NC

Charles Savrda
Auburn University

Michael Smith
American Geological Institute

Earth's Fluid Spheres

Chet Bolay
Cape Coral High School
Cape Coral, FL

Steven Dutch
University of Wisconsin

Virginia Jones
Bonneville High School
Idaho Falls, ID

Laurie Martin-Vermilyea
American Geological Institute

Joseph Moran
University of Wisconsin

Mary-Russell Roberson
Durham, NC

Bruce G. Smith
Appleton North High School
Appleton, WI

Michael Smith
American Geological Institute

Earth's Natural Resources

Chuck Bell
Deer Valley High School
Glendale, AZ

Jay Hackett
Colorado Springs, CO

John Kemeny
University of Arizona

John Kounas
Westwood High School
Sloan, IA

Laurie Martin-Vermilyea
American Geological Institute

Mary Poulton
University of Arizona

David Shah
Deer Valley High School
Glendale, AZ

Janine Shigihara
Shelley Junior High School
Shelley, ID

Michael Smith
American Geological Institute

Earth System Evolution

Julie Bartley
University of West Georgia

Lori Borroni-Engle
Taft High School
San Antonio, TX

Richard M. Busch
West Chester University
West Chester, PA

Kathleen Cochrane
Our Lady of Ransom School
Niles, IL

Cathey Donald
Auburn High School, AL

Robert A. Gastaldo
Colby College

William Leonard
Clemson University

Tim Lutz
West Chester University

Carolyn Collins Petersen
C. Collins Petersen Productions
Groton, MA

Michael Smith
American Geological Institute

Content Reviewers

Gary Beck
BP Exploration

Phil Bennett
University of Texas, Austin

Steve Bergman
Southern Methodist University

Samuel Berkheiser
Pennsylvania Geologic Survey

Arthur Bloom
Cornell University

Craig Bohren
Penn State University

Bruce Bolt
University of California, Berkeley

John Callahan
Appalachian State University

Sandip Chattopadhyay
R.S. Kerr Environmental Research Center

Beth Ellen Clark
Cornell University

Jimmy Diehl
Michigan Technological University

Sue Beske-Diehl
Michigan Technological University

Neil M. Dubrovsky
United States Geological Survey

Frank Ethridge
Colorado State University

Catherine Finley
University of Northern Colorado

Ronald Greeley
Arizona State University

Michelle Hall-Wallace
University of Arizona

Judy Hannah
Colorado State University

Blaine Hanson
Dept. of Land, Air, and Water Resources

James W. Head III
Brown University

Patricia Heiser
Ohio University

John R. Hill
Indiana Geological Survey

Travis Hudson
American Geological Institute

Jackie Huntoon
Michigan Tech. University

Teresa Jordan
Cornell University

Allan Juhas
Lakewood, Colorado

Robert Kay
Cornell University

Chris Keane
American Geological Institute

Bill Kirby
United States Geological Survey

Mark Kirschbaum
United States Geological Survey

Dave Kirtland
United States Geological Survey

Jessica Elzea Kogel
Thiele Kaolin Company

Melinda Laituri
Colorado State University

Martha Leake
Valdosta State University

Donald Lewis
Happy Valley, CA

Steven Losh
Cornell University

Jerry McManus
Woods Hole Oceanographic Institution

Marcus Milling
American Geological Institute

Alexandra Moore
Cornell University

Jack Oliver
Cornell University

Don Pair
University of Dayton

Mauri Pelto
Nicolas College

Bruce Pivetz
ManTech Environmental Research Services Corp.

Stephen Pompea
Pompea & Associates

Peter Ray
Florida State University

William Rose
Michigan Technological Univ.

Lou Solebello
Macon, Gerogia

Robert Stewart
Texas A&M University

Ellen Stofan
NASA

Barbara Sullivan
University of Rhode Island

Carol Tang
Arizona State University

Bob Tilling
United States Geological Survey

Stanley Totten
Hanover College

Scott Tyler
University of Nevada, Reno

Michael Velbel
Michigan State University

Ellen Wohl
Colorado State University

David Wunsch
State Geologist of New Hampshire

Pilot Test Evaluator

Larry Enochs
Oregon State University

Pilot Test Teachers

Rhonda Artho
Dumas High School
Dumas, TX

Mary Jane Bell
Lyons-Decatur Northeast
Lyons, NE

Rebecca Brewster
Plant City High School
Plant City, FL

Terry Clifton
Jackson High School
Jackson, MI

Virginia Cooter
North Greene High School
Greeneville, TN

Monica Davis
North Little Rock High School
North Little Rock, AR

Joseph Drahuschak
Troxell Jr. High School
Allentown, PA

Ron Fabick
Brunswick High School
Brunswick, OH

Virginia Jones
Bonneville High School
Idaho Falls, ID

Troy Lilly
Snyder High School
Snyder, TX

Sherman Lundy
Burlington High School
Burlington, IA

Norma Martof
Fairmont Heights High School
Capitol Heights, MD

Keith McKain
Milford Senior High School
Milford, DE

Mary McMillan
Niwot High School
Niwot, CO

Kristin Michalski
Mukwonago High School
Mukwonago, WI

Dianne Mollica
Bishop Denis J. O'Connell
High School
Arlington, VA

Arden Rauch
Schenectady High School
Schenectady, NY

Laura Reysz
Lawrence Central High School
Indianapolis, IN

Floyd Rogers
Palatine High School
Palatine, IL

Ed Ruszczyk
New Canaan High School
New Canaan, CT

Jane Skinner
Farragut High School
Knoxville, TN

Shelley Snyder
Mount Abraham High School
Bristol, VT

Joy Tanigawa
El Rancho High School
Pico Rivera, CA

Dennis Wilcox
Milwaukee School of Languages
Milwaukee, WI

Kim Willoughby
SE Raleigh High School
Raleigh, NC

Field Test Workshop Staff

Don W. Byerly
University of Tennessee

Derek Geise
University of Nebraska

Michael A. Gibson
University of Tennessee

David C. Gosselin
University of Nebraska

Robert Hartshorn
University of Tennessee

William Kean
University of Wisconsin

Ellen Metzger
San Jose State University

Tracy Posnanski
University of Wisconsin

J. Preston Prather
University of Tennessee

Ed Robeck
Salisbury University

Richard Sedlock
San Jose State University

Bridget Wyatt
San Jose State University

Field Test Evaluators

Bob Bernoff
Dresher, PA

Do Yong Park
University of Iowa

Field Test Teachers

Kerry Adams
Alamosa High School
Alamosa, CO

Jason Ahlberg
Lincoln High
Lincoln, NE

Gregory Bailey
Fulton High School
Knoxville, TN

Mary Jane Bell
Lyons-Decatur Northeast
Lyons, NE

Rod Benson
Helena High
Helena, MT

Sandra Bethel
Greenfield High School
Greenfield, TN

John Cary
Malibu High School
Malibu, CA

Elke Christoffersen
Poland Regional High School
Poland, ME

Tom Clark
Benicia High School
Benicia, CA

Julie Cook
Jefferson City High School
Jefferson City, MO

Virginia Cooter
North Greene High School
Greeneville, TN

Mary Cummane
Perspectives Charter
Chicago, IL

Sharon D'Agosta
Creighton Preparatory
Omaha, NE

Mark Daniels
Kettle Morraine High School
Milwaukee, WI

Beth Droughton
Bloomfield High School
Bloomfield, NJ

Steve Ferris
Lincoln High
Lincoln, NE

Bob Feurer
North Bend Central Public
North Bend, NE

Sue Frack
Lincoln Northeast High
Lincoln, NE

Rebecca Fredrickson
Greendale High School
Greendale, WI

Sally Ghilarducci
Hamilton High School
Milwaukee, WI

Kerin Goedert
Lincoln High School
Ypsilanti, MI

Martin Goldsmith
Menominee Falls High School
Menominee Falls, WI

Randall Hall
Arlington High School
St. Paul, MN

Theresa Harrison
Wichita West High
Wichita, KS

Gilbert Highlander
Red Bank High School
Chattanooga, TN

Jim Hunt
Chattanooga School of Arts
& Sciences
Chattanooga, TN

Patricia Jarzynski
Watertown High School
Watertown, WI

Pam Kasprowicz
Bartlett High School
Bartlett, IL

Caren Kershner
Moffat Consolidated
Moffat, CO

Mary Jane Kirkham
Fulton High School

Ted Koehn
Lincoln East High
Lincoln, NE

Philip Lacey
East Liverpool High School
East Liverpool, OH

Joan Lahm
Scotus Central Catholic
Columbus, NE

Erica Larson
Tipton Community

Michael Laura
Banning High School
Wilmington, CA

Fawn LeMay
Plattsmouth High
Plattsmouth, NE

Christine Lightner
Smethport Area High School
Smethport, PA

Nick Mason
Normandy High School
St. Louis, MO

James Matson
Wichita West High
Wichita, KS

Jeffrey Messer
Western High School
Parma, MI

Dave Miller
Parkview High
Springfield, MO

Rick Nettesheim
Waukesha South
Waukesha, WI

John Niemoth
Niobrara Public
Niobrara, NE

Margaret Olsen
Woodward Academy
College Park, GA

Ronald Ozuna
Roosevelt High School
Los Angeles, CA

Paul Parra
Omaha North High
Omaha, NE

D. Keith Patton
West High
Denver, CO

Phyllis Peck
Fairfield High School
Fairfield, CA

Randy Pelton
Jackson High School
Massillon, OH

Reggie Pettitt
Holderness High School
Holderness, NH

June Rasmussen
Brighton High School
South Brighton, TN

Russ Reese
Kalama High School
Kalama, WA

Janet Ricker
South Greene High School
Greeneville, TN

Wendy Saber
Washington Park High School
Racine, WI

Garry Sampson
Wauwatosa West High School
Tosa, WI

Daniel Sauls
Chuckey-Doak High School
Afton, TN

Todd Shattuck
L.A. Center for Enriched Studies
Los Angeles, CA

Heather Shedd
Tennyson High School
Hayward, CA

Lynn Sironen
North Kingstown High School
North Kingstown, RI

Jane Skinner
Farragut High School
Knoxville, TN

Sarah Smith
Garringer High School
Charlotte, NC

Aaron Spurr
Malcolm Price Laboratory
Cedar Falls, IA

Karen Tiffany
Watertown High School
Watertown, WI

Tom Tyler
Bishop O'Dowd High School
Oakland, CA

Valerie Walter
Freedom High School
Bethlehem, PA

Christopher J. Akin Williams
Milford Mill Academy
Baltimore, MD

Roseanne Williby
Skutt Catholic High School
Omaha, NE

Carmen Woodhall
Canton South High School
Canton, OH

Field Test Coordinator

William Houston
American Geological Institute

Advisory Board

Jane Crowder
Bellevue, WA

Arthur Eisenkraft
Bedford (NY) Public Schools

Tom Ervin
LeClaire, IA

Mary Kay Hemenway
University of Texas at Austin

Bill Leonard
Clemson University

Don Lewis
Lafayette, CA

Wendell Mohling
National Science Teachers Association

Harold Pratt
Littleton, CO

Barb Tewksbury
Hamilton College

Laure Wallace
USGS

AGI Foundation

Jan van Sant
Executive Director

The American Geological Institute and EarthComm

Imagine more than 500,000 Earth scientists worldwide sharing a common voice, and you've just imagined the mission of the American Geological Institute. Our mission is to raise public awareness of the Earth sciences and the role that they play in mankind's use of natural resources, mitigation of natural hazards, and stewardship of the environment. For more than 50 years, AGI has served the scientists and teachers of its Member Societies and hundreds of associated colleges, universities, and corporations by producing Earth science educational materials, *Geotimes*–a geoscience news magazine, GeoRef–a reference database, and government affairs and public awareness programs.

So many important decisions made every day that affect our lives depend upon an understanding of how our Earth works. That's why AGI created *EarthComm*. In your *EarthComm* classroom, you'll discover the wonder and importance of Earth science by studying it where it counts—in your community. As you use the rock record to investigate climate change, do field work in nearby beaches, parks, or streams, explore the evolution and extinction of life, understand where your energy resources come from, or find out how to forecast severe weather, you'll gain a better understanding of how to use your knowledge of Earth science to make wise personal decisions.

We would like to thank the AGI Foundation Members that have been supportive in bringing Earth science to students. These AGI Foundation Members include: American Association of Petroleum Geologists Foundation, Anadarko Petroleum Corp., The Anschutz Foundation, Apache Canada Ltd., Baker Hughes Foundation, Barrett Resources Corp., Elizabeth and Stephen Bechtel, Jr. Foundation, BP Foundation, Burlington Resources Foundation, CGG Americas, Inc., ChevronTexaco Corp., Conoco Inc., Consolidated Natural Gas Foundation, Devon Energy Corp., Diamond Offshore Co., Dominion Exploration & Production, Inc., EEX Corp., Equitable Production Co., ExxonMobil Foundation, Five States Energy Co., Geological Society of America Foundation, Global Marine Drilling Co., Halliburton Foundation, Inc., Kerr McGee Foundation, Maxus Energy Corp., Noble Drilling Corp., Occidental Petroleum Charitable Foundation, Ocean Energy, Optimistic Petroleum Co., Parker Drilling Co., Phillips Petroleum Co., Santa Fe Snyder Corp., Schlumberger Foundation, Shell Oil Company Foundation, Southwestern Energy Co., Texas Crude Energy, Inc., Unocal Corp. USX Foundation (Marathon Oil Co.).

We at AGI wish you success in your exploration of the Earth System and your Community.

Michael J. Smith
Director of Education, AGI

Marcus E. Milling
Executive Director, AGI

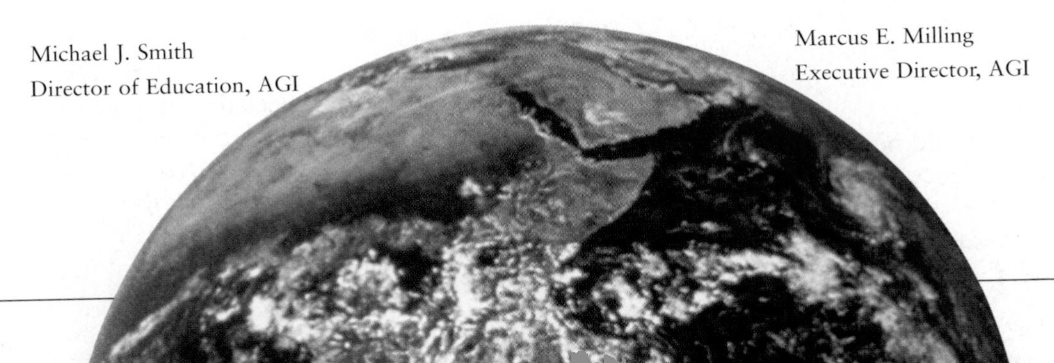

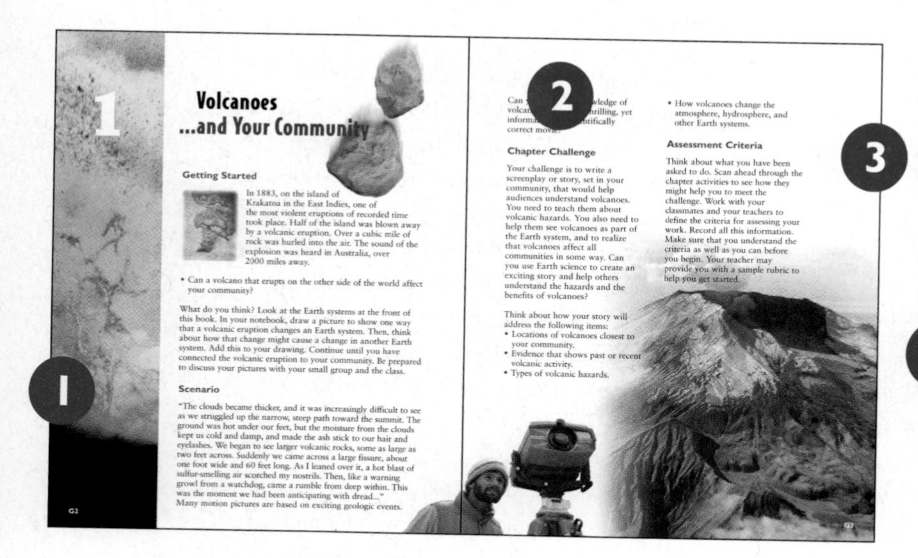

1

Volcanoes ...and Your Community

Getting Started

In 1883, on the island of Krakatoa in the East Indies, one of the most violent eruptions of recorded time took place. Half of the island was blown away by a volcanic eruption. Over a cubic mile of rock was hurled into the air. The sound of the explosion was heard in Australia, over 2000 miles away.

- Can a volcano that erupts on the other side of the world affect your community?

What do you think? Look at the Earth systems at the front of this book. In your notebook, draw a picture to show one way that a volcanic eruption changes an Earth system. Then, think about how that change might cause a change in another Earth system. Add this to your drawing. Continue until you have connected the volcanic eruption to your community. Be prepared to discuss your pictures with your small group and the class.

Scenario

"The clouds became thicker, and it was increasingly difficult to see as we struggled up the narrow, steep path toward the summit. The ground was hot under our feet, but the moisture from the clouds kept us cold and damp, and made the ash stick to our hair and eyelashes. We began to see larger volcanic rocks, some as large as two feet across. Suddenly we came across a large fissure, about one foot wide and 60 feet long. As I leaned over it, a hot blast of sulfur-smelling air scorched my nostrils. Then, like a warning growl from a watchdog, came a rumble from deep within. This was the moment we had been anticipating with dread..."
Many motion pictures are based on exciting geologic events.

G2

Can y... ...wledge of volcan... ...hrilling, yet informa... ...ntifically correct movie?

Chapter Challenge

Your challenge is to write a screenplay or story, set in your community, that would help audiences understand volcanoes. You need to teach them about volcanic hazards. You also need to help them see volcanoes as part of the Earth system, and to realize that volcanoes affect all communities in some way. Can you use Earth science to create an exciting story and help others understand the hazards and the benefits of volcanoes?

Think about how your story will address the following items:
- Locations of volcanoes closest to your community.
- Evidence that shows past or recent volcanic activity.
- Types of volcanic hazards.
- How volcanoes change the atmosphere, hydrosphere, and other Earth systems.

Assessment Criteria

Think about what you have been asked to do. Scan ahead through the chapter activities to see how they might help you to meet the challenge. Work with your classmates and your teachers to define the criteria for assessing your work. Record all this information. Make sure that you understand the criteria as well as you can before you begin. Your teacher may provide you with a sample rubric to help you get started.

G3

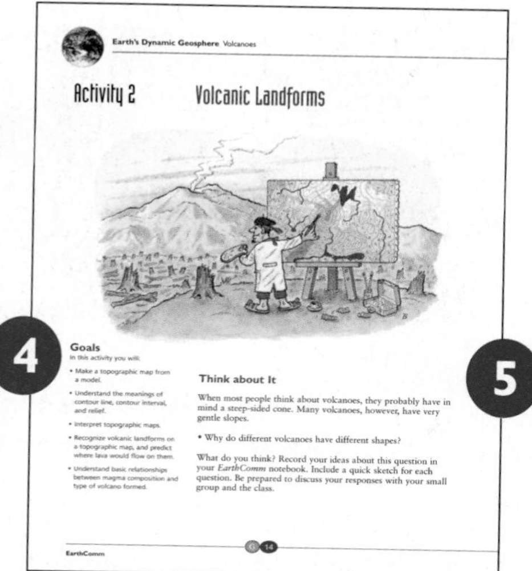

Earth's Dynamic Geosphere Volcanoes

Activity 2 Volcanic Landforms

Goals

In this activity you will:
- Make a topographic map from a model.
- Understand the meanings of contour line, contour interval, and relief.
- Interpret topographic maps.
- Recognize volcanic landforms on a topographic map, and predict where lava would flow on them.
- Understand basic relationships between magma composition and type of volcano formed.

Think about It

When most people think about volcanoes, they probably have in mind a steep-sided cone. Many volcanoes, however, have very gentle slopes.

- Why do different volcanoes have different shapes?

What do you think? Record your ideas about this question in your *EarthComm* notebook. Include a quick sketch for each question. Be prepared to discuss your responses with your small group and the class.

EarthComm G 14

1. Scenario
Each chapter begins with an event or situation in the Earth system that has happened or could actually take place.

2. Challenge
This feature challenges you to solve a problem based in your community. You will need to use the Earth science knowledge you gain from working through the chapter.

3. Assessment Criteria
Before you begin, you and your classmates, along with your teacher, will explore exactly how you will be graded. You will review the criteria and expectations for solving the challenge, and make decisions about how your work should be evaluated.

4. Goals
At the beginning of each activity you are provided with a list of goals that you should be able to achieve by completing your science inquiry.

5. Think about It
What do you already know? Before you start each activity you will be asked one or two questions to consider. You will have a chance to discuss your ideas with your group and your class. You are not expected to come up with the "right" answer, but to share your current understanding and reasoning.

6. Investigate
In EarthComm you learn by doing science. In your small groups, or as a class, you will take part in scientific inquiry by doing hands-on experiments, participating in fieldwork, or searching for answers using the Internet and reference materials.

7. Reflecting on the Activity and the Challenge
Each activity will help you meet the chapter challenge. This feature gives you a brief summary of the activity. It will help you relate the activity that you just completed to the "big picture."

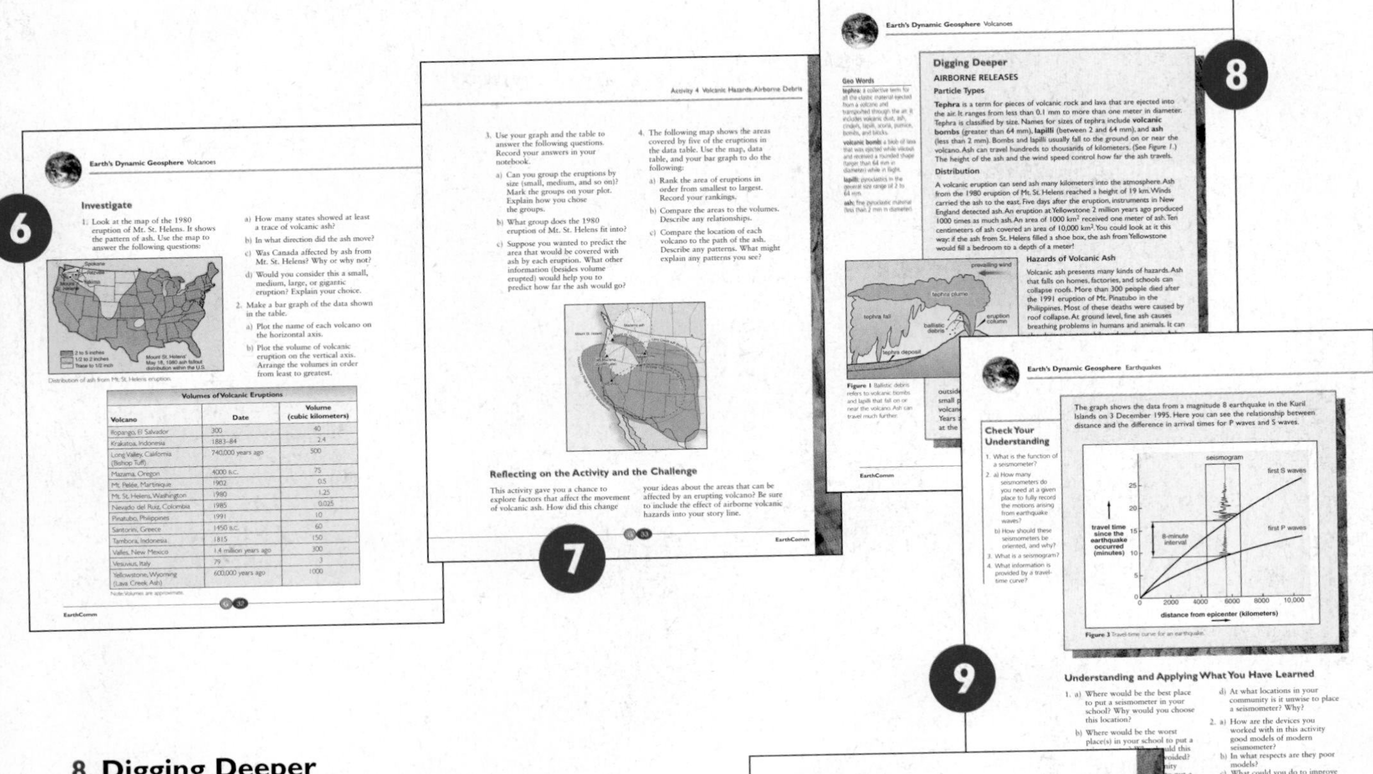

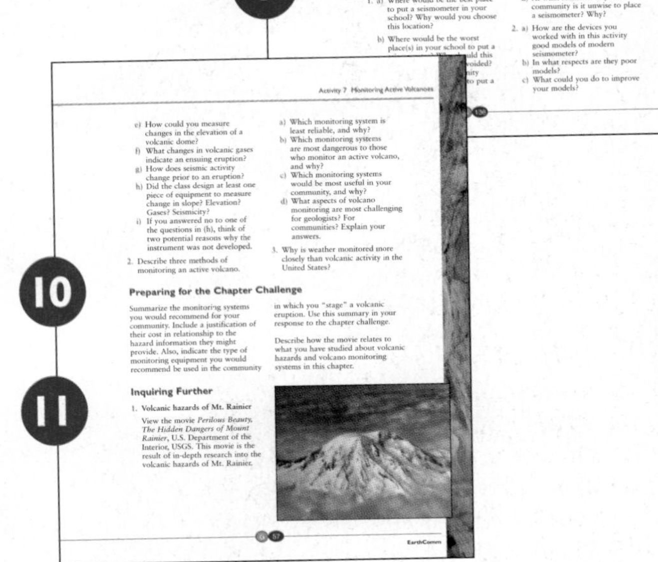

8. Digging Deeper

This section provides text, illustrations, and photographs that will give you a greater insight into the concepts you explored in the activity. Words that may be new or unfamiliar to you are defined and explained. "Check Your Understanding" questions are included to guide you in your reading. Key terms are highlighted as "Geo Words."

9. Understanding and Applying What You Have Learned

Questions in this feature ask you to use the key principles and concepts introduced in the activity. You may also be presented with new situations in which you will be asked to apply what you have learned.

10. Preparing for the Chapter Challenge

This feature suggests ways in which you can organize your work and get ready for the challenge. It prompts you to combine the results of your inquiry as you work through the chapter.

11. Inquiring Further

This feature provides lots of suggestions for deepening your understanding of the concepts and skills developed in the activity. It also gives you an opportunity to relate what you have learned to the Earth system.

EARTH SYSTEMS

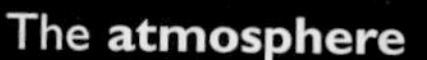

The **atmosphere**
is the gaseous envelope that surrounds the Earth and consists of a mixture of gases composed primarily of nitrogen, oxygen, carbon dioxide, and water vapor.

The **biosphere**
is the life zone of the Earth and includes all living organisms, including humans, and all organic matter that has not yet decomposed.

The **cryosphere**
is the portion of the climatic system consisting of the world's ice masses and snow deposits. This includes ice sheets, ice shelves, ice caps and glaciers, sea ice, seasonal snow cover, lake and river ice, and seasonally frozen ground and permafrost.

The **geosphere**
is the solid Earth that includes the continental and oceanic crust as well as the various layers of the Earth's interior.

The **hydrosphere**
includes the water of the Earth, including surface lakes, streams, oceans, underground water, and water in the atmosphere.

EarthComm®

Earth System Science in the Community

Earth's Dynamic Geosphere

Chapter 1: Volcanoes...and Your Community

Chapter 2: Plate Tectonics...and Your Community

Chapter 3: Earthquakes...and Your Community

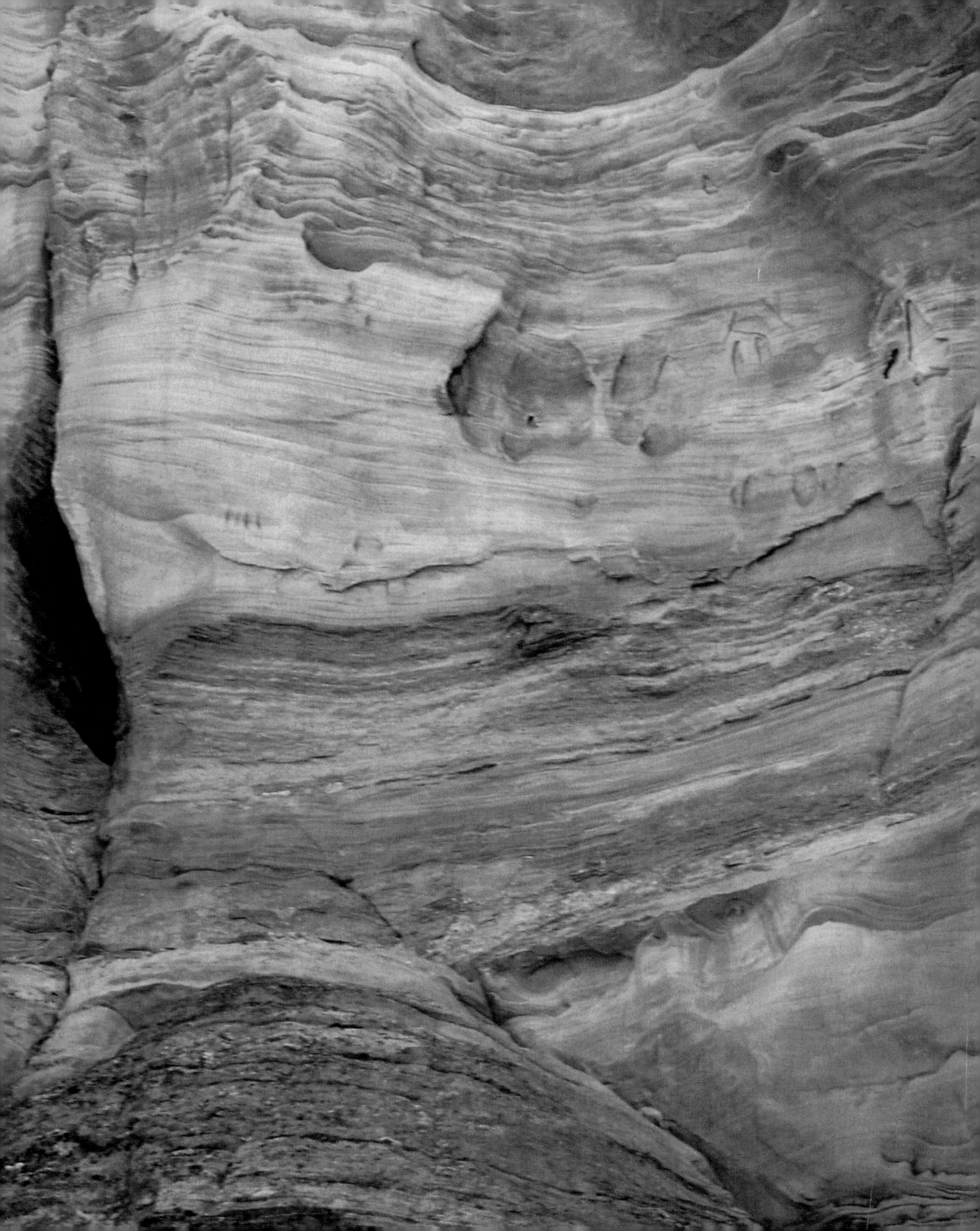

1

Volcanoes... and Your Community

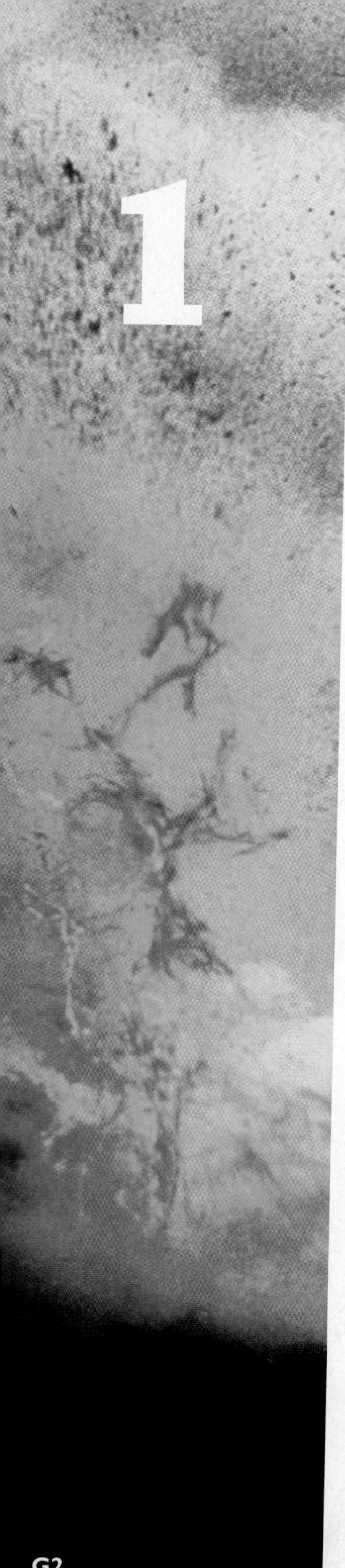

1

Volcanoes ...and Your Community

Getting Started

In 1883, on the island of Krakatoa in the East Indies, one of the most violent eruptions of recorded time took place. Half of the island was blown away by a volcanic eruption. Over a cubic mile of rock was hurled into the air. The sound of the explosion was heard in Australia, over 2000 miles away.

- Can a volcano that erupts on the other side of the world affect your community?

What do you think? Look at the Earth systems at the front of this book. In your notebook, draw a picture to show one way that a volcanic eruption changes an Earth system. Then, think about how that change might cause a change in another Earth system. Add this to your drawing. Continue until you have connected the volcanic eruption to your community. Be prepared to discuss your pictures with your small group and the class.

Scenario

"The clouds became thicker, and it was increasingly difficult to see as we struggled up the narrow, steep path toward the summit. The ground was hot under our feet, but the moisture from the clouds kept us cold and damp, and made the ash stick to our hair and eyelashes. We began to see larger volcanic rocks, some as large as two feet across. Suddenly we came across a large fissure, about one foot wide and 60 feet long. As I leaned over it, a hot blast of sulfur-smelling air scorched my nostrils. Then, like a warning growl from a watchdog, came a rumble from deep within. This was the moment we had been anticipating with dread..."
Many motion pictures are based on exciting geologic events.

Can you use your knowledge of volcanoes to make a thrilling, yet informative and scientifically correct movie?

Chapter Challenge

Your challenge is to write a screenplay or story, set in your community, that would help audiences understand volcanoes. You need to teach them about volcanic hazards. You also need to help them see volcanoes as part of the Earth system, and to realize that volcanoes affect all communities in some way. Can you use Earth science to create an exciting story and help others understand the hazards and the benefits of volcanoes?

Think about how your story will address the following items:

- Locations of volcanoes closest to your community.
- Evidence that shows past or recent volcanic activity.
- Types of volcanic hazards.
- How volcanoes change the atmosphere, hydrosphere, and other Earth systems.

Assessment Criteria

Think about what you have been asked to do. Scan ahead through the chapter activities to see how they might help you to meet the challenge. Work with your classmates and your teachers to define the criteria for assessing your work. Record all this information. Make sure that you understand the criteria as well as you can before you begin. Your teacher may provide you with a sample rubric to help you get started.

Activity 1 Where are the Volcanoes?

Goals

In this activity you will:

- Find the latitude and longitude of volcanoes nearest your community when given a map of historically active volcanoes.
- Search for and describe patterns in the global distribution of volcanoes.
- Make inferences about possible locations of future volcanic activity.
- Understand that most volcanism occurs beneath the ocean.
- Understand that map projections distort regions near the poles and eliminate some data.

Think about It

Volcanoes are one of nature's most feared, yet spectacular activities.

- Can volcanoes form anywhere on Earth? Why or why not?

What do you think? Record your ideas about this question in your *EarthComm* notebook. Be prepared to discuss your responses with your small group and the class.

Investigate

1. "Thought experiments" are experiments that scientists dream up and then run in their imagination, rather than in a real laboratory. They are a useful way to develop new ideas and insights into scientific problems. Here's a thought experiment for you to do, to help you understand the Mercator map projection. It wouldn't even be too difficult to do this in real life, if you could obtain the right materials.

   Visualize a large, see-through plastic ball. Poke holes on opposite sides, and stick a wooden dowel or a chopstick through to make the North Pole and the South Pole. Install a bright light directly at the center of the ball, somehow. With a felt-tipped pen, draw a fake continent on the ball. Make the continent extend from near the Equator to near the North Pole. Now wrap a clear sheet of stiff plastic around the globe, to make a tight-fitting cylinder that's parallel to the Earth's axis. See the figure on the right for how to arrange this. Turn the light on, and observe how the border of your continent projects onto the plastic cylinder. Trace that image on the cylinder with the felt-tipped pen. Unwrap the cylinder from the globe, and lay it flat on the table. You now have a map with what's called a Mercator projection of your continent!

   a) Describe how the image of your continent is changed in shape (distorted) when it is projected onto the cylinder.

   b) If you drew a short east–west line with a certain length near the southern end of your continent, and another east–west line with the same length near the northern end of your continent, how would the lengths of the lines compare when they are projected onto the cylinder map?

   c) If you drew a short north–south line with a certain length near the southern end of your continent, and another north–south line with the same length near the northern end of your continent, how would the lengths of the lines compare when they are projected onto the cylinder map?

   d) How would the image of a continent that is centered on the North Pole project onto the cylinder map?

   e) Which part of your map shows the least distortion?

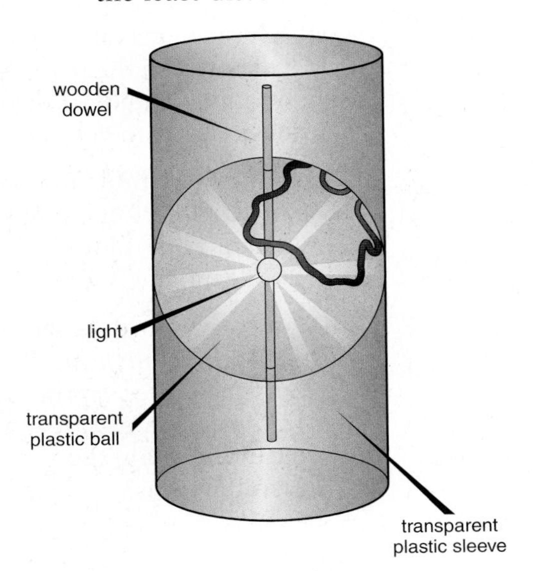

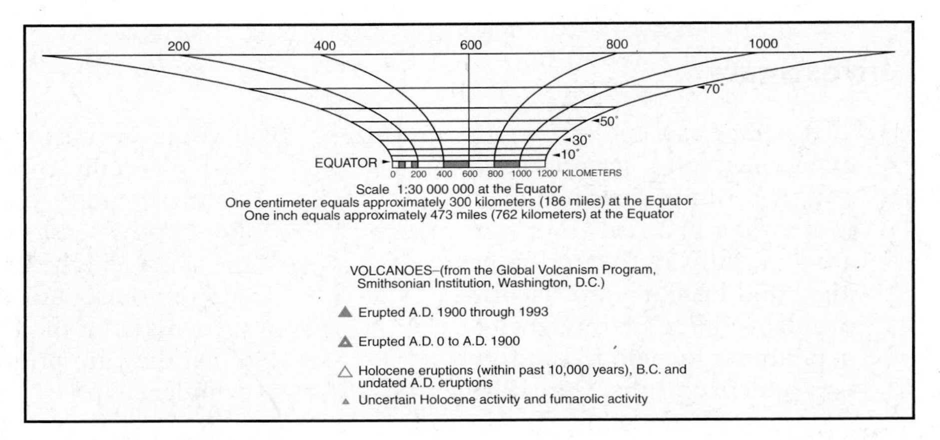

2. Obtain the USGS map called *This Dynamic Planet*. Look at the map key, also shown above, to learn the meaning of the various symbols and how to use the map scale.
   a) What do each of the four kinds of triangles represent?
   b) What do the solid red lines represent?
   c) Describe how the scale of the map changes with latitude.
   d) Does the map cover the entire Earth? Why or why not?
3. For each time interval of volcanic activity shown on the map, find the latitude and longitude of three volcanoes closest to your community.
   a) Make a data table to record your results. When complete, the data table should list 12 volcanoes.
   b) Compare your data with that of other groups in your class. Did your class agree on the locations of the nearest historically active volcanoes? How did you resolve any differences?
4. Obtain a copy of a world map. Use this map to summarize any patterns in the global distribution of volcanoes.
   a) When volcanoes follow a linear pattern, draw a thick line on the world map. Use the string of volcanoes within the Aleutian Islands and southern Alaska as an example.
   b) For the red lines that appear on the USGS map, draw thin lines on your copy of the map. See the examples in the Pacific Ocean near Oregon and Washington.
   c) Where volcanoes are less concentrated, outline (circle) the area that they cover. Try to be as accurate as possible. See the group of volcanoes in the Cascades as an example.
5. When your map is complete, answer the following questions in your notebook:
   a) Are most volcanoes found in random places or do they show a trend or pattern? Explain.

b) Does the USGS map show volcanoes that have not erupted during the last 10,000 years?

c) Does the USGS map show eruptions after 1993, or new volcanoes?

d) Does the USGS map show any volcanoes associated with the red lines in the ocean basins?

e) What information does the map give about the size or hazard of the volcanoes?

f) Suppose that tomorrow a volcano forms somewhere in the United States. Could it form in or near your state? Support your answer with evidence from this activity.

g) What are some limitations of the evidence you used?

Reflecting on the Activity and the Challenge

By looking at a world map of recent volcanic activity you found patterns in the data. This helped you to make inferences about the possible location of the next volcanic eruption in the United States. The data you looked at are incomplete. This may limit the conclusions you can draw. However, you now have some knowledge that will help you decide where in the U.S. you might "stage" a volcanic eruption.

Digging Deeper

THE GLOBAL DISTRIBUTION OF VOLCANOES

Volcanoes beneath the Sea

The USGS map *This Dynamic Planet* shows historical volcanic activity throughout the world. It tells a story about how our dynamic planet releases its internal storehouse of energy. No single source of data tells the whole story, but a map is a great place to begin.

On average, about 60 of Earth's 550 historically active volcanoes erupt each year. Geologists have long known that volcanoes are abundant along the edges of certain continents. The presence of volcanic rocks on the floors of all ocean basins indicates that volcanoes are far more abundant under water than on land.

Geo Words

mid-ocean ridge: a continuous mountain range extending through the North and South Atlantic Oceans, the Indian Ocean, and the South Pacific Ocean.

rift valley: the deep central cleft in the crest of the mid-oceanic ridge.

magma: naturally occurring molten rock material. generated within the Earth

All of the Earth's ocean basins have a continuous mountain range, called a **mid-ocean ridge** extending through them. These ridges, over 80,000 km long in total, are broad rises in the ocean floor. They are usually in water depths of 1000 to 2000 m. *Figure 1* shows a vertical cross section through a mid-ocean ridge. At the crest of the ridge there is a steep-sided **rift valley**. **Magma** (molten rock) from deep in the Earth rises up into the rift valley to form submarine volcanoes. These volcanoes have even been observed by scientists in deep-diving submersibles. All of the floors of the oceans, beneath a thin layer of sediments, consist of volcanic rock, so we know that volcanoes form all along the mid-ocean ridges, at different times at different places. At a few places along the mid-ocean ridges, as in Iceland, volcanic activity is especially strong, and volcanoes build up high enough to form islands.

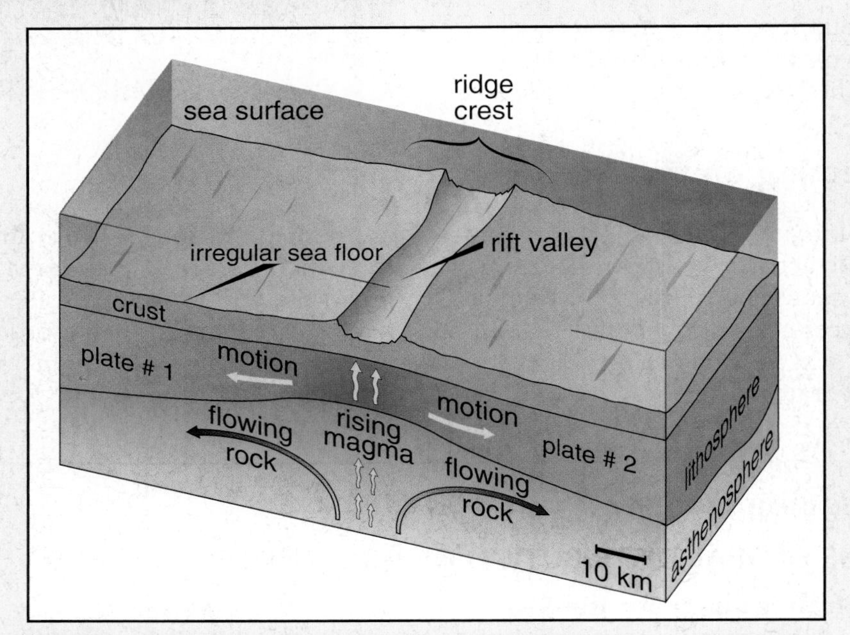

Figure 1 Cross section through a mid-ocean ridge.

Volcanoes on Land

Volcanoes that erupt on land are much more dangerous than volcanoes beneath the ocean. Eruptions along the western edge of the United States have formed the Cascades volcanic mountain range. They also form island chains, like the Aleutians in Alaska. Volcanoes like these are common in a narrow belt all around the Pacific Ocean. Geologists call this the "Ring of Fire." A famous example of an eruption along the Ring of Fire was the dramatic eruption of Mt. St. Helens in Washington in 1980.

Ring of Fire

Around the edges of the Pacific Ocean, the plates of the Pacific Ocean slide down beneath the continents. Look at *Figure 2* to see an example. The Nazca Plate, moving eastward from the East Pacific Ridge, slides down beneath the west coast of South America. The plate is heated as it sinks into the much hotter rocks of the deep Earth. The heat causes fluids, especially water, to leave the plate and rise into overlying hot rocks. The added water lowers the melting point of the solid rock. If enough water is added the rock melts and magma is formed. The magma rises upward, because it is less dense than the rocks. It feeds volcanoes on the overlying plate. Nearly four-fifths of volcanoes on land form where one plate slides beneath another plate.

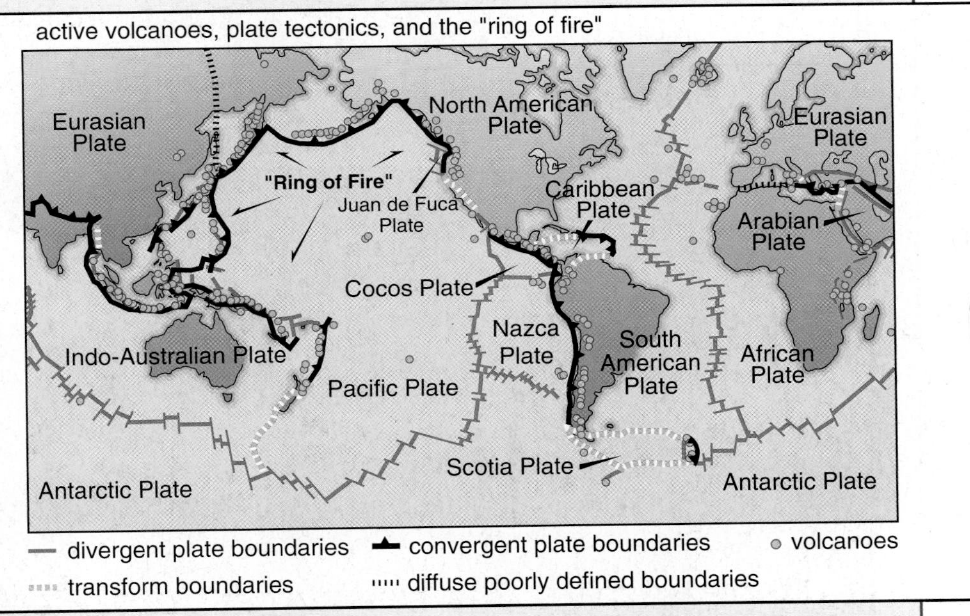

Figure 2 The plates of the Earth, and the "Ring of Fire" around the Pacific. The circles show active volcanoes.

Volcanoes Formed by Rifting on the Continents

Volcanoes in the East African rift valley form where two parts of the African continent are moving apart from each other. The process is very similar to what happens at mid-ocean ridges. The continental plate is stretched and broken. One of the breaks becomes the main one, and opens up to form the rift valley, as shown in *Figure 3*.

Geo Words

lava: molten rock that issues from a volcano or fissure.

hot spot: a fixed source of abundant rising magma that forms a volcanic center that has persisted for tens of millions of years.

map projections: the process of systematically transforming positions on the Earth's spherical surface to a flat map while maintaining spatial relationships.

Mercator projection: a map projection in which the Equator is represented by a straight line true to scale, the meridians by parallel straight lines perpendicular to the Equator and equally spaced according to their distance apart at the Equator, and the parallels by straight lines perpendicular to the meridians and the same length as the Equator. There is a great distortion of distances, areas, and shapes at the polar regions.

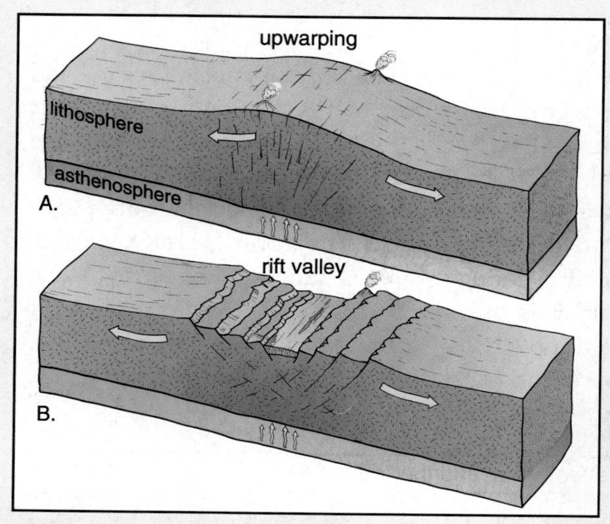

Figure 3 Formation of a rift valley on a continent.

In the United States, continental rifting long ago formed the rocks that make up the tall cliffs on the western bank of the Hudson River. These rocks formed when magma intruded the crust during this rifting. The rocks are seen for more than 80 km along the bank of the Hudson River and can be up to 300 m thick! Other evidence of magma formed during this rifting is found in many states along the East Coast.

Figure 4 Mount Kilimanjaro is a famous example of volcanism at a continental rift. Many other volcanoes in the East African rift valley have erupted in historic times.

Volcanoes at Hot Spots

Volcanoes discussed so far occur near the edges of plates. However, a small percentage of volcanoes occur in the interior of a plate. The Hawaiian Islands, shown in *Figure 5*, are an example. Studies of volcanic rock show that the islands get older to the northwest. Only the youngest island, the "big island" of Hawaii, has active volcanoes.

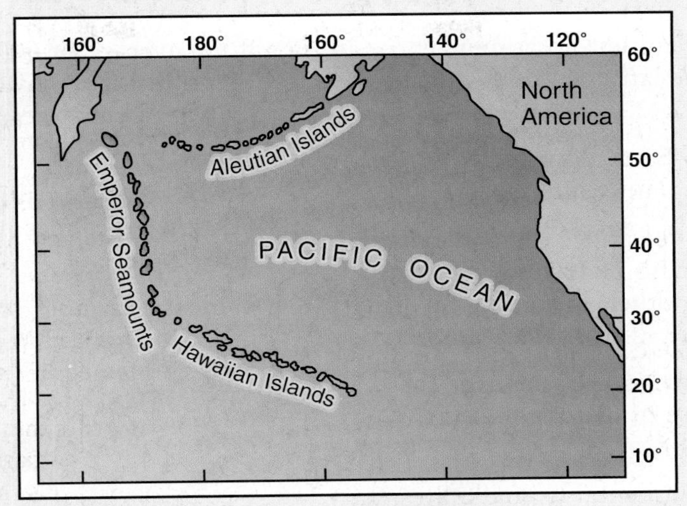

Figure 5 The Hawaiian island chain and the Emperor seamount chain.

Here's how geologists explain the pattern of the Hawaiian Islands. Deep beneath Hawaii, there is a fixed source of abundant rising magma, called a **hot spot**. As the Pacific Plate moves to the northwest, away from the East Pacific Ridge, it passes over the fixed hot spot. Magma from the hot spot punches its way through the moving plate to form a chain of islands. The sharp bend in the chain was formed when the direction of movement of the plate changed abruptly at a certain time in the past. Far to the northwest the chain consists of **seamounts**.

Map Projections

There is always a big problem in drawing a map of the world, because you have to try to show the curved surface of the Earth's globe on a flat sheet of paper. Many different ways of doing this, called **map projections**, have been developed, but they all have some kind of distortion. The USGS map uses a **Mercator projection**. As you move away from the Equator, the map becomes more and more distorted. For example, it makes all lines of latitude look like they are equal in length. This makes it difficult to measure distances on the map. Another problem is that the USGS map stops at 70° north and south latitude, because of the Mercator projection. This keeps you from seeing all of the data. For example, the USGS map cuts off the mid-ocean ridge north of Iceland. The scale of the map also presents a problem. The larger the area covered by the map, the less detail the map can show. In this case, the triangular symbols that represent volcanoes often overlap in areas with many volcanoes. This makes them difficult to count.

Geo Words

seamount: a peaked or flat-topped underwater mountain rising from the ocean floor.

Check Your Understanding

1. What evidence do geologists have that volcanoes occur on the ocean floor?
2. What is the Ring of Fire, and where is it located?
3. Where do most volcanoes on land form?
4. How are rift valleys formed?
5. What are hot spots? Provide an example of a hot spot on Earth.
6. Why does the horizontal scale of a Mercator projection increase with latitude?

Understanding and Applying What You Have Learned

1. What difficulties did you have finding the latitude and longitude of volcanoes?
2. Where on Earth do most volcanoes occur? Explain your answer.
3. Are most volcanoes on land caused by the Earth's plates moving away from each other or moving toward each other? Explain your answer.
4. In your own words, describe the likely cause of historically active volcanoes in:
   a) The continental United States.
   b) The Aleutian Islands and southern Alaska.
   c) The Hawaiian Islands.
5. Based on your results from this investigation, list the five states that you think are most likely to experience the next volcanic eruption. Explain each choice.
6. Of the average of 60 volcanoes that erupt in any given year, how many are likely to erupt along the Ring of Fire?
7. Why did the Mercator projection not show volcanoes near the Earth's poles?
8. Do most volcanoes on land occur in the Northern Hemisphere or the Southern Hemisphere? Explain why you think this is so.

Preparing for the Chapter Challenge

Think about how you can help the audience understand why you chose the probable location of the volcanic eruption for your story. Explain the map that you made for this activity. Note the volcanic eruptions that are closest to your area. Explain where most volcanoes occur in the United States. You should also note where they have not happened recently.

Inquiring Further

1. **Eruptions near your community**

   Find out more about the historical eruptions at the volcanoes nearest your community. The Volcano World web site lists hundreds of historically active volcanoes. (Consult the AGI *EarthComm* web site for current addresses.) Your data table of latitudes and longitudes will help you to identify them.

2. **Volcanoes and the water on Earth (the hydrosphere)**

   Research to find answers to the following questions, and any other questions which you have formed:

   - How do volcanoes at mid-ocean ridges affect the temperature of seawater?
   - How do volcanoes change the chemistry of seawater?
   - How does seawater affect the composition of the volcanic rock that is formed at the mid-ocean ridge?
   - Would volcanoes affect a small body of seawater, such as the Red Sea, the same way as a large ocean like the Atlantic?
   - Can a change in the volume of volcanic rock formed at mid-ocean ridges change sea level?

Activity 2 Volcanic Landforms

Goals

In this activity you will:

- Make a topographic map from a model.
- Understand the meanings of contour line, contour interval, and relief.
- Interpret topographic maps.
- Recognize volcanic landforms on a topographic map, and predict where lava would flow on them.
- Understand basic relationships between magma composition and type of volcano formed.

Think about It

When most people think about volcanoes, they probably have in mind a steep-sided cone. Many volcanoes, however, have very gentle slopes.

- Why do different volcanoes have different shapes?

What do you think? Record your ideas about this question in your *EarthComm* notebook. Include a quick sketch for each question. Be prepared to discuss your responses with your small group and the class.

Investigate

1. Use a piece of paper and tape to make a model of a volcano. The model should be small enough to fit into a shoebox.

2. Draw horizontal curves on the model at regular heights above the table. To help you draw the lines, attach a strip of stiff cardboard at right angles to a centimeter ruler at the 1-cm mark, as shown in the diagram. Hold the ruler upright on the table, with the "zero" end down, and move it around the model so that the cardboard strip is near the surface of the model. Make a series of small dots on the model at this 1-cm height, and then connect the dots to form a horizontal curve. Repeat this with the cardboard strip attached at the 2-cm mark. Continue increasing the height above the table by 1 cm until you reach the top of the model.

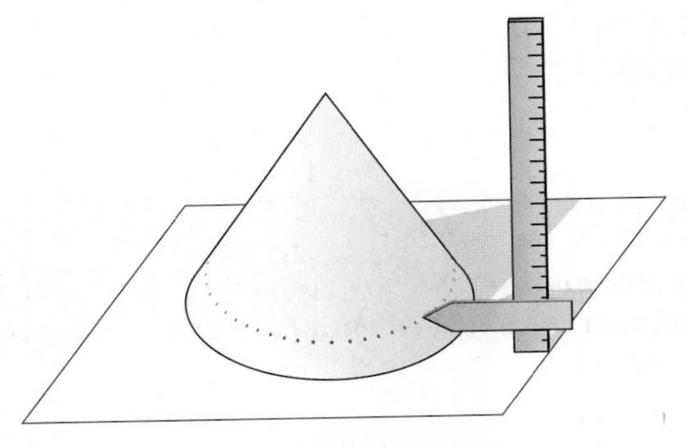

3. Place the model into a shoebox.

4. Clip an overhead transparency onto a clear clipboard. Lay the clipboard on the box.

5. Look straight down into the top of the shoebox at the lines you drew on the mountain. With a grease pencil or marker, trace the lines onto the transparency. Be sure to keep looking straight down, whenever you're tracing the lines! Also, it might help to keep one eye closed. These are contour lines, or lines of equal elevation above sea level.

6. Remove the transparency from the box. Write an elevation on each line of the transparency. Let each centimeter in height on the model represent 100 m in elevation on the map. The numbers should increase toward the center of the transparency.

7. Compare your map to the map of Mt. St. Helens shown below. Answer the following questions in your notebook and on your map:

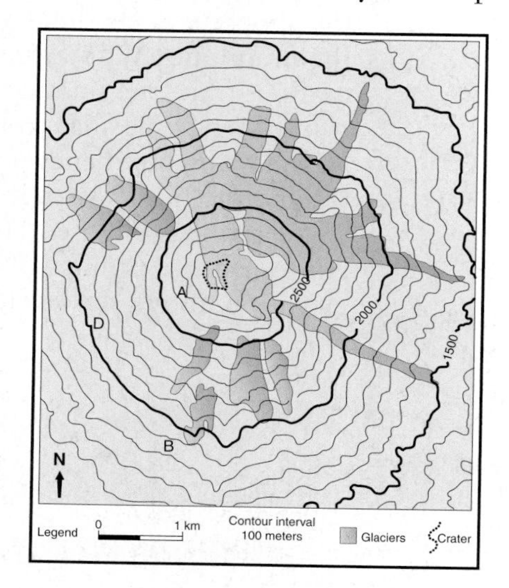

a) Describe two similarities between the maps.

b) Note the legend on the map of Mt. St. Helens. Add a legend to your map. Include a scale, north arrow, and contour interval.

c) What do the shaded regions on the map of Mt. St. Helens represent?

d) Why do the shaded regions cross the contour lines at right angles?

e) Which part of Mt. St. Helens is steeper, the slope between 1500 m and 2000 m, or the slope between 2000 m and 2500 m? Explain your answer.

f) What are the lowest and highest elevations on the map of Mt. St. Helens? What is the difference in elevation between these two points?

g) Note the locations A, B, C, and D on the map. If lava erupted at point A, would it flow to point B, C, or D? Explain why.

Reflecting on the Activity and the Challenge

You made a map from a model of a volcano. The map showed lines of equal elevation. The lines are contour lines, and the map is a topographic map. You can use topographic maps to predict volcanic hazards. Gravity pulls the lava erupted from volcanoes downhill. A topographic map shows the paths the lava might take. It might also help you to guess whether a certain region has volcanoes.

Digging Deeper

TOPOGRAPHY OF VOLCANIC REGIONS

Topographic Maps

Topographic maps have **contour lines.** These are curves that connect all points at the same elevation. The **contour interval** is the difference in elevation between adjacent contour lines. A **topographic map** shows how steep or gentle a slope is. It also shows the elevation and shape of the land. **Relief** is the difference in elevation between the highest and lowest points on the map.

The following are some important points to consider when interpreting topographic maps:

- Contour lines never cross (but two or more can run together, where there is a vertical cliff).
- The closer together the contour lines, the steeper the slope.
- Contour lines for closed depressions, such as a volcanic crater, are marked with "tick marks" (short lines at right angles to the contour line) pointing downslope, into the depression.
- On most topographic maps, every fifth contour line on a map is darker and its elevation is always marked.

Magma Composition

Volcanoes are often pictured as cone-shaped mountains. However, volcanoes come in many shapes and sizes. Ice, wind, and rain can change the shape of a volcano, both between eruptions and after the volcano becomes dormant. A large eruption or giant landslide can remove the top or side of a volcano. The chemical composition of magma can have an even greater effect on the shape the volcano takes as it forms.

Magma is a mixture of liquid, melted rock, and dissolved gases. The most abundant chemical elements in magma are silicon and oxygen. As the magma cools, minerals form. Silicon and oxygen are the building blocks of the most common minerals, called silicate minerals, that form from magmas. One silicon atom and four oxygen atoms become tightly bonded together to form an ion, called the silicate ion. These combine with ions of other elements, mainly aluminum, iron, calcium, sodium, potassium, and magnesium, to form silicate minerals.

When geologists make a chemical analysis of an **igneous rock**, a rock that formed when molten materials became solid, they express the results as percentages of several "oxides," such as SiO_2, Al_2O_3, or CaO. In one

Geo Words

contour lines: a line on a map that connects points of equal elevation of the land surface.

contour interval: the vertical distance between the elevations represented by two successive contour lines on a topographic map.

topographic map: a map showing the topographic features of the land surface.

relief: the physical configuration of a part of the Earth's surface, with reference to variations of height and slope or to irregularities of the land surface.

igneous rock: rock or mineral that solidified from molten or partly molten material, i.e., from magma.

Geo Words

silica: material with the composition SiO_2.

shield volcano: a broad, gently sloping volcanic cone of flat-dome shape, usually several tens or hundreds of square miles in extent.

way, this is a fake, because real oxide minerals are a very small part of most igneous rocks. It's just a generally accepted practice. Because silicon and oxygen are the most abundant elements in magmas, the "oxide" SiO_2, called **silica**, is the most abundant "oxide." The percentage of silica in magma varies widely. This is important to know for two reasons. First, magmas rich in silica tend to have more dissolved gases. Second, silica content affects how easily magma flows. Magmas that are rich in silica do not flow nearly as easily as magmas that are poor in silica. Because of this, silica-rich magmas are more likely to remain below the Earth's surface, at shallow depths, rather than flowing freely out onto the surface. These two factors combine to make eruptions of silica-rich magmas likely to be dangerously explosive. Here's why: As the magma rests below the surface, the dissolved gases gradually bubble out, because the pressure on the magma is much less than it was down deep in the Earth, where the magma was formed. It's just like what happens when you pour a can of soda into a glass: the carbon dioxide dissolved in the soda gradually bubbles out of solution. Unlike your soda, however, the magma is so stiff that the bubbles can't readily escape. Instead they build up pressure in the magma, and that often leads to a catastrophic explosion. The table in *Figure 1* shows how magma properties relate to magma composition.

Properties of Magma as They Relate to Magma Composition

Magma Property	Magma Composition (percent silica content)		
	Low Silica	**Medium Silica**	**High Silica**
Silica content (% SiO_2)	~ 50	~ 60	~ 70
Viscosity	lowest	medium	highest
Tendency to form lava	highest	medium	lowest
Tendency to erupt explosively	lowest	medium	highest
Melting temperature	highest	medium	lowest
Volume of an eruption	highest	medium	lowest

Figure 1 Adapted from *Earth Science*, 7th Edition, Tarbuck and Lutgens, 1994

Types of Volcanic Landforms

When low-silica magma erupts, lava tends to flow freely and far. If it erupts from a single opening (vent) or closely spaced vents, it forms a broad **shield volcano**, as shown in *Figure 2*.

Figure 2 Volcanoes such as these are called shield volcanoes because they somewhat resemble a warrior's shield. They are formed when low-silica magma erupts.

Figure 3 The eruption of low-silica magma along long, narrow openings in the Columbia Plateau flowed over a vast area. The result was a broad lava plateau that makes up the cliffs.

Silica-rich magmas are far less fluid. They often stop moving before they reach the surface. If they do reach the surface, they ooze slowly, like toothpaste squeezed out of a vertical tube. The thick, stiff lava forms volcanic domes with steep slopes, as shown in *Figure 4*. If the volcano's vent gets plugged, gases cannot escape and pressure builds up. The pressure can be released in a violent eruption that blasts pieces of lava and rock (pyroclastics) into the atmosphere.

Geo Words

composite cone (stratovolcano): a volcano that is constructed of alternating layers of lava and pyroclastic deposits.

caldera: a large basin-shaped volcanic depression, more or less circular, the diameter of which is many times greater than that of the included vent or vents.

Figure 4 Silica-rich magma does not flow readily and often forms a volcanic dome such as the one shown in this photograph.

A **composite cone**, as shown in *Figure 5*, forms by many eruptions of material with medium or high silica content. They erupt violently when pressure builds up in the magma. After the explosion, gooey (viscous) lava oozes out of the top. The volcano becomes quiet. Over time, pressure may build up and repeat the cycle. Composite volcanoes are tall and have steep slopes because the lava does not flow easily.

Figure 5 Composite cones include the beautiful yet potentially deadly Cascades in the northwestern United States (Shasta, Rainier, Mt. St. Helens, etc.).

When a very large volume of magma is erupted, the overlying rocks may collapse, much like a piston pushing down in a cylinder. The collapse produces a hole or depression at the surface called a **caldera**, shown in *Figure 6*. A caldera is much larger than the original vent from which the magma erupted.

Figure 6 Calderas are deceptive volcanic structures. They are large depressions rather than conical peaks. Oregon's Crater Lake, formed nearly 7000 years ago, is an example of this type of volcano.

Check Your Understanding

1. Explain in your own words the meaning of a contour line, contour interval, relief, and topographic map.
2. Arrange corn syrup, water, and vegetable oil in order of low to high viscosity.
3. What is the silica content of magma that has a low viscosity?
4. Why do silica-poor magmas produce broad volcanoes with gentle slopes?
5. Why does high-silica magma tend to form volcanic domes with steep sides?
6. How is a caldera formed?

Understanding and Applying What You Have Learned

1. What is the contour interval on the topographic map of Mt. St. Helens?
2. Sketch a contour map of a volcano that shows:

   a) a gentle slope
   b) a steep slope
   c) a nearly vertical cliff
   d) a crater or depression at the top
3. Imagine that your paper model was a real volcano. Lava begins to erupt from the top. Shade your topographic map to show where a stream of lava would flow. Explain your drawing.
4. For the volcanoes shown in *Figures* 2 and 5, sketch a topographic map. Show what the volcano would look like from above. Apply the general rules for interpreting topographic maps. Include a simple legend.

5. Use the topographic map below, or obtain a topographic map of your state or region, to answer the questions.
   a) Record the contour interval, and the highest and lowest elevations. Calculate the relief.
   b) Identify areas that look like the volcanic landforms you explored in this activity. Describe possible paths of lava flows.

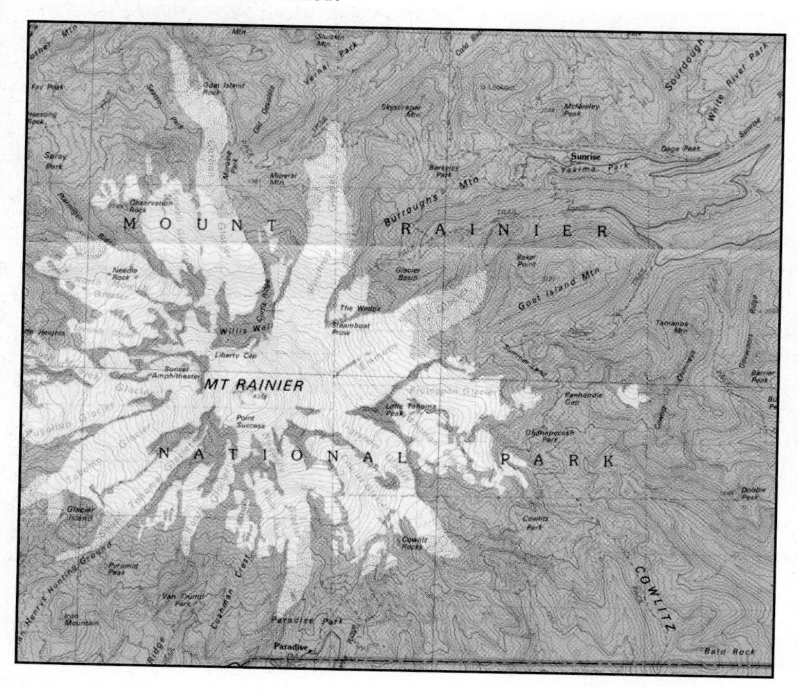

Preparing for the Chapter Challenge

In your story or play decide how you will convey to the audience the importance of using topographic maps to identify volcanic landforms. Indicate how the maps can also help geologists predict the paths of lava.

Inquiring Further

1. **Cascade volcano in your community**

   Build a scale model of a Cascade volcano and a scale model of your community. To do so, find a topographic map of a Cascade volcano. Trace selected contours on separate sheets of paper. Cut and glue each contour level onto pieces of card or foam board. Stack the board to make a three-dimensional model. Do the same using a topographic map of your community. Make sure that the scales of the maps match.

Be cautious when cutting foam board.

Activity 3 Volcanic Hazards: Flows

Goals

In this activity you will:

- Measure and understand how volume, temperature, slope, and channelization affect the flow of fluid.
- Apply an understanding of factors that control lava flows, pyroclastic flows, and lahars (mudflows).
- Apply understanding of topographic maps to predict lahar flow (mudflow) patterns from a given set of data.
- Describe volcanic hazards associated with various kinds of flows.
- Become aware of the benefits of Earth science information in planning evacuations and making decisions.
- Show understanding of the nature of science and a controlled experiment.

Think about It

Only one person in the entire city of St. Pierre, on Martinique in the Caribbean, survived the hot ash and rock fragments that swept over the city from the explosive eruption of Mt. Pelée in 1902. He was a prisoner in a dungeon deep underground.

- How do volcanoes affect the biosphere?

What do you think? Record your ideas about this question in your *EarthComm* notebook. Be prepared to discuss your responses with your small group and the class.

Investigate

Part A: Area of Lava Flow

1. Suppose a volcano produces twice the amount of lava than in a prior eruption. Write a hypothesis based upon the following question: What is the relationship between the volume of an eruption and the size of the area it covers?

   a) Record your hypothesis in your notebook.

2. Check your hypothesis to see if it could be disproved. A hypothesis must be a prediction that can be falsified. The statement "Some stars will never be discovered," cannot be disproved. Therefore, it is not a hypothesis.

3. In this investigation, you will use liquid soap to simulate flow during a volcanic eruption. Volcanic flows include lava, gases, and mixtures of solid particles and gases.

   a) In your notebook, set up a data table. The table should help you record the relationship between the volume of liquid soap and the surface area that the soap covers. You will do trials with 0.5, 1, 2, 4, 8, and 16 cm^3 (cubic centimeters) of liquid soap.

4. Place an overhead transparency of a square grid on a flat surface.

5. Pour 0.5 cm^3 of liquid soap onto the transparent graph paper.

6. When the soap stops flowing, measure the area of the flow.

   a) Record the area of the flow in your data table.

7. Wipe the surface clean. Repeat the trials using 1, 2, 4, 8, and 16 cm^3 of liquid soap.

   a) Record your data in your table. Look for patterns.

8. Develop a hypothesis and design a test for one of the following questions related to the flow of fluids. Remember that during scientific inquiry, you can return to the materials or your data and revise your procedures as needed.

   - What effect does temperature have on resistance to flow (viscosity)?
   - What happens to fluid when slope changes from steep to gentle?
   - What effects would you see if fluids moved through narrow channels?

   a) Write down your hypothesis.

   b) Record your procedure in your notebook.

   c) Describe the variables you investigated.

9. Present your procedure to your teacher for approval. Then run your test.

   a) Record your data.

   b) Summarize your conclusions.

   c) Was your hypothesis correct?

Heat sources can cause burns. Hot objects and liquids look like cool ones. Feel for heat at a distance before touching.

Clean up any spills immediately. Liquids being used can cause floors and equipment to be sticky or slippery.

Part B: Travel Time of Lahars

1. Examine the table of expected travel times of lahars (mudflows) triggered by a large eruption of Mt. St. Helens. The values in the table come from computer simulations and actual behavior of mudflows in the 1980 eruption.

Expected Travel Times for Lahars Triggered by a Large Eruption of Mt. St. Helens (USGS)

Distance (via river channels) from Mt. St. Helens (km)	Estimated travel time (hours:minutes)	
	North Fork Toutle River	South Fork Toutle River, Pine Creek, Muddy River, Kalama River
10	0:37	0:11
20	1:08	0:30
30	1:37	0:54
40	2:16	1:21
50	2:53	1:49
60	3:27	2:20
70	3:48	2:53
80	4:43	3:31
90	6:36	4:18
100	8:50	5:12

2. Convert the travel times into minutes.
   a) Record the times in your notebook.
3. Make a graph of travel time (in minutes on the vertical axis) versus distance (in kilometers on the horizontal axis) for both data sets.
   a) Plot both data sets on the same graph.
   b) Connect the data points so that you can compare the data.
   c) Calculate an average velocity for mudflows along each fork of the Toutle River.
4. Answer the following questions in your notebook:
   a) Which area (North Fork or South Fork) is more likely to have a steeper gradient? Use the results of your investigation in **Activity 2** to support your answer.
   b) Explain the evidence in your graphs that suggests the gradients are not constant?
   c) Based on the information in the table, explain whether or not you think that a community located 50 km from Mt. St. Helens along either of these river valleys would have time to evacuate in the event of an unexpected massive eruption.

Reflecting on the Activity and the Challenge

In this activity, you found that temperature, volume, channels, and slope affect the flow of liquids. Analyzing data from a computer model, you predicted the flow of volcanic fluids down river valleys near Mt. St. Helens. You can now describe the volcanic hazards associated with various kinds of flows and factors which affect the flows. In your movie you may wish to locate a town in the path of a potentially dangerous flow.

Geo Words

lava flow: an outpouring of molten lava from a vent or fissure; also, the solidified body of rock so formed.

viscosity: the property of a fluid to offer internal resistance to flow.

pyroclastic flow: a high-density mixture of hot ash and rock fragments with hot gases formed by a volcanic explosion or aerial expulsion from a volcanic vent.

Digging Deeper

FLOW-RELATED HAZARDS

Lava

Lava flows are streams of molten rock that come from vents and fissures in the Earth's crust. Lava flows destroy almost everything in their path. However, most lava flows move slowly enough for people to move out of the way. Slope and cooling affect the flow of lava. Lava flows faster on a steeper slope. As lava cools, it flows less and less easily. The term **viscosity** is used to describe a fluid's resistance to flow or internal friction. As lava cools, it becomes more viscous.

Figure 1 Lava tubes form when the surface of a flow cools and crusts over, but the interior of the flow is still fluid.

Lava that is low in silica is less viscous. (See *Figure 1* in the previous activity that shows the properties of magma as they relate to magma composition.) Flows of low-silica lava can travel tens of kilometers from the source. Sometimes it sets up an internal "plumbing system." The surface may cool, crust over, and insulate the interior. This keeps the lava at a higher temperature as it moves away from the source. Evidence of this is found in the lava tubes, as shown in *Figure 1*, found in flows of low-silica lava. When lava breaks out of the leading edge of a flow, the lava can drain out. A hollow tube remains behind.

Basalt flows can move at speeds of up to 10 km/h (kilometers per hour) on steep slopes. On a shallow slope, basalt flows typically move less than one kilometer per hour. Basaltic lava flows confined within channels or lava tubes can travel at speeds of 45 km/h. Basaltic lava flows can cover a considerable area. The largest lava flow in recent history occurred in 1783 at Laki in Iceland. Lava erupted from the Laki fissure covered 500 km^2, an area roughly equal to 100,000 soccer fields.

Since the start of the eruption in 1983, lava flows erupted from the Kilauea volcano in Hawaii and entered communities repeatedly. The flows destroyed more than 180 homes, a visitor center in a national park, highways, and historical and archaeological sites. The village of Kalapana was buried in 1990 by 15–25 m of lava erupted during a period of seven months. See *Figure 2*.

Figure 2 The former village of Kalapana was buried by lava flows.

It is sometimes possible to control the flow of lava. In 1973, lava flows at Heimaey, Iceland threatened to cut off a vital harbor. Citizens sprayed water onto the lava from ships in the harbor. This stopped the flow. Lava flows can also be diverted away from populated areas. Workers must carve a new channel or pathway through the landscape for the lava to follow.

Andesitic lava is cooler and has a higher silica content than basaltic lava. It moves only a few kilometers per hour. Andesitic lava rarely flows beyond the base of the volcano. Dacitic and rhyolitic lavas are even higher in silica and are even more viscous. Their lava usually forms steep mountains called lava domes, which extend only short distances from the vent.

Pyroclastic Flows

Topography plays a role in two other types of volcanic flows: pyroclastic flows and lahars. **Pyroclastic flows** are high-density mixtures of hot ash and rock fragments with hot gases. Pyroclastic flows occur in explosive eruptions. They move away from the vent at speeds up to 350 km/h. They often have two parts. A lower flow of coarse fragments moves along the ground. A turbulent cloud of ash rises above the lower flow. Both parts ride upon a cushion of air. This enables the material to move

Geo Words

lahar: a wet mixture of water, mud, and volcanic rock fragments, with the consistency of wet concrete, that flows down the slopes of a volcano and its river valleys.

rapidly. The more dense material follows the topography in a twisting path downslope. Pyroclastic flows are extremely dangerous. They destroy everything in their path. The pyroclastic flow produced by the Mt. St. Helens eruption, as shown in *Figure 3*, was impressive, but it was small compared to pyroclastic flows in prehistoric times.

Figure 3 Flow of pyroclastic materials from Mt. St. Helens destroyed everything in its path.

Lahar

A **lahar** is a wet mixture of water, mud, and volcanic rock fragments, with the consistency of wet concrete, that flows down the slopes of a volcano and its river valleys. Lahars can carry rock debris ranging in size from clay, to gravel, to boulders more than 10 m in diameter.

Eruptions may trigger lahars. Heat from the eruption may melt snow and ice, or the eruption may displace water from a mountain lake or river. Lahars sometimes form when the erupted material dams the mountain's drainage, causing a lake to form. The lake may spill over the loose volcanic material and send water and debris down valley. Lahars are also formed when rain soaks the loose volcanic debris during or after an eruption, causing it to start to flow. As a lahar flows downstream, it poses a risk to everyone in the valley downstream. When a lahar finally comes to a stop, it can bury an entire village under many meters of mud.

Figure 4 On the left side of the photograph, the dark region extending down the side of Mt. St. Helens is an example of a lahar flow.

Check Your Understanding

1. Name two factors that influence the viscosity of a lava flow.
2. Describe two ways in which lava flows can be controlled.
3. What is a pyroclastic flow?
4. What is a lahar?
5. How are lahars formed?
6. Explain how topography influences volcanic flows.

Understanding and Applying What You Have Learned

1. How does the volume of an eruption affect the area? Describe any mathematical pattern in your data.

2. When the Mauna Loa volcano erupted in 1984, lava flowed toward Hilo, Hawaii. It is an excellent example of how scientists used their understanding of the factors that control the flow of lava to predict where lava would flow and decide whether to evacuate residents. The map shows the path of a series of lava flows from Mauna Loa. Each flow is given a letter (A through G) in the order it happened.

   a) Look at flow D on the map. What is the elevation of the top of flow D, and what is the elevation of the Kulani Prison?

   b) How close did flow D get to the prison?

   c) Do you think that the prison was put on alert?

   d) Look at flow E on the map. The flow was channeled. Do you think it moved swiftly or slowly? Explain.

   e) Lava from flow E crossed an important road. It headed straight for the city of Hilo. The lava then broke through walls of the channel. What do you think happened to the width of the flow after it broke through the channel? How do you think this changed the speed of the flow?

3. Refer back to the reading that described the lava flow at Heimaey, Iceland.

   a) Why did spraying the lava flow with water slow it down?

   b) This was a very unusual circumstance. What factors made this effort successful?

4. Why might a lahar (mudflow of volcanic debris and water) affect a community more severely than a lava flow?

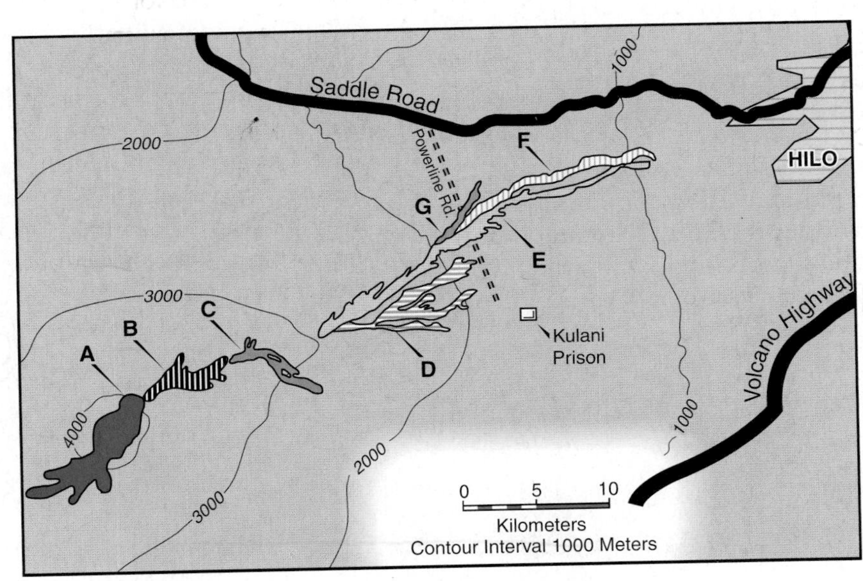

Source: USGS

Preparing for the Chapter Challenge

Think about what you have learned about volcanic flows. Prepare a one-page information sheet to raise awareness of how flows affect communities. Focus on three or more of the following:

a) How local topography controls where lava would flow.

b) Major roads that must be protected to ensure evacuation.

c) Natural and developed areas most likely to be affected.

d) Areas least likely to be affected, and why.

e) Living things that would not escape advancing flows.

f) Ways that flows might be controlled (diverting the flow, using water, and so on).

Consider how you can creatively work this information into your story line.

Inquiring Further

1. **Research a famous lava flow**

   Search the web for information about the Columbia River basalt group in the northwest. Prepare a report to the class about the members of this famous basalt group in relation to largest, longest, thickest, cooling characteristics, effects on ancient topography, and cause.

2. **Lava and the biosphere**

   How have lava flows at Mauna Loa and Kilauea volcanoes affected Hawaiian communities? How does the lava that enters the Pacific Ocean in Hawaii affect coastal ecosystems? What kinds of organisms develop and thrive at the "black smokers" along mid-ocean ridges? Research the 1783 Laki fissure flow in Iceland. It was 40 km long and covered 500 km^2. How did it affect vegetation and livestock?

3. **Lava and the cryosphere**

   What happens when lava erupts from an ice-capped or snow-capped volcano? This is an issue in the Cascade volcanoes. Mt. Rainier, which overlooks Seattle, has 27 glaciers. Some insights might be gained from exploring the recent eruption at Grimsvotn in Iceland.

Activity 4 Volcanic Hazards: Airborne Debris

Goals

In this activity you will:

- Understand why ash from volcanic eruptions can affect a much larger region than lava, pyroclastic flows, or lahars.
- Define tephra and describe some of the hazards it creates.
- Interpret maps and graph data from volcanic eruptions to understand the range in scale of volcanic eruptions.
- Understand that the explosive force of a volcano is not the only factor that determines its potential to cause loss of life and property.
- Interpret maps of wind speed and direction to predict the movement of volcanic ash.

Think about It

Volcanic ash put into the stratosphere from the great eruption of Krakatoa in Indonesia (see **Getting Started** on page two) caused spectacular sunsets all around the world for many months.

- Could material from a volcanic eruption ever reach your community? Explain your ideas.

What do you think? Record your ideas about this question in your *EarthComm* notebook. Be prepared to discuss your responses with your small group and the class.

Investigate

1. Look at the map of the 1980 eruption of Mt. St. Helens. It shows the pattern of ash. Use the map to answer the following questions:

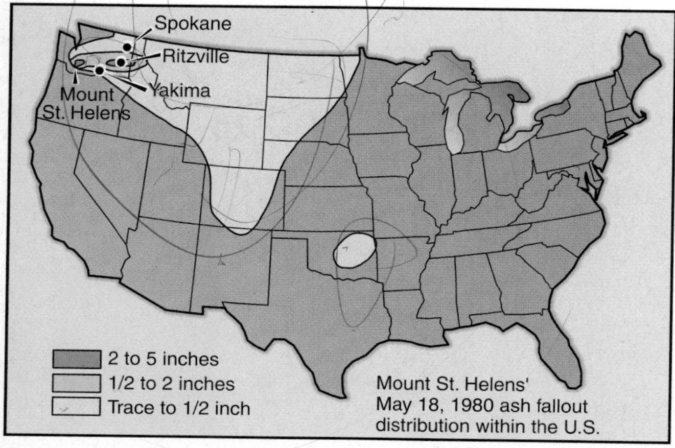

Distribution of ash from Mt. St. Helens eruption.

   a) How many states showed at least a trace of volcanic ash?

   b) In what direction did the ash move?

   c) Was Canada affected by ash from Mt. St. Helens? Why or why not?

   d) Would you consider this a small, medium, large, or gigantic eruption? Explain your choice.

2. Make a bar graph of the data shown in the table.

   a) Plot the name of each volcano on the horizontal axis.

   b) Plot the volume of volcanic eruption on the vertical axis. Arrange the volumes in order from least to greatest.

Volumes of Volcanic Eruptions

Volcano	**Date**	**Volume (cubic kilometers)**
Ilopango, El Salvador	300	40
Krakatoa, Indonesia	1883–84	2.4
Long Valley, California (Bishop Tuff)	740,000 years ago	500
Mazama, Oregon	4000 B.C.	75
Mt. Pelée, Martinique	1902	0.5
Mt. St. Helens, Washington	1980	1.25
Nevado del Ruiz, Colombia	1985	0.025
Pinatubo, Philippines	1991	10
Santorini, Greece	1450 B.C.	60
Tambora, Indonesia	1815	150
Valles, New Mexico	1.4 million years ago	300
Vesuvius, Italy	79	3
Yellowstone, Wyoming (Lava Creek Ash)	600,000 years ago	1000

Note: Volumes are approximate.

3. Use your graph and the table to answer the following questions. Record your answers in your notebook.
   a) Can you group the eruptions by size (small, medium, and so on)? Mark the groups on your plot. Explain how you chose the groups.
   b) What group does the 1980 eruption of Mt. St. Helens fit into?
   c) Suppose you wanted to predict the area that would be covered with ash by each eruption. What other information (besides volume erupted) would help you to predict how far the ash would go?

4. The following map shows the areas covered by five of the eruptions in the data table. Use the map, data table, and your bar graph to do the following:
   a) Rank the area of eruptions in order from smallest to largest. Record your rankings.
   b) Compare the areas to the volumes. Describe any relationships.
   c) Compare the location of each volcano to the path of the ash. Describe any patterns. What might explain any patterns you see?

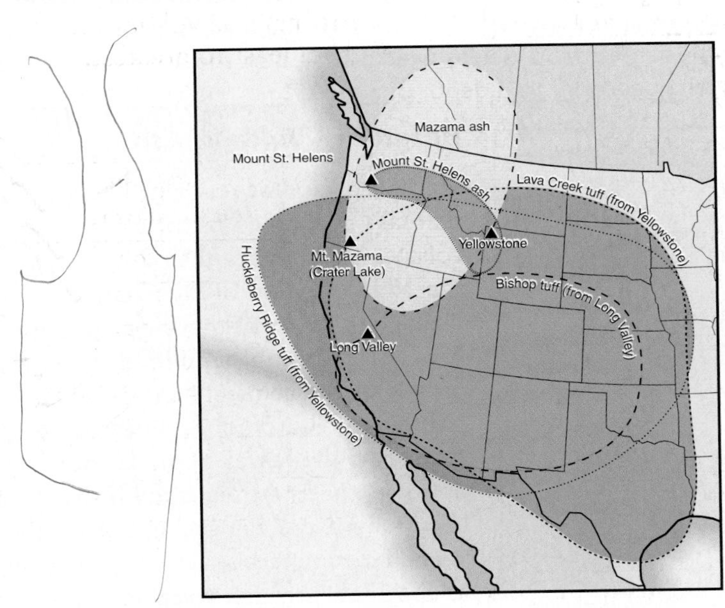

Reflecting on the Activity and the Challenge

This activity gave you a chance to explore factors that affect the movement of volcanic ash. How did this change your ideas about the areas that can be affected by an erupting volcano? Be sure to include the effect of airborne volcanic hazards into your story line.

Geo Words

tephra: a collective term for all the particles ejected from a volcano and transported through the air. It includes volcanic dust, ash, cinders, lapilli, scoria, pumice, bombs, and blocks.

volcanic bomb: a blob of lava that was ejected while viscous and received a rounded shape (larger than 64 mm in diameter) while in flight.

lapilli: pyroclastics in the general size range of 2 to 64 mm.

ash: fine pyroclastic material (less than 2 mm in diameter).

Digging Deeper

AIRBORNE RELEASES

Particle Types

Tephra is a term for pieces of volcanic rock and lava that are ejected into the air. It ranges from less than 0.1 mm to more than one meter in diameter. Tephra is classified by size. Names for sizes of tephra include **volcanic bombs** (greater than 64 mm), **lapilli** (between 2 and 64 mm), and **ash** (less than 2 mm). Bombs and lapilli usually fall to the ground on or near the volcano. Ash can travel hundreds to thousands of kilometers. (See *Figure 1*.) The height of the ash and the wind speed control how far the ash travels.

Distribution

A volcanic eruption can send ash many kilometers into the atmosphere. Ash from the 1980 eruption of Mt. St. Helens reached a height of 19 km. Winds carried the ash to the east. Five days after the eruption, instruments in New England detected ash. An eruption at Yellowstone 2 million years ago produced 1000 times as much ash. An area of 1000 km^2 received one meter of ash. Ten centimeters of ash covered an area of 10,000 km^2. You could look at it this way: if the ash from St. Helens filled a shoe box, the ash from Yellowstone would fill a bedroom to a depth of a meter!

Figure 1 Ballistic debris refers to volcanic bombs and lapilli that fall on or near the volcano. Ash can travel much further.

Hazards of Volcanic Ash

Volcanic ash presents many kinds of hazards. Ash that falls on homes, factories, and schools can collapse roofs. More than 300 people died after the 1991 eruption of Mt. Pinatubo in the Philippines. Most of these deaths were caused by roof collapse. At ground level, fine ash causes breathing problems in humans and animals. It can also damage automobile and truck engines. Ash that coats the leaves of plants interferes with photosynthesis. Ash injected higher into the atmosphere can damage aircraft. In the last 15 years, 80 commercial aircraft have been damaged as they flew through volcanic ash. The only death outside the immediate area of Mt. St. Helens occurred from the crash of a small plane that was flying through the ash. Ash that falls on the slopes of a volcano poses great risk. When soaked by rain, loose ash can form lahars. Years after the eruption, lahars remain a source of concern to communities at the base of Mt. Pinatubo.

Figure 2 Volcanic ash landing on buildings can result in death. Volcanic ash is also a hazard to airplanes on the ground as well as in the air.

Volcanic Explosivity Index

The table in *Figure 3* is a scale of eruption magnitude. The scale is known as **Volcanic Explosivity Index**, or VEI. The VEI is based on the volume of erupted material and the height it reaches. The size of an eruption depends upon several factors. Two important factors are the composition of the magma and the amount of gas dissolved in the magma. The viscosity of a magma depends on two things: the temperature of the magma, and its chemical composition. The higher the silica content of magma, the more viscous it is. The more viscous it is, the more likely it is for gas pressure to build. High-silica volcanoes, like Yellowstone, erupt extremely violently, but on a scale of tens or hundreds of thousands of years. Volcanoes with intermediate silica contents, like Mt. St. Helens, commonly produce violent eruptions with a frequency of hundreds or thousands of years. Silica-poor magmas, like those erupted at Kilauea, feed less explosive eruptions that occur more often.

Volcanic Explosivity Index (VEI)

VEI	Plume Height	Volume	How often	Example
0	<100 m	1000's m^3	Daily	Kilauea
1	100–1000 m	10,000's m^3	Daily	Stromboli
2	1–5 km	1,000,000's m^3	Weekly	Galeras, 1992
3	3–15 km	10,000,000's m^3	Yearly	Ruiz, 1985
4	10–25 km	100,000,000's m^3	10's of years	Galunggung, 1982
5	>25 km	1 km^3	100's of years	St. Helens, 1980
6	>25 km	10's km^3	100's of years	Krakatoa, 1883
7	>25 km	100's km^3	1000's of years	Tambora, 1815

Figure 3

Geo Words

Volcanic Explosivity Index: the percentage of pyroclastics among the total products of a volcanic eruption.

It might seem that the number of deaths caused by an eruption should always increase as the VEI increases. The table of VEI and deadliest eruptions in *Figure 4* shows that this is not the case. For example, mudflows after the 1985 eruption of Nevado del Ruiz (Colombia) killed more than 25,000 people. This was the worst volcanic disaster since Mount Pelée in 1902. However, both eruptions had a VEI below five. Of the seven most deadly eruptions since 1500 A.D., only Tambora and Krakatoa erupted with greater explosive force (VEI above 5).

Volcanic Explosivity Index (VEI) of the Deadliest Eruptions since 1500 A.D.

Eruption	Year	VEI	Casualties
Nevado del Ruiz, Colombia	1985	3	25,000
Mount Pelée, Martinique	1902	4	30,000
Krakatoa, Indonesia	1883	6	36,000
Tambora, Indonesia	1815	7	92,000
Unzen, Japan	1792	3	15,000
Lakagigar (Laki), Iceland	1783	4	9000
Kelut, Indonesia	1586	4	10,000

Figure 4

Tambora erupted in 1815. It had a VEI of 7. Pyroclastic flows streamed down its slopes. Ash rose 44 km. About 150 km^3 of ash were erupted (about 150 times more than the 1980 eruption of Mt. St. Helens). A caldera formed when the surface collapsed into the emptying magma chamber. The Tambora caldera is 6 km wide and 1.1 km deep. Tephra fall, a tsunami (a giant sea wave caused by the explosion), and pyroclastic flows killed about 10,000 people. More than 82,000 people died from famine. It is thought that the ash shortened the growing season.

Check Your Understanding

1. Review any words from the **Digging Deeper** sections of the previous activities that are used in this section but may still be unfamiliar to you. Briefly explain the meaning of each of the following terms: lahar, pyroclastic flow, caldera.
2. In your own words explain the meaning of tephra and how volcanic bombs, lapilli, and ash relate to tephra.
3. Name two factors that can affect the distance that volcanic ash can travel.
4. How does the silica content of magma affect how explosive a volcano can be?
5. a) What does VEI represent?

   b) Is VEI on its own a good indicator of the dangers involved with a volcanic eruption? Explain your answer.

Understanding and Applying What You Have Learned

1. In your own words, compare the sizes of the areas affected by lava, pyroclastic flows, and ash falls.
2. Is volcanic ash a concern only in the western United States? Explain your answer.
3. Why do eruptions in Hawaii differ from the Mt. St. Helens eruption?
4. Look at your list of the three volcanoes that are closest to your community. Go to the AGI *EarthComm* web site to find out how to simulate the eruption of one of your three volcanoes on the Internet. Simulate the eruption of one of your three volcanoes.
   a) Print out and describe the paths of the ash from the simulation.
   b) What do the maps tell you about the prevailing wind directions for your community?
   c) Do the prevailing wind directions change seasonally in your area? If yes, how would this affect the pattern of ash fallout?

Preparing for the Chapter Challenge

In **Activity 1** you were asked to describe to your audience the place that you thought a volcanic eruption was most likely to occur. Did you leave the impression in your story that this was the only area to be affected by the eruption? Think about how your study of the movement and hazards of volcanic ash has changed your ideas. Be sure to include this information in your story.

Inquiring Further

1. **Make a model of tephra transport**

   Build a model of a volcano that has exploded (Mt. St. Helens). Run a tube up through the vent of the volcano. Mix a small amount of baby powder with some sand. Use a funnel to pour the sand mixture down the other end of the tube. Attach a bicycle pump to pump the sand out of the volcano. Use a fan or hair dryer to simulate winds. Devise a method to outline the distribution of material when there is no wind, weak wind, and strong wind. Compare how far the sand travels and how far the baby powder travels. Consider the factors of particle size, wind speeds, wind direction, and topography. As part of your **Chapter Challenge** you may wish to include a presentation and explanation of your model.

Use eyewear whenever non-water liquids and/or particles such as sand are used. If this activity is done indoors use a large, clear area. Clean the area well when you are done. Sand on the floor can be slick.

Activity 5 Volcanoes and the Atmosphere

Goals

In this activity you will:

- Measure the amount of dissolved gas in a carbonated beverage.
- Understand that volcanoes emit gases such as water vapor, carbon dioxide, and sulfur dioxide.
- Describe how volcanoes are part of the hydrosphere and water cycle.
- Demonstrate awareness of how volcanoes can affect global temperatures.
- Recognize that volcanoes are part of interactive systems on Earth.

Think about It

Following the eruption of Tambora in Indonesia in 1815, snow fell in New England during each of the summer months that year!

- What else escapes from a volcano besides lava, rock, and ash?

What do you think? Record your ideas about this question in your *EarthComm* notebook. Be prepared to discuss your responses with your small group and the class.

Investigate

1. Use a can of your favorite carbonated soft drink to explore the quantity of gas that can be dissolved in a liquid under pressure.
   a) How many milliliters (mL) of liquid are in the can of soda?
   b) Predict how many milliliters of gas (carbon dioxide) a can of soda contains. Record and explain your prediction.
2. Obtain these materials: heat source, 1-liter Pyrex® beaker, water, rubber tubing (about 50 cm), smaller beaker or bottle, plastic container (shoebox size), modeling clay, safety goggles.
3. Devise a way to use the materials to measure the gas that escapes from the can of soda. Note: You will need to heat the soda after you have opened it. To do this safely, put the can in a water bath (container of water) and heat the water bath.
   a) Draw a picture of how you will set up your materials.
   b) Write down the procedures you will follow. Include the safety precautions you will take.
4. After your teacher has approved your design, set up your materials. Run your experiment.
   a) Record your results.
   b) How do your results compare to your prediction?
   c) Describe anything that might have affected your results.

Plan your activity carefully in detail to avoid potential hazards.

Reflecting on the Activity and the Challenge

You worked with a material that resembled volcanic products. When you opened the can of soda, you lowered the pressure inside the can. This allowed carbon dioxide (the dissolved gas) to come out of solution. Dissolved gases emerge from the Earth's interior in much the same way.

Digging Deeper

VOLCANOES AND THE ATMOSPHERE

Volcanic Gases

Gases that escape in greatest abundance from volcanoes are water vapor, carbon dioxide, hydrogen chloride, and nitrogen. These and certain other gases have played an important role in the Earth system throughout the long span of geologic time, and they continue to do so at the present time.

The atmosphere of the Earth early in its history contained abundant carbon dioxide but no oxygen. After primitive algae made their appearance partway through Earth history, the carbon dioxide emitted by volcanoes was gradually converted to oxygen by photosynthesis.

Carbon dioxide is more dense than air and sometimes accumulates in a low spot near a volcanic eruption. High concentrations of carbon dioxide are hazardous, because they cause people and animals to suffocate.

Water vapor is an essential component of the Earth system. It is especially important for human communities, because it sustains life. When you think of the water cycle, do you think of volcanoes? Volcanoes release abundant water vapor. Most of the Earth's surface water seems to have been released from the Earth's interior by volcanoes throughout the Earth's history.

Some volcanoes emit sulfur dioxide gas in great abundance. Sulfur dioxide combines with water vapor and oxygen to form sulfuric acid. The sulfuric acid is washed out of the atmosphere by rain, over large areas downwind of the eruption. Rain that contains sulfuric acid, and certain other acids as well, is called acid rain. It is produced not only by volcanoes but also by power plants that burn coal containing sulfur. Acid rain damages plants both on land and in lakes.

Volcanoes and Climate Change

Volcanoes illustrate the complexity of Earth's systems, because the gases from volcanic eruptions can contribute both to global cooling and to global warming.

How do volcanoes affect climate? If the Earth system were simple, the task of answering that question might be easy. Suppose that volcanic activity is the independent variable. This is the variable that, when changed, causes a change in something else (the dependent variable). In a simple model, climate would be the dependent variable. You could plot volcanic activity over time and compare it to temperature (an aspect of climate that can be measured) over time. Temperature changes that follow volcanic events would allow you to make inferences about the effects of eruptions on climate. However, the Earth system is complex. Records of climate and volcanic activity are imperfect. Some volcanic products should warm the atmosphere (carbon dioxide, a greenhouse gas). Others should cool the climate (dust, which reduces sunlight). The task of understanding climate change is obviously very complicated. The evidence at hand, however, suggests that major volcanic eruptions can lower the average temperature of the Earth's surface by a few tenths of a degree Celsius for as long as a few years.

It is often thought that volcanic eruptions increase or cause rainfall near or downwind of the eruption. Volcanoes put dust into the air. Water droplets in clouds form around small dust particles. Eruptions can also heat the local atmosphere. This should increase convection, or circulation, of the atmosphere. Finally, some volcanic eruptions release great quantities of water vapor needed to form clouds and rain. However, a number of studies show that an increase in rainfall is rare after an eruption. The major eruption of Krakatoa in 1883 did not increase rainfall, and it occurred during the wet (monsoon) season. It seems that conditions in the atmosphere near a volcanic eruption have to be just right for rainfall to increase just because of the eruption.

Enormous quantities of sulfur dioxide gas from a volcanic eruption can be put all the way into the stratosphere (the upper layer of the atmosphere, above the weather). It then slowly reacts with water to form tiny droplets of sulfuric acid, less than a thousandth of a millimeter in diameter. Unlike in the troposphere (the lowest layer of the atmosphere), these sulfur dioxide droplets are not affected by the water cycle. They stay suspended in the stratosphere for as long as a few years. The sulfur dioxide droplets, as well as the large quantities of very fine volcanic ash particles that also reach the stratosphere during major volcanic eruptions, reflect sunlight and are thought to cause the global cooling that is often observed for a few years after a major volcanic eruption. For example, following the eruption of Tambora in Indonesia in 1815, many areas in the United States and Canada had unusually cold summer weather. In New England, 1815 was called the "year without a summer."

Check Your Understanding

1. What gases escape from volcanoes?
2. Why does the emission of carbon dioxide pose a threat near volcanic eruptions?
3. How are volcanoes connected to the water cycle?
4. a) How is acid rain formed?

   b) Are volcanoes the only source of acid rain?
5. Do volcanic eruptions increase or decrease the temperature of the Earth? Explain your answer.

Understanding and Applying What You Have Learned

1. Think about the air you are breathing. How much of it came from some distant volcano?
2. If a volcano erupted huge amounts of ash, would you expect global temperatures to go up or down? Why?
3. If warm air rises, why would hot gases from a volcano be a threat to people in the valley below? (Hint: think about volume's effect in your work with the lava flow lab.)
4. If a system consists of many parts that affect each other, how are volcanoes part of systems on Earth?

Preparing for the Chapter Challenge

Use the information in this activity to argue that volcanoes have affected virtually every community. Consider ways in which you can include these arguments in your story line.

Inquiring Further

1. **Cascades eruptions**

   Examine the figure showing the eruptions of Cascade volcanoes during the last 4000 years.

   a) Which volcano has been most active? Which volcano has been least active? Explain.

   b) What three volcanoes do you think are most likely to erupt next?

   c) Visit the AGI *EarthComm* web site for a link to the USGS Cascades Volcano Observatory web site. Find out about their monitoring efforts.

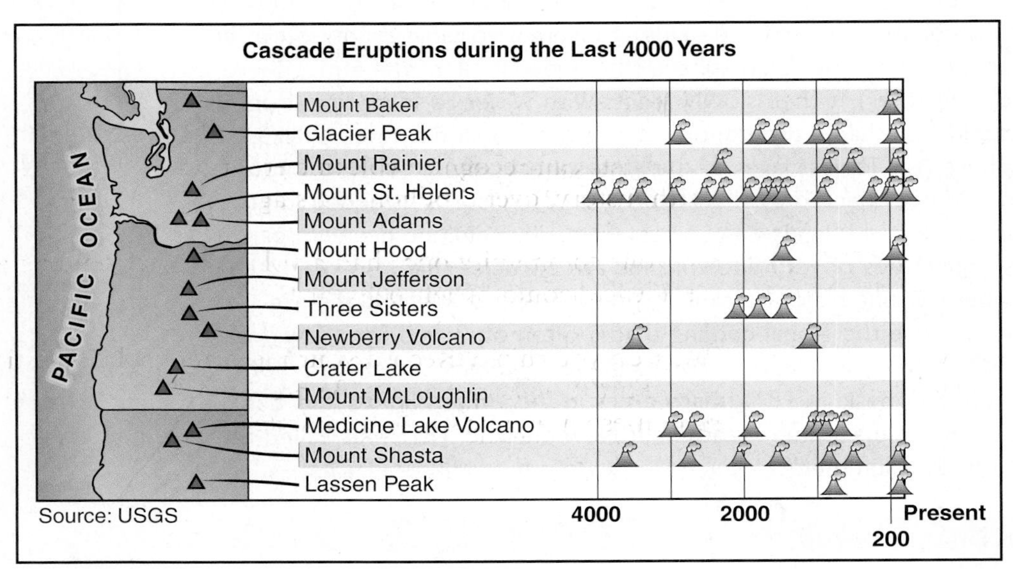

Activity 6 Volcanic History of Your Community

Goals

In this activity you will:

- Demonstrate awareness of the knowledge used to construct geologic maps.
- Examine and identify several common igneous rocks.
- Identify the distribution of active volcanoes on one map and rock types on a given geologic map
- Recognize that volcanic rocks indicate a history of volcanic activity.

Think about It

Geologists can recognize volcanic eruptions dating back to early Earth history, over 3 billion years ago.

- If your community once had a volcanic eruption long ago, what evidence would be left behind?

What do you think? Record your ideas about this question in your *EarthComm* notebook. Be prepared to discuss your responses with your small group and the class.

Investigate

Part A: Rock Samples

1. Examine a set of rocks collected near a volcano.

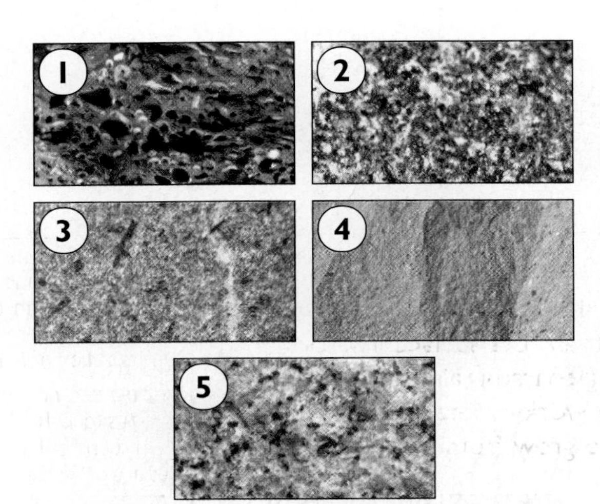

   a) Record the sample number and briefly describe the physical characteristics of each rock.

   b) Sort them into different groups by color (light, medium, dark).

   c) Sort them again by texture (fine grains, large grains, mixture of sizes of grains, bubble holes).

2. Examine a set of rocks collected in your community.

   a) Observe each rock, noting the size of the grains, color, and any other distinguishing features. Briefly describe each rock.

   b) Note which ones you think might be volcanic.

   c) Are all the rocks in the collection originally local? Explain.

Soft rock material can come off rock samples as grit. Avoid contact with eyes.

Part B: Geologic Maps

1. Examine the geologic map of your state or region. Find your community on the map. An example of a geologic map is shown.

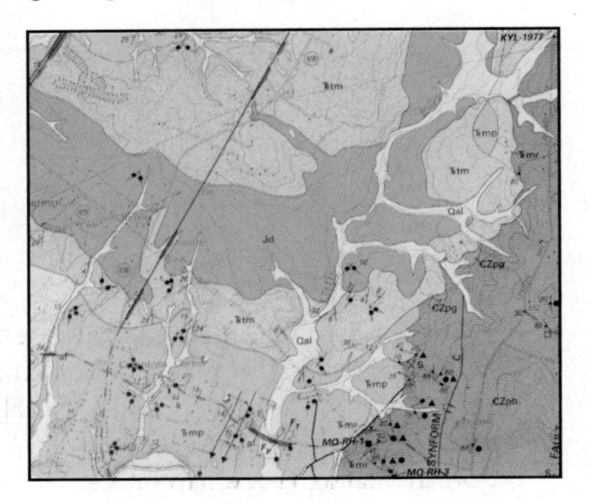

2. Look for patterns in the distribution of map colors (different rock types).

3. Compare the names of the rocks in the formations on the map with the names of rocks found in a reference book or field guide. Formation names followed by Basalt, Andesite, Dacite, Rhyolite, Ash, or Tuff indicate volcanic origin. Also, use any rock columns provided with the map and look for symbols that indicate volcanic rocks.

   a) Record in your notebook the geologic age of any volcanic rocks you find on the map.

4. Using a measuring device and the map scale, determine how far your community is from the nearest volcanic rock unit.

   a) Record this distance in your notebook.

Reflecting on the Activity and the Challenge

You have seen evidence that rocks formed by volcanic eruptions look different from those formed in other ways. They also vary greatly in their chemical composition. You can try to use what you have learned about the ages of the rocks in your area to decide on the probability of a volcano erupting in your community. Consider how you can share this knowledge with your audience.

Digging Deeper

IGNEOUS ROCKS

Introduction

Igneous rocks crystallize from cooling magma and lavas. Some igneous rocks form from magma that has cooled slowly beneath Earth's surface. The slow cooling allows crystals to form and grow, yielding coarse-grained igneous rocks such as granite. We know from laboratory work on melted rock that it takes extremely long times for large crystals to grow from a cooling magma. Such slow cooling can happen only deep within the Earth. These **intrusive igneous rocks**, also called **plutonic igneous rocks**, are made of crystals large enough to be seen with the naked eye. The sizes and shapes of bodies of intrusive igneous rocks vary greatly, from human scale to whole mountain ranges. Because intrusive igneous rocks form underground, they can only be seen where uplift and erosion have removed the overlying rocks.

Other igneous rocks, called **extrusive igneous rocks**, form from magma that is brought to the Earth's surface. Magma at or near the Earth's surface cools more rapidly, and crystals do not have time to grow to a large size. Sometimes, lava cools so fast that no crystals have a chance to form. Instead, the lava forms a kind of glass, called obsidian, as shown in *Figure 3*. Extrusive rocks are only one kind of volcanic rock. Remember from an earlier activity that pyroclastic volcanic rocks are also important.

Figure 1 Granite is one kind of intrusive igneous rock formed by slow cooling of magma below Earth's surface.

Geo Words

intrusive igneous rock (plutonic igneous rock): igneous rock formed at considerable depth by the crystallization of magma.

extrusive igneous rock: an igneous rock that has formed by eruption of lava onto the surface of the Earth.

Usually, the color of igneous rocks reflects the composition of the magma from which they form. Rocks from magmas high in silica (rhyolite and granite) tend to be lighter in color because their minerals are lighter in color, such as quartz, muscovite, and feldspar. These minerals are relatively poor in magnesium and iron, which are chemical elements that tend to make minerals dark.

Figure 2 Scoria is an extrusive igneous rock with a frothy texture.

Igneous rocks from magmas low in silica (basalt and gabbro) are darker in color, because they contain a large percentage of dark-colored minerals like pyroxenes, amphiboles, and dark micas. These minerals are relatively rich in magnesium and iron.

Figure 3 Obsidian is an extrusive igneous rock with a glassy texture.

Volcanic rocks of intermediate composition are mixtures of light and dark minerals that give the rock an intermediate color. A good example is andesite. It is named after the Andes Mountains, a mountain chain in South America that has many volcanoes.

In summary, igneous rocks crystallize from melted rock. They are divided into two types:

- Intrusive (plutonic) rocks are coarse-grained (>1 mm) and composed of crystals large enough to be seen with the naked eye. This implies slow cooling at depth.
- Extrusive (volcanic) rocks are fine-grained (<1 mm) and are composed mainly of crystals that are too small to be seen without a magnifying glass. This implies rapid cooling at or near the surface. Some may even be glassy, and many are filled with bubble holes, called vesicles.

Classification of Igneous Rocks

Color		Light	Intermediate	Dark	Dark
Mineral composition		quartz (≥5%) plagioclase feldspar potassium feldspar iron-magnesium rich minerals (≤15%)	quartz (<5%) plagioclase feldspar potassium feldspar iron-magnesium rich minerals (15-40%)	no quartz plagioclase feldspar (~50%) no potassium feldspar iron-magnesium rich minerals (~40%)	nearly 100% iron magnesium rich minerals
Texture	Crystals >10 mm	granite pegmatite	diorite pegmatite	gabbro pegmatite	
Texture	Crystals 1–10 mm	granite	diorite	gabbro	peridotite
Texture	Crystals <1 mm	rhyolite	andesite	basalt	
Texture	Glassy	obsidian		obsidian	
Texture	Frothy	pumice		scoria	

Figure 4

Rocks with Interesting Textures

Pumice, shown in *Figure 5*, feels like sandpaper. It is often light enough to float in water. Pumice forms when the gases inside the lava effervesce (bubble off) as pressure is released, just the way bubbles form when a can of soda, under pressure, is opened. The lava cools into a rock that is mostly tiny holes, with only very thin walls of rock between the holes. Much of the material blown out at first around Mt. St. Helens was pumice.

Figure 5 Pumice sample from Mt. St. Helens.

Obsidian is a glassy rock that was cooled so quickly from lava that crystals did not form. A broken surface of obsidian is usually smooth and shiny, just like a broken piece of glass (which it is!).

Tuff is a rock composed of pyroclastic material deposited from explosive eruptions. Some tuffs consist of ash that was put into the atmosphere by the eruption and then fell back to the surface downwind of the volcano. These tuffs may be lightweight, soft, and fragile, if they have not been buried deeply by later deposits. Other tuffs form from pyroclastic flows. These tuffs are often welded tightly together by the heat of the pyroclastic flow.

Scoria is a crusty-looking rock filled with holes made by gas bubbles trying to escape from the lava as it solidified. It is usually red or black, depending on the degree of oxidation of iron. Scoria is heavier to lift than pumice, because it doesn't have as great an abundance of holes. If you get to visit a cinder cone that is being mined, look at the cinders in the center and compare them to the cinders on the flank. The central cinders will be more oxidized and red and the outer samples darker and more like the basalt associated with the cone.

Types of Volcanic Rocks (based on silica content)

Basalt is dark and fine-grained. It is a low-silica rock. It is the most common rock on Earth and makes up the floors of all the oceans. Basalt covers about 70 percent of the surface of Earth. Basalt underlies the sediments on the floors of all the deep oceans of the Earth. Because the deep oceans occupy about 60 percent of the Earth's surface, basalt is the most common rock near the Earth's surface. On the continents, however, sediments and sedimentary rocks are much more common than basalt near the Earth's surface.

Andesite is a rock that is intermediate in silica content. It is also intermediate in color, typically gray, green, or brown. Andesite is typical of volcanic rocks from around the "Ring of Fire" in the Pacific Ocean. The ash from Mt. St. Helens was andesitic.

Dacite is a silica-rich rock that is between a rhyolite and an andesite in composition, somewhat like the difference between blue and violet in the color spectrum. It is lighter in color but not quite as silica rich as a rhyolite. It, too, is fine-grained but may have some visible crystals.

Rhyolite is usually light in color and has a fine-grained background. It forms from viscous lava that is high in silica. One of the identifying aspects of rhyolite is that it has visible quartz crystals that are not often seen in other volcanic rocks. These quartz crystals form because of (and are evidence of) rhyolite's high silica content. Perhaps the most famous rhyolite is the rock found in Yellowstone National Park. It came from one of the most massive volcanic eruptions known in Earth history.

Check Your Understanding

1. How can you distinguish intrusive igneous rock from extrusive igneous rock?
2. What is the difference between the way intrusive and extrusive igneous rock is formed?
3. Which chemical elements tend to make minerals dark?

Understanding and Applying What You Have Learned

1. Examine the sketch of the volcano. Some rocks shown on the sketch are extrusive, some are intrusive, and others are not igneous. Label a copy of the sketch to point out the following:
   a) Two locations of intrusive igneous rocks.
   b) Two locations of extrusive igneous rocks.
2. Consider what you know about volcanic hazards.
   a) Why is it important to study rocks?
   b) What information can be learned by looking at the geologic evidence in your community?

Cross section of a composite volcano. Rocks that form from the material shown in this diagram include extrusive igneous (erupted out of the volcano), intrusive igneous (cooled and crystallized inside the volcano or elsewhere beneath the surface), and non-igneous (metamorphic or sedimentary rock).

Preparing for the Chapter Challenge

You have now completed a variety of studies about volcanoes and volcanic rocks. Think about what you have learned about the relationships between rock types and volcanoes. Use the information you have learned about the distribution of rocks in your region to write a paragraph about the potential for a volcano to affect your community. Design a creative way to include geologic evidence about past eruptions in your movie script.

Inquiring Further

1. **Volcanoes as natural resources**

   Volcanoes provide energy to some communities. The Geothermal Education Office explains how heat associated with volcanic activity can be converted into electrical energy.

   Volcanoes also provide materials that are used for a variety of applications. Visit the Volcano World web site and compile a list of metals and other materials mined from volcanoes.

   Consult the AGI *EarthComm* web site for current addresses.

2. **Simulating gases in igneous rock**

   How does all the gas in pumice form?Acquire some bread dough (from the store or make from a recipe). Note the characteristics of the dough. Cut an end off to examine the interior. Set the dough out for a while and watch what happens. If you bake it, look at the interior again. It is not exactly the same as a volcano because the bubble holes in lava come from gases evolving out of solution, but it will give you the idea of how the gases are formed.

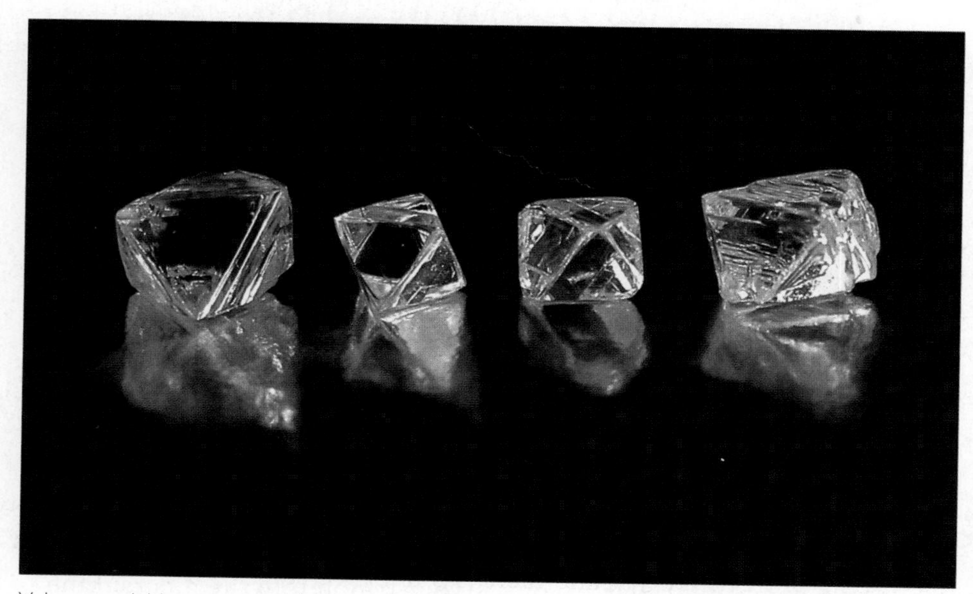

Volcanoes yield many materials that we use. Diamonds are just one example. Can you think of any others?

Activity 7 Monitoring Active Volcanoes

Goals

In this activity you will:

- Understand some of the changes that occur prior to volcanic eruptions.
- Describe volcanic monitoring systems.
- Design and build an instrument to monitor changes which occur prior to volcanic eruptions.

Think about It

Many volcanologists (geoscientists who study volcanoes) have been killed while observing active volcanoes, when the activity of the volcano increases unexpectedly.

- How would you be able to tell that a volcano was about to erupt?

What do you think? Record your ideas about this question in your *EarthComm* notebook. Be prepared to discuss your responses with your small group and the class.

Investigate

1. Read the report issued from the Montserrat Volcano Observatory. The report was released one day before an eruption.

2. Identify all the evidence that signaled an impending eruption.

   a) In your notebook, write down each kind of event that signals volcanic activity.

Montserrat Volcano Observatory Daily Report 3/25/99 to 3/26/99

A slight change in the nature and level of seismicity was observed during the period under review. There were nineteen (19) earthquakes, nineteen (19) rockfalls and three (3) regional events. An increase in small earthquakes that are thought to be associated with rock fracturing due to dome growth was noted. About one hundred and thirty-four (134) of these small events were recorded on the Gages and Chances Peak Seismographs. The southern and eastern EDM [electronic distance measurement] triangles were measured today. The very small changes in the elevation of the land measured were consistent with the recent trends. Visual observations were made early this morning from the helicopter in excellent viewing conditions and subsequently from Chances Peak. The dome continues to steam and vent gas from various locations. The focus of activity has shifted from the two previously active areas (near Farrell's and to the north of Castle Peak) to two new areas located on the eastern and western sectors of the dome. Rockfalls may continue to occur in the area north of Castle Peak. The rockfalls in this area have been the source of the recent ash clouds.

3. As a group, select one of the changes that occurs prior to an eruption. Discuss how you might design an instrument (or a model of an instrument) that would monitor the change. The instrument should be one that you could transport to an observation site. You should also be able to construct it from readily available materials. To learn more about volcano monitoring techniques, read the **Digging Deeper** section: "Volcano Monitoring."
   a) Sketch your design in your notebook.
   b) Label the parts of the instrument and list the materials from which it will be built.
4. Write a brief manual for the instrument. Be sure to specify the following:
   a) What the instrument does.
   b) Where on or near the volcano to place it.
   c) How to position it on the site.
   d) How to operate it.
   e) How to record observations.
5. Exchange your manual with another group and construct the instrument they have designed. If possible, take it to a test site and follow the procedures to take and record observations.

Reflecting on the Activity and the Challenge

Understanding how volcanoes are monitored can help you appreciate how scientists strive to keep people informed and safe from volcanic hazards. You looked at the aspects of the design of instruments that made them easy to understand, construct, and operate as well as the relative costs and savings associated with developing and maintaining monitoring systems. You can now share with your audience the general advantages and disadvantages of different monitoring systems.

Perhaps you could have a scene in your movie showing a debate in which a town council must decide whether or not to purchase monitoring equipment.

Geo Words

seismology: the study of earthquakes and of the structure of the Earth.

Digging Deeper

VOLCANO MONITORING

The technology that geologists use is likely to be more sophisticated and expensive than the instruments you designed and constructed. However, they follow the same principles and observe the same changes. In emergencies, particularly in remote regions or in underdeveloped countries, simple monitoring techniques may provide the response time needed to save lives and property. Since the United States Geological Survey (USGS) cannot afford to monitor all volcanoes, on-site and local monitoring information can provide clues to the need for more sophisticated equipment.

Only 25% of the world's active volcanoes are monitored. Most potentially active volcanoes are not monitored due to a lack of funding. In response to this problem, the USGS developed a mobile volcano monitoring system. This system allows scientists to quickly install monitoring equipment and assess the potential hazard when a volcano becomes restless. The USGS sent this system to Mt. Pinatubo, Philippines in 1991 to help the Philippine Institute of Volcanology and Seismology monitor their volcano. (**Seismology** is the study of earthquakes and the structure of the Earth.) The lessons learned from monitoring volcanoes around the world help USGS scientists further prepare for eruptions in the United States. This system is not capable of stopping nor can it be designed to stop any natural occurrence. Its purpose is to provide forecasts to local populations and agencies for the health and welfare of their communities. Everyone must realize that the forces involved in a volcanic eruption cannot be stopped and must be allowed to take their course.

Figure 1 Many potentially active volcanoes around the world are not monitored due to a lack of funding.

Mt. Pinatubo erupted cataclysmically in June 1991. Fortunately for the residents of the island, monitoring prevented property losses of at least $250 million and saved an estimated 5000 lives. While 300 lives were lost and some $50 million were spent preparing for and monitoring the eruption, the estimated savings in human life far outweighed the costs associated with the monitoring program.

Volcano monitoring involves the recording and analyses of measurable phenomena such as ground movements, earthquakes, variations in gas compositions, and deviations in local electrical and magnetic fields that reflect pressure and stresses induced by the subterranean magma movements. To date, monitoring of earthquakes and ground deformation before, during, and following eruptions has provided the most reliable criteria in predicting volcanic activity, although other geochemical and geophysical techniques hold great promise.

Most of the commonly used monitoring methods were largely pioneered and developed by the Hawaiian Volcano Observatory (HVO). The HVO began monitoring in 1912. It has been operated continuously by the U.S. Geological Survey since 1948. Years of continuous observations of Kilauea and Mauna Loa, two of the world's most active volcanoes, have led to new instruments and measurement techniques. This work has increasingly been used in the study of other active volcanoes the world over. Moreover, early major advances in seismic research at HVO contributed significantly to subsequent systematic investigations of earthquake and related crustal processes, now conducted as part of the U.S. Geological Survey's Earthquake Hazards Reduction Program.

The volcanic plumbing and reservoir system beneath Kilauea can be pictured schematically as a balloon buried under thin layers of sand and plaster. When magma is fed into the reservoir (analogous to air filling a balloon), the internal pressure increases, and the sand-plaster surface layers are pushed upward and outward in order to accommodate the swelling or inflation. The net effects of such inflation include the steepening of slope of the volcano's surface; increases in horizontal and vertical distances between points on the

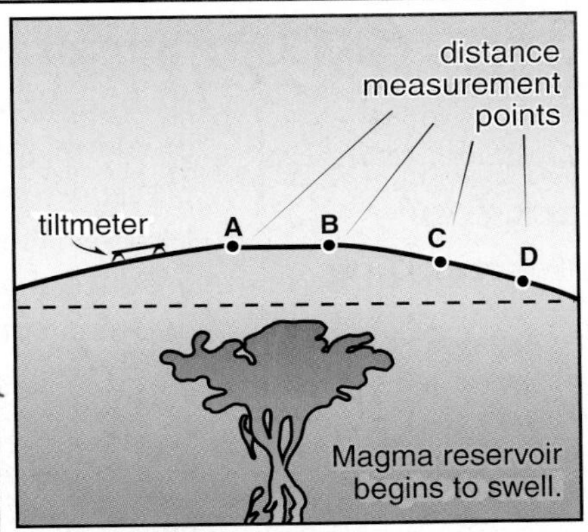

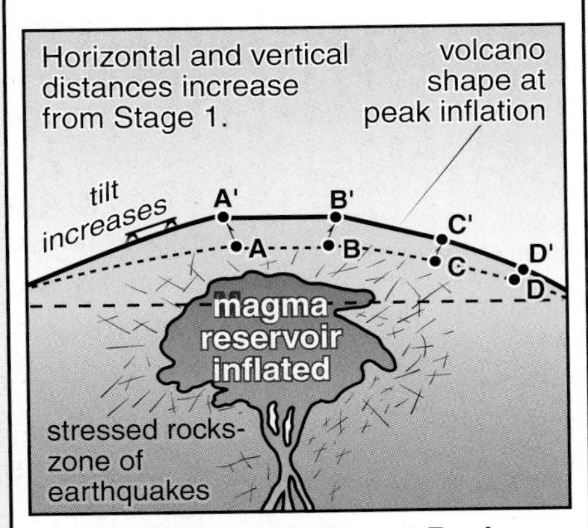

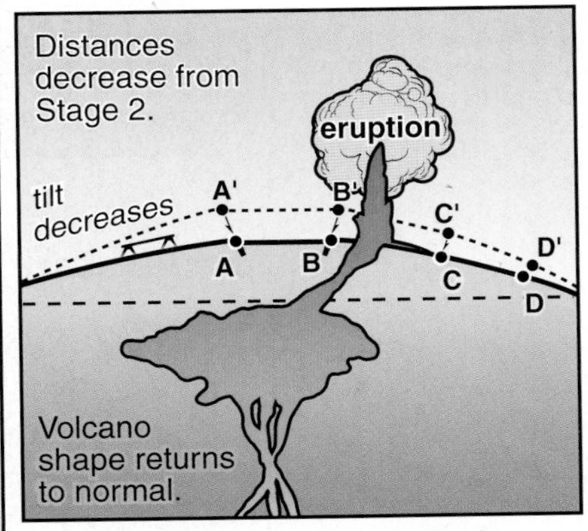

Figure 2 A schematic representation of changes in a volcano prior to and following an eruption.

surface; and, in places, the fracturing of rock layers stretched beyond the breaking point. Such rupturing of materials adjusting to magma-movement pressures results in earthquakes. A shrinking or rapidly draining reservoir to feed surface eruptions (analogous to deflating or popping the balloon) would produce the opposite effects: flattening of slopes, reduction in distances between surface points, and decrease in earthquake frequency.

Changes in slope in a real volcano can be measured precisely by various electronic mechanical "tiltmeters" or field tilt surveying techniques. Tiltmeters can detect the change in slope of a kilometer-long board raised by the thickness of a dime placed under one end. Similarly, an instrument that uses a laser beam can measure minute changes in horizontal distances. Tiny changes in vertical distances can be measured by making a series of precise leveling surveys. Such changes can be easily detected to a precision of only a few parts per million. The notion of one part per million can be compared to putting one drop of Kool-Aid® in 16 gallons of water. The frequency, location, and magnitude of earthquakes generated by magma movement can easily and accurately be determined by data obtained from a properly designed seismic network. The Hawaiian Volcano Observatory has recently expanded its seismic networks to 45 stations to continuously record the earthquake activity of Kilauea and Mauna Loa.

The monitoring of earthquakes and changes in volcano shape is not sufficient for predicting eruptions, however. The proper analysis of data requires a basic understanding of the prehistoric eruptive record and behavior of a volcano. Such glimpses into a volcano's prehistoric past are critically important because historic records extend over too short a time to permit the making of reliable predictions of future behavior.

Check Your Understanding

1. Can monitoring equipment be used to prevent a volcano from erupting? In your answer explain what monitoring equipment can do.
2. What does monitoring a volcano involve?
3. What external evidence indicates that a reservoir in a volcano is being "fed" with magma?
4. Is a tiltmeter a sensitive measuring device? Explain your answer.

Understanding and Applying What You Have Learned

1. Use the following questions to analyze your experiences designing, devising written plans, understanding the plans of other groups, and using monitoring instruments. In your notebook, record your responses.
   a) What aspect of your design would have worked well?
   b) What aspect of another group's design would have worked well?
   c) How useful do you think such devices would be in an emergency where expensive monitoring instruments were not readily available?
   d) How does elevation and slope of a volcano change prior to an eruption?

e) How could you measure changes in the elevation of a volcanic dome?
f) What changes in volcanic gases indicate an ensuing eruption?
g) How does seismic activity change prior to an eruption?
h) Did the class design at least one piece of equipment to measure change in slope? Elevation? Gases? Seismicity?
i) If you answered no to one of the questions in (h), think of two potential reasons why the instrument was not developed.

2. Describe three methods of monitoring an active volcano.
 a) Which monitoring system is least reliable, and why?
 b) Which monitoring systems are most dangerous to those who monitor an active volcano, and why?
 c) Which monitoring systems would be most useful in your community, and why?
 d) What aspects of volcano monitoring are most challenging for geologists? For communities? Explain your answers.

3. Why is weather monitored more closely than volcanic activity in the United States?

Preparing for the Chapter Challenge

Summarize the monitoring systems you would recommend for your community. Include a justification of their cost in relationship to the hazard information they might provide. Also, indicate the type of monitoring equipment you would recommend be used in the community in which you "stage" a volcanic eruption. Use this summary in your response to the **Chapter Challenge**.

Inquiring Further

1. **Volcanic hazards of Mt. Rainier**

   View the movie *Perilous Beauty, The Hidden Dangers of Mount Rainier*, U.S. Department of the Interior, USGS. This movie is the result of in-depth research into the volcanic hazards of Mt. Rainier.

   Describe how the movie relates to what you have studied about volcanic hazards and volcano monitoring systems in this chapter.

Earth Science at Work

ATMOSPHERE: *Air Traffic Controller*
Air traffic controllers are responsible for the safe passage of aircraft in their airspace.

BIOSPHERE: *Produce Manager*
Customers expect to find an ample supply of fresh produce throughout the year.

CRYOSPHERE: *Geohydrologist*
Some geologists are involved in the study of global water, its properties, circulation, and distribution. Mountain rivers and streams would be one area that they would investigate.

GEOSPHERE: *Commodities Supplier*
Some industries rely on an ample supply of sulfur in their manufacturing processes.

HYDROSPHERE: *Fisheries Worker*
In the fishing industry, income depends on the quality and the quantity of the catch.

How is each person's work related to the Earth system, and to volcanoes?

2

Plate Tectonics... and Your Community

Plate Tectonics ...and Your Community

Getting Started

Plate tectonics is a scientific theory. A scientific theory is a well-tested concept that is supported by experimental or observational evidence. It explains a wide range of observations. The theory of plate tectonics explains the formation, movement, and changes of the outer, rigid layer of the Earth, called the lithosphere.

- How do you think the Earth's lithosphere formed?
- How do you know that the Earth's lithosphere moves?
- List some ways that the Earth's lithosphere changes.

What do you think? Write down your ideas about these questions in your notebook. Be prepared to discuss your ideas with your small group and the class.

Scenario

A middle-school science teacher in your community has been asked to teach a unit on plate tectonics to her class and has asked for your help. To her students, the fact that they are riding on Earth on a floating plate makes about as much sense as riding on a magic carpet. The students feel that because plate tectonics is "just a theory" that scientists have "made up," there is no need for them to learn about it, or try to understand it. Also, they believe it is totally irrelevant to their lives.

The teacher hopes that her students might be more likely to listen to local high school students. She would like you to convince her class of the significance of plate tectonics in their lives. She wants you to explain to her class the key science concepts behind the theory. She thinks that it would help her students if they could see how research, evidence, modeling, and technology support plate tectonics.

Can the EarthComm students meet this challenge and help a middle-school class to understand why the theory of plate tectonics is important to learn about and understand?

Chapter Challenge

Think about how you can use the theory of plate tectonics to help middle-school students understand scientific theories. You might want to prepare a PowerPoint™ presentation, a web page, or a three-panel poster display. Your project will need to address:

- Evidence for movement and changes in the geosphere over time.
- The flow of matter and energy in the Earth.
- The nature of the Earth's interior.
- How plate tectonics accounts for the features and processes that geoscientists observe in the Earth.
- How the theory of plate tectonics developed.

Assessment Criteria

Think about what you have been asked to do. Scan ahead through the chapter activities to see how they might help you meet your challenge. Work with your classmates and your teachers to define the criteria for judging your work. Record all this information. Make sure that you understand the criteria as well as you can before you begin. Your teacher may provide you with a sample rubric to help you get started.

Activity 1 Taking a Ride on a Lithospheric Plate

Goals

In this activity you will:

- Determine the direction and rate of movement of positions within the plate on which your community is located, using data from the Global Positioning System and a computer model.
- Predict the position of your community in the near future, and "retrodict" its position in the recent past, by extrapolating from data already collected.
- Recognize that the rate and direction of plate motion is not necessarily constant.
- Describe several lines of evidence for plate motion.

Think about It

The motion of anything (you, your automobile, a lithospheric plate, or the Milky Way Galaxy!) has to be described in relation to something else.

- How can you locate your position on the Earth's surface?
- How would you be able to determine whether your position on the Earth has moved?

What do you think? Record your ideas about these questions in your *EarthComm* notebook. Include sketches as necessary. Be prepared to discuss your responses with your small group and the class.

Investigate

Part A: Data from the Global Positioning System

1. Data from Global Positioning System (GPS) satellites will help you find out if the position of your community has changed over time. The map shows measurements of movements at GPS recording stations in North America. Each station has a four-character symbol. Arrows show the rate and direction of motion of the Earth's surface at that station. Longer arrows indicate faster motion than shorter arrows. The motions shown are relative to the GPS frame of reference, which you can think of as "attached" to the Earth's axis of rotation.

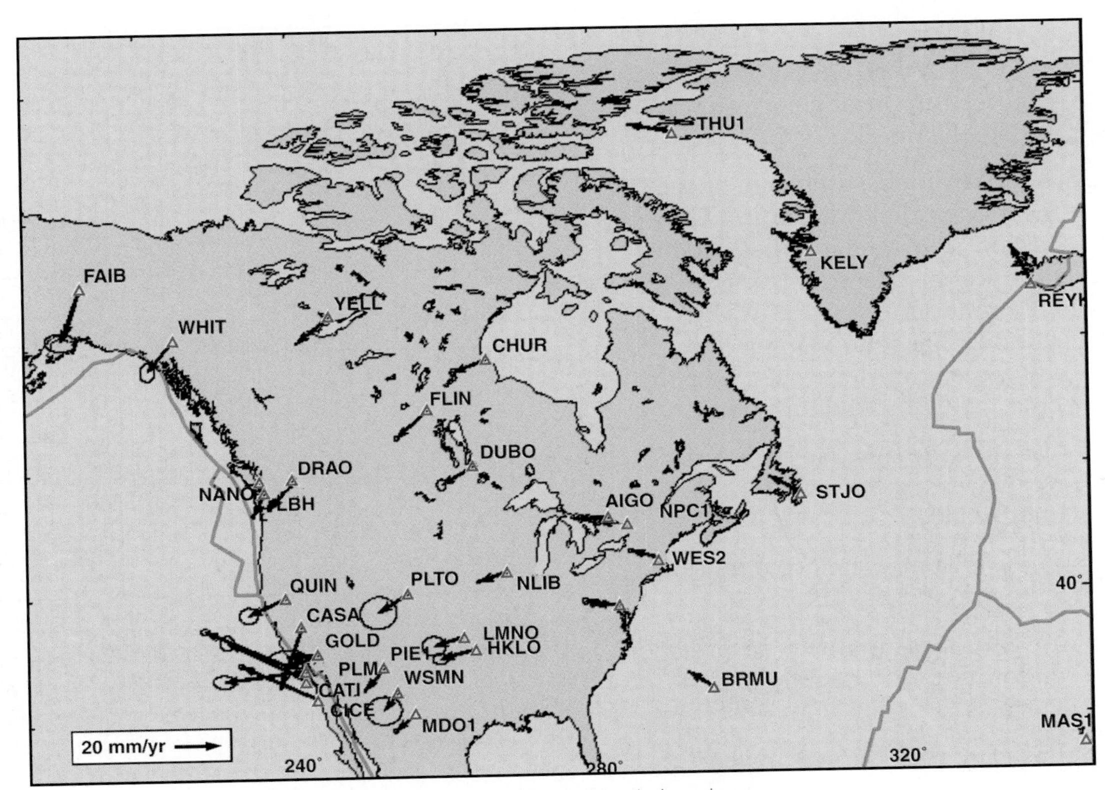

Measurements of movements at GPS recording stations in North America.

a) Find the WES2 station (in the northeastern United States). How do you know that the WES2 station has moved over time?

b) In what compass direction is the WES2 station moving? Be specific.

c) The arrow in the lower left corner of the map is a scale. It shows the length of a "20 mm/yr" arrow. Is the WES2 station moving more than or less than 20 mm/yr? Explain.

d) Are all stations on the map moving at the same speed? Explain.

e) Are all stations on the map moving in the same direction? Explain.

f) What is the general or average direction of movement of North America?

2. A series of measurements of the location and elevation of a GPS station over time is called a GPS time series. The graph shows the GPS time series for the WES2 station. The solid sloping lines on the three graphs are the "best-fit" lines through the data points.

Use the map and the time series to answer the following questions:

a) How many years of data does the time series show?

b) Were measurements recorded continuously or only at certain times? Explain your answer.

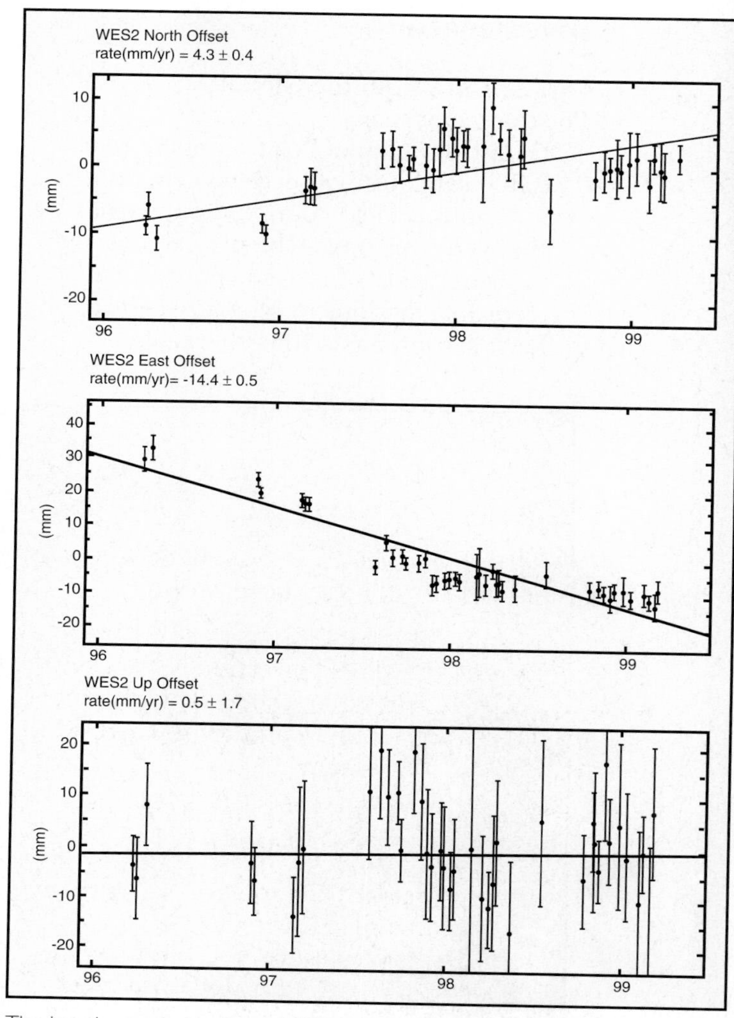

The location and elevation of GPS station WES2 over time. The vertical lines above and below each point are called "error bars". They show the uncertainty in the measurement. They tell you that the real value might lie anywhere within the error bar.

3. The top graph shows movement of the station to the north or south. Northward movement is indicated by positive values, and southward movement is shown by negative values. Find the calculation above the top graph. How many millimeters per year did WES2 move? Convert this value to centimeters per year. In which direction did it move?

a) Record the rate (in cm/yr) and the direction of motion in your notebook.

4. The middle graph shows movement to the east or west. Eastward movement is indicated by positive values, and westward movement is shown by negative values. Find the calculation above the middle graph. How many millimeters per year did WES2 move? Convert this value to centimeters per year. In which direction did it move?
   a) Record the rate (in cm/yr) and direction of motion in your notebook.
5. The bottom graph shows the movement up or down.
   a) Has the WES2 station always stayed at the same elevation? Explain.
6. Do the speed and direction of motion of WES2 shown in the graphs match the direction and length of the arrow shown on the map?
7. Look at the "best-fit" line in the top and middle graphs.
   a) Did the WES2 station move at a constant speed since 1996? Explain your answer.
   b) What additional data would you need to decide whether the differences between the measured data points and the best-fit straight line are due to the overall motion of the plate or are caused by processes in the local area around the WES2 station?
8. Obtain a GPS time series for a station nearest your community.
   a) Record the directions and rates of motion in cm/yr for the station nearest your community.

Part B. Data from a Computer Model

1. Computer models that use geologic data also provide information about the changes in position of your community over time. To use this model, you will need to know the latitude and longitude of your school in decimal format.

   Find your school (or another familiar place) on your local topographic map. Latitude and longitude are used to identify a position on the Earth's surface. Latitude is a measure of location in degrees, minutes, and seconds north or south of the equator. Therefore, it is found on the left or right side of the map. Longitude is a measure of location in degrees, minutes, and seconds east or west of the Prime Meridian, which passes through Greenwich, England. Therefore, it is found on the top or bottom of the map.
   a) Record the latitude and longitude of the position you chose in degrees, minutes, and seconds. (These "minutes" and "seconds" are *not* the same as the familiar minutes and seconds of time! They describe positions on a circular arc.)
   b) Convert the latitude and longitude values to a decimal format. An example for you to follow is provided on the following page.

Example:

42° (degrees) 40' (minutes) 30" (seconds) north latitude
Each minute has 60".
30" divided by 60" equals 0.5'.
This gives a latitude of 42° 40.5' north.
Each degree has 60'.
40.5' divided by 60' equals 0.675°.
The latitude in decimal format is 42.675° north.

2. Obtain a world outline map showing lithospheric plates, similar to the one shown.

   a) Place a dot on the map to represent the location of your community and label it with an abbreviation.

   b) In your notebook, record the name of the plate your community lies within. Record the name of a plate next to your community.

3. Visit the Relative Plate Motion (RPM) Calculator web site. (See the *EarthComm* web site for more information.) The RPM calculator determines how fast your plate is moving relative to another plate that is assumed to be "fixed" (non-moving). At the web site, enter the following information:
   - The latitude and longitude of your community (decimal format).
   - The name of the plate on which your community is located.
   - The name of the "fixed" reference plate adjacent to your plate. Use the African Plate as the reference plate.

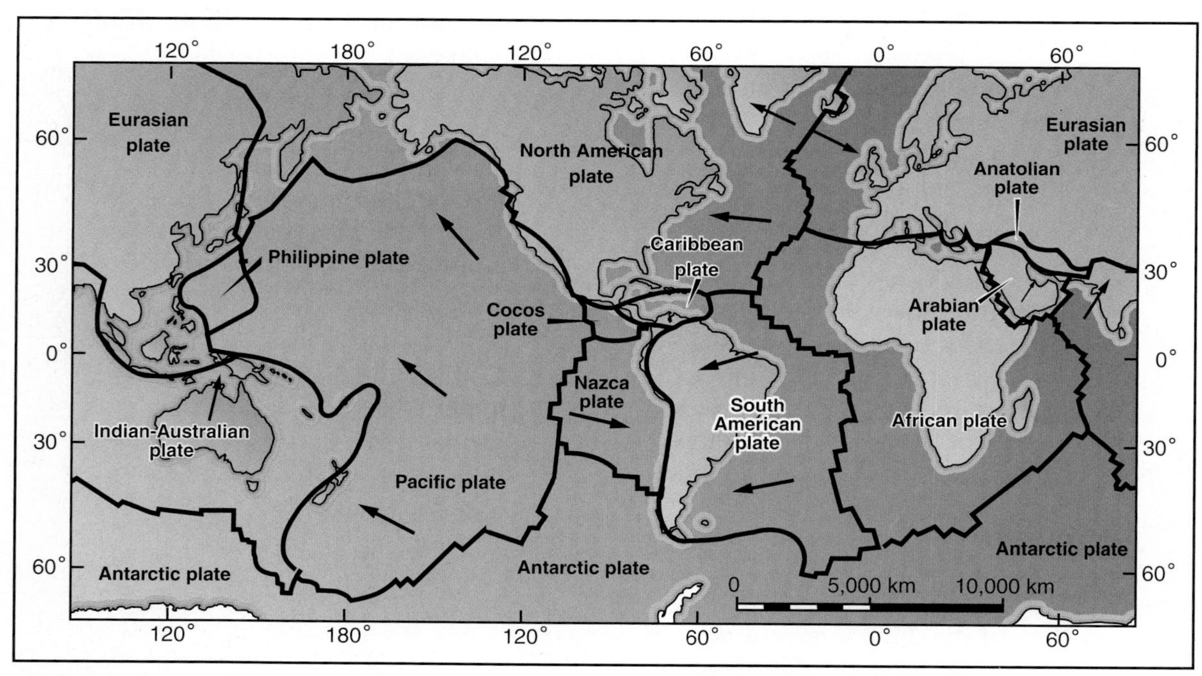

World map of major lithospheric plates. Arrows show the relative motions of the plates relative to the African Plate, which happens to be moving slowest relative to the Earth's axis of rotation.

Once you have entered the data, run the model. Print the results. (A sample printout for the location of station WES2 is provided for you.) Record the following information in your notebook:

a) The rate of movement of the plate on which your community is located (in centimeters per year).

b) Its direction of motion. (Note that direction is given in degrees, starting from 0°, clockwise from north. For example, 90° is directly east, 180° is directly south and 270° is directly west.)

c) In your own words, describe the motion of your plate over time.

d) How do the results from the computer model compare to those obtained from GPS data?

NUVEL-1A Calculation Results
Calculation results are as follows:

Relatively fixed plate = Africa
Relatively moving plate = North America
Latitude of Euler pole = 78.8 degree
Longitude of Euler pole = 38.3 degree
Angular velocity = 0.24 degree/million years
Latitude inputted = 42.364799 degree
Longitude inputted = –71.293503 degree
Velocity = 2.11 cm
Direction = 283.39 degree

e) What data does GPS provide that the plate motion calculator does not?

Reflecting on the Activity and the Challenge

So far, you have learned that when you use GPS data collected over time, you can find the speed and direction of movement of your plate. You can also use data from a computer model to show that your community is moving as part of the movement of a large plate. You are riding on a piece of the Earth's lithosphere! Your maps show how your community, along with the plate on which it rides, has moved over time. You can also predict its future position. You have now gathered some evidence to help you explain to the middle school students that their community is in fact moving.

Digging Deeper

MEASURING THE MOTION OF LITHOSPHERIC PLATES

The Interior Structure of the Earth

Refer to *Figure 1* as you read this section. The thin, outermost layer of the Earth is called the **crust**. There are two kinds of crust: continental and oceanic. The continental crust forms the Earth's continents. It is generally 30–50 km thick, and most of it is very old. Some continental crust has been dated as old as four billion years! The geological structure of the continental crust is generally very complicated, as you will learn in a later chapter. In contrast, the oceanic crust is only 5–10 km thick, and it is young in terms of geologic time. All of the oceanic crust on the Earth is younger than about 200 million years.

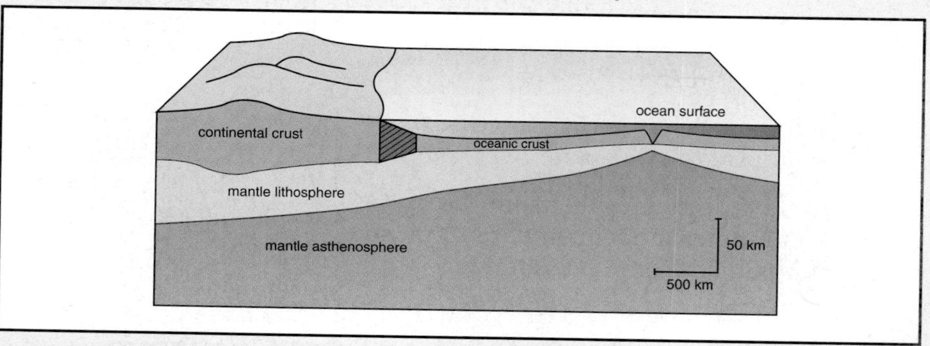

Figure 1 A schematic cross section through the outer part of the Earth. Note that the vertical and horizontal scales are very different. The diagram has a lot of "vertical exaggeration". If the diagram had been drawn without distortion, all of the layers would look much thinner. The boundary between continental crust and oceanic crust, shown by the shaded box, will be described later in the chapter.

Beneath the Earth's crust is the **mantle**. The rocks of the mantle are very different in composition from the crust, and the boundary between the crust and the mantle is sharp and well defined. The uppermost part of the mantle, which is cooler than below, moves as a rigid block, carrying the crust with it. The upper rigid part of the mantle, together with the crust, is called the **lithosphere**. The Earth's plates are composed of the lithosphere. At greater depths, the mantle is hot enough that it can flow very slowly, just like a very stiff liquid. That part of the mantle is called the **asthenosphere**.

In terms of composition and origin, the crust and the mantle are very different, but in terms of how they move, they behave in the same way. On the other hand, the lithosphere part of the mantle is the same in composition as the asthenosphere part of the mantle, but in terms of how they move, they behave very differently.

Geo Words

crust: the thin outermost layer of the Earth. Continental crust is relatively thick and mostly very old. Oceanic crust is relatively thin, and is always geologically very young.

mantle: the zone of the Earth below the crust and above the core. It is divided into the upper mantle and lower mantle with a transition zone between.

lithosphere: the outermost layer of the Earth, consisting of the Earth's crust and part of the upper mantle. The lithosphere behaves as a rigid layer, in contrast to the underlying asthenosphere.

asthenosphere: the part of the mantle beneath the lithosphere. The asthenosphere undergoes slow flow, rather than behaving as a rigid block, like the overlying lithosphere.

Measuring Plate Motions with GPS

The **Global Positioning System (GPS)** consists of 24 satellites that orbit the Earth at a height of 20,200 km. Receivers at stations on Earth (such as WES2 in Westport, Massachusetts) use the signals from satellites to calculate the location of the station. Geoscientists have set up a network of targets all over the world in order to monitor the movement of lithospheric plates. Steel spikes pounded into the ground (preferably embedded in solid rock) make up the targets, as shown in *Figure 2*. A high-precision GPS receiver is then mounted on a tripod and positioned directly above the target. The targets are revisited over a period of months or years. The receiver measures the distance to four or more GPS satellites and then uses stored data on satellite locations to compute the location of the target. Changes in horizontal and vertical positions can be detected within several millimeters.

Figure 2 A GPS receiver mounted in rock is used to measure changes in the elevation of this volcano.

GPS data collected at stations all over the world confirm that the surface of the Earth is moving. However, GPS time series show data for only the last several years. GPS is a new technology, and a global network of GPS stations has existed for less than a decade. How do we know that the surface of the Earth has been moving for a longer period of time? The answer to this question comes from the study of rocks.

Sea-Floor Spreading

The computer model at the Plate Motion Calculator web site uses several sources of geologic data. One source comes from the study of the magnetism of rocks that make up the sea floor. All magnets and

Geo Words

Global Positioning System (GPS): a satellite-based system for accurate location of points on the Earth.

materials that have magnetism have a north and south direction, or magnetic polarity. In the middle of the 20th century, geoscientists noted that they could group rocks by their magnetic polarity. Rocks with normal magnetic polarity match that of the Earth's magnetic field (the north end of the rock's "compass needle" points toward magnetic north.) The other group has magnetic minerals with reversed polarity(the north end of the rock's compass needle points south).

It was known that as lava cools to form **basalt** (an iron-rich volcanic rock that makes up the ocean floor), its iron minerals (such as magnetite) become magnetized and "lock in" the polarity of the Earth's magnetic field. Beginning in the 1950's, scientists began noting patterns in the magnetism of rocks on the ocean floor, as shown in *Figure 3*. The alternating belts of higher and lower than average magnetic field strength were of normal and reverse polarity, respectively.

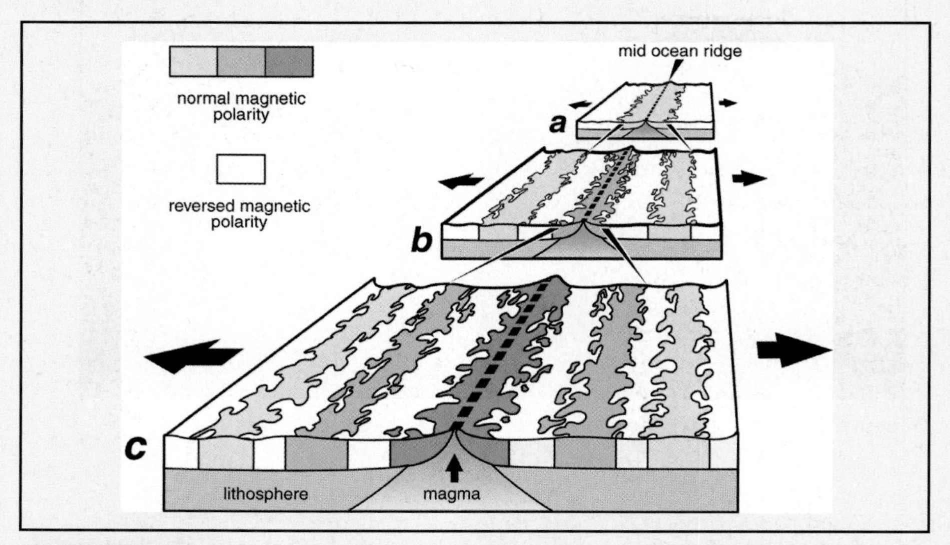

Figure 3 The formation of magnetic striping. New oceanic crust forms continuously at the crest of the mid-ocean ridge. It cools and becomes increasingly older as it moves away from the ridge crest with sea-floor spreading: a. the spreading ridge about 5 million years ago; b. about 2 to 3 million years ago; and c. present day.

In 1963, two scientists, F.J. Vine and D. H. Matthews, proposed the revolutionary theory of sea-floor spreading to explain this pattern. According to their theory, the matching patterns on either side of the **mid-ocean ridge** could be explained by new ocean crust forming at the ridge and spreading away from it. As ocean crust forms, it obtains the polarity of the Earth's magnetic field at that time. Over time, the strength of the Earth's magnetic field changes. When new ocean crust forms at the center of the spreading , it obtains a new kind of magnetic polarity. Over time, a series of magnetic "stripes" are formed.

Since the theory of sea-floor spreading was proposed, core samples of volcanic rock taken from the ocean floor have shown that the age of the rock increases from the crest of the ridge, just as the theory predicts. What's more, by measuring both the age and magnetic polarity of rocks on land, geologists have developed a time scale that shows when the magnetic field has reversed its polarity. Because the magnetic striping on the ocean floor records the reversals of the Earth's magnetic field, geoscientists can calculate the average rate of plate movement during a given time span. These rates range widely. The Arctic Ridge has the slowest rate (less than 2.5 cm/yr), and the East Pacific Ridge has the fastest rate (more than 15 cm/yr). The computer model in the plate motion calculator uses spreading rates from ocean ridges throughout the world to compute plate motion.

Geologic data is also used to find the direction of movement of the plate. Surveys of the depth of the ocean floor, mainly since the 1950s, revealed a great mountain range on the ocean floor that encircles the Earth, as shown in *Figure 4*. This mid-ocean ridge zig-zags between the continents, winding its way around the globe like the seams on a baseball. The mid-ocean ridge is

Figure 4 Map of the world ocean floor. The crest of the mid-ocean ridge system is shown as a broad light blue line throughout the ocean floor. The flanks of the mid-ocean ridges slope gradually down to the deeper part of the oceans, nearer the continents.

Geo Words

basalt: a kind of volcanic igneous rock, usually dark colored, with a high content of iron.

mid-ocean ridge: a chain of undersea ridges extending throughout all of the Earth's ocean basins, and formed by sea-floor spreading.

not straight; it is offset in many places (*Figure 5*). The offsets are perpendicular to the axis of the ridge. When combined with knowledge that the ocean floor is spreading apart at mid-ocean ridges, geologists realized that the offsets are parallel to the direction the plates are moving. By mapping the orientations of these offsets, and entering this data into the computer model, the plate motion calculator is able to generate the directions of plate motions. Comparisons between GPS measurements and results from geologic computer models show very good agreement, within 4%.

Figure 5 This map shows the network of fractures along the mid-ocean ridge in the eastern Pacific Ocean floor. Because the ocean floor spreads away from both sides of the ridge, the fractures indicate the direction of plate motion.

Check Your Understanding

1. What is the difference between the lithosphere and the asthenosphere?
2. What does the abbreviation GPS stand for?
3. From where does a GPS receiver get its signal?
4. Why are GPS data not enough to confirm that the Earth's surface has been moving for many years?
5. What has caused the "zebra pattern" in the rock of the ocean floor?
6. What is the significance of the patterns of offsets along mid-ocean ridges?

Understanding and Applying What You Have Learned

1. Describe the direction and the rate of movement for the plate on which you live.
2. Examine the scale of the USGS topographic map of your community. Given the rate of plate motion in your community, estimate the minimum number of years that it would take for a change in the location of your school to be detected on a topographic map.

3. How does GPS provide evidence that the surface of the Earth moves over time?
4. Describe at least one advantage of using GPS technology to gather evidence of plate motion.
5. How do studies of the magnetism of rocks on the sea floor provide evidence that the surface of the Earth moves over time?
6. What evidence examined in this activity suggests that the direction and rate of motion of plates is not constant?
7. Look at the world map in the **Investigate** section showing the relative motion of plates. This map shows how plates move relative to each other.
   a) Look at the names of the plates. On what basis does it appear that the plates were named?
   b) Write down the name of your plate and all the plates that border it. Describe the motion of your plate relative to all the plates that border it.
   c) How might the differences in motion of these plates affect the Earth's lithosphere?

Preparing for the Chapter Challenge

In your notebook, write a paragraph to convince the middle-school students that your community is moving as part of the movement of a much larger segment of the Earth's lithosphere. Describe the evidence used to make this determination.

Inquiring Further

1. **Technology used to detect plate motions**

   Explore how GPS allows plate movement to be measured. Excellent web sites that describe how GPS works can be found on the *EarthComm* web site.

2. **Investigating scales of motion**

   Plate motion is extremely slow. Make a list of other things you know about (or have heard about) that move or take place slowly. Possible examples include growth of fingernails, grass, tree height, tree-trunk diameter, and so on. Find out how fast they move. Compare the rate of these motions to the rate of movements of plates.

3. **Study animations of plate motions**

   Visit the *EarthComm* web site for the address of animated images of the motions of lithospheric plates. Describe how the motions shown in the animations match your analysis from this activity.

Activity 2 Plate Boundaries and Plate Interactions

Goals

In this activity you will:

- Classify and label the types of movement at plate boundaries, using a world map that shows relative plate motion.
- Identify the distribution of plates by means of the world map of relative plate motions.
- Describe the present plate-tectonic setting of your community, and infer possible past plate-tectonic activity based on your knowledge.

Think about It

New plates are created at certain places on Earth, and existing plates are consumed at certain other places. The total surface area of the Earth stays the same, so the creation of new plates has to be exactly equal to the consumption of existing plates.

- Where do you suppose you would have the most "interesting" ride on a plate? Would it be at the center, on a leading edge, on a trailing edge, or somewhere else on the plate?

What do you think? Record your ideas about this question in your *EarthComm* notebook. Include sketches as necessary. Be prepared to discuss your response with your small group and the class.

Investigate

Part A: Observing Plate Motions and Plate Interactions

1. Obtain the equipment shown in the following diagram.
2. Use the equipment to model a steady sea-floor spreading and subduction, as follows: One student holds the two rolled-up dowels in one place, loosely, so that they can turn but not shift their position. Another student holds the stapled piece of 2 x 4 lumber "continent" and pulls it away from the rolled-up dowels. A third student holds the dowel and piece of 2 x 4 lumber "subduction zone" at the other end loosely in place. A fourth student pulls the paper strip from under the piece of 2 x 4 lumber "subduction zone." Be sure to unroll the paper strips at the same rate, so that the numbers of the stripes stay matched up as they appear.
   - a) What does the area at the rolled paper strips on the dowels represent?
   - b) What does the section of paper between the dowels and the continental lithosphere (the piece of 2 x 4 lumber) represent?
   - c) What happens to the length of this section of paper as the dowels are unrolled?
   - d) As the dowels are unrolled, what happens to the width of the section of paper between the dowels and the subduction zone (the other piece of 2 x 4 lumber)?

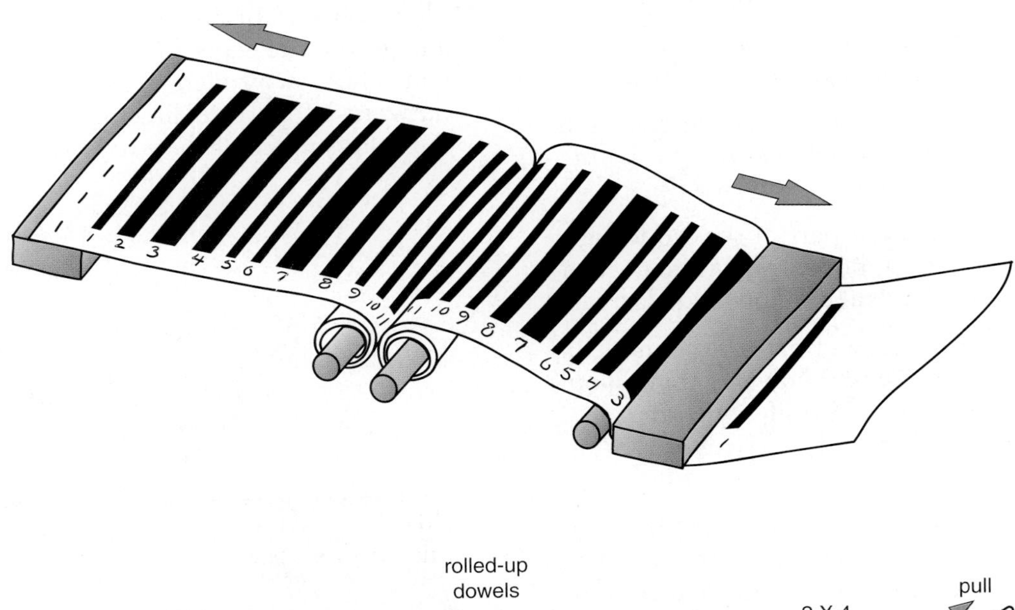

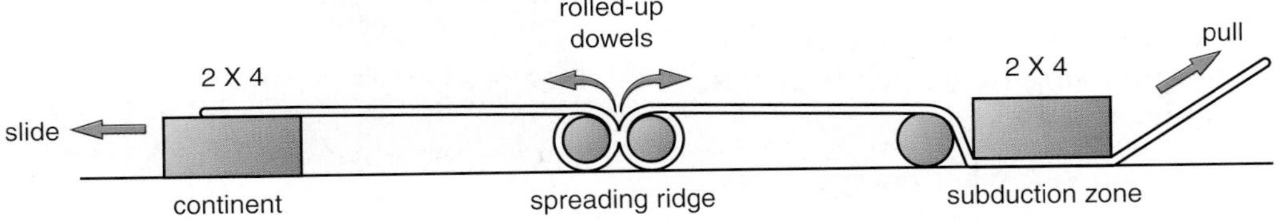

3. Use the equipment to model a collision of a spreading ridge and a subduction zone, as follows: Begin with the materials arranged in the same way at the end of **Step 1**. While two students pull the paper strips to unwind the two rolled-up dowels, the student holding those dowels slides them slowly toward the subduction zone. The student operating the subduction zone needs to make sure that the stripes appearing at the spreading ridge continue to have their numbers matched up.
   a) What happens to the length of the strip of paper between the dowels and the "continent" side in this situation?
   b) What happens to the length of the strip of paper between the dowels and the "subduction" zone?
   c) At what "place" does the spreading ridge eventually arrive?
4. Think about the following questions, and write a brief answer to each in your notebook.
   a) In the first part of the modeling (**Step 2**), how long will the ocean on the "subduction" side last?
   b) In the second part of the modeling (**Step 3**), what do you think would happen in real life when the spreading ridge arrives at the subduction zone?
   c) In the second part of the modeling (**Step 3**), how would the ocean on the "continent" side change after the spreading ridge arrives at the subduction zone?
   d) In both cases, what do you think would happen in real life if the continent became blocked in its movement away from the spreading ridge by something happening on the other side of the continent?

Part B: How Transform Faults Are Formed

1. When the theory of plate tectonics was young, back in the 1970s, there was an experiment to see how divergent plate boundaries are formed. Scientists might describe this experiment as "beautiful" or "elegant." It was small enough to be done on a table top. You can run it as a "thought experiment." A diagram of the setup is shown below. Use a shallow square pan. Fill it with melted wax, and heat the bottom of the pan to keep the wax melted. Cool the surface of the wax with a fan, so that it crusts over to form "lithosphere." Install roller cylinders along opposite edges of the top of the pan. These two rollers can be rotated outward in opposite directions to pull the solid surface of the wax apart.

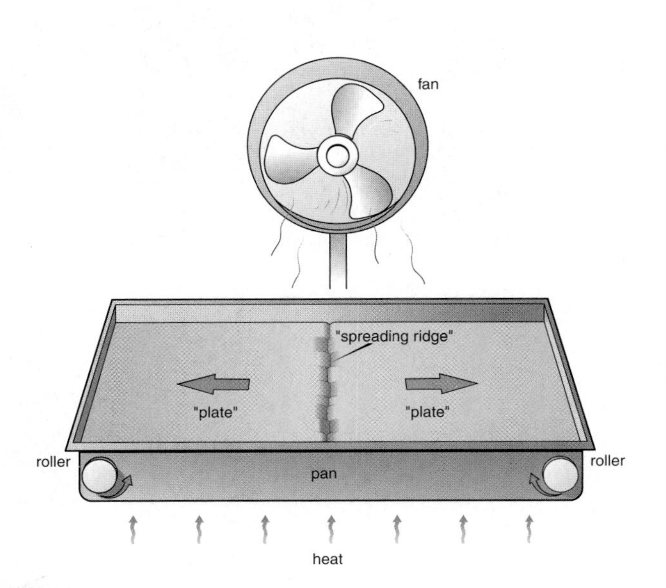

a) Predict what will happen to the crust of wax when the rollers are rotated. Assume that everything is adjusted right (composition of the wax; heating at the bottom; cooling at the top; speed of the rollers).

b) If a break were to occur in the surface wax, what do you think would happen to the liquid wax below?

2. In the actual experiment, the "lithosphere" broke along a crack across the middle of the pan. As the two "plates" on either side of the crack were formed, new "lithosphere" formed as the liquid wax welled up into the crack and solidified. The amazing thing about the experiment was that the original crack was formed with exactly the same pattern of ridge-crest segments and transform faults that can be seen in the real mid-ocean ridges!

a) Were your predictions correct? How do you explain any differences between what happened and your predictions?

b) What does the experiment suggest about the age of the transform faults compared to the mid-ocean ridges?

Part C: Plate Boundaries on World Maps

1. Look at the world map on page G66 in the previous activity, which shows the relative motion of the plates. Observe what it shows about how plates move relative to each other. Answer the following:

a) Name two plates that are moving toward each other (colliding/converging).

b) Name two plates that are moving apart (diverging).

c) Name two plates that are sliding past each other.

2. Use a blank world map to make a map that shows the three major types of plate boundaries.

a) On the map, color the boundary line that separates two converging plates. Do not outline both of the plates completely; highlight only the boundary between the two plates.

b) Using two other colors, highlight the divergent plate boundaries, (where plates are moving away from one another) and the plate boundaries, where plates slide past one another (called transform boundaries). Make a key that shows this color code.

Part D: The Plate-Tectonic Setting of Your Community

1. Describe the plate-tectonic setting of your community. Refer to your world map and the map *This Dynamic Planet* (United States Geological Survey) in your work.

a) How far is your community from the nearest plate boundary?

b) What type of plate boundary is it?

c) How might your community change its position relative to plate boundaries in the future?

Reflecting on the Activity and the Challenge

You have seen that plates can interact in three different ways: they can converge, diverge, or move parallel to each other. You have gained some experience in recognizing the three kinds of plate boundaries on world maps. You are better prepared for a later activity, on the different kinds of landforms that develop at the different kinds of plate boundaries, and how earthquakes and volcanoes are caused at or near plate boundaries. You are now in a better position to help middle school students understand one of the key concepts of plate tectonics: the material of the ocean floors is always much younger than the oceans themselves!

Geo Words

divergent plate boundary: a plate boundary where two plates move away from one another.

Digging Deeper

THE DYNAMICS OF PLATE BOUNDARIES

Types of Plate Boundaries

Plate boundaries are geologically interesting areas because they are where the action is! Geologists use three descriptive terms to classify the boundaries between plates: 1) divergent boundaries are where two plates move away from each other; 2) convergent boundaries are where two plates move toward each other; and 3) transform boundaries are where two plates slide parallel to each other.

Divergent Plate Boundaries

You have already learned some things about **divergent plate boundaries** in Activity 1, because mid-ocean ridges are divergent plate boundaries. The mid-ocean ridges are places where mantle asthenosphere rises slowly upward. As it rises, some of the rock melts to form magma. Why does melting happen there? To understand that, you need to know that the melting temperature of rock decreases as the pressure on the rock decreases. As the mantle rock rises, its temperature stays about the same, because cooling takes a long time. However, the pressure from the overlying rock is less, so some of the rock melts. The magma then rises up, because it is less dense than the rock. It forms volcanoes in the central valley of the mid-ocean ridge. Geoscientists in deep-diving submersibles can watch these undersea volcanoes! Because of the great water pressure in the deep ocean, and also the cooling effect of the water, these volcanoes behave differently from volcanoes on land. The lava oozes out of cracks in the rocks, like toothpaste out of a tube. Some of the magma stays below the sea floor and crystallizes into rock there. All of these new igneous rocks, at the sea floor and below, make new oceanic crust, which then moves away from the ridge crest. This would be a good time to go back to **Activity 1** and review the material on sea-floor spreading.

In the investigation you modeled how the "continent" moved farther and farther from the "spreading ridge." Look back again at the world map of lithospheric plates in Activity 1. In both the North Atlantic Ocean and the South Atlantic Ocean, there is no plate boundary along the coastlines on either side of the ocean. That tells you that the Atlantic Ocean is getting wider as time goes on. Why? Because new lithosphere is being created all the time at the mid-ocean ridge but is not being consumed at the edges of the continents. Does that make you wonder what would happen if you could make time run backward and watch the ocean shrink? At some time in the past, there was no Atlantic Ocean!

A new ocean begins when hot mantle material begins to move upward beneath a continent. Geoscientists are still not certain about why that happens. The lithosphere of the continent bulges upward and is stretched sideways. Eventually it breaks along a long crack, called a rift. See *Figure 1* for what a newly formed **rift valley** looks like. Magma rises up to feed volcanoes in the rift. As the rift widens, the ocean invades the rift. A new ocean basin has now been formed, and it gets wider as time goes on.

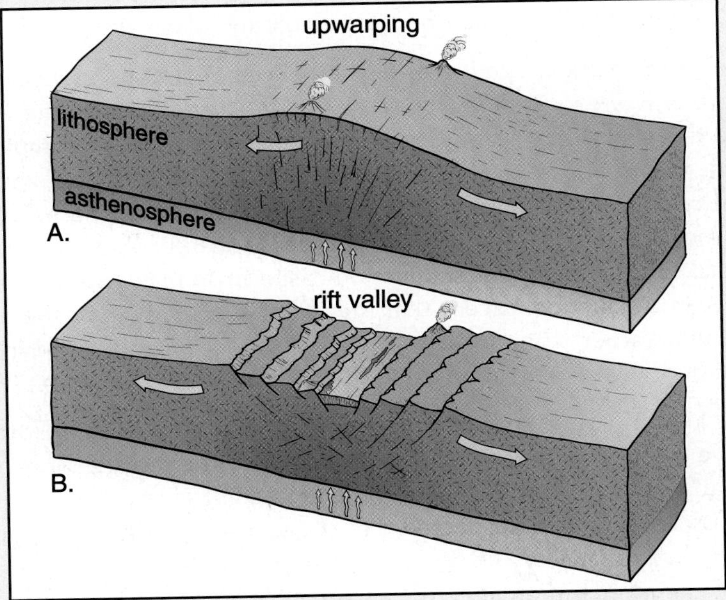

Figure 1 The formation of a rift valley on a newly rifted continent.

Geo Words

rift valley: a large, long valley on a continent, formed where the continent is pulled apart by forces produced when mantle material rises up beneath the continent.

Convergent Plate Boundaries

At a **convergent plate boundary**, two plates are moving toward each other. Your common sense tells you that one of them has to go under the other. (Would it surprise you to hear that common sense is important to a scientist, even though sometimes common sense can fool you?) There are three kinds of places where this happens. These are described below. In a later activity you will learn much more about the landforms that are produced at convergent plates boundaries, and how earthquakes and volcanoes are associated with convergent plate boundaries.

In some places, two oceanic lithospheric plates are converging. There are good examples along the western edge of the Pacific Ocean. Look back at the world map of lithospheric plates in Activity 1. The Pacific Plate and the Indo-Australian Plate are moving toward one another in the South Pacific, and the Pacific Plate and the Philippine Plate are moving toward one another in the Western Pacific. Look at the lower part of *Figure 2*. One plate stays at the surface, and the other plate dives down beneath it at some angle. This process is called **subduction**, and so these plate boundaries are called subduction zones. Where the downgoing plate first bends downward, a deep trench is formed on the ocean floor. These trenches are where the very deepest ocean depths are found. As the plate goes down into the mantle asthenosphere, magma is produced at a certain depth. The magma rises up to the ocean floor to form a chain of volcanic islands, called a volcanic island arc.

Other subduction zones are located at the edges of continents. Look at the upper part of *Figure 2*. The west coast of South America, where the Nazca Plate and the South American Plate are converging, is a good example. In places like this, the downgoing plate is always oceanic lithosphere and the plate that remains at the surface is always continental lithosphere. That's because the continental lithosphere is less dense than the oceanic lithosphere. Ocean–continent subduction is similar in many ways to ocean–ocean subduction, except that the volcanic arc is built at the edge of the continent, rather than in the ocean. The Andes mountain range in western South America is an example of a continental volcanic arc.

The third kind of convergent boundary is where two continental lithospheric plates have collided with each other. Think back to the investigation. You were asked what would happen when the "spreading ridge" arrives at the "subduction zone." If you suspected that the spreading ridge would go down the subduction zone, never to be seen again, you were right! If you had continued the investigation after the spreading ridge had disappeared, what would have happened? As the plate that was on the other side of the

Geo Words

convergent plate boundary: a plate boundary where two plates move toward one another.

subduction: the movement of one plate downward into the mantle beneath the edge of the other plate at a convergent plate boundary. The downgoing plate always is oceanic lithosphere. The plate that stays at the surface can have either oceanic lithosphere or continental lithosphere.

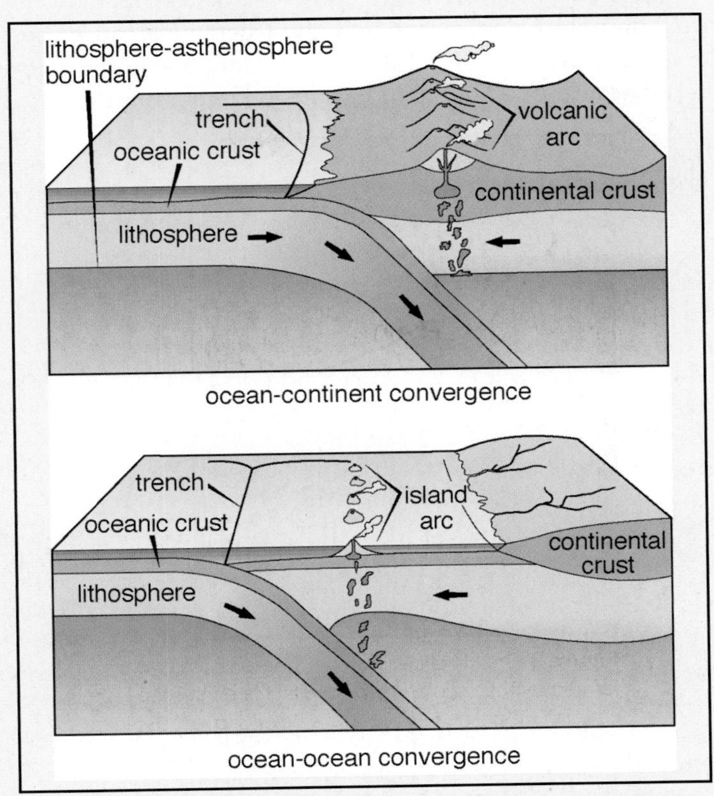

Figure 2 Cross sections through subduction zones. The upper cross section shows subduction of an oceanic lithospheric plate beneath a continental lithospheric plate (ocean–continent subduction). The lower cross section shows subduction of an oceanic lithospheric plate beneath another oceanic lithospheric plate (ocean–ocean subduction).

spreading ridge was consumed down the subduction zone, the continent on the other side of the ocean would have moved closer and closer to the subduction zone. This is how two continents can come together at a subduction zone. Remember that continental lithosphere is much less dense than the mantle, so continental lithosphere cannot be subducted. The subduction stops! The continent that was coming along toward the subduction zone keeps working its way under the other continent, for hundreds of kilometers, until finally the friction between the two continents is so great that plate movement stops. The zone where two continents have met and become welded into a single continent is called a **suture zone**. There is only one good example on today's Earth: the Indian Plate has collided with the Eurasian Plate and is still working its way under it.

Geo Words

suture zone: the zone on the Earth's surface where two continents have collided and have been welded together to form a single continent.

Geo Words

transform plate boundary: a plate boundary where two plates slide parallel to one another.

transform fault: a vertical surface of slippage between two lithospheric plates along an offset between two segments of a spreading ridge.

See *Figure 3*. There is a good reason why the Tibetan Plateau is the largest area of very high elevations in the world: the continental lithosphere is much thicker there because one continent has moved under another.

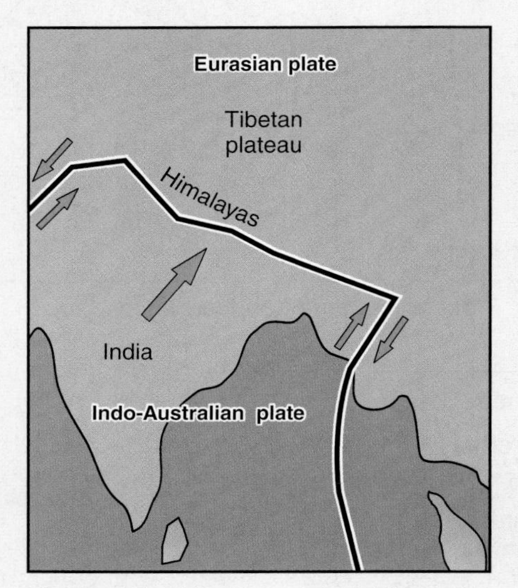

Figure 3 The Himalayas and the Tibetan Plateau were formed by the collision of the Indian Plate with the Eurasian Plate. The Indian Plate is being shoved horizontally underneath the Eurasian Plate, so the continental crust in the Himalayas and the Tibetan Plateau is much thicker than normal.

Transform Plate Boundaries

At **transform boundaries**, plates slide past one another. The surface along which the plates slide is called a **transform fault**. As you saw in Activity 1 in the material about mid-ocean ridges, transform faults connect the offsets along mid-ocean spreading ridges. Most are short, but a few are very long. The most famous transform fault forms the boundary between the North American Plate and the Pacific Plate, in California. It is several hundred kilometers long. You can see from the map in *Figure 4* that the San Andreas Fault connects short segments of spreading ridges at its northern and southern ends.

The movement along the transform fault is limited to the distance between the two segments of ridge crest. *Figure 5* is a sketch map of a mid-ocean ridge, showing segments of the ridge crest offset by transform faults. Between Points 1 and 2, Plates A and B are sliding past each other.

The transform fault extends only between Points 1 and 2. To the left of Point 1, Plate A is in contact with itself along the east-west line, and to the right of Point 2, Plate B is in contact with itself along the east-west line. Since the plates are rigid, there is no slipping movement along the east-west lines, which are like "dead" transform faults!

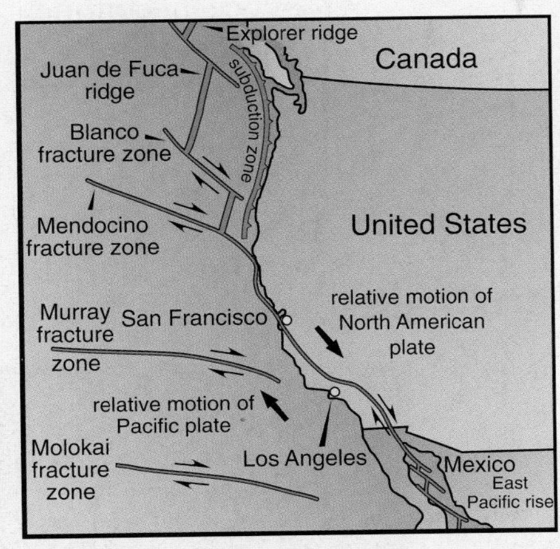

Figure 4 Plates showing a transform boundary.

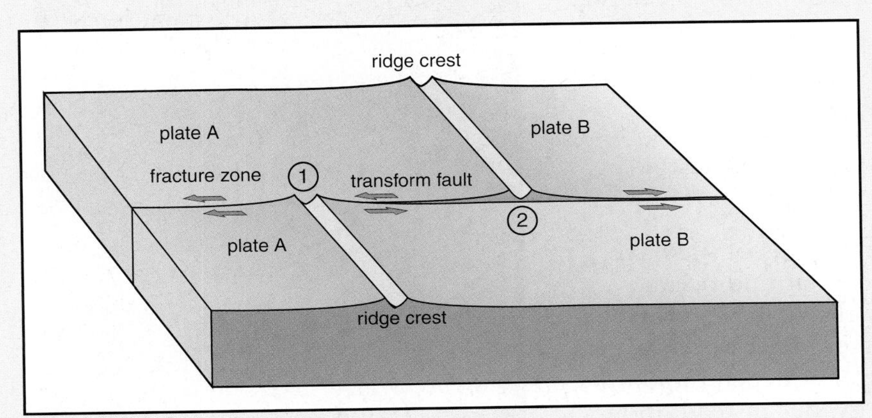

Figure 5 Sketch map of a mid-ocean ridge, showing segments of the ridge crest offset by transform faults.

Check Your Understanding

1. Name the three types of boundaries between lithospheric plates.
2. How and where are rift valleys formed?
3. How can ocean basins change in size?
4. Convergent plate boundaries can be in three different settings. What are they?
5. Describe subduction.
6. Why is it that transform faults can be used to figure out the directions of plate movements? Why can't subduction zones also be used for that?
7. What happens when two continents collide along a convergent plate boundary?

Understanding and Applying What You Have Learned

1. Design an investigation, with the materials like the ones you used in this activity, to model the creation of a new ocean basin.
2. Identify on a world map of lithospheric plates one example of each of the following settings:
   a) an established divergent boundary;
   b) a young divergent zone (continental rift zone);
   c) an ocean–ocean convergent boundary;
   d) an ocean–continent convergent boundary;
   e) a continent–continent collision zone:
   f) the interior of a plate;
   g) a transform plate boundary.

Preparing for the Chapter Challenge

Reflect on your answers to the above. Write a short summary (one or two paragraphs) describing what you have learned so far about how plate movements and their interactions at plate boundaries can change the arrangement of continents and ocean basins on the Earth. Be prepared to include this summary in your **Chapter Challenge.**

Inquiring Further

1. **Evolution of the Biosphere at Mid-Ocean Ridges**

   Recently, new forms of life have been discovered at mid-ocean ridges. They thrive in the presence of superheated, mineral-rich water. This life does not depend upon the Sun for energy, but instead upon the energy and matter from Earth's interior. How has life evolved in such environments? For further information check the *EarthComm* web site.

Black smokers form at the mid-ocean ridges.

Activity 3 What Drives the Plates?

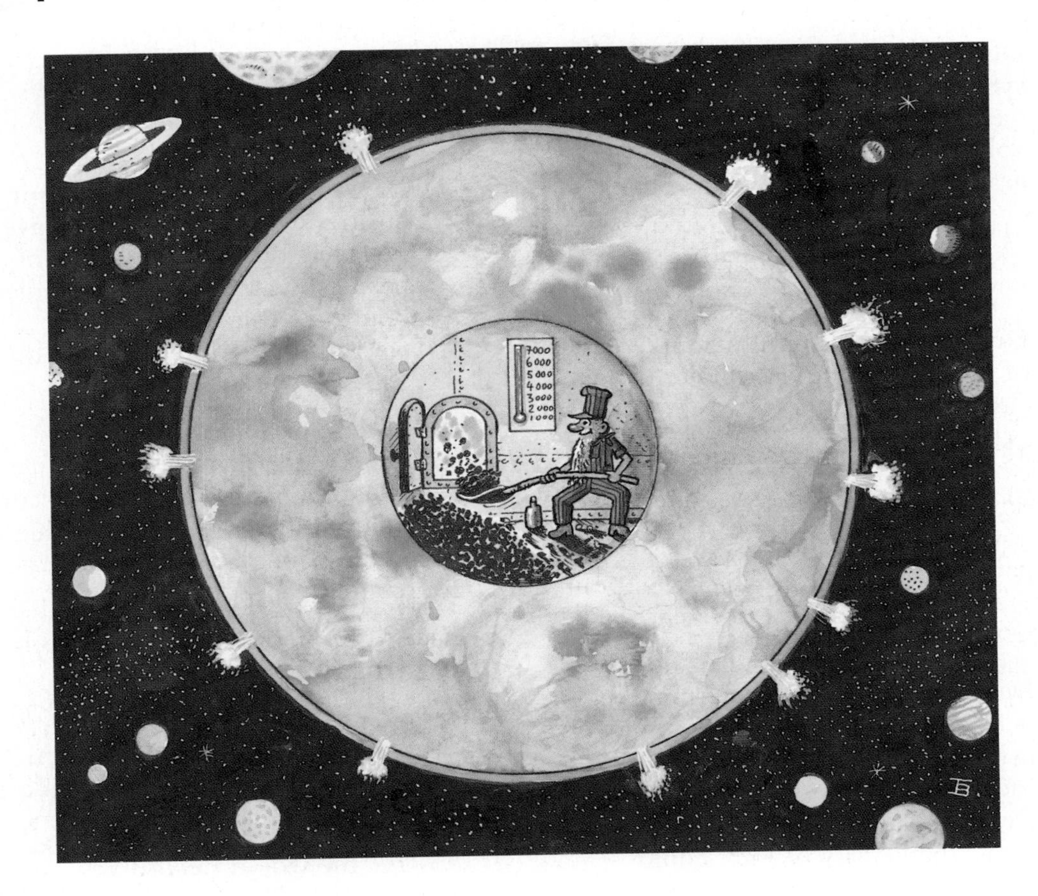

Goals

In this activity you will:

- Calculate the density of liquids and compare their densities with their position in a column of liquid.
- Observe the effects of temperature on the density of a material.
- Examine natural heat flow from within The Earth.
- Understand the results of uneven heating within The Earth.
- Understand the causes of the movement of lithospheric plates.

Think about It

Geoscientists are still uncertain about the most important forces that drive the plates.

- What causes the movement of the lithospheric plates?

What do you think? Record your ideas about this question in your *EarthComm* notebook. Be prepared to discuss your responses with your small group and the class.

Investigate

Part A: Effects of Density on the Position of Material

1. Obtain 30 mL each of water, pancake syrup, and vegetable oil. Suppose you were to carefully pour a small volume of each liquid into one graduated cylinder or clear tube.
   a) Predict what you think will happen. Sketch and explain your prediction.
2. One at a time, carefully pour 10 mL of each liquid into a cylinder or clear tube.
   a) Record your observations.
   b) Do your observations support your predictions?
   c) Does the order in which you pour the liquids make a difference in what you observe?
3. Develop a method to determine the density of each of the three liquids using a graduated cylinder, 10 mL of each liquid, and a balance scale. Density is mass per unit volume. Thus, the density of each liquid equals the mass of liquid (in grams) divided by the volume (10 mL).
   a) Write down your procedure for finding the density of each liquid.
   b) Make a data table to record your measurements and calculations for each liquid.
   c) After your teacher has approved your procedure, determine the density of each liquid.

Follow your teacher's safety advice about using a heat source. Hot corn syrup can cause burns. Clean up spills immediately.

4. Compare your calculations with your observations in **Step 2.**
   a) Describe how the densities you calculated explain what you observed.
   b) If layers of materials of different densities within the Earth behave like layers of liquids of different densities, what would you predict about the position of the rock layers of different densities in the Earth?

Part B: Effects of Temperature on Density of a Material

1. Pour about a 5 cm thick layer of corn syrup into a Pyrex® beaker or wide aluminum pan. Place the pan on a heat source.
2. Place three pieces of balsa wood on the syrup.
   a) Predict what you think will happen to the wood as the corn syrup is heated. Record your ideas in your notebook.
3. Observe the wood. Record any changes every 5 min for 20 to 30 min.
   a) Use diagrams to record the changes you observe.
   b) Do your observations support your predictions? What do you think caused the results you observed?

sandstone

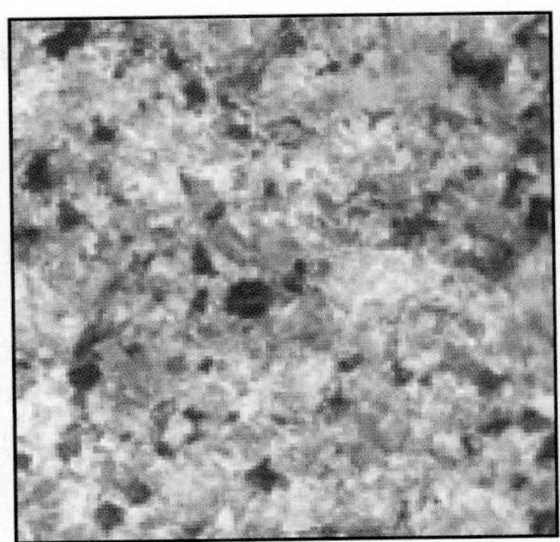
granite

basalt

Part C: Density of Earth Materials

1. Collect samples of rock from your community and also obtain samples of granite, basalt, and sandstone.
2. If you can, predict qualitatively the density of the samples. Which sample appears to be least dense? Which appears to be most dense?
   a) Record your predictions in your notebook.
3. Develop a method to find the density of each rock sample using the sample, water, a graduated cylinder, and a balance scale. Density is mass per unit volume. Thus, the density of each rock equals the mass of rock (in grams) divided by the volume of rock (in cubic centimeters). Note that 1 mL = 1 cm^3.
   a) Write down your procedure for finding the density of each rock sample.
   b) Make a data table to record your measurements and calculations for each rock.
   c) After your teacher has approved your procedure, determine the density of each rock sample.
4. Compare your calculations with your predictions.
   a) How does the density of the rock from your community compare with the densities of granite, sandstone, and basalt?

Part D: Forces Causing Subduction of Lithospheric Plates

1. Partly fill a large, rectangular tub with warm water. Wait until any tiny air bubbles have disappeared. The water has to be perfectly clear.

2. Very slowly and carefully, put a few ounces of liquid dish detergent in the water and mix it slowly and carefully with a mixing spoon. If any soap bubbles or foam remain on the water surface, scrape them off with a damp sponge.

3. Cut a piece of the vinyl plastic to be about six inches wide and about twelve inches long. Trim a flat, clear-plastic ruler with the scissors to be the same width as the plastic sheet. (The ruler should sink in water.) Tape the ruler to one end of the plastic sheet.

4. Dip the ruler end of the plastic sheet into the water to a depth of about 1 cm. Immediately place the plastic sheet on the water surface. Do this by holding the ends up, and letting the sagging middle part of the sheet touch the water surface first, to avoid trapping air bubbles under the sheet. Observe what happens. Repeat this step as many times as you need to make careful observations.

   a) Record your observations. Include a description of the motion of the plastic sheet in the water.

   b) What is the force that makes the plastic behave as it did?

   c) How does this demonstration show what happens in a subduction zone?

Reflecting on the Activity and the Challenge

You have seen evidence that liquids of varying densities will form layers in a container with the densest liquid on the bottom. You have also shown by modeling that a solid floating on a liquid seems to move away from a source of heat. You have seen evidence that different rocks are likely to have varying densities. Finally, you have made a direct observation of one of the main forces that cause subduction. These investigations will help you understand the Earth's interior and the flow of matter and energy in the Earth. You should now be able to explain why lithospheric plates can float and what might cause them to move.

Keep work area clean and dry.

Have paper towels ready for the wet plastic that is taken out of the tub.

Digging Deeper

THE EARTH'S INTERIOR STRUCTURE

Evidence for Earth's Layered Structure

Density (mass per unit volume) refers to how concentrated the mass (atoms and molecules) in an object or material is. Less dense material tends to rise upward and float on more dense material. Here are some everyday examples: a less dense solid floats in a more dense liquid; a more dense solid sinks to the bottom of a less dense liquid; a less dense liquid floats on a more dense liquid. Rocks in the Earth's crust (oceanic crust consists mainly of basalt; continental crust consists mainly of less dense rocks like granite) are less dense than the rocks of the underlying mantle. The crust "floats" on the more dense interior material.

Several kinds of evidence reveal that density varies within the Earth. Laboratory experiments in high-pressure apparatus show that rocks deep in the Earth are more dense than the same rocks when they are at the surface. The weight of the overlying rock puts pressure on rock below, making it more dense. The most dense material should be at the center of the Earth, where the pressure is greatest.

A second line of evidence comes from calculations of the average density of the Earth. You cannot put the Earth on a balance scale to find its mass, but the mass can be found indirectly using Newton's Law of Gravitation. According to that law, the gravitational force (F) between any two objects in the universe can be expressed this way:

$$F = \frac{gm_1m_2}{d^2}$$

where m_1 and m_2 stand for the masses of two objects,

d stands for the distance between them, and

g stands for the gravitational constant (known from experiments).

Because the Earth exerts a certain force on a body (like you) with a certain mass m_1 on the Earth's surface, some 6400 km from its center, the known values can be substituted into the equation and the mass of the Earth (m_2) can be calculated. Dividing the mass of the Earth by its volume gives an average density of the Earth (in metric units) of 5.5 g/cm^3. The density of the rocks commonly found at the surface (granite, basalt, and sandstone) is much lower. The average density of surface rocks is 2.8 g/cm^3. The density of

Geo Words

density: the mass per unit volume of a material or substance.

Geo Words

core: the solid, innermost part of the Earth consisting mainly of iron.

the Earth's interior must be much greater than 2.8 g/cm^3 for the entire Earth to average 5.5 g/cm^3. This is partly due to the effect of compression, but also partly because the material in the Earth's core is mostly iron, which is much more dense than rocks, even when it is not under great pressure.

The speed of earthquake (seismic) waves within the Earth generated by earthquakes also provides convincing evidence about the properties of rock in the Earth. Scientists have learned that these waves travel faster the deeper they are in the Earth. It's known from laboratory experiments that earthquake (seismic) waves that travel at 4.8 km/s at the surface travel at 6.4 km/s at a depth of 1600 km. The reason why the speed of seismic waves increases downward in the mantle is complicated. In the laboratory, scientists use special equipment to measure the speeds of seismic waves in different rocks. They can determine how the speed of seismic waves changes with changes in temperature, pressure, and rock type.

Studies also show that change in density is not uniform with depth. Instead, there are distinct jumps or changes in density. By studying the changes in the speed of earthquake (seismic) waves as they pass through the Earth, scientists have concluded that the Earth's interior structure is layered. The thickness and the composition of the layers are shown in the table in *Figure 1*.

Layer	Thickness (km)	Composition	Temperature (°C)	Density (g/cm^3)
Continental crust	30–60	Granitic silicate rock (>60% silica)	20–600	~2.7
Oceanic crust	5–8	Basaltic silicate rock (<50% silica)	20–1300	~3.0
Mantle	2800	Solid silicate	100–3000	~5
Outer core	2150	Liquid iron-nickel	3000–6500	~12
Inner core	1230	Solid iron-nickel	7000	~12

Figure 1 The composition of the layers of the Earth.

Using the evidence they have observed, geoscientists divide Earth into four main layers: the inner **core**, the outer core, the mantle, and the **crust**, as shown in *Figure 2*. The core is composed mostly of iron. It is so hot that the outer core is molten. The inner core is also hot, but under such great pressure that it remains solid. Most of the Earth's mass is in the mantle. The mantle is composed of iron, magnesium, and aluminum silicate minerals. At over 1000°C, the mantle is solid rock, but it can deform slowly in a plastic manner. The crust is much thinner than any of the other layers, and is composed of the least dense rocks.

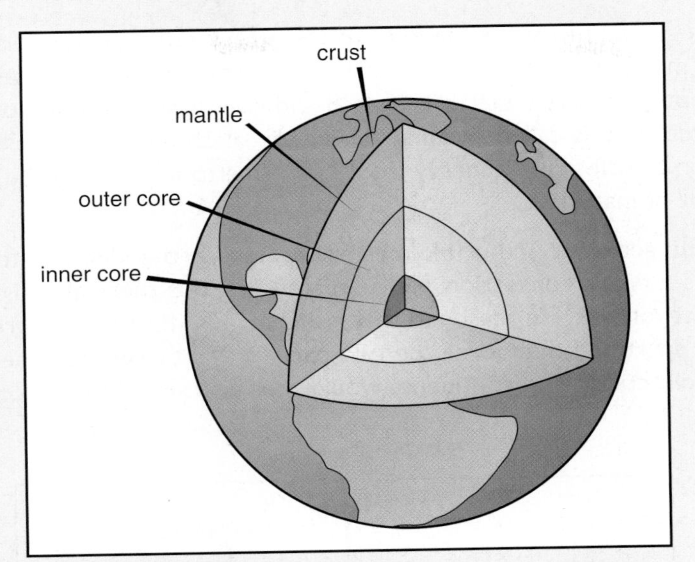

Figure 2 Schematic diagram showing the layered structure of the Earth's interior.

Geo Words

thermal convection: a pattern of movement in a fluid caused by heating from below and cooling from above. Thermal convection transfers heat energy from the bottom of the convection cell to the top.

The Flow of Matter and Energy within the Earth

The temperature of the Earth increases with depth. This can be observed directly in mines and in oil wells. Sources of the Earth's internal heat include the decay of radioactive elements, the original heat of Earth's formation, and heating by the impact of meteorites early in Earth's history. The Earth can be thought of as a massive heat engine. The transfer of heat from Earth's interior to its surface drives the movements of the Earth's crust and mantle.

Temperature affects the density of materials. Hot-air balloons show this effect well. When the air inside a balloon is heated it expands (increases in volume). The mass of the balloon stays the same, but the volume increases. When the ratio of mass to volume drops, the density drops. Therefore, heating makes the balloon less dense than the surrounding air. The hot-air balloon begins to rise. Similarly, as rocks in the interior of the Earth are heated enough, their density decreases. The less dense rock rises slowly over time, unless the rocks are too rigid to allow flow.

In the activity, you heated corn syrup and observed the movement of balsa wood. Why did the balsa wood move? The answer lies in the process of **thermal convection**. Heating lowers the density of the corn syrup at the bottom of the container. This causes it to rise. As the corn syrup approaches the upper surface, it flows to the side, making room for more corn syrup rising from below. As it moves to the side, it cools. As it cools, it becomes more dense, and it sinks back to the bottom of the container. At the bottom of the container it is heated and rises again. This kind of density-driven circulation is called thermal convection, as shown in *Figure 3*. Thermal convection transfers heat energy from one place to another by the movement of material.

In 1929, the geologist Arthur Holmes elaborated on the idea that the mantle undergoes thermal convection. He suggested that this thermal convection is like a conveyor belt. He reasoned that rising mantle material can break a continent apart and then force the two parts of the broken continent in opposite directions. The continents would then be carried by the convection currents.

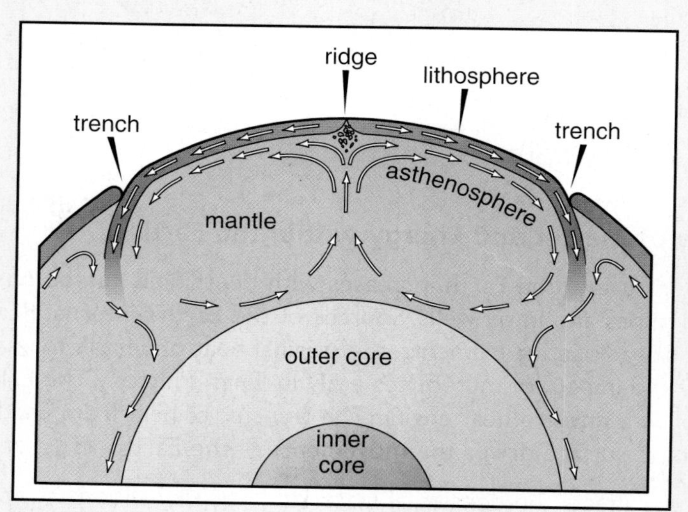

Figure 3 One possible pattern of thermal convection in the Earth's mantle. Convection cells like this might provide at least some of the driving force for the movement of lithospheric plates.

According to this hypothesis of mantle convection, material is heated at the core–mantle boundary. It rises upward, spreads out horizontally, cools, and sinks back into the interior. These extremely slow-moving convection cells might provide the driving force that moves the lithospheric plates (see *Figure 3*). Material rises to the surface at places where lithospheric plates spread apart from one another. Material sinks back into the Earth where plates converge. Although the idea was not widely appreciated during Holmes' time, mantle convection cells became instrumental in the development of the theory of plate tectonics.

Mantle convection can't be observed directly, the way you could have observed convection in the corn syrup if you had put some tiny marker grains in the syrup. Geoscientists are sure that the mantle is convecting, but they are still unsure of the patterns of convection. The patterns probably don't look much like what is shown in *Figure 3*! Geoscientists now think that the lithospheric plates themselves play a major part in driving the convection, rather than just being passive riders on top of the convection cells. Do you remember from Activity 1 that the mid-ocean ridges are broad, and they slope gradually down to the deep ocean nearer the continents? That means that the plates on either side of the ridge crest slope downward away from the ridge crest, and they tend to slide downhill under the pull of gravity! In this way, they help the convection cell to keep moving, instead of the other way around. Also, you probably know that most materials expand when they are heated and shrink when they are cooled. As the plates in the ocean cool, they become more dense than the deeper mantle because they have almost the same composition but they are not as hot. They sink into the mantle of their own accord, just as in Part D of the investigation. In that way they help to keep the convection cell moving.

Check Your Understanding

1. How can the density of the Earth be calculated?
2. How does the density of the Earth provide evidence that the interior of the Earth is denser than the surface?
3. Name three main layers of the Earth.
4. Why is the inner core of the Earth solid, even though it is hot?
5. How are convection currents set up?
6. What part of the Earth's interior layers are in motion due to density differences?

Understanding and Applying What You Have Learned

1. Look at the map of lithospheric plates near South America and the relative "horizontal" motion between these plates.
   a) At point A, the two plates are moving away from each other. What is happening between them?
   b) At point B, two plates are moving toward each other. What happens as they continue to push toward each other if they have:
      (i) Different densities
      (ii) The same density

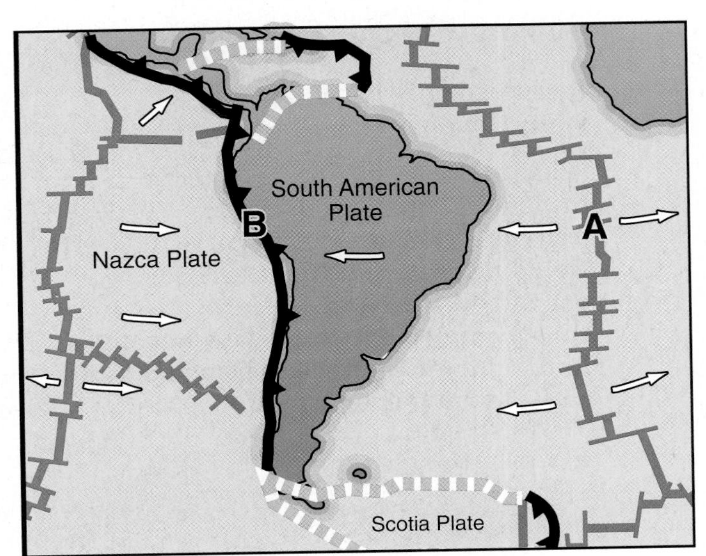

Map of the Nazca and South American plates.

2. Draw two pictures side by side. Make one the experiment with corn syrup and balsa wood. Make the other the Earth's interior structure. Show where heating and cooling occur, and use arrows to indicate the movement of material (the flow of matter and energy in both systems). Label the parts in each diagram and show how they correspond to each other.
3. What evidence is there at the Earth's surface for unequal heating somewhere within the Earth?
4. List some natural processes that occur when heat from the Earth's interior is transferred to the surface.
5. Use your understanding of density to calculate the missing values in the table below:

Object	Mass (g)	Volume (cm^3)	Density (g/cm^3)
Iron	41.8		7.6
Quartz	39.75	15.0	
Gold		8.0	19.3

Preparing for the Chapter Challenge

Reflect on your answers to the above. Write an essay that describes the Earth's interior structure and the flow of matter and energy within the Earth. Refer to the evidence you examined and the models you explored. Sketch and label a drawing or two to illustrate the main ideas. Be prepared to include this summary in your chapter report.

Inquiring Further

1. **Investigating Driving Forces for Plate Motions**

   What questions do you have about the driving forces behind plate tectonics? Develop a plan that would help you find an answer to one of your questions. Record your plan in your notebook. What further information might help you answer your questions?

Activity 4 Effects of Plate Tectonics

Goals

In this activity you will:

- Use maps to examine the distribution of earthquakes and volcanoes to the location of plate boundaries.
- Explain the location, nature, and cause of volcanic arcs in terms of plate tectonics.
- Explain the location, nature, and cause of hot spots.
- Explain how plate-tectonic processes have caused continents to grow through geologic time.
- Explain how plate-tectonic processes produce landforms.
- Explain how plate tectonics can affect the interior of a continent.

Think about It

Rocks high in the Himalayas, almost 8000 m (29,028 ft) above sea level, contain fossils of marine animals.

- Why are most high mountain ranges located at or near plate boundaries?

What do you think? Record your ideas about this question in your *EarthComm* notebook. Include sketches as necessary. Be prepared to discuss your response with your small group and the class.

Investigate

1. Compare the distribution of volcanoes and earthquakes shown on the following map with the map of crustal plates you created in **Activity 2.**
   a) Describe any differences between the distribution of volcanoes along plate boundaries and within plate interiors.
   b) Describe any differences between the distribution of earthquakes along plate boundaries and within plate interiors.
2. To model the rise of magma through the Earth, fill a tall, transparent jar almost up to the brim with honey. Put a very small volume of vegetable oil (about 5 mL, one teaspoon) on top of the honey. Screw on the lid. Try not to leave any air in the container. Turn the container upside down quickly and set it on a tabletop. Make your observations and answer the following questions:
   a) What does the honey represent?
   b) What does the vegetable oil represent?
   c) Describe and explain the behavior of the vegetable oil.
   d) What do you think are the similarities and differences between the behavior of this model and the rise of magma in the Earth?

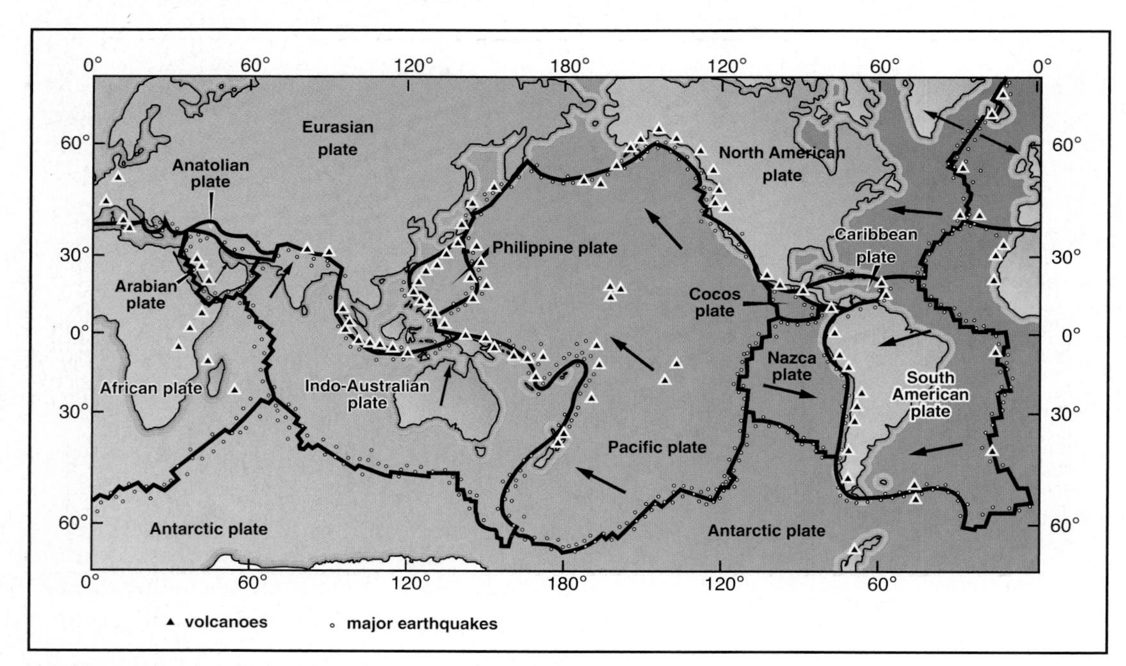

World map showing the location of volcanoes and large earthquakes.

Be sure the lid is on tight before the jar is turned over. Clean up spills immediately.

3. The diagram shows a cross section with two subduction zones, a spreading ridge, and a continental rift. Note the two zones where an oceanic plate is being subducted (plunged) under another plate. Volcanoes are common in a zone that is located a certain distance away from the trench where the subducted plate first bends downward.

   In your group, develop as many hypotheses as you can think of for why the volcanoes occur, and why they are located where they are. Here are four important facts you might need to use or think about in developing your hypotheses:

   (i) The downgoing plate is cooler than the mantle.

   (ii) The composition of the downgoing plate is slightly different from the composition of the mantle. (Why?)

   (iii) The oceanic crust on the downgoing plate contains water, which was added to the igneous rocks when they formed.

   (iv) Friction generates heat.

   a) Record your group's hypothesis.

   b) In a class discussion, compare the hypotheses you developed in your group with those of other groups. Were their hypotheses different from yours and still seemed reasonable?

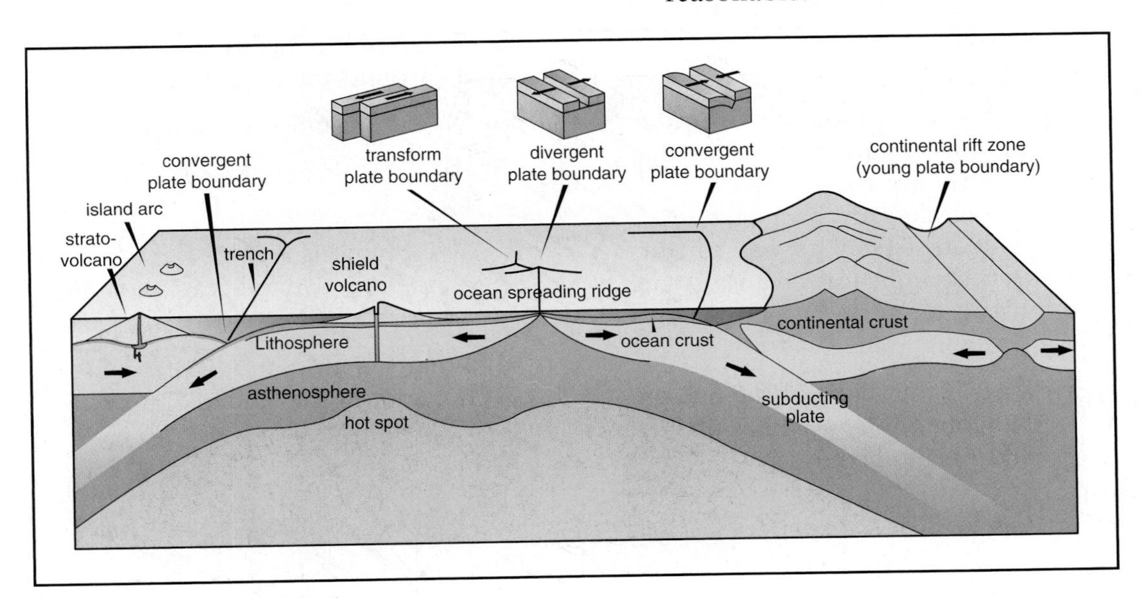

Cross-section with two subduction zones.

4. Look again at the diagram of the cross section with two subduction zones. Use the cross section to answer the following questions:
   a) Under what two types of plates is the oceanic lithosphere being subducted?
   b) What differences between oceanic volcanic arcs and continental volcanic island arcs can you see or infer from the cross section?
   c) On a copy of the map of volcanoes and earthquakes, circle two continental volcanic arcs and three oceanic volcanic arcs.
   d) In which part of the world are most volcanic arcs located? What does that suggest about the plate-tectonic setting of that part of the world?
5. To understand why volcanic arcs are called arcs, look at the Andes Mountains on a topographic map. The Andes are topped with volcanoes that are part of a volcanic arc. They appear to run along a straight line. Run a string or thread along their length on a globe.
   a) What is the shape of the line on the globe? Why are lines of volcanoes called arcs?
   b) Using the cross section of the subduction zones, explain why few volcanoes occur very far inland within a continent.
   c) If volcanic rock is found far inland within a continent, what is one possible reason why it is there?
6. One reason why volcanic rock might be located in the interior of a plate is illustrated on the cross section of the subduction zones where a hot spot is shown in the middle of one of the oceanic plates.
   a) Where does it appear that the hot spot originated?
   b) Is it related to subduction?
   c) Where does the hot spot begin to produce a pool of magma?
7. The map on the next page shows hot spots around the world:
   a) Where are most hot spots?
   b) Are they clustered or randomly located?
   c) What famous area of the continental United States sits over a hot spot?
   d) What sits atop another famous hot spot in the United States?

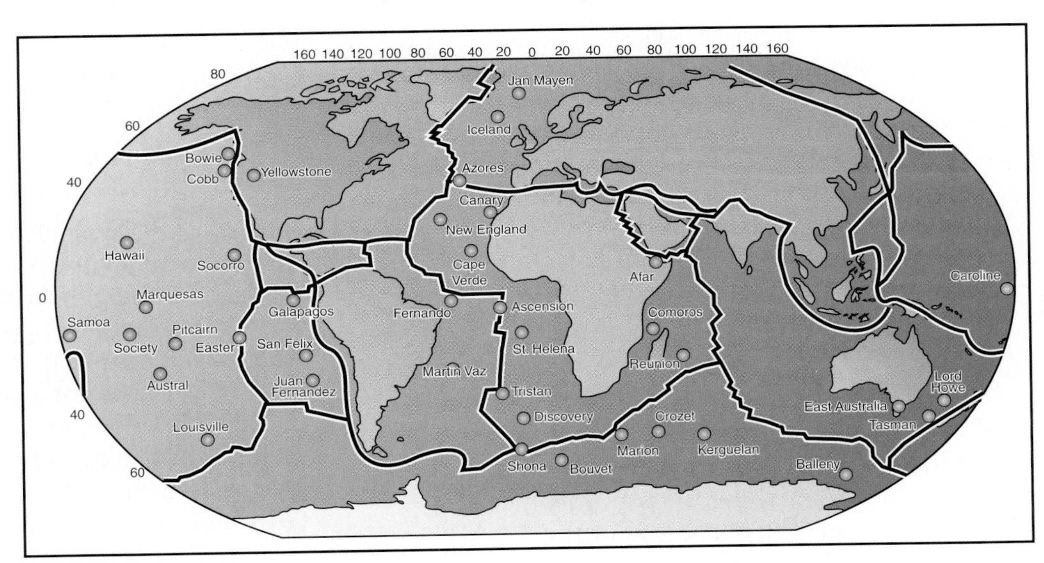

Map of hot spots around the world.

8. To model the growth in size at a subduction zone, set up pieces of lumber and a plastic sheet as shown in the diagram. With a table knife, spread a thin (about 2 mm) layer of cream cheese on the part of the plastic sheet that rests on the upper surface of the 2 x 6. Try to keep the layer as even in thickness as you can. Spread a layer of the cheese spread, of about the same thickness, over the layer of cream cheese.

   Have one student hold the piece of 2 x 4 in place, pressing down on it gently. Another student very slowly pulls the loose end of the sheet of plastic out from under the 2 x 4. Observe what happens to the layer of cream cheese and cheese spread as "subduction" proceeds.

   a) In your notebook, write a description of the process you observe. Be sure to note how the body of material at the subduction zone changes its shape, as well as its volume through time.

Keep work area clean. Have damp paper towels ready to wipe up spills and clean hands as required.

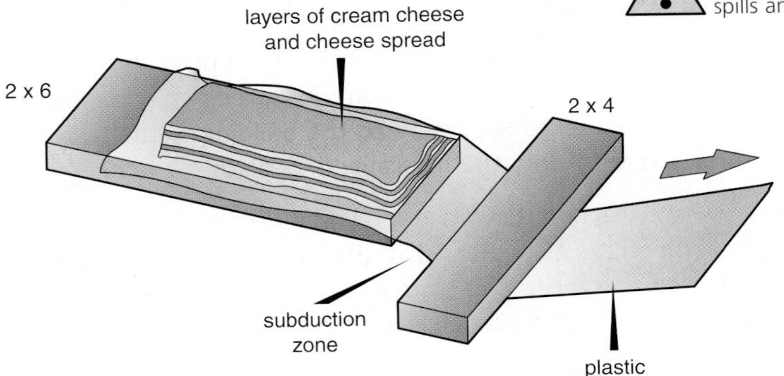

9. After all of the sheet of plastic has been consumed down the subduction zone, make a cut through the accumulated cream cheese and cheese spread, to see the internal structure of the new material at the subduction zone. The cut should be vertical, and parallel to the direction of subduction. Clear away all the material from one side of the cut and examine the face of the cut.
   a) In your notebook, draw a sketch of what the cut looks like. Show the topographic profile of the top of the material, and also any internal structures you observe.
   b) Is there a limit to the height of the "mountain range" that is formed at the subduction zone?
   c) Did any of the cream cheese and cheese spread go down the subduction zone (under the 2 x 4)? In a real subduction zone, what would happen to such material? (This is not a perfect model of what happens to oceanic sediment at a subduction zone, but it gives you the flavor of it. Geoscientists love to make puns.)

Reflecting on the Activity and the Challenge

In this activity you have observed features on the Earth's surface that are created as a result of crustal plate movements and activity within the geosphere. You have also seen that plate-tectonic processes can cause plates to grow in size. Finally, you have seen that plate movements can cause earthquakes that can be associated with specific plate boundaries. You will need to consider these surface features and disturbances created by plate motions in order to complete your **Chapter Challenge.**

Geo Words

plate tectonics: the field of study of plate motion.

Digging Deeper

"BUILDING" FEATURES ON EARTH'S SURFACE

Plate Tectonics

You can now see why the field of study of plate motion is called **plate tectonics**. Tectonics comes from the Greek word *tekton*, which means builder. Plate tectonics refers to the building of the features on Earth's surface due to deformation caused by plate movements.

You have learned that plate movements create mountain ranges, trenches, and rift valleys at or near plate boundaries. Also, there is a clear relationship between volcanoes and plate boundaries, and between earthquakes and plate boundaries. This is particularly evident around the rim of the Pacific Ocean, where the subduction of oceanic plates around much of the rim results in volcanic arcs and earthquakes.

Oceanic Trenches

The deepest valleys on Earth are in the ocean, where they can't be seen except from special deep-diving submersibles. Where an oceanic plate is subducted under another plate, it bends downward as it enters the subduction zone. The valley that is formed above the zone of bending is called a trench. Oceanic trenches are very deep. Many are deeper than 10,000 m, which is twice the average depth of the deep ocean. The word "valley" is a bit misleading, because trenches are wide, and their side slopes are not very steep. You can easily spot the locations of trenches in most world atlases, because they are shown with the darkest blue shading on maps of the world's oceans. Trenches are common in many places in the western Pacific, where there is ocean–ocean subduction. There is a long trench along the west coast of South America, where the Nazca Plate is being subducted under the continent.

Volcanoes at Plate Boundaries

You know already that volcanoes are common along mid-ocean ridges, where basalt magma rises up from the asthenosphere to form new oceanic crust. Volcanoes are also common along subduction zones, where they form volcanic arcs. At a depth of 200 to 300 kilometers, magma is produced above the subducted plate, and rises toward the surface because it is less dense than the surrounding rock. At first it was thought that the magma was produced as rock near the top of the downgoing plate and was heated by friction, but geoscientists are now convinced that the melting is for a different reason. When the oceanic crust is first produced, at the mid-ocean ridges, a lot of water is combined with certain minerals in the igneous rocks. As the pressure and temperature increase down the subduction zone, this water is driven off, and it rises upward from the plate. It's known that the melting temperature of the mantle rock above the plate is lowered when water is added to it. This causes some of the mantle rock to melt. This is a good way to explain why melting doesn't start until the plate has reached a certain depth down the subduction zone, and then stops at a slightly deeper depth. The "Ring of Fire" around the Pacific Ocean is caused by this melting at subduction zones all around the Pacific.

Hot Spots

Not all volcanoes are associated with mid-ocean ridges and subduction zones. Hot spots, which originate at the boundary between the mantle

and the outer core, are narrow plumes of unusually hot mantle material. These plumes rise up through the mantle and melt the rock at the base of the lithosphere, creating pools of magma. This magma then rises to the surface, resulting in hot spot volcanoes.

Some hot spots are located under continents. The hot spot producing the hot springs at Yellowstone National Park is an example. One theory suggests that the bulge created by a hot spot may initiate the rifting of a continent. It is thought that a hot spot lies below, and is responsible for, the Rift Valley of Africa. There is also evidence suggesting that the New Madrid fault, which runs down the Mississippi River Valley, may represent an aborted rift zone originally created by a series of hot spots. The largest series of earthquakes in the United States, outside of Alaska, occurred on the New Madrid fault in the early 19th century, ringing bells as far away as Philadelphia and causing the Mississippi River to run backwards for a short time! In this way, plate tectonics can even affect areas that are within the heart of a continent.

Mountains at Plate Boundaries

Most of the great mountain ranges of the world are located near convergent plate boundaries. When someone says "mountain ranges" to you, which of them do you think of? The Alps in Europe, the Himalayas in southern Asia, the Andes in South America, and the coastal mountain ranges in western North America are some examples. Mountain ranges like those are built in mainly two ways. You already know that the magma that is generated above the subducted plate rises up to form a chain of volcanoes. Much of the magma remains below the surface and cools to form large underground masses of igneous rock called batholiths. The combination of volcanoes at the surface and batholiths deep in the Earth adds a lot of new rock to the area above the subduction zone, and makes the elevation of the land much higher. Also, many subduction zones experience compression, when the two plates are pushed together by plate movements elsewhere. In places like that, great masses of rock are pushed together and stacked on top of one another in complicated structures, to form high mountains. This happens also where two continents collide with each other, as in the Himalayas.

Growth of Continents at Subduction Zones.

During the long travels of an oceanic plate from a mid-ocean ridge to a subduction zone, a hundred meters or more of oceanic sediment is deposited on the top of the plate. In Part 6 of the investigation, you saw how a lot of this sediment is scraped off and added to the edge of the other plate. This material, which is deformed into very complicated structures, is turned into rock by heat and pressure. It becomes a solid part of the other plate. When material is added to the edge of a continent in this way, the continent grows larger at its edge. Continents also grow as the igneous rock of volcanoes and batholiths are added to the continent above the subduction zone, as described above. The growth of a continent along its edge in these ways is called **continental accretion**. This has been going on through geologic time, making the continents larger and larger.

Earthquakes and Plate Tectonics

As plates move past each other at plate boundaries, they don't always slide smoothly. In many places the rocks hold together for a long time and then slip suddenly. You will learn more in Chapter 3 about how earthquakes are caused in this way. Earthquakes along mid-ocean ridges are common because of movement along the transform faults that connect segments of the ridge crest. Only where transform faults are on land or close to land, as in California, are these earthquakes likely to be hazardous. As you saw in Activity 2, the famous San Andreas fault in California is a transform fault. Earthquakes at subduction zones and continent–continent collision zones are a bigger problem for human society, because these areas are so common on the Earth, especially around the rim of the Pacific and along a belt that stretches from the Mediterranean to southeast Asia. Earthquakes in subduction zones happen at depths that range from very shallow, near the trench, to as deep as hundreds of kilometers, along the subducted plate. Earthquakes in continent–continent collision zones happen over wide areas as one continent is pushed under the other.

Geo Words

continental accretion: the growth of a continent along its edges.

Check Your Understanding

1. Why is plate tectonics a suitable name for the study of plate motion. Explain.
2. What geographic features would you expect to see at plate boundaries.
3. How do geoscientists suggest that "hot spots" are related to plate tectonics.
4. In your own words explain the process of continental accretion.

Understanding and Applying What You Have Learned

1. Review your work with earthquakes and volcanoes one more time.
   a) Summarize where most earthquakes are located compared to plate boundaries.
   b) Summarize where most volcanoes are located compared to plate boundaries.
2. Although most earthquakes and volcanoes are associated with plate boundaries, they are not always located directly along the boundaries. Considering boundaries between oceanic and continental plates:
   a) Why are volcanoes usually found on the continental side of a plate boundary?
   b) Why are earthquakes usually found on the continental side of a plate boundary?
3. Many volcanoes and earthquakes are found far from modern plate boundaries. Write a paragraph giving one idea you think might explain how at least some of them have formed. Be sure to point out examples by describing their location.
4. Make a list of the various plate-tectonic settings where mountain ranges are likely to be produced. For each item on the list, draw a cross section that shows the mountain range and how it relates to the plate-tectonic setting. For each item, give an example from somewhere in the world.

Preparing for the Chapter Challenge

In this activity you have learned that plate tectonics has an influence on every part of the world. Surface features and movements caused by plate tectonics are not only at plate boundaries, but also at locations far from modern plate boundaries. Write a paragraph, based on what you now know of movements or features in your region that indicate plate-tectonic activity (whether it is nearby or far away). Carefully list your evidence so you can present it to the middle school class.

Inquiring Further

1. **Plate tectonics and the local climate**

   Distant mountain ranges and plateaus created by plate tectonics can affect air flow in many ways, affecting local climate and thus vegetation, soil, wildlife, and drainage patterns. Research and report on how your local region has been affected directly or indirectly by plate tectonics. You may even wish to include some of this research in your **Chapter Challenge.**

Activity 5 The Changing Geography of Your Community

Goals

In this activity you will:

- Use several present-day distributions of minerals, rock formations, and fossils to help figure out the distribution of continents.
- Construct a map showing the position of continents 250 million years ago by reversing the present direction of plate motion.
- Recognize a convergence of presently widely scattered minerals, rock formations, and fossils when all the continents were part of Pangea.
- Compare present average community motions with that of the past 250 million years, by calculating the average yearly rate of motion over the last 250 million years.
- Describe the context in which the hypothesis of continental drift was proposed and why it was subjected to criticism.
- Show that your community has moved through different ecological regions over time.

Think about It

The plates forming the Earth crust can be compared to pieces of newspaper torn from the same page.

- How would you be able to decide if the pieces all came from the same page?
- How could you convince someone else that the pieces came from the same page?

What do you think? Record your ideas about these questions in your *EarthComm* notebook. Include sketches as necessary. Be prepared to discuss your responses with your small group and the class.

Investigate

1. Begin your work individually. Obtain four copies of the diagram, showing the outline of the continents at sea level, as well as the boundary between the continental crust and the oceanic crust. The diagram also shows the location of rock formations, mountain ranges, and fossil plants and animals. Cut out the continents on the first sheet along the edges of the continental shelves, which in most places are close to the boundaries betwen the oceanic lithosphere and the continental lithosphere.
   a) Why cut the pieces at the boundaries between the continental and oceanic crust?
   b) In which ecological region is your community today: tropics, subtropics, mid-latitudes, subpolar, or polar?
   c) Coal deposits originated in the swamps of tropical forests. Are the coal deposits shown on the map in the tropics today?
   d) Where do you find mountains similar in structure to the Appalachian Mountains?
   e) Where do you find rock formations similar to those in South America?
   f) Glossopteris is an extinct seed fern that had leaves like ferns of today. It produced seeds too large to travel by air or float on water. Where are fossils of these ferns located today?
   g) Mesosaurus is an extinct freshwater reptile that thrived during the Triassic Period (245 to 208 million years ago). Where are fossils of this reptile found today?
2. Rearrange the cut pieces on a blank piece of paper as the continents now appear and tape them in place.
   a) Label the outlines "Present."
   b) Draw a border around the map.
   c) Sketch in and label the Equator and latitude lines at 30° and 60° north and south.
   d) Title the map "Present."

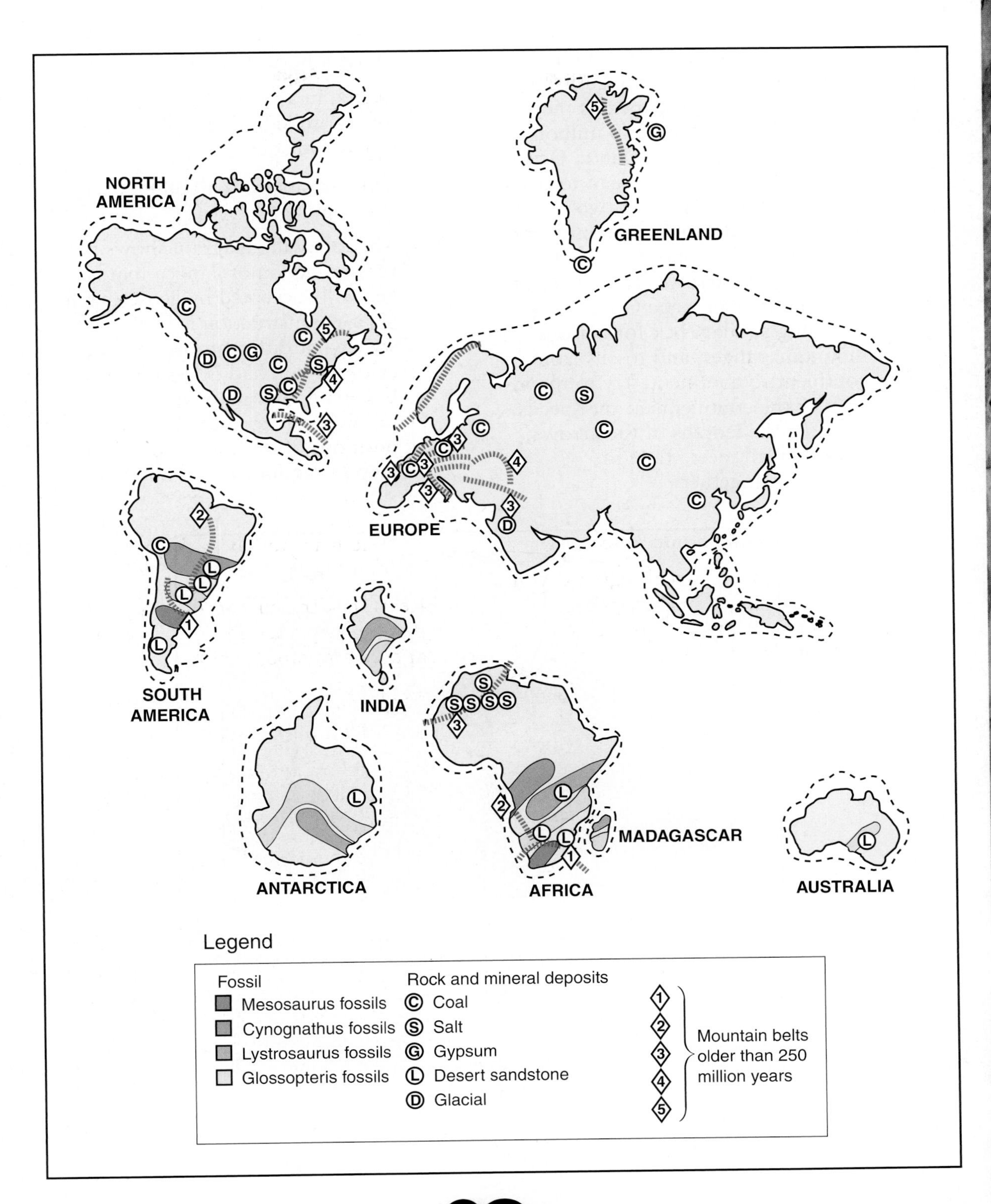
NORTH AMERICA
GREENLAND
EUROPE
SOUTH AMERICA
INDIA
ANTARCTICA
AFRICA
MADAGASCAR
AUSTRALIA
Legend
Fossil
Mesosaurus fossils
Cynognathus fossils
Lystrosaurus fossils
Glossopteris fossils
Rock and mineral deposits
C Coal
S Salt
G Gypsum
L Desert sandstone
D Glacial
1 2 3 4 5 Mountain belts older than 250 million years

3. Cut out the continents from the second sheet. Try to arrange them on another piece of paper, as they would have appeared 250 million years ago, before the Atlantic Ocean and the Indian Ocean began to open. You can do this by using two methods: (a) moving each continent in the direction opposite of that shown by the arrows on the map of plate motions on page G63; (b) matching similar rock formations, mountain ranges, and fossils from continent to continent. Try to move each of the continents at the speeds given by the lengths of the arrows, until they all meet. Tape the continents together.
    a) Draw a border around the map.
    b) Sketch in and label the Equator and latitude lines at 30° and 60° north and south.
    c) Title this map, "250 million years ago."
    d) The following diagram shows the reconstruction of Pangea that is generally accepted by geoscientists. Your reconstruction is likely to be somewhat different, because the evidence you had is less detailed. Compare your map with the following map, and adjust the positions of the continents on your map as required.

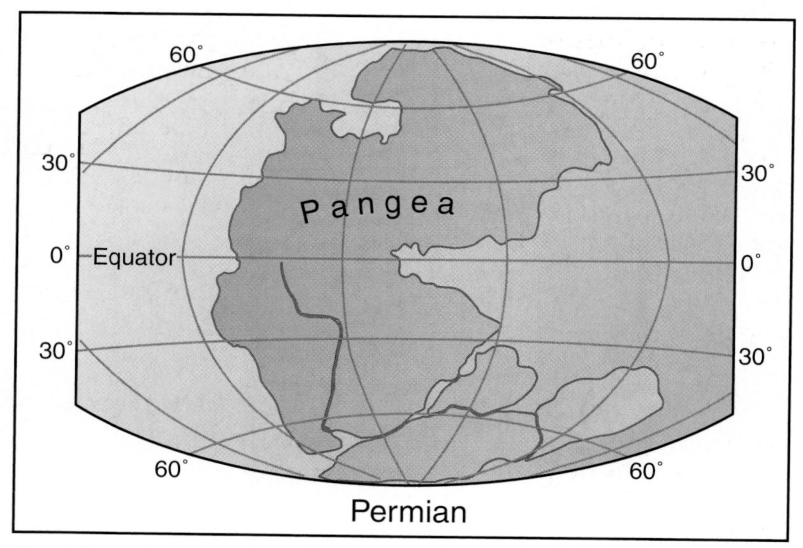

Generally accepted reconstruction of Pangea, in the Permian period of geologic time, 225 million years ago.

4. Use your adjusted map, "250 million years ago" to answer the following questions:
   a) Which two continents fit together best?
   b) Why do you think the continents do not fit together exactly?
   c) From the map of Pangea, what can you say about the latitude and longitude of your community 250 million years ago?
   d) In what ecological region was your community 250 million years ago?
   e) Many coal deposits were created before 280 million years ago in the tropics. Where were they 250 million years ago? Does this make sense? Explain your answer.
   f) Do the Appalachian Mountains line up with other mountain ranges that they resemble?
   g) Do rock formations in South America line up with other formations that they resemble?
   h) How does Glossopteris appear to have migrated to its present fossil distribution, since its seeds could not be carried by the wind or float on water?
   (i) How does Mesosaurus appear to have migrated to its present fossil locations, since it could not swim in the salty ocean?

5. Cut out the continents from the third sheet, in the same way as before. Arrange them on a new piece of paper, as they might appear 250 million years in the future. You can do so by starting with the present distribution of the continents and then moving each in the direction and at the speed shown by the arrows on the map of plate motions. Remember that some plates will be subducted under others.
   a) What will fill the spaces between the continents in the future?
   b) What will happen to the Mediterranean Sea? What will be created in southern Europe?
   c) Where will the southern coast of California be in 250 million years?
   d) In what latitude and in which ecological region might your community lie in 250 million years?
   e) How might the change in ecological region affect your community?
   f) Why might your prediction regarding the future location of your community and continent be in error?

Reflecting on the Activity and the Challenge

You have seen that by moving the continents "in reverse," in directions opposite to their present movement, you can make the continents fit together fairly well as a single continent. You have also seen that features like rock formations, mountain ranges, and fossil plants and animals that are similar but are now separated by wide oceans are brought together when the continents are assembled into the single large continent of Pangea. You have gotten some idea about how far a particular place on a continent might have moved in the 250 million years since Pangea broke apart.

Digging Deeper

DEVELOPMENT OF THE PLATE TECTONICS THEORY

In this activity, you examined some of the evidence that supports the idea that the continents of the Earth have moved during geologic time. Two features of the Earth were the subject of intense study in the late 1800s—the discovery of similar fossils on continents that are now separated widely by oceans, and the origin of mountain ranges. Both played a part in the early stages of the development of theory of plate tectonics.

In the late 1800s, an Austrian geologist named Eduard Suess (1831–1914) tried to solve a basic geological question: how do mountain ranges form? He based his model of mountain formation on some of the same principles that you explored in this chapter. Suess stated that as the Earth cooled from a molten state, the more dense materials contracted and sank toward the center, and the least dense materials "floated" and cooled to form the crust. He then speculated that mountain ranges formed from the contraction and cooling of the Earth. He likened this to the way that an apple wrinkles and folds as it dries out and shrinks.

Suess went on to explain the origins of oceans, continents, and the similarities of fossils on different continents now separated by oceans. In his model, during the cooling process, parts of the Earth sank deeper than others, forming the ocean basins. Suess claimed that certain parts of the sea floor and continents could rise and sink as they adjusted to changes in the cooling earth. This led him to propose land bridges between continents. Suess coined the term Gondwanaland for a former continent made up of central and southern Africa, Madagascar, and peninsular India. These areas all contained similar fossils that were hundreds of millions of years old. According to Suess, the land bridges allowed various animals and plants to migrate and spread without crossing an ocean.

Although other geologists proposed different models to explain mountains, oceans, and fossils, all generally agreed that the Earth's crust moved up and down, but not very far sideways. Land bridges were often cited as allowing various kinds of organisms to move between continents now separated by oceans. According to Suess and others, the land bridges sank into the ocean long ago and no longer exist.

Not all geologists accepted the theory of a contracting Earth. In 1912, the German geoscientist Alfred Wegener (1880–1930) proposed the hypothesis of continental drift. He saw a variety of problems with the contraction theory. One difficulty was the severe compression of the Alps. The Alps are a young mountain range. Rock layers in the Alps are severely folded and stacked up on top of one another, indicating a great horizontal shortening of original distances, as shown in *Figure 1*. Wegener thought that contraction could not produce such great shortening of the Earth's crust. He also thought that contraction should produce uniform "wrinkles" in the Earth, not narrow zones of folding. The discovery of radioactive heat in Wegener's time also provided evidence against cooling. Heat from radioactive decay in the Earth would work against the cooling and contraction process.

Figure 1 Wegener used the severe compression of the Alps as evidence to support his hypothesis of continental drift.

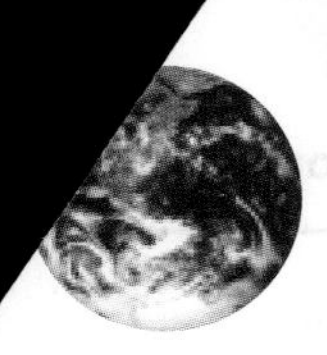

Geo Words

Pangea: Earth's most recent supercontinent which was rifted apart about 200 million years ago.

According to Wegener, about 200 million years ago, a huge supercontinent called **Pangea** (Greek for *all land*) broke into separate continents that moved apart. Wegener claimed that compression at the leading edge of the moving continent led to the formation of mountains. Wegener's hypothesis allowed him to explain the different ages of the different mountain belts. He claimed that the timing of the breakup was variable, with some parts of Pangea separating earlier than others. His evidence included the puzzle-like fit of the continents and the similarity of rocks, geologic structures, and fossils on opposite sides of the Atlantic Ocean. Wegener's hypothesis eliminated the need for (now sunken) land bridges that once connected widely separated continents.

But how did continents move? Wegener thought that the material beneath the Earth's lithosphere acts like a slow-moving fluid. If this is true for vertical movements, it should also be true for horizontal movements. To visualize Wegener's argument, think about a piece of candy taffy or "silly putty." At the right temperature, taffy that will shatter when struck with a hammer will deform by flowing rather than by breaking when a force is applied slowly and constantly. Although other geologists saw folded mountains as evidence of contraction, Wegener saw folded mountains as evidence of horizontal compression caused by movement of the continents. The presence of folded mountains convinced Wegener that forces within the Earth are powerful enough to move continents. A quote from Wegener summarizes his ideas about the way that all the geological evidence "fit together":

"It is just as if we were to refit the torn pieces of a newspaper by matching their edges and then check whether the lines of print run smoothly across. If they do, there is nothing left to conclude but that the pieces were in fact joined this way. If only one line was available to the test, we would still have found a high probablility for the accuracy of fit, but if we have *n* number of lines, this probability is raised to the *nth* power."

The reaction to Wegener's hypothesis was mixed. Some scientists accepted his arguments. Others argued that it would be impossible for continents to "plow through" the ocean floor (see *Figure 2*). Most geoscientists rejected Wegener's hypothesis. At an international meeting in 1926 devoted to the discussion of continental drift, only a handful of scientists were sympathetic to Wegener's ideas. One scientist raised 18 different arguments against Wegener's evidence! Although his evidence for drift was strong, the mechanism he proposed for the drift of the continents was inadequate.

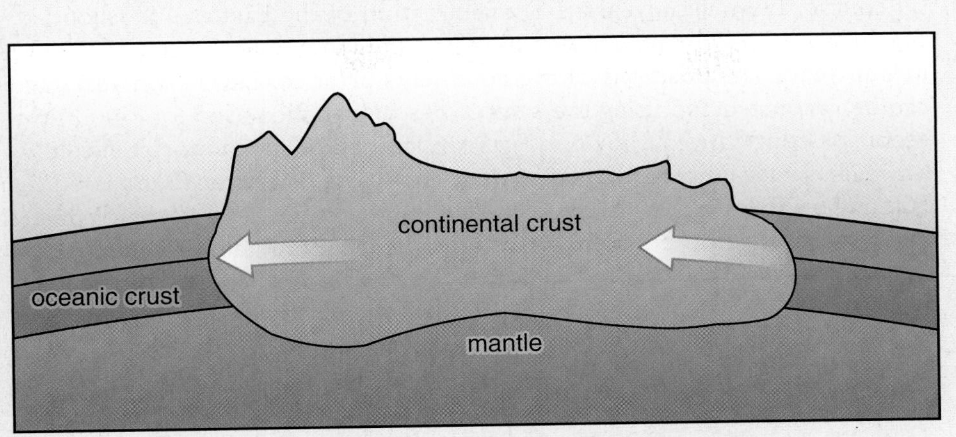

Figure 2 Wegener's proposal that continents plowed through oceanic crust was not accepted by many other geologists.

Convincing evidence began to emerge after World War II, as the sea floor was explored and mapped extensively. By the late 1960s, the theory of plate tectonics had been developed based on many types of evidence. Today, this evidence is considered so abundant and convincing that almost all geoscientists accept the theory. Much of the evidence that Wegener used to support his hypothesis supports plate tectonics. However, new evidence has emerged that provides a more plausible mechanism for the movement of the Earth's lithosphere.

Using evidence such as magnetic striping on the ocean floor (described in Activity 1), ages of ocean-floor basalts, outlines of continental plates, and the locations of similar fossils and rock types on widely spaced continents, geologists have reconstructed the record of the breakup of the supercontinent Pangea. Pangea started to break up about 200 million years ago, as continental rifts (divergent zones) began to open and oceanic crust began to form. As Pangea continued to be rifted apart, oceanic crust formed between the northern continents (called Laurasia) and the southern continents (called Gondwana). New ocean floor also was formed between Antarctica and Australia and between Africa and South America. India started to separate from Antarctica and travel northward.

The maps shown in *Figure 3* summarize what has been reconstructed as the breakup of Pangea, from 225 million years ago to the present. As you can see, continents that are now connected were not always that way, and continents that are now widely separated once were part of the same land mass.

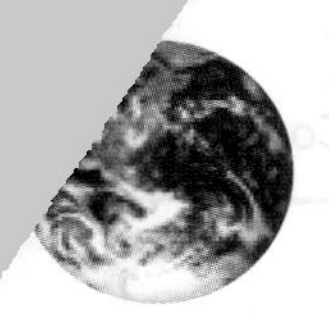

Geo Words

supercontinent: a large continent consisting of all of the Earth's continental lithosphere. Supercontinents are assembled by plate-tectonic processes of subduction and continent–continent collision.

Of course, 225 million years is a small fraction of the Earth's 4.5 billion year history. There may be rocks in your community much older than 225 million years. The positions of the continents prior to 225 million years ago can be reconstructed using the same types of evidence used for reconstructing Pangea, shown in *Figure 4*. The task is much more difficult, however, since the oldest oceanic crust geologists have ever found is only 200 million years old. Thus, the evidence for the earlier geography of the Earth must be gathered from the continents. Old mountain belts such as the Appalachians of North America and the Urals (which separate Europe from Asia) help locate ancient collision zones between continents of the past. Rock types and fossils provide evidence for the locations of ancient seas, glaciers, mountains and ecological regions. Continents like Pangea, which consist of all of the Earth's continental lithosphere in one single piece, are called **supercontinents**. Geoscientists are fairly sure that there was at least one earlier supercontinent before Pangea, and maybe others as well. The cycle of assembly, breakup, and reassembly of supercontinents is called the Wilson Cycle.

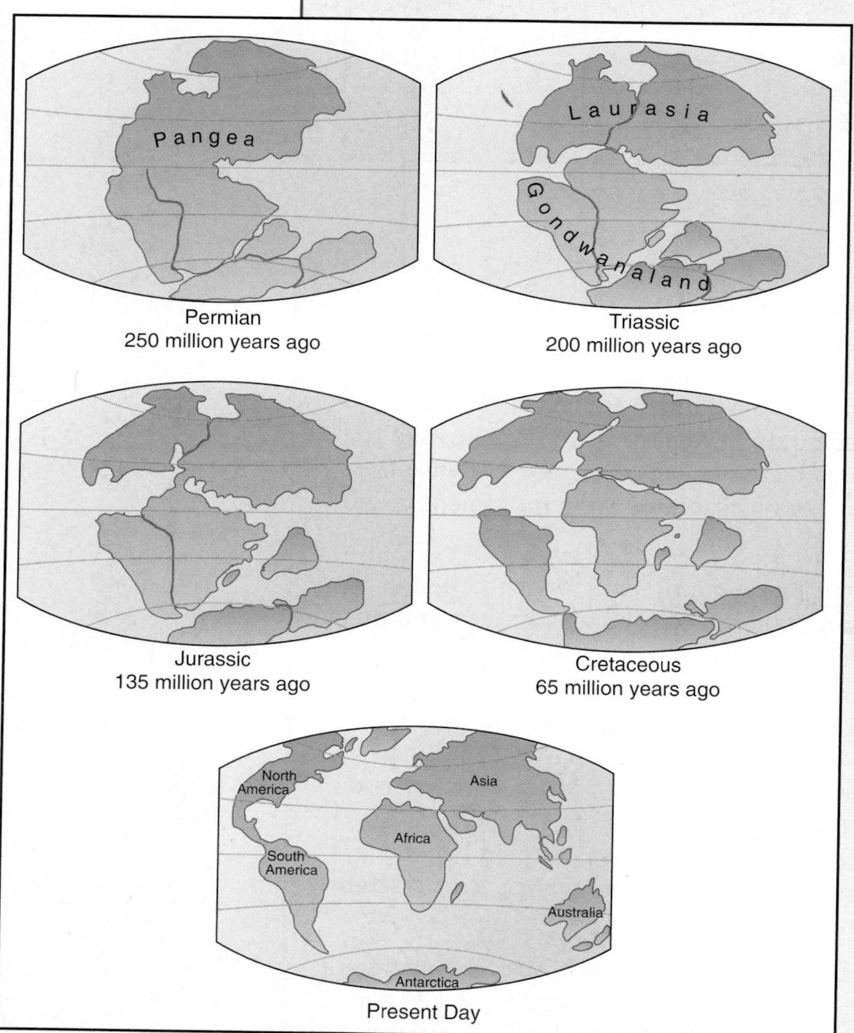

Figure 3 The breakup of Pangea.

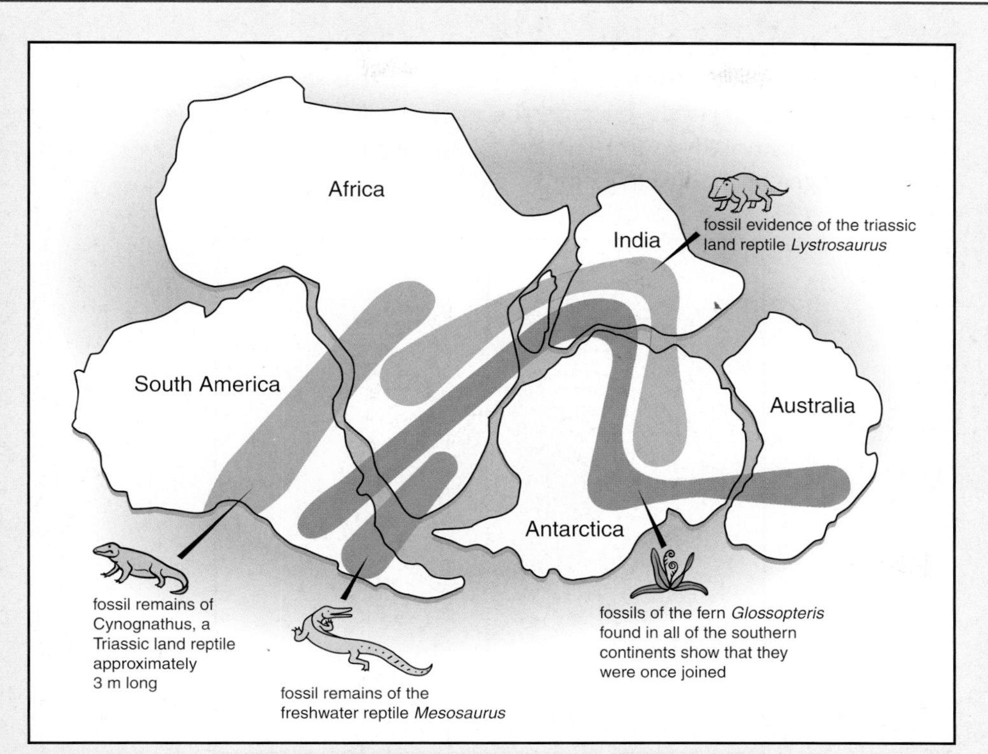

Figure 4 Evidence used to reconstruct Pangea.

Paleomagnetism

You learned in Activity 1 that the mineral magnetite "locks" the Earth's magnetic field into its atomic structure as it cools. Geoscientsts collect rock samples containing magnetite and measure the past magnetism the rocks record (called **paleomagnetism**). They do this by putting the sample in a special room that is arranged so that the present Earth's magnetic field is canceled out. The Earth's magnetic field has the same pattern that would be observed if there were a giant bar magnet inside the Earth, lying along the Earth's axis of rotation. There isn't really a big magnet in the Earth; the magnetic field is thought to exist because of movements of liquid iron in the Earth's core. *Figure 5* shows how the lines of the Earth's magnetic field are arranged. The angle that the magnetic field lines make with the Earth's surface changes from the Equator to the poles. Near the Equator the lines are nearly horizontal, and near the poles they are nearly vertical. This means that the paleomagnetism of a rock sample can tell you the latitude of the sample when it formed, called the paleolatitude. Measurement of

Geo Words

paleomagnetism: the record of the past orientation and polarity of the Earth's magnetic field recorded in rocks containing the mineral magnetite.

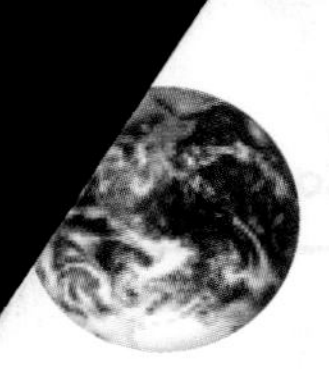

paleolatitudes is one of the things geoscientists use to reconstruct past supercontinents like Pangea. The big problem is that there is no way of measuring paleolongitude, because the magnetic field lines are always oriented about north–south! That's why no longitude lines are shown on the map of Pangea in the investigation.

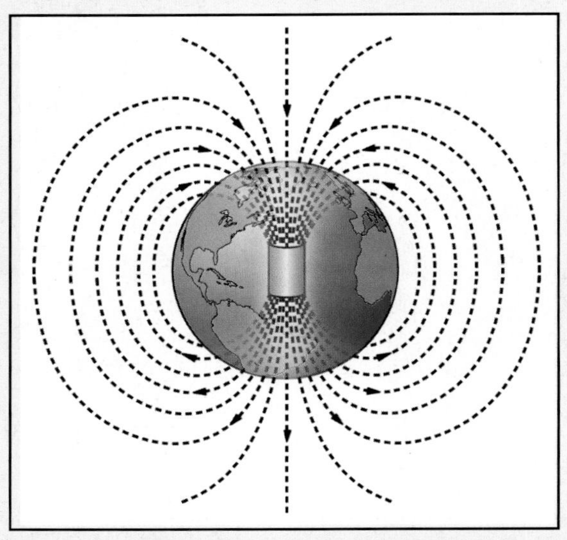

Figure 5 The diagram shows how the lines of Earth's magnetic field are arranged.

Check Your Understanding

1. How did Suess explain the formation of mountain ranges?
2. What evidence was found to contradict Suess's proposal that the Earth is cooling and shrinking?
3. What evidence did Wegener use to support his theory of the breakup of Pangea?
4. How did Wegener propose that the continents move horizontally?
5. How was fossil evidence used to reconstruct Pangea?

Understanding and Applying What You Have Learned

1. Geoscientists often try to figure out paleogeography (the geography of land and sea in the geologic past) using the clues given on your continent puzzle pieces. What additional evidence would you need to be more confident about your "225 Million Years Ago" map?
2. Paleoclimatology is also used to show how continents were connected in the past. What type of climate data might have been helpful to you in making your "225 Million Years Ago" map?
3. Why was the theory of continental drift questioned when it was first proposed by Alfred Wegener?
4. What discoveries helped scientists begin to accept the idea that parts of the Earth's lithosphere move? Why were the more modern clues not available in Wegener's time?

5. New scientific theories often take many years to be accepted by the scientific community. Explain why this is so, using the theory of plate tectonics as an example.
6. The theory of plate tectonics is now accepted by almost all geoscientists. The theory overcomes the objections many scientists had to the idea of continents moving around the globe. How does plate tectonics explain the seeming movement of continents through rigid oceanic crust?
7. Describe what has happened to the lithosphere under the Atlantic Ocean during the last 200 million years. What has happened to the lithosphere under the Pacific Ocean? How does this information support the theory of plate tectonics?

Preparing for the Chapter Challenge

In your notebook, write a brief essay describing how the arrangement of continents and oceans in the geologic past can be figured out, and what are the limitations to doing this.

Try to devise a way of animating the movements of all of the continents from the time of the breakup of Pangea to the present time, so that the middle school students can watch the movements with their own eyes. Write up your ideas in your notebook. Compare your ideas with those of others in the class.

Inquiring Further

1. **History of science**

   The history of the development of the theory of plate tectonics is a fascinating one. A very important piece of evidence that supported plate tectonic theory was the discovery of paleomagnetism in ocean bottom basalts. How was this paleomagnetic evidence of sea-floor spreading discovered?

   Write down in your notebook at least one additional question you have about the geologic history of your community. How would you go about gathering information to answer these questions? Write your ideas in your notebook.

2. **Plate tectonics and the Earth system**

   Write an essay explaining how Earth systems would change if plate tectonics were to "stop." You might begin with something directly connected to plate tectonics, such as volcanism or mountain building. For example, *"If plate tectonics were to cease, then global volcanism..."*

Earth Science at Work

ATMOSPHERE: *Meteorologist*
In predicting weather patterns, meteorologists must take into account the geological features of the surrounding land.

BIOSPHERE: *Paleontologist*
Geoscientists study the fossils of plants and animals and their relationship to existing plants and animals in order to "reconstruct" the geologic past.

CRYOSPHERE: *Tour Guides*
Some of the most spectacular scenery can be found on and off the shores of Alaska. Tourists are interested in the geography of the areas they visit.

GEOSPHERE: *Civil Engineer*
Transportation networks, whether within a city, or within the country, are a vital lifeline in today's society. The public relies on safe roads and bridges.

HYDROSPHERE: *Harbor Master*
Harbor masters and marina operators are responsible for the vessels that are tied up at their docks. Millions of dollars worth of boats could be housed in a single marina.

How is each person's work related to the Earth system, and to plate tectonics?

3
Earthquakes...
and Your Community

Earthquakes ...and Your Community

Getting Started

An average of about 350,000 earthquakes are detected on Earth each year.

- What causes an earthquake?
- How can an earthquake affect your community?

What do you think? In your notebook, draw a picture that shows a side view of the Earth's crust. Show what causes an earthquake, and how an earthquake can affect a community. Write a caption to explain your drawing. Be prepared to discuss your picture and ideas with your small group and the class.

Scenario

The following description of the 1906 San Francisco earthquake was written by author Jack London:

"Within an hour after the earthquake shock, the smoke of San Francisco's burning was a lurid tower visible a hundred miles away.

...There was no opposing the flames. There was no organization, no communication. The earthquake had smashed all the cunning adjustments of a twentieth century city. The streets were humped into ridges and depressions, and piled with the debris of fallen walls. Dynamite was lavishly used, and man crumbled many of San Francisco's proudest structures himself into ruins, but there was no withstanding the onrush of the flames. The troops were falling back and driving the refugees with them. From every side came the roaring of flames, the crashing of walls, and the detonations of dynamite."

Can you use your knowledge of earthquakes to develop a plan to ensure that your community would be prepared to deal with an earthquake?

Chapter Challenge

In many states of the U.S., damaging earthquakes can happen at any time. If you live along the Pacific Coast, in a mountain belt, or in the central Midwest, they may even be a very great hazard. Your community has asked your school to assess the earthquake hazard in your area and to find ways to reduce any damage. You have been asked to present these plans to the public as a public service message (video or audio) and in a brochure.

Think about how your message and brochure will address the following items:

- Educate the public about why earthquakes occur.
- Explain how earthquakes transmit energy.
- Explain the effects associated with earthquakes.
- Suggest ways to reduce the damage caused by earthquakes.

Assessment Criteria

Think about what you have been asked to do. Scan ahead through the chapter activities to see how they might help you prepare your brochure. Work with your classmates and your teacher to define the criteria for judging your work. Record all this information. Make sure that you understand the criteria as well as you can before you begin. Your teacher may provide you with a sample rubric to help you get started.

Activity 1 An Earthquake in Your Community

Goals

In this activity you will:

- Generate and describe two types of waves.
- Determine the relative speeds of compressional and shear waves.
- Simulate some of the motions associated with earthquakes.
- Infer the origin of earthquakes and the mechanism of transfer of seismic wave energy.

Think about It

In some earthquakes, the ground shakes back and forth so much that small objects are overturned, hanging objects and doors swing, and pictures are knocked out of plumb.

- If you have experienced an earthquake, describe your most vivid memory. If you have not experienced an earthquake, what would you expect to see, feel, and hear?

What do you think? Record your ideas about this question in your *EarthComm* notebook. Be prepared to discuss your response with your small group and the class.

Investigate

Part A: Rupture and Rebound

1. Obtain two "L"-shaped wooden blocks that have a slot cut in their short lengths, as shown in the diagram. Place the blocks so that the slots line up. Put a thin piece (less than 1 mm) of Styrofoam® into the two slots so that it connects the two blocks.

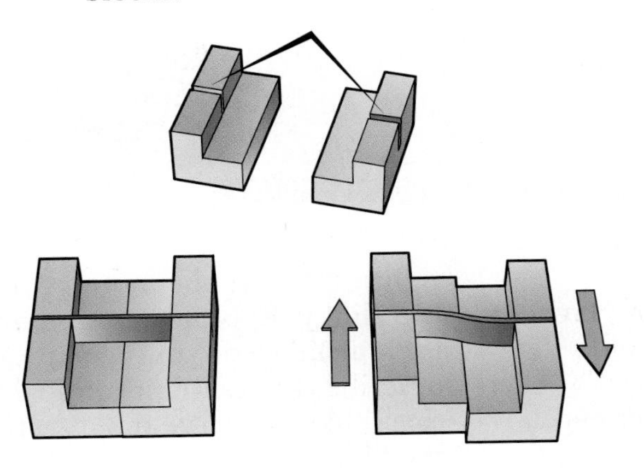

2. Put on your safety glasses. Holding the two blocks together, move the blocks parallel to each other, but in opposite directions. Do this very slowly. Gradually increase the offset between the blocks without breaking the Styrofoam.

   a) Record and sketch your observations in your notebook.

3. Continue to increase the offset until the Styrofoam snaps. Watch carefully what happens to the Styrofoam as it ruptures.

   a) Record and sketch your observations in your notebook.

   b) Did you feel a vibration in the wood? When?

Part B: A Thought Experiment on Rupture and Friction

1. The experiment with the Styrofoam strip shows some important things about the cause of earthquakes, but it's not a very realistic model of rocks. "Thought experiments" are experiments that scientists dream up and then run in their imagination, rather than in a real laboratory. They are a useful way to develop new ideas and insights into scientific problems. Here's a thought experiment that gets you closer to the real thing:

   Suppose you could make a "fake rock" that is just like a real rock, except that its strength is so much less that you can easily break it with your hands.

   Hold the two sides of the rock as you did to break the Styrofoam. As you gradually increase the sliding (shearing) force on the rock, it bends, just like the Styrofoam, but because it is much stiffer, like a real rock, you probably wouldn't be able to see the bending. Eventually the rock breaks along a line, and the two sides spring back to their original shape. "Seismic" waves are generated.

Styrofoam should not splinter. Do not substitute other material.

a) If you started to apply a shearing force again, with the two halves still pressed together, what do you think you would observe? Why would you need to exert some force to make the two halves slide past one another, even though they are already cut by a fracture?

b) Imagine pouring some liquid containing a bit of glue down into the fracture and waiting for the glue to set. The rock regains some of its strength. What happens when you again apply a greater and greater "sliding" force to the "healed" block of rock? Do you expect the force needed for rupture to be greater than originally, less, or the same? Why?

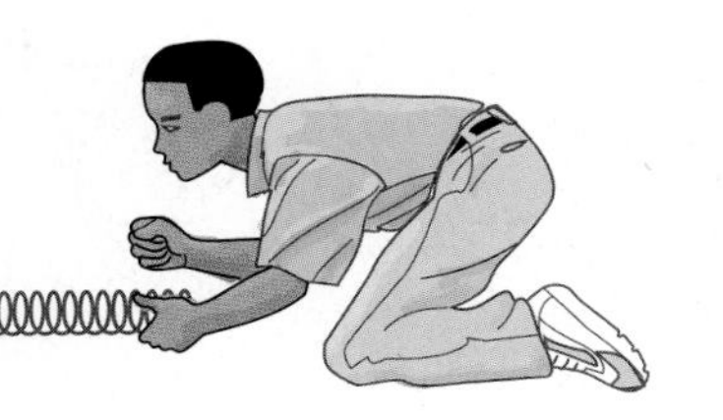

Part C: Vibration

1. Place a Slinky® on the floor. Have one person in your group hold one end of the Slinky and a second person hold the other end. Back away from each other so that the Slinky stretches out about 5 m long.

2. Have one person hold the end of the Slinky in a fist and then hit the back of the fist with the other hand. The palm should be facing the Slinky. Observe the motion of the Slinky. Repeat the pulse until each member of your group can describe the motion of the Slinky. Observe the direction the Slinky moves compared to the direction in which the pulse is moving.

   a) Record your observations.

3. With the Slinky stretched out about 5 m, have a person at one end quickly jerk the end of the Slinky back and forth (left to right). Observe the motions of the Slinky. Repeat the pulse until each member of your group can describe the resultant motion of the Slinky. Observe the direction the Slinky moves compared to the direction in which the pulse is moving.

   a) Record your observations.

4. To compare the two types of vibrations (wave motion) that you observed in the Slinky, stretch out two Slinkys along the floor, about 5 m. Starting at the same end at the same time, have one student strike their fist while the other student jerks the Slinky back and forth. Observe what happens. Try the movements several times until you are confident in your observations.

   a) Which of the two wave types arrives at the other end first (which one is faster)?

⚠ A stretched Slinky can move unpredictably when released. Spread out so that you can work without hitting anyone. Release the stretched Slinky gradually.

Reflecting on the Activity and the Challenge

In the experiments, you observed rupture, energy release, and energy transmission. These are the main processes in the occurrence of earthquakes.

In the experiment with the Styrofoam, you gradually applied a force to a solid material (the Styrofoam strip). The force caused the strip to bend. Bending like that is called elastic deformation. If you had removed the force before the strip broke, the strip would have returned to its original shape. Rocks in the Earth's crust behave in the same way. The material broke when the force exceeded the strength of the material. This instantly released the energy you had stored in the material by applying a force to bend it. In your model, you felt the sudden release of energy during rupture as a vibration in the wood. As the Styrofoam strip broke, its ends "jumped" a short distance in opposite directions, to straighten themselves out again.

In the thought experiment with the weak, fake rock, the force you applied caused the rock to rupture and slide along a fault plane. Friction along the fault plane made it necessary for you to apply a non-zero force to get the two halves to slide past one another even after they were cut by the fault. After the fault was partly healed, some force was still needed for the rock to slip along the fault plane again.

Energy can be transmitted from one place to another without permanent movement of the material. In the experiment with the Slinky, the energy of motion you put into the solid (the Slinky) by shaking it at one end was transmitted away from the source, without the Slinky changing its position after the waves had passed. Earthquake waves similar to the ones that you modeled carry the energy of the earthquake for long distances as they travel through the Earth.

Digging Deeper

WHAT IS AN EARTHQUAKE?

Earthquakes

An **earthquake** is a sudden motion or shaking of the Earth as rocks break along an extensive surface within the Earth. The rock masses on either side of the fault plane slip past one another for distances of as much as ten meters during the brief earthquake. The rocks break because of slowly built-up bending. The sudden release of energy as rock ruptures causes intense vibrations called **seismic waves** or earthquake waves.

Geo Words

earthquake: a sudden motion or shaking in the Earth caused by the abrupt release of slowly accumulated strain.

seismic (earthquake) waves: a general term for all elastic waves in the Earth, produced by earthquakes or generated artificially by explosions.

Geo Words

fault: a fracture or fracture zone in rock, along which the rock masses have moved relative to one another parallel to the fracture.

shear strength: the shear force needed to break a solid material.

elastic rebound: The return of a bent elastic solid to its original shape after the deforming force is removed.

friction: the force that resists the motion of one surface against another surface.

focus: the point of an earthquake within the Earth where rupture first occurs to cause an earthquake.

epicenter: the point on the Earth's surface directly above the focus of an earthquake.

fault scarp: the cliff formed by a fault that reaches the Earth's surface.

Geoscientists explain the occurrence of earthquakes in the following way. A **fault** is a surface between two large blocks or regions of rock, along which there has been rupture and movement in the past. Faults are very common in the rocks of the Earth's crust. Large-scale forces within the Earth's crust push the fault blocks in opposite directions. Most of these forces are caused by the movements of the Earth's plates.

As the forces gradually build up over time, the blocks are bent on either side of the fault, the same as with the Styrofoam strip. The region of bending can extend for very long distances away from the fault. All rocks have a shear strength. The **shear strength** of a rock is the force that is needed to break the rock when it is acted upon by forces in two opposite directions. Eventually the forces overcome the shear strength of the rock, and the rock breaks along the fault plane. The blocks then suddenly slip for some distance against each other to undo the bending, and stored energy is released. The straightening movement is called **elastic rebound**.

Usually, the rocks in a fault zone have already been ruptured by earlier earthquakes. Why don't they just slip continuously as force is applied? The answer is that in some places, they do slip continuously. In most places, however, the fault becomes "locked" and doesn't move again for a long time. There are two reasons for this. One reason is that there is a lot of **friction** along the fault plane, because the rock surfaces are rough and are pressed together by the great pressure deep in the Earth. You can see for yourself how effective this rock friction is, by gluing sandpaper to two wooden blocks and then trying to slide the sand-papered surfaces past one another while you squeeze the blocks together. The other reason is that new minerals tend to be deposited along the fault by slowly flowing water solutions. This new mineral material acts as a "cement" to restore some of the shear strength of the rock.

Earthquakes usually occur at some depth below the surface. The place in the Earth along the fault where rupture occurs is called the earthquake **focus**, as shown in *Figure 1*. The **epicenter** is the geographic point on the Earth's surface directly above the focus. Once a fracture starts, it spreads rapidly in all directions along the fault plane. It often reaches the Earth's surface. Where it does, the motion on the fault can cause a sharp step in the land surface, called a **fault scarp**. Fault scarps can be as much as a few meters high. Horizontal motions along a fault can cause roads or fences to be offset, by as much as 10 to 15 m.

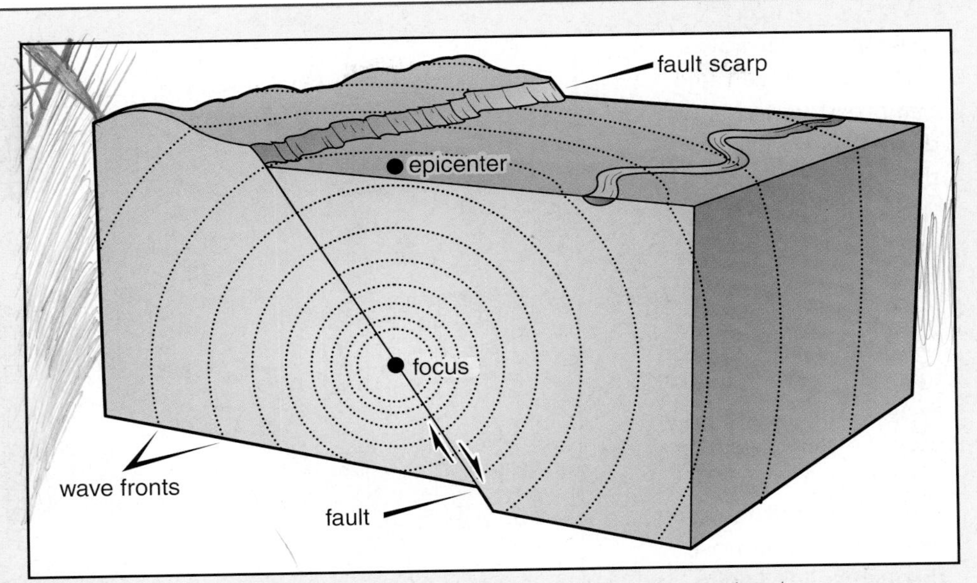

Figure 1 The relationship between the focus and the epicenter of an earthquake.

Earthquake or Seismic Waves

When an earthquake occurs by rupture along a fault, the elastic energy of bending is released and seismic waves spread out in all directions from the focus. Earthquakes produce several kinds of seismic waves. The different kinds of waves travel through rocks at different speeds, and each kind of wave causes a different kind of motion in the rock as it passes by. The various kinds of waves arrive at some distant point on the Earth at different times, depending on their relative speed and their path though the Earth. (See *Figure 2*.)

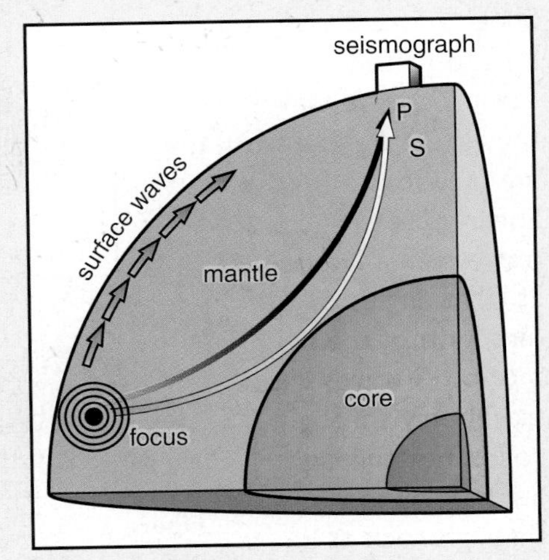

Figure 2 Earthquakes produce several types of seismic waves.

Geo Words

primary wave (P wave): a seismic wave that involves particle motion (compression and expansion) in the direction in which the wave is traveling.

secondary wave (S wave): a seismic wave produced by a shearing motion that involves vibration perpendicular to the direction in which the wave is traveling. It does not travel through liquids, like the outer core of the Earth.

surface wave: a seismic wave that travels along the surface of the Earth.

Check Your Understanding

1. What is an earthquake?
2. Explain how seismic waves are generated by an earthquake.
3. What is the relationship between the focus and the epicenter of an earthquake?
4. Use a diagram to describe the differences between P waves, S waves, and surface waves.
5. Rank P waves, S waves, and surface waves in order from fastest to slowest.

Compressional waves (*Figure 3a*) cause rapid compression and expansion of rock as they pass through the Earth. As the waves pass, the rock material is moved back and forth in the direction of wave motion. Compressional waves are the first to reach a location away from the focus, so they are called **primary waves**, or just P waves. P waves are similar to sound waves. They can move through solids, liquids, and gases. They move through solid rock at a speed of about five kilometers per second, or about fifteen times the speed of sound in air.

Shear waves arrive at a location after compressional waves, so they are called **secondary waves**, or just S waves. Shear waves (*Figure 3b*) move rock material at right angles to the direction of their motion. S waves can travel only through solids, not through fluids. They move through rock at a speed of about three kilometers per second.

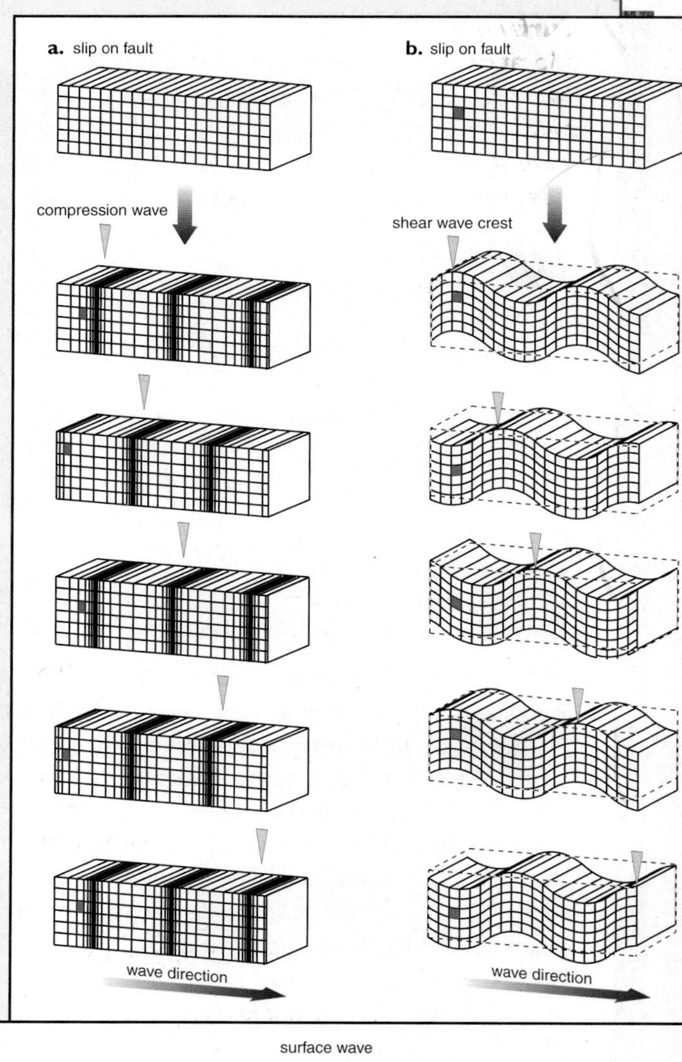

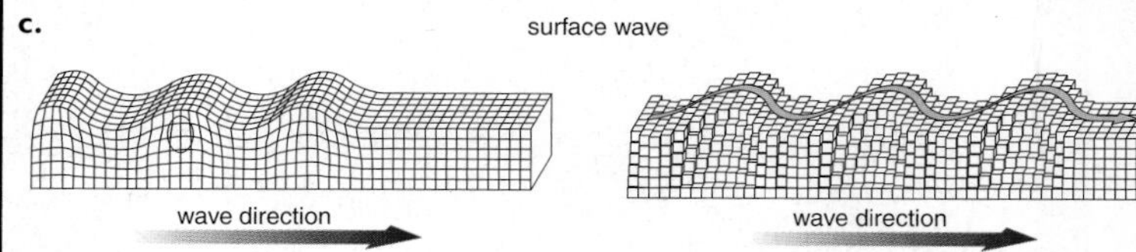

Figure 3 This diagram shows how **a.** primary (compressional), **b.** secondary (shear), and **c.** surface waves move through the Earth.

Surface waves, which travel along the Earth's surface, are the last to arrive at a location. They travel slower than S waves. There are two kinds of surface waves. One kind creates an up-and-down rolling motion of the ground, very much like a wave on a water surface (*Figure 3c.i*). The other kind of surface wave shakes the ground sideways (*Figure 3c.ii*). Surface waves usually cause the most movement at the Earth's surface, and therefore the most damage.

Understanding and Applying What You Have Learned

1. What kinds of motion would you expect to feel in an earthquake?
2. What effects might earthquake motions have on buildings, roads, and household furnishings?
3. Of the types of earthquake waves discussed in this section, which do you think are the most dangerous? Why?
4. Many people have some common beliefs about earthquakes. One of these is that earthquakes occur more frequently in areas of warm climates.
   a) How would you design an investigation that might test this idea?
   b) Do you have information available to you that either supports or contradicts this idea?
   c) Write a short paragraph either supporting or refuting this belief.
5. What other ideas about earthquakes did you have before doing these activities that were either supported or contradicted by what you have learned through your investigation? Describe your original ideas and how they were either confirmed or refuted.
6. Some faults are frequently active and produce numerous small earthquakes. Other faults are rarely active but produce large earthquakes. Based on the activities you completed, propose factors that might influence the number and size of earthquakes produced by a fault.
7. In the rupture activity, you provided the energy needed to break the styrofoam. Use this idea to describe why earthquakes reveal that Earth is a dynamic planet.

Preparing for the Chapter Challenge

Write a background summary for the brochure you will prepare for your **Chapter Challenge.** Include a concise, simple, but accurate explanation for the cause of earthquakes, how they transmit energy, and how different types of seismic waves move. Be sure to address any common beliefs that you may know to be false. This section should be no longer than one page. Include diagrams as appropriate.

Inquiring Further

1. **Forming questions to investigate**

   Write down other questions you have about the causes of earthquakes and their effects. How would you go about gathering information to answer these questions? Write your ideas in your notebook.

2. **Using seismic waves to study the Earth's interior**

   How do we know about the structure and composition of the interior of the Earth? Study of the distribution and effects of earthquakes, and especially of the transmission of seismic waves, has enabled geoscientists to develop many answers. To learn more about the details of the Earth's structure revealed by the study of seismic waves, visit the *EarthComm* web site.

3. **Using seismic waves to explore for oil and gas**

   Understanding the behavior of seismic waves allows geoscientists to use them as tools to study deep layers of the Earth. Find out how exploration geologists use seismic waves to draw inferences about the layers of sedimentary rock in which they find oil and gas deposits. Consult the *EarthComm* web site.

4. **Earth science careers**

   Do you think you would like to study earthquakes for a career? To see what a seismologist does at work, visit the *EarthComm* web site.

Marine seismic vessel.

Activity 2 Detecting Earthquake Waves

Goals

In this activity you will:

- Construct a simple seismometer.
- Record motion in two dimensions and also within a fixed time frame.
- Understand how seismometers record earthquake waves.
- Recognize P waves, S waves, and surface waves on seismograms.
- Read a graph to determine the distance to the earthquake focus.
- Locate the epicenter from time–distance graphs.

Think about It

Some of the energy released by an earthquake takes the form of seismic waves. Surface waves are responsible for most earthquake destruction.

- What specific observations would you want to make to study an earthquake?
- How could you detect and record the arrival of earthquake waves: P waves, S waves, and surface waves?

What do you think? Record your ideas about these questions in your *EarthComm* notebook. Be prepared to discuss your responses with your small group and the class.

Investigate

1. Attach one end of a spiral spring or thick rubber band to a small, heavy weight (a non-lead fishing-line sinker would work well). Attach the other end of the spring or rubber band to the bottom of a rectangular, open-sided storage box, such as a small milk crate. Turn the box upside down so that the weight is suspended and hanging freely.

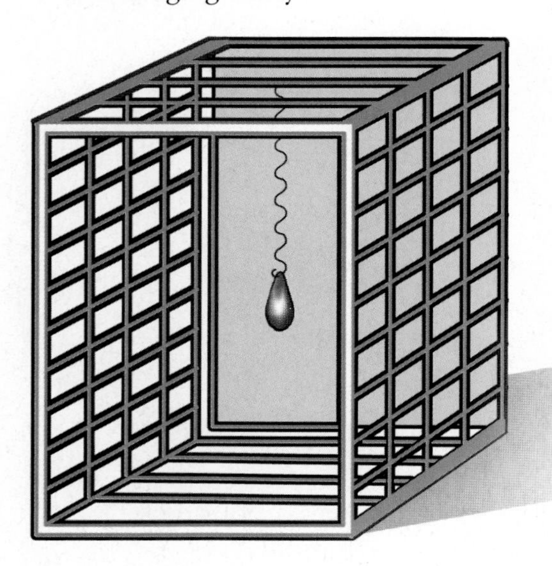

2. Move the frame of the box rapidly back and forth (horizontally). Now move the box vertically up and down. Move it back in forth in one direction, then back and forth in the other direction.

   a) In your notebook, write a careful description of what you observe.

   b) Are the motions you generate similar to the motions produced by the Slinky in **Activity 1**, or different?

   c) How would you describe the motions of the weight in comparison to the motions you impart to the box?

Be sure the spring and weight are securely fastened to the crate.

3. Obtain a piece of heavy paper or light cardboard and a very soft pencil or thin felt-tipped marker. Hold the marker firmly in place above the paper so that its tip is just touching the paper. This can be done by having the member of your group with the steadiest hand hold the marker in place above the paper. Have another group member move the paper under the marker in order to write the word "Earthquake" in cursive. Move only the paper, not the pen.

   a) Record what you observe and how this writing is achieved.

4. Drag the paper across the table smoothly toward you (with the tip of the marker touching the paper). Then pull the paper toward you again, but this time jiggle it back and forth perpendicular to the direction in which you're pulling.

   a) What does the line look like when the paper was pulled smoothly toward you?

   b) What does the resulting line look like when the paper was jiggled?

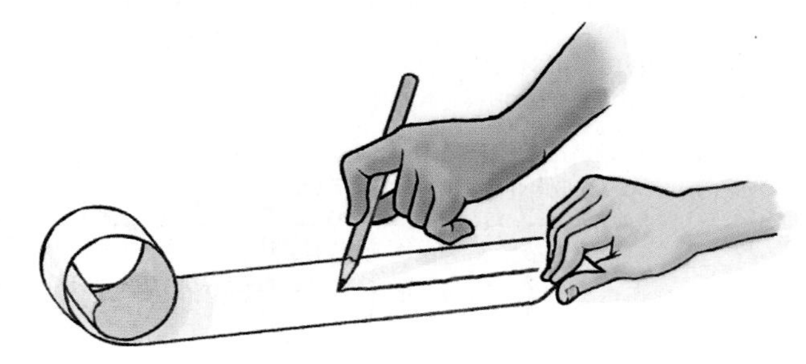

5. Repeat step 4, but this time use a timer or the second hand on a watch to record the time it takes to pull the paper through. Use a roll of adding machine paper this time so you have a strip of paper a meter or so long. Have a third person make a little mark on the edge of the paper strip every second as you move the strip along.
6. Examine a record of a real earthquake as shown in the figure below.
   a) Is the size (height) of the recorded wave the same for the entire duration shown on the seismogram?
   b) Is the shape of the recorded wave the same for the entire duration shown on the seismogram?
7. Make or obtain a copy of the diagram showing a real earthquake.
   a) Label the arrival of the P wave and the S wave.
   b) How much time separates the arrival of the two waves?
   c) Use the diagram and the difference in arrival times to determine the distance from the focus to the seismometer.

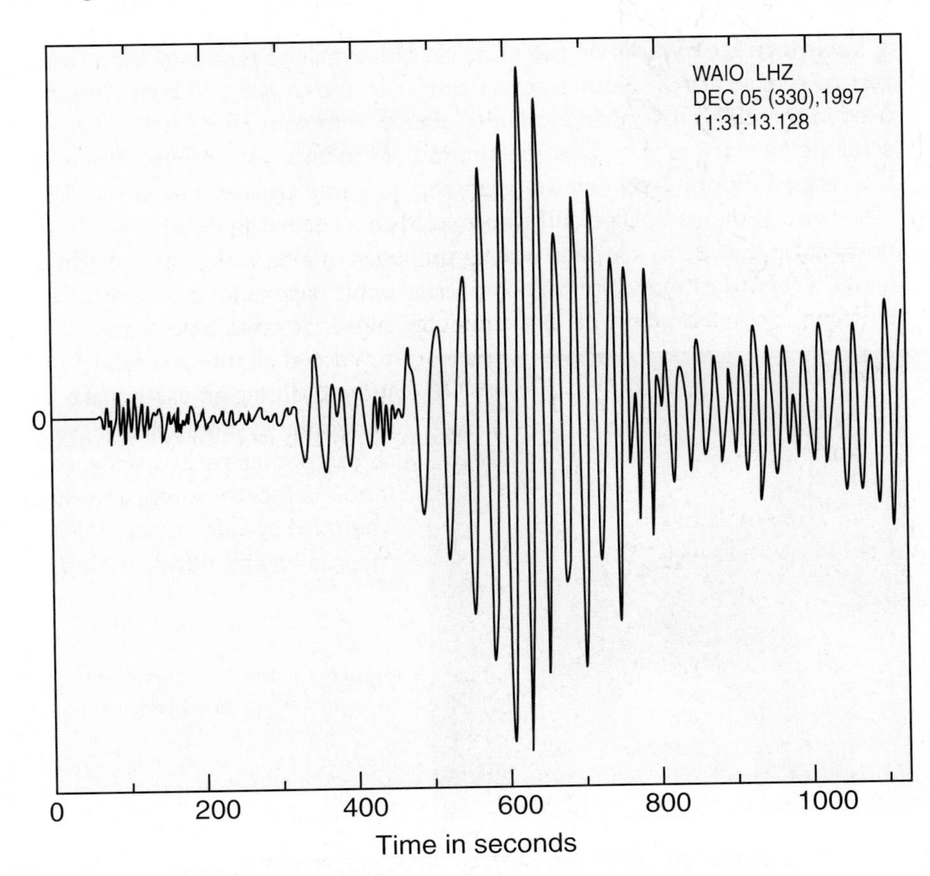

Reflecting on the Activity and the Challenge

You constructed a simple instrument that recorded passing vibrations. Your instrument used a stationary mass (like the fishing-line sinker) to provide a fixed reference against which vertical and horizontal movement could be compared. By marking time on your record, you could determine the "arrival" of the waves you generated. What you created is a very simple seismometer. A seismometer is a device for measuring shaking. You also made a seismogram, or record of shaking. Geologists use similar (but more complex) instruments to record passing earthquake waves. Time is also marked on their records. This allows the arrival times of P waves, S waves, and surface waves to be determined.

Digging Deeper

RECORDING EARTHQUAKE WAVES

Seismometers

A **seismometer** works on the principle of inertia, the tendency for a mass at rest to remain at rest. Seismometers similar to the device you built were first used in the 1800s. They had a cylinder coated with soot and a stylus that scratched a mark as it registered vibrations. In today's instruments (*Figure 1*), the relative motion between the mass and its frame creates an electric signal. The signal is then amplified and transmitted to a recording destination. The destination may be a) ink pens writing on paper, b) a narrow beam of light that leaves a record of the vibrations on photographic paper, c) a device that records a magnetic signal on tape, or d) a computer screen. It takes three seismometers to record all the motions of the ground during an earthquake. Two horizontal cylinders at right angles to each other record sideways motions (north–south, east–west). The third cylinder is vertical. It records up and down motions.

Geo Words

seismometer: an instrument that measures seismic waves. It receives seismic impulses and converts them into a signal like an electrical voltage.

Figure 1 In the 1960s, a worldwide network of seismometers was developed to verify nuclear test-ban treaties. When a nuclear device is tested, seismometers around the world record the seismic waves that result from the blast.

Instruments used to detect earthquakes also record any motion of the ground to which they are attached. These motions can be natural (earthquakes, landslides) or caused by humans (large trucks, passing airplanes and helicopters, blasting during construction, nuclear bombs).

Interpreting Seismograms

A **seismogram** is a written or mechanically produced record of earthquake waves. *Figure 2* shows a seismogram recorded at Dallas, Texas. Note the separation of P waves and S waves on the seismogram. This seismogram was recorded about 1600 km from the earthquake's focus. If it had been recorded near the focus, the two waves would appear much closer together. All the waves are produced during the rupture. As distance from the focus increases, the separation and arrival times between the wave types increase because they travel at different speeds.

Geo Words

seismogram: the record made by a seismometer.

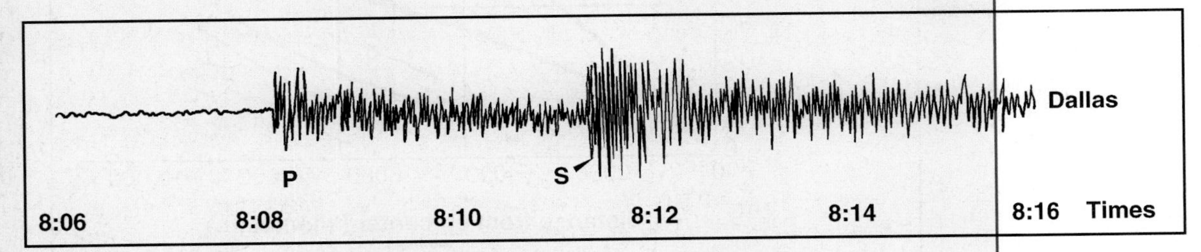

Figure 2 Seismogram recorded in Dallas, Texas.

P waves travel the fastest through the Earth, so they arrive first at a distant station. S waves arrive shortly after. Waves that arrive after the direct P waves complicate the seismogram. Various reflected and refracted P and S waves bounce off (are reflected at) layers of Earth's interior and eventually reach the station.

Using Travel-Time Curves

P and S waves travel at different speeds , so they arrive at different times at a seismological station. The difference in their arrival times increases with the distance from the focus. Travel-time curves (see *Figure 3* on the next page) show this relationship.

Check Your Understanding

1. What is the function of a seismometer?
2. a) How many seismometers do you need at a given place to fully record the motions arising from earthquake waves?
   b) How should these seismometers be oriented, and why?
3. What is a seismogram?
4. What information is provided by a travel-time curve?

The graph shows the data from a magnitude 8 earthquake in the Kuril Islands on 3 December 1995. Here you can see the relationship between distance and the difference in arrival times for P waves and S waves.

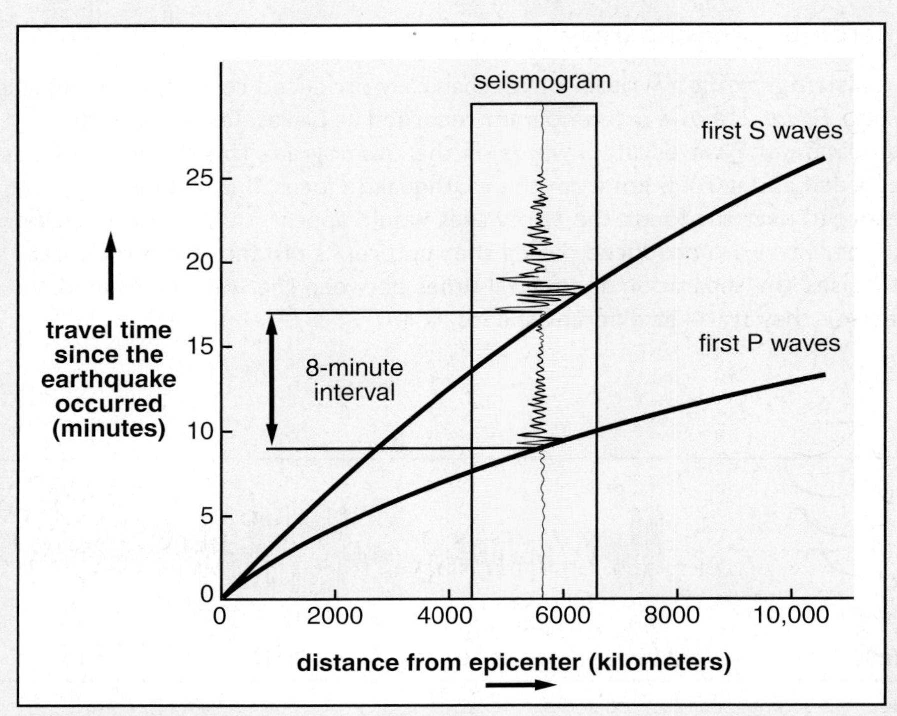

Figure 3 Travel-time curve for an earthquake.

Understanding and Applying What You Have Learned

1. a) Where would be the best place to put a seismometer in your school? Why would you choose this location?
   b) Where would be the worst place(s) in your school to put a seismometer? Why should this (these) location(s) be avoided?
   c) Where in your community might be a good place to put a seismometer? Why?
   d) At what locations in your community is it unwise to place a seismometer? Why?
2. a) How are the devices you worked with in this activity good models of a modern seismometer?
   b) In what respects are they poor models?
   c) What could you do to improve your models?

3. What advantages would be gained by having more seismometers at a particular location?
4. Not all vibrations of the Earth are made by natural earthquakes.
   a) Write in your notebook as many things as you can think of that could cause strong vibrations of the Earth's surface.
   b) How might you be able to distinguish seismograms of "natural" earthquakes from "human-made" earthquakes?

Preparing for the Chapter Challenge

Write a summary of this activity for your brochure. Include a concise, simple, but accurate description of how to detect earthquake waves and locate epicenters. Also record where to place a seismometer at your school (or somewhere else in your community). Explain why you chose this location.

Inquiring Further

1. **History of science**

   The study of earthquakes has a fascinating history. Humans have always felt the effects of earthquakes. Early civilizations interpreted the shaking of the Earth in different ways. How have the methods used to study earthquakes changed over time? Use electronic or print resources to prepare a report.

In 132 A.D., a Chinese scholar named Chang Heng made one of the earliest known devices used to record the occurrence of an earthquake.

2. **Recent seismic activity**

   Visit the *EarthComm* web site for a description of how to conduct an on-line investigation into seismic waves that travel complicated paths within the Earth.

3. **Virtual earthquakes**

   Find out about the Virtual Earthquake web site at the *EarthComm* web site. Practice using seismographs to find an earthquake epicenter. Simulate an earthquake in the region of your choice. Print out a record of your results. Include the seismograms and the map showing the epicenter location, but do not do the magnitude activity at this time.

Activity 3 How Big Was It?

Goals

In this activity you will:

- Rank the effects of earthquakes.
- Map the intensity of earthquakes.
- Interpret a map of earthquake intensity to infer the general location of the epicenter.
- Identify geologic materials that pose special problems during earthquakes.
- Explain how the magnitude of an earthquake is determined.

Think about It

Earthquakes can be so small that people can't feel them. They can be so great that the seismic energy of a magnitude 8 earthquake would supply one day's worth of electrical energy for the entire U.S.

- What factors would you look at to "measure" the size of an earthquake?
- What factors are most important for different uses? What measurements would interest each of the following: a seismologist? a city planner? a homeowner? you?

What do you think? Record your ideas about these questions in your *EarthComm* notebook. Be prepared to discuss your responses with your small group and the class.

Investigate

Part A: Measurement of Earthquake Effects

1. The reports listed below describe the effects of an earthquake felt in many cities in the eastern U.S.

   a) In your group, rank the effects in order of intensity. Assign numerical values to your ranking. Be prepared to explain your ranking system.

Newspaper data:

Detroit, MI: "Did you feel the earthquake? Last night a slight earth tremor was felt in this area."

Pittsburgh, PA headline: "Items Broken By Earthquake. Small Earthquake Awakens Many."

Syracuse, NY: "Mayor's 10th-floor office chandelier sways in the earthquake."

Baltimore, MD: "Citizens dash outside as earthquake brings down plaster in many homes."

Philadelphia, PA: "Doors and windows rattle as earthquake strikes."

Cleveland, OH: "Earthquake felt on top of new 23-story department store."

Roanoke, VA: "Pedestrians report parked cars rocked back and forth by earthquake."

Charleston, SC: "Almost no one here notices earthquake."

New York City: "Skyscraper offices sway in earthquake."

Washington, DC: "Chimneys tumble, new prefabricated buildings collapse in violent earthquake felt here."

Richmond, VA: "Furniture moved about by earthquake, but no major damage reported."

Winston-Salem, NC: "Patients in hospital report that building shakes in earthquake."

Atlanta, GA: "Earthquakes reported north of here apparently miss Atlanta."

Indianapolis, IN: "Few people here feel recently reported earthquake."

2. Use the data and your ranking to map the intensity of earthquake effects on the map of the Eastern United States.

   a) Draw several curves on the map connecting points or regions of equal intensity.

   b) Label each curve with the intensity value it represents.

   c) Attach your intensity scale.

3. Use the map to answer the following questions:

   a) Describe any pattern you observe.

   b) What can you infer about the probable location of the epicenter of the earthquake? Provide a reason.

   c) What problems did you have finding the epicenter?

   d) What other information do you think would have helped you to locate the earthquake more accurately?

   e) How would you describe your measurement scale? Is it qualitative or quantitative?

   f) What property or properties of the earthquake were you measuring? Explain.

4. Share your results with the class.

   a) How do the rankings of earthquake effects compare?

   b) How close were the estimates of the epicenter of the earthquake?

Part B: Measurement of Earthquake Wave Amplitude

1. Visit the *EarthComm* web site to find the location of the Virtual Earthquake web site. Simulate a new earthquake.

   a) Follow the directions to calculate the magnitude of the earthquake.

   b) How does the amplitude (height) of a seismic wave change when the size of an earthquake changes?

   c) How would you expect the amplitude as recorded on a seismogram to change as you get farther from the epicenter?

Reflecting on the Activity and the Challenge

To reduce earthquake risks, the potential effects of earthquakes should be studied. You made an intensity scale by ranking the effects of earthquakes on people and structures. You used the distribution of intensities to find the general area where an earthquake started. On the Virtual Earthquake web site, you learned that the amplitude of seismic waves on a seismogram is used to measure magnitude. For most earthquakes, both intensity and magnitude decrease as you move farther away from the epicenter. Intensity requires human observers. Magnitude requires a seismometer.

Digging Deeper

DESCRIBING EARTHQUAKES

Earthquake Intensity

The effects of an earthquake on the Earth's surface, including people and buildings, are an indication of its intensity. Intensity scales are based on certain key responses to the shaking of an earthquake. Examples include people awakening, damage to brick and stone structures, and movement of furniture. The intensity scale currently used in the United States is called the Modified Mercalli Intensity scale. There is no mathematical or quantitative basis for the scale. It is simply a ranking based on observed effects, as with the activity you just completed.

Earthquake intensity is a measure of the actual effects at a certain location. This makes an intensity value more meaningful than magnitude values to a non-scientist. The maximum observed intensity often occurs near the epicenter, but there are important exceptions to this rule. During the 1985 Mexico City earthquake and the 1989 Loma Prieta earthquake, the areas with maximum intensity were not nearest to the epicenter. In the Mexico City earthquake, the epicenter was hundreds of kilometers away, yet areas within the city experienced much higher intensities.

Modified Mercalli Scale of Earthquake Intensity

Value	Description of Effects
I	• not felt, except rarely by a few
II	• felt by few, especially on upper floors of buildings • delicately suspended objects may swing
III	• felt indoors by some • vibration similar to a passing truck • not always recognized as an earthquake
IV	• felt indoors by many, outdoors by few • awakens some sleeping people • dishes, windows, and doors rattle, walls creak • standing cars rock • hanging objects swing
V	• felt indoors by mostly everyone, outdoors by many • awakens most sleeping people • some dishes break, windows and plaster walls crack • small unstable objects overturned • hanging objects and doors swing considerably, pictures knocked out of plumb • some liquid spilled from full containers

continued on next page

Modified Mercalli Scale of Earthquake Intensity ***(continued)***

Value	Description of Effects
VI	• felt by everyone • general excitement and some fear • slight damage to poorly built structures • considerable amount of glassware and windows broken • some furniture overturned, and some heavy furniture moved • pictures and books fall from walls and shelves • some fallen plaster and damaged chimneys • small bells ring
VII	• everyone frightened, some have difficulty standing • negligible damage to well-designed, well-built structures, slight to moderate in well-built ordinary structures, considerable in poorly built or designed structures • weak chimneys broken • large church bells ring • trees and bushes shaken, water in lakes and ponds disturbed, some stream banks collapse
VIII	• general fear, approaching panic • slight damage to structures designed to withstand earthquakes, considerable damage to ordinary structures, great damage in poorly built or designed structures • heavy furniture overturned • chimneys and monuments topple • sand and mud ejected from the ground • changes in flow of wells and springs
IX	• general panic • considerable damage to structures designed to withstand earthquakes, well-designed buildings shifted off of foundations, great damage and partial collapse of substantial buildings • ground noticeably cracked • some underground pipes cracked
X	• severe damage to well-built wooden structures, most masonry structures and foundations destroyed, well-built brick walls cracked • bridges severely damaged or destroyed • ground severely cracked • considerable landslides from steep slopes and river banks • sand and mud on beaches shifted • water splashes over banks of canals, rivers, and lakes • underground pipes broken • open cracks and wavy patterns in concrete and asphalt • railroad tracks slightly bent

continued on next page

Modified Mercalli Scale of Earthquake Intensity *(continued)*	
Value	**Description of Effects**
XI	• few structures remain standing • bridges destroyed • broad fissures in the ground • Earth slumps and landslides in soft, wet ground • sand and mud-charged water ejected from the ground • underground pipelines completely out of service • railroad tracks greatly bent
XII	• total damage to all works of construction • numerous rock and landslides, river banks slump • waves seen on the ground • objects thrown up in the air

The Effect of Local Geologic Conditions on Intensity

Often, although not always, seismic waves increase in amplitude when they pass from solid bedrock to softer material near the Earth's surface, like sand, mud, or landfill material. The physical processes that cause this change are complicated, but you can get some idea of them by thinking about how a soft gelatin-like material responds to a source of shaking, compared to a much more rigid material like metal or rock. An earthquake that shook Mexico City in 1985 showed how local geologic conditions influence the intensity (and damage) of an earthquake. It caused about $4 billion in damage and killed at least 8000 people. The earthquake epicenter was about 300 km from Mexico City. Soft sand and clay deposits from an old lake bed under part of the city amplified (increased) the ground motions by a factor of 75 times. The amplification of shaking caused selective damage to tall buildings. Nearby structures on bedrock were relatively undamaged.

Another geologic process that affects earthquake intensity is liquefaction. Liquefaction is the temporary change of water-saturated soil and sand from a solid to a liquid state. Areas like the Marina district in San Francisco experienced very high intensities during the Loma Prieta earthquake. Studies revealed liquefaction of the wet landfill on which the district was built. The earthquake was centered 80 km south of the city. Nearby parts of the city built on hard bedrock did not experience intensities as high as in the Marina district.

Earthquake Magnitude

Earthquake magnitude is a measure of the amplitude of the seismic waves recorded on a seismogram. Charles F. Richter, a seismologist at the California Institute of Technology, developed the first magnitude scale in the 1930s. His basic idea was to observe the maximum amplitude recorded on a seismogram at a known distance from an earthquake. Earthquakes could be ranked quantitatively by size or strength. The amplitude measured is the "swing" of the stylus. The wider the swing, the "stronger" the vibrations, and therefore the stronger the earthquake. You modeled this in Activity 2.

The original Richter scale was developed only for shallow earthquakes detected in southern California by a certain type of seismometer. Since then it has been modified for more general use. Modern measurements of magnitude are still based on the amplitude of the recorded waves. However, more sophisticated methods are used today. Depending on which magnitude a seismologist is using to calculate the magnitude of an earthquake, the various magnitudes can vary by one unit or more.

An important aspect of magnitude scales is that they are logarithmic, based on powers of 10. This means that seismic wave amplitudes increase by 10 times for each unit of the scale. For example, the measured amplitude in a magnitude 6 earthquake is 10 times the measured amplitude in a magnitude 5 earthquake.

Check Your Understanding

1. Why is the intensity value of an earthquake more meaningful than magnitude to a non-scientist?
2. Use the Modified Mercalli Scale to determine the intensity of the earthquake for each of the following observations:
   a) "I was awakened from a deep sleep and observed the door to the bedroom swinging back and forth."
   b) "I thought it very unusual to see a small wooden decoy vibrate on the table beside my chair."
   c) "I noticed waves on the pond, and could hear the church bells ring."
3. Is the greatest intensity of an earthquake always found at the epicenter? Explain your answer.
4. What geological conditions influence the intensity of an earthquake?
5. What does an earthquake magnitude scale measure?

Understanding and Applying What You Have Learned

1. The Magnitude/Intensity Comparison table below compares the magnitude and intensity scales. The intensities listed are those typically observed at locations near the epicenter of earthquakes of different magnitudes. If intensity and magnitude measure different characteristics of earthquakes, how can such a chart be compiled? Why isn't it like comparing apples to oranges?

Magnitude/Intensity Comparison	
Magnitude	**Intensity**
1.0–3.0	I
3.0–3.9	II–III
4.0–4.9	IV–V
5.0–5.9	VI–VII
6.0–6.9	VIII–IX
7.0 and higher	X or higher

2. What is the highest intensity/magnitude that you would consider exciting to experience, but not dangerous? Explain your reasoning.

3. The following table shows the number of earthquakes per year of a given magnitude.

   Use the table to answer the following questions:

   a) Roughly how many earthquakes occur in a given year?
   b) Do humans feel most earthquakes?
   c) How many earthquakes did you hear about in the last year?
   d) What were their approximate magnitudes?
   e) Does this generate a bias in your perception of the number of earthquakes that happen per year and their size? Explain your answer.

Number of Earthquakes per Year of Given Magnitudes		
Description	**Magnitude**	**Number per Year**
Great earthquake	Over 8.0	1 or 2
Major earthquake	7.0 to 7.9	18
Destructive earthquake	6.0 to 6.9	120
Damaging earthquake	5.0 to 5.9	800
Minor earthquake	4.0 to 4.9	6000
Smallest usually felt by humans	3.0 to 3.9	49,000
Detected but not felt	2.0 to 2.9	300,000

Gutenberg, B, and Richter, C.F., 1954, *Seismicity of the Earth and Associated Phenomena*, Princeton University Press, Princeton, NJ.

Preparing for the Chapter Challenge

As a group, put together a text box for your brochure explaining how earthquakes are measured. You may use diagrams or charts. The box should contain a simple but accurate explanation of the two methods. It should also explain the advantages and limitations of each method. Remember that this will only be a sidebar for your brochure. It will help people to understand the meanings of terms you use in the rest of the brochure.

Inquiring Further

1. **Reporting earthquakes**

   Does your community experience frequent earthquakes? Maybe you would like to help seismologists when an earthquake happens in your community. The Earthquake Felt Report Form allows you to contribute your intensity observations directly to seismologists so that they can construct isoseismal maps. Visit the *EarthComm* web site page now to find the address that will let you know the kinds of observations you need to detect and record.

2. **Determine the intensity of an earthquake from a description**

   Do you know someone who has experienced an earthquake? If so, ask the individual to describe the experience to you. Compare the person's descriptions to the Mercalli Intensity Scale. Are the descriptions of earthquake effects in the Mercalli intensity scale consistent with the person's experience? Use the scale to rate the intensity of an earthquake they experienced. If they recall the approximate magnitude of the earthquake they experienced, how well does the magnitude/intensity chart match their experience?

3. **Investigate earthquake measurement**

   Write down other questions you have about the ways earthquakes are measured. How would you go about gathering information to answer these questions? Write your ideas in your notebook. Visit the *EarthComm* web site for suggestions of useful web sites to explore.

Activity 4 Earthquake History of Your Community

Goals

In this activity you will:

- Recognize patterns in the global distribution of earthquakes.
- Interpret maps and research written information to determine the earthquake history of the community and region.
- Examine correlations between faults and earthquakes on a regional and community scale.
- Assess the likelihood of future earthquakes in the community.
- Interpret graphical data to examine long-term trends in the number of earthquakes in the United States.

Think about It

Earthquakes occur all over the world, every day, but not many are strong enough for people to feel them.

- Is there a general pattern to where earthquakes occur on Earth?
- Has an earthquake occurred in your community? Could it occur in your community?

What do you think? Record your ideas about these questions in your *EarthComm* notebook. Be prepared to discuss your responses with your small group and the class.

Investigate

1. In your group, take a close look at the USGS map *This Dynamic Planet*. You may wish to use the map below instead. Discuss the following questions within your group and record your ideas.

   a) How would you describe the pattern of earthquakes around the Earth?

   b) Are earthquakes concentrated in any particular areas on the Earth's surface? If so, what other phenomena and features correspond to these areas?

   c) Are patterns different for ocean and continental areas?

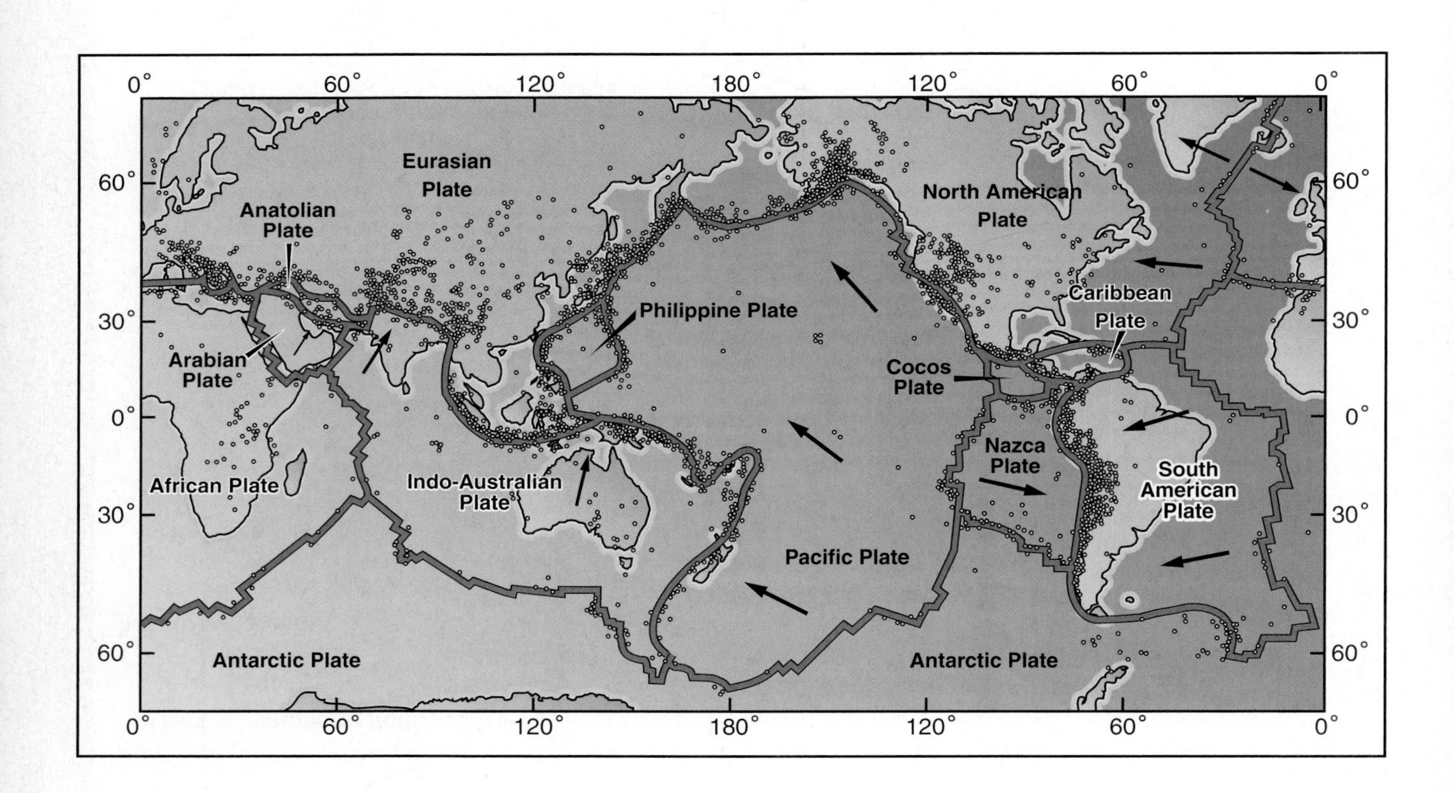

2. The next figure shows two maps of California. One map shows the San Andreas Fault system and other large faults. The other map shows recent earthquakes for a period of a week.

   a) Describe the distribution and arrangement of earthquakes in California.

   b) Do earthquakes correlate with identified large faults?

   c) Do earthquakes occur in areas without identified large faults?

   d) In your small group, discuss what a fault is and what causes one to occur. Record your ideas.

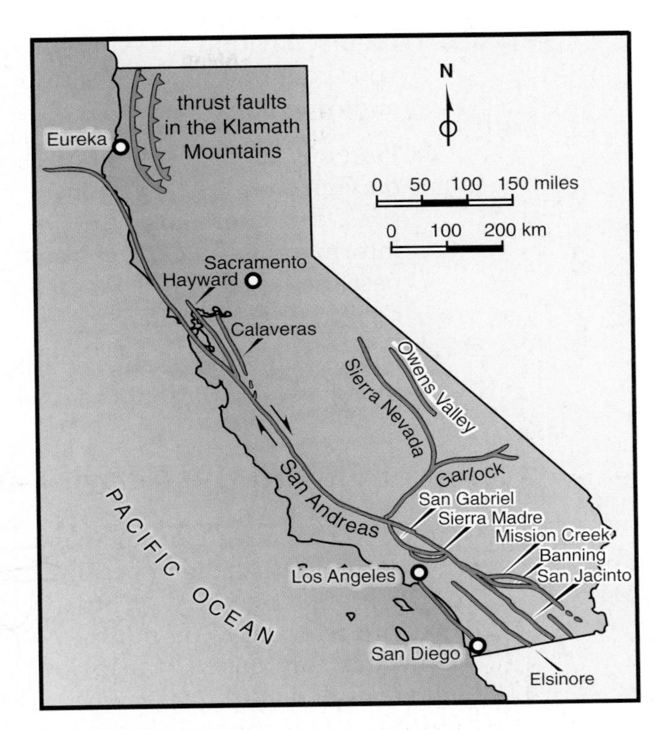

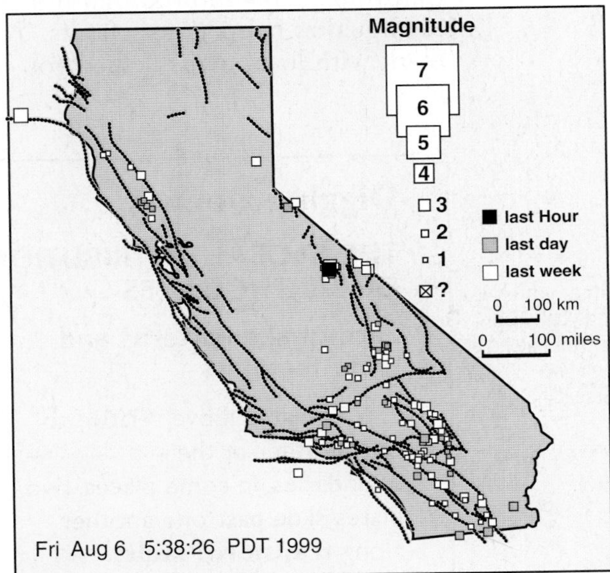

3. Obtain a map of the world and a geologic map of your state (with latitude and longitude marked). If possible, obtain a geologic map of your region (with latitude and longitude marked). Use electronic and print resources to answer the following questions. Visit the *EarthComm* web site for suggested Internet sites to visit and explore.

   a) Plot notable earthquakes of the world on the world map.

   b) Plot important earthquakes of the United States and Canada on the world map (or a map of North America).

   c) Write down the criteria you used to determine if an earthquake was notable or important.

d) Plot earthquakes that have occurred within 200–500 km of your community on your state and/or regional map.

e) Find out about, list, and mark on your maps the locations, sizes, dates, and any other information you can get about past earthquakes in your state, region, or community.

f) Share with other groups so that each has information about worldwide, continental, regional, and state earthquake distributions.

4. Look at the geologic map of your community or state. Locate faults on the map.

   a) Describe the relationship between the location of faults and the locations of earthquakes in your community or state.

Reflecting on the Activity and the Challenge

You found that most earthquakes occur along linear belts in oceans. You also found scattered or broad bands of earthquakes on most continents. You looked at maps of fault zones and earthquakes in California. This showed that most earthquakes happen near faults. Your work with local maps helped you to look for local patterns and relationships. Compiling information on past earthquakes helped you to think about the potential risk of earthquakes in your state, region, and community. This will help you explain to the public the risk of an earthquake in your area, as well as the magnitude of this risk.

Geo Words

transform fault: a vertical surface of slippage between two lithospheric plates along an offset between two segments of a spreading ridge.

Digging Deeper

THE GLOBAL DISTRIBUTION OF EARTHQUAKES

Earthquake Patterns and Plate Tectonics

Earth's plates move relative to one another at their boundaries. In some places, two plates slide past one another along **transform faults**. Earthquakes are common along transform faults, like the

San Andreas Fault in California. In other places, plates move away from each other or toward each other. These motions also cause forces in the rocks near the plate boundaries. When the forces build up to be greater than the strength of the rocks, the rocks break, causing an earthquake. Thus, you should expect to see many earthquakes near plate boundaries. The concentration of earthquakes along plate boundaries is very high.

The depths of earthquake foci also match the types of boundaries. Shallow-focus earthquakes occur at mid-ocean ridges and transform faults. At **subduction zones**, where one plate dives beneath another to great depths in the Earth's mantle, earthquakes range from shallow-focus ones to very deep-focus ones.

Areas of Risk in the United States

Risk is the impact of natural **hazards** on people. The size of the natural hazard, how often they occur, how close they are to people, and population density affect risk. Certain locations in the United States have had large earthquakes in recorded history. This puts them at higher risk from earthquakes again than at other places.

The map in *Figure 1* shows earthquake risks for the United States. The map is based mainly on earthquake history. The areas at highest risk are near plate boundaries. Southern Alaska is near a subduction zone. California has a very long transform fault (the San Andreas Fault), which extends from north of San Francisco all the way to the Gulf of California, in Mexico. Large earthquakes have also happened far from plate

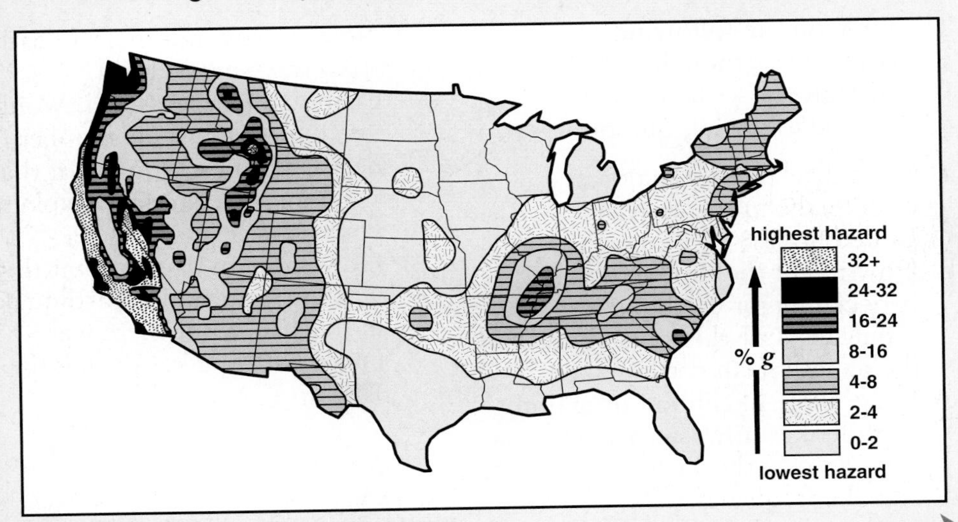

Figure 1 Earthquake risk for the United States.

Geo Words

subduction zone: a long, narrow belt in which one plate descends beneath another.

risk: the potential impact of a natural hazard on people or property.

hazard: a natural event, like an earthquake, that has the potential to do damage or harm.

Check Your Understanding

1. Why do scientists use the distribution of earthquakes as evidence to support the theory of plate tectonics?
2. In general, where is the highest risk of earthquakes?

boundaries. This suggests a zone of weakness and/or high stress in the crust. These earthquakes are not as easy to understand as the ones that happen at plate boundaries. Keep in mind that the forces produced by plate motions are transmitted all the way across the plates. Here's an example of that from everyday life: when you squeeze a brick from its ends or try to pull it apart, the forces you apply act throughout the brick, not just at its ends. Some areas within plates have unusually weak rocks, for a number of geologic reasons. In these areas, the forces that build up within the plates cause the rocks to break, causing an earthquake. Here are some examples: New Madrid, Missouri; Charleston, South Carolina; and parts of southern Quebec.

The legend of the map is in units of percentage of one *g*, the acceleration due to gravity, which is how fast an object gains speed when it is dropped. You have felt "*g* forces" yourself, when an airplane makes a sharp turn or when an elevator or an automobile stops suddenly. You would feel them if you were caught in an earthquake, too! Shaking during an earthquake causes an object that's attached to the ground, like a building, to move back and forth. How fast its velocity changes as it moves back and forth can be described in units of *g*.

Understanding and Applying What You Have Learned

1. The table on the following page shows major earthquakes in the United States. Use the data to address the following questions:
   a) When was the last major earthquake in the eastern United States?
   b) How might the time interval between major earthquakes influence your thoughts about the danger from earthquakes?
   c) What is the likelihood of major earthquakes in western states compared to the eastern states?
   d) Has the frequency of earthquakes in the western states been increasing over time? Explain.
   e) Note the low number of deaths from the New Madrid earthquake in Missouri. Would you expect the same number of deaths if an earthquake of that size took place today? Explain your answer.
   f) How can you explain that there was only one major earthquake in the 1700s but there were 15 major earthquakes during the 1900s?

Major Earthquakes in the United States			
Year	**Nearest City/ Epicenter**	**Richter Magnitude**	**Number of Deaths– Comments**
1755	Boston/Cape Ann, MA	6	Buildings damaged
1811 1812	Memphis, TN/New Madrid, MO	7.8	8 in 3 separate earthquakes
1812	San Juan Capistrano mission San Gabriel, CA	7	40 – roof of church caved in
1857	Los Angeles/Ft. Tejon, CA	8.0	1 – San Andreas Fault
1868	San Jose, CA	7.0	30 – Hayward Fault
1872	Bishop, CA/Lone Pine, CA	8.0	27 – Sierra Nevada Fault
1886	Charleston, SC	6.8	100 – Liquefaction of soil
1925	Santa Barbara, CA	6.3	13 – Unnamed offshore fault
1933	Long Beach, CA	6.3	120 – inspired codes of construction of schools
1934	Salt Lake City/Kosmo, UT	6.6	0 – Wasatch Fault System
1949	Seattle/Puget Sound, WA	7.1	8 – Area of occasional earthquakes
1952	Bakersfield/Kern County, CA	7.7	12 – White Wolf Fault
1954	Reno/Dixie Valle, NV	7.1	0 – Epicenter in rural area
1959	Bozeman/Hebgen Lake, MT	7.3	28 – In landslide caused by earthquake
1964	Anchorage/ Prince William Sound, AK	9.2	131 by tsunami and landslides; tsunami kills 11 in CA
1971	Los Angeles/San Fernando Valley, CA	6.4	65 – buildings and highway bridges collapse
1975	Kalapana, Hawaii	7.2	2 – tsunami damage
1983	Coalinga, CA	6.5	1 – older buildings destroyed
1987	Whittier, CA	5.9	8 – $358 million damage
1989	San Francisco – Oakland/Loma Prieta, CA	7.1	62 – Most in overpass collapse; over $6 billion damage
1991	Arcadia/Sierra Madre, CA	6	2 – $18 million damage
1992	Yucca Valley and Big Bear Lake, CA	7.4 and 6.5	2 – over 170 injured; Extensive ground cracking in remote area

2. Refer to the *This Dynamic Planet* map to answer these questions:
   a) Are all linear belts of earthquakes found with volcanoes?
   b) Do patterns depend on whether or not the earthquakes happen on continents or under oceans?
   c) Do you think this map is complete? Explain.
   d) How does seismic risk correlate with the edges of the continents?
3. The distribution of major earthquakes in the eastern states appears to be random. Unlike the western states, there are no linear belts. How might this make preparing the public for future earthquakes more difficult?

Preparing for the Chapter Challenge

Write a background summary for the brochure for your **Chapter Challenge**. Discuss the earthquake history of your state and community. Note any major earthquakes. Also, note the frequency of earthquakes that have been felt, and the maximum magnitude the public should prepare for. Make your summary concise, easy to understand, and accurate. It should fit on one page. Include maps and diagrams as needed.

Inquiring Further

1. **Earthquakes of magnitude 7 or greater**

   Use electronic or print resources to answer the following questions. Visit the *EarthComm* web site for suggested electronic resources.

   a) What is the maximum number of earthquakes with magnitude 7 or greater that occurred in one year from 1900 to 1989?
   b) On average, how many earthquakes of this size happen in a given year?
   c) Describe any patterns that you see in the data.
   d) Can you suggest any natural forces that might cause the observed variation in the number of earthquakes over time? Explain.

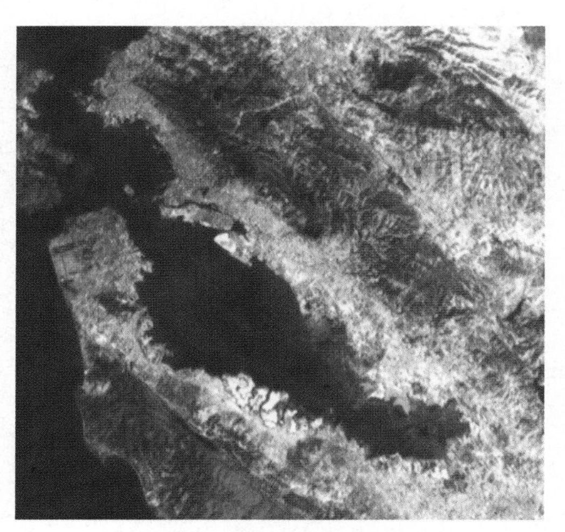

Can you find the San Andreas (transform) fault in this Landsat image of the San Francisco Bay Area?

Activity 5 Lessening Earthquake Damage

Goals

In this activity you will:

- Assign earthquake intensity based on observed effects.
- Describe the effects of passing earthquake waves.
- Recognize the secondary effects of earthquakes.
- Relate variations in intensity to the nature of the underlying geologic material.
- Identify places in your community with high and low risk from earthquakes.
- Outline steps to increase personal safety during an earthquake.

Think about It

In Japan, earthquake dangers are very great. All Japanese schools have regular earthquake drills, just as you have fire drills in your school.

- What damage do passing earthquake waves cause?
- How can you protect yourself in an earthquake?

What do you think? Record your ideas about these questions in your *EarthComm* notebook. Be prepared to discuss your responses with your small group and the class.

Investigate

1. At 4:48 a.m. on November 29, 1975, a magnitude 7.2 earthquake rocked the Island of Hawaii and neighboring islands. Construct a map of the earthquake intensities (as you did in **Activity 3**), using the observations listed below, the intensity scale, and a map of the islands.

Island of Hawaii

Hilo: Extensive damage to downtown area. Minor cracks in road, water pipes, concrete walls, plaster, and floors. Minor cracks and floor-to-wall separations a few millimeters wide in steel-reinforced concrete structures at the hospital, several schools, and libraries. In some of these buildings 5 to 10 mm vertical drops in some floor sections. Breaks in swimming pools. Shelved items in markets fell or tumbled over. Glass windows shattered. Breakage of chinaware. Collapse of stonewalls. Houses shifted from foundations. Cracking and crumbling of brick fireplace chimneys.

Hawaii Volcanoes National Park and Volcano: Extensive ground cracking caused heavy road damage. Water tanks damaged. Water lines broke. Some fireplace chimneys collapsed. At the Hawaiian Volcano Observatory, violent ground motion lasted half a minute, with many loose objects moved or turned over. Numerous rock falls in the crater.

Kurtistown: Damage to seven residential homes: cracked concrete steps, house and garage moved off foundation, chinaware broken from falling from cupboards.

Hawaii Paradise Park: One house shifted off foundation; cabinets toppled off walls.

Kalapana: A wood frame house shifted 1 m off the foundation.

Mountain View: Foundation of water tank cracked, Plexiglas cracked, television set shifted off stand and fell to floor, rock wall damaged.

Pahoa: Three homes moved from foundation. Other damage included a broken water line and a collapsed water tank.

Naalehu: The foundation of one ranch house cracked, and the roof was damaged.

Pahala: Doors distorted and house moved from concrete foundation. Furniture and stereo fell.

Laupahoehoe: Landslides along steep road cut along coastal road between Laupahoehoe and Honomu. Loose objects fell off shelves in homes.

Keaau and Pepeekeo: Moderate ground shaking and unstable objects moved.

Kailua-Kona: Loose objects fell off shelves. Strong shaking. Rock falls in Kealakekua.

Honokaa: Landslides on coast road. Loose objects fell off shelves.

Kohala: Some loose objects moved. Shaking felt by many people.

Island of Maui

Wailuku: Hanging objects moved slightly. Not recognized as an earthquake at first (thought to be a passing truck).

Island of Oahu

Honolulu: Felt indoors by a few sensitive people.

Island of Kauai

Lihue: Felt indoors by a few sensitive people.

a) Mark the places of equal intensity on your map. Seismologists call maps showing areas of equal intensity isoseismal maps.

b) List three effects directly related to the earthquake.

c) List three secondary effects, events triggered by the earthquake, noted in the above descriptions.

d) On which island was the earthquake located? In which quadrant of the island was the earthquake located?

e) Do(es) the area(s) of highest intensity correspond to the earthquake epicenter? If not, speculate why.

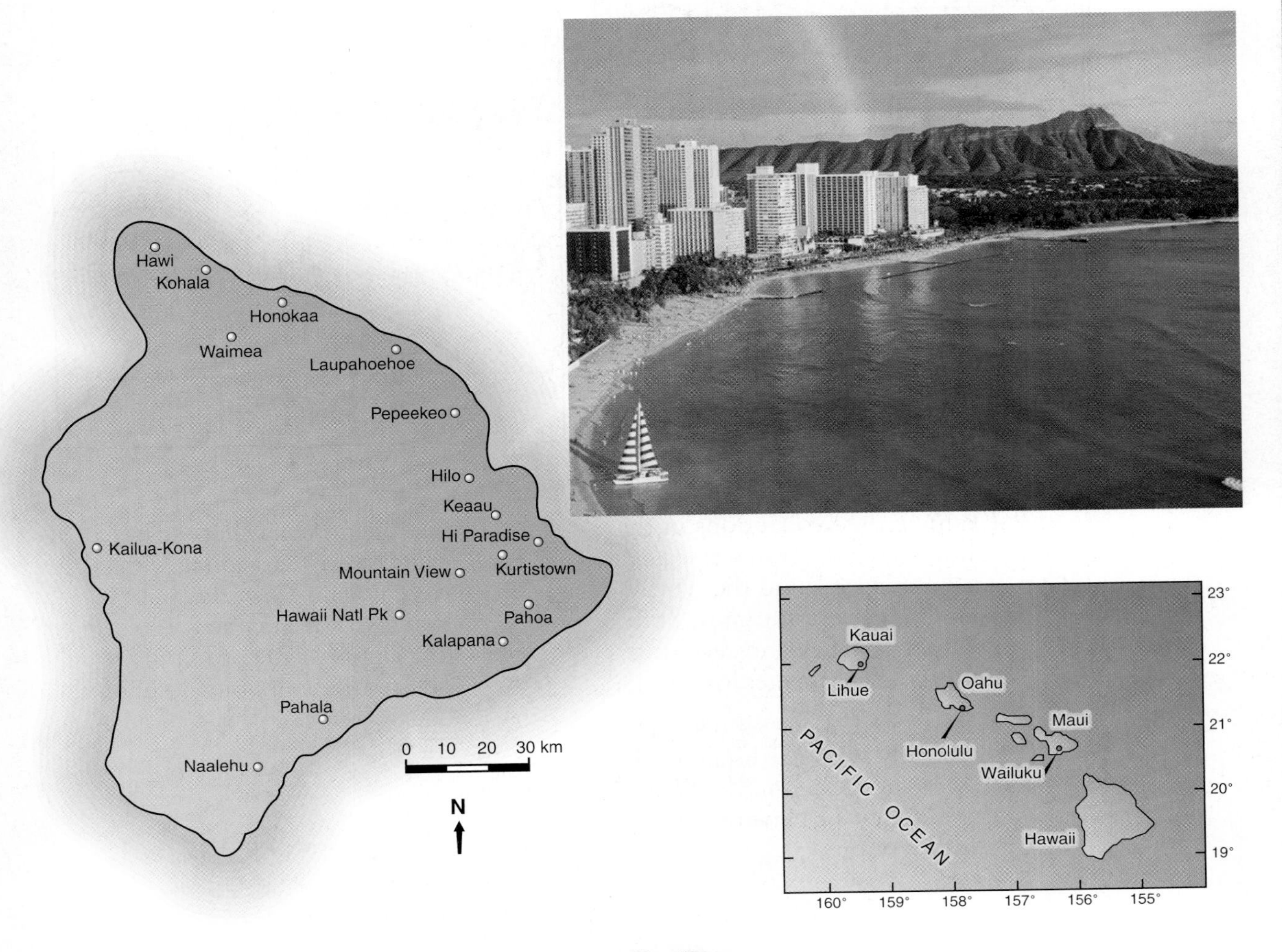

2. The figure below shows the location of three seismometers in San Francisco and the seismograms recorded at each station for a magnitude 4.6 aftershock of the Loma Prieta earthquake.
   a) What is the underlying geology at each seismometer?
   b) Which station(s) experienced low-amplitude waves during the aftershock?
   c) Which station(s) experienced high-amplitude waves during the aftershock?
   d) Describe any correlation between underlying geology and amplitude of seismic waves?

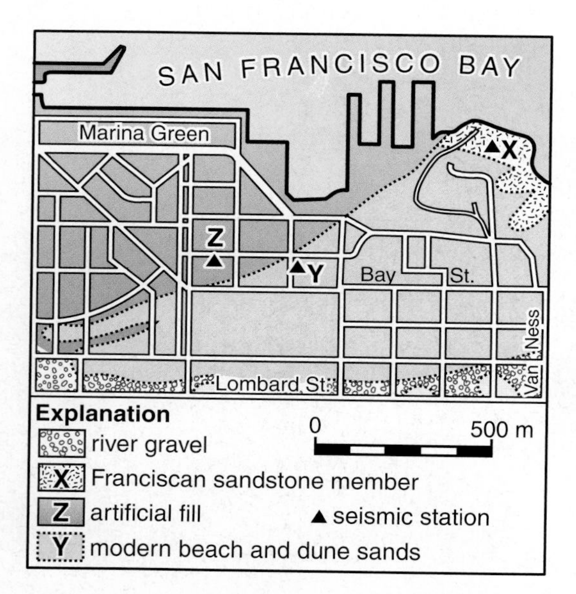

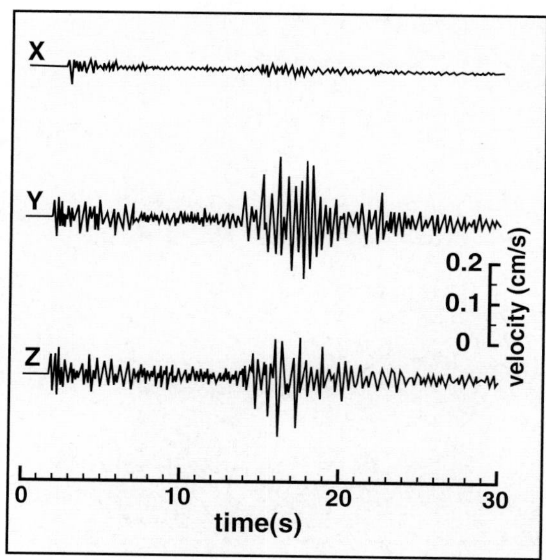

3. Examine the descriptions in the following chart of major earthquakes around the world, including a description of both geologic changes and resulting destruction (damage, loss of life).
   a) List phenomena associated directly with earthquakes that were not noted in the San Francisco or Hawaii descriptions.
   b) List phenomena that were triggered by earthquakes that were not noted in the San Francisco or Hawaii descriptions. Record your observations in your notebook.

Recent Seismic Events of Special Interest (IRIS)			
Event	**Magnitude**	**Geologic Changes**	**Destruction**
1. Northern Iran (5/97 & 6/97)	7.3	Landslides, rare sequence of large earthquakes	1567 killed, 2300 injured, 50,000 homeless, extensive damage
2. Windward Islands (4/97)	6.7	One of the largest known earthquakes in or near Trinidad or Tobago	None reported
3. S. Xinjiang, China (7/96)	5.2	Nuclear weapons test	None reported
4. Sakhalin Island (5/95)	7.5	None noted	1989 killed, 750 injured
5. Kobe, Japan (1/95)	6.8	Surface faulting for 9 km with horizontal displacement of 1.2 to 1.5 m, soil liquefaction	5502 killed, 36, 896 injured, 310,000 homeless, severe damage
6. Kuril Islands (10/94)	8.3	Tsunami with heights up to 346 cm	10 killed or missing, extensive damage throughout islands
7. Northern Bolivia (1/94)	8.2	At 637 km depth, the largest deep earthquake, first earthquake from this part of S. America to have been felt in N. America, including Canada	Several people killed
8. Northridge, CA (1/94)	6.8	A maximum uplift of 15 cm occurred, many rock slides, ground cracks, soil liquefaction	60 killed, 7000 injured, 20,000 homeless, severe damage
9. Southern India (9/93)	6.3	Large intraplate earthquake devastation	9748 killed, 30,000 injured, extreme devastation
10. Republic of South Africa (5/93)	3.8	Mine collapse	Several people killed
11. Flores Region (12/92)	7.5	Tsunami run-up of 300 m with wave heights of 25 m on Flores, landslides, ground cracks	2200 killed or missing
12. Switzerland (11/92)	3.7	Accidental explosion of a munitions cavern	6 killed
13. Northern Colombia (10/92)	7.3	Explosion of a mud volcano, landslides, soil liquefaction, small island emerged from the Caribbean Sea off San Juan de Uraba	10 killed, 65 injured, 1500 homeless
14. Landers, CA (6/92)	7.6	Surface faulting along a 70-km segment with up to 5.5 m of horizontal displacement and 1.8 m of vertical displacement	1 killed, 400 injured, substantial damage

continued on next page

Recent Seismic Events of Special Interest (IRIS) ***(continued)***			
Event	**Magnitude**	**Geologic Changes**	**Destruction**
15. Northern India (10/91)	7.0	2 events 1.6 s apart, landslides, 30 m deep crack	2000 killed 1800 injured, 18,000 buildings were destroyed
16. Luzon, Philippines (7/90)	7.8	Landslides, soil liquefaction, surface faulting	1621 killed, 3000 injured, severe damage
17. Western Iran, (6/90)	7.7	Landslides	40,000–50,000 killed, 60,000 injured 400,000 homeless, extensive damage
18. Loma Prieta, CA (10/89)	7.1	Maximum intensity in part of Oakland and San Francisco, landslides, soil liquefaction, small tsunami at Monterey	60 killed, 3757 injured, $5.6 billion damage
19. Macquarie (1/95)	8.2	Small tsunami along coasts of Tasmania and in Australia. Largest oceanic earthquake ever recorded	None reported
20. Turkey-USSR Border (12/88)	6.8	Surface faulting 10 km in length with a maximum throw of 1.5 m	25,000 killed, 19,000 injured, 500,000 homeless, $6.2 billion damage

4. Look into the municipal building code in your community (contact your town or city government office).
   a) Does part of the code address earthquake risks?
   b) What parts of the code are applicable to reducing earthquake risk, both primary (due to Earth shaking) and secondary (aftereffects)?
   c) Has the code been upgraded in recent years?
5. Examine a detailed map of your community. A USGS 7.5-minute topographic quadrangle would work well.
   a) Assign risk factors to the highway system, access to hospitals, the power grid, police and fire services, and other community services that would come into play in the case of an earthquake catastrophe.
   b) In case of an earthquake, what kinds of damage might occur that could cause even more damage to the community and its occupants than the shaking of the Earth itself?
   c) Report your findings and your ideas about building codes and community safety to the class.

Reflecting on the Activity and the Challenge

You learned that ground motion (shaking) happens with every felt earthquake. Ground motion is the leading cause of damage to materials and buildings. Your **isoseismal map** showed that the epicenter and the area of maximum intensity do not always coincide. The effects of an earthquake depend in large part on the rocks and soil in a region. The motion from earthquakes can trigger landslides and tsunamis, and break gas lines and water lines. You also looked at local structures that might be at risk from earthquakes. You now have the information to provide recommendations on ways to reduce damage from an earthquake.

Digging Deeper

REDUCING EARTHQUAKE HAZARDS

Direct Hazards

The main effect of earthquakes is shaking of the ground as seismic waves pass through an area. The main result of ground shaking is the collapse of buildings. Motion along a fault can break power lines, pipelines, roads, bridges, and other structures that cross the fault.

Indirect Hazards

Fire is a secondary hazard in cities. Fire can cause much more damage than ground movement during an earthquake. In the 1906 San Francisco earthquake, 70% of the damage was due to fire. Fires occur when ground motion breaks fuel lines, tanks, and power lines. Often, water lines are also disrupted or broken. This reduces the water available to fight the fires. One of the ways to reduce the risk of heavy fire damage is to place many valves in the water and fuel pipelines. If one part of the pipeline is damaged, those pipes can be isolated from the system.

Landslides are another serious secondary effect of earthquakes. Earthquakes can trigger the failure of unstable slopes. The best way to minimize this hazard is not to build in areas with unstable slopes. Because buildings already exist in such areas, it is difficult to reduce the risk.

When an undersea or nearshore earthquake occurs, a **tsunami**, also called a seismic sea wave, can be generated. Tsunamis are like the ripples that form when you throw a rock in a pond, except very much larger. Tsunamis form when a large area of the ocean floor rises or falls suddenly in an earthquake. This causes waves to move away from the area. In the open ocean, the waves have very long wave lengths (greater than 500 km), but heights of only a

Geo Words

isoseismal map: a map showing the lines connecting points on the Earth's surface at which earthquake intensity is the same.

tsunami: a great sea wave produced by a submarine earthquake (or volcanic eruption or landslide).

Check Your Understanding

1. What are the direct hazards of an earthquake?
2. Why are fires able to cause extensive damage after an earthquake?
3. What is a tsunami?
4. Why is it difficult to prepare against the destruction caused by a tsunami?

meter or so. The waves are so long and so low that ships at sea can't tell they are passing by! If you have been to the seashore, you probably have noticed that ordinary ocean waves get much higher and then break as they move into shallower water. The same thing happens with tsunami waves, only more so. As they come onshore, the waves build to heights as great as 30 m.

One of the difficulties in preparing for tsunamis is their great speed. They move very fast (1000 km/hour) over very long distances in the ocean. On average, there are two destructive tsunamis in the Pacific basin each year. An early warning system now monitors sea level around the Pacific. A tsunami can take several hours to travel across the Pacific. If a tsunami is detected, its estimated time of arrival is sent to areas in danger so that people can be evacuated. This early warning system has had many successes, but also some failures, since it was begun in 1948. It works well for areas far from the earthquake, but it is not very effective for areas close by, because the waves move so fast.

Personal Safety in an Earthquake

The Federal Emergency Management Agency provides safety rules for earthquakes. Earthquakes strike suddenly, violently, and without warning. Identifying potential hazards ahead of time and advance planning can reduce the dangers of serious injury or loss of life from an earthquake.

Understanding and Applying What You Have Learned

1. In a major earthquake, where in your school and in your community would you be safest? What places pose the greatest risk from the effects of an earthquake? Explain why you selected these locations.
2. If you live in an earthquake-prone area, your school may also have regular earthquake drills.

   If your school has a regular drill, examine the drill and evaluate its appropriateness.

   If your school has no earthquake drill plan, describe how you would develop such a plan for your school. Think about the following questions:

   - Where will students go when the shaking stops?
   - How will this be accomplished?

   Write down your ideas as a sequence of steps.

Preparing for the Chapter Challenge

Develop an overall community response and evacuation plan for earthquakes. Take into account the kinds of damage that may block roads, injure people, and collapse buildings. Note any steep slopes that might generate landslides during an earthquake.

Be sure to address the following issues:

- If a community mitigation (making less severe) plan already exists, how well does it address the important issues? Are there places where the current plan can be improved?
- Are there parts of your community where the physical conditions of the ground itself creates special risks of earthquake damage?
- Are there other areas where people and structures are more likely to be safer?

Inquiring Further

1. **Forming questions to investigate**

   Write down other questions that you have about ways to reduce the severity of earthquakes in your communities and around the world. How would you go about gathering information to answer these questions? Write your ideas in your notebook.

2. **The Pacific Tsunami Warning Center**

   The Pacific Tsunami Warning Center has provided successful advance warnings of potentially dangerous tsunami that may affect areas distant from the original earthquake. Find out about this tsunami warning system.

   a) Record in your notebook your ideas and any information you find on tsunamis and tsunami prediction.

   b) Why is it possible to predict a tsunami, but not an earthquake, in time to save lives?

   c) If your community is near the ocean, find out how well your community is prepared to deal with a tsunami warning.

Activity 6 Designing "Earthquake-Proof" Structures

Goals

In this activity you will:

- Determine factors that influence the stability of buildings.
- Build models to assess the behavior of buildings during ground motion.
- Design reinforcements to structures and re-test their structures.
- Compare real structural failures to failure of their models.
- Suggest structural improvements for local buildings.
- Identify types of motion that lead to structural failure.
- Identify geologic conditions that enhance ground motion.
- Measure the natural frequency of different objects.
- Learn that natural frequency is a control on the stability of structures.

Think about It

During earthquakes, some buildings stand up well to the shaking while others collapse immediately.

- What influences the extent of damage to a building during an earthquake?
- Where is the safest place to be in a building during an earthquake?

What do you think? Record your ideas about these questions in your *EarthComm* notebook. Be prepared to discuss your responses with your small group and the class.

Investigate

Part A: Modeling the Response of Buildings to Shaking

1. Obtain the materials you need for this investigation. These include wooden dowels (about 4 cm diameter would be good to start with, but perhaps others that have a larger or smaller diameter as well), large sheets of heavy-gauge mat board or corrugated cardboard, and map pins. You may want to find other materials of your own to test some of the models you develop further. Read the steps carefully so that you are familiar with the suggested methods.

2. On a table that can easily be shaken sideways, build a model using the dowels as pillars and sheets of mat board. Construct a model of a 4-story parking garage, a tall square building, or an overhead freeway. Do not fasten these together first. Assemble them just by stacking.

3. Design your investigation so that you will be testing one variable at a time, if possible.

   a) Record which variable you will investigate.

4. Once your structure is complete, shake the table very gently for 5 s and observe whatever happens to the structure. Then shake it for 10 s, 20 s, and 30 s.

   a) Record what you observe each time.

Work in an unobstructed area. As the structures topple, pieces will fall unpredictably. Do not use material that can break or cause injury as they fall. Keep feet and hands clear of falling objects.

5. Repeat your test and vary the "magnitude" of shaking (keeping the time and direction the same). Also, repeat the test and vary the direction of shaking, by moving to a different position around the table (keeping the "magnitude" and time the same). Observe what happens.

   a) Record your results.

   b) Was your building more or less resistant than you thought?

   c) How was the stability of your building affected by the violence of shaking? By the duration of the shaking?

   d) Does it matter from which direction the vibration comes?

6. Try other shapes of buildings, similar to buildings in your community, and use other kinds of materials if you wish to create more "realistic" models.

   a) Record your procedure and results carefully so that your experiment could be repeated and verified by others.

   b) Is there a difference between tall and short structures?

   c) Is there a difference between narrow and wide structures?

7. You should make your structures collapse eventually, because what you want to know is just how resistant the structures are to shaking. To determine the limits of your structures, you must continue the experiments until the structures fail.

Rebuild the structures you tested, but this time construct them so that they will not collapse, even with prolonged shaking. Identify materials like push pins, string, additional cardboard, tape, and paper binders that may help to reinforce your structure. Once your group has discussed several options, test them. Try to subject the rebuilt structures to as many of the types of wave motion you learned about in **Activity 2** as you can.

a) How did you add strength to your buildings?

b) Which methods of reinforcement worked the best?

8. The photographs show actual damage that happened to a bridge and buildings in real earthquakes.

   a) How does the damage you observe in the photos resemble the kind of collapse and damage that you observed in your models?

   b) For each case, can you suggest the type of force (compression, extension, or shear) that caused the structure to fail?

   c) For each case, can you tell if the underlying geologic materials played a role in increasing the intensity of waves or causing damage?

Part B: Modeling the Response of Buildings of Various Heights to Shaking

1. Obtain the following materials: two square pieces of wood board, each about 10 cm on a side; a 30-cm (10") thin metal strip; a small C-clamp; and a lump of modeling clay, about half the size of a fist (about 100 gm).
2. Attach the two pieces of wood together with nails or wood screws to make an "L"-shaped wooden base. Set it on a table with one side horizontal and the other side upright. Clamp the metal strip blade vertically to the top of the upright piece of wood, as shown in the diagram.
3. Mold the modeling clay into a ball, and push it onto the top of the metal strip. Measure the distance from where the blade is clamped, up to the center of the ball.

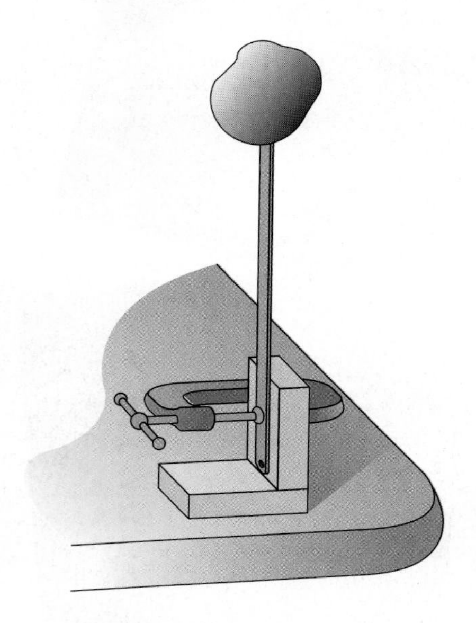

Stand back of the swinging metal strip. Avoid pulling back the metal strip to extreme bends. It is possible for all or part of the clay ball to come off. Position yourself to avoid being hit.

   a) Set up a table to record the measurements for this activity.

   b) Record the initial position of the clay ball.

4. Hold or clamp the wooden base to a table. Pull the ball to one side and let it go. You are modeling how a tall building sways back and forth when it is disturbed by the ground motion of an earthquake, or even by a strong gust of wind. Measure the period of the oscillation by timing how long it takes for several oscillations and then dividing the time by the number of oscillations.

   a) Record your measurements.

   b) Repeat your measurements for shorter buildings, by pushing the ball down the metal strip to three or four different heights. Record any observations and measurements.

   c) Plot a graph with oscillation period on the vertical axis and the height of the ball on the horizontal axis.

   d) How does the oscillation period depend on the height of the ball?

5. Put the ball back at the top of the metal strip. Slide the wood base on the table in a regular back-and-forth motion. Measure the period of the sliding by timing how long it takes for several cycles.

   a) Estimate and record how far the ball moves away from the upright "neutral" position during one oscillation cycle. It's easier to do this if you prepare a centimeter grid on a piece of poster board and then staple the grid onto the wooden base so that you can view the swaying of the ball in front of the grid.

b) Repeat the last step for several different sliding periods, from very slow (less than one cycle per second) to as fast as you can easily measure. Record your measurements.

c) Plot a graph with sliding period on the horizontal axis and estimated ball displacement on the vertical axis. Draw a smooth curve through your data points.

d) What is the shape of the curve?

e) What is the relationship between the sliding period that produces the greatest displacement and the natural period of the oscillation, which you measured earlier?

6. Use the results of your investigation to answer the following questions:

a) Will earthquakes affect all buildings equally?

b) Are large earthquakes required to damage tall buildings?

c) If you had to design a tall building to withstand an earthquake, how might you use the results of this investigation to help you?

7. Within your groups, examine the data you have gathered during the two parts of the activity investigations and draw some conclusions about structures and their ability to withstand earthquake shaking.

a) Record your conclusions. Have representatives from each group report to the class.

b) Record the similarities and differences among the group results, and suggest possible explanations for these differences.

Reflecting on the Activity and the Challenge

In this activity, you used various materials to make models of buildings. With these models, you examined the influence of design, support, and size on the stability of a building. You also saw that magnitude, duration, and direction of shaking are important influences. By rebuilding some structures, you recognized that improvements could be made to add stability. By examining damage from past earthquakes you gained insights about the interaction between building design and earthquake forces. You used a model to explore the relationship between height and stability and calculate the natural oscillation period of an object. You can use this information as part of your public service message.

Digging Deeper

BUILDINGS AND GROUND MOTION

Introduction

Severe earthquakes can cause major damage to buildings in any community. The problems are made worse by the loss of electric power, gas, and communication lines. Damage to roads often disrupts transportation. One of the problems is that it is difficult to build realistic small models that will behave like full-scale structures during an earthquake. Not all earthquakes produce the same ground motion. Areas with different underlying Earth materials react differently to the same shaking. However, certain design elements will help buildings and other structures to withstand ground motion.

Duration of Shaking

The duration of an earthquake can affect the extent of damage. For example, buildings made of reinforced concrete may withstand a 20-s main shock, yet collapse in a smaller-magnitude earthquake that lasts longer.

Direction of Motion

Many earthquakes in the San Fernando Valley of California caused horizontal displacement and ground motion. Structures that had collapsed in the past were rebuilt to withstand this kind of motion. The motion of the 1994 Northridge earthquake, however, had a strong vertical component. Many of the rebuilt structures and freeways collapsed again.

Underlying Earth Materials

Underlying Earth materials have a strong influence on the motions of structures during an earthquake. In the San Francisco earthquake of 1906, buildings on filled land (loose, wet soil) suffered much more damage than buildings built on bedrock. The same pattern was repeated in the 1989 Loma Prieta earthquake. Damage was high in the Marina district (Activity 5). This same part of San Francisco was hit hard in 1906. Buildings in Mexico City were affected greatly by the earthquake of 1985, although the epicenter was far away. Mexico City is built on a basin filled with weak layers of volcanic ash and gravel, sand, and clay. Acapulco, which was much closer to the epicenter, suffered less damage, because it stands firmly on bedrock. It's not easy to specify exactly why this happens. Materials like soils and sediments, which are much less rigid than bedrock, respond to the passage of seismic waves with much greater wave amplitudes. You can think of it this way: rocks and gelatin are both elastic solids, and both will vibrate when you hit them, but you know that the gelatin will shake more than the rock!

Resonance

Sometimes, particular structures are affected by ground motion when others are not. All structures have a natural period of oscillation or swaying, as you determined in the investigation. When the shaking of the ground during an earthquake happens to be close to the natural oscillation period of a building, the swaying of the building is at its maximum and contributes to building collapse. When the oscillation that is imposed on the building is nearly the same as the natural oscillation period of the building, the swaying is greatly intensified. This is called **resonance**. You all know about resonance from when you were little kids on a swing. The swing has a natural period of oscillation, so when you "pump" to go higher, you pumped at about the same period as the natural period. Pumping at a different period got you nowhere!

The 1985 earthquake in Mexico City caused severe damage or destruction to about 500 buildings. Ground vibrations were amplified by the vibrational properties of tall buildings. This caused 10 to 14-story buildings to sway even more (at a period of 1 to 2 s) and resulted in damage to many structures. Nearby shorter and taller buildings were not damaged.

Materials

Building materials also make a difference. Stone, brick, wood, concrete, and adobe all have different responses to the forces they experience during an earthquake. Building design must take into account not only the structural elements but also the material from which the structure is made.

Geo Words

resonance: a condition in which a vibration affecting an object has about the same period as the natural vibration period of the object

Check Your Understanding

1. In addition to damage to buildings, what other problems are associated with earthquakes?
2. Name five factors that affect the amount of damage caused by an earthquake.
3. Why were structures rebuilt to withstand earthquakes still heavily damaged during the 1994 Northridge earthquake?
4. Why was the Marina district of San Francisco so severely damaged during the 1989 Loma Prieta earthquake?
5. Why is earthquake risk greater on filled land with soft ground than on solid bedrock?
6. In the 1985 earthquake of Mexico City why were some buildings heavily damaged or destroyed, while surrounding shorter and taller buildings were not?

Understanding and Applying What You Have Learned

1. What are some of the major problems involved in getting your community to prepare for earthquake damage?
2. What modifications could be made to buildings of your community?
3. What types of scientists and professionals should be consulted in earthquake planning?
4. Suggest the kinds of areas where you think that buildings should not be constructed.
5. Which communities should prohibit the construction of tall buildings? Explain your reasoning.

Preparing for the Chapter Challenge

As a group, develop a set of recommendations to make to your town council and school board about how to minimize damage in your community from a major earthquake. Address the design of new structures and possible changes to older structures in your community. Write these recommendations as a list to include in your brochure. Be sure to address the following issues:

- suggestions for new building developers,
- suggestions for homeowners, and
- suggestions for local government and businesses.

Inquiring Further

1. **Earthquake engineering**

   The National Information Service for Earthquake Engineering provides an excellent overview of numerous structures as well as photographs of earthquake-resistant design. Visit the *EarthComm* web site for more information on how to locate this data.

   Compare their structures to the ones you built early in the activity. The service also has a description of shake tables, the large platforms used to test structural designs during simulated earthquake vibrations.

Earth Science at Work

ATMOSPHERE: *Firefighter*
The public depends on the availability of adequate firefighting equipment and the quick response time of the emergency personnel.

BIOSPHERE: *Insurance Agent*
Insurance agents need to consider the risks involved in insuring a building, and the lives and property of its owners.

CRYOSPHERE: *Ski Instructor*
Professional ski instructors arrange for helicopter skiing trips into remote mountain areas not accessible by roads. They are responsible for the safety of their clients.

GEOSPHERE: *Architect*
When designing a building, an architect must carefully calculate the forces that will act on the structure.

HYDROSPHERE: *Dam Maintenance Worker*
Dams are one of the engineering marvels of this world. Their maintenance is vital to their operation.

How is each person's work related to the Earth system, and to earthquakes?

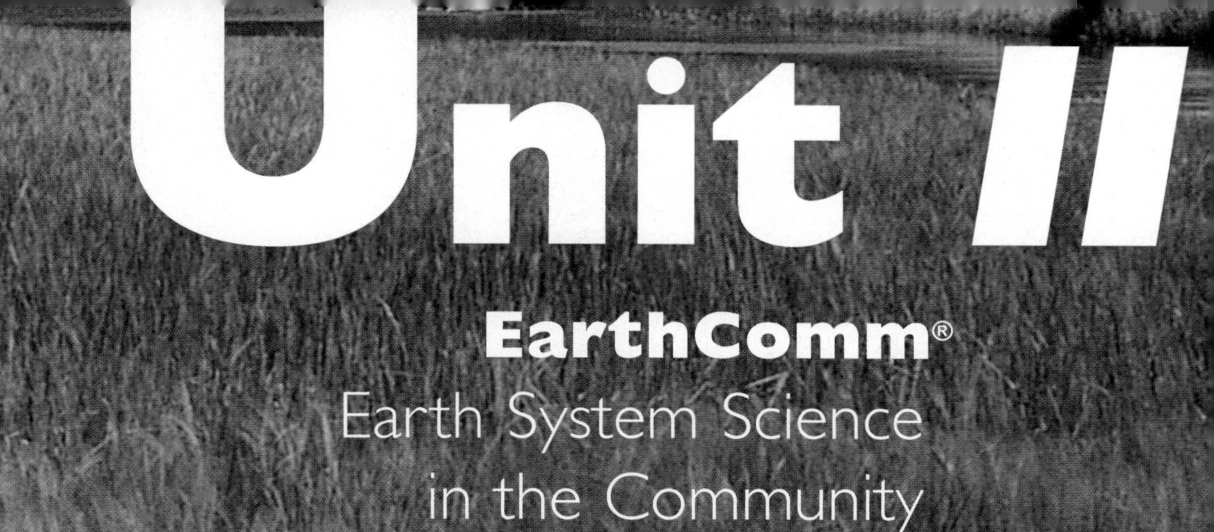

Unit II

EarthComm®

Earth System Science in the Community

Understanding Your Environment

1

Bedrock Geology ...and Your Community

Bedrock Geology ...and Your Community

Getting Started

Except for a few sudden, dramatic events in your lifetime, you will probably notice very little change in the physical features of the Earth around you. However, the rock record in your community will likely tell a long and complicated story. Therefore, geologists must understand the processes that form and shape rocks in order to "read the story" and reconstruct the past. Reconstructing the past is important, because unraveling the past can help you understand how the planet works.

- What are the three major types of rocks, and how do they form?
- How old are the oldest rocks on Earth?
- How old are the oldest rocks in your community, and how did they get there?

What do you think? Write down your ideas about these questions in your *EarthComm* notebook. Be prepared to discuss your ideas with your small group and the class.

Scenario

The Bed and Breakfast Association in your community would like to provide their guests with geological information about the region in which they are located. They know that visitors come to the area to explore the surroundings. They have asked the *EarthComm* students in your school for help. They need your study and report of the geologic history of your community to provide to tourists. Can you use your understanding of rocks to teach the public about the geologic history of your state? Can you develop a brochure or guide which the Bed and Breakfast Association in your community can actually use? Perhaps you might wish to work with other schools in your area or your state to develop a report that can be used over a wider region.

Chapter Challenge

Your challenge is to write a report about the geologic history of your area to be used by the Bed and Breakfast Association in your community. The report should:

- Give an accurate summary of the geologic history of your state.
- Help people understand the rocks and sediments in your state and community.
- Describe principles of Earth science as they relate to geologic change.
- Explain how your local geologic history fits into the geologic history of the United States and North America.

Here are some ideas for you to think about. Choose one that is most practical for your class and your community.

Road Log: Geology seen along a highway (or series of highways).
Trail Guide: Geology seen along a hiking or biking trail.
Park Brochure: Geology of local or state parks.

Assessment Criteria

Think about what you have been asked to do. Scan ahead through the chapter activities to see how they might help you to meet the challenge. Work with your classmates and your teachers to define the criteria for assessing your work. Record all this information. Make sure that you understand the criteria as well as you can before you begin. Your teacher may provide you with a sample rubric to help you get started.

Activity 1 Sedimentary Rocks and the Geologic History of Your Community

Goals

In this activity you will:

- Identify and classify several sedimentary rocks.
- Describe how the three main types of sedimentary rocks form.
- Understand that sedimentary rocks are divided into groups based on how they form.
- Infer the environment in which sediment was deposited when you are given a sedimentary rock.
- Understand that classification helps scientists organize the natural world into smaller, workable components.

Think about It

Sedimentary rocks, which are made of sediment, cover about three-fourths of the Earth's land surface.

- How does sediment "turn into" sedimentary rock?

What do you think? Record your ideas about this question in your *EarthComm* notebook. Include a quick sketch. Be prepared to discuss your responses with your small group and the class.

Investigate

Part A: Making Models of Sedimentary Rocks

1. Mudstone
   - Spread some wet mud in a pan.
   - Set it out in the sun undisturbed until all the moisture has evaporated from the mud.
2. Rock Salt
   - Add salt to a container of warm water until salt will no longer dissolve.
   - Pour a few millimeters of the water into a shallow plate, dish, or pan.
   - Let the water evaporate overnight. Do not disturb the setup until all the water has evaporated.
3. Sandstone
   - Make a mixture of half water and half white craft glue.
   - Combine this mixture with a handful of sand in a small container. Pour off any excess liquid.
   - Line a small bowl or beaker with waxed paper and pour in the sandy mixture.
   - Let it stand undisturbed until all the water has evaporated, which may take several days.
4. Conglomerate
   - Make a mixture of half water and half white craft glue.
   - Combine this mixture with a handful of sand, gravel, and clay in a small container. Pour off any excess liquid.
   - Line a small bowl or beaker with waxed paper and pour in the sandy mixture.
   - Let it stand undisturbed until all the water has evaporated, which may take several days.
5. Sediment deposition
   - Pour a mixture of clay, silt, sand, and gravel in water into a clear, unbreakable container.
   - Close and shake the container, then let it stand.
   - Observe the container over the next several days.

   a) Describe what you observe immediately, by the end of class, and the next day.
6. Examine the rock samples that you made. These samples are models of sedimentary rocks.

   a) Draw a labeled diagram of each sedimentary rock that you made.

Clean up spills immediately.
Take care not to get sediments into your eyes.
Wash your hands after you have completed the activity.

Part B: Observing Sedimentary Rocks

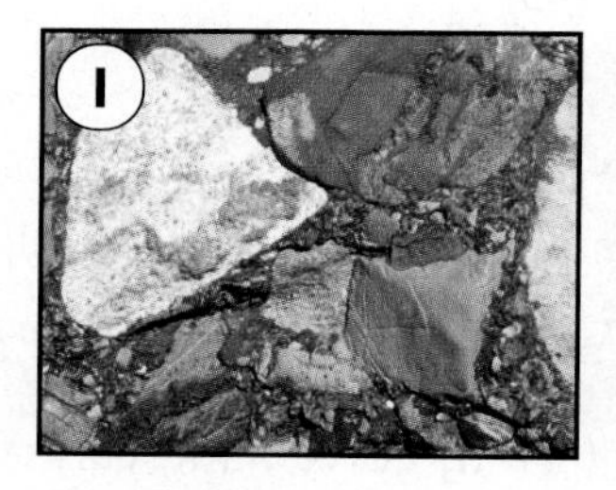

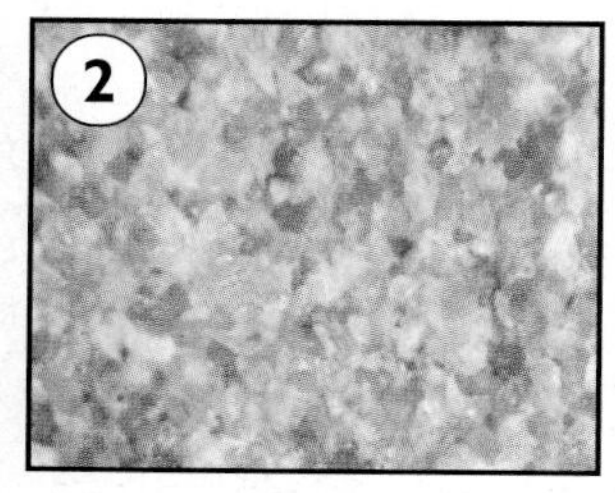

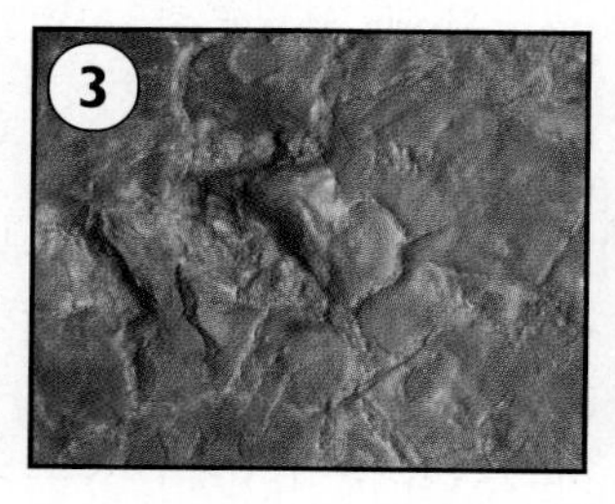

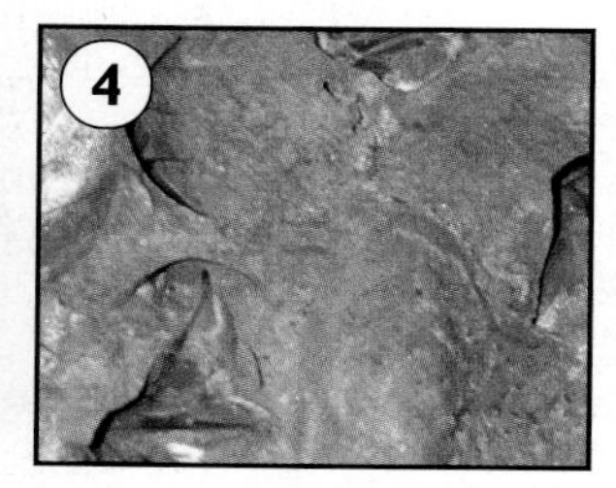

1. Obtain a set of sedimentary rocks. Common sedimentary rocks include limestone, dolomite, mudstone, sandstone, siltstone, shale, conglomerate, rock salt, and coal. Carefully observe and describe the rock samples.
   a) Make a data table to record your descriptions of each sedimentary rock. Note any distinguishing features.
   b) The three major sedimentary rock types are described below. Based on your descriptions, determine the sedimentary rock type of each rock sample.

Sedimentary Rock Type	Description
Clastic	Fragments of rocks and minerals that have been physically transported and deposited and then converted into rock.
Organic	Remains of plants and animals that have been converted into rock.
Chemical	Direct precipitation of minerals from a solution.

Part C: Sedimentary Rocks of Your Community

1. Examine the legend of a geologic map of your community or local area or state. A geologic map shows the distribution of bedrock at the Earth's surface. The bedrock shown on the map might be exposed at the surface, or it might be covered by a thin layer of soil or very recent sediment. Every geologic map has a legend that shows the kinds of bedrock that are present in the map area. The legend also shows the rock bodies or rock units that these rocks belong to, and their geologic age. You will learn more about rock units in Activity 4 of this chapter.

   Working with your group, interpret the data on the geologic map by answering the following questions:
   a) Are any sedimentary rocks described in the legend? If yes, write down the rock types, the names of the rock units they belong to, and their locations.
   b) What are the most common sedimentary rocks in your area?
   c) Which is the oldest sedimentary rock unit?
   d) Which is the youngest sedimentary rock unit?

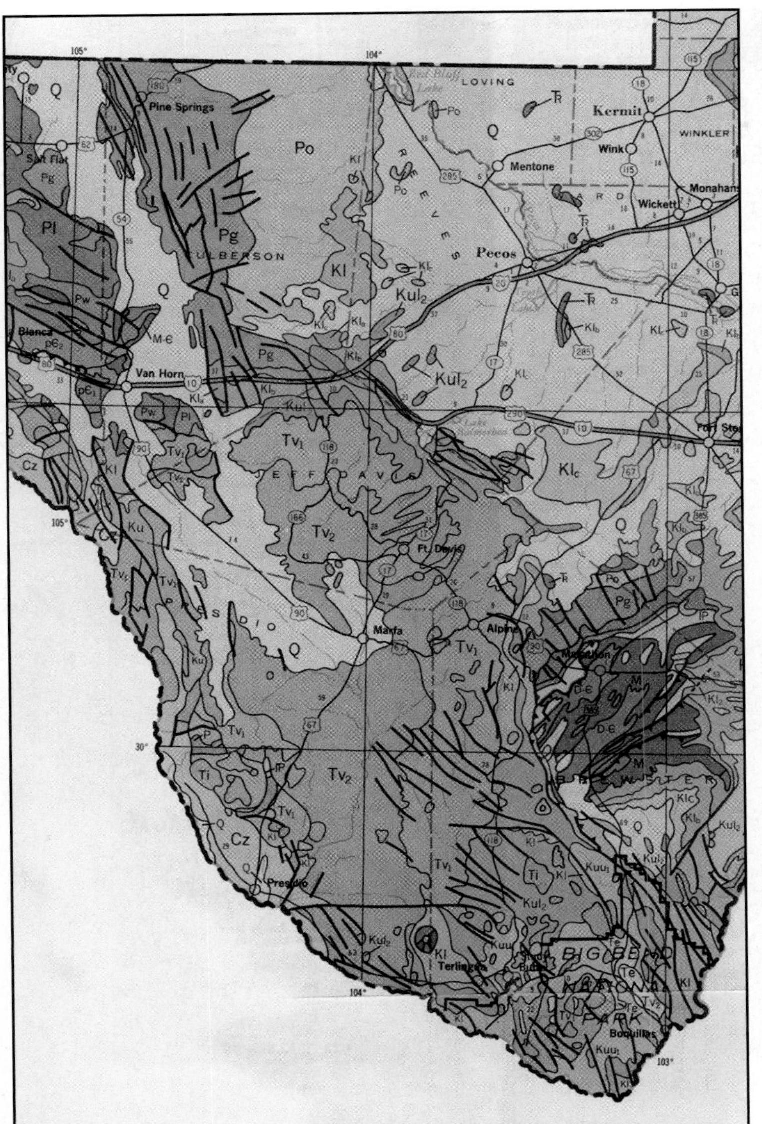

GENERALIZED CHART OF TIME AND ROCK UNITS WEST TEXAS

ERA	SYSTEM	OUTCROP COLUMNAR SECTION
CENOZOIC	Cz QUATERNARY	
	Ti TERTIARY	
MESOZOIC	CRETACEOUS	
	JURASSIC	
	TRIASSIC	
PALEOZOIC	PERMIAN	
	PENNSYLVANIAN	
	MISSISSIPPIAN	
	DEVONIAN	
	SILURIAN	M-6
	ORDOVICIAN	
	CAMBRIAN	
PRECAMBRIAN		

Conglomerate
Sandstone
Shale
Marl
Limestone
Dolomite
Chert
Anhydrite
Salt
Volcanic Rocks
Granite
Metamorphosed Rocks

Reflecting on the Activity and the Challenge

In this activity, you made models of sedimentary rocks and used these models to help you identify sedimentary rocks. You also interpreted sources of information about sedimentary rocks in your area. Being able to identify sedimentary rocks will help you understand how and where sedimentary rocks are formed. This information will help you to understand the geologic history of your community and help you to complete your Chapter Challenge.

Digging Deeper

SEDIMENTARY ROCKS IN THE EARTH'S CRUST

Distribution of Sedimentary Rock

Except for a thin layer of soil and very young sediments at the Earth's surface, the Earth's crust is made of solid **bedrock**. The crust consists of a very wide range of rock types, but **sedimentary rocks** are by far the most common rock type in the upper part of the crust. If you could somehow remove the thin layer of soil and sediment from the top of the crust and look at the exposed bedrock, about three-quarters of it would be sedimentary rock. Over large areas of the continents, sedimentary rocks form layers, as shown in *Figure 1*.

Figure 1 The Grand Canyon is a striking example of layering in sedimentary rocks.

In areas near the ocean, sedimentary layers usually indicate that those areas were below sea level at certain times in the past. In the middle of a continent, the presence of sedimentary layers generally means one of two things. The area might have been topographically low relative to nearby mountain ranges, which supplied **clastic sediments** to cover the low area. The other possibility is that the area was covered by a shallow sea in the past. Either clastic sediments or chemical sediments can be deposited in a shallow sea, depending upon the nature of the nearby land areas and the chemistry of the seawater.

Geo Words

bedrock: solid rock that is connected continuously down into the Earth's crust, rather than existing as separate pieces or masses surrounded by loose materials.

sedimentary rock: a rock, usually layered, that results from the consolidation or lithification of sediment, for example a clastic rock like sandstone, or a chemical rock like rock salt, or an organic rock like coal.

clastic sedimentary rock: a sedimentary rock made up mostly of fragments derived from pre-existing rocks and transported mechanically to their places of deposition.

Clastic Sedimentary Rocks

Clastic sedimentary rocks are made of fragments, called **clasts**, that are eroded from other rocks. Conglomerate, sandstone, siltstone, mudstone, claystone, and shale are clastic sedimentary rocks. Clasts are classified according to their size. The smallest clasts, too small to see without a microscope, are called clay, and clasts with sizes between clay and sand are called silt. Claystone consists of clay-size particles, and siltstone consists of silt-size particles. Mudstone consists of a mixture of silt-size and clay-size particles. When a claystone or mudstone breaks into small, flat chips, it is often called a shale. Sandstone is made of sand-size particles. Conglomerate is made of gravel-size particles, which range from small pebbles to large boulders. The size of the particles usually reflects the strength of the medium that carried the sediment. Pieces of gravel are much larger than tiny clay particles, so faster flows of water are needed to move them from where they originate to where they are deposited.

Chemical Sedimentary Rocks

When water cannot hold all of the material that is dissolved in it, some of the material comes out of solution as solids. This process, which is called **precipitation**, commonly happens when some or all of the water evaporates, or when the water is cooled. **Chemical sedimentary rocks** consist of materials that have precipitated from ocean water or lake water, as shown in *Figure 2*.

Figure 2 Evaporation of rainwater produces salt flats, as in Death Valley.

Geo Words

clast: an individual fragment of sediment produced by the physical disintegration of a larger rock mass.

precipitation: the process of forming solid mineral constituents from a solution by evaporation.

chemical sedimentary rock: a sedimentary rock formed by direct chemical precipitation of minerals from a solution.

Geo Words

organic sedimentary rock: a sedimentary rock consisting mainly of the remains of organisms.

In Part A of the investigation, you modeled the deposition of a chemical sediment by allowing a saturated saltwater solution to dry. The salt crystals that formed precipitated out of the solution. Limestone is the most common chemical sedimentary rock. It consists of the mineral calcite, a calcium carbonate mineral with the formula $CaCO_3$. Some of the calcium carbonate is precipitated directly out of sea water, and some of it is precipitated by marine animals to make their shells. Dolomite is another common chemical sedimentary rock. It consists of the mineral dolomite (with the same name as the rock!), with the chemical formula $CaMg(CO_3)_2$. Gypsum ($CaSO_4 \cdot 2H_2O$), and halite (NaCl), also called rock salt, are also precipitated out of solution. They form when evaporation concentrates the dissolved material enough for it to precipitate out of solution. Areas where intense evaporation is most likely to happen are those with arid (dry) climates.

Organic Sedimentary Rocks

Some sedimentary rocks are made of organic materials. Coal is the best example. Coal forms when plants in swamps with rich vegetation die and are buried by the remains of later plants. Over time, the plant material is compacted so much by the weight of overlying sediment that it is turned into rock. The first material to form is called peat. Peat, shown in *Figure 3*, which has not yet been buried deeply, is used by humans for fuel and for agriculture. With time and greater compaction, peat is converted to lignite ("brown coal") and, with further compaction, bituminous coal ("soft coal"). Approximately 35 m of original plant matter is compacted to form one foot of bituminous coal. The most deeply buried coal is called anthracite ("hard coal").

Figure 3 In Ireland, peat harvested from bogs is often used as a source of fuel.

Sedimentary Environments

Sedimentary rocks are formed from sediments that are deposited in various environments at the Earth's surface. Limestone, for example, is usually deposited in a shallow ocean. Sandstone can also be deposited in a shallow ocean, but it can also form in a beach environment, a desert environment, or a river environment. Coal is usually formed in swamps. A sedimentary rock can therefore tell you something about what the environment was like when and where the sediment was deposited. Each rock tells a story about the geologic environment in which it formed. It may not be easy, however, for geologists to read that story!

Sedimentary Rocks and Climate

Sedimentary rocks can give you information about past climates. For example, sandstone that was deposited as desert sand dunes records a time when the area was dry and lacking protective vegetation. Limestones suggest deposition in warm, shallow oceans. Coal forms in tropical to subtropical climates. Ancient coal is found in Antarctica. This suggests that climate has changed over time in the Antarctic.

How Sediment Becomes Rock

In many places where sediment deposition continues for a long time, the sediments become buried deep below the Earth's surface. The pressure on the sediment increases, causing the particles to be pressed together. Also, water solutions filtering up through the pore spaces of the sediment from deeper in the Earth tend to precipitate cementing material around the sediment particles. These processes of **compaction** and **cementation** cause the sediment to be converted into a solid sedimentary rock. In Part A of the activity, your mixture of glue and water modeled the natural cementation process. Clastic sediments usually turn into solid rock after

Geo Words

compaction: the reduction in bulk volume or thickness of fine-grained sediments owing to increasing weight of overlying material that is continually being deposited.

cementation: the process by which sediments are converted into rock by the precipitation of a mineral "cement" among the grains of sediment.

Check Your Understanding

1. What does the presence of sedimentary rock layers reveal about sea level or past topography in a region?
2. Why is gravel more likely to be found on a river bottom than on a lake bottom?
3. The top of Mt. Everest is made of limestone. What does this suggest about how the topography of that area has changed through time?
4. Rock salt is mined throughout the Great Lakes region. What does this suggest about the past climate of this area?

many hundreds, or even thousands, of meters of burial. Chemical sediments, on the other hand, can become sedimentary rocks with very shallow burial of meters to a few hundreds of meters.

Classifying Sedimentary Rocks

The great variety of sedimentary rocks stems from the variety of environments where sediments are deposited. Classification, or the grouping or ordering of objects by similar features, is meant to make it easier for people to think about and discuss what they are investigating. Classification schemes reflect the features that the observer considers to be important. When you looked at the sedimentary rocks, you made judgments about how to put them into groups. Each person or group that makes a classification system decides which features form the basis of classification. For example, you might have chosen color, texture, roundness of grains, or other features. Did you find differences in your groups or between groups? If you did, you would have experienced exactly what geologists experienced when they developed classification systems for sedimentary rocks. The classification of sedimentary rocks described in this activity (clastic, chemical, and organic) is about the simplest scheme that can be used. Sedimentary geologists have developed much more detailed classifications that stem from this simple one. The main features that are used in such classifications are the composition of the sediment particles and the size of the sediment particles.

Understanding and Applying What You Have Learned

1. In your own words, explain how the three main types of sedimentary rocks form.
2. From your knowledge of sedimentary rocks, label the following interpretations of depositional environments as true or false. Explain your answers.
   a) Coal and peat form from the same material.
   b) Limestone indicates that a shallow sea once covered an area.
   c) The presence of sandstone indicates that the area was once a shoreline.
   d) Rock salt indicates that a region once had an arid climate.
   e) Claystone is deposited by fast-flowing streams.
3. Look at the three rock samples shown in the photographs or provided to you by your teacher.
   a) What is the name of each of the sedimentary rocks?
   b) Describe a possible depositional environment in which each formed.
   c) How did the models you made help you identify these rocks?

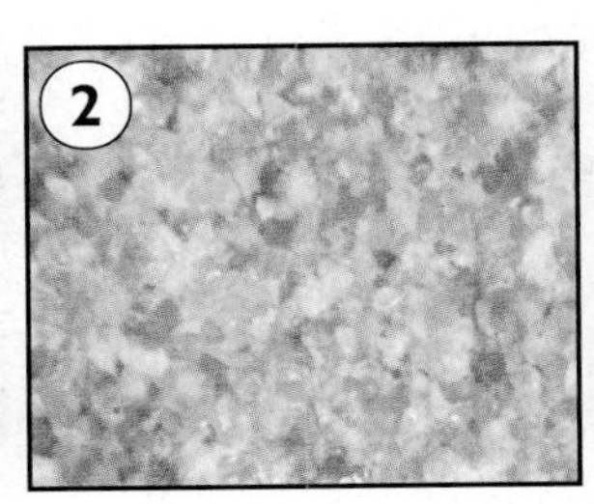

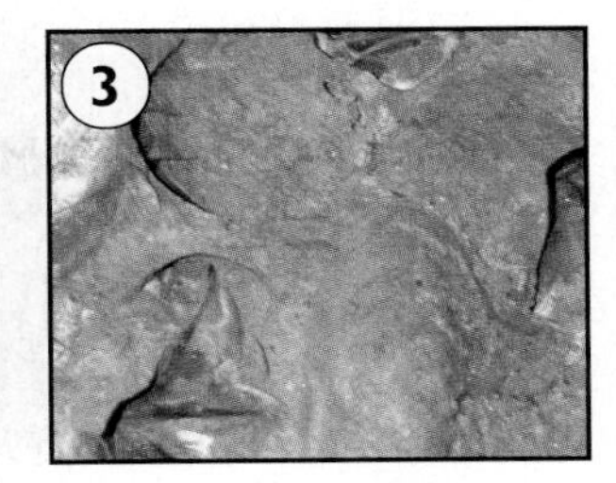

Preparing for the Chapter Challenge

In the area you have selected for your Chapter Challenge, describe the type of environment in which each sedimentary rock might have formed. Consult a local geologic map for information. Describe how the geologic history of the landform or location has changed over time as evidenced by the sedimentary record.

Inquiring Further

1. **Sedimentary rocks and energy resources**

   What sedimentary environments lead to the formation of oil and natural gas? Investigate the types of rocks associated with the successful mining of oil and natural gas.

Activity 2 Igneous Rocks and the Geologic History of Your Community

Goals

In this activity you will:

- Identify several igneous rocks using a rock chart.
- Describe how the two main types of igneous rocks form.
- Understand that igneous rocks are classified based on how they form.
- Use a geologic map and legend to search for evidence of past igneous rock formation.
- Understand that classification helps scientists organize the natural world into smaller, workable components.

Think about It

Igneous rocks cool and crystallize from a molten rock (magma).

- How are these rocks similar to, and different from, the sedimentary rocks that you have seen?

What do you think? Record your ideas about this question in your *EarthComm* notebook. Include a quick sketch. Be prepared to discuss your response with your small group and the class.

Investigate

Part A: Working with Igneous Rocks

1. Examine the photographs of the igneous rocks shown, or a set of igneous rock samples that you are provided.

   a) List some ways you can divide these igneous rocks into groups or categories.

2. Separate the samples into the categories you have decided to use. You might sort them in more than one way.

   a) List the rocks that you place in each category.

   b) Describe the difficulties you experienced trying to categorize them in each way you used.

3. Compare your classification system with the categories used by another group.

   a) Add any categories to your list they used that you had not thought about.

4. Refer to the chart of The Classification of Igneous Rocks on page U20.

   a) Use the chart to name each igneous rock sample.

   b) How do geologists classify igneous rocks?

   c) Describe the similarities and differences between your classification scheme and that of geologists.

   d) What is an advantage to classifying rocks into different groups?

5. Magma cools faster at the Earth's surface (for example, after a volcanic eruption) than it does when it cools below the ground. The faster the cooling and crystallization, the smaller the crystals. Observe the rock samples (or photographs) and the chart of igneous rocks to answer the following questions:

   a) Does rhyolite form at or below the Earth's surface? Explain.

   b) Does gabbro form at or below the Earth's surface? Explain.

Part B: Evidence of Igneous Rocks in Your Community

1. A geologic map, such as one shown in Activity 1, page U7, shows the rocks and sediments at the Earth's surface. Each color or symbol on the map stands for a type and/or age of rock. Geologic maps have a legend. Colors and symbols in the legend explain the types and/or ages of rock shown on the map. Examine the geologic map of your community or region.

   a) Are there any igneous rocks described in the legend? If so, write down a list of the rock type, locations, and ages (in millions of years) if possible. Make a data table to record your observations. If there are numerous igneous rocks in your community, limit your data table to about five different examples.

   b) What are the most common igneous rocks in your area?

   c) Many igneous rocks are very resistant to weathering, and thus erode more slowly than other kinds of rocks. When igneous rock is surrounded by softer rock, a distinct elevated landform may develop. Locate an elevated or prominent landform in your community or region (choose a familiar hill, mountain, rock exposure, or cliff). Is the landform made of igneous rock?

Reflecting on the Activity and the Challenge

Learning how to classify rocks by their physical and/or chemical properties will help you use rock charts to identify the rock specimens. Being able to identify igneous rocks and understanding how they form will help you determine the geology of the area as you work on your Chapter Challenge.

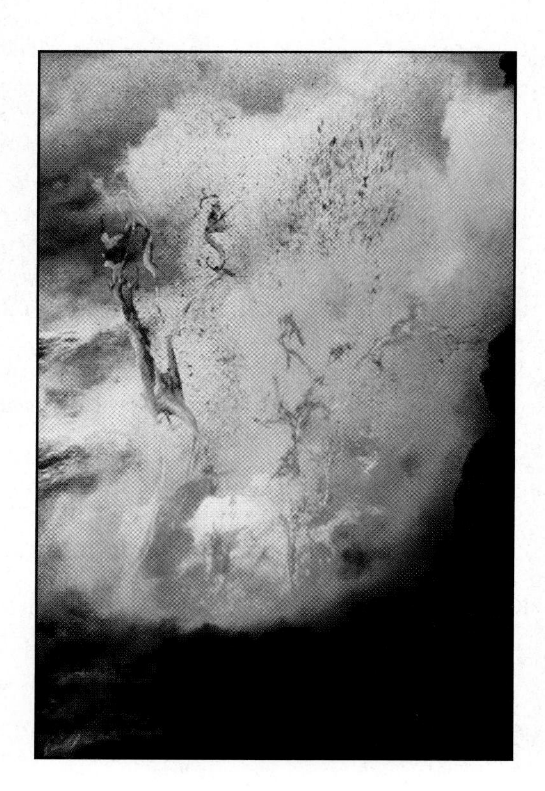

Digging Deeper

IGNEOUS ROCKS

The Nature of Igneous Rocks

All igneous rocks are made of interlocking crystals of minerals that cool and crystallize out of **magma** (molten rock). The interlocking nature of the crystals makes igneous rocks very resistant to **physical weathering** and **erosion**. **Minerals** are the building blocks of igneous rocks (and all other rocks as well). Minerals are compounds of usually several chemical elements. Each mineral has a particular chemical composition and crystal structure. Each mineral has a chemical formula, which expresses the proportions of the various chemical elements in its composition. Although there are literally thousands of kinds of minerals in the Earth's crust, only six kinds of minerals are common in igneous rocks (fortunately for beginning students of igneous rocks!). These are quartz, feldspars, micas, pyroxenes, amphiboles, and olivines. These are all called **silicate** minerals, because their basic structure consists of very tightly bonded units consisting of silicon and oxygen (called silica) that are bonded less strongly to various other atoms. Of the six kinds, all but quartz are listed in the plural form (-s), because the details of their compositions can vary widely even though the basic nature of the mineral is the same. For example, there are two kinds of feldspars (plagioclase and potassium feldspar), with slightly different structures and very different chemical compositions. And there are two kinds of micas (muscovite and biotite), again with slightly different structures but very different chemical compositions.

Magma, Lava, and Igneous Rock

Igneous rocks are formed from cooling of magma. Suppose that you could drill a hole very deep into the Earth. You would find that the Earth's temperature initially rises by about 30°C with every kilometer of depth. This rate of increase slows down at deeper depths. At a depth of 100 to 350 km, the temperature is high enough for large volumes of rock to melt and form magma at certain times and places. Nearly all substances expand when they are heated. When rock is melted into magma, its volume increases by about 10%. This makes the magma less dense than the surrounding rock. Like a hot-air balloon that rises through less dense surrounding air, magma rises toward the Earth's surface. (See *Figure 1*.) Some magmas cool and solidify into igneous rock before they reach the surface. The igneous rock that forms in that way is called **intrusive igneous rock** (because the magma "intrudes" into solid rock that was already there). In some places, magma reaches the surface before it solidifies into igneous rock.

Geo Words

igneous rock: a rock that solidified from molten or partly molten material, i.e., from magma.

magma: naturally occurring molten rock material, generated within the Earth, from which igneous rocks have been derived through solidification and related processes.

physical weathering: the processes of weathering by which rock is broken down by physical forces or processes, including gravity, water, ice, wind, or human actions at or near the Earth's surface.

mineral: a naturally occurring inorganic, solid material that consists of atoms and/or molecules that are arranged in a regular pattern and have characteristic chemical composition, crystal form, and physical properties.

erosion: the wearing away of soil or rock by weathering, mass wasting (downhill movement of material under the influence of gravity), and the action of streams, glaciers, waves, wind, and underground water.

silicate: a compound whose basic structure consists of very tightly bonded units consisting of silicon and oxygen (called silica) that are bonded less strongly to various other atoms.

intrusive igneous rock: igneous rock formed at considerable depth by the crystallization of magma.

Geo Words

lava: magma that reaches the Earth's surface.

extrusive igneous rock: igneous rock that has erupted onto the surface of the Earth.

Magma that reaches the surface is called **lava**. Rock that is formed when lava cools is called **extrusive igneous rock** (because the lava is "extruded" onto the Earth's surface, like toothpaste from a tube). As you will see, the appearance of an igneous rock reveals whether or not it formed below or at the Earth's surface.

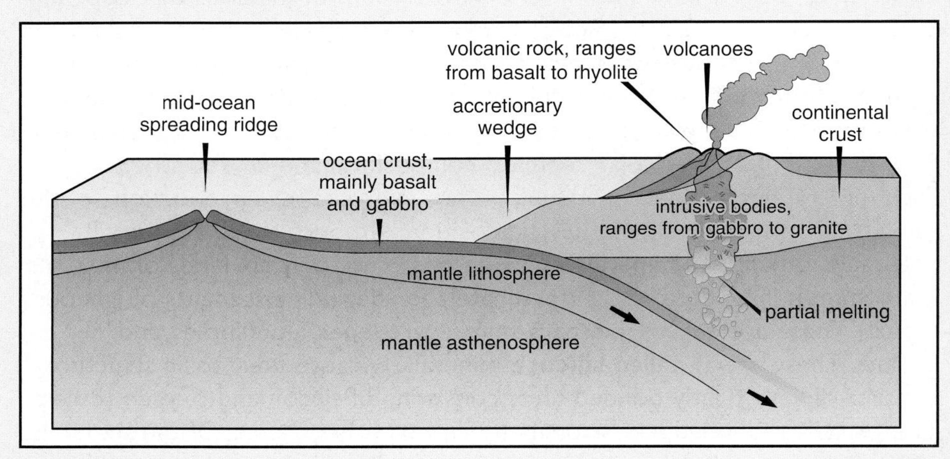

Figure 1 Cross section of mantle and crust.

Classifying Igneous Rocks: Texture

The crystal size of an igneous rock depends very strongly on how fast the magma cools. When magma cools very slowly, only a small number of crystals are formed in a given volume of the magma, but they have plenty of time to grow to be large. The resulting igneous rock is coarse-grained, with mineral grains that are usually several millimeters, or even a few centimeters, in size, as shown in *Figure 2*. On the other hand, when a magma is extruded at the Earth's surface and cools very rapidly, a large number of crystals are formed in a given volume of the magma. However, there is not enough time for them to grow to be large. The resulting igneous rock is very fine-grained with minerals grains that are usually too small to be seen without a magnifying glass. If the lava cools even more quickly, a glassy textured rock called obsidian can form. Obsidian forms when magma cooled so quickly that no crystals have time to form.

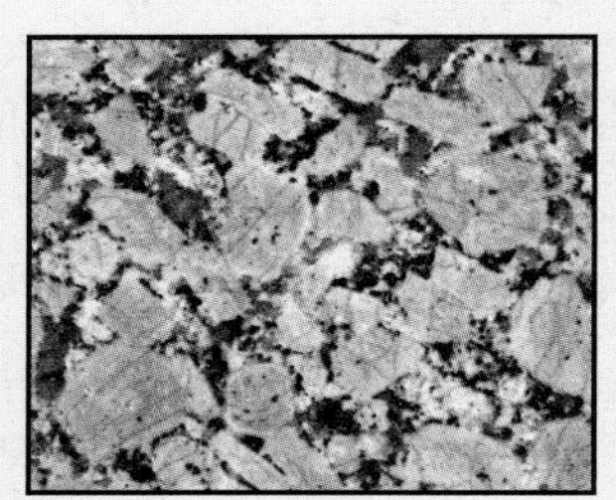

Figure 2 Granite with coarse-grained texture.

Figure 3 Granite with medium-grained texture.

Classifying Igneous Rocks: Chemical and Mineral Composition

The color of an igneous rock is determined mainly by the chemical composition, and therefore the mineral composition, of the rock. Recall the common igneous rock-forming minerals. Quartz, potassium feldspar, and muscovite mica are light in color. Igneous rocks with high percentages of these minerals tend to be light in color. They are the most common minerals in igneous rocks of the continental crust. Pyroxenes, amphiboles, plagioclase feldspar, biotite mica, and olivines are darker in color. Igneous rocks with high percentages of these minerals tend to be dark in color.

Igneous rocks that consist mostly of minerals like quartz, potassium feldspar, and muscovite mica, which are rich in silica (silicon and oxygen) but poor in iron and magnesium, are lighter in color, because these minerals are usually white, light gray, or pink. These rock types, whether intrusive or extrusive, are associated with places where **lithospheric plates** are moving together and magma is formed. Magmas rich in silica do not flow very easily and usually cool before they reach the Earth's surface to form granite. Granites found at the Earth's surface today formed below the surface long ago and have been exposed by uplift and erosion. If the same magma reaches the surface, it cools quickly to form an extrusive igneous rock called rhyolite.

Igneous rocks that contain minerals that are rich in iron and magnesium (olivines, amphiboles, pyroxenes, and biotite mica) are dark in color, typically black to dark green. One extrusive igneous rock of this kind, basalt, is the most common rock on the Earth's surface, because it is the major rock of the oceanic crust. Basalt is formed where lithospheric plates are spreading apart or where magma is rising through a mantle hot spot. These rocks are common in the Hawaiian Islands and Iceland. Gabbro is an intrusive igneous rock that contains minerals rich in iron and magnesium. It is the coarse-grained counterpart of basalt. It is common deep in the oceanic crust.

Some igneous rocks are intermediate in chemical composition. These rocks are made of a mixed content of minerals that contain iron and magnesium. These rocks are therefore also intermediate in color. Andesite, an extrusive rock, and diorite, the corresponding intrusive rock, are examples. Andesite (named for the Andes Mountains, where it is abundant) and diorite often form where an oceanic lithsopheric plate is being subducted beneath a continental lithospheric plate. Water rising up into the mantle from the down-going plate causes some of the mantle rock to melt. As the magma rises up through the continental plate, it melts some of the continental rocks, causing it to have an intermediate composition.

Geo Words

lithospheric plate: a rigid, thin segment of the outermost layer of the Earth, consisting of the Earth's crust and part of the upper mantle. The plate can be assumed to move horizontally and adjoins other plates.

Classification of Igneous Rocks					
Color		Light	Intermediate	Dark	Dark
Mineral composition		quartz (≥5%) plagioclase feldspar potassium feldspar iron-magnesium rich minerals (≤15%)	quartz (<5%) plagioclase feldspar potassium feldspar iron-magnesium rich minerals (15-40%)	no quartz plagioclase feldspar (~50%) no potassium feldspar iron-magnesium rich minerals (~40%)	nearly 100% iron magnesium rich minerals
Texture	Crystals >10 mm	granite pegmatite	diorite pegmatite	gabbro pegmatite	
	Crystals 1–10 mm	granite	diorite	gabbro	peridotite
	Crystals <1 mm	rhyolite	andesite	basalt	
	Glassy	obsidian		obsidian	
	Frothy	pumice		scoria	

Check Your Understanding

1. In your own words, describe the difference between an intrusive igneous rock and an extrusive igneous rock.
2. How do the two main types of igneous rocks form?
3. Explain the relationship between the mineral composition of an igneous rock and the color of the rock.
4. Explain how the texture of an igneous rock reveals how the rock formed.

Explosive Volcanic Eruptions

Some magmas, especially magmas that are silica-rich, have a high content of dissolved gases, like water vapor and carbon dioxide. When these magmas rise up to near the Earth's surface, the dissolved gases tend to bubble out of the magma, because the pressure is so much lower than deep in the Earth. Sometimes the pressure is released suddenly by a violent explosive volcanic eruption, as happened at Mt. St. Helens in the Pacific Northwest in 1980. The products of such an eruption are pieces of mineral grains and broken igneous rock, called volcanic ash. In one sense, the rock formed from volcanic ash is a sedimentary rock because it is formed by the deposition of material. However, because it came directly from a volcano it is usually considered to be an igneous rock. Pumice is a volcanic rock that consists mainly of bubble holes, with only thin walls between the holes. Because of its very low density, pumice floats on water!

Figure 4 Pumice sample from Mt. St. Helens.

Understanding and Applying What You Have Learned

1. Use the photographs of the rocks shown, or obtain several new samples of igneous rocks. Use the rock chart to answer the following questions:
   a) Is the rock light, intermediate, or dark in color?
   b) Is the rock glassy, or does it have fine crystals, or coarse crystals?
   c) Is the rock intrusive or extrusive?
   d) Name each rock.

2. Examine the geologic map of your community and the list of igneous rocks that you generated in Part B of the investigation.
   a) Did the igneous rocks in your community or area form underground or at the Earth's surface? Explain.
   b) Focus in on the area that you have selected for your chapter report. Describe any evidence of igneous rocks.

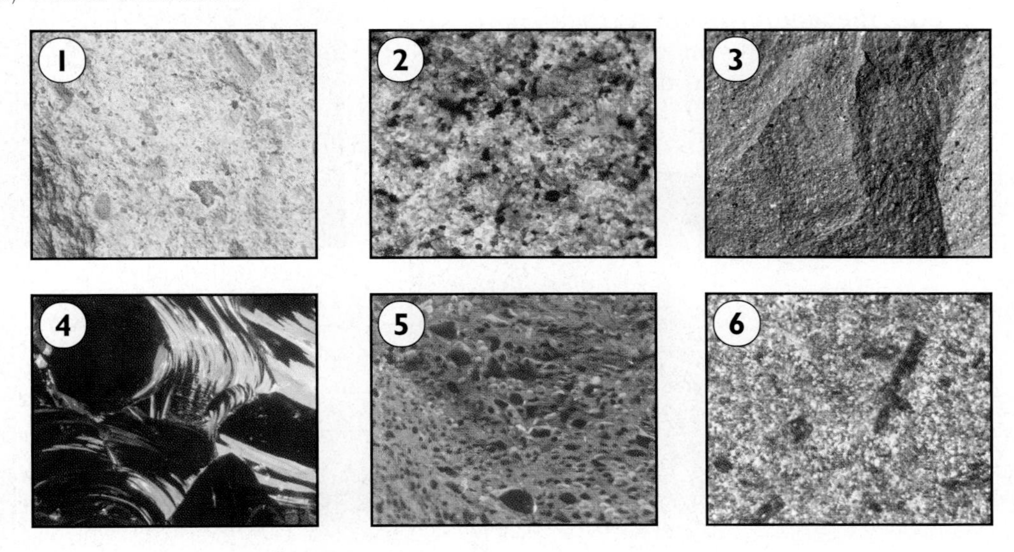

Preparing for the Chapter Challenge

Using the information you have gathered about the igneous rocks in your community or state, make inferences about past plate motion that affected your community or area. Name the igneous rock types that helped you make your decisions.

Inquiring Further

1. **Igneous rocks and famous landscapes**

   Investigate one of the following:

   - Ship Rock, New Mexico
   - Sierra Nevada Batholith, Yosemite National Park, California
   - Devil's Postpile National Monument, California

   From what igneous rock is the famous landform made? What does the landform and its rock composition tell you about the geological history of that place?

Devil's Postpile National Monument, California.

Ship Rock, New Mexico.

Sierra Nevada Batholith, Yosemite National Park, California.

Activity 3 Metamorphic Rocks and Your Community

Goals

In this activity you will:

- Identify and classify several metamorphic rocks using a rock chart.
- Describe two agents of metamorphism.
- Use a geologic map to search for evidence of past metamorphism in a community.
- Understand that properties of materials can change over time.

Think about It

Metamorphism is the "magic" that transforms one rock into a new form.

- What factors are responsible for changing a rock from one form to another?
- Where does metamorphism occur?

What do you think? Record your ideas about these questions in your *EarthComm* notebook. Provide a sketch. Be prepared to discuss your responses with your small group and the class.

Investigate

Part A: Modeling Deformation during Metamorphism

1. Obtain a ream (500 sheets) of paper, or an old, very thick (at least 3 cm) telephone book or catalog.
2. On the side of the stack of sheets, draw a large circle. Then draw a straight line through the center of the circle and parallel to the sheets, and another straight line perpendicular to the sheets. See the diagram below.
3. Change the shape of the stack of sheets or pages by sliding them parallel to one another so that the stack "leans sideways." Change in the shape of an object is called deformation. The kind of deformation you are producing here is called shear. If you use a ream of paper rather than a book or catalog, you will be able to make the stack lean farther (in other words, you will be able to make it deform more).
   a) How does the shape of the circle change when you deform the stack?
   b) How does the line parallel to the sheets change when you deform the stack?
   c) How does the line perpendicular to the sheets change when you deform the stack? Record your observations in your notebook along with a sketch of the stack before and after deformation.
4. What do you think would happen to a rock if it is sheared in the same way as the ream of paper or the book?
   a) Record your conclusions in your notebook. Compare your conclusions with those of the other groups, and discuss any differences in your conclusions.

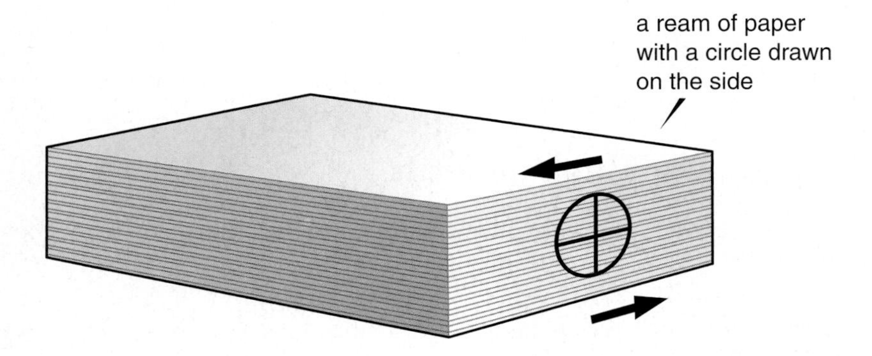

Part B: Classifying Metamorphic Rocks

1. Examine the photograph of the collection of metamorphic rocks (or obtain rock samples) and the chart on the following page.
   a) What properties do geologists use to classify metamorphic rocks?
   b) Use the metamorphic rock chart to identify your rock samples (or the pictured rocks).

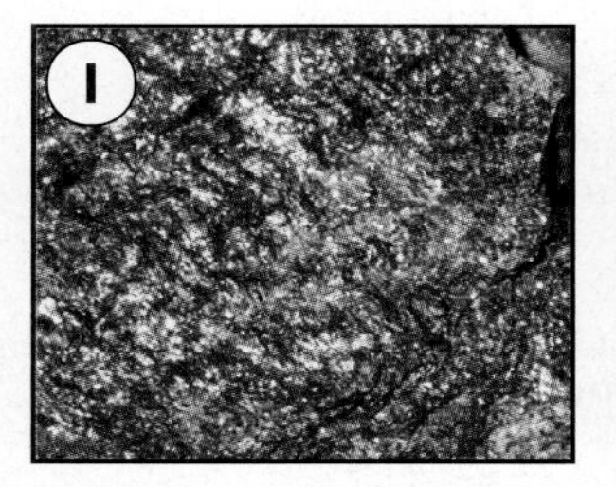

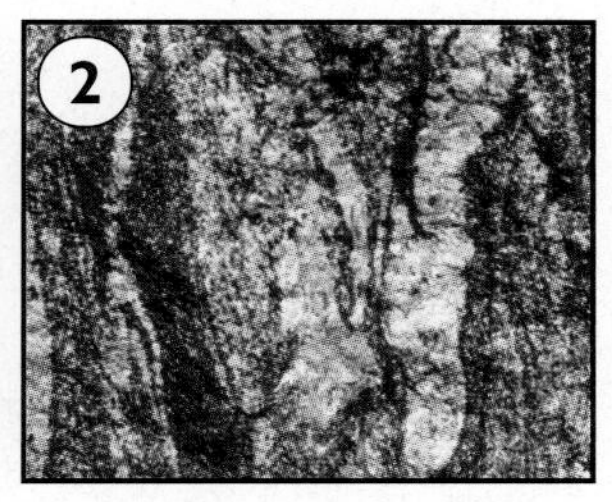

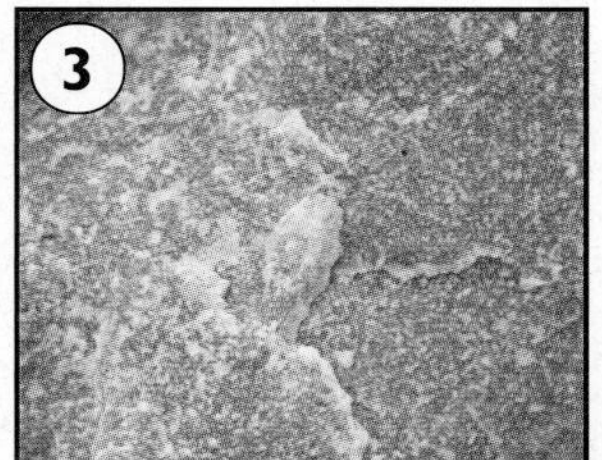

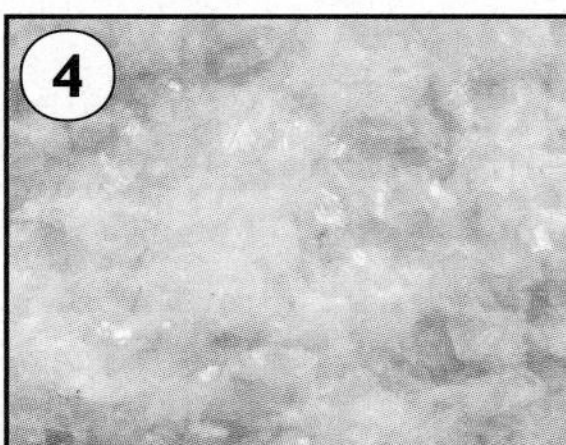

Part C: Evidence of Metamorphic Rocks in Your Community

1. Examine the geologic map of your community or region.
   a) Are any metamorphic rocks described in the legend? If so, make a list of the rock type, locations, and ages (in millions of years). Record your observations in a data table. If there are numerous metamorphic rocks in your community, limit your data table to several different examples. Be sure to include any evidence found in the area that you have selected for your Chapter Challenge.
   b) What are the most common metamorphic rocks in your area?

Classification of Metamorphic Rocks			
Texture	**Rock Name**	**Description**	**Rock before Metamorphism**
Strongly foliated: rocks in which platy minerals are arranged to be approximately parallel, causing the rock to split easily along parallel planes.	Slate	Very fine-grained, usually dark, splits easily along parallel planes.	Mudstone, claystone, shale.
	Phyllite	Fine-grained, usually dark, splits easily along parallel planes: often crinkled or folded. Not as fine-grained as slate.	Mudstone, claystone, shale.
	Schist	Medium-grained to coarse-grained, with parallel alignment of platy mineral grains like micas.	Mudstone, claystone, shale, some volcanic rocks.
Weakly foliated or nonfoliated: rocks without abundant platy mineral; the rocks do not split easily along parallel planes.	Gneiss	Medium-grained to coarse-grained rock, often with alternating layers of light and dark minerals.	Granite, rhyolite, some sandstones, some volcanic rocks.
	Marble	Usually light-colored, composed of calcite crystals.	Limestone
	Quartzite	Usually light-colored, composed of quartz crystals.	Quartz sandstone
	Greenstone	Dark green, fine-grained rock made of various minerals rich in iron and magnesium.	Basalt
	Amphibolite	Dark-colored, medium-grained to coarse-grained rock with abundant amphibole minerals.	Basalt

Reflecting on the Activity and the Challenge

In this investigation, you saw how materials can be deformed by shear. You also examined samples of metamorphic rocks and searched for evidence of metamorphic rocks in your community. Understanding how rocks are changed will help you interpret the geological history of your area and complete your Chapter Challenge.

Digging Deeper

Formation of Metamorphic Rocks

Sedimentary rocks and igneous rocks can be metamorphosed (turned into a **metamorphic rock**) if they are subjected to high temperatures and/or pressures. The changes occur while the rock is still solid, before the temperature becomes so high that part of the rock melts. If the temperature becomes too high, part of the rock melts to form a magma, which later cools to form an igneous rock. The basic idea behind metamorphism is that a crystal of a mineral can grow only in a certain range of temperature and pressure. If a mineral crystal in a rock is subjected to the high temperatures and pressures of metamorphism, it is likely to be outside of its range of temperature and pressure, and it is changed into crystals of one or more different minerals. This is why the minerals in a metamorphic rock are usually very different from the minerals in the original rock. A few common minerals, however, like quartz and calcite, are "just as happy" under metamorphic conditions as they are near the Earth's surface. When a limestone is metamorphosed, the calcite continues to exist, but the crystals grow to be much larger, and all evidence of the original features of the limestone, like fossils, are destroyed.

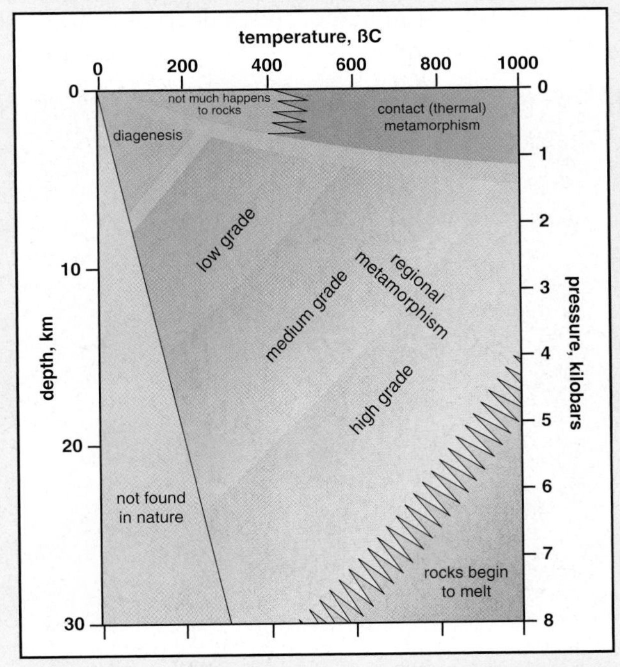

Figure 1 Diagram to explain regional metamorphism.

Geo Words

metamorphic rock: rock that has been changed (metamorphosed) into a different rock type, without actually melting, by an increase in temperature and/or pressure, and/or the action of chemical fluids.

Geo Words

fault: a fracture or fracture zone in rock, along which rock masses have moved relative to one another parallel to the fracture.

regional metamorphism: a general term for metamorphism affecting an extensive region.

By using special pressurized furnaces in the laboratory, geoscientists have learned a lot about the conditions of temperature and pressure that are characteristic of metamorphic minerals. The kinds of metamorphic minerals in a rock can then tell a story of what the temperatures and pressures in the Earth were where the rock was metamorphosed.

The temperature of a rock can be increased in two ways. Rocks can become buried deeper and deeper in the Earth. This can happen either by deposition of a very thick layer of sediment on top of the rock, or by Earth movements that shove very thick masses of rock on top of the rock by movement along **faults**. As the rock is buried, its temperature gradually increases. That is because the temperature in the Earth increases with depth. Enormous volumes of rock can be metamorphosed in this way by deep burial. This is the most important kind of metamorphism. It is called **regional metamorphism**, because large regions of the Earth's crust can be affected in this way. The temperature of a rock can also be increased if a body of magma is put into place near the rock. As the magma cools, the surrounding rock is heated, and this can metamorphose the rock, as shown in *Figure* 2. If the igneous intrusion is small, only a thin layer of the surrounding rock is metamorphosed, but very large intrusions can metamorphose the surrounding rock for thousands of meters away from the intrusion. The intensity of metamorphism decreases gradually away from the contact with the intrusion.

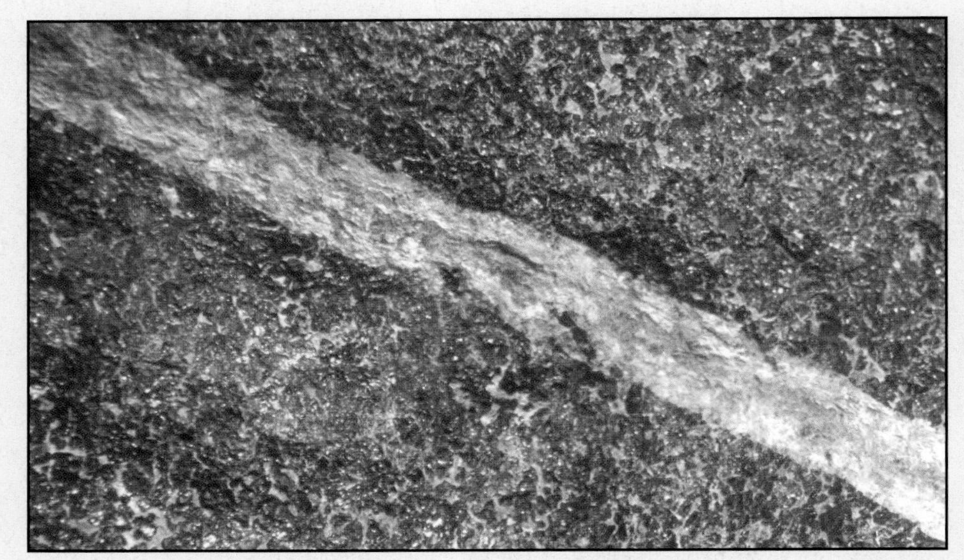

Figure 2 When an igneous rock intrudes another rock, the intense heat of the intrusion can result in metamorphism of the surrounding rock. This is known as contact metamorphism.

Deformation in Metamorphism

Extreme deformation (change in the shape of a material) is common during regional metamorphism. In Part A of the investigation you modeled the deformation of a rock by shearing. The same thing happens, although usually even more so, when rock is sheared by forces within the Earth. This is especially common where one lithospheric plate slides down beneath another. You saw in the investigation that when a material is sheared, lines or planes within it become more nearly parallel. This is called **transposition**. In many metamorphic rocks, all kinds of features and structures are "smeared out" by transposition to become nearly parallel planes. The layering you see in a metamorphic rock may not have anything to do with layering in the original rock! Forces within the Earth can also stretch or compress the rock. In some metamorphosed conglomerates, the pebbles are stretched into the shape of cigars.

Geo Words

transposition: the process by which lines or planes within a material become more nearly parallel when they are sheared.

foliation: the tendency for a metamorphic rock to split along parallel planes.

Foliation in Metamorphic Rocks

Figure 3 This gneiss is an example of a strongly foliated metamorphic rock.

Sedimentary rocks like claystone, mudstone, and shale contain a high percentage of very fine flakes of mica minerals. These rocks become metamorphosed first to slate, then to phyllite, and then to schist, as the intensity of metamorphism increases. As you saw in the classification table earlier in this activity, all of these rocks tend to split easily along parallel planes. This is because the mica minerals in the rock have grown to be

parallel to one another, causing weakness in the direction parallel to the planes of the mineral grains. This parallel growth develops for a combination of two reasons. First, the mica minerals grow with their planes perpendicular to the direction of greatest force on the rock. Second, when the rock is sheared the mica grains tend to become parallel, as you learned earlier. The tendency for a metamorphic rock to split along parallel planes is called **foliation.** Foliation, as shown in the photograph in *Figure 3* is a very prominent feature of many metamorphic rocks.

The Protoliths of Metamorphic Rocks

The rock from which a metamorphic rock was formed is called the **protolith** of the metamorphic rock. Both sedimentary and igneous rocks are commonly the protoliths of metamorphic rocks. Geoscientists who study metamorphic rocks are always interested in trying to figure out what the protolith was. Sometimes that is easy; for example, a quartzite probably started out as a quartz sandstone, and a marble probably started out as a limestone. Sometimes, however, it is very difficult to guess the protolith.

Geo Words

protolith: the rock from which a metamorphic rock was formed.

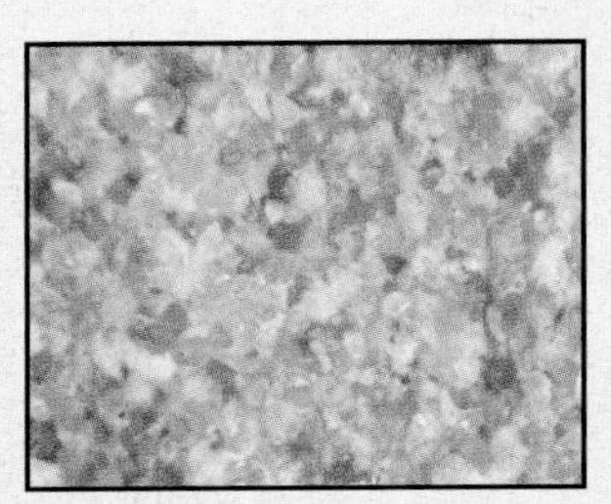

Figure 4a Quartz sandstone.

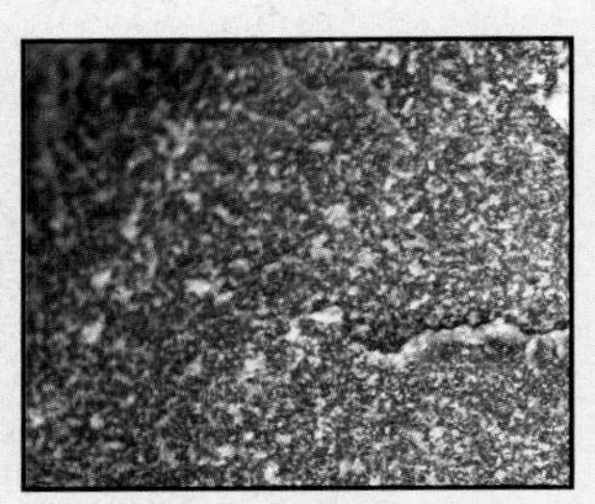

Figure 4b Quartzite.

Check Your Understanding

1. Describe in your own words two sources of heat that lead to metamorphism.
2. Why do temperature and pressure increase with depth in the Earth?
3. Why is the mineral composition of a metamorphic rock usually different from the mineral composition of the protolith (the original rock)?

Understanding and Applying What You Have Learned

1. Why is foliation more likely to occur during mountain building than through the contact of rock with magma?
2. Why are some metamorphic rocks foliated but others lack foliation?
3. Calculate the mass (kg) of a column of rock one meter square (m^2) and 100 m deep. Assume that the density of the rock is 2700 kg/m^3. Repeat this for a depth of 500 m, 1 km, 2 km, 4 km, 8 km, and 16 km. Graph your results. Use your results to describe the relationship between pressure and depth in the Earth. (Remember that 1 km = 1000 m.)
4. Look at the photographs of the metamorphic rocks, or the samples of metamorphic rocks provided by your teacher.
   a) Are the rocks foliated or non-foliated?
   b) What are the names of these metamorphic rocks?
   c) How did these rocks form?
5. Examine the geologic map of your state and your list of metamorphic rocks found in the region covered by the map. Look at the ages of the metamorphic rocks in your area. Did you find evidence of more than one period of metamorphism? Discuss reasons why you might see more than one time period of metamorphism.
6. In your own words, describe how metamorphic rocks demonstrate the principle that the properties of materials can change over time. Discuss crystal size, foliation, and hardness.
7. Suppose you found gneiss and/or schist in a region that currently has no mountains. How would you use this evidence to describe to a friend that there used to be mountains in that region?

Preparing for the Chapter Challenge

You have learned how metamorphic rocks are formed. You have also learned that looking at the features in a rock can provide clues about the history of the rock. The history of the rock will help you understand how the land's surface has changed. You have also gathered information about metamorphic rocks from the geologic map. Using this information, you can add to your understanding of the geological history of your community. Write a one or two paragraph description about the events and evidence for metamorphism in your state and in your local area.

Inquiring Further

1. **Metamorphism in the United States**

   Research the history of the formation of metamorphic rocks in the Appalachian Mountains.

2. **Metamorphism and mineral resources**

   A third major type of metamorphism is that caused by the movement of heated solutions of mineral-rich groundwater. The groundwater is heated by bodies of hot magma. Investigate how hydrothermal alteration leads to the formation of deposits of valuable minerals like gold, silver, and copper.

Activity 4 Rock Units and Your Community

Goals

In this activity you will:

- Understand that rocks are arranged in the Earth's crust as well-defined bodies or units.
- Identify the general shapes of rock units based on rock type.
- Read and interpret geologic maps and geologic cross sections.

Think about It

Rock types exposed at the Earth's surface generally vary from place to place, but over large areas the rock type is often about the same.

- If you start out at an outcrop of bedrock and see a particular kind of rock, how far would you have to walk until the rock type changes?

What do you think? Record your ideas about this question in your *EarthComm* notebook. Be prepared to discuss your responses with your small group and the class.

Investigate

Use the geologic map and cross section of Georgia and Alabama to answer the following questions:

1. a) Read the legend. What kinds of sedimentary rocks are shown on the map and cross section?

   b) What environment do you think was present during the deposition of the rocks in this area?

   c) On the cross section, how does the thickness (vertical extent) of the rock units compare to the distance covered (lateral extent) by the rock units?

   d) Sedimentary rocks are originally deposited in flat, horizontal layers. Given this fact, why do you think that some of the sedimentary layers in the cross section are tilted?

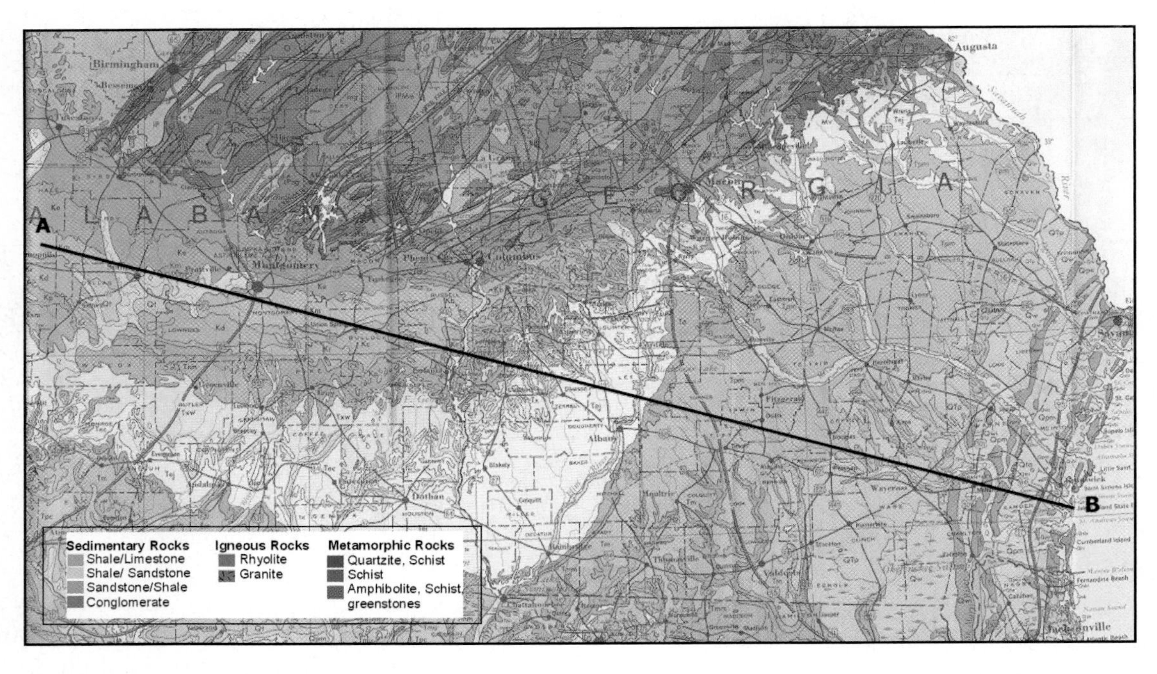

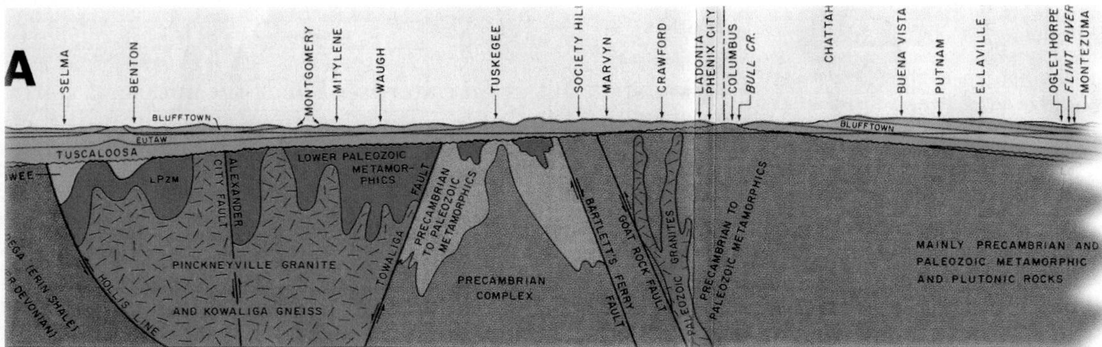

e) On the cross section, follow the units named Eutaw and Tuscaloosa (green layers). Are the units continuous? Why or why not?

2. a) What kinds of igneous rocks are shown on the map and cross section?

   b) Are the igneous rocks shown intrusive or extrusive?

   c) Compare the map and the cross section. Are the igneous rocks that you see in the cross section visible on the map? Why or why not?

   d) Compare the rock units labeled Pinckneyville Granite, Paleozoic Granite, and Jurassic–Triassic Intrusives. How do their sizes vary?

   e) Look at the Paleozoic Granite, and the Jurassic–Triassic Intrusives. How are they different? How are they similar? How do they differ from the Pinckneyville Granite?

   f) Describe the relationship between the Pinckneyville Granite, Paleozoic Granite, and Jurassic–Triassic Intrusives and the surrounding sedimentary rocks.

3. a) What kinds of metamorphic rocks are shown on the map and cross section?

   b) How does the placement of the metamorphic rock units compare with the surrounding rock units?

Reflecting on the Activity and the Challenge

In this activity, you examined how rocks are arranged below the Earth's surface. You learned that rocks are not randomly arranged but occur in distinctive bodies called rock units, and that the size and the shape of rock units vary greatly, depending on the type of rock and how it was formed. Understanding how rocks are arranged in the Earth's crust will help you to decipher the geological history of your community and help you to complete the Chapter Challenge.

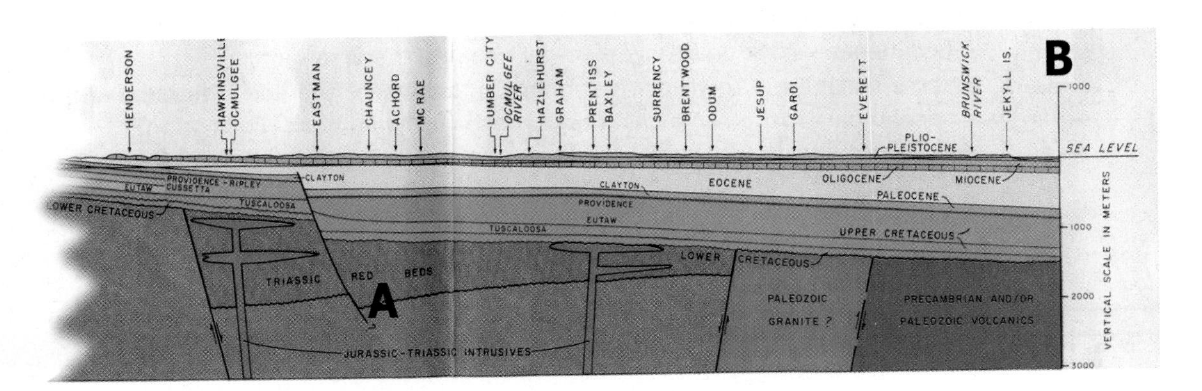

Digging Deeper

ROCK UNITS

What are Rock Units?

Geo Words

rock unit: a body of rock that consists dominantly of a certain rock type, or a combination of types.

The number of different rock types in the Earth's crust is enormous, but if you examine exposed bedrock on the Earth's land surface, you would find that the rocks are of about the same type over large areas. You might walk for hundreds or thousands of meters, or even for some tens of kilometers, and find about the same rock type. This is because rocks, whether sedimentary, igneous, or metamorphic, are originally formed in large volumes by a specific process. The rock bodies that are formed during the same process are called **rock units**. Geologists, when they work in the field, try to recognize or identify such rock units. The change from one rock unit to another is usually abrupt, across some surface or narrow zone of change. This surface or zone is called a contact, and it is what geologists plot on their geologic maps. You saw many examples of contacts between rock units in the activity. Rock units vary greatly in their size and shape, depending on the processes that formed them.

Sedimentary Rock Units

Early geologists believed that sediments were laid down in uniform sheets over large areas of the Earth, a concept referred to as "layer cake" geology. Geologists now understand that, at any given time, different kinds of sediments are deposited in different places, sometimes very close together. Different sedimentary rock units, which might consist of conglomerate, sandstone, shale, or limestone, can be exposed in the same area of the land surface. However, simply because they represent different rock units, you should not assume that they are all of different ages. In many places, such units can be of the same age, having been deposited in different environments at the same time. Usually, however, as you saw in the cross section, sedimentary rock units have a "width" (lateral extent) that is much greater than their thickness. They typically range in thickness from tens of meters to thousands of meters, and their lateral extent can be hundreds of kilometers. Sedimentary rock units have two-part names: the first part is the name of some place like a town or river or mountain where the unit is found, and the second part is the name of a rock type, or just the word "Formation."

Igneous Rock Units

You learned in Activity 2 that igneous rocks are formed in two ways: by cooling and crystallization of magma below the Earth's surface, to form intrusive igneous rocks, and by volcanic activity (extrusion of lava, or explosive eruption of pyroclastic material). Intrusive igneous rock units vary greatly in size and shape, depending on how the magma is emplaced in the surrounding rock. **Sills** are sheets of igneous rock that intruded along layers of sedimentary rocks, and **dikes**, shown in *Figure 1*, are sheets of igneous rock intruded along fractures that cut through any existing rock. **Batholiths** are large masses of intrusive igneous rock with irregular shapes. (See *Figure 2*.) The reason why units of intrusive igneous rocks can be seen exposed today is that the land surface has slowly been worn down by weathering and erosion, to expose the once deeply buried igneous rock body. Volcanic igneous rock units are very much like sedimentary rock units in their size and shape, because they are also spread over the land surface in broad layers. Igneous rock units are given names in much the same way as sedimentary rock units.

Figure 1 Dikes and sills cut through existing rocks.

Metamorphic Rock Units

As you learned in Activity 3, any rock can be metamorphosed. Therefore, metamorphic rock units vary greatly in composition. The shapes of metamorphic rock units are more complicated than those of sedimentary and igneous rock units. The reason is that metamorphism is usually accompanied by large-scale changes in the shape of the rock units by processes like folding and faulting, which you will learn about in a later activity. Metamorphic rock units are usually named in the same way as sedimentary rock units. In some areas, sedimentary or volcanic rock units change slowly and continuously into metamorphic rock units, because the intensity of metamorphism usually changes slowly from place to place. You should not be surprised if you saw on a geologic map that the Smithtown Limestone changed gradually from one place on a geologic map to the Smithtown Marble in another area of the map!

Figure 2 The Sierra Nevada batholith is a massive intrusive structure in California.

Geo Words

sills: sheets of igneous rock that intruded along layers of sedimentary rocks.

dikes: sheets of igneous rock intruded along fractures that cut through any existing rock.

batholiths: large masses of intrusive igneous rock with irregular shapes.

Check Your Understanding

1. Why is the shape of metamorphic rock units usually much more complicated than that of igneous or sedimentary rock units?
2. What is the change from one rock unit to another called?
3. What is the difference between a sill, a dike, and a batholith?
4. What kinds of rocks are laid down in nearly horizontal layers?

Understanding and Applying What You Have Learned

1. If only some of the rocks are visible at the Earth's surface, how do geologists construct cross sections?
2. Examine your state geologic map and geologic cross sections.
   a) Can you tell what rock types are present in the area from the size and shape of the units? Explain.
   b) What rocks are present on the Earth's surface?
   c) What rocks are present below the Earth's surface?
   d) What is the relative order of the rock units in your community?
   e) Is one layer always found beneath another?
   f) How do the sizes of the rock units in your community compare with one another? Which unit is the thickest? Which occupies the most surface area?

Preparing for the Chapter Challenge

Using your understanding of rock units, describe the rock units near your community. Make certain to include a description of the size, shape, and orientation (horizontal, tilted, folded, etc.) of each unit as well as a description of how the units are arranged relative to one another.

Inquiring Further

1. **Making sedimentary formations**

   Sift two batches of sand of different colors into the ends of a water-filled roasting pan. Drain the water. Make a drawing of the structure from above. This is your "geologic map." While the sand is still damp, cut the "deposit" in half. Make a drawing of the side view. This is your "cross section." Describe what you see. How do the "rock units" you have created relate to each other?

2. **Making plutons**

   Create a series of layers using colored clay. Create a crack, or fissure, in the block of clay. Inject caulk, toothpaste, or cake frosting into the fissure. Cut the clay block horizontally, at a distance below the surface that will reveal the upper part of the "intrusion." Make a drawing of the block from above. This is your "geologic map." Then cut the lower part of the clay block in half vertically. Make a drawing of the block from the side. This is your "cross section."

Activity 5 Structural Geology and Your Community

Goals

In this activity you will:

- Describe the relationship between fault movement and the forces that cause this motion.
- Understand that Earth movements can create faults and folds.
- Understand that models help scientists understand how things work.

Think about It

Marble is metamorphosed limestone, a rock that is deposited in nearly horizontal layers.

- What would happen if you tried to use a powerful machine to fold a marble bench?
- How are rocks able to fold naturally without first breaking?

What do you think? Record your ideas about these questions in your *EarthComm* notebook. Be prepared to discuss your responses with your small group and the class.

Investigate

Part A: Making a Model of Folds

1. Mold a large lump of soft craft clay into the shape shown in the figure. The block should be about 10 cm long and about 8 cm wide, as shown in the diagram below. This is your "base block."

2. Put another lump of clay onto a sheet of waxed paper and roll it out with a rolling pin or a wooden dowel into a sheet about 6 mm thick. Drape the sheet over your base block, and trim the edges.
3. Repeat step 2 with lumps of clay of different colors until you have five or six layers. As you place the layers on the block, shape them slightly with your fingers so that they keep almost the same shape as the top of your base block. You now have a block of folded rock layers.
4. Use your hands to stretch a length of dental floss or fishing line tightly, and use it to slice through your folded block. Make two slices: one should be straight down through the block, parallel to its long dimension and off to one side, and the other should be horizontal, through the middle of the folds, as shown in the diagram below.

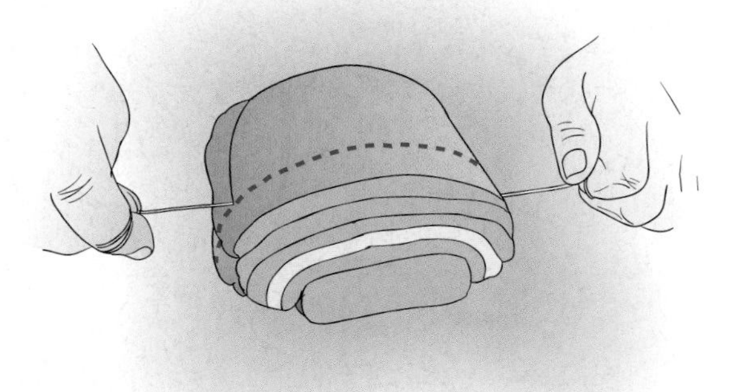

   a) In your notebook, sketch what the folds in your block look like on the faces of each of the cuts you made. The vertical cut is what real folded rocks look like in cross-section view, and the horizontal cut is what real folded rocks look like in map view.

Part B: Making a Model of Faults

1. Obtain foam blocks that have been cut into two pieces at an angle, as shown in the photograph. Use colored pencils or pens to create at least three horizontal layers on the sides of your square.

2. Place the two cut sides of the Styrofoam together, face to face. The plane where the two cut surfaces are in contact is a model of a fault plane. Slowly push the two pieces together, so that the upper piece slides upward relative to the lower piece.

   a) Draw a side view and a top view of what happens. Use arrows to show the direction of force. Use a different type of arrow to show how the colored layers moved in relation to one another along the fault plane.

3. Start over again, with the two pieces of Styrofoam face to face. Slowly slide the lower piece upward relative to the upper piece.

   a) Draw a side view and a top view of what happens. Again, use arrows as above to indicate the direction of force and the movement of layers.

   b) What is the difference in the way the Styrofoam blocks moved in these two trials?

4. Return the Styrofoam to the original position. Move the Styrofoam pieces so that they slide sideways past each other.

   a) Draw a picture of the pieces. Use arrows as before to indicate the direction of force and the movement of layers.

Part C: Interpreting Structure Using Geologic Maps and Cross Sections

1. Geologic maps have special symbols to indicate the locations of faults and folds. Many maps also have one or more cross sections drawn. A cross section shows how the rocks are deformed and makes it easier to infer the forces that caused the deformation. Use a copy of the geologic map and cross section on the following page to complete the following:

   a) Color the cross section and the map. Use a different color for each of the five rock layers A through E.

   b) Sediments are almost always deposited in nearly flat, horizontal layers. What evidence suggests that the rock layers in this region were deformed by forces within the Earth?

   c) Were the faults produced by compression (pushing forces), tension (pulling forces), or shear (sideways forces) in the rock layers? Explain.

   d) Are the folds in the rock layers consistent with your answer above? Explain.

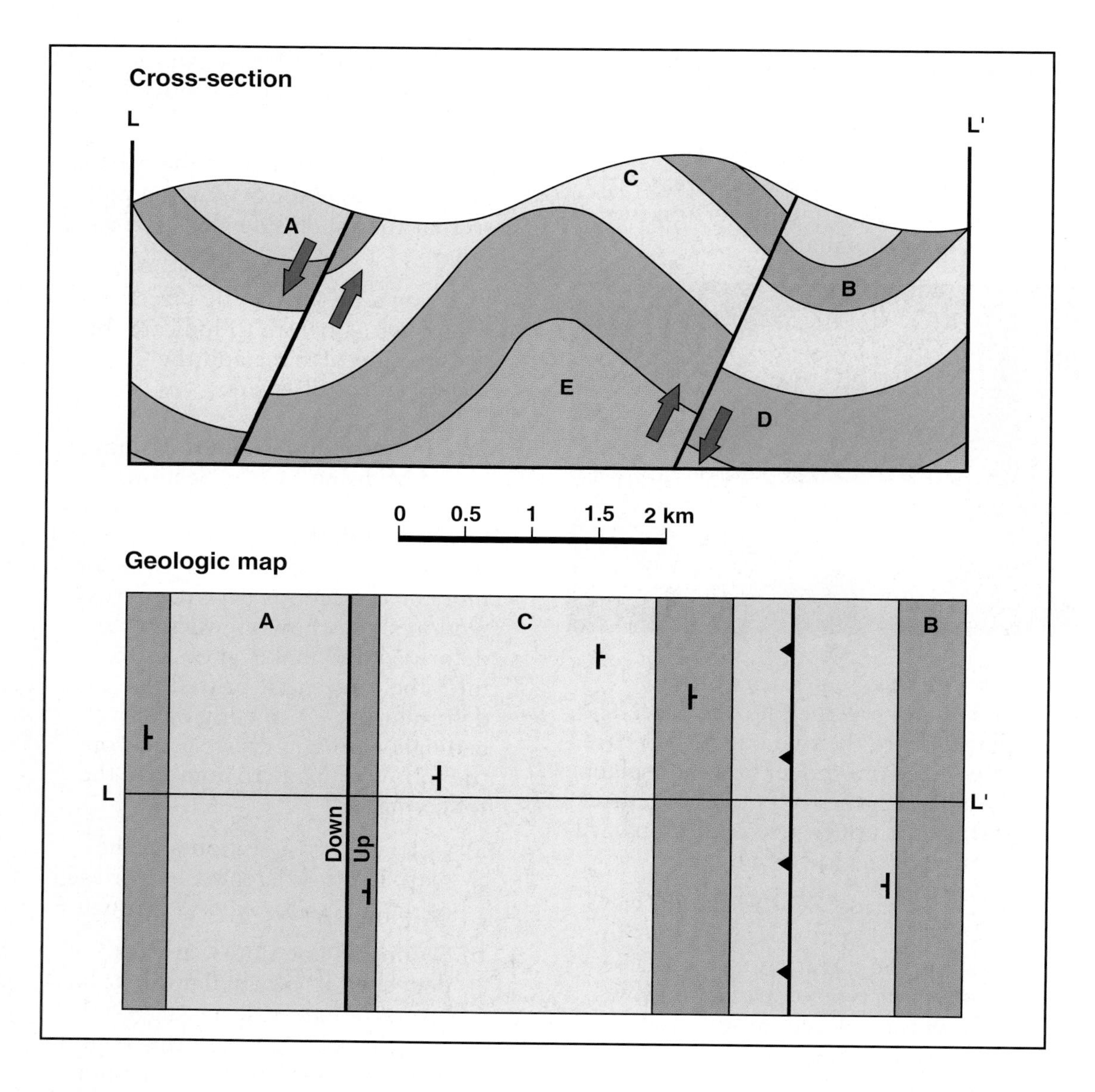

Reflecting on the Activity and the Challenge

In this activity, you modeled the way that forces on rocks can cause the rocks to bend. If the forces on the rocks are large enough, they can cause the rock to fold or fracture. Sometimes forces can move broken pieces up or down. Understanding that changes within the Earth's crust can cause folds and fractures will help you determine the geological history of your community.

Digging Deeper

FORCES IN THE EARTH'S CRUST

Types of Forces

Earlier in this chapter you learned that sedimentary rocks form from sediments that are originally laid down in almost flat, horizontal layers. In many places on Earth, however, sedimentary rocks, and other kinds of rocks as well, are not found in a horizontal position. Instead, they are tilted at some angle to the horizontal. Sometimes they are tilted so much that the layers are vertical! The tilting of the layers is a sign that they have been deformed by forces acting within the Earth. The forces can cause the rocks to become folded, and they can cause the rocks to fracture and then to slip along the fracture surfaces. A fracture surface along which rocks slip is called a **fault**. Forces within the Earth that cause folding and faulting are created by the movement of the Earth's lithospheric plates. The plates move very slowly (2 to 20 cm/yr), but over time spans of hundreds to thousands of years, great forces build up in the Earth's crust. These forces are transmitted for long distances through the crust, so folding and faulting can happen not only near plate boundaries but also in the interiors of the continents, far from plate boundaries.

Three different kinds of forces can cause rocks to deform: **tension forces, compression forces, and shear forces**. Suppose you are holding a solid, rectangular block between your hands. When you try to pull the block apart at its ends, you are exerting a tension force. When you push the ends of the block together, you are exerting a compression force. When you hold two opposite edges of the block and try to move them in opposite directions, you are exerting a shear force. In all three cases, if the force you exert is greater than the strength of the solid material, it deforms, either by fracturing or just changing its shape without actually breaking. The forces that are created in the Earth's crust by the movement of lithospheric plates are often great enough to deform the rocks.

Folding and Faulting

What determines whether a rock is faulted or folded? It's partly a matter of temperature. At low temperatures, as in the upper parts of the Earth's crust, the temperature of rocks is relatively low. At low temperatures rocks are brittle, and they tend to deform by fracturing. At higher temperatures, as in the lower parts of the Earth's crust, the temperature of rocks is relatively high. When forces are exerted on rocks at high temperatures, the rocks tend to deform by changing their shape continuously rather than by faulting.

Geo Words

fault: a fracture or fracture zone in rock, along which the rock masses have moved relative to one another parallel to the fracture.

tension force: a force that tends to pull material apart.

compression force: a force that tends to push material together.

shear force: a force that tends to make two masses of material slide past each other.

Geo Words

normal fault: a fault formed by tension forces that cause the body of rock above the fault plane to slide down relative to the body of rock below the fault plane.

reverse fault: a fault formed by compression forces that cause the body of the rock above the fault plane to slide upward relative to the body of rock below the fault plane.

thrust fault: a reverse fault in which the fault plane is nearly horizontal.

strike-slip fault: a fault formed by horizontal shear forces that cause the bodies of rock on either side of the fault plane to slide past each other horizontally.

Figure 1 Intense heat and pressure folded these previously horizontal sedimentary layers.

Folding is an example of how rocks can change their shape continuously without breaking, as shown in *Figure 1*. Whether a rock is folded or faulted is also a matter of time. If forces build up very fast, rocks are more likely to fracture, but if forces build up very slowly, the rocks are more likely to change their shape without breaking.

Faults

When a fault is formed by tension forces, which cause the rocks of the crust to be pulled apart, the body of rock above the fault plane slides down relative to the body of rock below the fault plane, as shown in *Figure 2(a)*. Faults of this kind are called **normal faults**. When a fault is formed by compression forces, the body of rock above the fault plane slides upward relative to the body of rock below the fault plane, as shown in *Figure 2(b)*. Faults of this kind are called **reverse faults**. When the fault plane is nearly horizontal, the faults are called **thrust faults** instead. When a fault is formed by horizontal shear forces, the bodies of rock on either side of the fault plane slide past each other horizontally, as shown in *Figure 2(c)*. Faults of this kind are called **strike-slip faults**.

Movement on faults usually occurs suddenly after a long time without any movement. The forces that cause the faulting build up very slowly, and when they become greater than the strength of the rock, the fault moves. That relaxes the forces, which then slowly build up again. Major faults typically move only every few centuries. Most earthquakes are caused by sudden movement on faults.

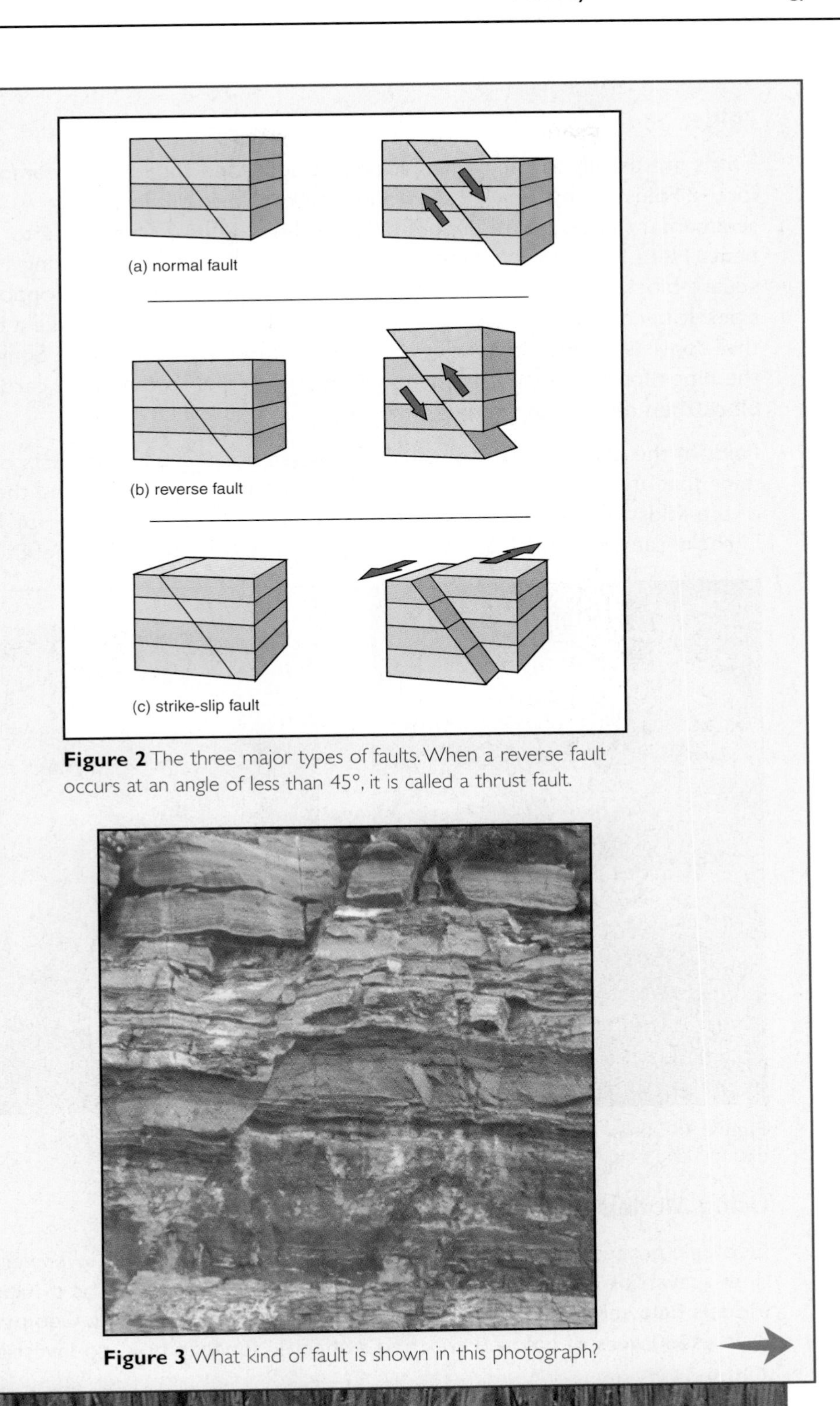

Figure 2 The three major types of faults. When a reverse fault occurs at an angle of less than 45°, it is called a thrust fault.

Figure 3 What kind of fault is shown in this photograph?

Geo Words

fold: a bend in a planar feature in rocks. A fold is usually a result of deformation.

syncline: part of the fold that is concave upward.

anticline: part of the fold that is convex upward.

Folds

Folds are usually formed when rocks are squeezed together by compressive forces. Folds are especially common when the rock is layered, like sedimentary rock, and the layers differ in their stiffness (resistance to being bent). Here is a "thought experiment" to show the effect of layering. Make a square block of soft, modeling clay and squeeze it together from opposite sides. It becomes compressed, but it doesn't form folds. Now make a block that consists of many thin layers of clay separated by index cards. Squeeze the block from opposite sides, in a direction parallel to the index cards. The block then deforms by folding.

Folds in the Earth's crust are usually arranged horizontally. The parts of the folds that are concave upward are called **synclines** (*Figure 4a*), and the parts of the folds that are convex upward are called **anticlines** (*Figure 4b*). Folds in rocks can be as small as centimeters or as large as many kilometers.

Figure 4a Sideling Hill in Maryland is an example of a syncline.

Figure 4b Anticline in Wyoming.

Using Models to Investigate Geologic Structures

Geologic processes like the folding and faulting of rocks work over very long time spans. The folding of rocks cannot be observed in nature as it happens. Models help scientists to understand the deformation of rocks. Geologists have used layers of colored wax and a special "squeeze box" to investigate the folding process. The box had a glass wall for viewing the wax from the side. One side of the box moved along threaded rods, creating a force on the wax as the walls closed in on one another. Wax was poured into the box one layer at a time. A knob on one end of the box allowed the wall of the box to be moved horizontally. This movement squeezed the wax from the ends. The model allowed scientists to vary the rate at which the force was applied and the thicknesses of the various layers. The model helped them to create folds that they saw in nature and make interpretations about the forces that created these folds.

Check Your Understanding

1. Describe tension, compression, and shear forces in your own words. You may wish to use a diagram.
2. What factors determine whether a rock will fault or fold?
3. Why do scientists work with models to understand folding and faulting?

Understanding and Applying What You Have Learned

1. Look at the photograph of faulted rock layers in Pennsylvania.
   a) Do the rocks appear to have been pulled apart, pushed together, or slid past each other to form this structure?
   b) What type of fault is this?
2. Look at the photograph of a fault found in California.
   a) Were the rocks pulled apart, pushed together, or slid past each other to form this structure?
   b) What type of fault is this?
3. Examine your state geologic map and geologic cross sections.
   a) How many faults do you see near your community?
   b) What types of faults do you see?
   c) How many folds do you see near your community?
   d) What types of plate motion does this suggest?

Pennsylvania

California

Preparing for the Chapter Challenge

Using your understanding of the plate motions associated with faults and folds, interpret the geological history of your area in terms of motion in the Earth's crust.

Inquiring Further

1. **Geologic structures in the National Parks**

   Look at a geologic map and cross section of the Grand Tetons in Wyoming. Research the types of structures, folds, and faults in this area. Interpret the geologic history of the area.

2. **Careers in Structural Geology**

   Use the *EarthComm* web site to find a structural geologist who lives and works in your state or region. Introduce yourself through a letter or e-mail message. Briefly describe the work that you are doing in your *EarthComm* classroom. Ask one or two questions that will help you learn about careers in structural geology. Examples include:

   - What makes structural geology so interesting to you?
   - What made you decide to become a structural geologist?
   - What do you think is the most challenging aspect of structural geology?
   - What do you enjoy most about your work?

Activity 6 Reading the Geologic History of Your Community

Goals

In this activity you will:

- Understand the basic principles used to determine the relative ages of rock units.
- Understand the nature and significance of unconformities and their role in deciphering geologic history.
- Interpret the geologic history of an area using the basic principles.

Think about It

Determining the ages of rock units relative to one another is, in a sense, similar to solving a puzzle.

- When you are studying rocks at an outcrop in your community, and you identify two different rock units, how can you tell which is older and which is younger?

What do you think? Record your ideas about this question in your *EarthComm* notebook. Include a quick sketch for each question. Be prepared to discuss your responses with your small group and the class.

Investigate

Part A: Basic Geologic Principles

1. Roll out three different colors of soft, craft clay: red, yellow, and blue. Place the red layer flat on the table. Place the yellow layer on top of it, followed by the blue layer.

   a) Which layer is the "oldest" (i.e., has been there the longest)? Which layer is the "youngest"?

2. The geologic cross section in Figure A shows a series of layers of sedimentary units. As you learned in Activity 1, sedimentary rocks are laid down in layers, much like the layers of clay in step 1.

   a) Which of the units in the cross section do you think is the oldest? Which unit do you think is the youngest? How do you know?

Figure A

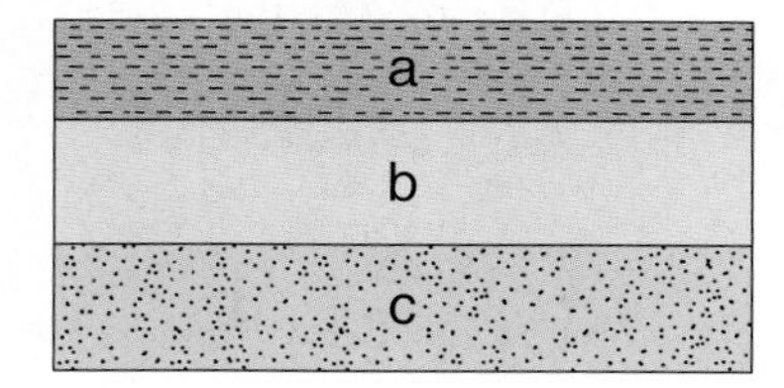

3. Take the clay layers from step 1.

   a) Sketch a side view of what you see.

   b) Now form the layers into folds, as you did in Activity 5. Sketch a side view of what you see.

   c) Sedimentary and igneous extrusive rocks are originally laid down in nearly horizontal layers. Why do you think that the layers are not horizontal? Number the cross sections in Figure B in the order in which they would occur.

Figure B

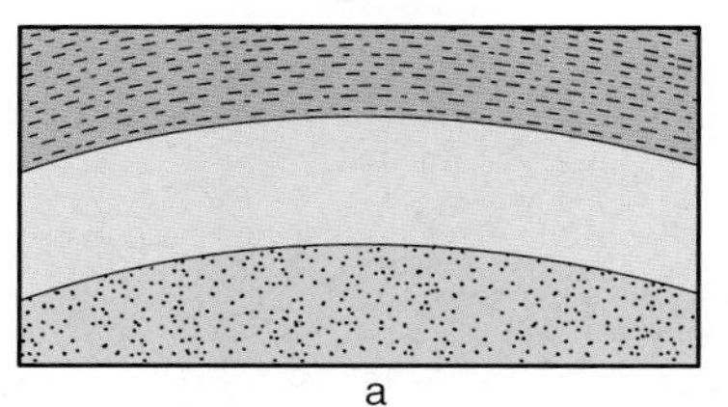
a

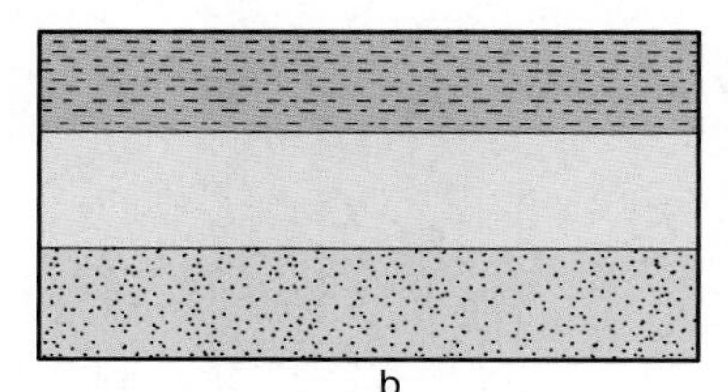
b

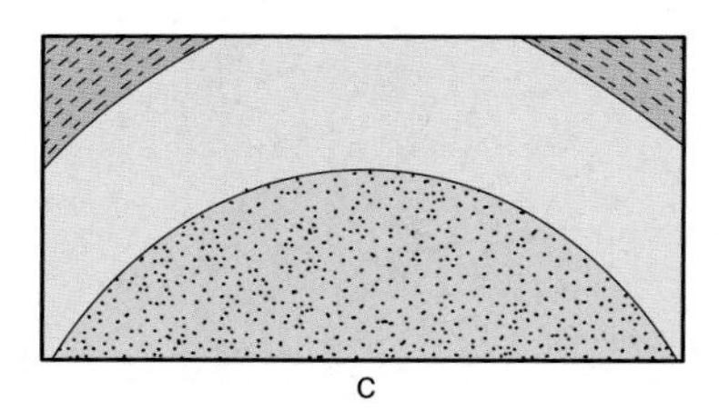
c

4. Flatten out the clay layers and again stack them into a block. Make a slanting cut through the block. Lift the lower side up relative to the upper side so that the red layer on the left matches up with the yellow layer on the right. Remember from Activity 5 that you have produced a normal fault.

   a) Sketch what you see.

b) Now, look at the two cross sections shown in Figure C. What is the youngest feature in each of the two cross sections above? How do you know?

Figure C

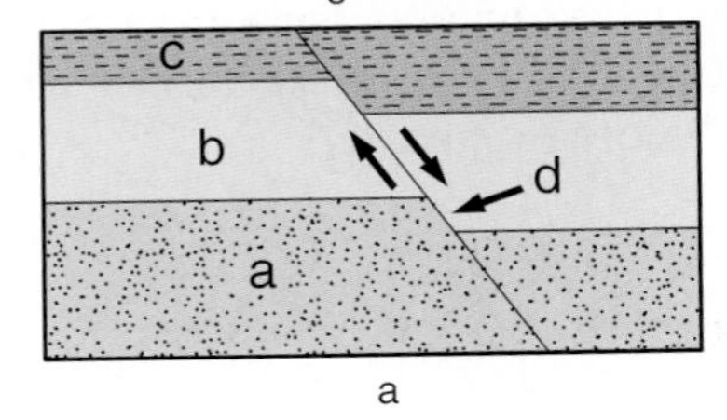

a

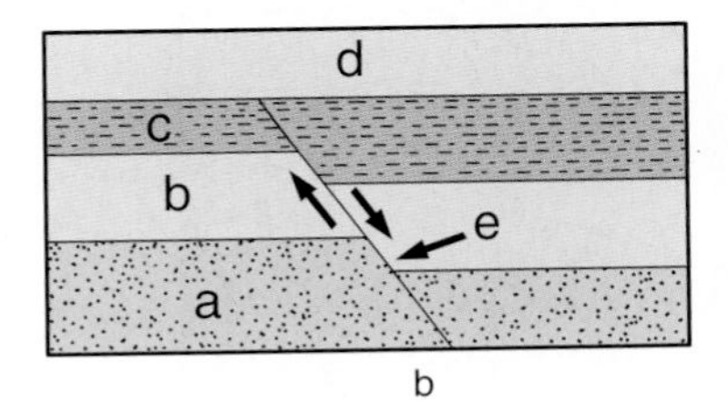

b

5. The following geologic cross section in Figure D shows a sedimentary rock unit A and an intrusive igneous rock unit B.

   a) From what you know about how intrusive igneous rock units form, which of these units do you think is older? How do you know?

Figure D

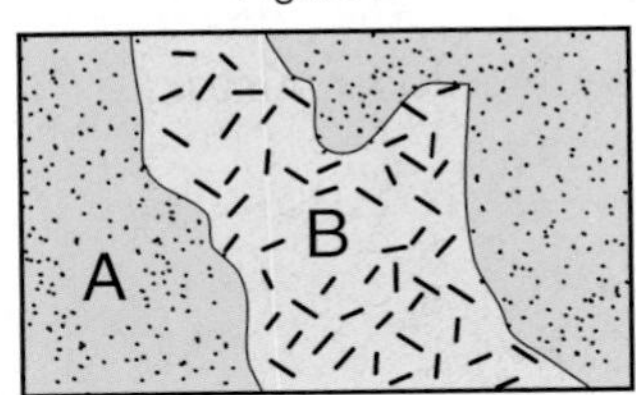

6. The rock units in the following cross section in Figure 6 have been assigned approximate age ranges.

   a) Are the ages continuous, or do you see any time gaps?

   b) Assume that these are sedimentary rocks that were formed as sediment was slowly deposited, layer upon layer. Can you think of an explanation for why there is a time gap in the record?

Figure E

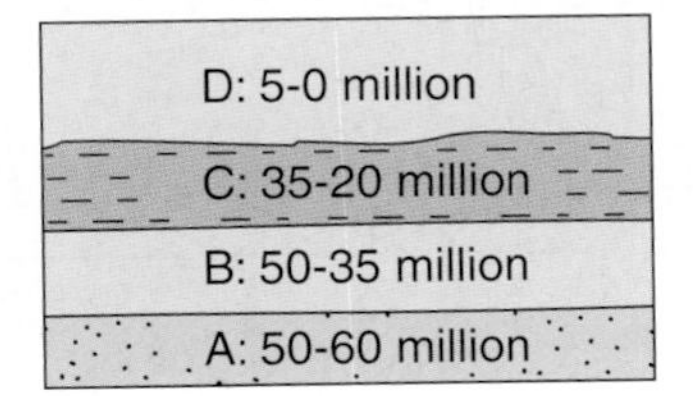

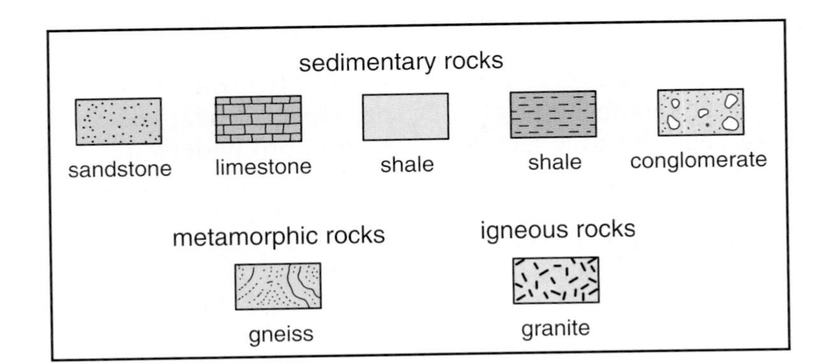

Part B: Using the Principles to Interpret Geologic History

1. The cross section shows several rock units in an area that has had a long and varied geologic history.

   a) Put the rock units and other geologic features marked with letters in the cross section in order of occurrence from earliest to latest. Start by asking yourself what was there first, and then work your way forward through time. You can think of this as a "geologic puzzle."

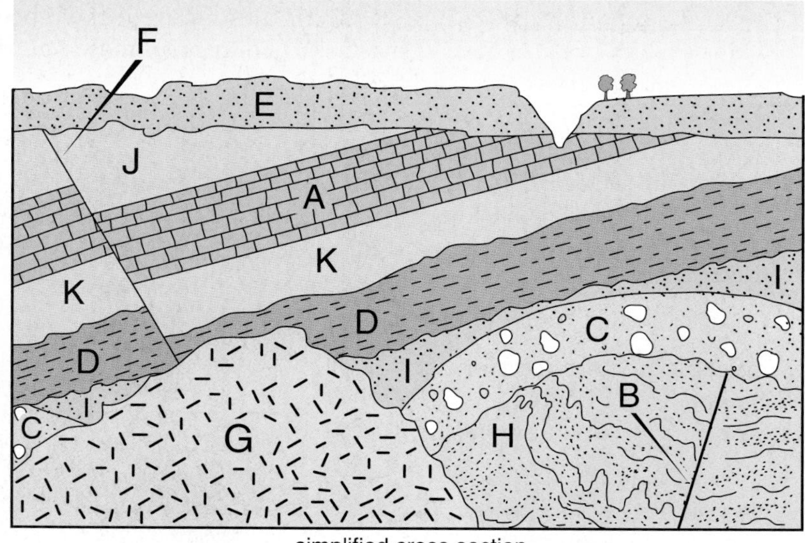

simplified cross section

Reflecting on the Activity and the Challenge

In this activity, you learned how geologists can use the relationships of rock units and geologic features such as folds or faults in order to interpret the geologic history of an area. You then applied basic principles to interpret the history of a geologically complex cross section. Being able to apply the basic principles of relative geologic time is essential to understanding the geologic history of your community and to helping you to complete the Chapter Challenge.

Digging Deeper

INTERPRETING GEOLOGIC HISTORY

Geologic Events and Processes

Most areas of the Earth's crust have had a long and often complex geologic history. Many kinds of geologic events can occur and many kinds of geologic processes can operate to shape the geologic history of an area. You have learned about many of these in previous activities. Here is a list of the most important ones: deposition, erosion, folding, faulting, uplift, subsidence, igneous intrusion, volcanism, and metamorphism.

The only processes in the list above that have not been mentioned up to now are **uplift** and **subsidence**. Local areas of the Earth's crust can be slowly raised (uplift) and lowered (subsidence) by large-scale forces acting within the Earth. Vertical changes in elevation can range from meters to kilometers. Much of uplift and subsidence is caused by the movement of the Earth's lithospheric plates, but it can also be caused just by changes in the temperature of the rocks. When rocks cool, they contract, and that causes subsidence. When rocks are heated, they expand, and that causes uplift. The degree of contraction and expansion is small, but because great thicknesses of rock are affected (kilometers to tens of kilometers), uplift and subsidence of the Earth's surface caused in this way can amount to hundreds of meters.

Basic Geologic Principles

You have learned that after geologists map an area of bedrock they usually construct one or more cross sections. They do that by projecting the rock units and other geologic features they see at the surface downward into the Earth. Once geologists have described the bedrock of the area by means of a map and one or more cross sections, they try to interpret the geologic history of the area. They do that by using several basic principles, which are listed below. Some of these principles might seem like just "common sense" to you, and in a way they are. When they were first developed long ago, however, they were revolutionary advances in how early geologists thought about the geologic record.

- **Principle of Superposition**: younger sedimentary and volcanic rocks are deposited on top of older rocks, as shown in *Figure 1*.
- **Principle of Original Horizontality**: sedimentary and volcanic rocks are laid down in approximately horizontal layers.
- **Principle of Lateral Continuity**: sedimentary and volcanic rocks are laid down in layers that are usually much greater in lateral extent than in thickness.

Geo Words

uplift: the process by which local areas of the Earth's crust can be slowly raised by large-scale forces acting within the Earth or the heating of rocks.

subsidence: the process by which local areas of the Earth's crust can be slowly lowered by large-scale forces acting within the Earth or the cooling of rocks.

Geo Words

unconformity: the contact between an earlier rock and younger sedimentary and/or volcanic layers.

- **Principle of Crosscutting Relationships**: if one rock unit or geologic feature cuts across another rock unit or geologic feature, it was formed later in geologic time. Here are two examples of this principle. If you see a rock unit cut by an igneous intrusion like a dike, you can be sure that the dike is younger than the rock unit. (See *Figure 2*.) If you see one or more rock units cut by a fault, you know that the fault is younger than the rock units. (See *Figure 3*.)

Unconformities

When you are trying to interpret the geologic history of an area, it is important to understand the concept of **unconformity**. All successions of sedimentary and volcanic rocks are deposited on some earlier rock surface. The contact between that earlier rock and the younger sedimentary and/or volcanic layers is called an unconformity. The significant thing about the unconformity is that for some period of geologic time, nothing except perhaps erosion was occurring on that surface. Some period of geologic time was not recorded at the unconformity. The "missing" time might be as short as thousands of years, but it is usually much longer: hundreds of thousands to many millions of years. At some unconformities, more than a billion years of Earth history is unrecorded!

Figure 1 In a series of rock layers, the oldest rocks are usually found on the bottom while the youngest rocks are on the top.

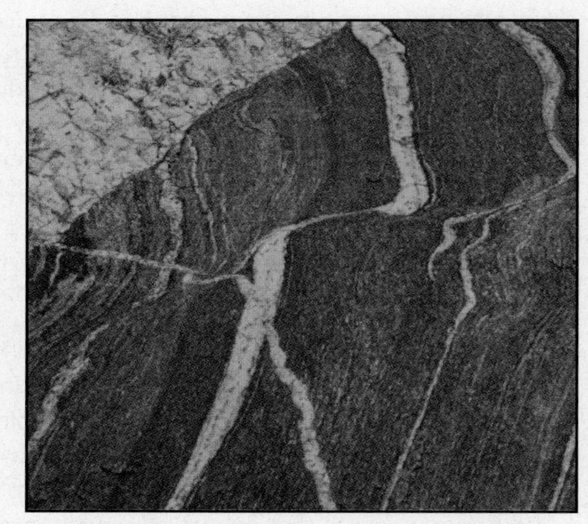

Figure 2 Which rock unit shown in the photograph is the youngest?

Figure 4 shows two common kinds of unconformity. In *Figure 4a*, a younger sedimentary succession is resting unconformably on an older sedimentary succession that was folded and then eroded down before conditions changed and more sediment was deposited. In *Figure 4b*, a younger sedimentary succession is resting unconformably on an intrusive body of granite. The granite was emplaced deep in the Earth, and then later erosion wore down the land surface to the level of the granite intrusion. Then conditions changed for some reason, and sediment was deposited on the previously eroded surface. You can see from these examples that recognizing an unconformity can be very helpful in interpreting geologic history.

Figure 3 After intruding into the surrounding granite, this dike was offset by a fault.

Figure 4a An unconformity in which the older, underlying rocks are at a different angle than the younger, overlying rocks is called an angular unconformity.

Figure 4b An unconformity developed when older igneous rocks were exposed to erosion before sedimentary rocks covered them is called a nonconformity.

Check Your Understanding

1. What causes uplift and subsidence?
2. How do unconformities form?
3. Define and explain two of the major principles used by geologists to interpret the rock record.

Understanding and Applying What You Have Learned

1. Can you think of a situation in which the Principle of Superposition does not work (i.e., when the oldest rocks are on top)?
2. Can the Principle of Original Horizontality be applied to metamorphic rocks? Why or why not?
3. Examine your state geologic map and geological cross sections.
   a) Which rock unit near your community is the oldest? Use the legend and the geologic time scale to estimate an age range for this rock.
   b) Which rock unit near your community is the youngest? Use the legend and the geologic time scale to estimate an age range for this rock.
   c) If there are any faults or folds in your area, when did they occur relative to the surrounding rocks?
   d) Is there evidence for any unconformities near your community? If so, use the legend and geologic time scale to estimate the length of time that is not recorded.

Age at base	Epoch	Period		Era	Eon
0.01	Holocene	Quaternary		Cenozoic	Phanerozoic
1.6	Pleistocene				
5	Pliocene	Neogene	Tertiary		
23	Miocene				
35	Oligocene	Paleogene			
57	Eocene				
65	Paleocene				
146		Cretaceous		Mesozoic	
208		Jurassic			
245		Triassic			
290		Permian		Paleozoic	
323		Pennsylvanian	Carboniferous		
363		Mississippian			
409		Devonian			
439		Silurian			
510		Ordovician			
570		Cambrian			
		Precambrian			Proterozoic
					Archean

Preparing for the Chapter Challenge

Using the techniques that you have learned, prepare a chronological list (oldest to youngest) of geologic events that have occurred in your community. Be sure to include the order in which rocks formed, as well as when events like folding or faulting may have occurred. Remember to ask yourself the question: "What was there first?"

Inquiring Further

1. **Dating techniques**

   In this activity you have learned how to tell the relative age of rocks from their relationships to each other. Geologists also use other techniques to determine the ages of rocks. Research the following dating techniques and describe how they are used and how they differ from the technique that you learned in this activity:

   - Biostratigraphy
   - Radiometric dating

Activity 7 Geology of the United States

Goals

In this activity you will:

- Determine the ages of the rocks of the United States.
- Understand the general geology of the United States.
- Become familiar with some of the major physiographic provinces in the United States.

Think about It

The oldest rocks that have been found on Earth up to now are almost 4 billion years old.

- If you were instantly transported to another part of the United States and given only a geologic map, how could you find your way home using only the local bedrock as your guide?

What do you think? Record your ideas about this question in your *EarthComm* notebook. Include a quick sketch for each question. Be prepared to discuss your responses with your small group and the class.

Investigate

Part A: Bedrock Geology of the United States

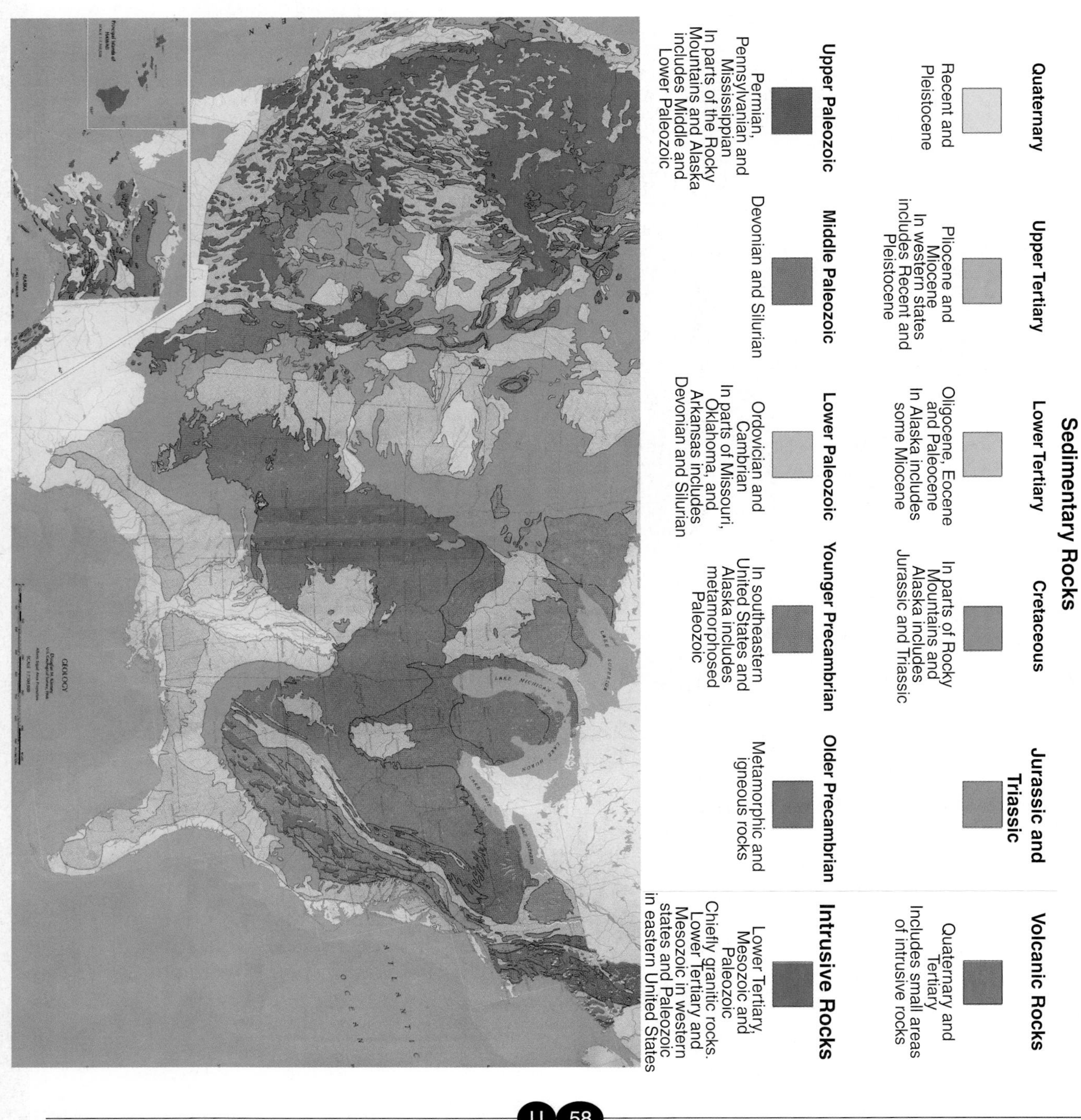

1. In your small group, start by examining a geologic map of the United States. The map shows bedrock by geologic age. Using the legend of the map, identify all of the areas of the United States with the following ages: Precambrian, Paleozoic, Mesozoic, and Cenozoic. For the subdivisions of these major time divisions refer to the geologic time scale. These major divisions of geologic time include all of Earth history, so all of the area of the United States falls into one or another of them.

   Like other skilled professionals (good examples are surgeons, engineers, carpenters, and auto mechanics), geologists try to think carefully about a project before they begin to work on it. Inspect the map very carefully before you begin your work. Try not to feel overwhelmed by all the detail you see on the map. You will see that in some parts of the United States, large regions have rocks that fall into only one of the major age divisions. In such regions the geology is likely to be fairly simple. In other regions the geology is very complicated, and there are many small areas with bedrock of very different ages.

2. Geologists often need to generalize so that they can see patterns that hide in masses of detail. In this exercise, you will need to generalize when you are working with regions of the United States with complex geology.

   a) Use the geologic map of the United States to estimate the position of boundaries between areas with bedrock of the following ages: Precambrian, Paleozoic, Mesozoic, and Cenozoic. Draw these boundaries on a blank map of the United States. The boundaries can be somewhat approximate in their positions. Use a pencil to draw the boundaries, as you will probably change your mind a number of times and need to erase some lines!

   b) Using colored pencils, color the areas of each of the major divisions of geologic time with the following colors:

      - Precambrian, red
      - Paleozoic, yellow
      - Mesozoic, blue
      - Cenozoic, green

   c) For complex regions don't try to duplicate the patterns that you see. Instead, simplify them by drawing a single region and coloring it with "candy cane" stripes of colors that correspond to the range of ages that are represented in that region. Geologists know that there is no single "right" answer. Some of you will decide to generalize more than others will.

3. After your group has finished the map, compare it with those of the other groups. As a class, discuss the differences among the maps created by the different groups.
   a) Record the major differences in your notebook.
4. Use the simplified geologic map of the United States that you have created to answer the following questions:
   a) Can you detect any system or regularity to the pattern of colors? In your notebook, describe what you see. What do you think might have caused such a pattern?
   b) In what regions did you have to make the greatest generalizations? Why do you think that is the case?
5. For each of the major time divisions, make a survey of the kinds of rocks they contain. To do this, you need to examine the various areas of the geologic map and then use the legend to get an idea of the general rock type present (sedimentary, intrusive igneous, volcanic igneous, or metamorphic).
   a) In your notebook, make a list of the most common rock types for each of the major geologic time divisions.
   b) Where are the oldest rocks in the United States? What kind of rock(s) are they?
   c) Where are the youngest rocks in the United States? What kind of rock(s) are they?

Part B: The Geology of the Physiographic Regions of United States

1. Below is a list of several of the major physiographic regions of North America.
   1 Eastern Coastal Plain
   2 Allegheny Plateau
   3 Valley and Ridge
   4 Blue Ridge Province
   5 Piedmont
   6 Central Lowland
   7 Great Plains
   8 Rocky Mountains
   9 Colorado Plateau
   10 Basin and Range
   11 Columbia Plateau
   12 Sierra Nevada
   13 Pacific Border
   a) Compare the geologic map of the United States and the map outlining the major physiographic regions of the United States. In your notebook, describe the geologic characteristics of each of the physiographic regions. Things to focus on are the rock types of the region, the ages of the rocks in the region, and the map pattern of the rock units in the region.
2. You can use what you have learned thus far about how different rock types form and how they are expressed at the Earth's surface to interpret the geologic history of the major physiographic regions of the United States. In your notebook, write a brief geologic history of each region. Things to focus on include how the rocks present may have formed, how they may have been exposed so that they are visible today, and where the regions lie geographically (on the coast, in the mountains, etc.).

Major Physiographic Regions of the United States

Reflecting on the Activity and the Challenge

In this activity, you learned about the geology and geologic history of the United States. You also learned about the major physiographic regions of the United States and how they relate to the geology of the United States. This knowledge will allow you to put the geologic history of your community into a larger context and will help you to complete the Chapter Challenge.

Digging Deeper

THE EARTH'S CONTINENTS

Each of the Earth's major continents (North America, South America, Eurasia, Africa, Australia, Antarctica) has an extremely long geologic history. Rocks as old as three billion years are found on all of the continents, and in some places rocks almost four billion years old are still preserved. In earlier activities in this chapter, you learned about the great variety of processes that shape the geology of the continents. These processes operate on time scales that are much shorter than the total span of geologic time. That is why the geology of the continents is so complex: the processes have had so long to operate.

Until the 1960s, most geologists thought that the Earth's continents have always stayed in the same place. Geoscientists now know that the continents have moved relative to each other. One continent can be split apart into two or more continents, and two or more continents can collide to form a single continent. The processes that cause these changes in continents are termed **plate tectonics**.

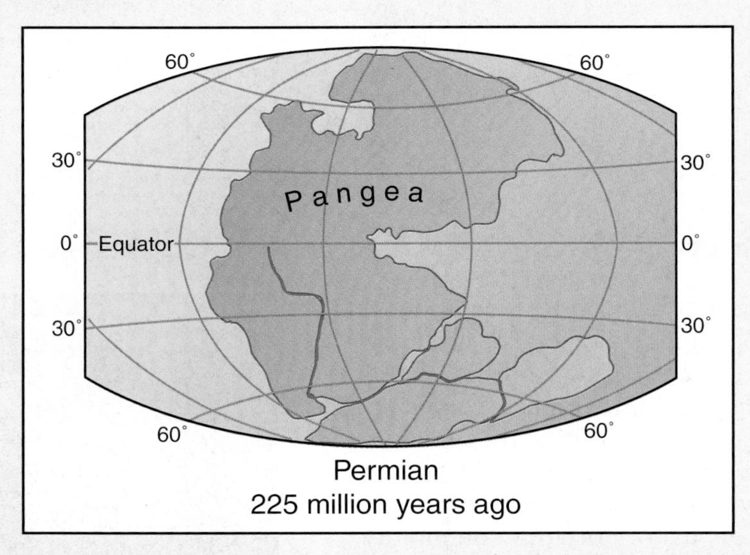

Permian
225 million years ago

North America has not always looked the way it looks today. About two hundred million years ago (just a short time ago, geologically speaking!), North America was part of a single giant continent, called **Pangea**. Then Pangea split apart, and the pieces drifted apart from one another to form today's arrangement of continents. The present outlines of North America are in a sense just an "accident" of how Pangea was split apart.

Geo Words

plate tectonics: the study of the movement and interactions of the Earth's lithospheric plates.

Pangea: Earth's most recent supercontinent, which was rifted apart about 200 million years ago.

continental accretion: the growth of a continent along its edges.

The Canadian Shield, seen in *Figure 1*, seems to be the nucleus of North America. These very old Canadian Shield rocks extend down into the northern United States, near Wisconsin and Minnesota. Throughout later geologic time, more rocks were added to North America around that old Precambrian nucleus, by a process called **continental accretion**. Continental accretion is the result of the movement and collision of the Earth's lithospheric plates. Before the development of the theory of plate

Figure 1 The Canadian Shield is composed of very old igneous rocks, and forms the nucleus to the North American continent.

Figure 2 The Appalachian Mountains are a continuous range stretching from southern Canada to Georgia.

tectonics, geoscientists were frustrated because they had no good way of accounting for how North America developed through geologic time.

Long, narrow mountain ranges are characteristic of many areas of the Earth's continents. There are two major mountain ranges in North America: the Appalachian mountain range in eastern North America, shown in *Figure 2*, and the Cordilleran mountain range in western North America. The geology of such mountain ranges is very complicated, because of the great variety of geologic processes associated with the collision of lithospheric plates: uplift, subsidence, erosion, deposition, deformation, metamorphism, igneous intrusion, and volcanism.

Between the major mountain ranges of North America, the crust is very stable. Folding, faulting, and igneous activity are uncommon. At some times in the geologic past, when sea level was relatively low, this large central area of North America underwent slow erosion, and at other times, when sea level was relatively high, the area received widespread thin layers of sediment eroded from the mountainous areas to the east and west. Can you imagine a time when more than half of North America was covered by a shallow sea, no deeper than a few hundred meters? It is difficult to doubt the existence of such a sea, when you can find fossils of marine animals like mollusks in flat-lying shales and limestones in places like Iowa!

Figure 3 What kinds of rocks do you think are found below the flat-lying landscapes of Kansas?

Check Your Understanding

1. Explain the process by which new material is added to a continent.
2. What is the name of the supercontinent that existed 200 million years ago?
3. Explain why the geology of continents is so complex.
4. What types of rocks would you expect to find in Nebraska?

Understanding and Applying What You Have Learned

1. Locate your community (or state) on the geologic map of the United States.
   a) How does the age of the rocks near your community (or state) compare with the age of the surrounding rock units?
   b) How does the size (surface area) of the rock units near your community compare with the size of other units in the United States? How do you explain the difference?
   c) How does the general shape of the rock units near your community compare with the general shape of other units in the United States? How do you explain the difference?
2. a) What are the most common rock types exposed at the surface in the United States?
   b) How might you account for your answer?
3. From the results you obtained in the investigation, do you think that the United States has grown, shrunk, or stayed about the same size through geologic time?

Preparing for the Chapter Challenge

Use the information that you have gained in this activity to insert your community's physiographic province into the geologic history the United States.

Inquiring Further

1. **Development of the Appalachian Mountains**

   Use the Internet to research the history of development of the Appalachian Mountains. When and how were the mountains formed?

Earth Science at Work

ATMOSPHERE: *Timber Industry*
The timber industry is concerned with the quality and quantity of wood that can responsibly be harvested. The type of soil as well as acid rain impact on the growth of a forest.

BIOSPHERE: *Carpenter*
People like to build their homes in many different places. The carpenter who frames the house must rely on a firm foundation on which to erect the structure.

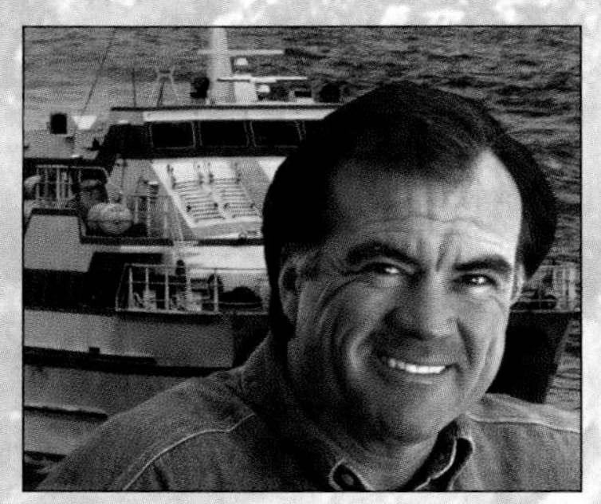

CRYOSPHERE: *Navigator*
The navigator on a vessel that sails in coastal waters must be aware of the hazards that lie below the surface of the ocean. The waters along the coastal northeast USA are particularly tricky to negotiate.

GEOSPHERE: *Geographic Information Systems Technician*
A GIS technician maintains data sets of bedrock geology. These may be accessed by many people for a large variety of purposes.

HYDROSPHERE: *Travel Agent*
Over five million people a year visit the Grand Canyon National Park in Arizona. Travel agents know that the Grand Canyon is unmatched in the spectacular views it offers to visitors.

How is each person's work related to the Earth system, and to bedrock geology?

2 River Systems ...and Your Community

2 River Systems ...and Your Community

Getting Started

The river near your community is part of a connected system of water that flows on and within the Earth. A change in one part of the river system can affect other parts of the system. What might happen to the river system if:

- A forest is cleared to construct a shopping center?
- A dam is built to make a reservoir?
- Chemicals leak from an underground storage tank located upstream from your community?
- Fertilizer is added to enrich the soil on farms?
- The region has unusually dry weather for several years?

What do you think? Select one of these situations and write a one-paragraph description of how different parts of the river system might be affected. Write down your ideas as clearly and with as much detail as possible. Draw a diagram to illustrate your ideas. Be sure to look at the diagram of the Earth Systems at the front of this book. Be prepared to discuss your descriptions with your small group and the class.

Scenario

Your town council has requested a plan for a riverside parkland site in your community. They realize that changing one part of a river system (such as paving a parking lot in a riverside park) can change other parts of the river system. The council wants to know the potential for flooding and erosion in and near the parkland. They want the design and use of the park to be safe and cost-effective. They also ask that the plan include an educational exhibit on the river system. The town council has invited the *EarthComm* students in your school to submit a riverside parkland plan.

Can you use your understanding of river systems to plan a riverside park in or near your community? Can you make a plan that explains how rivers change the land and how any changes caused by building the park will affect the river?

Chapter Challenge

Your challenge is to plan a riverside park in or near your community. Your plan should:

- Briefly outline how the park will benefit the community.
- Describe the parkland site and river with a scaled map.
- Explain the potential for flooding at the parkland.
- Explain how stream erosion occurs in the parkland.
- Explain how the river "works" and how it fits into the larger river system.
- Explain how the development and use of the parkland will affect the river system within and beyond your community.

Assessment Criteria

Think about what you have been asked to do. Scan ahead through the chapter activities to see how they might help you to meet the challenge. Work with your classmates and your teachers to define the criteria for assessing your work. Record all this information. Make sure that you understand the criteria as well as you can before you begin. Your teacher may provide you with a sample rubric to help you get started.

Activity 1 High-Gradient Streams

Goals

In this activity you will:

- Use models and real-time streamflow data to understand the characteristics of high-gradient streams.
- Identify characteristics of high-gradient streams.
- Calculate stream slope or gradient.
- Identify areas likely to have high-gradient streams.
- Identify possible hazards and benefits of a high-gradient stream on a community.

Think about It

Look at the two different streams shown in the illustrations.

- How are the two streams different?
- Could both streams be located in the same geographic area?

What do you think? Record your ideas about these questions in your *EarthComm* notebook. Be prepared to discuss your responses with your small group and the class.

Before you begin it would be a good idea to cover tables with newsprint or other material to facilitate effective cleanup. Keep paper towels handy for cleanup.

Investigate

Part A: Investigating High-Gradient Streams Using a Stream Table

1. To model a high-gradient stream, set up a stream table as follows:
   - Cover the bottom of a stream table with a layer of sand about 2.5 cm thick.
   - Using additional sand, make high mountains separated by narrow river valleys at the upper end of the stream table.
   - Using pieces of toothpicks or small blocks, set up communities of "buildings" in the stream valleys and on the hillsides and hilltops.
   - Prop up the stream table about 30 cm to create a steep slope. You may need to support the lower end to prevent it from sliding.
   - Be prepared to drain, bail, or recycle the water that accumulates at the lower end of the stream table.

Stream table setup for a high-gradient stream.

2. Turn on a water source with a low rate of flow. Observe and record the changes in the stream valleys and hillsides. You may wish to make a "before and after" video or photo to record your observations of the stream table.

   a) Which parts of the landscape are most prone to erosion, the steeply sloping parts or the gently sloping parts?

Be ready to turn off the flow of water at any moment. Mass wasting (sand slide) is possible.

b) Where is sediment deposited?

c) Where does water flow fastest, and where does it flow slowest?

d) Where is the largest volume of water flowing in the stream and where is the smallest?

3. Turn off the water and rebuild your landscape and "community."

If toothpicks were used be sure to retrieve them from the sand. Wash your hands after handling the sand.

Part B: Stream Gradients

1. Streams and rivers always flow downhill. The gradient, or slope, of a stream or river expresses the loss in elevation of the stream or river with distance downstream. Obtain one or more topographic maps that cover your community and nearby areas. Identify the stream or river nearest to your school. Find two adjacent contour lines that cross the river. Note the contour interval. It may be 5, 10, 20, or 40 feet, or it might be in meters instead.

   a) Record the contour interval as change in elevation.

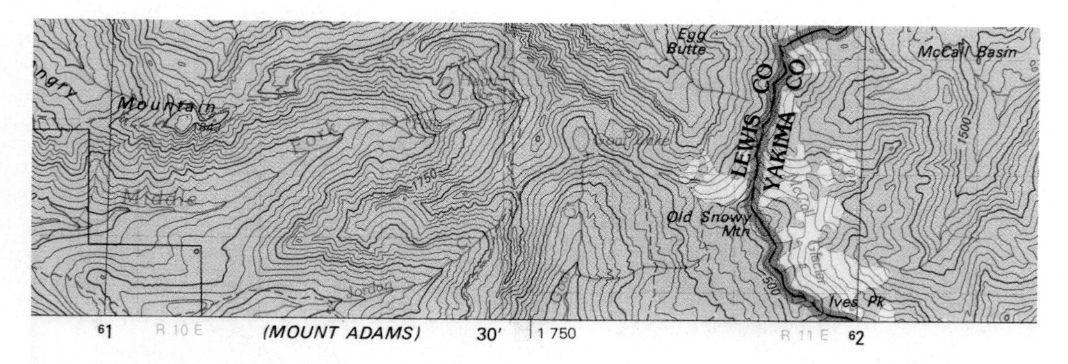

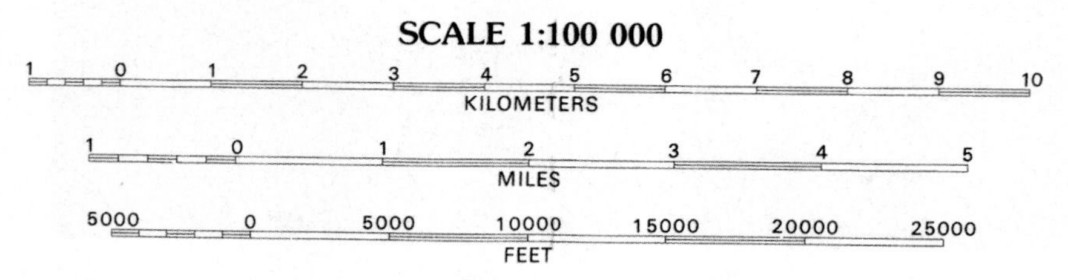

CONTOUR INTERVAL 50 METERS
NATIONAL GEODETIC VERTICAL DATUM OF 1929
To convert meters to feet multiply by 3.2808
To convert feet to meters multiply by 0.3048

2. Use a piece of string to measure the horizontal distance between the two points along the river where the contours lines cross the river. Use the scale on the map to convert this distance on the map to miles or kilometers on the ground.

   a) Record this value.

3. Use the two values obtained above to calculate the stream slope, or gradient, in feet per mile or in meters per kilometer. To do this, divide the change in elevation by the horizontal distance between the two points. For example, if the river drops two meters over a horizontal distance of four kilometers, the gradient of the river is one-half meter per kilometer. The gradient can also be expressed as just a number, by using the same units of measurement for both the vertical drop and the horizontal distance. In the example above, the gradient would be two meters divided by 4000 meters, or 0.0005.

   a) Record the gradient of the stream.

4. Study the data for the Mississippi River system in the table below.

   a) Search for patterns in the data that would enable you to characterize how a river changes over its course. For example, using the data, complete the following sentence: "As the distance from the sea decreases, floodplain width..." Write down two more sentences that describe patterns or relationships in the data.

Data on the Mississippi River System
Discharge, Floodplain Width, Distance to the Sea, and Elevation for Various Locations on the Mississippi River and Its Tributaries, Late Summer 1998

Location	Discharge (ft^3/sec)	Floodplain Width (miles)	Distance to Sea (miles)	Elevation (feet)
Shoshone Canyon Wyoming, Shoshone River	350	0.02	2925	5200
Cody, Wyoming, Shoshone River	600	0.2	2900	4800
Culbertson, Montana, Missouri River	10,000	1.65	2400	1900
Jefferson City, Missouri, Missouri River	46,000	1.8	900	530
Kimmswick, Missouri, Mississippi River	110,000	5.3	750	380
Natchez, Mississippi, Mississippi River	~140,000	32	180	30

b) Use the data to make a graph showing one of the patterns that you have described above.

c) Calculate the stream gradients (in feet per mile) between the following segments of the Mississippi:

i) Between Kimmswick, Missouri, and Natchez, Mississippi.

ii) Between Shoshone Canyon, Wyoming, and Cody, Wyoming.

d) Describe the relationship between stream gradient, elevation, and discharge.

The Missouri River near Culbertson, Montana

The Mississippi River near Natchez, Mississippi

5. Copy the data table below into your *EarthComm* notebook. Use a topographic map of your area to fill in parts (a) to (j) in the table for your local stream.

Comparison of Characteristics of Local and High- and Low-Gradient Streams			
Characteristic	**Local Stream**	**High-Gradient Stream**	**Low-Gradient Stream**
a) Difference in elevation (ft) between highest and lowest points in study area			
b) Stream gradient (ft/mile)			
c) Steepness of valley walls (steep, moderate, gentle)			
d) Channel shape (straight, curved, meandering)			
e) Channel width (miles)			
f) Floodplain width (miles)			
g) Area of land available for farming in valley			
h) Number of tributaries within four miles			
i) Rapids or waterfalls present?			
j) Could a large boat travel upstream here?			
k) Basin area (ft^2)			
l) Current discharge (ft^3/s)			
m) Current stream velocity (ft/s)			
n) Minimum discharge (ft^3/s)			
o) Maximum discharge (ft^3/s)			

6. Go to the *EarthComm* web site to find the USGS web site that gives data on the discharge of rivers in the United States. Use the data at the web site to fill in the discharge (or flow, in ft^3/s), the drainage area (ft^2), and stream velocity (calculate using discharge and drainage area) of your local river. Use the location of the data that is closest to your school. If your river is not listed, use data for the closest river. Complete parts (k) to (o) for your local stream.
7. Look at the state or regional topographic or shaded relief maps to determine where the gradient of your river is greatest. Note this location in your *EarthComm* notebook. Fill in sections (a) to (o) in the column labeled "High-Gradient Stream" in the table for this location, as you did for your local river.
8. Use your completed data table to answer the following:
   a) Compare the width of the floodplain in your local area and in the high-gradient area.
   b) Compare the stream velocity in the two areas.
   c) Compare the current discharge of your local stream to the maximum and minimum discharges. How do you account for the differences between the numbers?

   (You will fill in the data for the river in a low-gradient area for the next activity.)

Reflecting on the Activity and the Challenge

In this activity, you used a stream table to explore the nature of erosion and deposition along rivers with steep gradients. You used a topographic map to calculate the gradient (the change in elevation with horizontal distance) of a stream near your school. You also searched for patterns and relationships between variables used to characterize a river along its course. Finally, you examined real-time data of streamflow in a river that flows near your community. These explorations will help you characterize the river system in your community.

Knowing the characteristics of the river system in your community will help you decide how best to make use of the river in the parkland you are planning. It will also help you to determine how your use of the land near the river will affect the river system, and how the river will affect any construction that you have proposed. You will need to include this information in your Chapter Challenge.

Digging Deeper

CHARACTERISTICS OF HIGH-GRADIENT STREAMS

You may have been uncertain about the difference between a **stream** and a **river**. Geoscientists use both words to describe a flow of water in a natural channel on the Earth's surface. The word "river" is usually used for a flow in a relatively large channel, and the word "stream" is usually used for a flow in a relatively small channel. Often, however, the word "stream" is used in a general way for all flows in natural channels, large and small. Very small streams are often called **brooks** or **creeks**.

The **gradient**, or slope, of a stream or river expresses the loss in elevation of the stream or river with distance downstream. High-gradient streams are usually located in the headwater areas of river systems. The **headwaters** are the areas of the river system that are farthest away from the mouth of the river. The headwaters are at the highest elevations in the river system. Slopes of the land surface are generally much steeper at the headwaters than in the lower parts of the river system, as shown in *Figure 1*.

The velocities of flow in high-gradient streams are high, sometimes greater than 3 m/s (10 ft/s). Because such streams are usually in the headwaters of the river system, however, they have not collected much water from upstream, and they are relatively small and shallow. Streams with high velocities and shallow depths exert very strong forces on the stream bottom. The reasons for that are complicated, and have to do with the dynamics of flowing water. High-gradient streams can move even very large particles on the streambed, even large boulders, especially during floods. During floods in some high-gradient streams, you can stand on the bank of the stream and hear a thunderous roar caused by boulders colliding with one another as they are moved by the stream.

Figure 1 The slope of the land at the headwaters of a river is generally very steep.

Geo Words

stream: a small or large flow of water in natural channels.

river: a relatively large flow of water in a natural channel.

brook: a term used for a small stream.

creek: a term used for a small stream.

gradient: the slope of a stream or river expressed as a loss in elevation of the stream or river with distance downstream.

headwaters: the areas of the river system that are farthest away from the mouth of the river.

Geo Words

downcutting: erosion of a valley by a stream.

floodplain: the area of a river valley next to the channel, which is built of deposited sediments and is covered with water when the river overflows its banks at flood stage.

stream discharge: the volume of water passing a point along the river in a unit of time.

Because high-gradient streams can exert large forces on the stream bed, they tend to erode their valleys rapidly. Erosion of a valley by a stream is called **downcutting**. Sometimes streams cut straight down to form canyons with vertical walls, but usually the valley is in the form of a "V" with steeply sloping sides. Weathering produces loose material on the valley slopes. That material then slides down or is washed down by rainfall to the stream. The stream carries the material downstream.

High-gradient streams cut their valleys vertically downward too rapidly for the valleys to widen out to form **floodplains**. In most high-gradient streams, the sloping sides of the valley come down very near the stream channel as shown in *Figure 2*. There is only a limited area of flat land available for farming in the valleys.

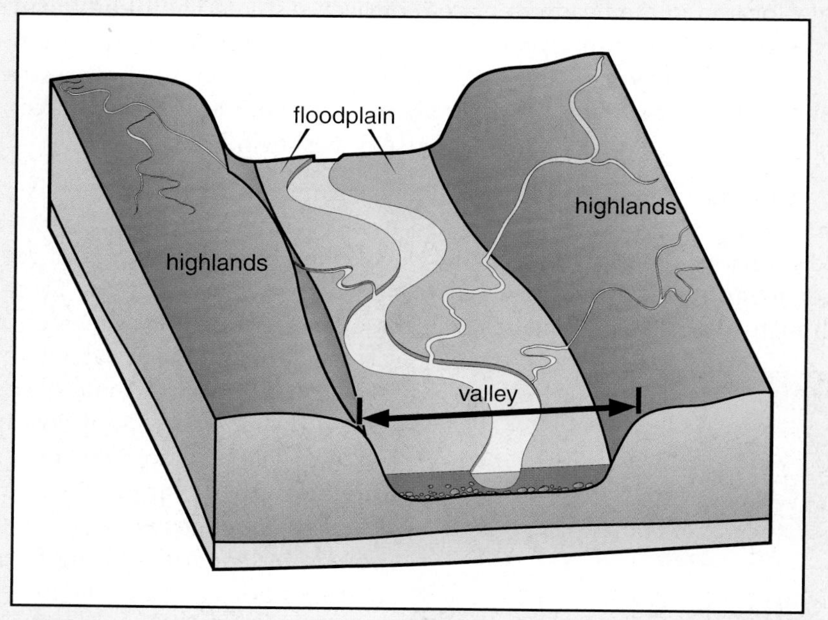

Figure 2 Illustration of high-gradient streams in the highlands and a low-gradient stream forming a broad valley below.

As you noted when you compared different points along the Mississippi River, high-gradient streams also tend to have low **stream discharge**. Stream discharge is the volume of water passing a point along the river in a unit of time. It is calculated by multiplying the cross-sectional area of the river channel by the velocity of the water. The cross-sectional area of the river channel is not easy to measure. You could imagine taking depth soundings all across the river from a bridge, and then plotting these depths on a graph to show the cross section, and then measuring the area of the cross section. The discharge is measured in cubic feet per second (often called "cusecs") or in cubic meters per second (often called "cumecs").

Stream velocity and stream discharge vary a lot over time. Dry periods cause a reduction of stream discharge, as revealed by the sample plot of stream discharge from the USGS real-time water data web site, shown in *Figure 3*. No rain fell during the week spanned by the plot of stream flow. Stream discharge fell in the first three days shown, then leveled off. Periods of increased rainfall or snowmelt cause stream discharge to increase, until the rain or snowmelt diminishes and the stream flow returns to normal levels. Over much of the colder and temperate areas of the United States, spring rains and snowmelt cause stream discharge to increase seasonally, occasionally resulting in major floods.

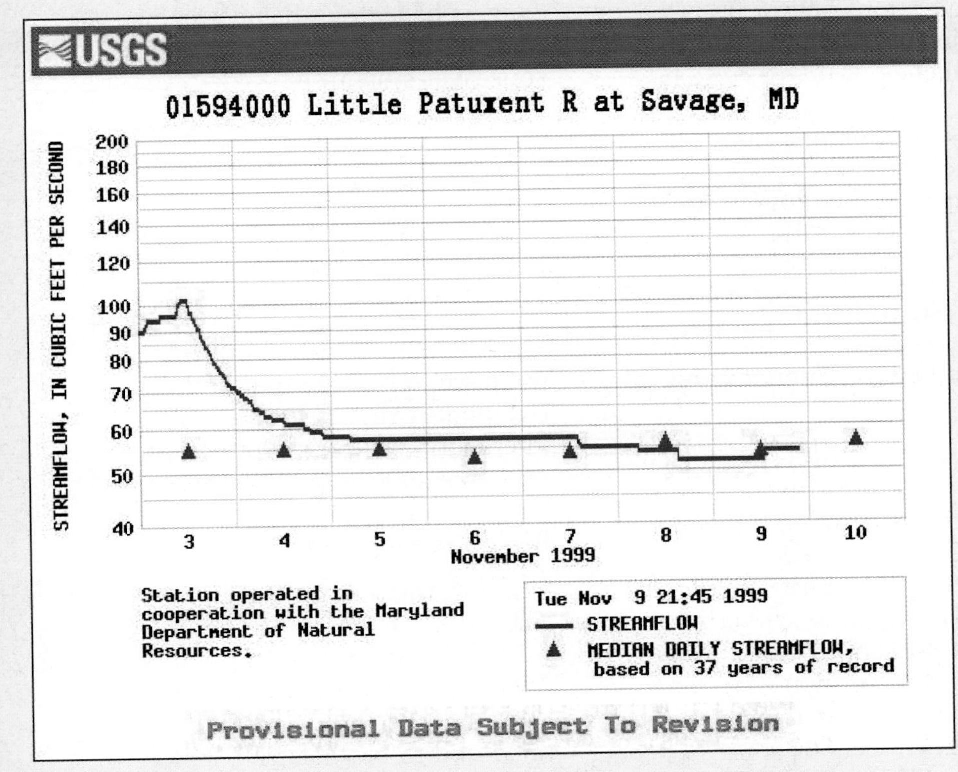

Figure 3 Plot of stream discharge versus time for the Little Patuxent River at Savage, Maryland.

Check Your Understanding

1. Why can high-gradient streams move large sediment particles?
2. What is stream discharge?
3. How does stream discharge change from high-gradient to low-gradient streams?
4. Why do high-gradient streams cause downcutting of their valleys?
5. What causes stream discharge to change over time?

Understanding and Applying What You Have Learned

1. Describe three characteristics of a high-gradient stream.
2. Is the major stream in your community a high-gradient stream? How do you know?
3. How does the elevation of your community compare to the elevation of areas around high-gradient streams?
4. What events would cause the velocity of your river to increase? To decrease?
5. Is there a time of year when a high-gradient stream is likely to pose a hazard to communities? Explain.

Preparing for the Chapter Challenge

Write a short paper in which you address the questions below. Be prepared to include this paper in your Chapter Challenge.

- What are some potential uses of high-gradient streams near your home or upstream in your river system?
- What are some of the potential dangers of high-gradient streams near your home or upstream in your river system?

Inquiring Further

1. **Interaction between humans and rivers**

   Many stories and novels have been written that focus on rivers, or on the interactions between humans and rivers, including *The Adventures of Huckleberry Finn*, by Mark Twain, *Siddhartha*, by Herman Hesse, and *A River Runs Through It*, by Norman Maclean. Write a story or essay that involves a river and members of your community. What you write does not have to be centered on the river, but it should involve some interaction between community residents and the river or stream.

2. **Big Thompson, Colorado flood**

   Find information on the Big Thompson, Colorado flood of July 1976 and describe how it is related to high-gradient streams and land use. What factors caused this flood to be so catastrophic?

Activity 2 Low-Gradient Streams

Goals

In this activity you will:

- Use models and real-time streamflow data to understand the characteristics of low-gradient streams.
- Understand how models can help scientists interpret the natural world.
- Identify areas likely to have low-gradient streams.
- Describe hazards of low-gradient streams.

Think about It

During the Mississippi River flood of 1993, stream gauges at 42 stations along the river recorded their highest water levels on record.

- What happens during a flood?

What do you think? Record your ideas about this question in your *EarthComm* notebook. Include a sketch of the water line (the line where the water surface meets the riverbank) during normal flow in the river and during a flood. Be prepared to discuss your responses with your small group and the class.

Investigate

Part A: Investigating Low-Gradient Streams Using a Stream Table

1. To model a low-gradient stream, set up a stream table as follows. Use the photograph to help you with your setup.
   - Make up a batch of river sediment by mixing a small proportion of silt with a large proportion of fine sand.
   - Cover three-fourths of the stream table with a layer of the river sediment at least 2.5 cm thick.
   - With your finger, trace a winding river between 0.6 cm and 1.3 cm deep in the sediment. Make several bends.
   - Using pieces of toothpicks or small blocks, set up communities of "buildings" along both the insides and outsides of river bends.
   - Prop up the stream table about 2.5 cm to create a very gentle slope.
   - Be prepared to drain, bail, or recycle the water that accumulates at the lower end of the stream table.

The stream table setup of a low-gradient stream. Pieces representing buildings and houses are placed in the sand on the inside and outside bends of the river.

Before you begin, review all safety cautions provided in Activity 1 regarding the stream table setup.

2. Using additional sediment, make landforms that you think are typical of areas with low-gradient streams. Refer to a topographic map for ideas.
3. Turn on a water source to create a gently flowing river. Observe and record the changes to your stream table model.
   a) Which parts of the landscape are most prone to erosion? To deposition?
   b) What shape does the river channel take?
   c) Describe all the areas where silt is being deposited. Describe all the areas where sand is being deposited.

d) Observe and sketch the distributary system that develops where the river enters the "ocean."

e) Increase the velocity of the river slightly. What happens?

f) Increase the velocity again. What happens?

g) What events might cause the velocity of your river to increase?

h) Would you expect the discharge to increase when the velocity of the river increases?

i) In general, which have larger discharges: high-gradient streams or low-gradient streams?

Part B: Characteristics of Low-Gradient Streams

1. Complete the data table you began in Activity 1, Part B.

   a) Look at the state, regional, or U.S. map to determine where the stream gradient for your river would be the gentlest. Note the location in your *EarthComm* notebook.

   b) Use the map to fill in rows (a) to (j) in the column labeled "Low-Gradient Stream."

   c) Use the USGS web site (consult the *EarthComm* web site) that gives data on the discharge of rivers in the United States to fill in parts (k) to (o) for the low-gradient stream.

   d) Why is this part of the river called a low-gradient river?

   e) Compare the width of the floodplain in the low-gradient area with the width of the floodplain in the high-gradient area of the previous activity.

   f) Compare the stream velocity in the low-gradient area and the high-gradient area.

   g) Compare the area of land available for farming. If there is a difference, why does it exist?

Reflecting on the Activity and the Challenge

This investigation helped you to realize that streams with lower gradients and larger discharges tend to have wider floodplains than streams with higher gradients and smaller discharges. Large, low-gradient rivers carry large amounts of sediment into lakes and oceans as they change the landscape and transport large volumes of water as part of the Earth's hydrosphere.

By comparing the river system in your community with both high-gradient and low-gradient streams, you will be able to better understand the characteristics of the river on which you have chosen to locate your parkland. Also, you need this information to explain how the river "works" and how it fits into the larger river system.

Digging Deeper

LOW-GRADIENT STREAMS

Meandering Streams

As anyone who has been to both the headwaters and the mouth of their river system knows, there are striking differences between high-gradient and low-gradient streams. The energy of high-gradient streams is focused mostly on downward erosion or downcutting, making steep, straight valleys with little or no floodplain. Low-gradient streams, on the other hand, erode sideways as well as downward, making wider and wider valleys, as shown in the photograph in *Figure 1*.

Figure 1 How does this stream differ from the one on page U77?

Typically, streams in the lower areas of a river system have lower gradients, wider channels, and wider floodplains than streams in the higher areas of river systems. The fact that the width of the valleys increases as discharge increases demonstrates that rivers erode the valleys that they occupy.

Low-gradient streams cut wide valleys because their channels tend to shift sideways. Most low-gradient streams do this by meandering. A **meandering stream** is a stream with a channel that curves or loops back and forth on a wide floodplain, as shown in *Figure 2*. Each curve is called a **meander bend** or meander loop. The velocity of the water is greatest on the outside of the meander bend, and thus erosion tends to occur there. In contrast, the

Geo Words

meandering stream: a stream with a channel that curves or loops back and forth on a wide floodplain.

meander bend: one of a series of curves or loops in the course of a mature river.

velocity is lower on the inside of the bend, so this is where sediment is deposited. Over time, erosion on the outside of the meander bend combined with the deposition on the inside of the meander bend causes the river to meander farther and farther sideways. As a result, a wider and wider valley is cut. The flat, low-lying valley bottom surrounding the channel is called the floodplain, because that is where water spreads when the river overflows its banks during floods. The floodplain is built of the sediments that the river has deposited during meandering, as well as sediments deposited during floods.

As each flood deposits some sediment on the inner side of the meander bend, a low ridge, usually no more than a meter or so high, is formed. The area of the floodplain on the inside of the meander bend shows a large number of these ridges, called **meander scars**. They reveal the earlier positions of the meander bend.

Geo Words

meander scars: low ridges on the part of the floodplain inside the meander bend caused by deposition of sediment on the point bar during a flood.

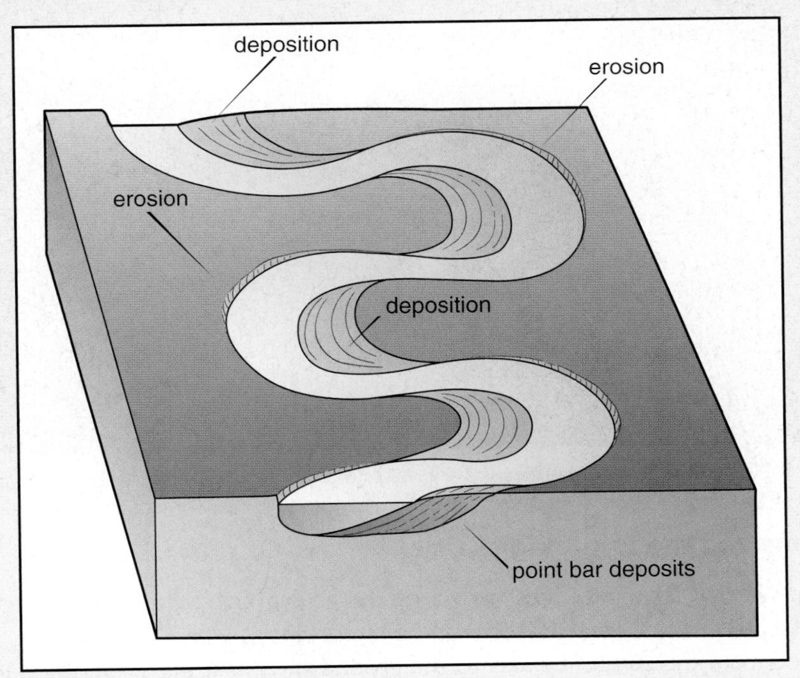

Figure 2 Illustration of a meandering stream. Notice that erosion occurs on the outside of the meander bend while deposition occurs on the inside.

As a meander bend grows wider, its neck usually also becomes narrower. Eventually, the meander bend is cut off during a flood as the water begins to flow across the narrow neck to follow a more direct route downstream. The more direct route is shorter and steeper than the longer route all the way around the meander. The river abandons the former meander bend. Soon

Geo Words

oxbow lake: a crescent-shaped body of standing water situated in the abandoned channel (oxbow) of a meander after the stream formed a neck cutoff and the ends of the original bend were plugged up by fine sediment.

afterward the ends of the abandoned bend are plugged with river sediment. The bend becomes a curved lake, called an **oxbow lake**. Later floods deposit sediment in the oxbow lake, until eventually it is filled in completely with sediment. Oxbow lakes, including those partly or completely filled with sediment, are common features on the floodplains of low-gradient streams. If you're ever in an airplane flying over a big meandering river (like the Mississippi or the Missouri), look out the window and you will see the patterns of meander bends and oxbow lakes, as shown in *Figure 3*. You might also be able to see meander scars, and even the faint outlines of former oxbow lakes, now filled with sediment.

Figure 3 Meander bends and oxbow lakes are characteristics of low-gradient streams. Continued plugging of the channel with fine sediment will eventually turn this meander into an oxbow lake.

Streams and the Hydrologic Cycle

The main factor that influences stream discharge is precipitation in the drainage area of a stream, but other factors can be important as well. Water can be removed from the stream by loss to the groundwater system, evaporation into the atmosphere, or diversion for municipal water supply or crop irrigation. Water can enter the stream from the groundwater system, from the melting of snow or glaciers, or from release of water from reservoirs.

The flow of water in streams is intimately connected to the groundwater system. Have you ever thought about why most rivers flow throughout the year, even during long periods when no rain falls to feed the river? Some of the rain that falls on the land runs off directly into streams, but some soaks into the soil and becomes groundwater. Groundwater flows slowly through porous underground sediments and rocks, called **aquifers**, until the aquifer intersects the ground surface, causing an outflow of water. Outflow from aquifers is a major source of water for many rivers, especially during periods of drought. Refer to the plot of the stream flow in Activity 1 on page U79. Rain had not fallen in the drainage basin of the river shown in the plot, yet water continued to flow in the stream. This is mainly the result of groundwater charging, or adding to, the stream. Groundwater that leaves an aquifer and flows into the bed of a stream is referred to as base flow. Because water generally flows much more slowly through rock and sediment than it does over the Earth's surface, base flow can charge a stream even long after precipitation has stopped.

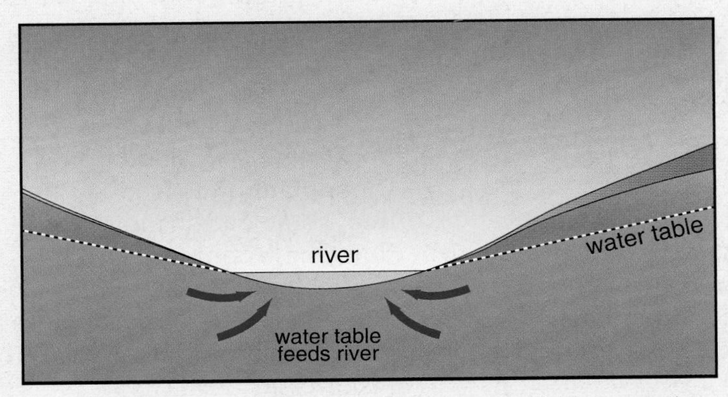

Figure 4 Groundwater flows through the aquifer to the stream. This prevents the stream from going dry during extensive periods of drought.

Hazards: Floods on Low-Gradient Streams

Flooding on low-gradient streams occurs when the stream channel cannot contain the discharge of water that is passing through it. Commonly, the discharge in low-gradient streams changes with the seasons, because seasonal changes cause changes in precipitation. For example, flooding is not common in Maine during the winter because most precipitation is in the form of snow, which remains on the ground surface. In the spring, however, when warm weather causes the snow to melt rapidly, much of the snowmelt flows directly into streams, or into groundwater systems which then feed the streams. All of the snowmelt in the upstream parts of the drainage basin eventually drains into the low-gradient streams in the downstream parts of the drainage basin, resulting in flooding. In such areas, the hazard of flooding is especially great during heavy rains in warm spring weather after a very snowy winter. In contrast, during the hot summer months precipitation is less abundant, and more water is lost to evaporation and growing vegetation, thus reducing the risk of flooding.

Geo Words

aquifer: a body of porous rock or sediment that is sufficiently permeable to conduct groundwater.

Check Your Understanding

1. How does meandering change the pattern of a stream channel in a low-gradient stream?
2. Why do low-gradient streams have a broad flood plain?
3. What types of sediment are carried and deposited by low-gradient streams?
4. What causes low-gradient streams to flood?

Understanding and Applying What You Have Learned

1. The stream gradient you measured in the preceding activity on high-gradient streams is really the gradient of the valley in which the river flows. If a stream meanders on its flood plain, is the gradient of the stream channel itself equal to, greater than, or less than the overall gradient of the valley? How might you measure the gradient of the stream channel, rather than the stream valley, from a topographic map?
2. Are the streams in your community generally high-gradient streams, low-gradient streams, or somewhere in between? Explain your interpretation.
3. Because they are physical barriers to travel, streams have been used as political boundaries throughout history. This includes boundaries between cities, counties, states, and countries.
   a) Do rivers serve as boundaries in your community? In your state?
   b) On a map of the United States, identify rivers that form the boundaries between states, between the United States and Mexico, and between the United States and Canada.
   c) How could meandering of a stream channel on its flood-plain affect the boundaries?
   d) How would communities react to changing of boundaries because of meandering of rivers?
4. Will a high-gradient stream or a low-gradient stream likely have a large population center near them? Why?
5. Which is the most likely to contain pollution, a high-gradient stream or a low-gradient stream? Why?
6. Low-gradient streams have wide, flat floodplains.
   a) List some advantages to locating a community on a floodplain of a river.
   b) List some disadvantages to locating a community on a floodplain of a river.
   c) Do you think the advantages outweigh the disadvantages, or vice versa?
7. Compare the hazards posed by low-gradient streams with the hazards posed by high-gradient streams.
8. Is there a time of year when a low-gradient stream poses a particular hazard to communities? Explain your answer.

Preparing for the Chapter Challenge

For a low-gradient stream, what changes and hazards would you expect to occur in a parkland site along the stream as part of the natural changes in the river system? Write a short paper in which you address this question.

Inquiring Further

1. **The floods of 1993 and 1997**

   Research the Mississippi and Missouri River floods of the summer of 1993, or the Red River flood in Grand Forks, North Dakota and East Grand Forks, Minnesota in the spring of 1997. What happened in cities on the floodplains? Pick a city that was affected by one of the floods and describe the impact of the flood. Was the city prepared for floods? What did the city do once it became clear that the river would flood? Was the city damaged? What has the city done to prepare for future floods?

Sandbags provide added protection against rising waters during a flood.

Activity 3 Sediments in Streams

Goals

In this activity you will:

- Describe and classify sediments according to particle size and shape.
- Describe what happens to sediments composed of different rock types as they are transported in streams.
- Understand the relationship between stream velocity and particle size.
- Understand the relationship between transport distance and particle size.

Think about It

Rocks the size of cars can be transported in streams during floods.

- What can you learn about a stream by looking at the materials in the streambed?
- How do streams change the material they carry?

What do you think? Record your ideas about these questions in your *EarthComm* notebook. Be prepared to discuss your responses with your small group and the class.

Investigate

Part A: Modeling the Breakdown of Sediment

1. Obtain four small pieces of gypsum and four small pieces of shale. Determine the total mass of the gypsum and the total mass of the shale.

   a) Record these masses.

2. Make a table in your *EarthComm* notebook in which you will display the following data for each piece: roundness, length (*a* axis), width (*b* axis), thickness (*c* axis), the ratio *b*/*a*, the ratio *c*/*b*, and shape. You will need to be able to record three measurements for each piece.

3. Determine the roundness of each piece according to the roundness chart shown.

   a) Record the data in your table.

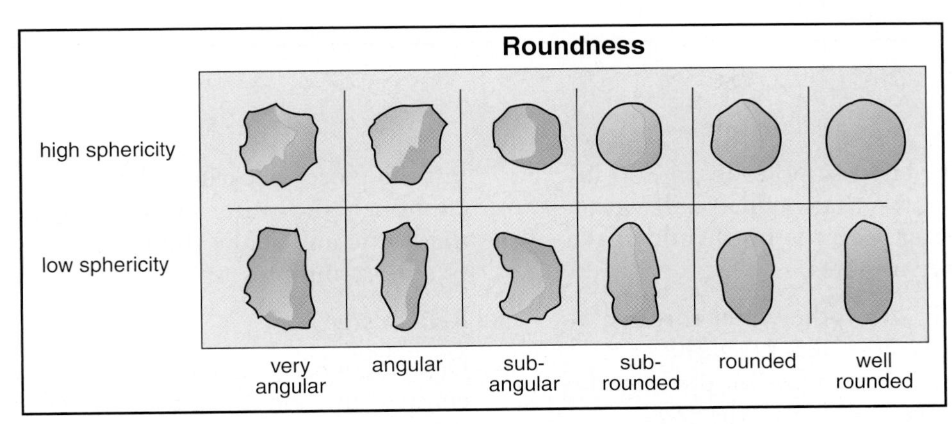

Chart showing how to estimate the roundness of sediments.

4. Determine the shape of each piece, recording all of the data in your table. Place each piece on a flat surface so that the longest axis is approximately horizontal. (Refer to the diagram on page U92.)

   a) Measure and record the longest axis. This is the *a* axis.

   b) Measure and record the horizontal axis that is perpendicular to the *a* axis. This is the *b* axis.

   c) Now measure and record the vertical axis that is perpendicular to the first two axes. This is the *c* axis.

   d) Compute and record the ratios *b*/*a* and *c*/*b*.

   e) Using these ratios, plot the location of each piece on a particle-shape graph with the ratio *b*/*a* on the vertical axis and the ratio *c*/*b* on the horizontal axis. Use the graph on the next page as a guide.

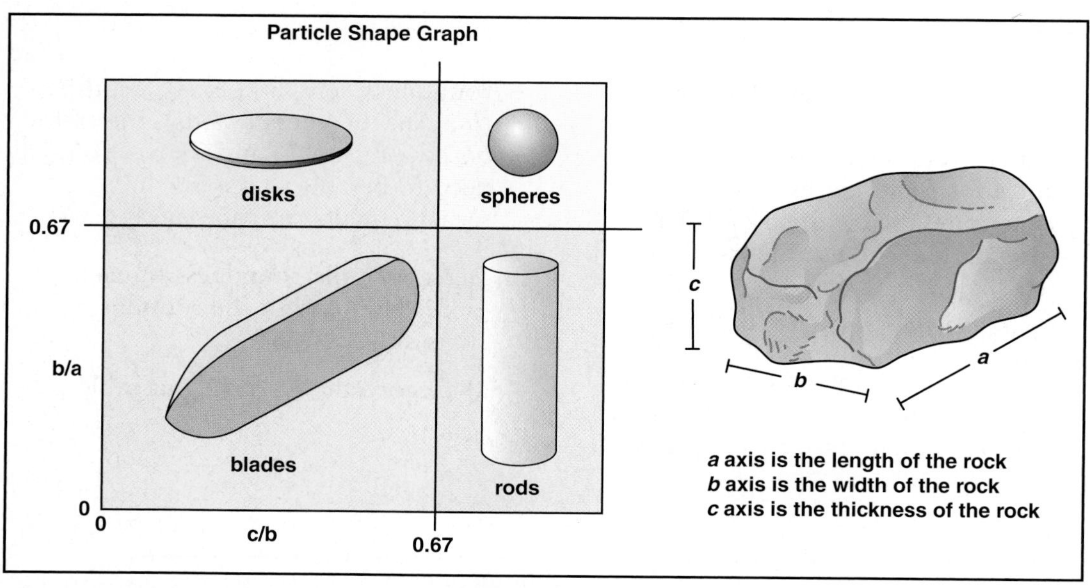

Chart showing how to estimate the roundness of sediments.

5. Place the eight samples in an 800-mL plastic bottle that is filled halfway with water. Cap the bottle and shake it for five minutes.
6. Carefully strain the water through a screen. Avoid spills. Place the material that remains on the screen on a paper towel. Dry the rock samples and find the total mass of the gypsum and the shale as before.
   a) Record the mass.
   b) Determine the roundness of the particles. Record this in your table.
   c) Measure the *a*, *b*, and *c* axes again. Record each measurement.
   d) Plot the location of each piece on a new piece of particle-shape graph paper.

Dry the outside of the bottle before shaking so it is not slippery.

7. Put the pieces back into the container, add water as before, cap the bottle and shake the mixture for five more minutes.
8. Repeat step 6.
9. Describe what you saw each time you emptied the container and analyzed the pieces. Consider the following:
   a) How did the mass, roundness, and shape change?
   b) What differences did you notice between the changes in gypsum versus shale?
   c) What type of material did you collect when you sieved the water?

Wash your hands after each part of the activity.

Part B: Analyzing Stream Sediments

1. Obtain 5–10 pieces of coarse sediment (at least a few centimeters in diameter) collected from a local river by your teacher.

   a) Describe the characteristics of the pieces. Identify the rock types, if possible.

2. Make a table in your *EarthComm* notebook in which you will display the following data for each piece: roundness, length (*a* axis), width (*b* axis), thickness (*c* axis), the ratio *b/a*, the ratio *c/b*, and shape. You will need to be able to record three measurements for each piece.

3. Determine the roundness of each piece using the roundness chart.

   a) Record the data on your table.

4. Determine the shape of each piece, recording all of the data on your table. Place each piece on a flat surface so the longest axis is roughly horizontal.

   a) Measure and record the *a*, *b*, and *c* axes.

   b) Determine and record the ratios *b/a* and *c/b*.

   c) Using these ratios, plot the location of each piece on particle-shape graph paper.

Part C: Measuring Sediment Sizes

1. Obtain sediment samples from either a local stream or your teacher. Find the mass of the sample.

   a) Record the mass.

2. Sieve the sediment sample using a set of sieves, or, if these are not available, a piece of plastic window screen from a hardware store. You have now separated your sample into at least two groups.

3. Dry the sediment. Find the mass of the groups and classify them by particle size (sand, silt, etc.).

   a) Record all of your data in a table.

4. You have grouped the sediments by particle size. Use your data to determine the percentages of each group of sediment size. Look at the largest and smallest particle sizes with a hand lens.

   a) Can you see any differences in roundness and sphericity? Write a short paragraph explaining your findings.

Part D: Using a Stream Table to Observe the Beginning of Sediment Movement in Streams

1. Obtain two thick wooden boards as long as the stream table and about 5 cm wide. Place the boards on edge along the center of the stream table, leaving a space about 5 cm wide between them, to form a channel. Place a wooden block 5 cm wide, 8 cm long, and 2 cm thick between the wooden boards near the upstream end of the stream table. See the photo for how to arrange the board and the block in the stream table.

2. Place a layer of fine sand 2 cm thick between the boards. Level the bed of sand so that it is nearly flat and at the same level as the wooden block between the boards.
3. From the water supply, run a small stream of water onto the stream table just upstream of the wooden block. The water will flow across the surface of the block and down the sand bed in the channel. Maintain a constant flow that is low enough so as not to disturb the sand.
4. Measure the velocity of the water flow in the channel. Do this by floating a tiny piece of cork on the water surface and timing how long it takes to move down the channel. Divide the downstream travel distance (in centimeters) by the travel time (in seconds) to obtain the velocity in centimeters per second. Check the sand bed to make sure that no sand is being moved by the water flow.
   a) Record the flow velocity in your notebook.
5. Increase the water supply slightly, and observe the sand bed closely for any sand movement. Measure the velocity of the flow again.
   a) Record the velocity in your notebook.
6. Repeat steps 4 and 5 until you notice that many of the sand grains are being moved by the flow.
   a) Record the flow velocity for which the sand is first moved. This is called the threshold velocity for sand movement.
7. Repeat the experiment using coarse sand instead of fine sand in the channel.
   a) Record all your data.

Clean up all spills immediately.
Wash your hands after the activity.

Reflecting on the Activity and the Challenge

In this activity you discovered that particles are changed in size and shape as they are agitated in water, and that particles with different compositions change at different rates. You also saw that stronger water flows are needed to move coarse particles than fine particles. Being able to apply these concepts to the sediments you find in the river in your community will help you understand the types of flow that have helped shape the river and carry sediments through your community. You will need this information to complete your Chapter Challenge.

Digging Deeper

SEDIMENTS IN STREAMS

Size Range of Sediments

Sediments come in a very wide range of sizes. To make communication easier, geoscientists have officially named several ranges of sediment size; see the table of sediment sizes below. To geoscientists, the words clay, silt, sand, and gravel mean something very definite. It is easy to measure the sizes of sand and gravel particles, but it is very difficult to measure the sizes of silt particles and, especially, clay particles.

Particle Size Classification of Sediments & Sedimentary Rocks			
	Sediment	**Particle Size**	
Gravel	Boulder	>256 mm	**Coarse**
	Cobble	256 – 2 mm	↑
	Pebble	64 – 2 mm	
	Sand	2 – 0.062 mm	
Mud	Silt	0.062 – 0.0039 mm	↓
	Clay	< 0.0039 mm	**Fine**

Sediment particles also vary greatly in their composition. Most gravel particles are pieces of rock. In most streams and rivers, sand and silt particles consist mainly of the mineral quartz, which is abundant at the Earth's surface and which is very resistant to chemical dissolution and mechanical abrasion. Depending on the source of the sand, however, several other minerals may also be common in sand sizes. Most clay-size particles consist of minerals, called clay minerals, that exist in the form of tiny plates or flakes.

Geo Words

suspended load: material that travels down a stream suspended in the water.

turbulence: the irregular motion of water.

eddies: swirling masses of turbulent fluid.

bed load: sediment particles that travel along the streambed, by sliding, rolling, and bouncing.

Transportation of Sediment by Streams

Sediment can be carried by streams in several ways, as shown in *Figure 1*. Sediment can be dissolved in water and carried along invisibly in a stream. Fine sediment particles, of clay and silt size, travel mostly while they are suspended in the water and "ride" along with the stream. This material is called **suspended load**. The suspended sediment is held up above the stream bed by the irregular motions of the water, called **turbulence**. To get a good idea of what turbulence in a stream looks like, watch steam or smoke coming out of a smokestack. You will be able to see the swirling masses of turbulent fluid, called **eddies**. On the other hand, very coarse sediment particles, of gravel size, travel mostly along the stream bed, by sliding, rolling, and bouncing. This material is called **bed load**. Sand is moved mostly as bed load when the stream flow is moderate, but as both bed load and suspended load when the stream flow is very strong. Whether a stream carries most of its sediment in suspension or as bed load depends both on the size of the sediment in the stream and on the velocity of flow in the stream.

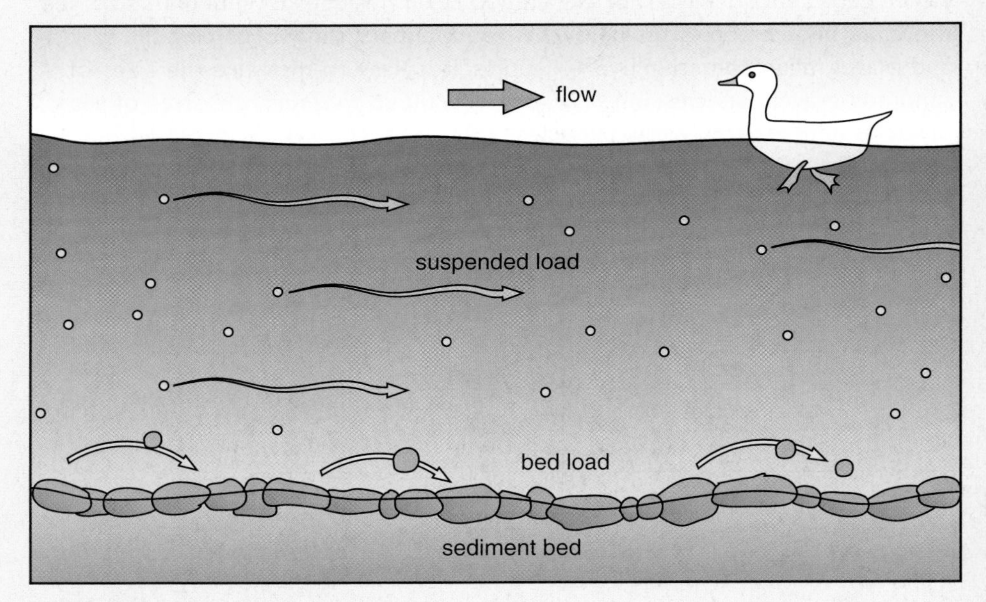

Figure 1 Sediment transport by a stream.

The flowing water in a stream exerts forces on sediment particles resting on the streambed, in the much the same way a stream exerts forces on you when you are standing or sitting in a shallow stream. Because the sediment particles are resting in little "pockets" between the particles underneath, a certain force is needed to move a given particle from its position on the bed. For each sediment size, a certain velocity of flow, called the **threshold velocity**, is needed to move some of the particles on the bed. As you saw in Part D of the activity, stronger flows are needed to move coarse sediment than to move fine sediment. This shown in the graph in *Figure 2*. Below the threshold curve, no sediment is moved, but above the threshold curve, the flow can move at least some of the sediment. In Part D of the activity, you identified two points on a graph like this! Another way of looking at the graph is that there is a maximum size of sediment particle that can be moved by a given velocity of flow.

Geo Words

threshold velocity: the velocity of flow that is needed to move certain particles along the bed of a stream.

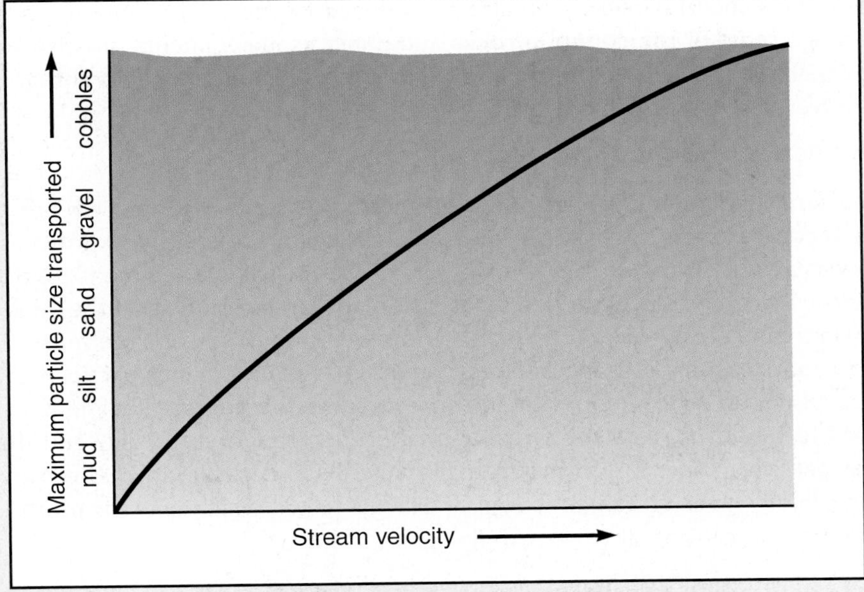

Figure 2 Graph showing relationship between stream velocity and maximum particle size transported.

How Streams Reduce the Sizes of Sediment Particles

Sediment particles in streams can become rounded as they bounce along the bottom of the stream and collide with other particles. The collisions chip the edges of the particles and grind them down. In general, the higher the flow velocity, the harder and more frequent are the collisions that break down the sediment. Smaller particles, such as sand and silt, are commonly picked up and carried in suspension. These sediments can "sandblast" the larger sediment particles they come in contact with. In this way, large particles that are not in constant motion can still be worn down and rounded. Also, powerful collisions between pieces of gravel during floods can break or split the gravel into small pieces.

Sediment particles composed of different types of rocks or minerals become rounded at different rates. Particles composed of softer minerals become rounded more rapidly than those composed of harder minerals. For example, limestone, which is composed of relatively soft calcite, becomes rounded more quickly than quartzite, which is composed of relatively hard quartz. Rocks that have layering or other planes of weakness also break down more quickly than rocks that are uniformly strong. For example, a layered rock like gneiss may break down more rapidly, because of its layering, than a nonlayered rock like granite.

Rock and mineral particles can also be reduced in size by dissolution, although most of the common rocks and minerals in sediments dissolve very slowly, if at all. Calcite is the only very common sedimentary mineral that dissolves fairly rapidly in streams.

Downstream Fining

Ordinarily, sediment particles in the upstream areas of a river system are much coarser than the particles in the downstream areas. This is known as **downstream fining**. It can have various causes. All of the sediment particles could be slowly reduced in size by abrasion and/or dissolution as they travel downstream. Most geoscientists think, however, that this is not the most important reason. Breakage of larger particles to make smaller particles is probably much more important. In some streams, the coarser sediment tends to be dropped by the stream and stored in the stream valley, and the finer particles move on downstream. That would also cause downstream fining. In any given stream, it is usually difficult to tell which effect is more important in causing downstream fining.

Figure 3 As you proceed downstream you will find that the sediments carried by the stream become finer and finer.

Geo Words

downstream fining: the decrease in sediment size downstream in a stream or river.

Check Your Understanding

1. Compare physical breakdown with dissolution.
2. What would baseball-size particles in a streambed indicate about the maximum velocity of the stream flow?
3. In your own words, describe what might happen to a large piece of granite as it is transported farther and farther downstream. What are the processes that would be acting on the piece?

Understanding and Applying What You Have Learned

1. From your data, what can you say about the relationship between the velocity of a river and the size of the sediment it carries?
2. What was the likely velocity of the river from which the following sediments were taken:
   a) Silt and clay?
   b) Fine sand?
   c) Large, rounded boulders?
3. One of the political leaders in your community has suggested making a "swimming hole" along a stream in your community. The politician proposes to dredge gravel from some part of the stream channel to make it sufficiently deep, then add sand to the banks and bottom. This politician maintains that this will be a low-budget, "natural" swimming hole. As the expert on sedimentation in your community's streams, do you agree with the politician? Explain your answer.

Preparing for the Chapter Challenge

Write a one-page paper in which you address how the sediments in a streambed help indicate how the flow in the stream might affect your community's use of the stream. Consider the following:

- How can you tell if the stream is subject to periods of high-velocity flow?
- Will the streamflow potentially affect the streamside park area?
- Will the streamflow potentially affect areas of the community beyond the park?

Inquiring Further

1. **Cleaning up sediment**

   Has a stream in your community ever flooded and deposited sediments on a road, athletic field, or parking lot? How did your community handle the cleanup? How much did it cost? What was done with the sediment?

2. **Sediment and living things**

   In what ways could the types of sediment in a streambed indicate the various plants and animals that could live there? Do plants and animals that live in streams use specific types of sediments? Would you find a different set of plants and animals in a mud-bed stream as opposed to a gravel-bed stream?

Activity 4 Rivers and Drainage Basins

Goals

In this activity you will:

- Interpret topographic maps to identify large and small streams within your community.
- Understand the nature of a drainage basin.
- Analyze maps to identify the drainage basin in which your community is located.
- Evaluate important interactions between communities and river systems.

Think about It

Look at the system of veins on a leaf. Pick a spot on a small vein near the edge of the leaf. Trace the vein until it joins the stem. Repeat this for another spot on the other side of the leaf.

- How is the system of veins on a leaf similar to and different from the network of streams and rivers that carry water into a larger river, like the Mississippi?

What do you think? Record your ideas about this question in your *EarthComm* notebook. Include a quick sketch. Be prepared to discuss your responses with your small group and the class.

Investigate

Part A: Local Stream Drainage

1. Use a topographic map of your community for the following exercises. (See the example in the text on page U102.) If you do not have a river or stream in your community, use a topographic map from a nearby community.

   Find a stream on the map that flows into or joins another stream.

   a) What do you notice about its size relative to the stream that it flows into?

   b) In which compass direction does it flow?

   c) Describe how the relative sizes of streams can be used to determine the direction in which a stream flows.

   d) Contour lines on a topographic map show elevation above sea level. What is the highest elevation along the course of the stream you chose? What is the lowest elevation? Record these values in your notebook.

   e) How can you use contour lines to determine the direction in which a stream flows?

2. Use a photocopy of the topographic map (or a clear overlay) to show the range of stream sizes in your community.

   a) Trace the pattern of streams on a copy of the map, or on the clear overlay. Devise a way to show small streams, medium streams, and the largest stream. Be prepared to explain your drawing.

   b) Count the number of streams of each size. Make a data table for your results.

   c) Describe one or two relationships between smaller streams and larger streams.

   d) Write a paragraph describing the pattern that is made by the rivers and streams in your community.

   e) Exchange your drawing and explanation with another group. In your notebook, explain any similarities and differences.

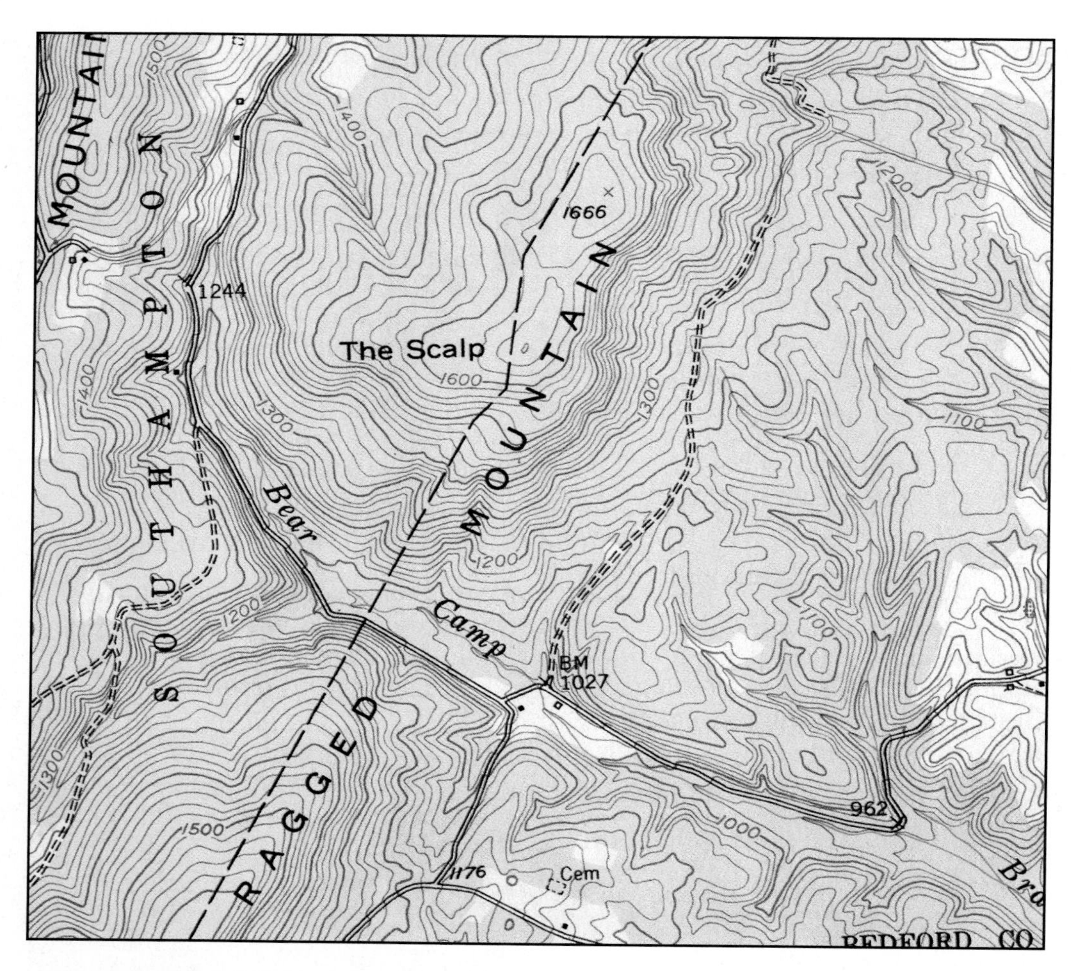

Part B: Regional Stream Drainage

1. Depending upon the region where you live, you will need some of the following: topographic maps (community and/or state), road maps or road atlases (your state and the United States), and a satellite image of the United States. A relief map of your state or region might be helpful as well. Be prepared to share these resources between groups.
   Look at the local topographic map. Find your school or a familiar landmark on the map. Imagine a rainstorm at your school. Consider the rainwater that does not evaporate, soak into the soil, or get swallowed by a thirsty animal.
   a) On a copy of the topographic map, or on a clear overlay, trace the path of a drop of water that falls on your school as it flows downhill from your school to the nearest stream. Keep in mind that water that falls on the ground follows a path downhill and always perpendicular to the contour lines. Where would a drop of water that fell on your school leave your community (or go off the map)?

2. Working with your group, figure out a way to outline boundaries of the area that drains into the stream you have chosen. Use either a photocopy of the map, a piece of tracing paper, or a clear plastic overlay. The area you have drawn is called a drainage basin.
   a) In your own words, summarize the meaning of a drainage basin.
3. Work with the local topographic map and maps that show elevation of larger regions.
   a) Locate where the rainwater that fell on your school flowed out of your community (or off the local topographic map). Follow the path of the water farther downstream. Name several cities that it passes.
   b) What is the ultimate destination of rainwater that landed on your school? Explain how you know.
   c) From what you have explored so far, explain why pollution that enters a stream near your school can affect a community many miles away within the same river system.

Reflecting on the Activity and the Challenge

In this activity, using different kinds of maps, you found streams of different sizes in your community and traced the water flow from higher elevations to lower elevations. Knowing the area over which the water in your community flows and how it changes both its own composition and the land features can help you determine boundaries of river systems and sources of streamflow. This information will help you in determining ways your community could affect the natural river processes as you work on the Chapter Challenge.

Geo Words

tributary system: a group of streams that contribute water to another stream.

trunk stream: a major river, fed by a number of fairly large tributaries; the main stream in a river system.

distributary system: an outflowing branch of a river, such as what occurs characteristically on a delta.

Digging Deeper

RIVER SYSTEMS

Parts of a River System

A river system is a network of streams that drain the surface water off a continent or part of a continent. A river system has three parts: a tributary system, a trunk stream, and a distributary system.

- A **tributary system** consists of many small streams that flow together into slightly larger streams, which flow into larger streams, which flow into even larger streams, as shown in *Figure 1*. Tributary systems are commonly found in mountainous areas.

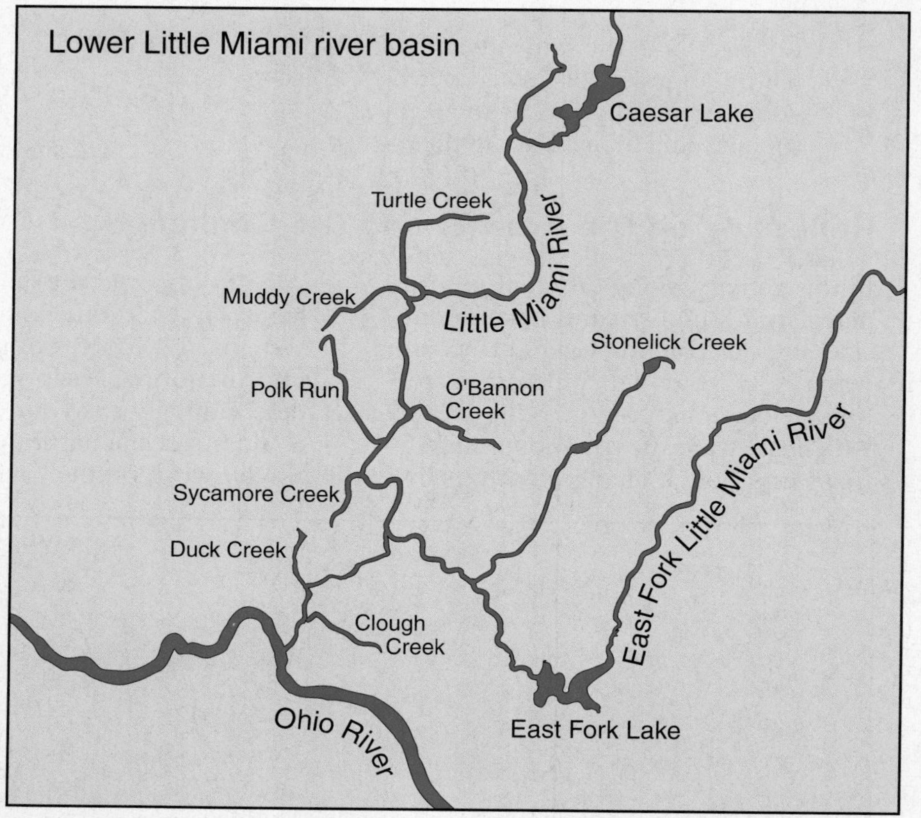

Figure 1 Map of a tributary system. How many tributary streams are shown in this map?

- A **trunk stream**, as shown in *Figure 2*, is a major river, fed by a small number of fairly large tributaries. The word trunk is used because of the tree-like drainage pattern made by most river systems.

Figure 2 A trunk stream is fed by many smaller streams.

- A **distributary system** consists of a number of small channels that branch off from the main river near the river's final destination, which is often a delta or large depositional feature. Distributaries deposit sediment and dissolved materials in the ocean, often in association with deltas, as shown in the photograph in *Figure 3*.

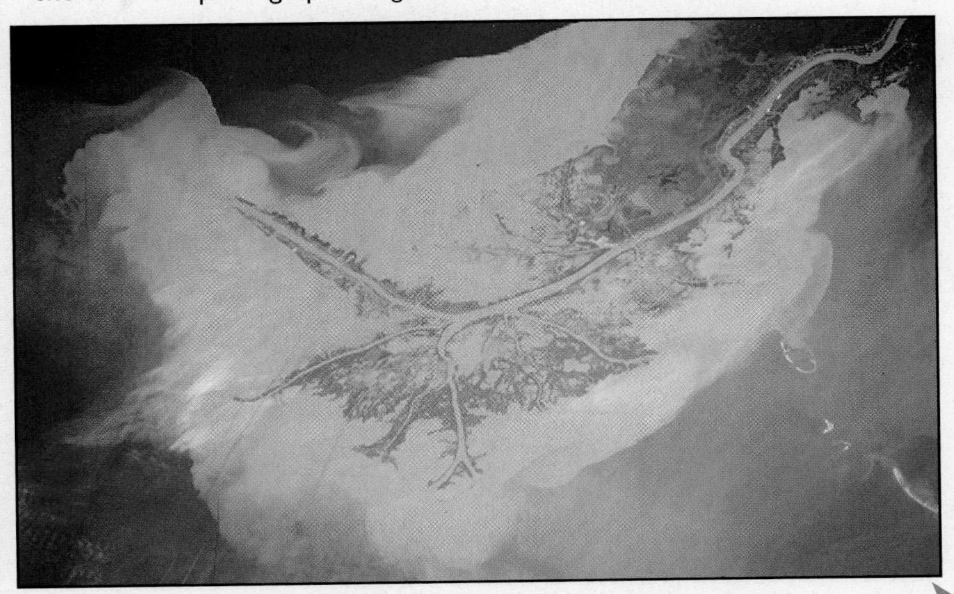

Figure 3 The Mississippi carries a large amount of sediment into the Gulf of Mexico.

Geo Words

drainage basin: the area from which all of the rain that falls eventually flows to the same final destination, usually the ocean.

All parts of all river systems have one thing in common: the water flows downhill. Rain that falls in the United States flows downslope to the Atlantic Ocean, the Gulf of Mexico, the Gulf of California (a part of the Pacific Ocean), or the Pacific Ocean. There are two exceptions: in northern Alaska, water flows into the Arctic Ocean, and in some areas of the western United States rivers flow into large depressions rather than into oceans. Some of the depressions are below sea level (Death Valley, for example, is more than 200 feet below sea level).

A **drainage basin** is the area from which all of the rain that falls eventually flows to the same final destination, usually the ocean. In the United States, there are drainage systems of different sizes. (See *Figure 4*.) In the northeast, the largest drainage basins are the Hudson, Connecticut, Delaware, and Potomac river systems, but even these are relatively small. The southeastern part of the United States is dominated by rivers that flow to the east and south off the high Appalachian Mountains. Some of these, like the Savannah River, flow into the Atlantic Ocean; others, like the Appalachicola River, flow into the Gulf of Mexico.

The largest river system in the United States is the Mississippi River, which enters the Gulf of Mexico downstream of New Orleans, Louisiana. It collects water from a huge area of the midsection of North America. Its giant tributaries include the Ohio River, including the Tennessee and Cumberland Rivers, the Missouri River, and the Arkansas River, as well as many branches of the Mississippi River.

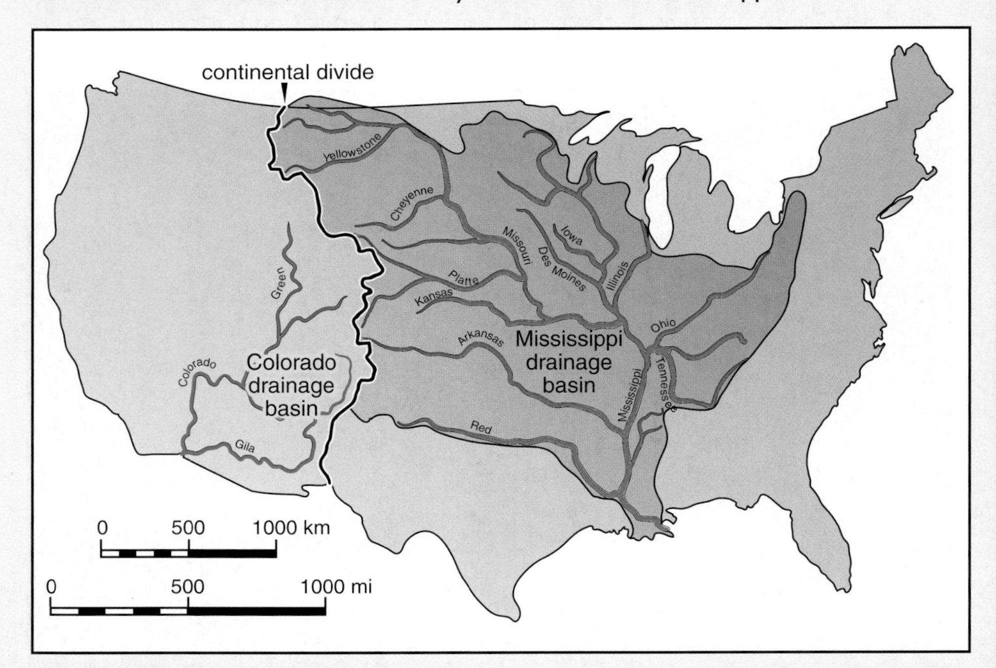

Figure 4 Map of the United States showing the major river systems with the Continental Divide.

Drainage Divides

Hilltops and mountains serve as boundaries between drainage basins. These boundaries are called **drainage divides.** Water falling on one side of a divide flows into one river system; water falling on the other side of the divide flows into a different river system. In this way, raindrops that fall within inches of each other on a mountaintop can end up thousands of miles away from each other! There are divides between streams of all sizes within a river system. Any hilltop or ridge causes some rainfall to flow in one direction and other rainfall to flow in another direction. In some cases, however, the rainfall might flow into two different tributaries that actually end up in the same larger stream farther down. The continental divide, which stretches north-south through the mountainous areas of the western United States, separates the United States into two major drainage systems, one emptying into the Pacific Ocean and the other emptying into the Atlantic Ocean. (See *Figure 4.*)

Geo Words

drainage divide: the boundary between adjacent drainage basins.

River Systems and Community Growth

Why are river systems important? Humans use river systems in many ways. Rivers provide a source of water for drinking, for domestic and industrial use, and for irrigation of farmlands. They are also used to wash away waste products like used chemicals from industrial processes and treated sewage. Throughout history, rivers have served both as giant water faucets and as giant sewers—two roles that are fundamentally incompatible. As recently as the late 1960s, several major cities in the United States allowed human waste to enter large rivers as part of their waste-disposal systems. Communities on all scales, from local to national, have recognized the problems and have worked to limit the use of rivers as waste-disposal systems. Yet accidental spills of industrial and human waste continue to occur every year.

People also use river systems for transportation. River transportation is relatively cheap. Barges carrying materials like gravel and coal move up and down the national river systems, particularly in the eastern and midwestern parts of the United States. In the northern United States, the St. Lawrence River flows from Lake Ontario into the Atlantic Ocean. A system of canals and locks connects the Great Lakes to the St. Lawrence, making it possible for goods to be shipped from inland ports like Duluth, Minnesota, Chicago, Illinois, and Detroit, Michigan, to the Atlantic Ocean and then to ports worldwide.

This dam literally "stops up" the flow of the river water, generating electricity in the process.

Rivers provide power. Since the colonial times, Americans have harnessed this power. In the 1700s and 1800s Americans used the energy of flowing water to move waterwheels that powered mills for cutting wood and for grinding corn and wheat. In the 20th century they built dams and hydroelectric power plants along rivers. A dam forms an artificial lake. Some of the water is allowed to run through conduits in the dam. As it moves down through the conduits, the water turns the blades of turbines, and the mechanical energy of the water is converted into electrical energy. Hydroelectric power plants are common in the United States, which has already harnessed much of its potential hydroelectric power.

Dams are also used to control water flow, and thus lessen the impact of flooding. To do this, the operators of the dam drop the level of the water behind the dam during dry periods, to make room for storage of water during heavy rains. The water held behind dams can also supply cities with water for domestic use, and agricultural areas with water for irrigation. However, dams disrupt the natural flow of rivers as well as the river ecosystems. Recognition of the negative as well as positive aspects of dams has sparked a national debate about dams.

A by-product of dams is lake formation.

Rivers also provide recreation. People are awed by waterfalls; they love the sound of rushing mountain streams; they crave the thrill of rafting, canoeing, or kayaking in a swift-flowing river; and they are calmed by a boat ride or a picnic on a river bank. Millions of Americans swim, fish, and boat in rivers and in the lakes created by dams along rivers.

Recreational uses of rivers include swimming, boating, and fishing.

In addition to providing humans with water, waste disposal, power, and fun, rivers change the surface of the Earth. Water moving downhill toward the ocean erodes bits of soil and rock and carries it downstream toward the coast. The process of picking up and transporting loose soil and rock lowers the level of mountains and gives the Earth's surface its shape. Even in deserts, where water is scarce, streams that flow after infrequent rainstorms are important agents in shaping the landscape, because there is little vegetation to hold the soil in place.

This desert landscape shows how rivers shape the land.

Check Your Understanding

1. Describe the three main parts of a river system.
2. What is a drainage divide?
3. Describe at least one benefit and drawback to building a dam on a river.

Understanding and Applying What You Have Learned

1. Describe the drainage basin in which your community is located.
2. How is your local river system part of a larger drainage basin in the United States?
3. Draw a diagram showing the aerial or map view of your concept of a river system and how it changes from upstream to downstream. Mark where your community fits in.
4. Examine a copy of the topographic map shown.
   a) On a copy of the map, draw arrows along the streams to show the direction of flow. Explain the basis for your interpretation.
   b) Outline the drainage basin shown on the map.

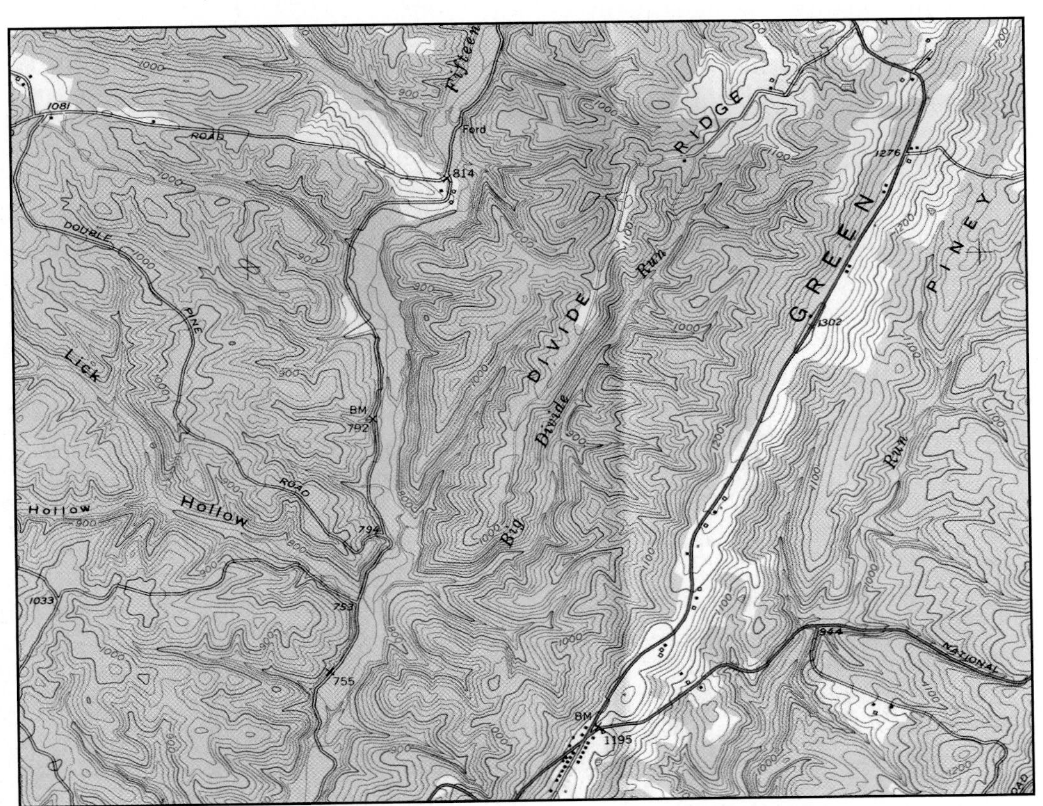

Preparing for the Chapter Challenge

Working with a copy of your local topographic map, select a potential site for your riverside parkland. Describe the location in terms of its position in the local river system relative to smaller and larger streams. Brainstorm with members of your group at least three uses for the riverside parkland in your plan.

Inquiring Further

1. **Water quality in your community river system**

   If water quality is a big issue in your community, do some in-depth research on the causes of the water-quality problems, the effects on your community, and the solutions that have been proposed to address the problems.

   - What are some of the different strategies being suggested to improve water quality?
   - What are the pros and cons of the different strategies? What course of action do you recommend?

2. **River pollution and ecosystems**

   How does water pollution affect ecosystems that depend on the river? Research a particular ecosystem in your community that has been affected by water pollution. Has anything been done to address the problem? What do you suggest?

3. **Dams and river systems**

   Research the controversies surrounding one of the following dams, some of which have been removed, some of which are slated to be removed, and some of which are still being debated. Include reasons for and against removal of the dam.

   - Edwards Dam, Kennebec River, Maine
   - Quaker Neck Dam, Neuse River, North Carolina
   - Kirkpatrick Dam (also known as Rodman Dam), St. Johns River, Florida
   - Glen Canyon Dam, Colorado River, Arizona
   - Lower Granite Dam, Snake River, Idaho
   - Elwha Dam and Glines Canyon Dam, Elwha River, Washington.

4. **Local river systems and wastewater treatment**

   - Where does the sewage from your city go?
   - Does sewage from your community enter the river system before or after treatment?
   - Which communities downstream would this affect?
   - What would happen to the drainage system and sewage system if it rained 10 cm or more in one day?
   - Are there any communities upstream of your community that might put sewage or pollutants into your river system? If so, what are they?

Check the *EarthComm* web site for assistance with **Inquiring Further** research.

Activity 5 Parkland Field Study

Goals

In this activity, you will:

- Interpret and describe how weathering, transportation, and deposition occur in your local stream.
- Describe your parkland site in terms of its relation to the river system and the water cycle.
- Develop a scaled drawing of your parkland site that shows its relationship to Earth systems and cycles.

Think about It

Weathering, transportation, and deposition occur in every stream, whether the stream is fast moving or slow moving, shallow or deep.

- What would you look for if you searched for these three processes at your parkland site?

What do you think? Record your ideas in your *EarthComm* notebook. Write down your ideas about how you would gather evidence of these processes in your local stream. Be prepared to discuss your ideas with your small group and the class.

Investigate

You will need to plan two investigations before going into the field.

Part A: Scaled Drawing
As a class, devise a way to make a scaled drawing of the parkland site. The drawing should accurately show the path of the stream and locations of prominent features in the park or along the stream (rock outcrops, large trees, tributaries, buildings, parking areas, and so on).

Part B: Stream Processes
In your group, plan how to identify, record, and analyze evidence of weathering, transportation, and deposition at your riverside parkland site. The questions below are suggestions. Feel free to generate your own questions for inquiry.

Once you have selected questions for inquiry, develop a hypothesis and a method of testing your hypothesis. What will you measure? How will you make measurements? How will you analyze the data you collect? Be sure to represent your stream processes on your scaled drawing. Share your plan with your teacher.

a) What is the range of sediment sizes carried by the stream?

b) Where is erosion most likely to occur along the stream?

c) Where is deposition most likely to occur along the stream?

d) Which dissolved solids does the stream contain?

e) Is there evidence that flooding events occur along this stream?

f) Which areas within the park are prone to flooding? Which areas are protected from flooding?

⚠ If you can't swim, do not wade in. If you will be wading in the stream, don't go barefoot or wade in above your knees. If you are wearing waders, don't allow any water in the tops of the waders. The weight of water-filled waders could quickly pull you under. Don't touch any animals you find, dead or alive.

Reflecting on the Activity and the Challenge

Three main processes work in every stream: weathering, transportation, and deposition. Your field study allowed you to observe how streams work. You made a scaled drawing to note where these processes operate. Understanding how streams work will help you to design a riverside park that is safe and educational.

Digging Deeper

RIVER SYSTEMS AS PART OF THE EARTH SYSTEM

Throughout this chapter you have learned how rivers are part of the hydrologic cycle. Water can exist in three different physical states at a riverside park (solid, liquid and gas), is stored in various **reservoirs**, and moves from one reservoir to another. In Earth systems science, the movement of matter or energy from one reservoir to another is known as a **flux**. Fluxes can be **inflows** (movement into a reservoir) or **outflows** (movement out of a reservoir).

Stream systems are dynamic; they are always changing. As fluxes change, so can the amount of water in the stream. Changes in the amount of water result in changes in stream flow. This affects the amount of weathering, transportation, and deposition in the stream. These changes, in turn, affect how the stream interacts with the four Earth systems: geosphere, hydrosphere, atmosphere, and biosphere. As you can imagine, changes in any part of a stream system cause other parts of the system to change.

The flow of water between different reservoirs within a stream system can be compared to a sink that is being filled with water. As you turn the handle on the faucet, you increase the flow of water into the sink (the reservoir). This makes the water level in the sink rise. One outflow of water is the drain. As the inflow into the sink increases, outflow down the drain increases. If water enters the sink faster than the sink can drain, the water level in the reservoir increases until the water spills over the sides and floods. Flooding over the sides is another outflow of water. As you may know from experience, this is likely to create other consequences and effects as well!

This drainage basin shows how accumulated water flows out as a stream.

Geo Words

reservoir: a natural or artificial storage place of water.

flux: the movement of water from one reservoir to another.

inflow: the amount of water entering a reservoir.

outflow: the amount of water leaving a reservoir.

A stream system behaves the same way, although it is a system with many more reservoirs (sinks), and many more fluxes (drains and faucets). Reservoirs in a stream system include the atmosphere, the biosphere, groundwater, surface water runoff, streams, and lakes. A period of heavy rainfall creates a greater inflow into reservoirs like groundwater and lakes. The system responds by filling the reservoirs and increasing their outflows. Increased outflow from these reservoirs becomes greater inflow into streams. If the stream is filled faster than it can drain, it floods. Effects of flooding include undercutting of stream banks, increased sediment transportation, and damage to vegetation both within and outside of the stream.

Natural changes in stream processes can also be subtle. For example, during summer months there is more vegetation to absorb water. Warmer temperatures cause an increase in evaporation. This can result in lower groundwater levels, less inflow to the stream, lower stream flow, and thus less sediment transport.

Humans can also alter stream processes. **Urbanization**, the process of converting natural lands to developed areas for human use, greatly affects stream processes. For example, paving of roads and parking lots limits the inflow of precipitation into groundwater reservoirs. Instead of being absorbed into the ground, most of the precipitation becomes surface water that flows into streams, which increases stream flow and sometimes causes flooding. Diverting stream water for irrigation and human use reduces stream flow. Adding contaminants to stream water, such as sewage and pesticides, can kill fish and vegetation.

Geo Words

urbanization: the process by which humans convert natural lands into areas developed for specific human uses. This can include building, paving, or other construction. It also includes partitioning, grading, and clearing areas of land that were previously in a natural state.

The convergence of a river and an urban environment in Minneapolis.

When creating any development along a stream, one must consider two questions:

- How will natural stream processes affect the development?
- How will development affect stream processes?

Scientists, engineers, city planners, and other professionals are often consulted to answer these questions. They look for evidence of natural hazards to any development, such as boulders or collapsed stream banks in a shallow streambed, which indicate episodes of flooding. They also evaluate how developments will alter processes both in and around the stream, such as how paved parking lots will increase surface runoff and inflow, as well as contaminant transport, into the stream.

Check Your Understanding

1. Explain flux in your own words. Refer to inflow and outflow in your answer.
2. List at least five natural reservoirs in a stream system.
3. Describe two natural changes in stream processes that can occur.
4. What effects does urbanization have on stream processes?

Understanding and Applying What You Have Learned

1. Characterize the hydrologic cycle at work at your parkland site. Use your scaled drawing to make a map and a cross section that shows:
   a) where water is stored in reservoirs within the stream system (reservoirs like groundwater, river water, atmosphere, biosphere, and so on);
   b) how water flows into and out of (inflows and outflows) the different reservoirs within the stream system.
2. Did you observe large rocks and boulders in your stream?
   a) How and when do they move down the streambed?
   b) Describe two different ways that running water changes the larger particles in the streambed.
3. Observe a sample of sediment from your stream.
   a) Describe the variety of particle sizes.
   b) Describe the variety of rock types in the stream.
   c) How do the rock types in the stream compare to the types of rocks in the local and regional area? Obtain a geologic map to make comparisons.
4. Was the water in your stream clear enough for you to see the bottom of the stream?
   a) If so, why? If not, why not?
   b) How would the removal of vegetation in the drainage basin or at the parkland site affect the clarity of stream water? Explain.
   c) How will the addition of man-made structures at your parkland affect the clarity of stream water? Explain.

5. Look at the flowchart that describes how human activities can affect streams. Choose one specific human activity or development that will be necessary to construct your parkland (like paving a parking lot, burying sewage lines for public bathrooms, clearing vegetation to make picnic areas, etc.). Describe how this activity or development could affect the stream. Will this effect influence how the stream interacts with any of the Earth Systems? If so, how?

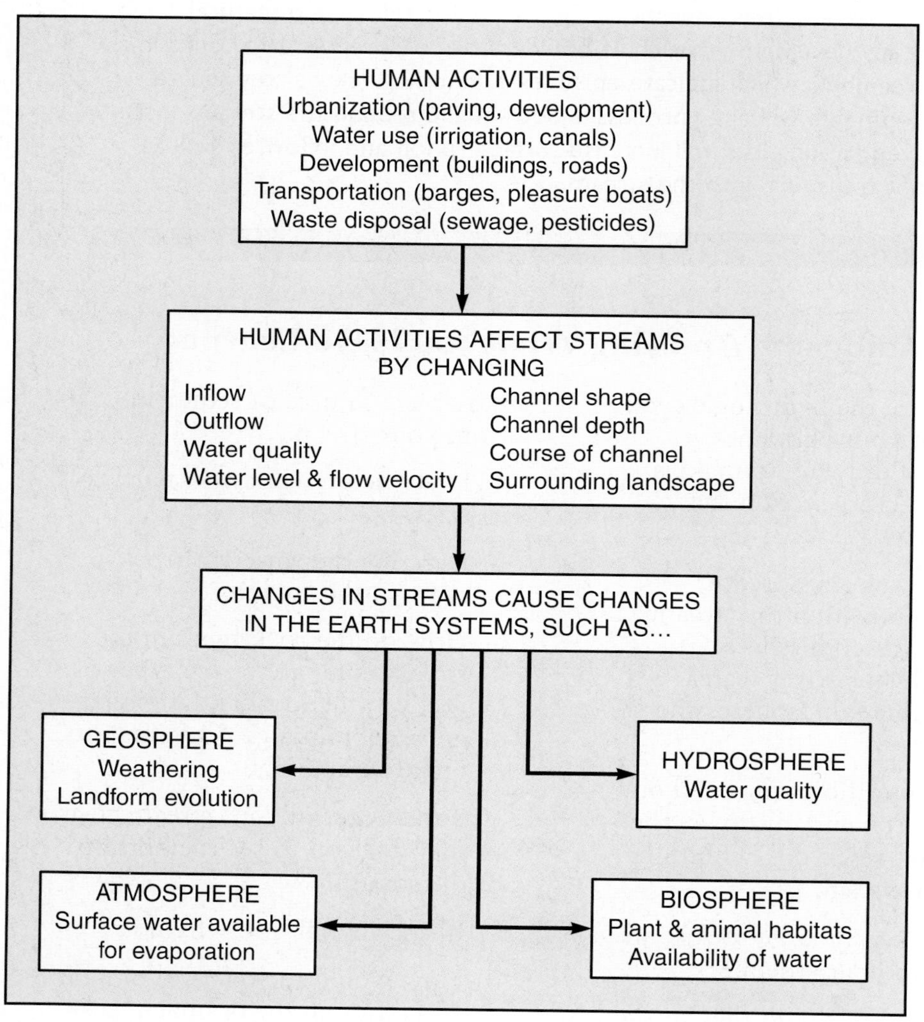

6. From what you have learned in this chapter, how would you expect your stream and the valley it flows through to change over time? Consider short-term and long-term changes.

Preparing for the Chapter Challenge

Make a new copy of your scaled drawing. Plan your parkland on the basis of your investigation into the stream and surrounding area. Describe the uses of the park and how each use might affect the stream. Be sure to note where your educational exhibit will be located and used.

Inquiring Further

1. **Parkland design**

   How are parklands planned and developed in your community? Which laws or regulations must be followed? What role do scientists, engineers, and other professionals play in the process of park design?

2. **Landscape architecture as a career**

   Landscape architects design outdoor spaces (parks, nature trails and landscapes around buildings, just to name a few) so that they are useful and scenic. In many cases, they must also be sensitive to the environment. Investigate the issues a landscape architect would encounter in planning a nature trail that runs from your parkland down along the stream. What kinds of structures would be constructed? How would these structures affect the environment?

3. **Construction practices**

   When a building or road is constructed, vegetation is stripped away, exposing the soil to erosion during rainstorms. Sediment that enters streams can be harmful to wildlife. Investigate the steps that developers at construction sites must take to minimize soil erosion and the amount of sediment that enters a stream.

Earth Science at Work

ATMOSPHERE: *Red Cross Worker*
Those who provide help and relief to communities during floods must carefully follow the analysis of weather data so that they are aware of imminent danger. They want to be prepared to act immediately when disaster strikes.

BIOSPHERE: *Farm Worker*
The livelihoods of many people depend upon a productive growing season. Farm workers rely on healthy crops produced on fields supplied with ample nutrients.

CRYOSPHERE: *Coast Guard*
The crew on an icebreaker clears inlets, harbors, and rivers on the Great Lakes to maintain as long a shipping season as possible. Millions of dollars of revenue can be lost if a boat is unexpectedly stranded by ice.

GEOSPHERE: *State Survey Geologist*
State survey geologists monitor stream and river flow for the public welfare. They study past evidence of flooding to make predictions about future events.

HYDROSPHERE: *Film Maker*
Many levels of government organize efforts to work towards a balanced and sustainable ecosystem in areas such as the Everglades in South Florida. Educating the public through videos and films is an important part of these efforts.

How is each person's work related to the Earth system river systems, and communities?

3

Land Use Planning ...and Your Community

3 Land Use Planning ...and Your Community

Getting Started

Your town or city has probably changed considerably since the first settlers arrived in the area. Their lifestyles and needs were likely different from yours. Imagine yourself among the group of pioneers who first began the settlement that later grew into your town or city.

- Why would you have chosen to settle in this area?
- What natural resources do you think the early pioneers found?

What do you think? Record your ideas about these questions in your *EarthComm* notebook. Be prepared to discuss your responses with your small group and the class.

Scenario

In recent years, the residents in your area have expressed an interest to the town council in establishing a number of new facilities in the community. Interest has been expressed in a sports complex, a theme park, an airport, a shopping center, a school or college, a housing community, a business complex, an industrial park, a hospital, a new facility for solid-waste management, a new water-treatment and sewage plant, a forested trail, and a picnic area.

Before making any decisions, the town council has decided to update the master plan for land use in your community. The members of council appreciate the need to maintain a favorable

living environment and at the same time protect the integrity of the area. They are looking for input from those who understand the importance of Earth systems to the development and growth of a community.

Chapter Challenge

Your challenge is to engage in a community planning project. To complete this project you will:

- Research the geological and climatological features, and the human use patterns and needs of your community.
- Use the information to develop a comprehensive land-use plan for the next 20 years.
- Include provisions for residential, commercial, industrial, and agricultural areas; water, sewage, and waste management; recreation; and the protection of natural resources and quality of life.
- Consider the impact of the plan on natural resources, including, but not limited to, water, air, and soil.
- Address any issues of human health and safety.
- Suggest modifications to the present land use.
- Assure that the plan adheres to local ordinances and land-use laws.

Assessment Criteria

Think about what you have been asked to do. Scan ahead through the chapter activities to see how they might help you to meet the challenge. Work with your classmates and your teacher to define the criteria for assessing your work. Devise a grading sheet for the assessment of the challenge. Record all this information.

Make sure that you understand the criteria and the grading scheme as well as you can. Your teacher may provide you with a sample rubric to help you get started.

Activity 1 Your Community's Water Resources

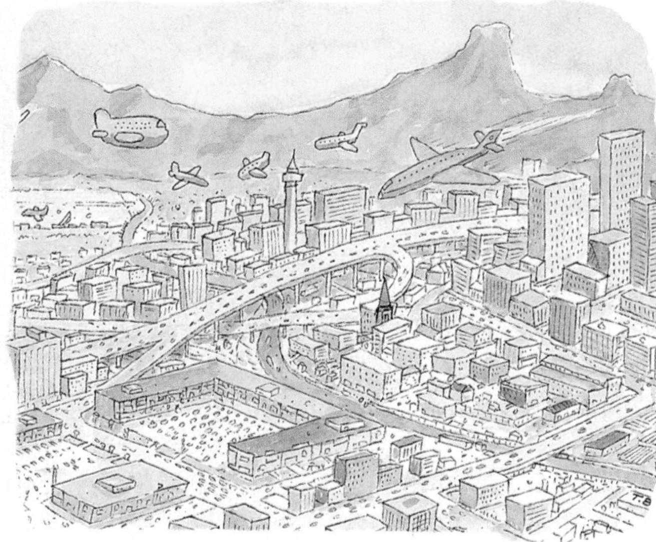

Goals

In this activity you will:

- Map the drainage basin(s) in which your community lies, and identify the drainage divides.
- Locate and identify water sources within the community.
- Understand the importance of protecting the water resources of a community.
- Understand the importance of knowing the location of a community's watersheds and drainage divides for land-use and planning purposes.

Think about It

Development within a community can change the flow of water on and within the land. It may also affect the quality of water resources.

- How do you think the water resources of your community have changed over time?

What do you think? Record your ideas about this question in your *EarthComm* notebook. Be prepared to discuss your responses with your small group and the class.

Investigate

1. Lay a piece of tracing paper or a clear plastic sheet on a topographic map that includes your community. (You may need to use more than one map to cover your entire community.) Carefully trace all of the major streams. We will refer to this as your local stream-network map.

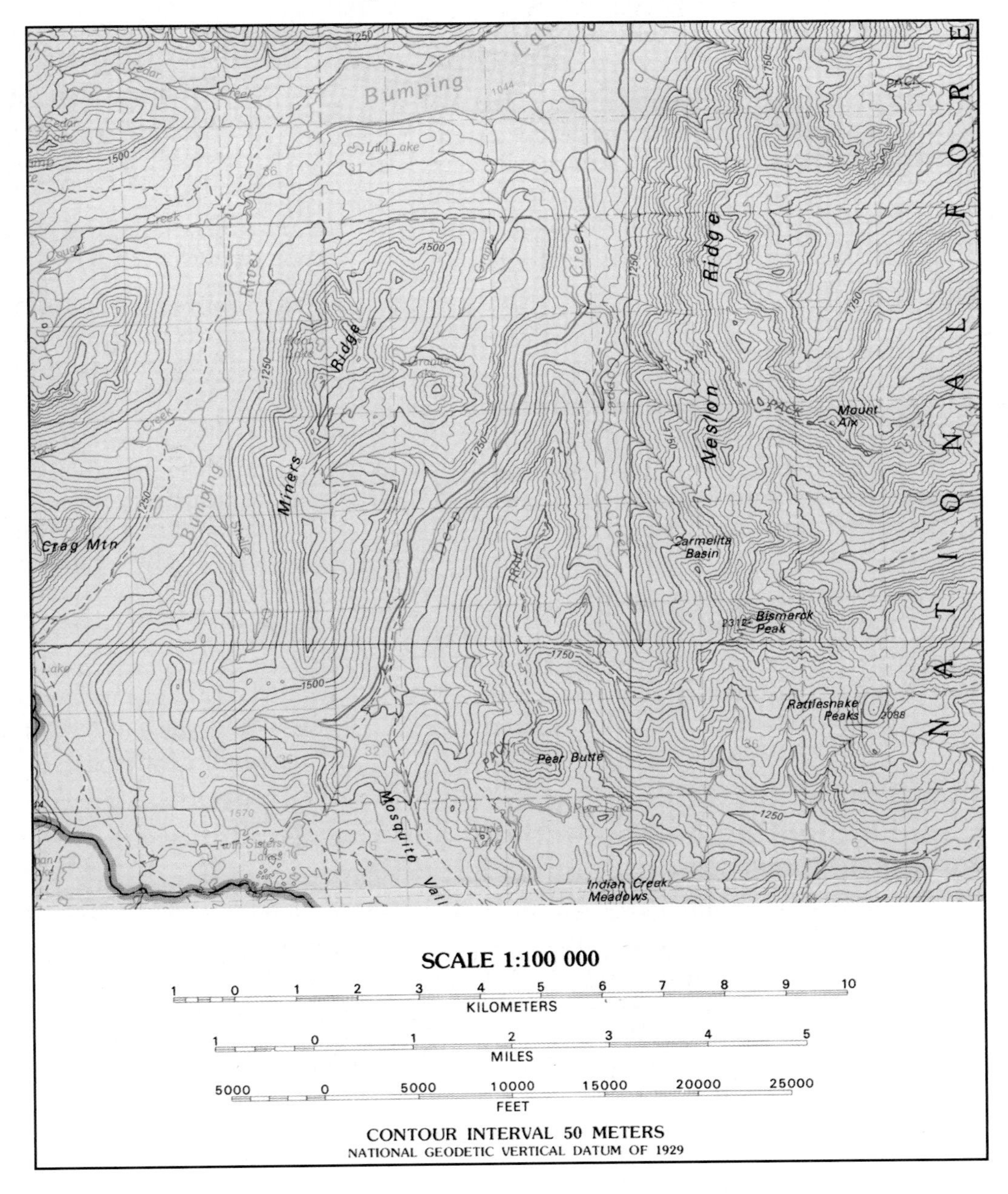

2. Find out the source of your community's drinking water. It may be a local river, reservoir (natural or man-made) or deep well, or it may be delivered by pipeline from a source many miles away.
   a) Record your community's source of drinking water in your *EarthComm* notebook.
3. If your community and its source of water are in different drainage systems, trace the drainage divide(s) (on your local stream-network map) between where your drinking water comes from and where your school is located.
4. Refer to your local stream-network map:
   a) Lightly shade over the land area from which all the streams drain into connecting streams within your community's drainage basin.
   b) Show any existing major land-use features such as factories, airports, agricultural sites, areas of dense population, schools, hospitals, water and sewage treatment facilities, or landfills. Devise a key for your map.
   c) Consider the activities that occur at each of the sites you identified. Divide the sites into:
      i) Sites that require clean water.
      ii) Sites that may pollute water sources.
   d) Locate specific sources of potentially clean water in and around your community and circle them. These sources could include large or small rivers, ponds, lakes, or reservoirs.
   e) On your map, draw two kinds of arrows:
      i) Arrows that go from water sources to nearby areas of land use that require clean water.
      ii) Arrows that go from areas of land use that create pollutants to nearby water sources. Keep in mind that water on the surface always flows downhill, perpendicular to local contour lines.
5. In your own words, explain how watersheds might influence where you would place a development that requires clean water.

Reflecting on the Activity and the Challenge

In this activity, you mapped the land-use patterns and water resources in and around your community. You also identified how different land uses affect water resources. Finally, you showed that watersheds can influence your community's water resources, and thus land use. Knowing where your community's watersheds are will help your town council decide how to distribute land use to take advantage of water resources, while protecting water resources from contamination.

Digging Deeper

WATER SOURCES AND LAND-USE PLANNING

The words **stream** and **river** mean basically the same thing, except that "river" is used for relatively large flows of water in natural channels, and "stream" is used for both small and large flows of water in natural channels. The terms **watershed**, **catchment**, and **drainage basin** all mean the same thing. They refer to the land area from which rainfall collects to reach a given point along some particular river. The farther down that river you go, the larger the land area that is included in the watershed. The idea is easiest to understand if you think about the point where a large river flows into the ocean. There is some well-defined area upstream that supplies the water that flows from that river into the ocean. Watersheds are of all sizes, from ones like the Mississippi-Missouri River drainage basin shown in *Figure 1*, which covers a large part of the continent of North America, to ones that cover only a very small local area, less than one square kilometer.

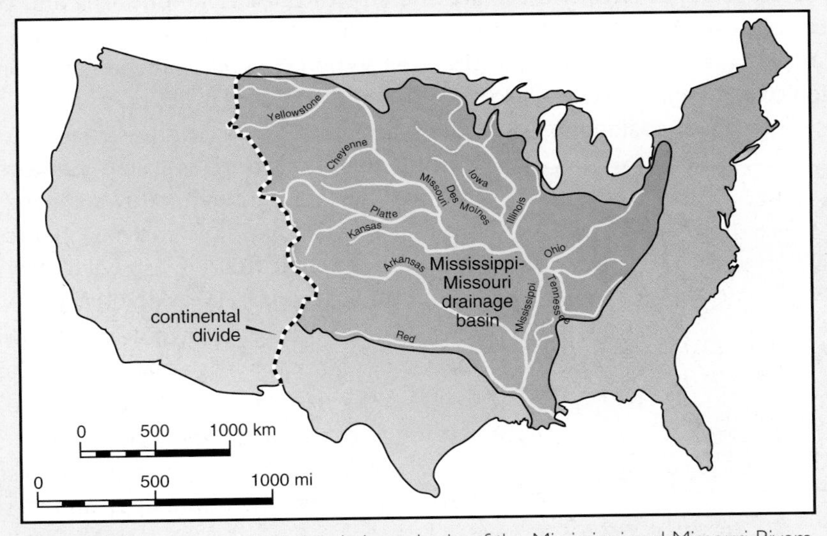

Figure 1 A large watershed: the drainage basin of the Mississippi and Missouri Rivers.

Except for the rainfall that falls directly into a stream or river, the water first flows over the land surface in thin, slow-moving sheets before reaching a stream. The water then collects into small streams, called **tributaries**, which drain into larger streams. These larger streams are tributaries to even larger streams. Eventually, all the water that falls over a particular watershed ends up in a single large stream, called a **trunk stream**. The collection of all of the streams, large and small, in a watershed is called a **drainage network**.

Geo Words

stream: a small or large flow of water in natural channels.

river: a relatively large flow of water in natural channels.

watershed (also, catchment, drainage basin): the land area from which rainfall collects to reach a given point along some particular river.

catchment (also drainage basin, watershed): the land area from which rainfall collects to reach a given point along some particular river.

drainage basin (also catchment, watershed): the land area from which rainfall collects to reach a given point along some particular river.

tributary: a small stream that drains into a larger stream.

trunk stream: the main stream in a watershed.

drainage network: the collection of all the streams in a watershed.

When you trace out all the streams of a drainage network on a map, the pattern looks somewhat like the branching pattern of a tree, with the tributaries represented by the little twigs and the larger branches, and the trunk stream represented by the tree trunk. (In fact, that's where the term "trunk stream" comes from!)

The line on the land surface that separates one drainage basin from another is called a **drainage divide.** You have probably heard of the "continental divide," which is the line, running approximately north–south in the high mountains of the western U.S., that separates areas that drain into the Pacific Ocean from areas that drain into the Atlantic Ocean. Every drainage basin, small or large, is bounded by drainage divides. In areas of steep topography, drainage divides are usually easy to locate. Locating drainage divides in areas with very gentle topography, however, is often much more difficult.

If you were asked to list the things that control how much water reaches a trunk stream, what would you put on the list? It would probably seem obvious to you that two of them are the size of the catchment area and the amount of precipitation in the catchment area. A less obvious thing is the infiltration capacity of the soils in the area. When rain falls on the land, some of it infiltrates (sinks directly into) the soil and some of it flows on the surface instead of soaking in. All of the water that flows on the surface rather than infiltrating is called **surface runoff.** At first the water flows across the surface in thin, slow-moving sheets, called **overland flow**. If you have ever been on a paved parking lot during a heavy rain, you might have noticed a thin sheet of water flowing at your feet. The same thing happens on the soil surface, although the flow is less regular. The overland flow eventually reaches a stream channel, where it flows deeper and faster.

The time it takes for runoff to reach some point in a stream system is also important for flood prediction. The steepness of the land surface is significant because water flows faster on steep slopes than on gentle slopes. Also, dense vegetation tends to slow the surface runoff in the stream channels themselves as well as in the land areas between stream channels. In a highly developed area, there are more houses, commercial buildings, roads, and parking lots. Precipitation falling on these surfaces cannot infiltrate; instead, it moves quickly to streams. The total amount of runoff reaching the trunk stream is also greater.

Geo Words

drainage divide: the line on the land surface that separates one drainage basin from another.

surface runoff: the part of the water that travels over the ground surface without passing beneath the surface.

overland flow: the flow of water on the Earth's surface in the form of thin, slow-moving sheets, rather than in localized channels.

Check Your Understanding

1. How does rain falling on the land surface reach a trunk stream?
2. How would you locate the position of the drainage divide that separates your watershed from the adjacent watershed?
3. Describe natural features that may help to slow surface runoff in your watershed.
4. Explain the difference between contaminants and pollutants in a stream system.

Water that passes over or through both natural and man-made surfaces may pick up solid or dissolved materials along the way, referred to as contaminants. When contaminants reach high concentrations that are dangerous or poisonous to people, animals, or plants they are called pollutants. Contaminants can be transported great distances by the tributaries and the trunk stream. In rural areas, runoff may pick up contaminants from agricultural operations or old septic tanks and leach fields. Large amounts of sediment washed from agricultural fields can also contaminate streams and reservoirs. In suburban and urban areas, contaminants can come from industrial operations, building materials, and landfills, as well as from chemicals and fuels that accumulate on paved surfaces.

Water-treatment plants remove contaminants (such as household chemicals and sewage) using both filtration and chemical treatments, usually in open reservoirs. During floods, water-treatment plants commonly close because heavy precipitation can cause the reservoirs to overflow, and then runoff can transport the abundant contaminants directly into the watershed area.

In planning a community, it is desirable to locate land uses that use or create contaminants away from important water resources. In areas where land uses already release contaminants into the watershed system, measures must be in place to minimize or avoid contamination of community water supplies.

Understanding and Applying What You Have Learned

1. Using your stream-network map, compare the locations of land uses that create pollutants to land uses that require uncontaminated water. Questions to address include:
   - How far away from water resources are land uses that create pollutants?
   - How far away from water resources are land uses that require uncontaminated water?
   - What natural features are present in and around your community that might help reduce water contamination?
   - What has your community done to prevent the contamination of water resources?

   Identify any land uses that pose a threat to the quality of your community's water resources.
2. From what you know about stream networks and drainage patterns, what factors would you consider when deciding on a location for a landfill? a hospital?
3. How could natural events bring about the contamination of water resources in your community's watershed?

4. One of the developers in your community believes that tenants in an apartment complex would like to have a scenic view. Therefore, the developer suggests building such a complex on the highest topographic feature in your community. The developer maintains that because the development is not industrial, and will be quite far from your community's water supply, the complex cannot adversely affect local water quality. From your knowledge of how land use can affect water resources, do you agree with the developer's analysis? Explain why or why not.

Preparing for the Chapter Challenge

Write a paper in which you address potential hazards to your community's water resources. Also describe measures your community has taken to protect its watershed area, and comment on how successful you think these measures will be in the future. Finally, make a set of recommendations on how to direct your community's growth in order to protect local water resources.

Inquiring Further

1. **Monitoring local issues**

   Keep a scrapbook throughout the year of newspaper articles that deal with water pollution, water treatment, safety of drinking water, development in watersheds, protection of watersheds, etc., in and around your community. Identify three specific issues, and consider how you would address these issues if you were:

   a) a politician
   b) a scientist

   What are some reasons why approaches used by politicians and scientists may differ? As a member of the community, in what ways could you make your opinions known to local politicians? To local scientists?

2. **Water and sewage treatment**

   Visit your community's water and sewage treatment plant to see firsthand how it operates. Find out what kinds of contaminants are removed at the facility. Also, find out where the water goes after it is treated.

3. **Problems in other communities**

   Search the web for articles discussing problems other communities may be experiencing in their watershed area. Determine if any of the problems you find could occur in your community's watershed area and, if so, how they could be avoided or lessened. (Consult the AGI *EarthComm* web site for suggestions.)

Activity 2 Urban Development and Air Quality

Goals

In this activity you will:

- Investigate how urban heat islands can affect local temperatures.
- Develop an understanding for factors that cause the urban-heat-island effect.
- Investigate ways to minimize the urban-heat-island effect.

Think about It

"The temperature today in Atlanta reached a sweltering 90°F. Nearby Athens reported in with a reading of 86°F."

Think of a time when you may have been in the heart of a city or out in the open countryside.

- What are the differences in the weather and/or air quality between the two places?

What do you think? Record your ideas about this question in your *EarthComm* notebook. Be prepared to discuss your responses to these questions with your small group and the class.

Investigate

Part A: Modeling the Urban-Heat-Island Effect

1. Measure equal masses of damp soil and rocks (about 1 kg of each, or enough to fill the container about 2/3 full). Place the damp soil in one container and the rocks in a second identical container.

2. Suspend a thermometer about three centimeters above the soil and another thermometer about three centimeters above the rocks. Make sure the thermometers are at the same height above both containers.

3. Place a light source, shining directly on the materials filling each container, an equal distance from both containers. Be careful not to shine the light directly on the thermometers, or to place the light too far away from the setup. Make sure each setup is identical except for the contents of the containers.

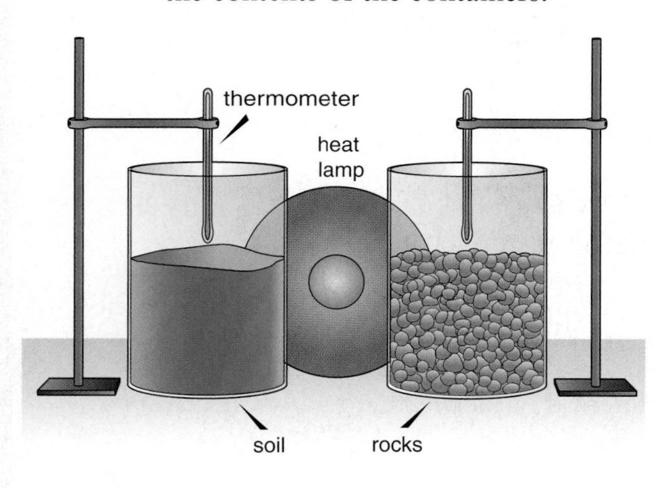

4. Read the temperature on each thermometer at 5-min intervals for a total of 25 min.

   a) Copy the chart shown below into your notebook.

   b) Record your data in the chart.

Light ON	Temperature	
Time (min)	Soil	Rocks
5		
10		
15		
20		
25		

5. Turn off the light and continue to read the temperatures at 5-min intervals for a total of 15 min.

   a) Copy the chart shown below into your notebook.

   b) Record your data in the chart.

Light OFF	Temperature	
Time (min)	Soil	Rocks
5		
10		
15		

6. Plot the data you collected in steps 4 and 5 on one graph showing temperature versus time with the light on and off.

Do not touch the hot lamp. Report any broken thermometers to your teacher.
Clean up any spills immediately.

7. Write a paragraph comparing the temperature data for the soil and rock samples.

   a) How long does the temperature difference between the two containers last after the lamp is turned off?

8. How might slightly higher temperatures affect weather patterns, plants, animals, and human residents of a community?

Part B: Changing the Urban-Heat-Island Effect

1. Repeat the investigation from Part A, but this time modify your setup. Use one or more of the following examples, or design your own variations to simulate conditions found in your own community.

   Possible modifications:

   - Use lighter-colored or darker-colored rocks.
   - Try materials other than rocks to represent other building materials, such as brick, concrete, pieces of wood, or pieces of man-made materials such as fiberglass or plastics.
   - Use containers made of different materials, such as wood, aluminum, or ceramic.
   - If you live in or near a desert or beach, fill one container with sand.
   - If you live near a large lake or ocean, fill one container with water.
   - Devise a model in which you include solid materials and a man-made lake in your containers.
   - Partially or completely cover the tops of the containers with a reflective coating (such as aluminum foil).
   - Partially or completely cover the tops of the containers with grass, leaves, or other plant material.

2. Analyze your data.

   a) How did your data change between Part A and Part B?

   b) What possible sources of error may exist in your models, and how would they affect your data?

3. Supplement your data by collecting temperature data from real cities. If you do not live in a city, pick a city near your community to study. You can find temperature records for many places in the United States from the following sources:

   - Local newspaper (go to the library for back issues).
   - Local TV meteorologist.
   - Local weather service.
   - GLOBE data base.
   - National Climatic Data Center. (Consult the AGI *EarthComm* web site for suggestions.)

Keep flammable materials away from the hot lamp.

Reflecting on the Activity and the Challenge

Changing one Earth system can have a dramatic effect on other Earth systems. In this investigation you modeled how the composition of the Earth's surface (part of the geosphere) can affect the atmosphere above it. Understanding how characteristics of the Earth's surface affect its ability to absorb and release heat helps urban planners recognize how development projects may affect Earth systems. You will need to consider how human activities can influence Earth systems as you prepare to advise the town council on development plans for your community.

Geo Words

urban: relating to a city.

urban-heat-island effect: the fact that many cities have higher average temperature than the surrounding countryside.

solar radiation: energy from the Sun available to be absorbed by all of the Earth systems.

Digging Deeper

URBAN-HEAT-ISLAND EFFECT

The **urban-heat-island effect** refers to the fact that many cities have higher average temperatures than the surrounding countryside, especially in the summer. Several factors cause this, the most obvious being that buildings, roads, and parking lots absorb more solar radiation than soil and vegetation do. Also, some of the **solar radiation** that falls on vegetation is used by the plants for life processes, rather than raising the temperature of the vegetation. This extra urban heating during the day keeps the city from cooling off as quickly as the country at night.

Figure 1 The numerous buildings and vast paved areas of Chicago absorb large quantities of solar radiation, causing temperatures in the city to be much higher than those of surrounding areas.

Another reason urban areas tend to be warmer is that rain in the city runs quickly off buildings and over paved areas into underground drainage networks rather than soaking into the ground as it does in undeveloped areas. As a result, after a rainstorm there is less surface and ground water remaining for **evaporation** in the city than there is in the country. Because water must absorb heat in order to evaporate, evaporation cools the air. Therefore, having less water available for evaporation in urban settings means that an important cooling process is less effective.

Heat is also generated in developed areas by car emissions, industrial operations, and heating and air-conditioning systems. Pollutants produced by cars and industry also act like a blanket to hold heat in the atmosphere over a city. The decrease in temperature as you move away from the center of a city is shown in the thermal data image *Figure 3*.

Figure 2 How would the temperature in this area differ from the temperature in Chicago? Why?

Geo Words

evaporation: the change of state of matter from a liquid (water) to a gas. Addition of heat is required for this change.

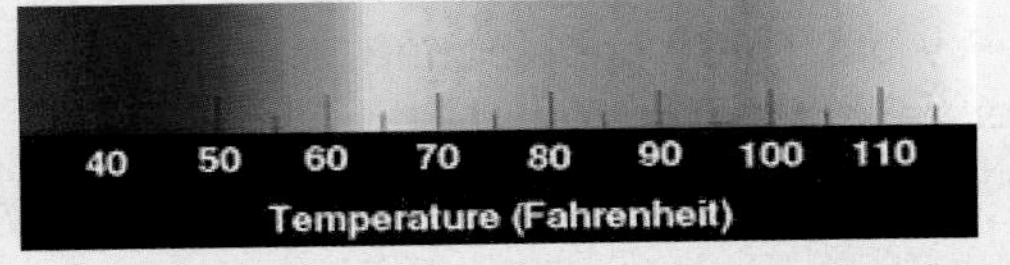

Figure 3 This thermal data image of Atlanta, GA. was taken during the day by a Landsat satellite.

Urban-heat-island effects cause people to use more energy for air conditioning in the summer. Using this extra energy increases the cost of electric bills and causes more pollution. Urban heat islands exist in winter as well as in summer, but then they act to moderate cold winter temperatures. Clear winter nights are almost always colder in areas outside the city, except when strong winds are blowing. Winds tend to even out temperature differences from place to place, by bringing cooler air into warm areas and vice versa.

Urban heat islands also can change local weather patterns by causing more rainstorms, and especially more thunderstorms. The hotter air over the city contributes to strong upward air currents that trigger thunderstorms. In addition, pollutants are actually small particles that can act as **condensation** nuclei that aid in the formation of raindrops.

The extra heat also increases ozone pollution. Ozone pollution is caused when emissions from cars react with warm summer sunshine. (Do not confuse ozone pollution in the lower atmosphere with the ozone hole in the upper atmosphere; they are two separate problems.) Ozone formation increases as the temperature increases. Ozone pollution is dangerous for people with asthma or other breathing problems, as well as the very young and very old. These people must stay inside on days that are predicted to exceed federal ozone standards.

Across the country, as many urban areas continue to grow, urban heat islands are getting bigger. Even in small communities, human development can measurably affect local temperatures. There are, however, ways to help lessen heat-island effects.

Areas of vegetation, especially large trees, can help to reduce the urban-heat-island effect. Vegetation does not absorb as much solar radiation as roads and buildings do. Trees shade roads and buildings from the Sun, reducing the amount of solar radiation absorbed. In addition, large volumes of water evaporate or are transpired from vegetation on most days, and this produces a cooling effect. Vegetation absorbs and/or traps pollutants, slowing their introduction into both the atmosphere and the hydrosphere. Plants provide additional benefits such as providing oxygen, slowing runoff (which can help reduce flooding), and filtering runoff (which can help reduce water contamination).

Careful selection of building materials can also reduce the urban-heat-island effect. Some materials absorb less solar radiation than others. For example, wood absorbs less solar radiation than brick. New materials that absorb less heat are constantly being developed. Also, lighter-colored materials absorb less radiation than darker materials. Covering structures with reflective materials reduces the amount of solar radiation that the structures absorb.

Geo Words

condensation: the change of state of matter from a gas (water vapor) to a liquid. Heat is released.

Check Your Understanding

1. What factors cause the urban-heat-island effect?
2. What kinds of building materials do you think contribute most to urban-heat-island effect?
3. During which time(s) of the day would you expect urban ozone pollution to be greatest? Why?
4. Describe how vegetation can be used to reduce the urban-heat-island effect.

Understanding and Applying What You Have Learned

1. How would the addition of a new shopping mall effect a community with an urban-heat-island problem?
2. How would the average temperature of a community be affected if an airport were built on an area formerly used as farmland?
3. What are some ways of lessening the urban-heat-island effect of city skyscrapers?
4. Many cities in the United States cite rush-hour commuting as a major cause of severe ozone pollution. What kinds of programs could city planners design and/or promote to help reduce this kind of pollution?
5. A developer in a large town wants to build a large office complex to house businesses. Although the town leaders support this plan because it will create jobs in their community, they are concerned that large developments may affect the local environment. What characteristics can the leaders suggest for this development in order to minimize heat-island effects?

Preparing for the Chapter Challenge

In your Chapter Challenge you will need to make recommendations on how to guide future growth in and around your community. Write a paragraph explaining what can be done to direct future growth in such a way as to avoid or minimize the urban-heat-island effect, as well as ozone pollution, in your community.

Inquiring Further

1. **Urban-heat-island effect in your community**

   Design an experiment to study the urban-heat-island effect in your community or in a city near your community. Pick a question that has occurred to you during this activity, such as *"Would reflective roof covers really help?"* or *"I wonder if there's been increased precipitation in my community because of an urban-island-heat effect?"* Make a plan for how you would go about answering the question. To be valid, your study should cover a span of at least five years.

2. **A model community**

   Make a model community (to scale) that has been designed to address and minimize the problems associated with urban-heat-island effects. Explain at least five features of the model.

Have your teacher approve your design before proceeding.

Activity 3 Flooding in Your Community

Goals

In this activity you will:

- Determine which areas of your community would be affected by floods of various levels.
- Gather and analyze rainfall data for the community for the past 50 years.
- Propose areas for development and areas not to be developed, using data on floods and rainfall.
- Understand the importance of floodplains in land-use planning.

Think about It

In some river valleys, such as the Nile, people have depended on flooding to enrich the soil each year. In many regions, however, floods are natural disasters.

- Have you ever experienced a flood or seen one on television? How did it occur? What did the flooded areas look like? What damage was caused?
- Do floods cause more damage in a city or in the country?

What do you think? Record your ideas about these questions in your *EarthComm* notebook. Be prepared to discuss your responses with your small group and the class.

Investigate

Part A: Mapping Floodplains in Your Community

1. Study a map of your state or county, and locate cities (or large towns) that are located on rivers.
   a) Create a table that lists the following: city names, population sizes, and river names.

2. Choose a section of the map to investigate. (Your teacher may choose to assign a section to your group.)
   a) Determine and record the elevation of the river along your section of the map.
   b) If the elevation of the river changes over the length of your section, record the elevation at several locations.

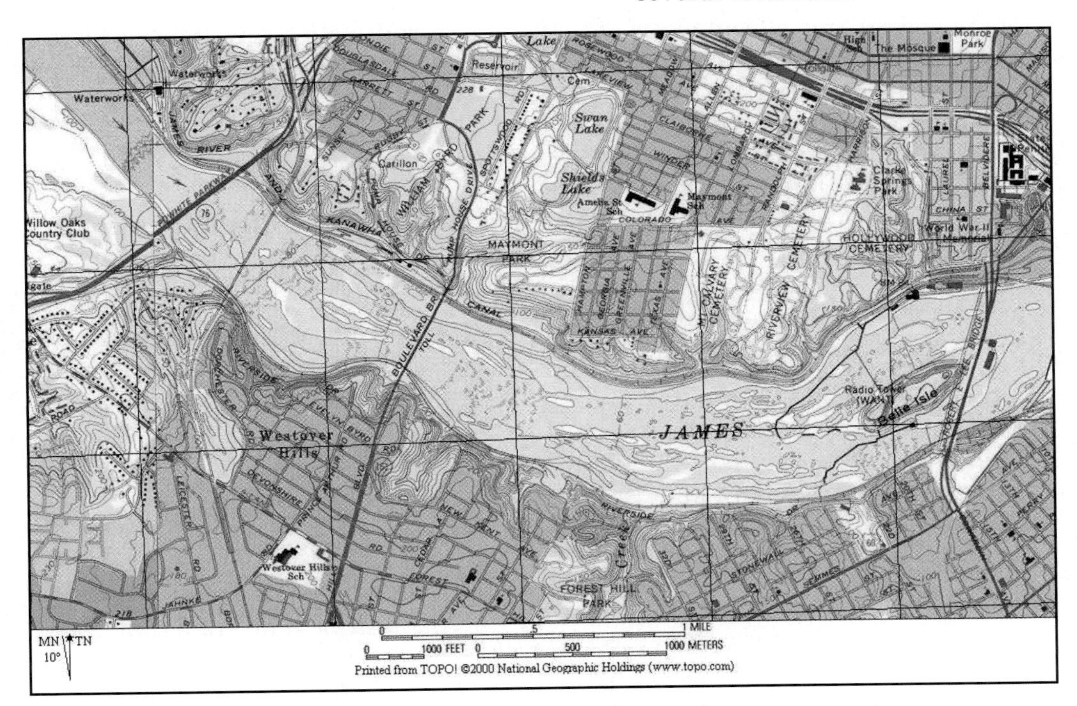

3. You will now make a floodplain risk map. Place a piece of tracing paper or a transparent plastic sheet on the map.
   a) Trace the river in your section with a blue marker or pencil.
   b) Using the contour lines on the map as a guide, draw a red contour line that represents elevations 10 feet above the elevation(s) of the river. This is the 10-foot flood line, which represents the position of the river's edge if it rose 10 feet above the normal elevation.
   c) Shade the area between the 10-foot flood line and the river with a red marker or pencil. This represents the area that would be under water during a 10-foot flood.

d) Draw a yellow contour line that represents elevations 25 feet above the elevation(s) of the river. Shade the area between the 10- and 25-foot flood lines yellow.

e) Draw a green contour line that represents elevations 50 feet above the elevation(s) of the river. Shade the area between the 25- and 50-foot flood lines green.

Part B: Tracking Flood Patterns in Your Community

1. Use the United States Geological Survey (USGS) Internet site to find out the frequency of 10-, 25-, and 50-foot floods along the section of the river you are studying. Use the *EarthComm* web site to help you access the information. If you don't have access to the Internet, call the USGS (1-800-ASK-USGS) to find out where the nearest library is that keeps USGS archival records.

   a) Make a table and graph in your notebook that shows how often and when your section of the river has flooded at 10-, 25-, and 50-foot levels.

2. Research precipitation levels for your community over the last 50 years. Possible sources of information are a local weather service office or television meteorologist, newspaper records, colleges, cooperative extension services, or the National Climatic Data Center.

   As a class, decide how best to chart or graph the data you have collected to show precipitation patterns.

   Once a plan has been decided upon, divide the data analysis between the student groups. Your teacher will be a resource for ideas on how large amounts of data can be sampled to be more easily studied.

   a) What cycles, patterns, or changes did you find in the historical rainfall data for your community? Examples of types of cycles that could be looked at are:

      - Yearly cycles. (Do some months routinely receive more or less precipitation than others?)
      - Drought cycles. (Are there patterns to the frequency at which drought occurred?)
      - Flooding cycles. (Are there patterns to the frequency at which flooding occurred?)
      - Drought vs. flooding. (Are droughts or floods more common?)
      - Overall pattern. (Did the amount of rain increase, decrease, or stay about the same over the 50-year period you studied?)

   b) On the basis of your analysis of the historical precipitation data, when would you expect the next major flood and/or drought to occur in your community?

   c) Draw a graph that shows the frequency of floods and/or droughts over the last 50 years, and what you predict to be the frequency for the next 50 years.

Reflecting on the Activity and the Challenge

Understanding which areas are likely to flood and having information on the frequency of floods will help you determine where land use may need to be restricted in order to prevent loss of property. Identifying areas where land should not be developed will aid you in your Chapter Challenge.

Digging Deeper

DEVELOPMENT AND FLOODPLAINS

Rivers have been, and continue to be, major sites of commerce, as shown in *Figure 1*. Rivers provide energy, a source of food and water, and a means for transporting raw materials and shipping out finished goods and products. In some areas a network of canals has been built off rivers to direct water to industrial operations. This harnesses the power of the water to turn and drive machines and tools for cutting, weaving, and transforming raw materials into manufactured goods.

Because rivers were the principal means for transporting materials and products in the past, most major industrial cities in the United States are on major waterways. In most cities, businesses and industries hold riverfront property. During the past 50 years, however, trucks, airplanes, and trains have transported most goods and materials. As a consequence, many cities have now abandoned the warehouses and factories on their waterfronts.

Figure 1 Situated on the Hudson and East Rivers, New York City is an example of a city that has benefited from its location on the water.

Many cities located along rivers are revitalizing old warehouses and factories, using them for new apartment complexes, shops, restaurants, and tourist attractions. But there is a risk in building along the edges of rivers: the occurrence of damaging floods. Before a city redevelops waterfront property, planners and developers need to be aware of the potential impact of floods.

Figure 2 Pittsburgh, Pennsylvania is an example of a city whose riverfront property, once dominated by factories, warehouses, and steel mills, has been redeveloped for other purposes.

The height of the water surface in a river channel, relative to sea level, at a given place along the river is called the **stage** of the river. During periods of normal flow in a river, water is confined to the channel of the river. When the stage of the river reaches what is called **flood stage**, water overtops the banks of the channel. The area of a river valley that is covered by water during a flood is called the **floodplain**, as shown in *Figure 3*. When water flows out onto the floodplain, it spreads out as a wide and shallow flow.

Figure 3 Deposition of fertile sediment on floodplains provides ideal farming conditions.

Geo Words

stage: the height of the water surface in a river channel, relative to sea level, at a given place along the river.

flood stage: the river stage (water level) at which a river rises above its banks and begins to cause a flood.

floodplain: any flat or nearly flat lowland that borders a river and which is covered by water when the river rises above the flood stage.

Because the flow is shallower, the friction between the flow and the ground plays a greater role than in the channel, so the flow across the floodplain is slower than in the river channel. The slower flow across the floodplain can't carry as much sediment in suspension as it did in the channel, so the floodwaters deposit much of sediment load (mainly sand and silt) across the floodplain. In many areas, floodplain sediments create fertile land that is good for farming. In areas that are not agricultural, cleaning up the sediment left by a flood is an expensive, labor-intensive job. (See *Figure 4*.)

Flooding is a natural process that occurs when a river system develops and evolves. Efforts by man to control this process along one section of a river can increase the effects of flooding along other sections of the river downstream. Sound land-use planning allows for the natural development of floodplains, and helps limit property damage potentially caused by flooding.

Figure 4 The flooding of the Red River of the North in Grand Forks, ND April, 1997, caused almost two billion dollars in property damage.

In many large rivers, especially along their lower courses as they approach the ocean, the land area through which the river flows is gradually subsiding relative to sea level. This causes the river to deposit some of the sediment it carries, so that it can maintain its same elevation relative to sea level. Most of this new deposition takes place in the channel itself, and also along its banks. As sediment is deposited along the banks, ridges called natural **levees** are formed. These natural levees stretch continuously along both sides of the river. With

Geo Words

levee: a natural or man-made embankment built along the bank of a river to confine the river to its channel and/or to protect land from flooding.

Geo Words

avulsion: a major change in the course of a river when the river breaks out of its levees during a flood.

headworks: an engineering structure built to control the flow of river water out of a river channel during a flood.

time, the river gets higher and higher above its floodplain. Eventually, during some especially large flood, the river breaks out of its levees and finds an entirely new and lower course across the floodplain. This is called **avulsion**. Not only does this result in a catastrophic flood on the floodplain; a city located somewhere downstream of the point of avulsion would be abandoned by the river!

At present, some of the lower Mississippi River in Louisiana flows out to the Gulf of Mexico along the Atchafalaya River, to the west of the main Mississippi. The U.S. Army Corps of Engineers has built an enormous structure, called a **headworks**, at the point along the Mississippi where the Atchafalaya branches off. The purpose of the headworks is to control how much water is diverted from the Mississippi.

Figure 5 Without a headworks, probably most of the Mississippi would by now be flowing down the Atchafalaya, leaving the city of New Orleans as a backwater city!

Check Your Understanding

1. Explain the meanings of river stage and flood stage.
2. What is a floodplain?
3. Under what circumstances are floodplain sediments useful, and when do they pose a problem?
4. What is a levee?
5. Describe how a downstream city could find itself abandoned by a river.

Understanding and Applying What You Have Learned

1. Describe at least one benefit and one drawback to placing artificial levees along a river that frequently floods.
2. Interpret the colors on your floodplain risk map as you would a traffic light—red means stop, yellow means proceed with caution, and green means safe. Using this map, list physical characteristics that could be used to identify land that is "high risk" for development.

3. What general recommendations for your community can you make about:
   a) How the river should be managed?
   b) How areas on the floodplain should be used?
4. Are there any current regulations in your community regarding floodplain development? If so, do you agree with them? If not, is it important that there should be?
5. Looking at your map and at the historical data on flooding in your community, make some recommendations for how floodplain land might best be used. Think about whether or not riverfront development is a good choice for your community. Be prepared to explain to your classmates the reasons behind your recommendations.
6. Using your floodplain risk map, recommend possible sites for the following developments: municipal park, retirement home for senior citizens, apartment complex, public swimming pool, man-made levees, hospital, university campus, gas station, single-family home, apple orchard, commuter parking lot, water-treatment plant, communications tower, lumber yard. Give reasons for your placements. If you think that some of these developments should not be located within the limits of your map, explain why.

Preparing for the Chapter Challenge

Provide an answer, two to four sentences long, for each of the following questions:

a) Where is flooding least likely to occur in and around your community?
b) Where is flooding most likely to occur in and around your community?
c) What factor(s) caused the last major flood in your community?
d) When can you expect the next major flood in your community?
e) How could future flood damage be limited in your community?

Inquiring Further

1. **Predicting future floods**

   Use your rainfall and flooding data to make predictions about whether the risk of flooding in your community will increase or decrease in the future. How sure can you be about your predictions? What might affect their accuracy?

Activity 4 Slope of Land in Your Community

Goals

In this activity you will:

- Calculate the angle of repose for different kinds of soils and other granular materials.
- Determine if any areas in your community have slopes that are too steep for safe development.
- Understand the importance of considering slopes in land development.

Think about It

The slope of the land and the nature of the materials beneath the surface must be considered when planning how to use the land in a community. Changing the slope of the land (or even the amount of vegetation on a slope) can have dangerous consequences.

- How would the slope of land control development in a community?
- How might changing the slope of the land create potential hazards for citizens (say, cutting through the land to build a road or housing project)?

What do you think? Record your ideas about these questions in your *EarthComm* notebook. Be prepared to discuss your responses with your small group and the class.

Investigate

Part A: The Slope of a Sand Pile

1. Slowly pour 500 mL of dry sand through a funnel onto a flat surface, such as your lab table, so that it makes a pile.
   a) Describe what happens to the sides of the pile as you pour the sand.
2. Hold a protractor upright (with the bottom edge held against the flat surface) and carefully begin to slide it behind the pile as shown in the diagram.
3. At the point where the curved upper edge of the protractor intersects the surface of the pile of sand, read the angle in degrees. This is the natural angle of the side (slope) of the pile. It is called the angle of repose. It is the steepest slope that can be formed in the material without slumping or sliding of the material down the slope.
   a) Record this angle in your *EarthComm* notebook.
4. Repeat step 3 several times.
   a) Record the measurement of the angle of the slope each time.
   b) Do you get the same angle each time? Why or why not?
   c) Why is it important to make this measurement several times?
5. Repeat steps 1, 2, and 3 using different amounts of sand.
   a) Record the measurement of the angle of the slope each time.
   b) Does the angle of the slope change? If so, how much?
6. Pour extra sand onto a pile of sand several times.
   a) Record the measurement of the angle of the slope each time.
   b) Does the angle of the pile change?

Clean up spills immediately. Cover desk with newsprint to facilitate cleanup. Wear safety goggles when pouring sand or other particles.

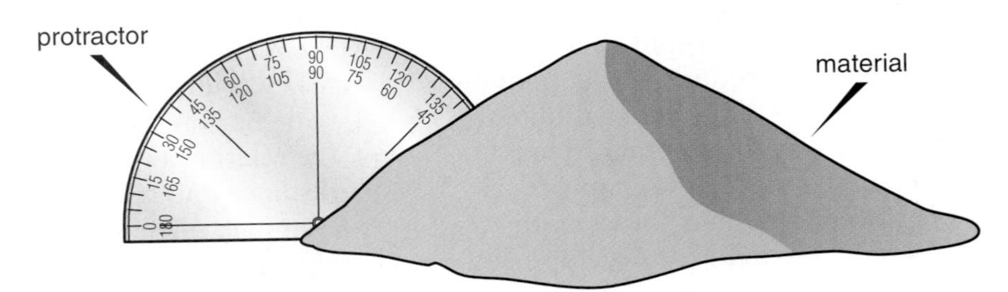

Part B: The Slope of Other Materials

1. Obtain some or all of the following materials (make sure they are dry): fine sand, coarse sand, clay or mud, gravel, silt, soil, table salt, granulated sugar.
   a) Predict what would happen if you repeated the activity in Part A using these materials, which have different sizes and shapes of particles. Record your prediction in your *EarthComm* notebook.

2. For each of the available materials, repeat the following procedures:
   - Place a handful of the materials in a dry container such as a can or jar.
   - Cover the container with a piece of cardboard.
   - Turn the container upside down onto a flat surface.
   - Lift the container very slowly. The material should form an inverted, cone-shaped pile.
   - Measure the angle of the slope of the pile.
   - Make three measurements for each material.

   a) Record your measurement on a copy of a chart similar to the one below.

   b) In your notebook, write a summary paragraph discussing conclusions you can draw from the data on your chart. Your paragraph should address how particle size relates to the maximum slope angle the particles will maintain.

Measurement of the Slope Angle of Different Materials				
Material	**Angle Measure of Slope**			**Average**
fine sand				
coarse sand				
clay				
gravel				
silt				
soil				

Part C: Characteristics of Slopes in Your Community

1. Obtain a topographic map of an area in or around your community that shows a variety of different slopes. If your community is relatively flat, use a map of another area that shows both slopes and areas of development.

2. Determine the contour interval on your map, either from the legend or from identifying the spacing between contour lines.

3. Determine the scale of your map. This may be expressed as a ratio, such as 1:24,000, which means one unit (of any measure of distance, such as inches or centimeters) on the map equals 24,000 units in the real world. There may also be a scale bar on the map that indicates the relationship between map distance and real distance (for example, one inch on the map equals one kilometer in the real world). You will need to use a scale that makes it easy to measure distances on your map.

4. Convert your horizontal scale to the same units that are used for the contour interval (probably feet or meters).

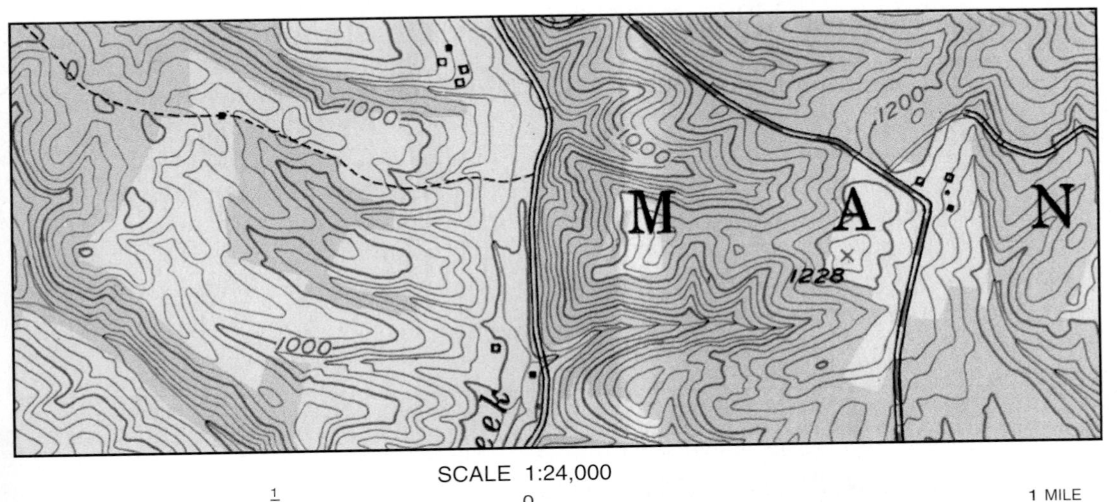

SCALE 1:24,000

1 ½ 0 1 MILE

1000 0 1000 2000 3000 4000 5000 6000 7000 FEET

1 .5 0 1 KILOMETER

CONTOUR INTERVAL 20 FEET

5. Choose a slope on your map, and record the following data in a table of your own design:
   a) Measure a specific horizontal distance perpendicular to the slope. Record the actual (not map) horizontal distance.
   b) Use contour lines to measure the change in elevation over that specific horizontal distance.
   c) Divide the change in elevation by the horizontal distance (make sure they are in the same units!), and then multiply by 100. This gives you the percent grade of the slope.
6. Repeat step 5 for several slopes on your map.
   a) Record all of the data in your table.
7. Create a second table that lists the location, percent grade, and characteristics for each slope. Some characteristics you could list include: kind(s) and density of vegetation; kind(s) of developments above, below, and on the slope; population density above, below, and on the slope; underlying geology; and surface deposits. Add any other characteristics that you think will be important when evaluating land use on and around slopes.
   a) Include this table in your notebook.

Reflecting on the Activity and the Challenge

In this activity you learned that materials of a given grain size will consistently pile up to a maximum slope angle. You also learned that you can calculate slope and percent grade from a topographic map, and can classify slopes on the basis of physical characteristics seen on maps.

Allowing for development on slopes in your community may require a detailed analysis of each slope in order to ensure human safety and protection of natural lands. These will be important issues to address as you work on your Chapter Challenge.

Geo Words

sediment: solid fragments or particles that are transported and deposited by wind, water, or ice.

unconsolidated material: sediment that is loosely arranged, or whose particles are not cemented together, either at the surface or at a depth.

lithification: the conversion of unconsolidated sediment into a coherent, solid rock.

angle of repose: the maximum slope or angle at which loose material remains stable, commonly ranging between 33° and 37° on natural slopes.

Digging Deeper

SLOPES AND MASS MOVEMENT

Angle Of Repose

Sediments are **unconsolidated materials**, which have not been **lithified**. A pile of sand is unconsolidated sediment, but sandstone is a rock. Mud is unconsolidated sediment, but shale is a rock. Unconsolidated materials cover solid bedrock in many environments including glaciated areas, soil horizons, deserts, beaches, lakes, rivers, and sand dunes.

Unconsolidated materials are far less stable than rock. Solid bedrock is stable at almost any slope angle. Unconsolidated sediments, however, are stable only up to a maximum slope angle, as seen in *Figure 1*. This maximum angle is called the **angle of repose**. If you add sand to a pile of sand with sides already at the angle of repose, the extra sand just slides down the sides. The angle cannot become any steeper without the sides collapsing. In general, the angle of repose for dry, unconsolidated sediments ranges from 33° to 37°. The angle of repose does not vary much with sediment size, but more angular (jagged) particles can maintain steeper slopes than more rounded particles.

Figure 1 Developers must take care not to build on slopes that exceed the angle of repose.

Mass Movement

An important factor to consider for your chapter challenge is the natural slope of the land, and the materials on and under these slopes. Sediments that were deposited by rivers or glaciers lie beneath many communities. If these deposits are sloping, you need to consider how stable the slope is before deciding to develop the area. The stability of a slope depends on a number of factors, including: kind and amount of vegetation; sediment composition, texture, moisture content, and underlying geology.

Under certain conditions, slopes can be modified to accommodate development. Examples of modifications are terraced slopes, retaining walls, and drainage channels at the top of slopes placed to reduce erosion.

Buildings, roads, and other structures built on slopes of any angle can be damaged or destroyed when **mass movement** occurs, as shown in *Figure 3*. Examples of mass movement include rockfalls, landslides, debris flows, debris avalanches, and creep (gradual rock and debris movement). The basic cause of these types of movement is the downward pull of gravity. Part of the pull of gravity acts parallel to the sloping surface. If that does not make sense to you at first, think about what happens to you when you stand on a slippery slope: gravity pulls you straight down the slope! This same downslope pull is always acting on the materials that underlie a sloping land surface. Under certain conditions, the downslope pull of gravity overcomes the strength of the material, and the material moves downslope. This downslope movement varies enormously both in speed (from so slow you can't see it happening, to tens of meters per second) and in volume of material (from single sediment particles, to cubic kilometers). The additional weight of water added to soil or sediment, as well as the lubrication that water provides between grains, significantly increases the potential for mass movement.

Figure 2 A small seaside community north of Santa Barbara, CA, felt the effects of mass movement. The slide of an unstable hillslope destroyed several homes and resulted in an evacuation of the area.

Geo Words

mass movement: downslope movement of soil, sediment, or rock at the Earth's surface by the pull of gravity.

In cold regions, cycles of freezing and thawing can trigger mass movements. As the water in soil or sediment freezes it expands, lifting the grains parallel to the slope. When the ice melts, the grains settle parallel to the slope, and then slide downhill (often with water providing lubrication) because of the pull of gravity.

In areas where the ground freezes to a depth of several feet in the winter, the top layers of soil are loosened during spring thaw while the bottom layers remain frozen and solid. The water-saturated upper layers then slide downhill. In areas where freeze-thaw cycles are frequent, building foundations and pipes carrying gas, water, or sewage must be placed below the freezing zone (3–4 feet deep in northern states) so they can resist damage from surface slides.

Vegetation can stabilize slopes. Trees, shrubs, bushes, and grasses can help to keep soil layers intact, depending on the depth of the roots. Most vegetated areas, however, are still subject to landslides if they become saturated with water.

Slopes in your community can be evaluated on topographic maps, because the maps indicate the elevation of the land with contour lines. The standard to which all elevations are compared is average sea level, and a contour line represents equal elevations, or heights, above sea level. Therefore, a 10-foot contour line connects all the points in a region that are 10 feet above sea level. There is a basic rule for drawing contour lines: contour lines can never cross, because two elevations cannot exist at the same location. The spacing of the contour lines is a measure of the steepness of the land. The closer the spacing of the contour lines, the steeper the slope they represent. A region showing great variation in elevation is referred to as having high **relief**, and a region showing relatively little variation is referred to as having low relief.

When you have to work with a slope on a topographic map, use the contour lines to measure the steepness of the slope (how much the land rises over a particular horizontal distance). Use the scale on the map to figure out the horizontal distance from one point on the slope to another point, measured perpendicular to the contour lines. Convert so that both the vertical change (the change in elevation) and the horizontal distance are expressed in the same units, usually in feet or miles, or in meters or kilometers. Divide the vertical change by the horizontal distance, then multiply by 100, to get what is called the **percent grade**.

Engineers usually measure the steepness of slopes in percent grade instead of in degrees. The percent grade is the change in elevation divided by the horizontal distance, multiplied by 100. For example, a slope where the land rises 20 feet over a horizontal distance of 100 feet has a 20% grade. In general, slopes exceeding a 20% grade cannot be safely developed. Those who are familiar with trigonometry will realize that the percent grade is just the tangent of the slope angle, multiplied by 100.

Geo Words

relief: the general difference in elevation of the land from place to place in some region.

percent grade: the ratio of the vertical and horizontal distance covered by a given slope multiplied by 100.

Check Your Understanding

1. What is the relationship between the size of sediment particles and the angle of repose? What is the relationship between the angularity of sediment particles and angle of repose?
2. Describe two ways in which slopes can be stabilized.
3. Describe three human activities that may make slopes unstable.

Understanding and Applying What You Have Learned

1. What range of slope angles would you consider safe for possible development? Why?
2. When considering whether a particular slope is safe for development, what else would you need to know in addition to the angle of repose of the unconsolidated surface material? List several important factors and how each affects development.
3. The development of roads and buildings typically involves moving and shaping the land. What basic guidelines should be followed when cutting a slope or piling loose material and creating a slope?
4. Why would a developer be motivated to build on a potentially unstable slope? In your opinion, what advantages would outweigh the dangers?
5. Specifically describe how slopes might have influenced your community's growth over the last:
   a) 5 years
   b) 20 years
   c) 50 years
6. Consider other communities you have visited or researched where slope influences development.
   a) Describe a community where slopes have limited development.
   b) Describe a community where slopes have been helpful for development.

Preparing for the Chapter Challenge

Write a short paragraph answering each of the following questions:

a) Which slopes in your community cannot be safely developed?
b) Which slopes in your community could be developed if they are modified?
c) What modifications would you suggest?
d) Which present developments in your community might be at risk from mass movements? How can these risks be minimized?

Inquiring Further

1. **Effect of water on mass movement**

   Repeat Part B of the activity using materials that have water added to them and see if your results change.

   - What do your results lead you to believe regarding slopes without vegetation during times of heavy rain?
   - What practices during times of heavy construction in a community does the information support?

2. **Slope-risk map of your community**

   Compile slope information into a risk map for your community using the "traffic light" colors of green = safe, yellow = proceed with caution, red = stop, do not proceed. Include a key that classifies what type of development may be of concern in each area.

3. **Underlying materials in your community**

   Determine what kinds of materials underlie different parts of your community by consulting geologic maps, local developers, and/or town officials.

   - Does your community lie on unconsolidated sediment or relatively solid bedrock?
   - If you found areas of unconsolidated sediment, are the sediments naturally occurring or were they deposited by human activity?
   - Is building on bedrock always safer than building on sediment? Explain why or why not.
   - In your own words, describe how the distribution of underlying materials has shaped your community's building patterns.

4. **Famous catastrophic mass movements**

   Go to the *EarthComm* web site to investigate famous examples of catastrophic mass movements that have affected communities. Conduct research on the mass movement. Answer these questions:

   - What happened?
   - How did the mass movement affect the community?
   - What factors led to the mass movement?
   - How might the event have been avoided?
   - What lessons were learned from the event?

Activity 5 Soil and Land Use in Your Community

Goals

In this activity you will:

- Collect, study, and describe local soils and develop a classification system for them.
- Understand how soils form, what a soil profile is, and the importance of soil as a natural resource.
- Understand the relationship between the physical characteristics of a soil and how the soil formed.
- Understand that soil characteristics may vary over time, and that these variations can greatly impact a community.
- Map the location of different soils in your community.

Think about It

During the 1930s, severe dust storms called "black blizzards" affected the midwestern states. This period in American history has been referred to as the "Dust Bowl" era.

- In what ways is soil part of the Earth systems (geosphere, atmosphere, hydrosphere, biosphere)?
- How is soil important in your life?

What do you think? Record your ideas about these questions in your *EarthComm* notebook. Be prepared to discuss your responses with your small group and the class.

Investigate

Part A: Classifying Soil Samples

1. Obtain a soil sample from your backyard, a farm, or a relative's house in the country. Bring the soil to school in a covered plastic container or zip-lock bag. Label the container with your name, where the soil came from, and the kind of place where the sample was taken (field, hilltop, riverbank, forest, etc.). Each student group obtains a sample of each soil available.

2. Work in your small group to study the soil samples available.

   a) Decide on and record a set of descriptive terms that your group will use to classify all of the samples.

   b) Make a table in your notebook that lists the different soil samples down rows and your set of descriptive terms across columns. Fill in the table, describing each sample with the set of terms.

   c) Develop a classification scheme for your soil samples. Write the name and definition for each type of soil identified by your classification scheme.

3. On a map of your community, show the locations where each soil sample came from.

   a) Label each point with the soil name your group came up with.

   b) If possible, draw boundaries between different soil types.

Part B: Determining Specific Characteristics of Soil

1. Characteristics of different soils make them appropriate for different uses. You can test soil for desired qualities such as:
   - how well it drains;
   - how well it absorbs and holds water;
   - how well it promotes plant growth;
   - stability during earthquakes;
   - strength (e.g., ability to support heavy structures).

   a) Design a test your group can perform on each soil sample that will identify a quality of your choosing. In your notebook, describe how you will perform the test. Submit this description to your teacher for approval.

   b) In your notebook, predict the results you will get when you test each soil sample.

2. Perform the test you designed on each soil sample.

   a) Display the results of your test in tables or graphs, and present the results to the class.

   b) Take notes on the results of the other groups' tests.

   c) As a class, discuss various test designs and results. In your notebook, summarize the results of all the tests.

Have your teacher approve your design before proceeding. Wear safety goggles when working with sand, soil, or other particles. Avoid contact with eyes. Wash hands when done.

Reflecting on the Activity and the Challenge

In this activity you learned that there are different kinds of soils, and that soil can be classified and mapped on the basis of its physical properties and where it is found. You also learned that you can test soil for various properties that might be desirable for specific uses. You will need to be aware of different ways to recognize and test for soil types in order to make sound land-use recommendations that will help you complete your Chapter Challenge.

Digging Deeper

SOIL

Classifying Soils

As you learned in your investigations, soil types can vary significantly. Soil classifications vary as well. One way to classify soils is by texture, which refers to the distribution of the sizes of the particles. Most soils are a mixture of gravel, sand, silt, and clay sizes, as well as organic materials. Texture controls many properties of soil, such as how fast water will drain through it, how much water it can hold, or how much it compacts under heavy loads.

Soil that contains approximately equal proportions of sand, silt, and clay is called **loam**. The soil texture diagram illustrated in *Figure 1* shows how soils are classified and named on the basis of the various percentages of grain sizes contained. Loam is excellent for growing plants because it does not drain water too rapidly or slowly and contains organic materials.

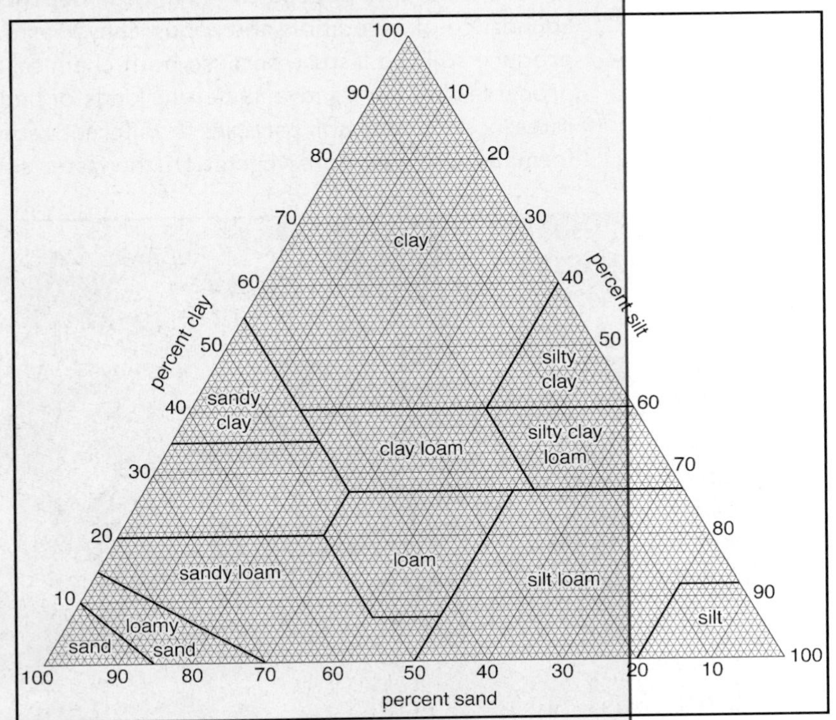

Figure 1 Soil texture triangle. Plotting the relative percentages of clay, silt, and sand in a soil sample allows for classification of the soil by texture.

Geo Words

loam: in general, a fertile, permeable soil composed of roughly equal portions of clay, silt, and sand, and usually containing organic matter.

Geo Words

weathering: the destructive process by which rocks are changed on exposure to atmospheric agents at or near the Earth's surface.

bedrock: solid rock that is connected continuously down into the Earth's crust, rather than existing as separate pieces or masses surrounded by loose materials.

Soil Formation

In general, soil is a relatively thin layer of rock, mineral fragments, and decaying organic material that covers most of the Earth's land surface. Soil forms when **weathering** and biological processes break down **bedrock** and organic material like dead plants. Physical weathering includes processes related to the freeze-thaw cycle, wind, rain, running water (e.g., rivers), temperature changes, and sunlight. Chemical weathering occurs when some of the minerals in the rock react with water and dissolved oxygen and acids and are converted into other minerals. The products of chemical weathering are mainly fine clay-size particles, along with dissolved chemicals. Some of the dissolved chemicals produced by weathering, like potassium and phosphorus, are important nutrients for plants. Biological processes include decomposition of organic matter by bacteria and other microorganisms, digestion by earthworms and other organisms, uptake of nutrients by roots, and addition of nutrients by decay of dead organic matter.

It can take from a few hundred to several hundred thousand years for a soil to form. The time needed to form a soil depends on climate, bedrock type, abundance of vegetation, and topography. Warm, humid climates tend to produce soil the fastest, because both chemical and physical weathering processes are very active. Different kinds of bedrock weather at different rates, contributing soil particles at different rates. Plants help facilitate soil formation, so the more vegetation, the faster soil tends to develop. Typically, there is little or no soil on steep mountain slopes because gravity and water transport the sediment to lower elevations as fast as it is produced. Valleys usually contain thick soil deposits, as do broad, flat areas.

Figure 2 Well-rounded rocks found at a beach or in a streambed are evidence for physical weathering.

Figure 3 Ice and snow can act to break down rocks to produce soil. What other weathering processes are at work in the photograph above?

Geo Words

soil horizon: a layer of soil that is distinguishable from adjacent layers by characteristic physical properties such as structure, color, texture, or chemical composition.

Soil Horizons

If you were to look at a vertical cross section of sediment from the surface down to a depth of several feet, you would see various layers of the soil, or what soil scientists call **soil horizons**, as shown in *Figure 4*. The top layer, called the A horizon, contains more organic matter than the other layers. This layer provides nutrients to plants and contains enormous numbers of insects, microbes, and earthworms. The next layer down, called the B horizon (or subsoil), is a transition layer between the layers above and below. It contains less organic material than the A horizon. In the lowest layer, the C horizon, partially broken-up bedrock is easily recognized, and organic material and organisms are scarce or absent. The thicknesses of the layers vary greatly from location to location, but these three layers are present in most soils.

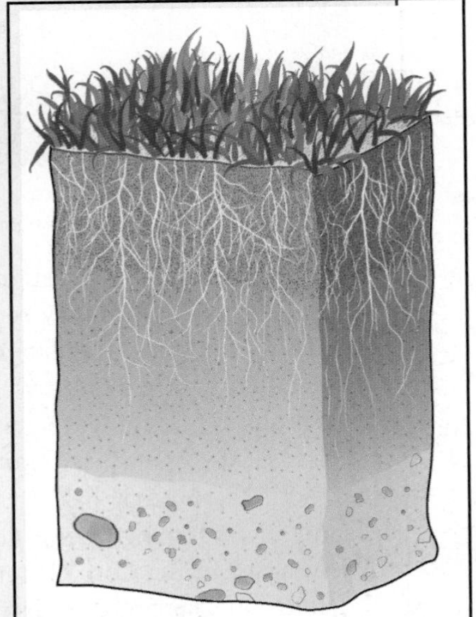

Figure 4 Can you see the three soil horizons in the diagram?

Soil as a Natural Resource

Soil is a vital natural resource because it allows humans to grow food crops. Because soil takes so long to form, it should be considered a non-renewable resource. Unfortunately, in many places around the world, including the United States, soil is being eroded away by wind and running water much faster than it is being formed. Soil is lost when rain or wind carries soil particles away from fields or construction sites that are left bare, without a cover of vegetation. Windbreaks (strategically placed walls or rows of trees or other plants) can protect soil from wind, and contour plowing and terracing can help to reduce soil erosion. Another strategy is to always grow plants in unused fields. This helps to hold the soil in place, and also adds valuable nutrients to the soil when the plants die and decompose.

Figure 5 The "Dust Bowl" provides a clear example of what can happen when soil is not considered in planning a community.

Some soils may be less suited to development than others. For example, poorly drained soils may require expensive drainage systems to protect buildings and property from flooding. Soils that contain many rocks and large boulders may be expensive to excavate and sift for development. During earthquakes, some water-saturated sandy or muddy soils undergo liquefaction. That is, they behave temporarily like a liquid and therefore cannot support structures. You may have modeled this yourself at the beach by jiggling wet sand and watching how it flows like a liquid for a short time before becoming firm again. In the Loma Prieta earthquake in California in 1989, much of the damage in Oakland and San Francisco was caused by liquefaction. Many buildings that were built over old, water-saturated landfill deposits collapsed as the shaking caused the soil below them to liquefy and flow.

Check Your Understanding

1. Describe three processes that are involved in physical weathering.
2. Draw and label a diagram that explains the major features of the three main soil horizons.
3. Describe two methods of preventing soil erosion.
4. Why are some soils less suited for development?

Understanding and Applying What You Have Learned

1. What are the different soil types in and around your community? For each different type, describe the following (organize your data in a table):
   a) appearance
   b) texture
   c) content (kind and amount) of organic matter
   d) other physical characteristics
   e) location(s) where it is found
   f) location(s) where it is being removed naturally or by human activity
2. Are there certain soils in your community that may be good or bad for agriculture or for development? Use data to support your answer.
3. Have you ever noticed changes in soils as you traveled? For example, you might notice the appearance of sandy soils as you get closer to the seashore, or the absence of soils as you enter a region with steep topography. From a trip you have taken, or in photographs you have seen of different regions, list some differences you may have noticed in soils.
4. Consider one soil type you have seen on a trip or in photographs. Describe how the characteristics of that soil can tell you something about the climate and geology of the region where it is found.

Preparing for the Chapter Challenge

From your research and the soil map of your community, make recommendations for land use as your community grows.

- Are there particular places that should be reserved for agriculture because the quality of the soil is very good?
- Are there particular places that should not be developed because the soil is unstable?
- On a map, draw boundaries to illustrate how soil type should guide land use.

Inquiring Further

1. **Soils in your community**

   a) Contact a state or local soil conservation agency, the cooperative extension at your state university, or your state geological survey to obtain a map of the soils in and around your community. Describe any correlation between soil type and current land use in your community.

   b) Analyze the physical characteristics of a particular soil type or sediment type found in your community. Questions you might answer include:

   - Why is the soil a certain color in your area but a different color in an adjacent area?
   - Why do deposits of sand and gravel tend to be found only at lower elevations?
   - Why will some sediments or soils liquefy during an earthquake?

   c) Describe a soil profile. To do this, you will need to find an area where you can observe 3–4 feet (about one meter) of fresh, vertically layered soil (such as a riverbank). In your notebook, draw what you see in detail. Include measurements of the various soil layers. Describe each layer as completely as you can, including observations such as color, texture, composition, grain size, and grain shape.

   d) Investigate soil erosion in your community? Write a report in which you describe the cause(s) of the problem and state what is being done to minimize damage. Offer your own suggestions for dealing with the problem. Include interviews with town officials, and/or local newspaper articles, if available.

   Consult the *EarthComm* web site for assistance with your research.

Activity 6 Surveying the Community on Land Use

Goals

In this activity you will:

- Use a process to build consensus from many diverse opinions.
- Develop a list of factors that are most important to consider when making land-use decisions for your community.
- Develop an appreciation for the importance and difficulty of considering different points of view when making land-use decisions for a community.
- Understand that land features and natural resources play an important role in a community's planning process.

Think about It

Have you ever been in a group that was trying to make a decision, but not everyone had the same opinion on what the correct decision was?

- How did the group finally agree on a decision?
- Did everyone get their way, or did some people have to change their minds?
- Was everyone happy with the final decision?

What do you think? Record your ideas about these questions in your *EarthComm* notebook. Be prepared to discuss your responses with your small group and the class.

Investigate

Part A: Reaching a Consensus in Class

1. On your own, develop a list of general factors that you think should be taken into consideration for land-use planning in your community. Try to think of as many factors as you can. Use your textbook, newspapers, and other resources to assist you in making the list. Some factors to consider include: landscape, geology, soil, land value, employment, underground conditions, natural beauty, preservation of natural environments, and mineral resources.

   Here are some examples of factors that may be important in a community:

   - If your community is located in a hilly area, then steep slopes might be an important factor in deciding how to use land.
   - If you live in a beautiful part of the country that profits from tourism, then protecting that natural beauty might be important.
   - If your community's water supply is polluted or in danger of becoming polluted, then watershed protection might be important.
   - If you live in a farming region, soil preservation might be important.
   - If you live in a valley where smog tends to get trapped, then air quality might be important.

   a) Enter your list in your *EarthComm* notebook.

2. Working in small groups, make one master list of factors by combining everyone's list.

   a) Record this master list in your notebook.

3. Looking at your group's master list, assign points to each factor listed based on this scale:

   1 = not very important

   2 = somewhat important

   3 = very important

   On your own, write down your point value for each factor on the master list. As a group, add up the total number of points given to each factor by all group members. Divide the total number given to each factor by the number of people in the group. This will give you a numerical ranking for each factor.

   a) Order and list the factors from the one receiving the highest ranking to the one receiving the lowest ranking. Copy this list, in order, into your notebook.

   b) Make one neatly written list to be duplicated so each person in the class can have a copy.

4. Obtain a copy of each group's ranking list. Cut up the lists so individual factors are separated. You now have a pile of pieces of paper, each with a factor and a ranking on it. On your own, separate the pieces of paper with a factor on each one into three piles:

   - very important
   - somewhat important
   - not very important

5. Have all students in the class put their pile of very important factors into one big pile. From this pile make a new ranked list of factors.

   a) Make one neatly written list to be duplicated so each person in the class can have a copy.

6. Obtain a copy of this new list. Cut up the list so individual factors are separated. You now have a pile of pieces of paper, each with a factor and a ranking on it. On your own, separate the pieces of paper with a factor on each one into three piles:

   - very important
   - somewhat important
   - not very important

   This time place a maximum of five factors in the very important pile.

   a) Make a list of the three factors that were put in the very important pile by the most students.

7. You now have a consensus list of the three items your class believes are the most important factors for determining land use in your community.

   a) Record this list in your *EarthComm* notebook.

Part B: Reaching a Consensus in the Community

1. Every student should take home three cards, each listing one of the three factors from the consensus list developed in your class (do not put ranking numbers on the cards).

2. Ask four adults and two students to assign a value to each of the three factors using the following scale:

   1 = not very important

   2 = somewhat important

   3 = very important

   a) Copy the format of the sample table below into your notebook, and record the ranking of each person.

3. Ask each person if there are any other factors he or she thinks are important to consider.

   a) Record their comments in your notebook.

4. In class, compile the data. Decide as a class whether or not to alter the class consensus list on the basis of strong opinions from the community.

Tally of Ranking							
Factors	**Adult 1**	**Adult 2**	**Adult 3**	**Adult 4**	**Student 1**	**Student 2**	**Average**
A							
B							
C							

Reflecting on the Activity and the Challenge

In this activity you used a survey process to develop a consensus, or agreement, about what your class and community believes are important issues to consider in land-use planning. You have seen firsthand that people can have very diverse opinions on a subject. A land-use plan is effective only if it addresses community needs and gains local support. This activity has given you information on which issues are important to your community, where there are areas of disagreement, and which issues you should specifically address in your chapter challenge.

Digging Deeper

REACHING A CONSENSUS ON LAND USE IN THE COMMUNITY

Because communities change over time, plans need to be continually developed and modified to avoid the potential for future land-use problems. Often, not all residents of a community agree on how to plan for changing conditions or future land use. Some individuals believe that encouraging development and population growth improves the local economy. Others believe that development should be restricted in order to limit growth and maintain the size of the community. Still others will want to ensure that policies do not restrict their individual land-use plans. Therefore, a community must reach a consensus on how present and future land use should be directed.

Figure 1 In an area such as New York City, high population density means that making good land-use decisions is even more crucial. Where does your trash go if you live in a big city?

Community planning can vary greatly from one community to another. Residents in urban areas have different needs and wishes than those in rural areas. Land-use regulations can also be divided within a single community, such as separating residential, business, and industrial districts.

In many areas of the world, population is increasing, which means more land is affected by human activities. If growth and development are unavoidable, then it is vital to consider how to direct growth and development. Unplanned growth can damage or destroy natural resources and the economic health and quality of life of a community. It is very important, therefore, to continually promote and implement sound land-use planning.

The first step toward establishing sound land-use planning is to determine the values and goals of members of the community. Next, these values and goals must be evaluated against available natural resources, environmental health, geologic constraints, and the economic health of the community. Sustainable development must be sensitive to both the environment and the economy. Development provides citizens with jobs, goods, and services. It also provides cities and towns with tax revenues that pay for basic services such as clean water, waste disposal, health care, and educational services for its citizens. At the same time, development efforts can consume natural resources, spread pollution and occupy lands that may have been useful for alternative uses.

Figure 2 What assumptions could you make about the values and goals of the people in this community? What environmental factors may have shaped those values?

Land-use planning must guide development before it begins, because it can be very difficult and expensive, or even impossible, to convert land back to its original condition, before development. With poor planning, or no planning, development can take place on any piece of land without consideration

Geo Words

terrain analysis: the process of interpreting a geographic area to determine the effects of the natural and man-made features on a planned activity.

for other uses for that land. With good land-use planning, the land in and around a community is analyzed so that different areas are assigned for uses that take advantage of the location and/or characteristics of those areas. **Terrain analysis** is the process of evaluating land for specific uses.

People who conduct terrain analyses look at physical, chemical, and even legal aspects of land use. These analyses generally involve making different maps to summarize various characteristics of areas of land. The maps are divided into grids, and each grid is classified as to how a specific characteristic could affect development plans. For example, a map showing the locations of geological faults could be used to ensure that hospitals are not built on land areas that are susceptible to earthquakes. The same map may suggest, however, that faulted land is suitable for farming. A map of soil types could show areas that are not appropriate for farming because of poor soil texture or fertility, but those areas might be ideal for development of industrial parks. A different map, however, might show that adjacent areas have housing developments, so building industrial operations nearby would not be in the best interest of the residents' health.

Other terrain-analysis maps might summarize factors like access to highways, extent of floodplains, risk of landslides, habitats for endangered species, water content of sediments, or areas contaminated by previous land use. Terrain-analysis maps are often printed on clear plastic overlays so they can be put on top of each other. This can give a picture of all the factors affecting a particular area.

Check Your Understanding

1. In your own words, describe what a consensus is.
2. Select two characteristics that could be used in terrain analysis and explain why you think they would be useful in evaluating land for various uses in your community.
3. Why would scenic beauty be important in land-use planning?

Understanding and Applying What You Have Learned

1. When you were ranking land-use factors with your class:
   a) Which land-use factors were most important to you?
   b) Did you disagree with any parts of the group consensus?
   c) Why is it important for a community to go through this potentially confrontational process?
   d) What are the benefits and drawbacks of using the survey technique that your class used in this activity?
2. Identify a controversy in your community that involves land use.
   a) What are the arguments given in support of each side?
   b) What kind of consensus agreement can you recommend that considers the interests of both sides?
3. Describe a few examples from your community where land-use decisions may not have been wise. Examples of questionable land use could include:
   - Building a house on a fault.

- Developing farms on clay-rich soil.
- Placing buildings below steep, possibly unstable slopes.
- Locating industrial operations next to water resources.
- Damaging forested areas to obtain natural resources.

4. Using grids and traffic light color assignments, develop a map to guide land use. In this method, you divide your map into grids and color-code each grid according to the following classification:

   - red = stop
   - yellow = proceed with caution
   - green = safe

   For example, if the factor you are evaluating is watershed protection, you would put red in grid areas near water resources to prevent contamination. You can use one clear overlay for each traffic-light-colored map of a single factor. Then layer all of the overlays over the basic map of the community to see a composite picture of factors that may dictate land use. You may wish to contact state geological and agricultural agencies, forest and wildlife agencies, private industries, and organizations to help you find information.

5. Have each small student group create one traffic-light-color overlay map of a single factor that would be important to consider when choosing a location for a hospital. Each group should propose three different locations for the hospital from their overlay map alone.

   a) Layer all of the group overlay maps.
   b) On the basis of all of the overlays, are there any areas in your community where a hospital should definitely not be located? If so, why not?
   c) Are there any proposed hospital locations that were suggested by every group? If so, what makes the location(s) appropriate for that kind of land use?
   d) If there is more than one appropriate location indicated, have all of the groups come to a consensus on the single best location for a hospital. Which group(s) had the strongest opinion on the location? Why?

Preparing for the Chapter Challenge

In this activity you found that many different factors and opinions must be considered when establishing land-use plans. How can local government officials help a community reach a consensus on future land-use plans? Write a short paper in which you address this question.

Inquiring Further

1. **Public survey of your list of important factors**

   Prepare and conduct a mail or e-mail survey of engineers, geoscientists, lawyers, developers, elected officials, and/or members of planning committees in your community. Ask these people to review and comment on your list of very important factors to consider when determining land use in your community. Ask them to add any important factors that you might not have considered.

2. **Land-use scrapbook**

   Interview one or more city planners in or near your community. Ask about how decisions about development and growth are made. Compile a scrapbook of newspaper articles to chronicle land-use planning issues and events in your community. You can include agendas or minutes from planning committee or town council meetings.

3. **Local and regional meetings**

   Attend a local or regional planning board, zoning board, city council, or other government meeting to see firsthand how government officials address land-use problems. Interview a government official after a meeting and ask what process the board or council uses to reach a consensus on difficult issues.

Washington, D.C. before and after the development of the Mall. What factors needed to be considered when determining how to use the land in Washington, DC?

Earth Science at Work

ATMOSPHERE: *Physicians*
Doctors treat many people with asthma and other breathing difficulties. These conditions are often aggravated by air pollution.

BIOSPHERE: *Environmental Protection Agency Ecologists*
EPA ecologists monitor and ensure safe development of land.

CRYOSPHERE: *Ski-Resort Operator*
Many ski resorts rely on the production of artificial "snow" to extend their skiing season.

GEOSPHERE: *Town Planner*
A town planner works with the local government to recommend ways that land should be used for the commercial, industrial, or recreational needs of a community.

HYDROSPHERE: *Public Utilities Worker*
A community relies on an abundant supply of clean water. Water quality must be continuously monitored to ensure the health of the residents.

How is each person's work related to the Earth system and to land use?

Unit III

EarthComm®

Earth System Science in the Community

Earth's Fluid Spheres

1

Oceans ...and Your Community

1 Oceans ...and Your Community

Getting Started

Millions of people in the United States live within 50 km of the oceans. Even if you live very far from an ocean, it still can affect your day-to-day life. Refer to the photographs, and then answer the following questions:

- What do you see in the photographs that might suggest that water from the ocean affects communities in the interior of a continent? Explain.
- What Earth systems affect the movement or circulation of ocean water? Explain.

What do you think? One way to begin to think about answers to these questions is to think about the Earth as a set of closely linked systems. Look at the descriptions of the Earth systems at the front of this book, then write down your answers to these questions. Be prepared to discuss your ideas with your small group and the class.

Scenario

All communities would like to know how and when changes in the local economy might occur. Natural disasters, financial crisis, a change in product

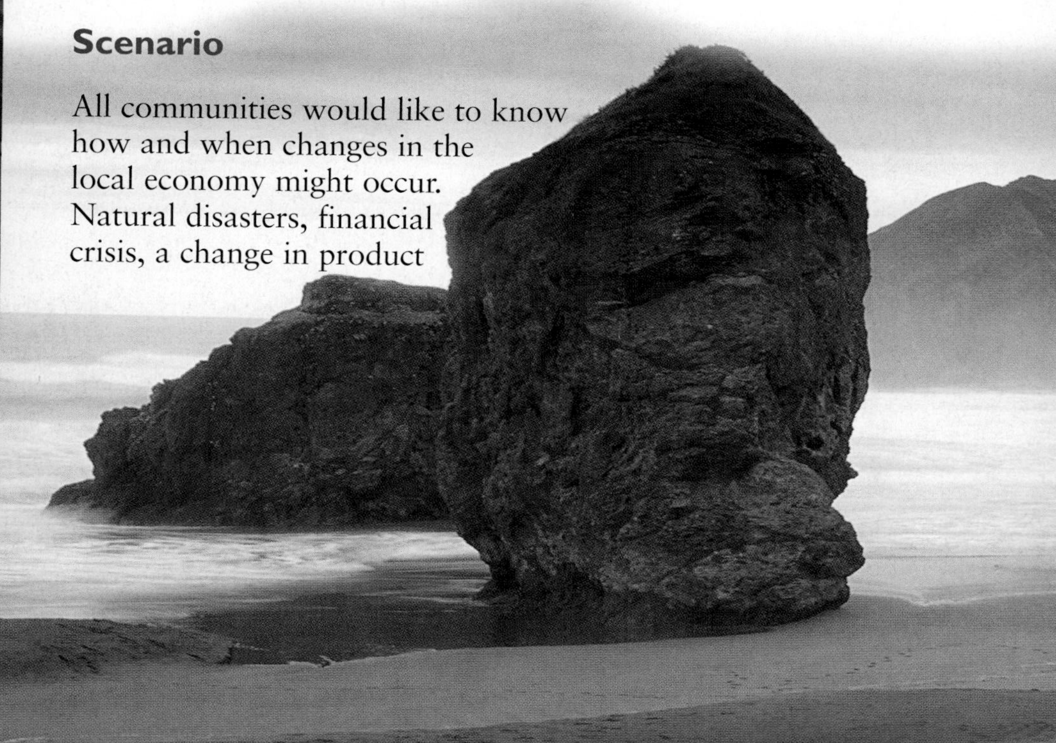

supply, or a variety of other factors could cause changes in the local economy. If communities know of changes in advance, they can better prepare themselves to minimize losses. Sometimes it is difficult to predict events that can affect a community. Sometimes it is easy. Sometimes it is not possible.
You have probably heard media reports about the effects of El Niño events. Leaders in your community want to know if an El Niño event could affect the climate and economy within your community. They have asked the *EarthComm* students in your class for help. They will host a statewide conference to discuss how to minimize the impact of El Niño, but only if your research reveals that El Niño events can actually affect your community. Can you help your community leaders to understand what an El Niño event is and how it affects communities?

Chapter Challenge

Your challenge is to help your community leaders decide whether to host a statewide conference on preparing for El Niño events. They have asked you to prepare a report that investigates the causes of ocean circulation and how they relate to El Niño and your community. Your report should:

- Explain surface circulation of the ocean and how it changes during an El Niño event.
- Explain the relationship between ocean circulation patterns during an El Niño event and atmospheric circulation.
- Describe any changes in weather or climate that have occurred or might occur in your community or state because of El Niño.
- Explain the relationship between ocean circulation patterns and the marine food chain and how this is affected by an El Niño event.
- Determine if El Niño–related changes in food production might affect your community and state.

Assessment Criteria

Think about what you have been asked to do. Scan ahead through the chapter activities to see how they might help you to meet the challenge. Work with your classmates and your teacher to define the criteria for assessing your work. Devise a grading sheet for the assessment of the challenge. Record all this information.

Make sure that you understand the criteria and the grading scheme as well as you can. Your teacher may provide you with a sample rubric to help you get started.

Activity 1 The Causes of Ocean Circulation

Goals

In this activity you will:

- Understand the effect of the wind on the movement of water at the ocean surface.
- Understand the effect of the Coriolis force on the movement of objects or materials when they are observed in a rotating system.
- Learn why the ocean is layered by temperature and density.
- Describe the meaning of pycnocline and thermocline in your own words.

Think about It

Ocean waters are constantly on the move. How they move influences climate and living conditions for plants and animals, even on land.

- Why does water move in the oceans?

What do you think? Record your ideas about this question in your *EarthComm* notebook. Be prepared to discuss your ideas with your small group and the class.

Investigate

Part A: Wind on the Ocean Surface

1. Prior to starting the investigation, predict the effect that wind speed has on water circulation.

   a) Record your hypothesis in your *EarthComm* notebook.

2. Place a fan on a tabletop. Set a pan on the tabletop so that its rim is level with the lower part of the fan, as shown in the diagram. Fill the pan almost full of water. Position the fan so that it blows across the water surface in the pan.

3. Turn the fan on the lowest speed, and let it blow across the water surface for a minute or two, long enough for the resulting water motion to settle into a steady state.

4. Put one drop of food coloring onto the water surface near the "upwind" edge of the pan.

5. Using a stopwatch, time how long it takes the mass of colored water to move from the edge of the pan to the center. Ignore the little waves generated by the wind on the water surface.

   a) Measure and record the distance traveled, and the time taken.

   b) Calculate the speed of the water by dividing the distance traveled by the time of travel. Record the speed in your notebook.

   c) In your notebook, describe the pattern of water motion in the pan that is revealed by the travel of the colored water. You can repeat the experiment by waiting until the coloring is uniformly mixed in the water, and then adding another drop of colored water at the surface.

6. Repeat Steps 3–5 for the medium speed of the fan, and then for the highest speed.

   a) Summarize your results.

   b) How do your results compare to the hypothesis that you developed at the start of the investigation? Explain.

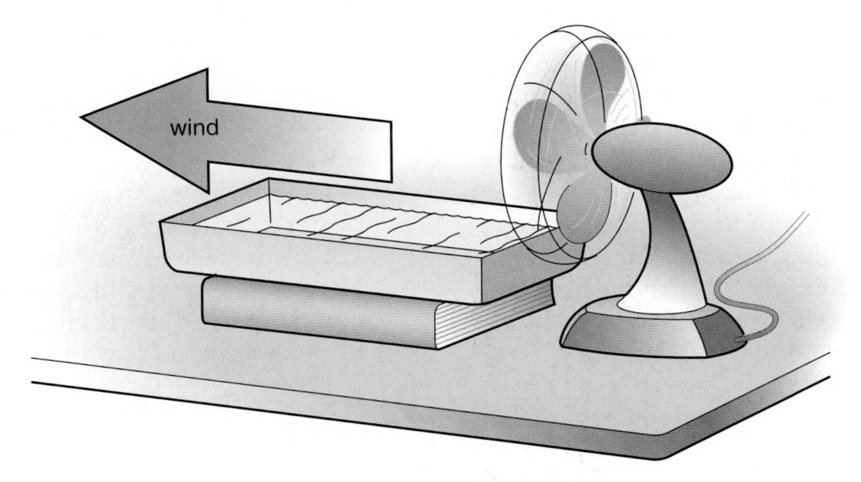

Be sure that the fan has an adequate guard around the blade so that fingers cannot be put near the blade. Do not put any items into the guard. Keep water away from the fan, especially the motor housing. Wipe up any spills immediately.
Wear eye protection. Wash your hands when you are done.

Part B: The Coriolis Effect

1. Read Steps 2–5 below. Before you do the investigation, predict the path that the pen will trace.
   a) Draw two rectangles on a sheet of paper. Indicate the direction of the posterboard rotation with an arrow. Record your predictions by drawing them in the appropriate rectangle.
   b) Explain your drawings. Why do you think that the pen will take the path that you predict?
2. At the chalkboard or a blank wall, one student holds a piece of posterboard against the wall with one thumb. The thumb should be pressed tightly enough against the center of the posterboard to prevent it from slipping down, but not so tightly that the posterboard cannot be rotated.
3. A second student stands next to one edge of the posterboard and rotates it slowly, and as nearly as possible at a constant speed, with a "hand-over-hand" motion. Try for a rotation rate of about one full turn in five seconds.
4. A third student stands next to the other edge of the posterboard. With a marker pen, he or she slowly draws a straight line on the posterboard while it is rotating. Important: the trick is to try to ignore the motion of the posterboard, and move the pen downward in a straight line relative to the wall or chalkboard beneath. It would be good to practice with the tip of a finger before using the marker pen.
   a) In your notebook record the shape of the line and the direction of rotation.

5. Turn the posterboard over, and repeat Steps 2–4 while rotating the posterboard in the opposite sense.
   a) Record in your notebook the shape of the line and the direction of rotation.
   b) How do your results compare with your predictions? Explain.
   c) How do you think the shape of the line would vary with:
      - the speed of rotation of the posterboard?
      - the speed of travel of the pen point?
   d) Which sense of rotation of the posterboard corresponds to the Earth's surface in the Northern Hemisphere, and which sense corresponds to the Earth's surface in the Southern Hemisphere?
   e) If time permits, test your ideas from 5(c).

Reflecting on the Activity and the Challenge

You observed that a wind blowing across a water surface creates a surface current that moves in the direction of the wind. You also observed that the wind-driven current affects only the surface layer. In the second part of the activity, you found that it was difficult to draw a straight line on a rotating sheet of posterboard. The Coriolis effect makes for some very strange phenomena on the Earth's surface, including ocean currents. Knowing this, would you feel comfortable predicting ocean circulation patterns just from the results of your investigations?

Digging Deeper

THE NATURE OF THE OCEANS

Oceans of the World

You are probably familiar with three large oceans on Earth: the Pacific Ocean, the Atlantic Ocean, and the Indian Ocean. Each of these three oceans is constricted at its northern end but wide open at its southern end, as shown in *Figure 1*. Each is connected at its southern end to a fourth major ocean, called the Southern Ocean or the Antarctic Ocean. The Southern Ocean is different from the other three because it is continuous all around the Earth instead of being bordered on the east and west by large continents.

The open ocean is mostly very deep: the average deep ocean depth is four or five kilometers. The bottom of the deep ocean is a place of inky blackness, eternal cold, and water pressures that are unimaginably large. In another sense, however, even the deep ocean is shallow relative to the surface area covered by the oceans. Suppose that an ocean like the Atlantic or Pacific were the size of a football field or a soccer field. You would be wading around in an ocean that is no deeper than your ankles! Keep that fact in mind as you work through the activities in this chapter.

The Warm Ocean and the Cold Ocean

The ocean is layered by temperature. There is an upper layer of warm water and a much thicker deep layer of cold water. At low latitudes, the change between these two layers is at a depth of a few hundred meters. Above that

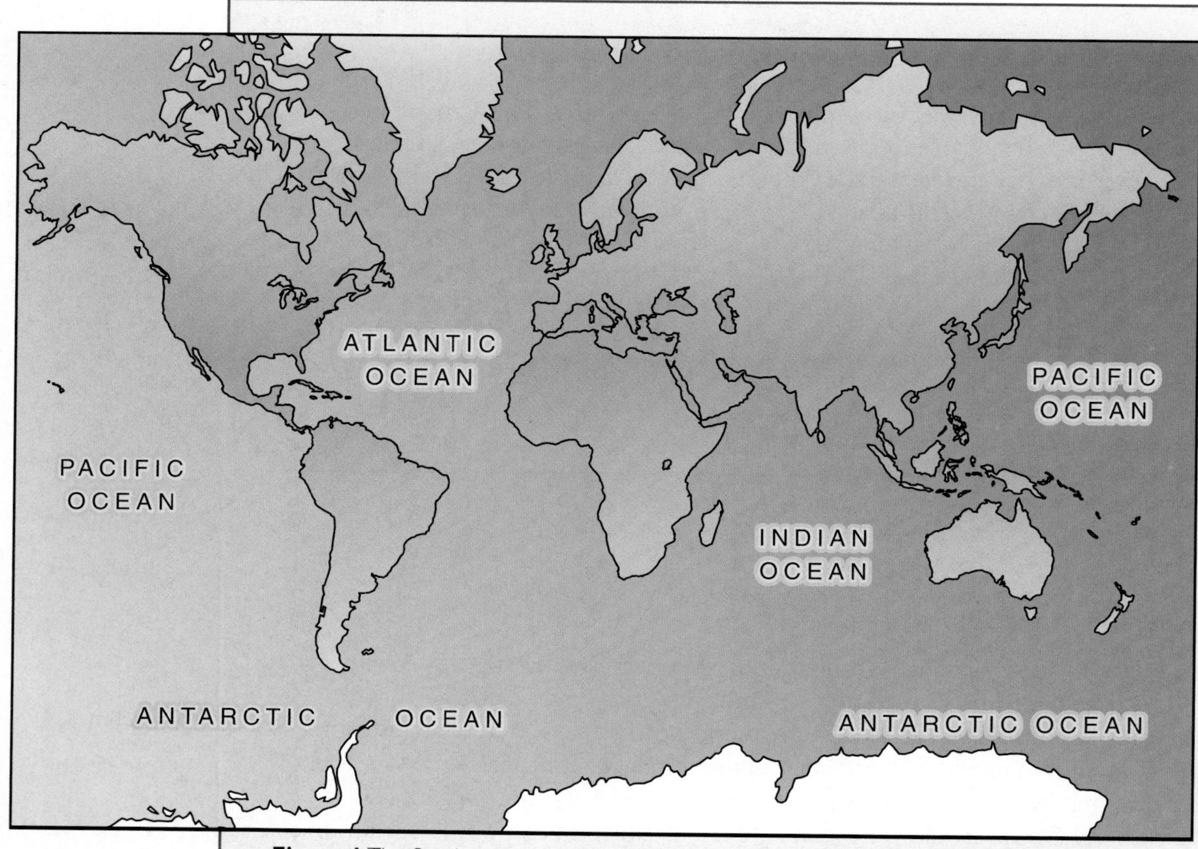

Figure 1 The Southern or the Antarctic Ocean is continuous all around the Earth.

Geo Words

thermocline: the zone of rapid change from warm water to cold water with increasing depth in the ocean.

pycnocline: a layer of water in the ocean characterized by a rapid change of density with depth.

depth the oceans are as warm as 25°C over large areas at low latitudes. Below that depth, the ocean is much colder. In the lowest parts of the ocean, the temperature is not much above 0°C, even at the Equator!

Between about 200 m and 1000 m below the surface, the temperature of the water decreases sharply with depth. This zone of rapid change from warm water to cold water with increasing depth is called the **thermocline**. It is a larger-scale version of the difference in temperature that you may have felt when treading water in a lake. Water around the upper part of your body might feel warm, while the water near your feet feels cooler. Because cold water is more dense than warm water, this transition zone is also called the **pycnocline** (*pycno-* means density). The reason that the pycnocline is also the thermocline is that the density of seawater depends mainly on the temperature of the water.

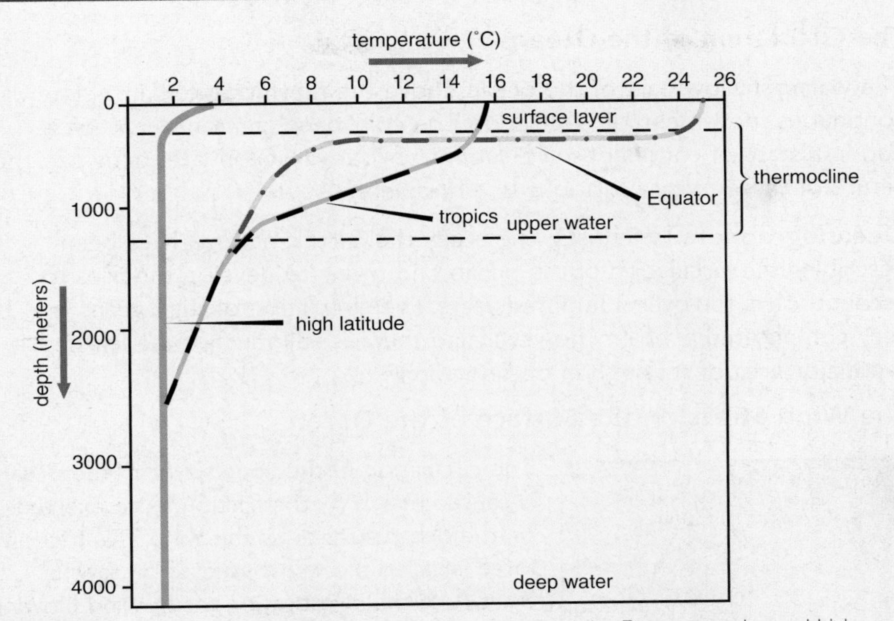

Figure 2 Sample temperature–depth profiles for waters at the Equator, tropics, and high latitudes (poles).

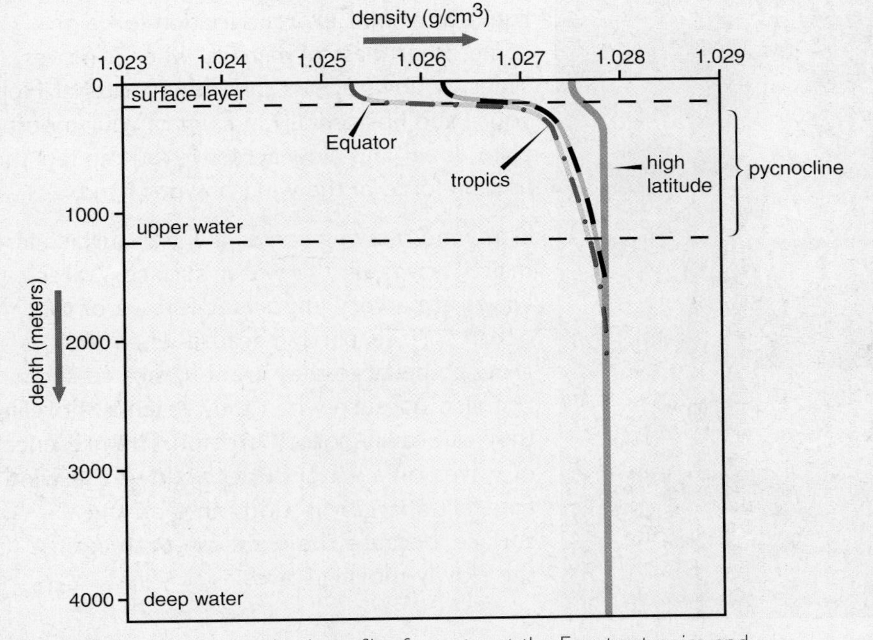

Figure 3 Sample density–depth profiles for water at the Equator, tropics, and high latitudes (poles).

Geo Words

oceanographers: scientists who study the Earth's oceans.

wind stress: the friction force exerted on the ocean surface by the wind.

The Circulation of the Oceans

The warm, shallow part of the ocean, above the thermocline, is in a state of continuous movement, or circulation. The cold, deep part of the ocean is also in a state of continuous circulation. As you will see in later activities, the nature of these two circulations is very different.

Oceanographers (scientists who study the Earth's oceans) have been describing the circulation of the ocean, and trying to develop theories to account for it, for over a hundred years. The circulation of the oceans is a very complicated problem, however, and there is still much to be learned. It is a major area of research in oceanography.

The Wind Stress on the Surface of the Ocean

The circulation of the upper layer of the ocean is caused mainly by the friction force exerted on the ocean surface by the wind. This friction force is called the **wind stress**. You saw in **Part A** of the investigation that a wind blowing across a water surface sets the water in motion. This should make good sense to you, because any material, whether it's a gas, a liquid, or a solid, exerts a friction force on some other mass of material when it moves, slides, or flows across that other material. Hold your hand horizontally in front of your mouth, palm down, and blow across it. You can feel the friction force of the wind on your hand.

The wind blowing across a water surface also makes waves on the water surface. Water waves are a very important feature of the ocean surface, partly because when they grow to be large they are a hazard to ships, and also because when they reach a shoreline they can cause coastal erosion. The presence of waves on a water surface allows the wind to exert a larger friction force on the surface, because the wind can push against the slowly moving waves.

Figure 4 Surface winds on the ocean can produce waves that are hazardous to boats.

The Coriolis Effect

One of the basic principles of physics is that a mass of material moves in a straight line at constant speed unless it is acted upon by some force. When water in the ocean is set into motion, it tends to move in a straight line. What makes this seemingly simple idea very complicated, however, is that the Earth is rotating under the water while it's moving in a straight line! You, the observers of the water movement, are rotating with the Earth, so it looks to you as though the water is moving in a curved path. This is what you saw in **Part B** of the investigation. The pen point was moving in a straight line, but to an ant rotating with the posterboard, it seemed to be moving along a curved path.

It takes a force to make a mass of material move in a curved path instead of a straight line. The force that seems to make the water or the pen point move in a curved path is called the **Coriolis force**. It's not really a force, but to someone rotating with the Earth or the posterboard, it seems like a force. Scientists use the word "counterintuitive" for things like this: your intuition tells you one thing, but the real-world situation shows you something very different. The overall circulation of the oceans depends in a very fundamental way on the Coriolis effect. You will see this in more detail in later activities in this chapter.

Geo Words

Coriolis force: the apparent force caused by the Earth's rotation which serves to deflect a moving body on the surface of the Earth to the right in the Northern Hemisphere and to the left in the Southern.

Check Your Understanding

1. Examine *Figures 2* and *3* on page F9. Compare and contrast the thermocline and the pycnocline. How are they similar? How are they different?
2. How does the wind cause surface currents in the ocean?
3. Why are waves an important feature of the ocean surface?
4. Describe the Coriolis effect in your own words.

Understanding and Applying What You Have Learned

1. Based on the observations you made in **Part A** of the investigation, do you think that waves always travel in the same direction as currents? Explain.
2. How would you expect waves to change over time when:
   a) wind speed increases?
   b) wind blows over a part of the ocean for a long time?
3. Refer to *Figures 2* and *3* (thermocline and pycnocline) to answer the following questions:
   a) What is the temperature of ocean surface water at the Equator, in the tropics, and at high latitudes?
   b) In what part of the ocean is the thermocline best developed? Least developed?
   c) In what part of the ocean is the pycnocline best developed? Least developed?
   d) Use your understanding of the relationship between temperature and density in the ocean to explain your answers to questions 3 (b) and (c).

Preparing for the Chapter Challenge

Use the information you have gained from this activity to write an introduction that explains the forces that drive ocean circulation. Think about the ultimate energy sources for these forces and try to think of how a change in these sources might affect ocean circulation.

Inquiring Further

1. **Careers in oceanography**

   Would you like to study the oceans for a career? To learn more about the different fields of oceanography and what a professional oceanographer does at work, visit the *EarthComm* web site.

2. **Harnessing wave power**

   Energy produced by moving water can be used to generate electricity, as at hydroelectric power plants located along rivers. The energy created by ocean waves exceeds that produced by rivers, but difficulties lie in efficiently harnessing this energy. Research the benefits and difficulties associated with generating electricity using the energy of ocean waves. Where is the potential for using ocean wind-power systems the greatest? To get started, visit the *EarthComm* web site.

3. **Scale diagram of the ocean floor**

   You were told in the **Digging Deeper** portion of this activity that oceans are shallow in comparison to their width. Investigate this claim by making a 1:1 scale profile of water depth across an ocean. A 1:1 scale profile is a cross section in which the horizontal scale is equal to the vertical scale. You will need a world map that indicates ocean bathymetry (water depth), a metric ruler, and graph paper. You may need to tape sheets of graph paper together.

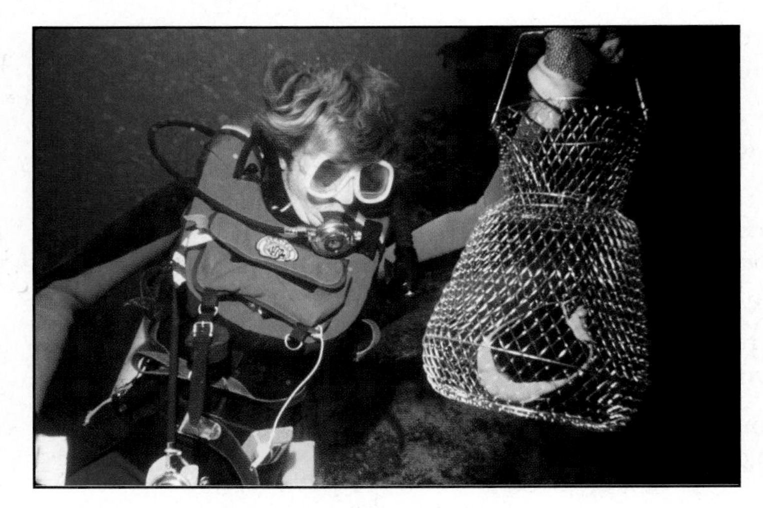

Activity 2 The Deep Circulation of the Ocean

Goals

In this activity you will:

- Understand how water temperature affects circulation within a body of water.
- Understand how salinity affects circulation within a body of water.
- Understand the meanings of ocean currents and water mass.
- Understand the nature and causes of water circulation in the deep ocean.
- Investigate convection cells and their relationship to deep ocean circulation.

Think about It

The circulation of oceans has been likened to a great conveyor. As you learned in **Activity 1**, the circulation of the surface layer of the ocean is caused mainly by wind stress.

- What makes deep ocean waters move?

What do you think? Record your ideas about this question in your *EarthComm* notebook. Be prepared to discuss your response with your small group and the class.

Investigate

Part A: Temperature and Circulation

1. Prepare "seawater" by adding 35 g of salt per liter of tap water. Stir until the salt is dissolved. Pour the contents into a transparent plastic container about the size of a shoebox. Continue adding one liter of seawater at a time until you fill the plastic container halfway with the seawater. This will represent your "ocean."
2. Predict what you think will happen if you add cold seawater to the ocean water. What do you think will happen if you add hot seawater to the ocean water?
   a) Record your predictions and explain the reasons behind your predictions.
3. Prepare 250 mL of cold seawater. Use refrigerated water or water that has been kept in a cooler with ice. Add a few drops of blue food coloring.
4. Prepare 250 mL of hot seawater. Use hot water from a tap. Add a few drops of red food coloring.
   a) Why is it important to keep the salt concentration of the ocean water in the plastic shoebox, the cold seawater, and the hot seawater equal?
5. Measure the temperature of the three different containers of seawater.
   a) Record these values in a data table.

Hot water should not be so hot that it does not present a hazard if touched. Clean up spills immediately.

6. Very slowly and carefully pour the hot and cold seawater in at opposite ends of the plastic box. Members of your group should be in a position to view the contents of the plastic box from the side. (Note: Sometimes seeing this effect is tricky. You might try slowly pouring the colder or hotter seawater onto a sponge floating in the plastic box.)
   a) Make a diagram of what was observed when the colored seawater was poured in. This should be a side view or cross section of the plastic box. Label the diagram clearly.
7. Use the observations to answer the following questions:
   a) How did your results compare to your predictions? Explain any differences.
   b) What do you think would happen if you made the hot water hotter and the cold water colder in this investigation? Explain.

Part B: Salt Content and Circulation

1. Fill a plastic container halfway with seawater (35 g of salt per liter of tap water). This will represent normal ocean water.
2. Work with members of your group to determine how to prepare solutions of different salt concentrations:
   - 250 mL of a solution that is 1/4 as salty as the normal ocean water.
   - 250 mL of a solution that is four times as salty as the normal ocean water.

a) Write down how you would make the solutions. When the teacher has approved your method, prepare the two solutions.

b) The solutions you make and add to the ocean water should be the same temperature as the ocean water. Why is it important to keep the temperature the same during this investigation?

3. Add a few drops of blue food coloring to the more saline solution and a few drops of red food coloring to the less saline solution.

4. Predict what will happen when each solution is added to the "ocean" water.

   a) Record your prediction. Be sure to include the reason for each prediction.

5. Slowly and carefully pour the two solutions at opposite ends of the container at the same time. Observe what happens.

   a) Make a diagram of what was observed when the colored seawater was poured in. This should be a side view or cross section of the plastic container. Label the diagram clearly.

6. Use your observations to answer the following questions:

   a) What would happen if the high-salinity water had been made even more saline and the low-salinity water had been made even less saline? Explain.

   b) Which do you think would sink below the other: warm salty water or cold fresh water? Explain your answer.

   c) How did your results compare to your predictions? Explain any differences.

Part C: "Thought Experiment" on Convection Cells

Scientists like to run "thought experiments" in their heads to get an idea about how a real experiment might work. Sometimes thought experiments lead to surprises, when the experimenter has ignored some important factor! Here is a two-part thought experiment for you to look at the importance of convection in deep-ocean circulation. (Convection is the circulation in a fluid from place to place that is caused by differences in fluid density.)

1. Imagine a simple tank, like an aquarium, that is filled with seawater. Now suppose that there is a source of heat beneath one end of the tank and a source of cooling beneath the other end of the tank, as shown in the diagram below.

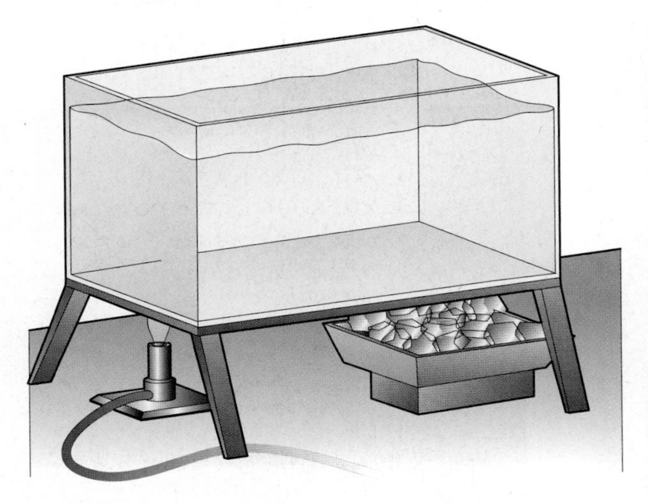

Do not attempt to try the experiment as a real setup.

a) What will happen to the warm water at the heated end of the tank?

b) What will happen to the water that flows at the surface of the tank at the cooled end?

c) Draw a diagram to show the convection cell that results, using arrows to indicate the direction of circulation.

d) What causes the circulation of the water?

2. Now imagine setting up the same tank, but this time putting the source of heating at the surface at one end of the tank and the source of cooling at the surface at the other end. Think of this as a very simple model of the ocean.

   a) At first, what happens to the cooled water at the surface?

   b) What will happen to the temperature of the water in the lower part of the tank as time goes by? (Ignore the freezing of freshwater ice near the source of cooling.)

   c) With time, what will happen to most of the water in the tank?

   d) With time, where will there still be some motion of water in the tank?

Reflecting on the Activity and the Challenge

You observed in this activity that colder water sinks below warmer water. Warm water tends to lie as a layer on top of cooler water. Water with high salt concentration sinks below water with lower salt concentration and flows across the bottom of its container. Similarly, fresh water or water that is less salty floats atop "normal" seawater. You should now be able to predict the circulation patterns created when changes in temperature and salt concentration are made to different parts of the water. This will help you understand and explain circulation patterns that occur in the ocean.

Digging Deeper

OCEAN CURRENTS

Seawater Temperature, Salinity, and Density

Just like what you observed in the investigation, water in the oceans can sink, rise, or flow horizontally relative to the surrounding water. The horizontal movements of ocean water are called **ocean currents**. Ocean currents carry matter and energy from one part of the ocean to another. They can be local, or they can be as broad as the oceans themselves. Some ocean currents, called surface currents, flow only in the surface layer of the ocean. Other currents flow in the deep ocean. The surface currents flow much faster than the deep currents, but the deep currents affect a much greater thickness of the ocean. Surface currents and deep currents have very different causes, as you saw in this activity and will see in the next.

The movement of deep ocean water is controlled by the density of the water. The density of seawater is determined by two factors: temperature and **salinity**. Both are important, although temperature has a much greater effect than salinity on seawater density.

Temperature – The Sun heats the Earth's surface unequally, because the Sun's rays strike the Earth from nearly overhead at low latitudes but at a very low angle at high latitudes. In winter, the Sun doesn't shine at all north of the Arctic Circle and south of the Antarctic Circle. The warm surface water at low latitudes stays at the surface because it is less dense than the deeper water. Surface seawater that is chilled at high latitudes, especially in winter, increases its density. When a mass of water is more dense than surrounding water, it sinks under the pull of gravity, like the cold water you observed sinking in the investigation.

Salinity – Ocean water contains dissolved salts. The concentration of dissolved salts in seawater, expressed as parts per thousand, is called salinity. The salinity of ocean water is approximately 35 parts per thousand (35 g of salt per 1000 g of ocean water). The higher the salinity of ocean water is, the more dense the water is. When the salinity of a mass of water is increased, the water sinks relative to surrounding masses of water. In some areas of the ocean, evaporation from the ocean surface is greater than precipitation onto the ocean surface. This increases the salinity, and the water tends to sink. In other areas of the ocean, precipitation is greater than evaporation. This decreases the salinity, and the water stays at the surface. The salinity of ocean water is also lowered where fresh river water flows into the ocean. At high latitudes, where the temperature of the air above the ocean is

Geo Words

ocean current: a predominantly horizontal movement of ocean water.

salinity: the concentration of dissolved salts in seawater, expressed as parts per thousand.

far below freezing, seawater freezes to form sea ice. The formation of sea ice removes fresh water from the ocean. This increases the concentration of salt in the water left behind. Because ocean water at high latitudes is also very cold, it is the most dense water in the world ocean.

Geo Words

water mass: a large region of water within the ocean with about the same temperature and salinity of water.

Water Masses

The ocean is a single body of water, but it is not the same everywhere. Both the temperature and the salinity of ocean water vary from place to place. Large regions of the oceans have water with about the same temperature and salinity. Such regions are called **water masses**. Oceanographers define water masses mainly by their temperature and salinity characteristics.

In the surface layer of the ocean, down to depths of 100 to 200 m, winds mix the water vertically. This mixing causes the temperature and salinity to be about the same throughout the layer. In most areas of the deep ocean, however, there is no vigorous vertical mixing, so individual water masses retain their distinctive properties for a long time.

Water masses are named for where they form. Antarctic Bottom Water and North Atlantic Deep Water are two examples. Antarctic Bottom Water is formed in the Southern Hemisphere near Antarctica by cooling of the near-surface water and increase in salinity by formation of freshwater ice. The water then sinks and flows northward away from Antarctica for thousands of kilometers along the ocean floor, all the way to high northern latitudes.

Figure 1 Why are the water masses formed off the coast of Antarctica so dense?

Both North Atlantic Deep Water and Antarctic Bottom Water play an important role in the world ocean circulation. North Atlantic Deep Water forms in the region around Iceland. It has an average salinity of 34.9 parts per thousand and a temperature of about 3°C. It takes as long as 1000 years for a parcel of North Atlantic Deep Water to sink, move south, circulate around Antarctica, and then move northward to the Indian or Pacific Oceans. Antarctic Bottom Water is the most dense of all deep ocean water masses. It has a salinity of 34.7 parts per thousand and a temperature of –0.5°C. It forms at the edge of the Antarctic continent and flows under all other water masses as it moves northward along the deepest part of the ocean bottom.

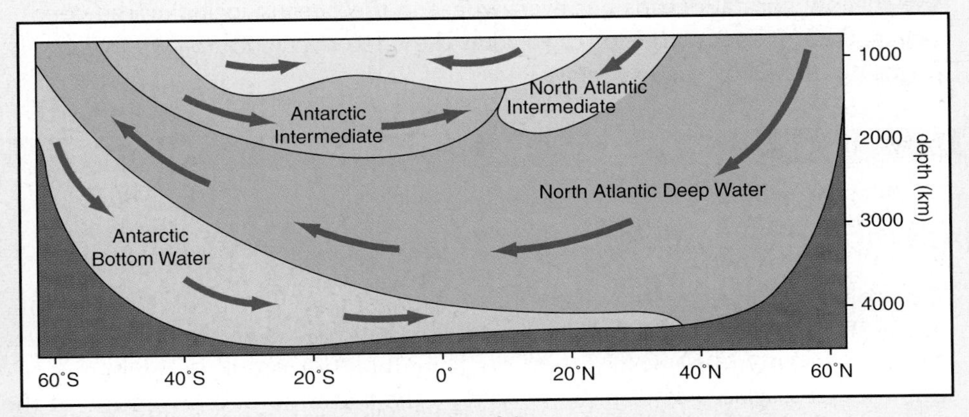

Figure 2 Circulation cell formed by movement of deep water masses.

What Causes the Deep Circulation?

The slow movements of the deep water masses form a gigantic and very complicated circulation cell, as shown in *Figure 2*. The cold waters that form near the surface at high northern and southern latitudes flow toward the Equator, and the deep waters gradually rise up toward the surface at low latitudes, where they are warmed and flow back to high latitudes to complete the loop. Oceanographers are not certain about the overall flow rate of this circulation, but it might be as high as 20 million cubic meters per second. That's 100 times the flow rate of the Amazon River, the largest river on Earth by flow rate! One cycle on this "conveyor belt" takes between 500 and 2000 years.

It might seem natural to you that the oceans, which are heated at the surface at low latitudes and cooled at the surface at high latitudes, form a

gigantic convection cell. You saw in **Part C** of the investigation, however, that a tank of water that is heated from the top at one end and cooled from the top at the other end does not form an active convection cell. Oceanographers have realized for a long time that another explanation is needed to account for the deep circulation of the oceans. What is needed is a way to mix the cold water in the deep ocean at low latitudes with the warmer water that lies above. Then, because warm water is less dense than cold water, the mixed water rises toward the surface, to complete the circulation cell. The problem is that oceanographers are still not sure about what causes the mixing. One idea that is becoming widely accepted is that the tides cause the mixing. The tides cause movement of water not just in the shallow coastal oceans but everywhere in the oceans, including the deep ocean. Does it seem strange to you that the deep circulation of the ocean might be caused by the moon?

Check Your Understanding

1. What are the two factors that determine the density of a water mass?
2. Thinking only about the effect of temperature, where would you expect water to be more dense—near the poles or near the Equator? Explain.
3. Explain what an ocean current is in your own words.
4. Why do individual water masses within the oceans retain distinctive physical properties for long periods of time?

Understanding and Applying What You Have Learned

1. The properties of different water masses in the Atlantic and Pacific Oceans are shown below, and simplified cross sections of each ocean are shown on the following page. Use your understanding of the factors that influence ocean circulation to fill in the names of the water masses on a copy of the figures.

Atlantic Ocean			
Water mass	**Temp (°C)**	**Salinity (ppt)**	**Density (g/cm³)**
Antarctic Bottom Water (AABW)	–0.5	34.7	1.0278
Antarctic Circumpolar Water (AACP)	0	34.7	1.0272
Antarctic Intermediate Water (AAIW)	4	34.2	1.0269
North Atlantic Deep Water (NADW)	3	34.9	1.0276

Pacific Ocean			
Water mass	**Temp (°C)**	**Salinity (ppt)**	**Density (g/cm³)**
North Pacific Intermediate Water (NPIW)	5	34.2	1.027
Antarctic Circumpolar Water (AACP)	0	34.7	1.0272
Antarctic Intermediate Water (AAIW)	4	34.2	1.0269
North Pacific Deep Water (NPDW)	2	34.6	1.0273

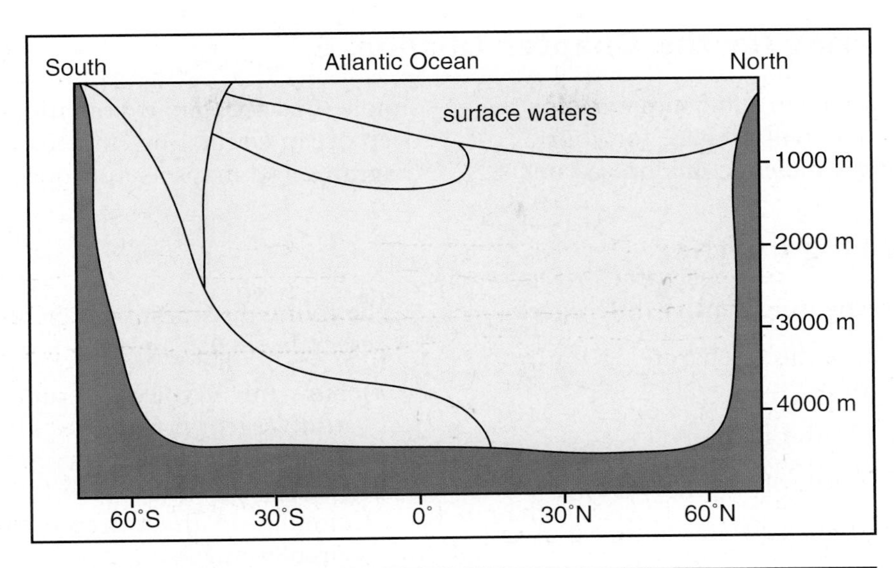

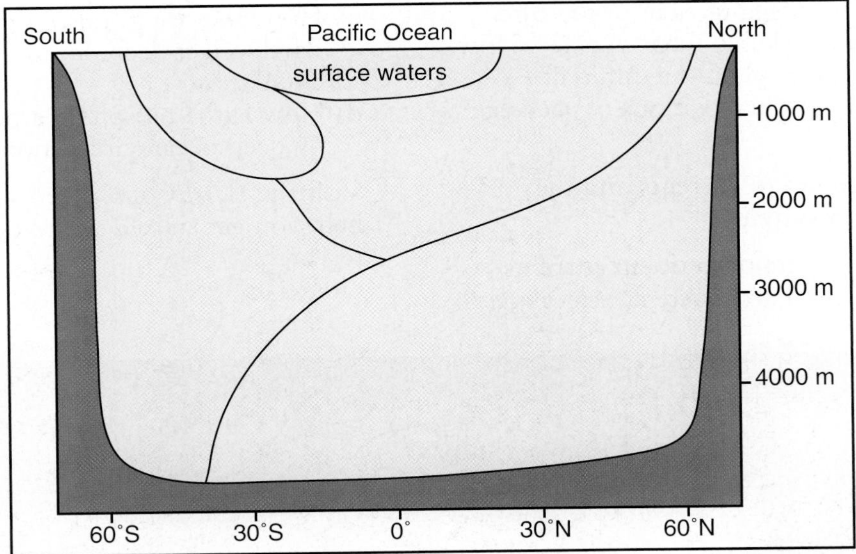

2. Temperature and salinity are the two primary factors controlling the density of a water mass.
   a) Propose several mechanisms that could cause these factors to change, thereby changing the density of the water mass.
   b) How would these changes affect ocean circulation patterns?
3. Does wind affect deep ocean circulation? Explain.
4. Explain how the tides may have an impact on deep water circulation in the oceans.

Preparing for the Chapter Challenge

Write a background summary for the report you will prepare for your **Chapter Challenge.** Include a concise, simple, but accurate explanation of deep ocean circulation. Include diagrams and maps as appropriate.

Inquiring Further

1. **Investigating density differences**

   Prepare three separate 250-mL solutions:

   - 250 mL of water
   - 250 mL of water with 8 g of salt
   - 250 mL of water with 4 g of salt

   Place an egg in each of the three liquids. Sketch what you see. How do you explain the difference between the positions of each egg?

2. **Deep ocean currents and the Earth system**

   Investigate deep ocean currents and the Earth system. Select one of the following questions for further research:

   a) How might volcanic eruptions at divergent boundaries affect deep ocean currents?
   b) How does atmospheric circulation affect deep ocean circulation?
   c) How does the rotation of the Earth affect deep ocean circulation?
   d) How is the biosphere affected by deep ocean circulation?

   Visit the *EarthComm* web site to help you get started with your research.

Activity 3 The Surface Circulation of the Ocean

Goals

In this activity you will:

- Understand the general paths of surface ocean currents.
- Determine the average speed of surface ocean currents in the Pacific Ocean.
- Understand how the winds drive surface ocean circulation.
- Understand how surface ocean circulation patterns change during an El Niño event.
- Understand how ocean circulation can influence climate.
- Understand how the Coriolis effect influences ocean circulation patterns.

Think about It

Horizontal surface currents move vast quantities of ocean water. Surface currents transfer heat from one region to another and moderate Earth's climate.

- How do you think Earth's climate would change if there were no surface currents?

What do you think? Record your ideas about this question in your *EarthComm* notebook. Be prepared to discuss your response with your small group and the class.

Investigate

Part A: Mapping Surface Currents

1. On May 27, 1990, 40,000 pairs of sneakers were washed overboard from a ship bound from Korea to Tacoma, Washington. In the months following this northern Pacific Ocean spill, beachcombers found shoes at the locations reported in the table below.
   a) Plot the data from the table on the map of the North Pacific.
   b) From the locations of the shoes that were found, draw arrows on your map to indicate the direction of the North Pacific current.

Day	Event	Latitude	Longitude	Distance since last discovery (miles)
1	Location of spill	48°N	161°W	—
190	Location of first shoe discovery	48°N	125°W	1649
249	Location of next shoe discovery	46°N	125°W	138
310	Location of next shoe discovery	45°N	124°W	84
340	Location of next shoe discovery	43°N	125°W	147
730	Location of next shoe discovery	20°N	157°W	2439
1200	Location of last shoe discovery	36°N	140°E	3931

2. Calculate the rate of "sneaker drift" for the first 190 days in miles per day.

   a) Record your calculations and rate in your notebook.

   b) What is the rate of drift southward along the coast between the Day 190 location and the Day 340 location?

3. Use your calculated rate of drift and the direction of the current.

   a) Place an X on a copy of the map where sneakers might be discovered in the future. Estimate the day of sneaker arrival. Write this number next to your X.

4. A few sneakers were found along the southern coast of Alaska. Look at a world map on the next page that shows ocean currents.

   a) Explain how this could be possible.

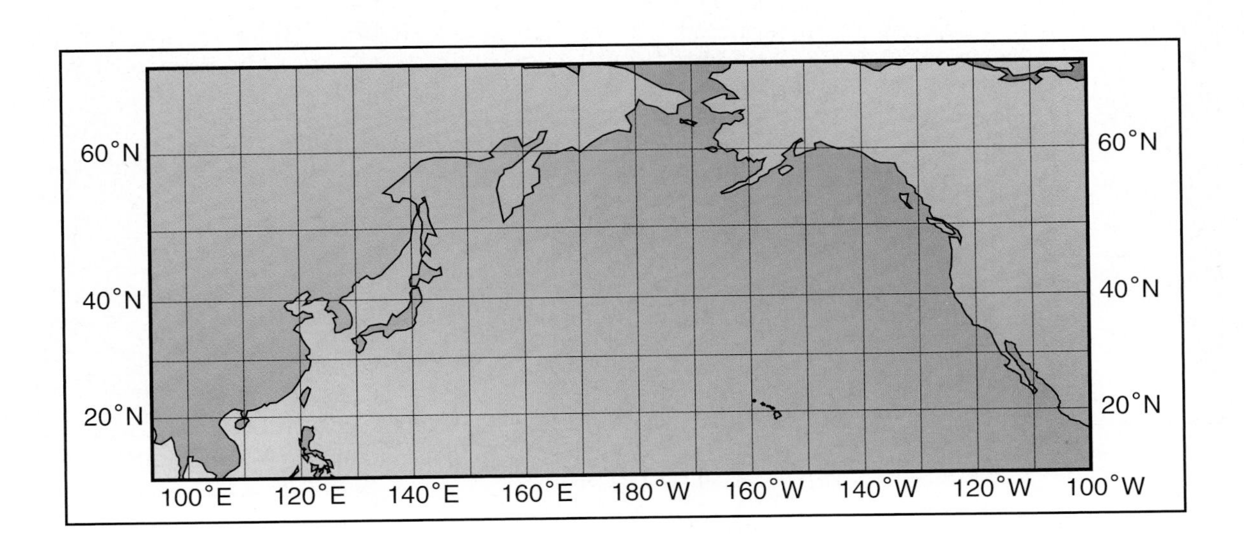

Part B: Tracing Wind and Surface Ocean Currents

1. Look at the map of surface ocean circulation.

   a) Which current(s) did you map out in the "sneaker spill" in **Part A**?

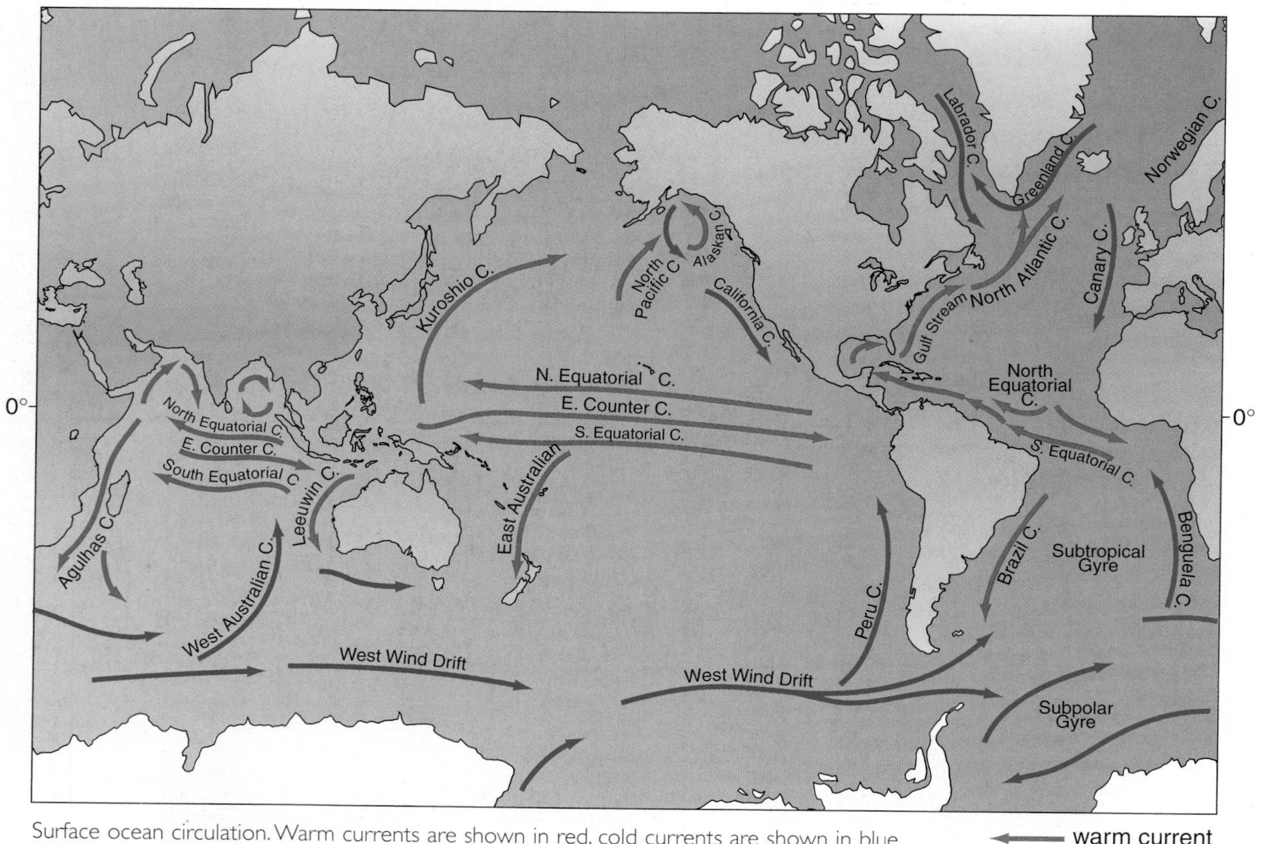

Surface ocean circulation. Warm currents are shown in red, cold currents are shown in blue.

2. Compare the map of normal wind patterns with the map of surface ocean circulation.

   a) How are the maps alike? How are they different?

   b) What would the map of ocean circulation look like if there were no continents?

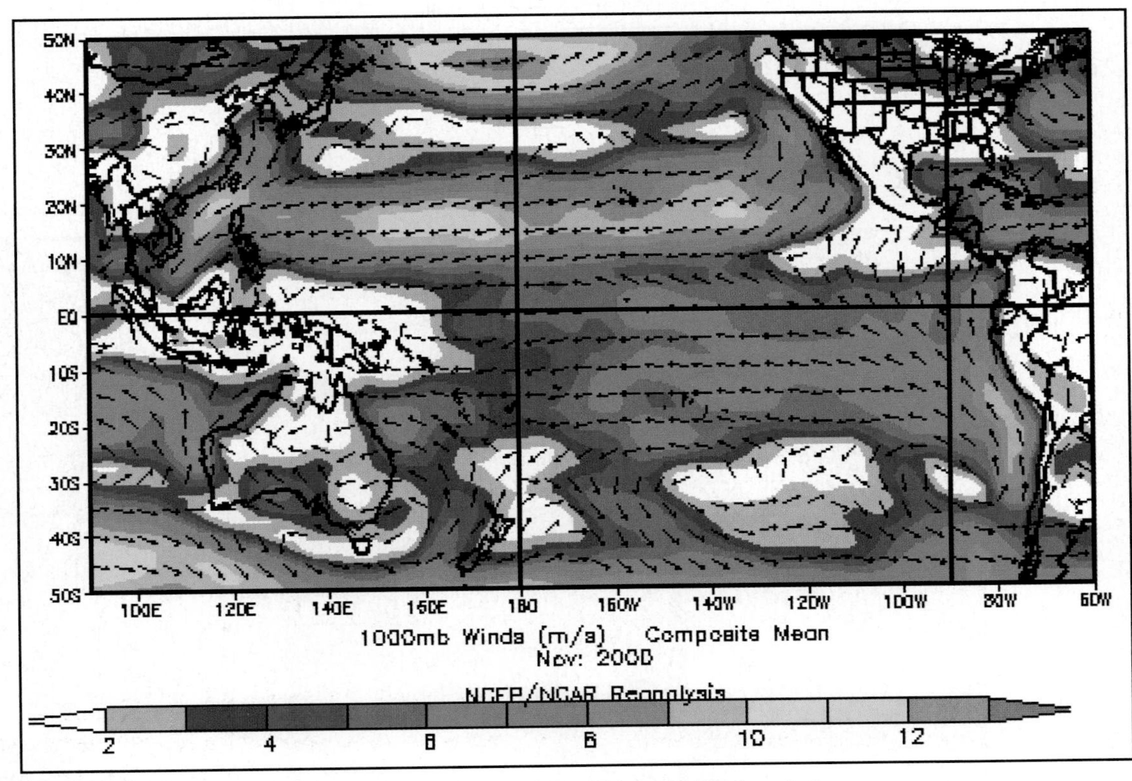

Map of vector wind direction for November 2000, a "normal" (non–El Niño) period.

3. An El Niño event is caused by a change in the ocean–atmosphere system around the tropical Pacific. As you will soon learn, this change can have important consequences for weather around the globe, including in your community. Below is a map of wind direction at the height of the 1997–1998 El Niño event. Compare the El Niño wind map with the map of normal wind patterns and answer the following questions:

   a) How do the two maps differ?

   b) Using what you know about how winds affect surface ocean circulation patterns, how do you think the pattern of surface ocean currents changes during an El Niño event? Indicate this on a copy of the map of surface ocean circulation.

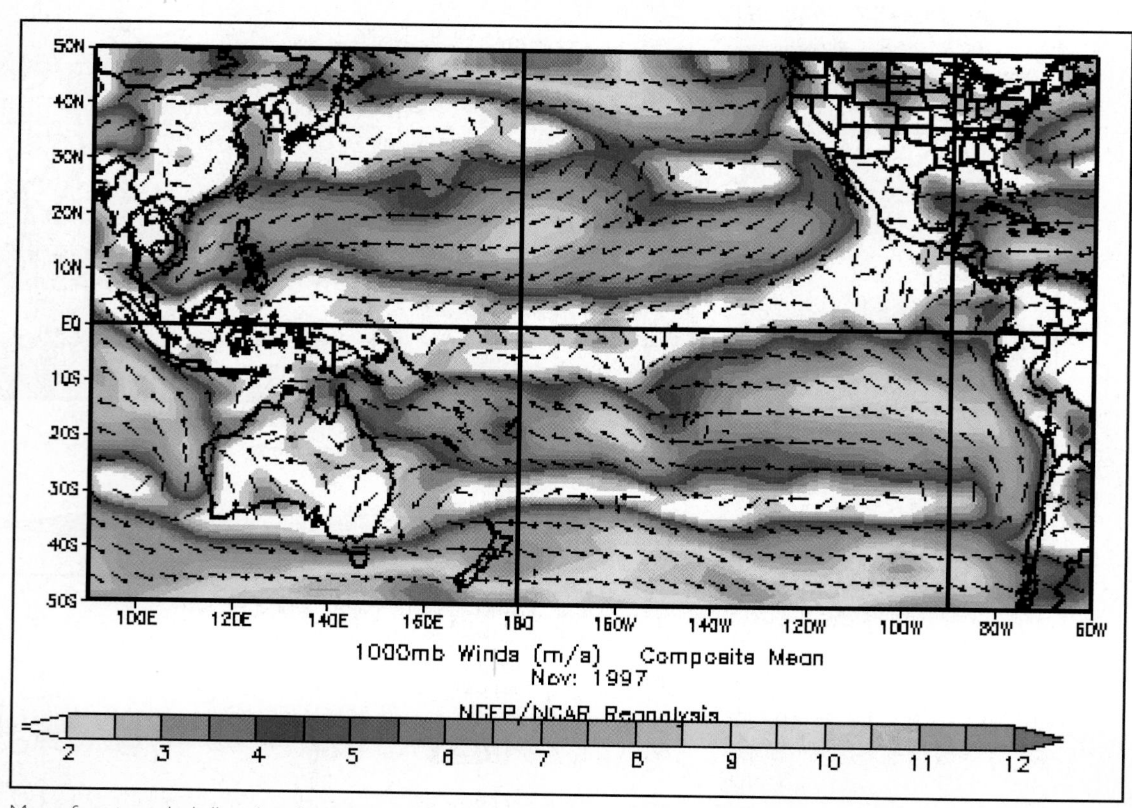

Map of vector wind direction for November 1997, an El Niño period.

Part C: Geostrophic Currents

1. The movement of water in the oceans under the influence of forces can seem very strange, because of the Coriolis effect. The figure below is a map of a horizontal surface with a body of matter resting on it. Suppose that this object is acted upon by only two forces. The blue force models a real force that always acts in the same direction, to push or pull the body toward the left on the map. The red force models the Coriolis force, which always acts at right angles (90°) to the direction of movement of the object.
    a) In which direction will the object start to move under the influence of these two forces of equal magnitude? Show this direction by a long, thin black arrow on a copy of the figure. If you have trouble with this, here's a hint: think about two people pulling in different directions with ropes on a big boulder resting on the ground.
2. On a blank sheet of paper, draw a fresh map like the one in the diagram, with the same black arrow you drew in **Step 1(a)**. Draw new arrows to show the blue force and the red force. Remember that the blue force always acts toward the left on the map, but the red force always acts at right angles to the direction of motion of the object.
    a) Now draw a new black arrow to show the direction of movement that is caused by these two forces.
    b) This direction is different from the one shown that you drew in **Step 1(a)**. How is this new direction different from the old direction?

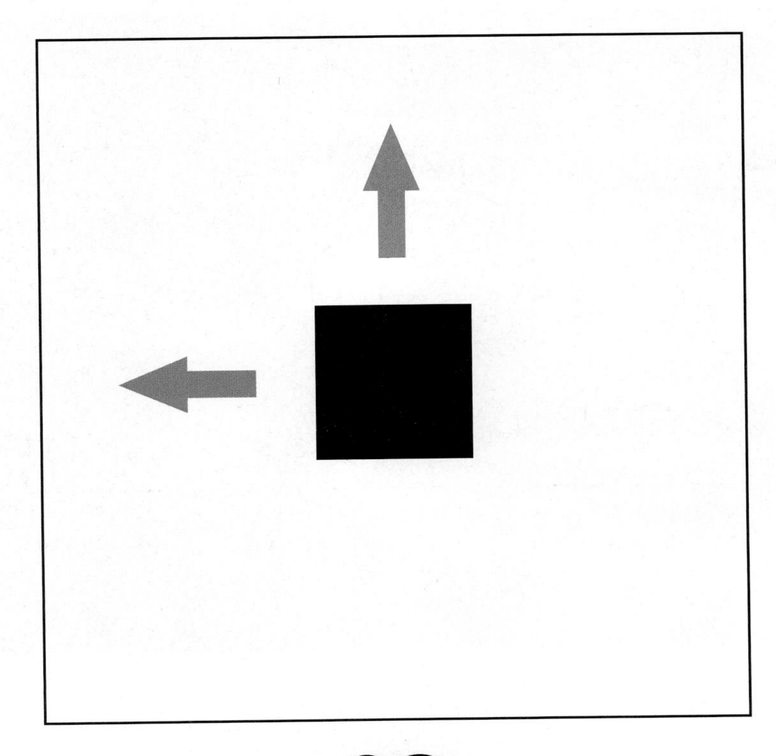

3. Using fresh sheets of paper, keep on making new maps that show the two forces and the new direction of movement, until the direction of movement stops changing. You may need to make several maps before you are convinced that the direction of movement has stopped changing.

   a) Keep a record of your drawings in your *EarthComm* notebook.

   b) What is the final direction of movement, after the direction stops changing?

   c) What are the directions of the two forces when the direction of movement stops changing?

Reflecting on the Activity and the Challenge

In this activity, you learned that when floating objects fall into the ocean, currents can move them for many hundreds or even thousands of miles. You saw that sneakers spilled overboard drifted from the site of the spill to the west coast of North America, then to Hawaii, and eventually to Japan. You also saw how wind determines the path of surface ocean currents and how the shift in wind during an El Niño event can change surface ocean circulation patterns. You learned that materials on the Earth's surface, like ocean water, can move in surprising directions because of the Coriolis effect. You can now explain how ocean circulation changes during an El Niño event.

Digging Deeper

Surface Currents

Currents at ocean depths down to 200 m encompass about 10% of the water in the oceans. These currents flow for long distances across the oceans. They form enormous circulation cells, called **gyres**. There is a gyre in each of the major ocean basins, as shown in *Figure 1*. These gyres circulate clockwise in the ocean basins of the Northern Hemisphere and counterclockwise in the ocean basins of the Southern Hemisphere. You can see from *Figure 1* that when the current in a gyre approaches a large land mass like a continent, it changes direction.

The gyres play a very important part in the Earth's climate, because they help to transport heat energy from low latitudes, where the Earth is warmed by the Sun, to high latitudes, where the Earth's surface loses heat to outer space. The poleward-flowing currents on the western sides of the gyres, like the warm Gulf Stream off the east coast of the U.S. or the Kuroshio Current off the east coast of Japan, are warm currents. The equatorward-flow currents on the eastern sides of the gyres, like the cold California Current off the west coast of the U.S. or the Peru Current off the west coast of South America,

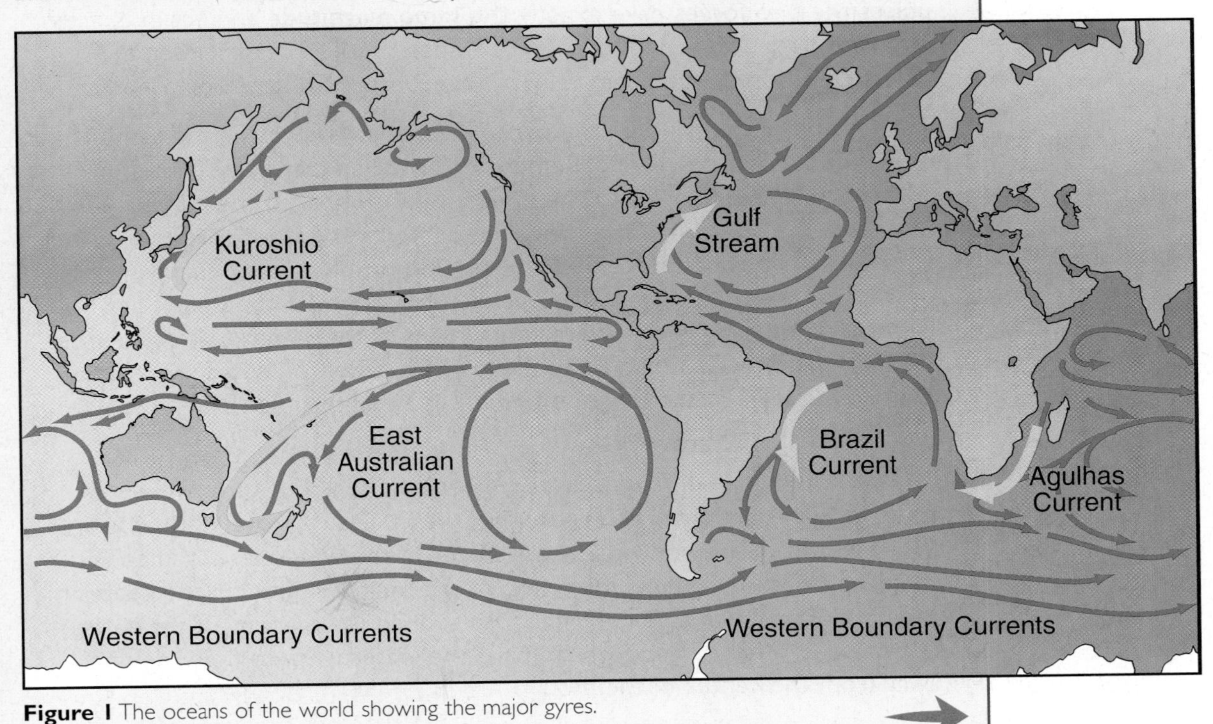

Figure 1 The oceans of the world showing the major gyres.

Geo Words

gyre: a circular motion of water in each of the major ocean basins.

Geo Words

Ekman drift: horizontal movement of ocean water that results from the balance between wind stress and the Coriolis force.

are cold currents. That's why it takes hardier souls to go swimming on the coast of northern California than on the coast of North Carolina!

The winds at low latitudes, in the tropics, are easterlies, meaning that they blow from the east. They are also called the "trade winds." The winds at mid-latitudes are less regular, but they blow mainly from the west. They are called the "prevailing westerlies."

It might seem obvious to you that the easterlies push water to the west and the westerlies push water to the east, to cause the currents in the gyres. It's true that the winds cause the currents, but only in an indirect way, not in the obvious way. To understand how the wind causes the currents, you need to know about surface **Ekman drift**.

Think back to **Part C** of the investigation. Water moves in the ocean because it is acted upon by forces. You can think of a force as a push or a pull on some object or material. All forces have a magnitude, but they also have a direction of action, so they can be shown by arrows with a length and a direction. When a force acts on an object, it causes the object to change its speed and maybe also its direction. If two different forces act on the object, they will cause the object to change its speed and direction unless the two forces have exactly the same magnitude and act in exactly opposite directions. The movement of water in the oceans under the influence of forces can seem very strange, because of the Coriolis effect.

The friction of the wind on the ocean surface always acts in the same direction, straight downwind. Remember, however, that the Coriolis force always acts at right angles to the direction of motion, so the current can't be exactly in the downwind direction. The outcome is a flow of surface water at an angle to the wind direction. The direction of the current changes downward from the surface. It turns out that, although the details are not easy to explain simply, the overall movement of the water is at right angles (90°) to the direction of the wind! The direction of the current is to the right of the wind in the Northern Hemisphere and to the left of the wind in the Southern Hemisphere.

The surface Ekman drift causes the easterly trade winds to move water towards higher latitudes and causes the westerlies to move water towards lower latitudes. This converging movement of the water results in an actual mound of water in the center of the ocean basin. Sea level in the center of the basin is as much as 2 m higher than around the margins of the basin—not much of a hill, but enough to make the water want to flow "downhill," away from the center of the mound.

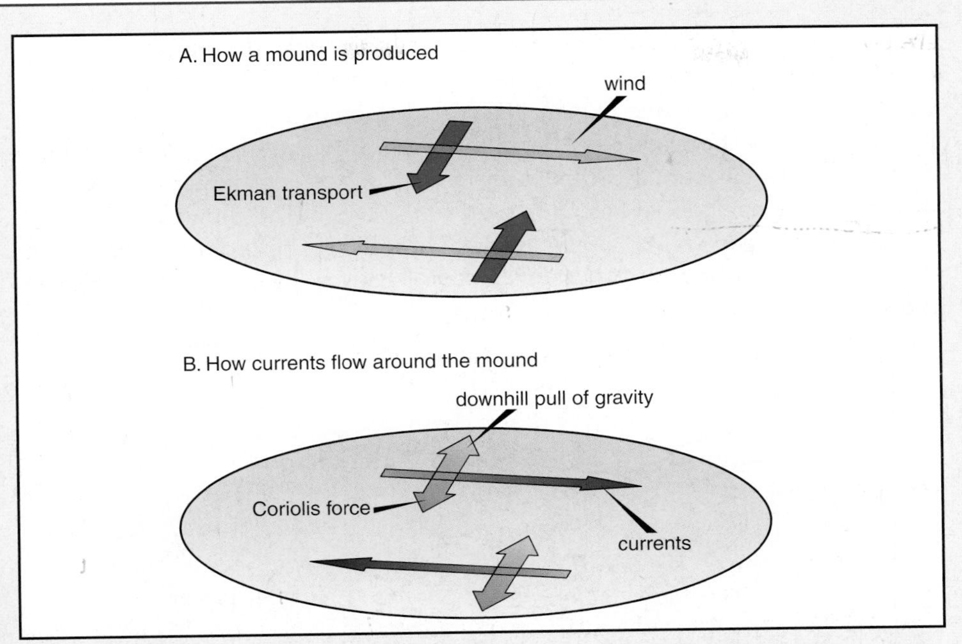

Figure 2 Surface Ekman transport creates mounds in the centers of the ocean basins.

Geo Words

geostrophic currents: in the ocean, a current in which the horizontal pressure force is balanced by the equal but opposite Coriolis force.

upwelling: the upward movement of ocean water from deep in the ocean to the surface.

Now you need to think back to **Part C** of the investigation once again. The Coriolis effect causes the water to move around the mound, rather than downhill away from the center! This is the real cause of the gyres. Currents that flow in this way are called **geostrophic currents**. Do these geostrophic currents make sense to you? Perhaps not, but remember that intuition or "common sense" does not always produce the right answer.

Not everything about surface currents is as strange as it probably seems to you by now. Very near the Equator the Earth's surface is parallel to the Earth's rotation axis, so the Coriolis effect disappears, and the rotation doesn't have any effect on the currents. The wind pushes the water straight ahead, and the direction of the current is in the same direction as the wind. You worked with surface currents of this kind in **Part B** of the investigation.

Upwelling

Most water movements in the ocean are horizontal currents. At some times and places, however, there are strong vertical movements as well. As the term suggests, **upwelling** is the upward movement of ocean water to the surface from deep in the ocean. As with so many things about water movements in the oceans, you again need to appeal to the Coriolis effect to understand upwelling.

Suppose that a wind blows over the ocean surface near the coastline, in the direction shown in *Figure 3*. In the preceding section of **Digging Deeper** you learned about surface Ekman drift. What does that tell you about how the wind moves the water? It pushes the water at right angles to the wind, in the offshore direction. As the water at the surface is pushed offshore, water from deeper down has to rise up to replace it. That upward movement of deep water is called coastal upwelling. The cross section shows how the wind pushes the water offshore and causes deeper water to rise to the surface near the coast.

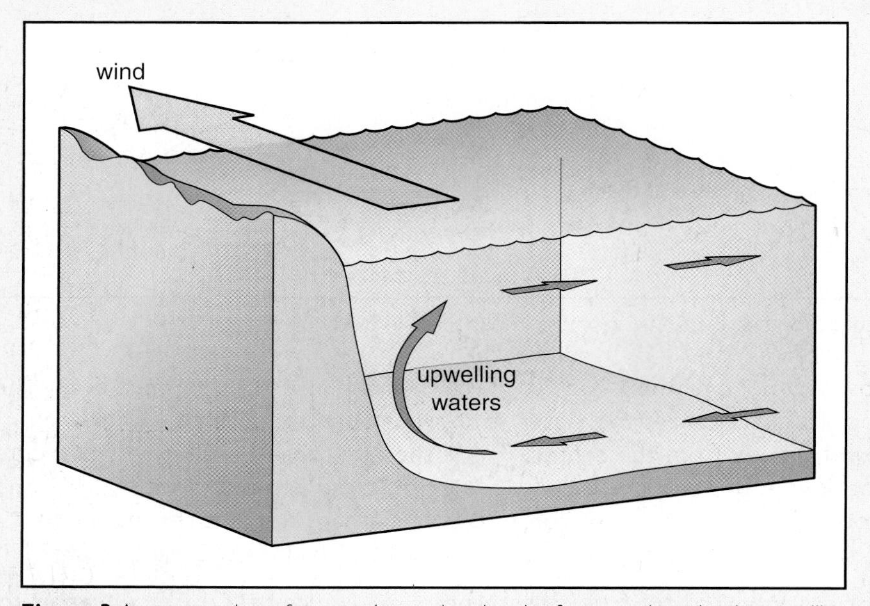

Figure 3 A cross section of a coastal area showing the forces and motion in upwelling.

Remember that everywhere in the oceans, even at low latitudes, the ocean is cold below the surface layer. Upwelling taps this colder water and brings it to the surface. This is one reason why a current like the California Current is so cold, even though the coastal area is not at high latitudes.

Very near the Equator, where the Coriolis effect vanishes, winds blowing directly offshore cause upwelling, in just the way your intuition might suggest to you. At higher latitudes, however, offshore winds move water only along coasts, not away from them.

When organisms living in the warm upper layers of the ocean die, their remains sink to lower layers of the ocean. These remains become nutrients that can be used by other organisms. This "rain" of nutrients enriches the deep ocean waters. Deep ocean circulation moves this enriched layer around the ocean basins. Warm water that has been at the ocean surface for a long time tends to be depleted of nutrients, because of the large population of near-surface plankton and larger marine animals. Upwelling moves the deeper water, which is rich in nutrients, to the surface. This upwelling of nutrient-rich water causes many coastal areas on the eastern sides of ocean basins to teem with marine life. Upwelling supports about half of the world's fisheries, even though these cool waters account for only 10% of the surface area of the world ocean.

What would happen if the wind in *Figure 3* blows in the opposite direction? Surface Ekman drift would cause water to be piled up along the coast. The water near the coast would then sink down under the added weight and be replaced by a surface flow of offshore water. That is the opposite of upwelling, and is called **downwelling**.

Geo Words

downwelling: the downward movement of ocean water from the surface

Check Your Understanding

1. What causes the circulation of the gyres in the ocean basins?
2. What are the major differences between eastern boundary currents in gyres and western boundary currents in gyres?
3. How do surface currents help to regulate the temperature of the planet?
4. How does upwelling promote the abundance of life in the surface waters of the ocean?

Understanding and Applying What You Have Learned

1. Refer back to the map of surface ocean circulation. Can you explain why the southern regions of the west coast of South America are colder than the northern regions of the east coast of North America at the same latitude?
2. Benjamin Franklin made the map of the warm Gulf Stream Current shown to the right. He based his map upon information reported by American ship captains. Why did Franklin believe that the map would be useful to sailors crossing the Atlantic Ocean?
3. Solely from what you have learned in this activity, how would you identify an El Niño event?
4. How and why would downwelling affect fisheries in coastal areas on the eastern sides of ocean basins?

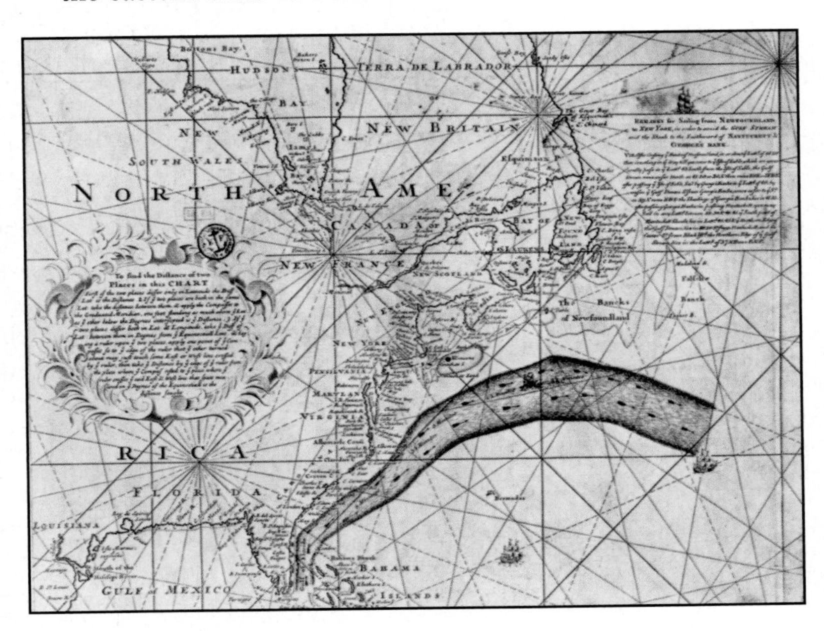

Preparing for the Chapter Challenge

Using what you have learned about surface currents, write a short essay about what drives surface currents. Include a discussion of the relationship between changing surface ocean circulation and atmospheric events and how ocean circulation can impact the climate of a region. You will refer to this essay when you are completing your **Chapter Challenge**.

Inquiring Further

1. **Surface ocean circulation and climate**

   Investigate the relationship between surface currents and climate. Consider researching the following:

   a) Compare the climate of England with the climate of a coastal area at a similar latitude on the western side of the North Atlantic Ocean.

2. **Seasonal Upwelling**

   Scientists can measure wind stress across the ocean water surface to make estimates of Ekman transports, which can then be used to calculate numbers called upwelling indices. These indices are a means for scientists to monitor the intensity of upwelling currents in a particular area. In general, the greater the upwelling index value, the more intense the upwelling. Negative upwelling index values imply downwelling (the opposite of upwelling).

   Go to the *EarthComm* web site to visit the Pacific Fisheries Environmental Laboratory web site. Here, you will find a map of the west coast of the United States. Click on the red data points to view graphs of average monthly upwelling index values for several years. Look at several of these graphs and answer the following questions:

   a) Are the patterns of upwelling the same all along the coast? Why or why not?
   b) Based on the data you viewed, would you say that upwelling currents are the strongest in the winter or in the summer? How would you explain this seasonal variation?
   c) How would the observed seasonal variation affect the distribution of organisms along the coast?

Activity 4 El Niño and Ocean Circulation

Goals

In this activity you will:

- Understand how sea surface temperatures vary during an El Niño event.
- Understand how to use data on sea surface temperatures to make inferences about changes in ocean circulation.
- Use remote-sensing data to determine the extent and duration of the 1997 El Niño event.

Think about It

Imagine that you and your friends have to make a nighttime parachute jump into unfamiliar waters.

- What are some ways of learning about the distance to the water surface, the depth of the water, and the water temperature, without actually touching the water?

What do you think? In your group, brainstorm some of the methods that could be used. Record your ideas in your *EarthComm* notebook. Be prepared to discuss your responses with your small group and the class.

Investigate

Part A: Sea Surface Temperature

1. Examine the two data sets on these pages. Think about how you would represent this data so that it is easier to understand and can be communicated to a wide variety of people. Discuss your ideas with members of your group. In your discussion, consider the following questions:

   a) What is the fundamental difference between the two data sets?

   b) How would you communicate this difference?

   c) Why is it important to communicate data clearly and efficiently?

   d) What patterns can you see? How would the patterns be easily recognizable?

   e) What is the highest temperature in the two data sets?

   f) What kind of scale or key would you develop to include all the data?

   g) If colors were used to represent the data, how many different colors would you use for this data set? (Too many might be confusing; too few might misrepresent the data.)

   h) What would the colors represent?

   i) Can the way data is represented affect the way data is interpreted?

Data Set 1. Sea surface temperatures (degrees Celsius) in the Equatorial Pacific in August during a normal (non–El Niño) year.

	125°E	137°E	149°E	161°E	173°E	175°W	163°W	151°W	139°W	127°W	115°W	103°W
11°N	29.0	29.1	29.1	28.9	28.5	28.0	27.4	27.5	27.4	27.4	27.5	28.1
9°N	29.0	29.1	29.1	28.9	28.5	28.1	27.7	27.7	27.6	27.3	27.2	27.6
7°N	29.0	29.2	29.2	28.9	28.6	28.1	27.9	27.7	27.4	27.0	26.7	26.9
5°N	28.9	29.3	29.2	28.9	28.6	28.1	27.9	27.4	27.1	26.5	26.0	25.9
3°N	28.8	29.3	29.2	28.8	28.5	28.0	27.9	27.1	26.7	25.9	25.1	24.7
1°N	28.6	29.1	29.1	28.7	28.2	27.9	27.8	26.8	26.2	25.3	24.3	23.5
1°S	28.2	28.9	29.0	28.5	28.2	27.8	27.7	26.8	26.0	25.0	23.8	22.6
3°S	27.7	27.9	28.9	28.5	28.2	27.8	27.7	27.0	26.2	25.1	23.7	22.4
5°S	27.5	26.8	28.6	28.5	28.4	27.9	27.8	27.2	26.4	25.3	23.9	22.5
7°S	27.3	26.6	28.1	28.4	28.5	28.1	27.8	27.4	26.5	25.5	24.1	22.8
9°S	27.2	26.4	27.3	28.2	28.2	28.2	27.9	27.4	26.5	25.4	24.3	23.0
11°S	27.0	26.3	26.6	27.8	27.8	28.2	27.8	27.3	26.4	25.3	24.2	22.9

Data Set 2. Sea surface temperatures (degrees Celsius) in the Equatorial Pacific in August during an El Niño year.

	125°E	137°E	149°E	161°E	173°E	175°W	163°W	151°W	139°W	127°W	115°W	103°W
11°N	28.9	29.2	29.0	29.2	28.8	28.4	28.0	27.7	27.9	27.8	28.2	28.6
9°N	28.7	29.1	28.9	29.1	29.0	28.5	28.4	28.1	28.2	28.0	28.1	28.2
7°N	28.6	29.0	28.8	29.1	29.2	28.7	28.7	28.3	28.3	28.0	27.8	27.8
5°N	28.5	28.8	28.8	29.1	29.3	28.9	28.8	28.3	28.2	27.7	27.3	27.2
3°N	28.2	28.6	28.8	29.1	29.4	29.1	28.8	28.0	27.9	27.2	26.6	26.2
1°N	27.8	28.4	28.8	29.2	29.6	29.2	28.8	27.7	27.6	26.7	25.8	25.2
1°S	27.4	28.3	28.7	29.2	29.7	29.2	28.8	27.7	27.5	26.4	25.3	24.5
3°S	27.0	27.7	28.5	29.2	29.7	29.2	28.8	27.9	27.6	26.3	25.2	24.3
5°S	26.8	26.1	27.9	29.0	29.6	29.2	28.7	28.0	27.6	26.3	25.3	24.3
7°S	26.6	25.7	27.1	28.7	28.9	29.0	28.6	28.2	27.6	26.3	25.4	24.3
9°S	26.6	25.6	26.5	28.2	28.4	28.8	28.3	28.2	27.4	26.3	25.3	24.1
11°S	26.4	25.6	25.8	27.5	28.0	28.3	28.0	28.0	27.2	26.2	25.1	23.7

2. Typically, red is used to represent the warmest temperature and purple or dark blue is used for the coldest temperature.

   a) Make a color scale to represent the sea surface temperatures given for the Pacific Ocean. You must have a color that can be used to plot every data point in the two data sets, from the lowest temperature to the highest temperature. If you wish, you may round the data to the nearest whole number. For example, 21.9 becomes 22, and 18.4 becomes 18. Draw your scale on two separate copies of the Pacific Ocean, similar to the one shown on the following page.

3. Use Data Set 1.

   a) Plot this data on a map of the Pacific Ocean. If available, a clear plastic ruler might be useful for plotting. Use the appropriate colored pencil for plotting.

   b) Where data points of the same color are located next to each other, lightly shade in that area with the corresponding colored pencil. When you are finished doing these for all the data points, you will have a map that is almost completely colored in the area between 11°N and 11°S.

   c) Give this map the title "Sea Surface Temperatures in August during a non–El Niño Year."

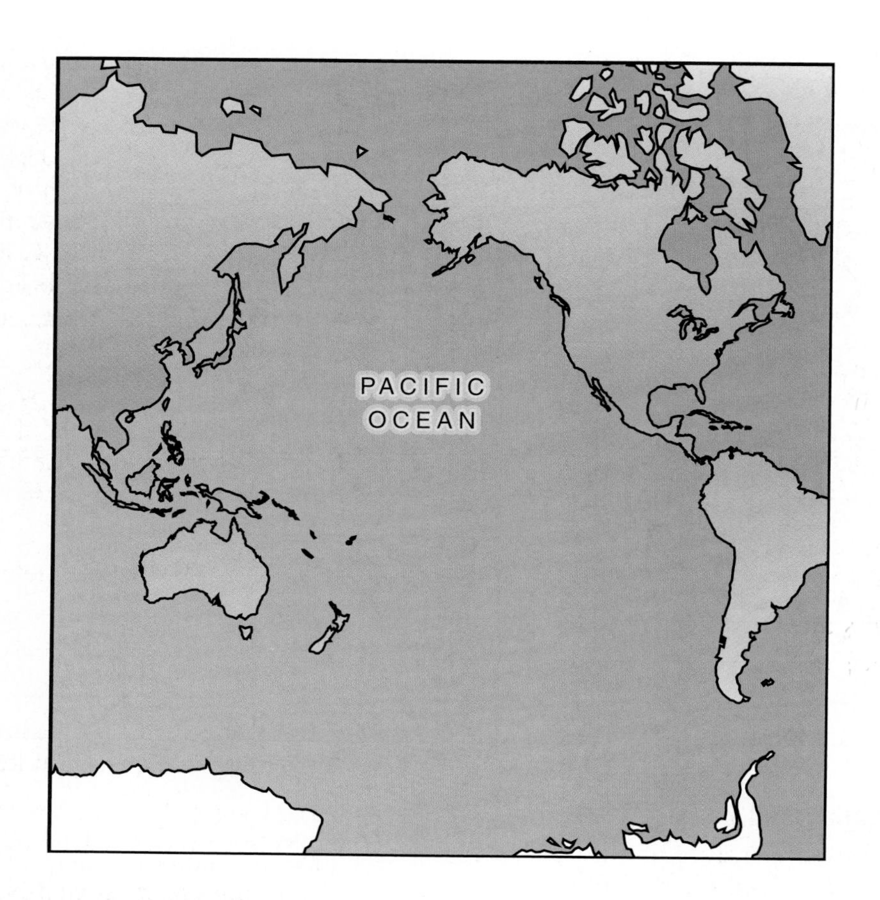

4. Use Data Set 2.
   a) Plot this data on a second copy of the map of the Pacific Ocean.
   b) Shade in the map and label it appropriately.
   c) Give this map the title "Sea Surface Temperatures in August during an El Niño Year."
5. When you have completed your map, answer the following questions:
   a) At what latitude and longitude do the warmest sea surface temperatures occur in August during a non–El Niño year?
   b) At what latitude and longitude do the warmest sea surface temperatures occur in August during an El Niño year?
   c) In general, what happens to sea surface temperatures in the equatorial Pacific Ocean during an El Niño event?
   d) During an El Niño event, what are the surface currents bringing to the eastern edge of the Pacific Ocean?
   e) In which direction are surface currents in the equatorial Pacific moving during an El Niño event?

f) Think carefully about where the surface ocean currents are pushing the water during an El Niño event. Which side of the equatorial Pacific Ocean would you expect to have higher sea level during an El Niño event?

g) From the differences that you noted in the data, would you infer that surface circulation is in the same direction during El Niño and non–El Niño years? Explain.

h) Are these maps easier to interpret than the data table?

i) In what ways are the colored maps inferior to the data table?

j) Is it important for scientists to have access to the map and the data table? Why or why not?

Part B: Extent and Duration of an El Niño Event

1. The following maps illustrate data collected by the TOPEX/Poseidon satellite, launched on behalf of NASA. TOPEX stands for Topography Experiment, and Poseidon was the Greek god of the sea. The TOPEX/Poseidon satellite actually measures sea level rather than water temperature. A number on one of your map sections is 10 cm higher or lower than the next. However, scientists have found when mapping the TOPEX satellite data that the highest sea levels match the warmest water, and the lowest are indeed the coolest. These bulges, or hills and valleys of water, are easily tracked.

   a) Color a copy of the maps using the color scheme found at the bottom of the maps. (Remember: the higher the number the warmer the temperature.)

2. After your maps are complete, answer the following questions:

   a) Which way did the warm water move?

   b) The El Niño was at maximum in November. How do you know?

   c) How long did it take for the El Niño to reach maximum?

   d) How long did it take El Niño to disappear?

   e) Look at the size of the area affected by El Niño. How does this compare to the size of your state?

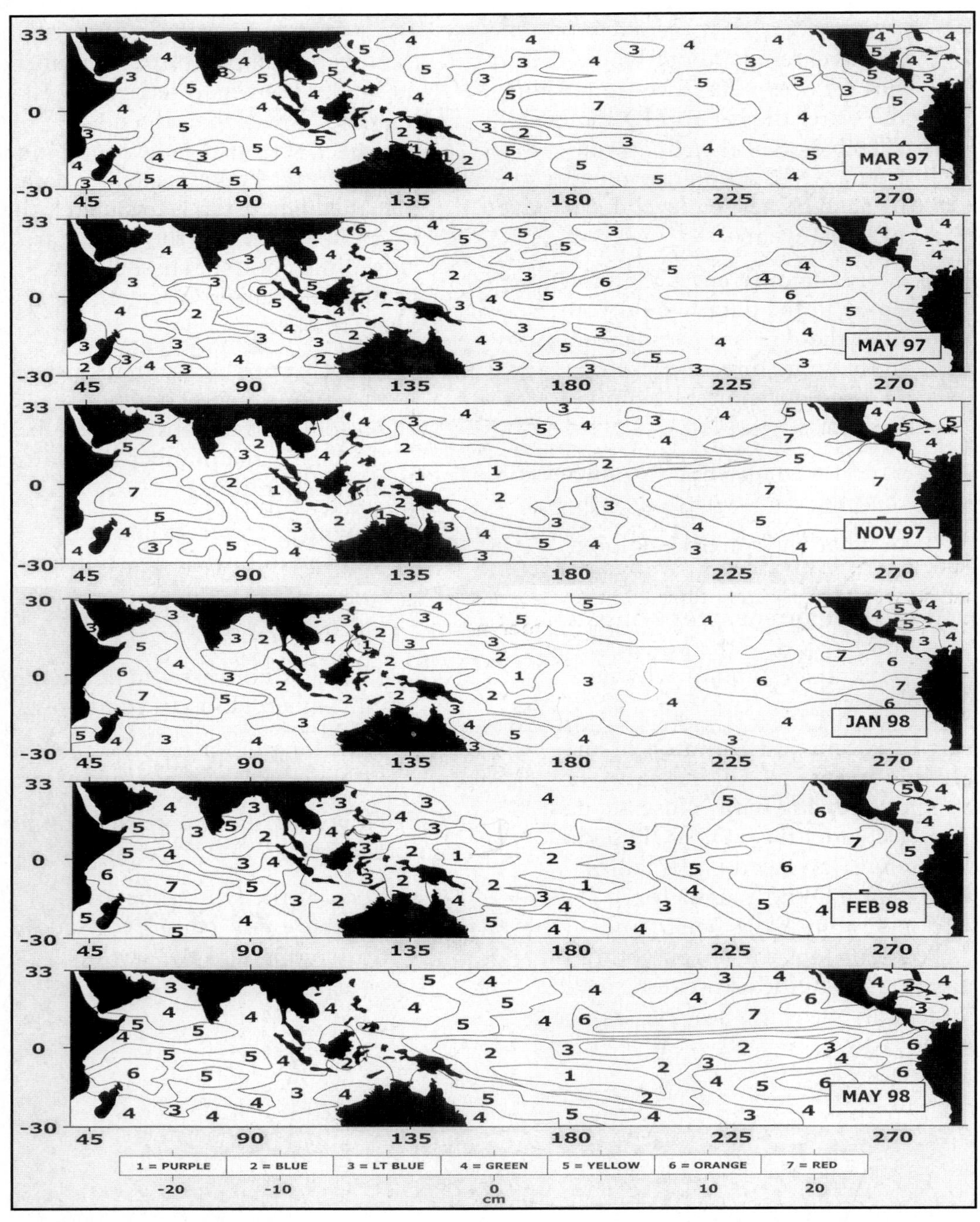

TOPEX/Poseidon data, March 1997–May 1998.

Reflecting on the Activity and the Challenge

In this activity you saw that the El Niño conditions changed the surface of the equatorial Pacific Ocean considerably compared to "normal" (non–El Niño) conditions. These conditions lasted for several months and affected the entire west coast of the United States, where El Niño pushed warm water especially close to Mexico and California. Indian Ocean temperatures were also increased, but water temperatures in Southeast Asia were cooler. This should help you understand why El Niño could affect the weather and climate. You will learn more about this in the next activity.

Digging Deeper

El Niño Conditions and Non–El Niño Conditions

During most of the time along the Peruvian coast of South America, a strong northwest-flowing current causes upwelling, bringing deeper water that is rich in nutrients up to the surface (*Figure 1*). These cold waters support a rich population of fish. An El Niño is a condition, lasting one to three years, when the sea surface temperatures in the eastern equatorial Pacific Ocean, off the coast of Peru, are much higher than during other times. During these times the fish population is much smaller, because cold, nutrient-rich waters are no longer brought up to the surface.

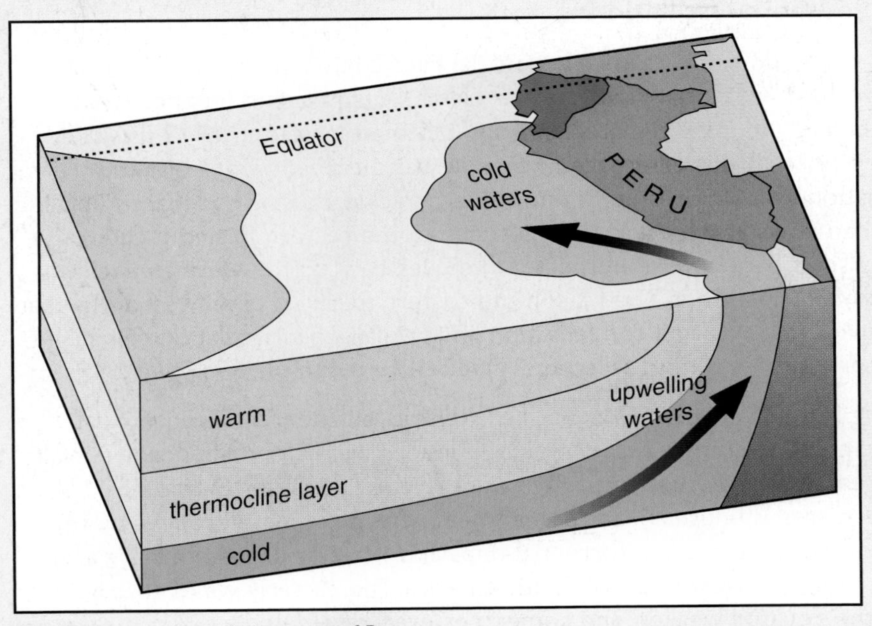

Figure 1 Upwelling off the coast of Peru.

Figure 2 The waters along the Peruvian coast of South America receive nutrients via upwelling.

El Niño conditions have been known to Peruvians for centuries. It has only been in the past 100 years, however, that geoscientists have recognized that an El Niño involves changes in ocean circulation, sea surface temperatures, and climate throughout the entire equatorial Pacific. This is an area that stretches a quarter of the way around the world.

An El Niño is usually called an "event," and it is contrasted with "normal" conditions in the equatorial Pacific. It's more realistic, however, to think of El Niño conditions and non–El Niño conditions as two different states of the ocean and the atmosphere, which alternate with each other irregularly as the years go by. An El Niño lasts one to three years, and non–El Niño conditions usually last about five to ten years. The equatorial Pacific is in the "El Niño state" about 20% of the time. Geoscientists still don't have a clear understanding of what causes the switch between non–El Niño conditions and El Niño conditions.

The Southern Oscillation

Most of the time, the eastern equatorial Pacific has a very dry climate, and the western equatorial Pacific, in the area of Australia and Indonesia, has a very humid and rainy climate. Long ago, a climatologist named G.T. Walker recognized that this difference in climate was a reflection of a gigantic circulation cell. Air rises up in the western Pacific, causing abundant rainfall, and then flows eastward at high altitudes to the eastern Pacific. It then slowly sinks back to low altitudes and moves back to the western Pacific as easterly surface winds. The reason the eastern Pacific is usually so dry is that the sinking air prevents condensation and rainfall. This circulation cell in the equatorial Pacific is now called the **Walker circulation**. See *Figure 3*.

During El Niño conditions, the atmospheric circulation in the equatorial Pacific changes drastically. The area of humid, rising air and abundant rainfall shifts eastward, to the central Pacific and even the eastern Pacific. The western area is unusually dry. Sometimes there is extreme drought there, and sometimes there are torrential rains and flooding in the normally arid areas along the west coast of South America. The easterly winds that blow near the Equator weaken, and sometimes even reverse.

Geo Words

Walker circulation: circulation cells within the equatorial atmosphere caused by differences in climate.

The Equatorial Pacific Ocean

During non–El Niño years, the easterly winds move warm surface water toward the western boundary of the Pacific. Sea level is then slightly higher along the western boundary of the Pacific, partly because of the push of the wind on the water but also because the water is much warmer. You know that water expands when it is raised to a higher temperature, and that raises the sea surface temperature slightly because of the expansion of the water in the upper layer of the ocean. Also, the thermocline (go back and review the material in Activity 1) is much deeper in the western Pacific than in the eastern Pacific. The reason that upwelling along the coast of South America causes the ocean surface to be so cold is that thermocline is very shallow, allowing the upwelling to tap cold water from below.

Figure 3 Normal (top) oceanic and atmospheric conditions in the equatorial Pacific, illustrating Walker circulation cells. During El Niño (bottom), shifts in atmospheric pressure result in a weakening of the trade winds and a shift in precipitation.

During El Niño years, the equatorial Pacific Ocean is very different. The weakening of the easterly winds causes warm water to move eastward from the western Pacific area, all the way to the west coast of South America. At the same time, the thermocline gradually deepens from west to east, in a kind of wavelike motion. Upwelling continues along the west coast of South America, but because the thermocline is now much deeper there, warm water instead of cold water is brought up to the surface.

Cause and Effect

Which comes first, the change in atmospheric circulation or changes in the equatorial ocean? The changes in the atmosphere and the ocean that are involved in the beginning of an El Niño event seem to develop at about the same time. There is therefore no easy way to decide which is the cause and which is the effect. Geoscientists are still not sure about this. One way of thinking about this is that the ocean and the atmosphere in the equatorial

Check Your Understanding

1. Why is the climate along the Peruvian coast of South America very dry during normal, non–El Niño times?
2. How do southerly winds along the coast of Peru cause upwelling there?
3. Why are the ocean surface waters along the Peruvian coast of South America unusually warm during an El Niño event, even though upwelling still operates?

Pacific form a coupled system. The system can exist in two different states, and perhaps some minor "trigger" can flip the system from one state to the other. Much more research is needed before geoscientists have a full understanding of El Niño.

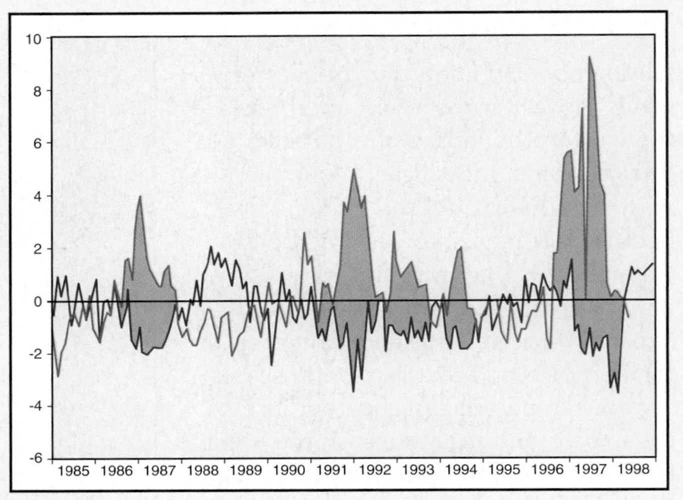

Figure 4 The red line gives sea surface temperature anomalies at Puerto Chicama, Peru, an indicator of El Niño. The blue line represents the Southern Oscillation Index (SOI) that is the difference in sea-level pressure between Darwin, Australia and Tahiti. Major El Niño events are shaded in.

Understanding and Applying What You Have Learned

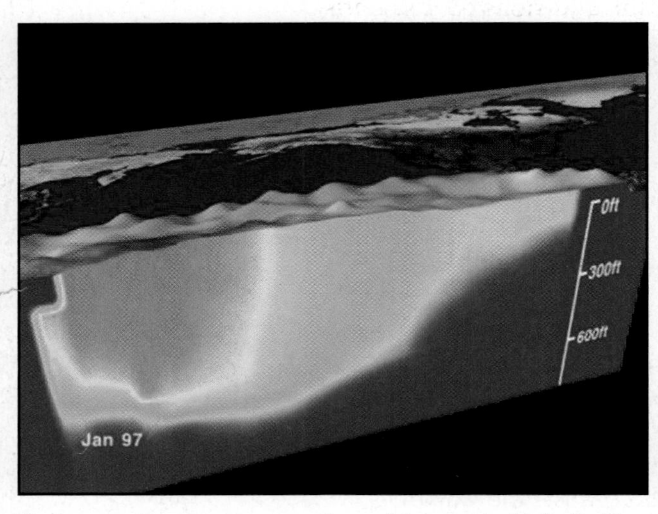

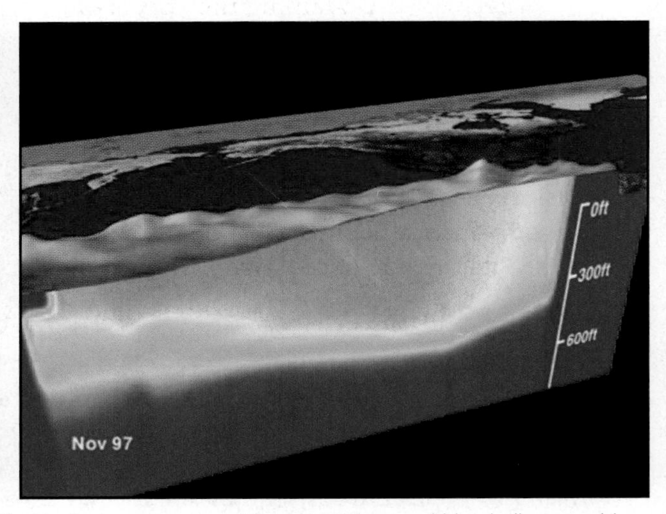

The images shown are cross sections of the equatorial Pacific Ocean. Red indicates warmer water temperature, and blue indicates colder water temperature. The January 1997 image represents "normal" non–El Niño conditions, while the November 1997 image illustrates El Niño conditions.

1. Under non–El Niño conditions:
   a) What happens to deep water at the eastern boundary of the Pacific Ocean (around 80° W longitude)?
   b) Are there places where deep water is exposed to conditions on the ocean surface?
   c) Are deep water circulation and surface water circulation interconnected in certain places? Why or why not?
   d) In which direction does the surface wind blow at the Equator?
   e) What explains the westward spread of cold surface water off the coast of North and South America?
2. Under El Niño conditions:
   a) How do deep circulation patterns change in the equatorial Pacific Ocean?
   b) Are the surface water temperatures along the eastern boundary of the equatorial Pacific Ocean warmer or colder during an El Niño event? Why?
   c) In which direction does the surface wind blow at the Equator?
   d) What happens to the location of the warm surface water during El Niño?

Preparing for the Chapter Challenge

Meet with your group and decide which area of your **Chapter Challenge** could be strengthened with more information. Keep in mind that you will be examining the effects of El Niño on weather, climate, and the food chain in upcoming activities. Visit the *EarthComm* web site to find Internet sites where you can pose your questions to a scientist.

Inquiring Further

1. **Technology used to study ocean–atmosphere interactions**

   How do scientists collect the data you studied in this activity? To learn more about the instruments scientists use to help them better understand El Niño events, visit the *EarthComm* web site to investigate the following El Niño research projects:

   - TOPEX/Poseidon
   - TAO (Tropical Atmosphere Ocean Project)

2. **La Niña**

   What is La Niña? What are the impacts of a La Niña event? How are they different from an El Niño event? Which do you think affects your community more?

Activity 5 Weather, Climate, and El Niño

Goals

In this activity you will:

- Map variations in temperature and precipitation in the United States during the 1982–1983 El Niño event.
- Understand that weather changes caused by this El Niño event had global consequences.
- Determine if the variation caused by El Niño occurred in your own state.
- Predict what effect a new El Niño event might have on your state's agriculture.

Think about It

The Earth system is complex. A change in one part of the system (the hydrosphere, for example) has an effect in other parts of the system (atmosphere, cryosphere, and so on).

- Suppose that evaporation increases over the ocean. How might the water that is added to the atmosphere cause a change in another part of the Earth system?
- Suppose that precipitation increased dramatically over land. How might the increased precipitation affect another part of the Earth system?

What do you think? Record your ideas about these questions in your *EarthComm* notebook. Be prepared to discuss your responses with your small group and the class.

Investigate

1. The table on the following page contains data for anomalies in temperature and precipitation for the United States during the 1982–1983 El Niño event. An anomaly is simply a departure from the long-term average. The data have been compiled by comparing average temperature and precipitation data from October 1982 to March 1983 (during an El Niño event) with the long-term temperature and precipitation averages from 1950 to 1995. Therefore, a positive variation (i.e., 2) means warmer than normal temperature during the El Niño event, whereas a negative variation (i.e., –2) means colder than normal.

   a) Plot the temperature variation data in *Table 1* on a map of the U.S.

2. Contour the data using an isotherm interval of 1°F. An isotherm is a line on a map that passes through points with equal temperature. Recall that in the data table a positive variation means warmer than normal temperature and a negative variation mean colder than normal. Thus, your map should show isotherms between –1 and +4.

   a) Draw the contours on your map. Drawing contours is not always easy, especially when data points are far apart compared to how rapidly the values change from place to place. If you have trouble drawing contours, ask your teacher for help.

3. Using the map, answer the following questions:

   a) Which areas of the map show the smallest temperature variation?

   b) Which areas of the map show the greatest temperature variation?

   c) What is the relationship between increasing latitude and temperature variation?

4. Use your map of the U.S., and the data in *Table 1*.

   a) Plot the data on precipitation variation from *Table 1*.

   b) Contour the data using a 5-inch isohyet interval. An isohyet is a line of equal precipitation. Note that a positive variation means wetter than normal and a negative variation mean dryer than normal. Thus, your map should show isolines that range from minus 5 inches to plus 20 inches.

Table 1: Anomalies for October 1982–March 1983 El Niño event. The data has been compared to 1950–1995 long-term average. No data are available for Alaska and Hawaii.

State	Temperature Anomaly (in degrees F)	Precipitation Anomaly (in inches)	State	Temperature Anomaly (in degrees F)	Precipitation Anomaly (in inches)
Alabama	0	8	Nebraska	2	4
Arizona	0	4	Nevada	0	3
Arkansas	1	6	New Hampshire	3	0
California	0	16	New Jersey	2	1
Colorado	1	1	New Mexico	–1	2
Connecticut	3	1	New York	3	–2
Delaware	2	1	North Carolina	1	6
Florida	0	10	North Dakota	4	4
Georgia	0	5	Ohio	4	–2
Idaho	1	2	Oklahoma	0	2
Illinois	3	4	Oregon	1	8
Indiana	4	–1	Pennsylvania	3	–3
Iowa	3	5	Rhode Island	2	2
Kansas	1	2	South Carolina	0	7
Kentucky	2	–5	South Dakota	3	3
Louisiana	0	15	Tennessee	2	–3
Maine	3	–1	Texas	–1	3
Maryland	2	–1	Utah	0	3
Massachusetts	2	3	Vermont	3	–1
Michigan	4	1	Virginia	1	0
Minnesota	4	4	Washington	2	5
Mississippi	0	14	West Virginia	2	–4
Missouri	2	2	Wisconsin	4	4
Montana	3	0	Wyoming	1	0

(Data adapted from NOAA-CIRES Climate Diagnostics Center)

5. Using the map, answer the following questions:
   a) The wettest areas occurred in which areas of the U.S.?
   b) The driest areas occurred in which areas of the U.S.?
   c) The greatest variation in precipitation (wettest) occurred in which state?
   d) The least variation in precipitation (driest) occurred in which state?

Reflecting on the Activity and the Challenge

In this activity you determined that during the El Niño event of 1982–83, most of the United States experienced temperatures that departed noticeably from the long-term average. This variation increased northward, with the greatest variation in the northeast and the least variation in the southwest. Precipitation variation also shows a definite pattern, with the wettest areas along the west and southeast coasts of the U.S. and the driest areas in the northeast. You are now aware of the relationship between an El Niño event and weather and climate.

Geo Words

weather: the atmospheric conditions at a particular time, day to day.

climate: the long-term average of weather in a particular region of the Earth, over years, decades, or centuries.

meteorologists: scientists who study the weather.

climatologists: scientists who study the Earth's climate.

teleconnections: the effects that climate change in one region of the Earth have on the climate in distant regions (*tele-* means far or distant, as in *tele*vision, "distant seeing").

Digging Deeper

EL NIÑO EVENTS AND CLIMATE

Weather and Climate

Weather is what the atmospheric conditions are like at a particular time, day to day. **Climate** is the long-term average of weather in a particular region of the Earth, over years, decades, or centuries. **Meteorologists** (scientists who study the weather) are fairly good at predicting the weather a few days in advance, but predicting the weather a month ahead is still very uncertain. Most **climatologists** (scientists who study the Earth's climate) are predicting that the Earth's climate will continue its warming trend at an increasing rate, probably because of the buildup of carbon dioxide in the atmosphere by burning of coal, oil, and natural gas.

There are also cycles of climate over times of several years, which are difficult to predict. The cycle of El Niño conditions and non–El Niño conditions in the equatorial Pacific is the clearest example of such short-term climate change. Climatologists who specialize in the study of El Niño are trying to develop ways of predicting the onset of an El Niño event long before it happens.

Teleconnections

The term **teleconnections** is used by climatologists for the effects that climate change in one region of the Earth have on the climate in distant regions (*tele-* means far or distant, as in *tele*vision, "distant seeing"). El Niño conditions in the equatorial Pacific can cause major changes in climate in distant regions of the Earth. Understanding such teleconnections is very difficult, because interactions within the atmosphere and between the atmosphere and the oceans are very complex. During El Niño years, the jet stream changes its pattern. Variations in the jet stream, which is a current of fast-moving air in the upper atmosphere, lead to changes in overall weather patterns. As shown in the map in *Figure 1*, the 1982–1983 El Niño event caused weather-related disasters on almost every continent.

The event was blamed for 1300 to 2000 deaths and more than $13 billion in damage to property and livelihoods. On September 24, in just 24 hours, sea surface temperatures along the coastal village of Paita, Peru shot up 7.2°F. There were also secondary problems caused by the 1982–1983 El Niño event. Encephalitis outbreaks occurred on the east coast of the U.S., attributed to a warm, wet spring (perfect for mosquitoes). Increases in snakebites were reported in Montana, as the hot, dry weather drove mice from high elevations downward in search of food and water, and the rattlesnakes followed. A rise in bubonic plague occurred in New Mexico

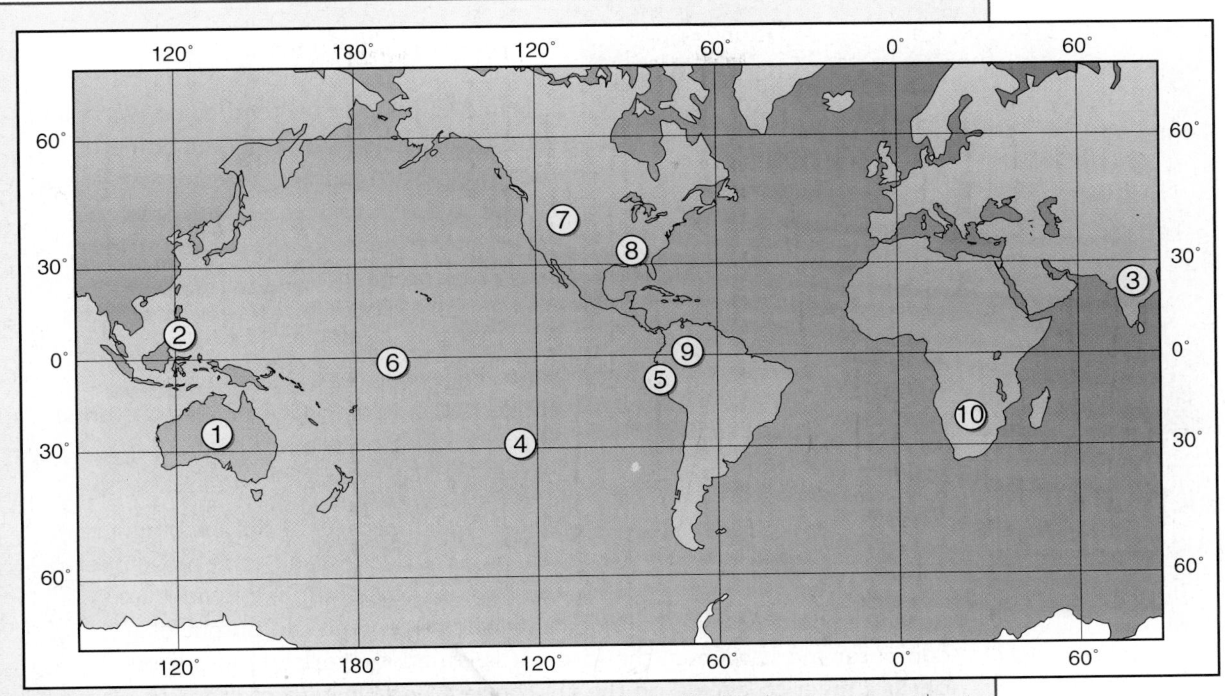

Figure 1 The impacts of the 1982–1983 El Niño were far-reaching, affecting almost every continent on the globe.

KEY
① Australia – drought and bush fires
② Indonesia, Philippines – crops fail, starvation follows
③ India, Sri Lanka – drought, fresh water shortages
④ Tahiti — six tropical cyclones
⑤ South America – fish industry devastated
⑥ Pacific Ocean – coral reefs die
⑦ Colorado River basin – flooding, mud slides
⑧ Gulf States – downpours cause deaths, property damage
⑨ Peru, Ecuador – floods, landslides
⑩ South Africa – drought, disease, malnutrition

as a cool, wet spring provided favorable conditions for flea-ridden rodents. There was an increase in shark attacks off the Oregon coast, because of unseasonably warm sea temperatures. There even was a rash of spine injuries in California, as weather-altered coastal sea floors fooled surfers. Some climatologists have connected El Niño to above-normal temperatures recorded in Alaska and northwestern Canada, and a reduction in the salmon harvest. In the eastern U.S., it was the warmest winter in 25 years.

Evidence of past El Niño events goes back for hundreds of years. This evidence includes historical accounts from early settlers and indirect evidence about climate from tree rings, sediment cores, ice cores, and coral reefs. El Niño events have occurred in the past and will occur in the future. The time between events varies, typically ranging from two to seven years. In addition, some El Niño events are strong and some are weak. Typically, an event lasts from 14 to 22 months, but it can be much longer or shorter.

Figure 2 Heavy rains in southern California associated with the 1997–1998 El Niño resulted in extensive damage, causing landslides and debris flows.

The country of Peru provides a good example of how El Niño forecasts can be valuable. El Niño years tend to be warm and thus unfavorable for fishing, and some of them have been marked by damaging floods along the coast. Fishermen welcome cold years, but farmers do not because these years have frequently been marked by drought and crop failures. Such cold years often come on the heels of strong El Niño years. Peruvians have reason to be concerned, not only about El Niño events but also about both extremes of the El Niño cycle.

Since 1983, forecasts of the upcoming rainy season have been issued each November on the basis of observations of winds and water temperatures in the tropical Pacific region and the output of computer models. Once the forecast is issued, farmers and government officials meet to decide on the appropriate combination of crops to plant. Rice and cotton, two of the primary crops grown in northern Peru, are highly sensitive to the quantities and timing of rainfall. Rice thrives on wet conditions during the growing season followed by drier conditions during the ripening phase. Cotton, with its deeper root system, can tolerate drier weather. Hence, a forecast of El Niño weather might induce farmers to sow more rice and less cotton than in a year without El Niño.

Check Your Understanding

1. Give two examples of how the El Niño event in the equatorial Pacific caused a change in another location.
2. Describe how predicting an El Niño is valuable to an economy.
3. What is the difference between climate and weather?
4. What evidence exists for past El Niño events?

Understanding and Applying What You Have Learned

1. Describe the effect that El Niño has on your state in terms of temperature and precipitation.
2. In your particular area of the country, which agricultural crops would be adversely affected by conditions like those of the 1982–83 El Niño?
3. List three industries that are dependent on long-term weather forecasting.
4. Ski areas in many parts of the country depend on snowfall.
   a) In which areas of the country would the ski industry be hit hardest from an El Niño like that of 1982–83?
   b) Which ski industry would have benefited from the 1982–83 El Niño?
5. How might El Niño affect the mosquito population?

6. Use the precipitation map you created to propose a hypothesis for why coastal areas had the greatest amount of precipitation.
7. What kind of weather event could have affected the additional amount of precipitation in Florida during the 1982–83 El Niño event?

Preparing for the Chapter Challenge

Make a chart showing the relationship between recorded El Niño events and yearly precipitation in your area. Your teacher may have this information available to you. If not, collect precipitation data from your local weather service or visit the *EarthComm* web site for web links to help you collect the data.

Inquiring Further

1. **El Niño and hurricanes**

   Research the relationship between El Niño and the occurrence of hurricanes in the Atlantic. Go to the *EarthComm* web site for help with your research.

2. **El Niño and crop prices**

   Investigate the fluctuations in wheat prices over the last 20 years. Can you find any relationships between crop prices and El Niño events?

3. **El Niño events and the Earth system**

   How does an El Niño event demonstrate the interrelationship between the hydrosphere and the atmosphere?

   How might severe weather associated with an El Niño event affect the biosphere? the geosphere?

Activity 6 El Niño and the Oceanic Food Chain

Goals

In this activity you will:

- Graph the annual Pacific fish catch from 1957 to 1983.
- Using your graph, identify past El Niño events and predict future events.
- Understand the relationship between upwelling, phytoplankton abundance, and fish catches in the equatorial Pacific.
- Use an Earth systems approach to consider how phytoplankton production is related to cycles of matter that could affect your state and community.

Think about It

The oceanic food chain is controlled by several factors, one of which is ocean circulation. Because humans are at the top of the food chain, changes in ocean circulation can affect what is available for you to eat.

- How might an El Niño event cause an increase in the cost of an anchovy pizza?

What do you think? Record your ideas about this question in your *EarthComm* notebook. Be prepared to discuss your responses with your small group and the class.

Investigate

1. The data in *Table 1* below show the annual fish catch in the eastern equatorial Pacific between 1957 and 1983.
   a) Use the data to construct a graph of fish catch over time.
2. Use your graph and the data in *Table 1* to answer the following questions:
   a) Describe how the fish catch changed during the period 1957–1983.
   b) In which years were fish catches the lowest?
3. From data on sea surface temperature, scientists have determined that over the period covered by the fish harvest data, three strong El Niño events and two moderate El Niño events occurred.
   a) If El Niño has a negative impact on fish harvest, during which years do you think El Niño events occurred?
4. Phytoplankton are tiny plants that float in the near-surface waters of the ocean. They represent the most important community of primary producers in the ocean, and hence form the base of the oceanic food chain. On the following page are images that illustrate the distribution of phytoplankton in the Pacific Ocean during an El Niño condition and during a non–El Niño condition.
   a) According to the key, what do the colors on the images represent?
   b) If an area had a high concentration of phytoplankton in the water, what color would you see on the maps?

Table 1: Annual Fish Catch in the Eastern Equatorial Pacific between 1957 and 1983

Year	Fish Catch (million tons)	Year	Fish Catch (million tons)	Year	Fish Catch (million tons)
1957	1	1966	17	1975	6
1958	2	1967	19	1976	7
1959	4	1968	20	1977	2
1960	6	1969	18	1978	3
1961	8	1970	24	1979	4
1962	11	1971	20	1980	2
1963	13	1972	8	1981	3
1964	18	1973	4	1982	2
1965	15	1974	7	1983	1

c) Do the red values represent areas you would look for anchovies?

d) What happened to the areas rich in phytoplankton during El Niño years? Why do you think that is?

e) From what you have learned about sea surface temperatures during El Niño events, would you say that phytoplankton prefer warm or cold waters? Can you think of a reason as to why this is so?

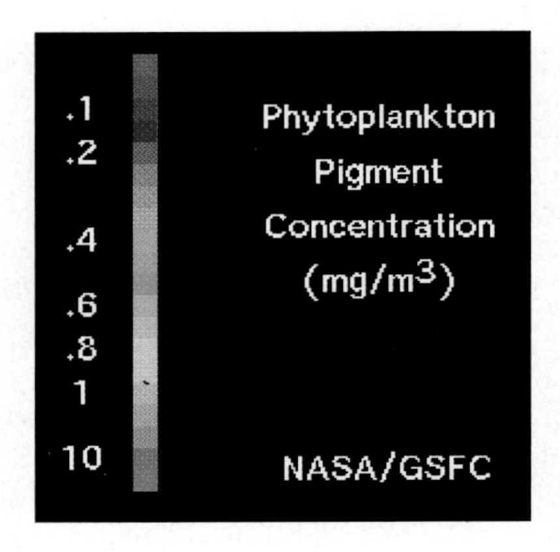

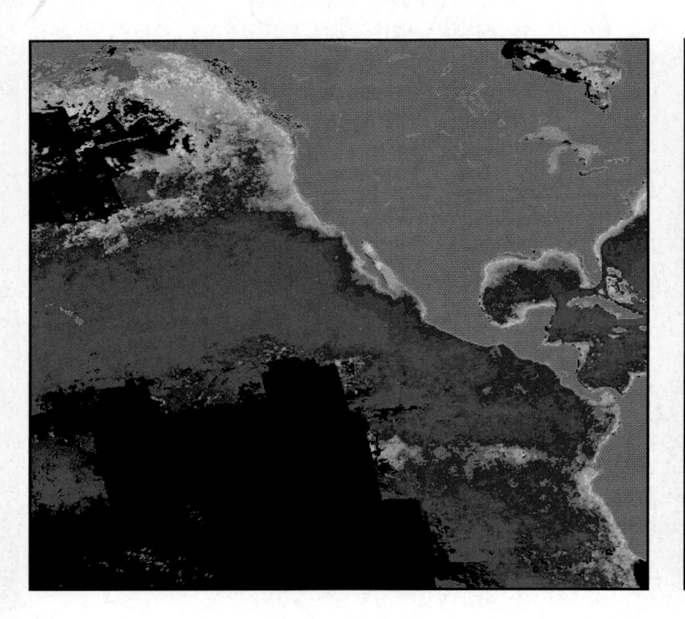

Reflecting on the Activity and the Challenge.

In this investigation, you graphed data describing fish catches in the equatorial Pacific Ocean. Using these data and the relationship between phytoplankton and upwelling, you made inferences about the impact of El Niño events on the fishing industry. You saw that the fish catches in the Pacific vary widely and that often the lowest catches are associated with strong El Niño events. These periods of low fish productivity can be related to low concentrations of phytoplankton. You can now speculate on how the effect of El Niño event on food production might have consequences in your community.

Digging Deeper

El Niño, Phytoplankton, and Global Geochemical Cycles

The Peruvian anchovy fishery is sustained by upwelling. As you learned in Activity 3, upwelling is caused by winds that create a surface current that flows offshore. As deep water rises to replace this water, nutrients are brought to the surface. The upwelled nutrients promote the growth of **phytoplankton**, which, in turn, provide food for **zooplankton** such as copepods, arrow worms, fish larvae, and other small marine animals. Other predators, including anchovies, prey upon these organisms. Marine birds like cormorants, pelicans, large fish, and squid feast on the anchovies, as do the fishing trawlers.

Figure 1 How are marine birds impacted by El Niño events?

This upwelling of nutrient-rich water causes the coastal areas on the eastern side of ocean basins to teem with life. During an El Niño event, however, the upwelling brings warm water rather than cold water to the surface, as you learned in **Activity 4**. The higher sea surface temperatures force the fish located in coastal areas to migrate north and south in search of cooler waters and food. Many fish, not able to migrate, die from lack of food or the increase in temperature. Those that are able to move north and south do not fare much better. These fish find themselves in unusually cold waters in which many cannot survive.

Figure 2 Large fish like sharks, near the top of the oceanic food chain, feel the impacts of El Niño events.

An increase in rainfall that accompanies El Niño events also has an effect on coastal species of fish. This results from an increase in turbidity (cloudiness) and a decrease in salinity. Torrential rains greatly increase river discharge, which brings with it large amounts of sediment and fresh water to the ocean.

Geo Words

phytoplankton: tiny plants that float in the near-surface waters of the ocean.

zooplankton: tiny animals that live in the near-surface waters of the ocean. They consume phytoplankton.

Check Your Understanding

1. Revisit the question asked at the start of this activity (Think about It). Revise your response on the basis of what you have learned.
2. In your own words, describe the oceanic food chain from solar energy to humans.

These factors have been shown to have some negative effect on the fish populations, but they do not seem to be as important as increased sea surface temperatures and reduction in upwelling.

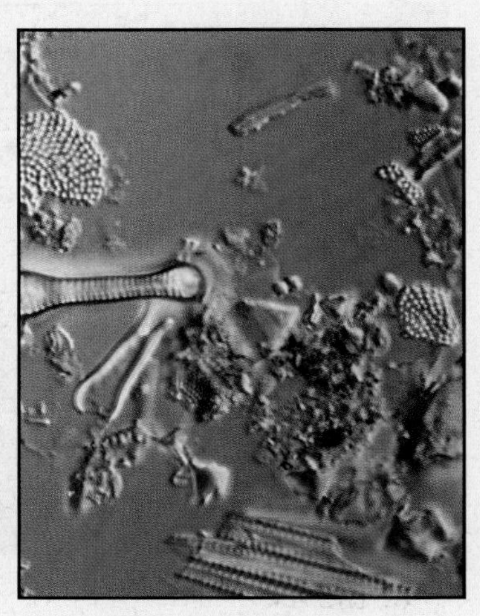

Figure 3 Microscopic organisms like diatoms are the first link in the marine food chain.

Organisms that float with the ocean current, but may or may not have some capability for weak movement, are known as plankton. Minute plants are known as phytoplankton. Zooplankton are tiny animals. Some, such as krill, spend their entire lives as zooplankton, whereas others are simply the larval stages of other animals, such as fish. Phytoplankton, which consist mainly of algae, convert the energy of the Sun into matter. Diatoms are a main component of the marine phytoplankton. Massive blooms of this organism can change the ocean from being crystal clear to having the consistency more like pea soup in a matter of days.

Besides acting as the first link in the food chain, phytoplankton are a critical part of ocean chemistry. The carbon dioxide in the atmosphere is in balance with carbon dioxide in the ocean. During photosynthesis, phytoplankton remove carbon dioxide from seawater and release oxygen as a byproduct. This allows the oceans to absorb additional carbon dioxide from the atmosphere. If fewer phytoplankton existed, atmospheric carbon dioxide would increase.

Figure 4 General diagram of the oceanic food chain. Note that organisms are not shown to scale.

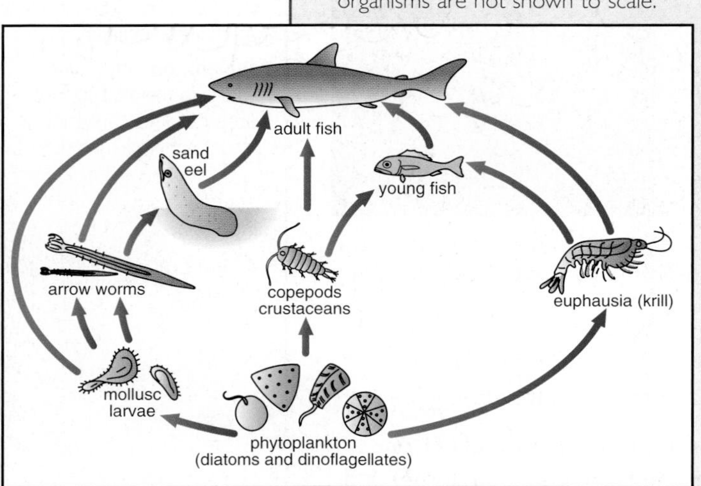

This is an important example of the complexity of the Earth system. To those who live far from the west coast of the United States, an event such as an El Niño seems remote. It is difficult to imagine how a change in ocean water temperature in the eastern equatorial Pacific Ocean might affect someone living in Washington, D.C. But the Earth's systems are connected. An El Niño event that reduces the abundance of phytoplankton can have global effects because it causes a change in the amount of carbon dioxide being removed from the atmosphere.

Understanding and Applying What You Have Learned

1. According to your data and graph, which was the stronger El Niño event: 1982–1983 or 1957? Explain your answer.
2. Think of all the steps involved and people employed to bring fish to your community. Which occupations and resources would be affected in your community if there is a poor fish catch?
3. Would you depend on fish data exclusively to determine when an El Niño event occurs? Why or why not?
4. Do you see a cyclic or predictable trend to the fish-catch data such that you might be able to predict the next El Niño event? If so, what is your prediction? Visit the *EarthComm* web site to check your prediction above with the latest data on El Niño events.

Preparing for the Chapter Challenge

Prepare a brief report giving a few examples of how natural events like floods or drought have affected the supply of staple foods. These effects could be availability, prices, or quality of the food. The price of orange juice is a good example to pursue. You can use this report as possible effects of El Niño on your community.

Inquiring Further

1. **El Niño and marine birds**

   Use the Internet to research the effect of El Niño on marine birds. Visit the *EarthComm* web site for links to get you started.

2. **El Niño and the Atlantic fishing industry**

   The Grand Banks in the North Atlantic are a world-renowned fishing area. Determine through research if El Niño events in the Pacific had an impact on fishing in the Atlantic.

Earth Science at Work

ATMOSPHERE: *Cinematographer*
When filming a movie, cinematographers must consider the setting and background as much as the actors.

BIOSPHERE: *Waitress*
Restaurants depend upon a reliable, daily harvest from the ocean.

CRYOSPHERE: *Highway Maintenance Worker* Snow clearing and removal are vital in many states.

GEOSPHERE: *Real Estate Agent*
Real estate agents work to meet the needs of the many people who appreciate an oceanview setting.

HYDROSPHERE: *Diver*
Using scuba gear or a deep sea diving suit, a diver works below the surface of the water to inspect, repair, install, and remove equipment and structures.

How is each person's work related to the Earth system, and to oceans?

Severe Weather ...and Your Community

2 Severe Weather ...and Your Community

Getting Started

Thunderstorms, tornadoes, or floods can happen almost anywhere. In some regions, these events occur very often, whereas in others they are unusual. However, at some time in your life you will probably experience some form of severe weather.

- What severe weather events are most common in your community?
- How do those events compare with severe weather events in other parts of the country?

What do you think? Write down your answers to these questions in your *EarthComm* notebook. Be prepared to discuss your descriptions with your small group and the class.

Scenario

An entertainment company wants to build a public arena in or near your community. The company proposes a circular arena that will seat 25,000 people. The arena will have a retractable roof that will allow for open-air rock concerts in good weather. Closing the roof will keep the company from having to cancel concerts in bad weather. Local government officials will approve the proposed arena if a plan is in place to ensure the safety of concert-goers. Of special concern is safety in severe weather. It takes an hour to close the roof of the building. Before and after concerts, roads leading to and

from the arena will be clogged with cars. Some of the roads may cross streams that may flood in severe weather. Will people be safe if severe weather develops? What warning signs will help the company decide whether to close the roof or cancel a concert? The entertainment company has asked the *EarthComm* students in your school to help them evaluate the potential for severe weather hazards in your community. Can you use your understanding of weather to explain how to recognize when severe weather threatens your region? You will also need to explain what to do during a weather-related emergency.

Chapter Challenge

Your challenge is to develop a report that will explain to the entertainment company how to recognize when severe weather hazards may be approaching, and what to do in case of such hazards. Your report will help the company to complete their final plans for the proposed arena. Without your report, the arena project will be dropped. Your report should address:

- The weather-related hazards that are likely to occur in your community.
- How often the hazards are likely to occur, and at what times of the year.
- How dangerous these hazards are and what kind of damage they can produce.
- What conditions might contribute to the development of these hazards.
- The warning signs that indicate when severe weather is developing.
- How citizens can be informed on short notice (in less than one hour) when weather-related hazards are expected.
- How citizens can prepare in advance for specific weather hazards.
- What safety procedures citizens should follow during and after severe weather events.

Assessment Criteria

Think about what you have been asked to do. Scan ahead through the chapter activities to see how they might help you to meet the challenge. Work with your classmates and your teacher to define the criteria for assessing your work. Devise a grading sheet for the assessment of the challenge. Record all this information.

Make sure that you understand the criteria and the grading scheme as well as you can. Your teacher may provide you with a sample rubric to help you get started.

Activity 1 What Conditions Create Thunderstorms?

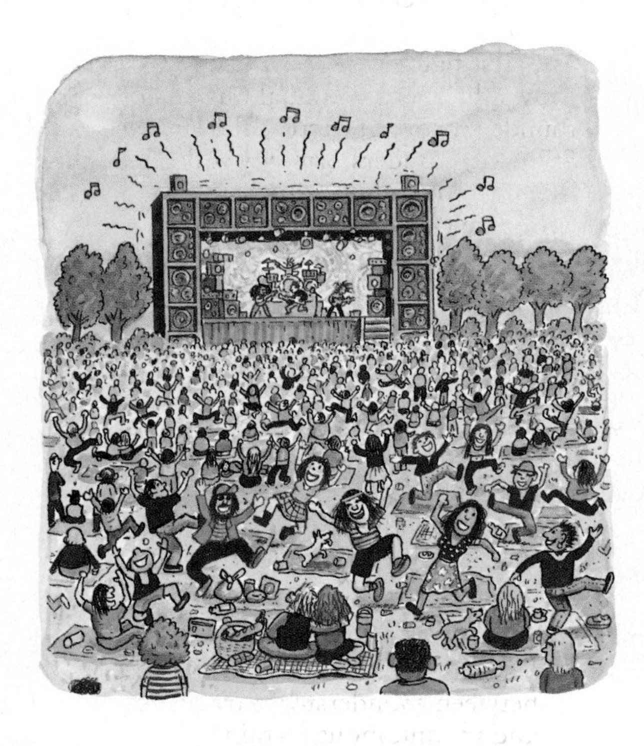

Goals

In this activity you will:

- Investigate how the density of air changes with change in air temperatures.
- Relate the frequency of occurrence of thunderstorms to the topography of an area.
- Relate the occurrence of thunderstorms to the temperatures in an area.
- Recognize areas in the United States where thunderstorms are relatively frequent.
- Learn how air masses at different temperatures interact.
- Understand why severe weather often occurs during specific seasons.

Think about It

At any given moment about 1800 thunderstorms are happening on Earth. This equates to 16 million thunderstorms per year worldwide.

- Where are thunderstorms most likely to occur in the United States? Why?
- How do thunderstorms form?

What do you think? Record your ideas about these questions in your *EarthComm* notebook. Include any sketches or diagrams. Be prepared to discuss your responses with your small group and the class.

Investigate

Part A: Temperature Effects on Air

1. Inflate two balloons to the same volume. Place one balloon in the refrigerator (or in an ice-filled cooler). Keep the other balloon at room temperature. After 30 minutes, remove the balloon from the refrigerator. Quickly measure the volumes of the two balloons.
   a) Record all the observations you made.
   b) Record the method you used to determine the volumes of the two balloons.
   c) Assume that no air escaped from either balloon. What do the volumes of the balloons tell you about the density of air in each balloon?
   d) How does air temperature affect the density of air?
2. Obtain a plastic clothes bag—the type used at a dry cleaner. Seal the hole at the top end of the bag with a small piece of masking tape. Tape the bottom of the bag so that it has an opening of about 15 cm. Using a hair dryer with its setting on "low," inflate the plastic bag. (Do not attach the bag to the hair dryer.) Now release the inflated bag.
   a) Record your observations in your notebook.
   b) Explain your observations. Refer to the density of air in your explanation.

If the balloon pops, collect all of the scraps and discard them.

Part B: The Meeting of Warm and Cold Air

1. The first figure on the following page shows the average annual number of thunderstorm days across the United States. A thunderstorm day is defined as a day when thunder is heard. Compare this to the second figure, a shaded relief map of the United States.
   a) On average, which places experience the most thunderstorms?
   b) On average, which places experience the fewest thunderstorms?
   c) How does the frequency of thunderstorms change from north to south over the eastern half of the nation?
   d) What is the relationship, if any, between thunderstorm frequency and mountainous terrain?
   e) How does thunderstorm frequency relate to proximity to large bodies of water?
   f) Compare the thunderstorm frequency on the East Coast and the West Coast. How do they compare?
   g) In your notebook, list all the factors that you think may influence thunderstorm frequency.
   h) Which of these factors are present in or around your community? Support your answer based on the frequency of thunderstorms in your community.

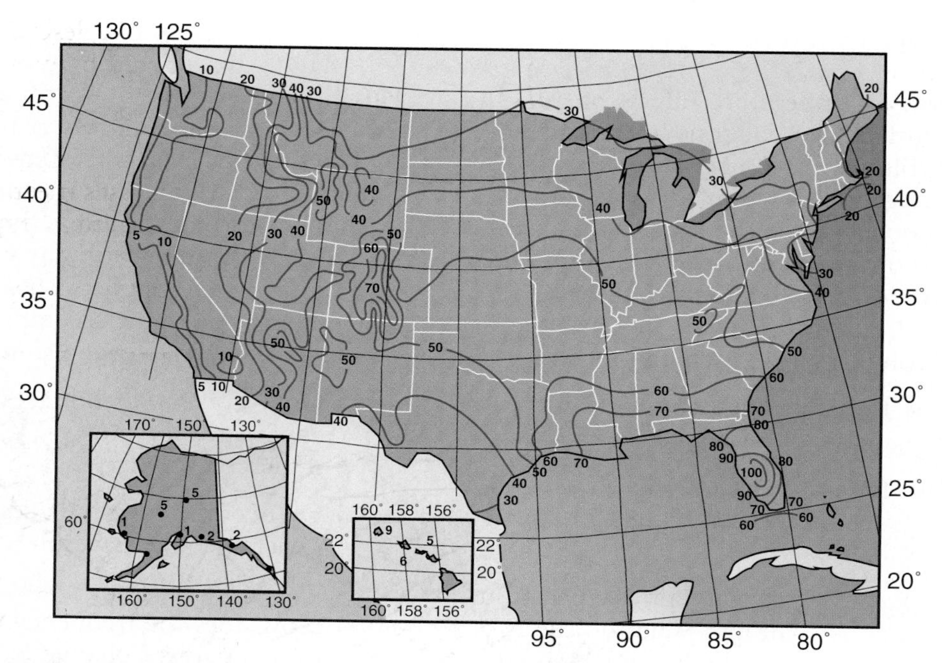

Average number of thunderstorm days across the United States.

Shaded relief map of the United States.

2. Below is a map of temperature observations reported at the same hour of March 8, 1998 at various weather stations in the United States. Place a clear transparency or sheet of tracing paper over the map.
   a) Draw a thick line to separate warm air masses from cold air masses.
   b) In your notebook, describe how you differentiated between warm and cold air masses.
   c) In your notebook, describe what you think happens when warm and cold air masses meet. Draw and label a diagram to explain your reasoning.

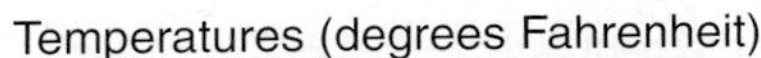

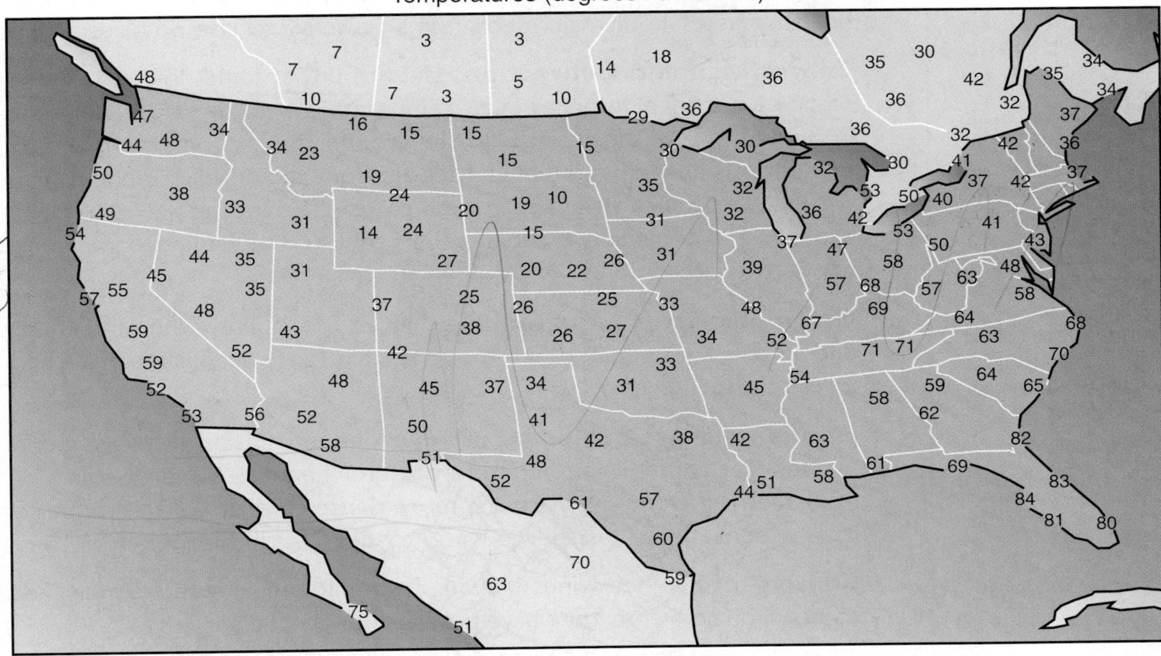

Map of temperatures (°F) across the United States on March 8, 1998.

Reflecting on the Activity and the Challenge

In this activity, you saw that a sample of air contracts, and therefore becomes denser, when it cools, even when the pressure acting on the sample is not changed. You also saw that warm air rises above cooler air. You noted that thunderstorms seem to be most frequent at lower latitudes over the central and eastern United States and in areas near mountainous terrain. You observed that thunderstorms are infrequent along the West Coast. Finally, you saw that warm and cold air masses meet in different regions of the United States. Understanding how air masses at different temperatures interact will help you understand how thunderstorms form. This, in turn, will help you identify regions that may be susceptible to severe weather.

Digging Deeper

HOW AND WHY WARM AIR RISES

As air warms, it becomes less dense than its cooler surroundings and rises. Warm air is less dense than cold air because molecules in warm air are more active than the molecules in cold air. (Air is composed of molecules of gases, mostly nitrogen and oxygen.) As the molecules move about with increasing speed, the spacing between individual molecules increases. As the spacing between molecules increases, density of the air decreases. Air masses can also be forced to rise when they flow over mountains or collide with other air masses. Four main mechanisms cause air to rise on a regional scale:

- ***Convective uplift:* Convection** refers to upward and downward motions of air caused by differences in air temperature (and therefore, differences in air density). As the Sun warms the ground, the ground warms the air immediately above it. Cooler, denser air from above sinks and forces the warmer, less-dense air at the ground to rise. The ascending warm air expands and cools and eventually sinks back to the ground, completing the convective circulation.
- ***Orographic uplift:* Orographic** refers to mountains. When the wind encounters a mountain range, the mountain range acts as a barrier, and forces the air upward.
- ***Frontal wedging:*** A **front** is a narrow zone of transition between air masses that contrast in temperature and/or humidity. When a cold air mass meets a warm air mass, the more dense cold air wedges or forces its way beneath the less dense warm air along a cold front.
- ***Convergence:*** When winds blowing from different directions meet head-to-head, or converge, they have nowhere to go but up.

What is a Front?

Figure 1 shows the "source regions" for the air masses that regularly move over North America. As indicated by the arrows in the figure, cold air masses usually flow southeastward and warm air masses flow northeastward. As air masses move out of their source region, they usually change, depending on the route they travel. For example, in winter a cold air mass is warmed as it moves southeastward from its snow-covered source region to the bare ground of the southern United States. This is one of the main reasons why winter temperatures do not fall nearly as low in Florida as they do in the Great Lakes region.

Geo Words

convection: the transfer of heat by vertical movements in the atmosphere as a result of density differences caused by heating from below.

orographic: relating to mountains.

front: a narrow zone of transition between air masses that contrast in temperature and/or humidity.

Air masses are classified by their temperature and humidity (or moisture content), as follows:

- Continental: relatively dry air masses that form over land.
- Maritime: relatively humid air masses that form over the ocean.
- Polar: cold air masses that form at high latitudes (northern Canada, for example).
- Tropical: warm air masses that form at low latitudes (Gulf of Mexico, for example).

Therefore, the general types of air masses include continental polar (cold and dry), continental tropical (warm and dry), maritime polar (cold and humid), maritime tropical (warm and humid) and arctic air (exceptionally cold and dry).

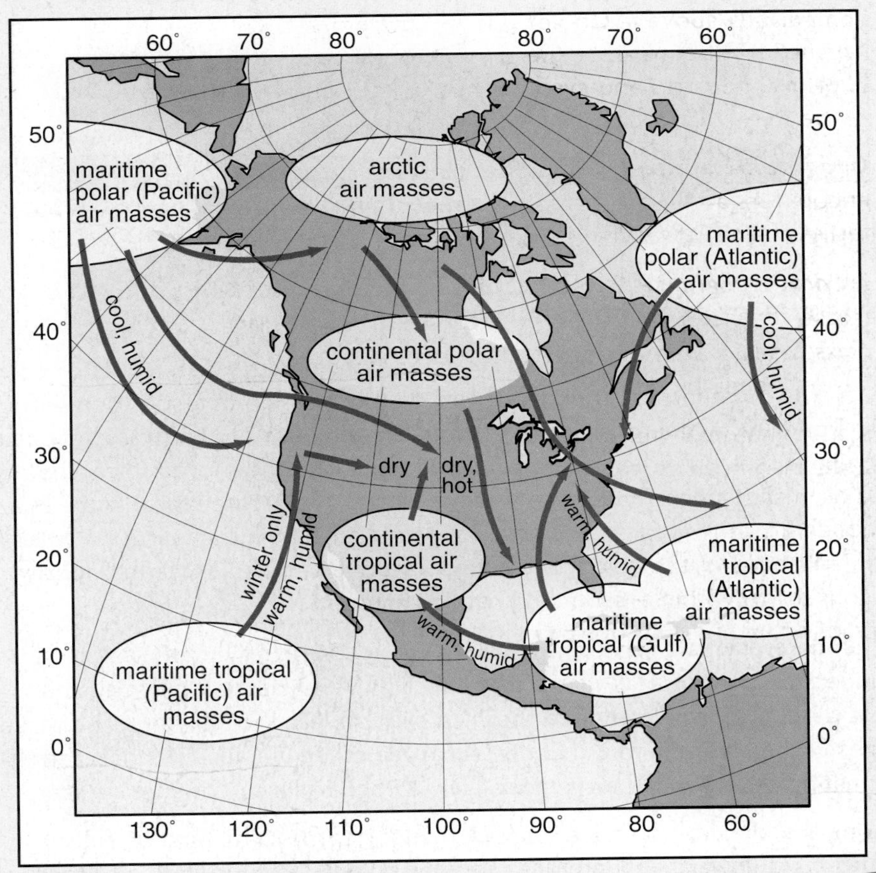

Figure 1 Source regions for air masses in North America.

When different air masses come together, the narrow zone of transition that forms between them is called a front. If the colder (or drier) air advances while the warmer (or more humid) air retreats, the transition zone is a cold front (*Figure 2*). If the warmer (or more humid) air advances while the colder (or drier) air retreats, the transition zone is a warm front.

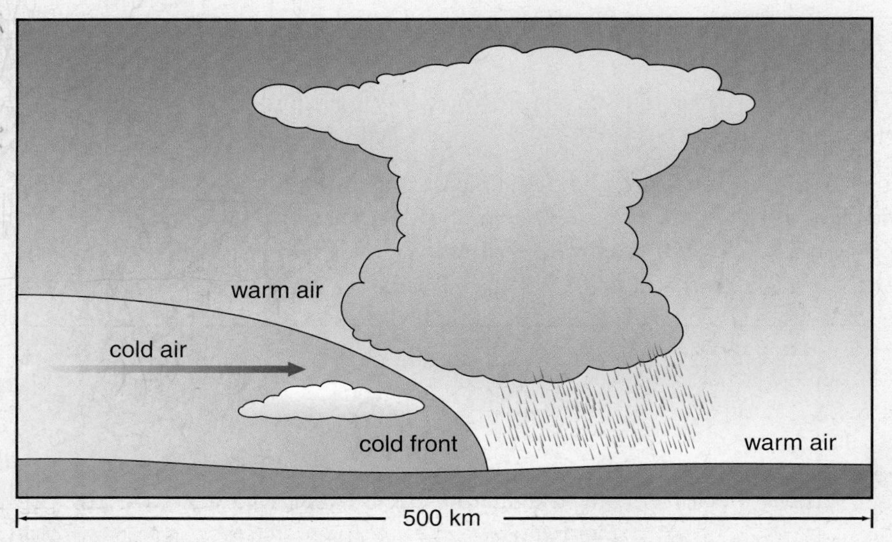

Figure 2 A cold front is the transition zone between cold, dry air and warm, humid air.

Development of a Thunderstorm

Thunderstorms can be caused by air rising along either warm fronts or cold fronts. The most intense thunderstorms usually develop in association with cold fronts. As a general rule, the greater the temperature contrast between two air masses on either side of the cold front, and the more rapidly the cold-air mass wedges under the warm-air mass, the greater the chance that a thunderstorm will form. The advancing cold air forces the warm air to rise along or ahead of the cold front. Thunderstorms can also be caused locally by convection within a warm and humid air mass.

A thunderstorm is a relatively small, short-lived weather system. A typical thunderstorm is so small that it may affect only one section of a city, and most thunderstorms complete their life cycle in less than an hour. The three stages in the life cycle of a thunderstorm are called cumulus, mature, and dissipating. Severe weather is most likely to occur during the mature stage.

During the initial or cumulus stage of thunderstorm formation, cumulus clouds build upward and laterally. Growing cumulus clouds have flat bases and towering tops, resembling cauliflower, as seen in *Figure 3*. Cumulus clouds develop where air ascends as an updraft (*Figure 4*). As the air reaches higher levels in the atmosphere, it expands, because the pressure is lower. The air cools as it expands, because some of the heat energy of the air is used up in the work of expansion. The cooling of the air causes some of the water vapor in the air to condense, producing clouds. No precipitation occurs during the cumulus stage of thunderstorm development. Most cumulus clouds do not develop into thunderstorms but all thunderstorms begin with cumulus clouds.

Figure 3 The flat bases and towering tops of growing cumulus clouds often represent the first stage of thunderstorm formation.

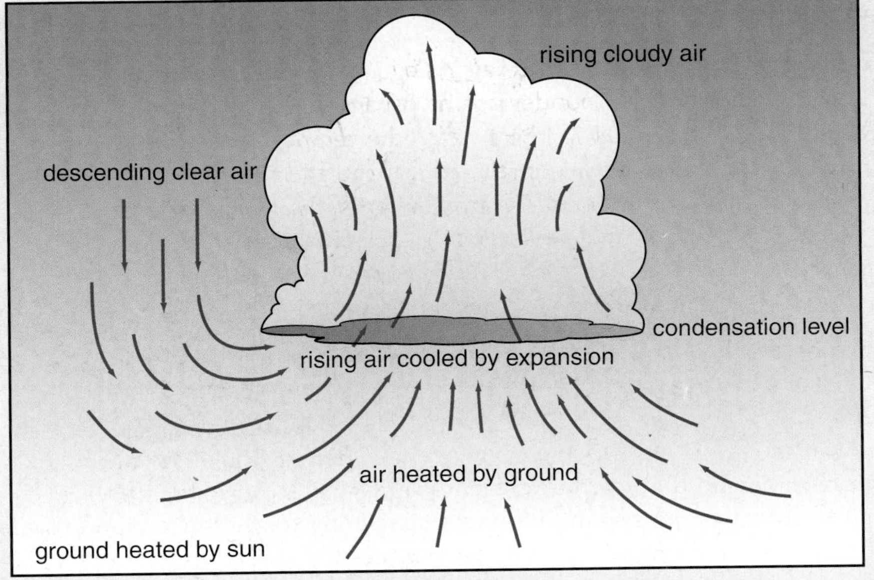

Figure 4 Surface heating and updrafted air result in the formation of clouds.

A thunderstorm that produces severe weather requires a very strong updraft that builds the developing cumulonimbus cloud to great altitudes. Most of the time an updraft due to convection alone is not strong enough to build a cumulus cloud into a cumulonimbus (thunderstorm) cloud. In order for this to happen, the updraft must be strengthened by some mechanism other than convection. Frontal wedging, orographic uplifting, or converging surface winds can strengthen an updraft and cause cumulus clouds to build vertically into thunderstorm clouds.

If an updraft is strong and persistent, the cumulus cloud becomes taller and taller and eventually produces rain, hail, or even snow. Once precipitation begins, the thunderstorm has entered its mature stage, and the cloud is called a cumulonimbus cloud (nimbus means rain). Falling raindrops (or hailstones) drag the surrounding air downward, forming a downdraft alongside the updraft. During the mature stage, precipitation is heaviest, lightning is most frequent, and hail and even tornadoes may develop. As a rule, the higher a thunderstorm builds into the atmosphere, the more likely it is to produce severe weather.

Figure 5 A thunderstorm reaches its mature stage with the onset of precipitation.

Precipitation and the downdraft eventually spread throughout the thunderstorm as the system enters its dissipating stage. Precipitation tapers off and ends, clouds dissipate, and the chances for severe weather decrease dramatically.

Check Your Understanding

1. In your own words, describe a front.
2. Draw a diagram that shows how an updraft forms due to:
   a) Convection.
   b) Orographic uplift.
3. Provide a definition of a thunderstorm.
4. Describe how an approaching cold front can promote the development of thunderstorms.
5. Describe the three stages of a thunderstorm.

Understanding and Applying What You Have Learned

1. In the central and eastern parts of the United States, what accounts for the general increase in thunderstorms as you go south?
2. In the United States, during which season(s) (spring, summer, winter, fall) do you expect the most thunderstorms to occur? Explain your answer.
3. Refer back to the maps used in this activity. Explain why there are so many thunderstorms in Louisiana and so few thunderstorms in the state of Washington.
4. Why is convective uplift likely to be strongest during the warmest time of day?
5. Central Florida's high thunderstorm frequency is related to its sea breezes, which often develop on hot afternoons. A sea breeze is a cool wind that blows inland from the ocean. In Florida, sea breezes push inland from both the Atlantic Ocean and Gulf of Mexico. Because Florida is nearly flat, the sea breezes push inland and converge over the central part of the state.
   a) Sketch a cross-sectional diagram that shows what happens where the sea breezes come together.
   b) How does the meeting of the sea breezes help explain central Florida's high thunderstorm frequency?
6. On a warm summer afternoon, why might you see scattered cumulus clouds? In other words, why might a cumulus cloud form in one place and not another?
7. A severe thunderstorm is located 100 km directly west of your community and headed toward the east. Is severe weather inevitable in your community? Explain your answer.

Preparing for the Chapter Challenge

You have been asked to develop a report that will explain how severe weather might affect your community. Local government officials will use your report to authorize construction of a concert arena. Review what you have learned about thunderstorms in this activity. In your notebook, outline several basic scientific concepts that will help local officials to understand the frequency and formation of thunderstorms in your community. Be sure to note how the topography and latitude of your community influence the development of thunderstorms. Include diagrams and sketches as needed to support your outline.

Inquiring Further

1. **Personal memories of thunderstorms**

   Ask members of your family or neighbors about the most memorable thunderstorms they ever witnessed in your community. (You may wish to describe a thunderstorm you have witnessed.)

   - What do they remember about the appearance, development, and how long the storm lasted?
   - What thunderstorm hazards or impacts have they experienced?
   - What was the worst thunderstorm they ever experienced?

2. **Keeping a weather log**

   Does your school have a weather station? If so, maintain a daily log of weather conditions. For example, record temperature, humidity, air pressure, wind speed and direction, and cloud cover at a specified time each school day. Download and print daily weather maps of your state or region from the Internet to track weather events over time.

3. **Severe weather events in your community**

   Research information on severe weather events that have affected your community in the past. Visit the *EarthComm* web site to help you get started with your research. You may even wish to include some of this information in your **Chapter Challenge** report.

Activity 2 A Thunderstorm Matures

Goals

In this activity you will:

- Investigate relationships between air temperature and water vapor.
- Observe and record changes in the shape and movement of clouds.
- Simulate cloud formation in a bottle.
- Describe cloud formation in terms of expansional cooling.
- Explain the conditions that cause thunderstorms to mature.
- Learn about conditions that may lead to the development of severe weather.

Think about It

Thunderstorm clouds can build tens of thousands of feet up into the atmosphere as they move over Earth's surface.

- What causes towering rain clouds to form?
- What causes such clouds to move from place to place?

What do you think? Record your ideas about these questions in your *EarthComm* notebook. Include a diagram showing towering cumulus clouds and any relationships that you can make between how they form and how they move. Be prepared to discuss your responses with your small group and the class.

Investigate

Part A: Characteristics of Clouds

1. Examine the three photographs shown below. The photographs are of the same mass of clouds, taken about 10 minutes apart. (Note that the distance from the cloud mass is increasing.)

   a) What changes can you see in the appearance of the clouds?

   b) How do you explain these changes?

   c) Draw a diagram in your notebook that shows how you think these changes occurred.

2. Look at the photograph below of an intense thunderstorm that has reached the mature stage. The photograph was taken some distance from the thunderstorm so that the full height of the cloud can be seen.

   a) In your notebook, draw a diagram of the shape of the clouds. Using arrows on the sketch, indicate the directions of the wind.

   b) Explain why you think the cloud has a relatively flat top.

3. Spend some time outside observing the sky.

   a) In your notebook, record wind direction, wind speed (light, moderate, strong), temperature, and precipitation. Also sketch the different kinds of clouds you see (if any).

   b) Describe any changes in the shape and location of clouds over time.

   c) Describe any differences in the appearance of clouds at different altitudes.

   d) Do any of the clouds look like they are producing precipitation? How do you know?

Part B: Observing the Behavior of Water Vapor

1. Half fill a 500-mL beaker with room-temperature water. Carefully observe the outside of the beaker while slowly adding ice cubes one by one.

   a) Record your observations in your notebook.

   b) Explain your observations. Consider how adding ice changed the situation.

2. Obtain two clean, transparent plastic 2-L bottles. Fill one bottle with very hot tap water and the other with cold tap water. Pour water out of both bottles until about 3 cm of water remains in each.

   Place an ice cube on the top of the open neck of each bottle. Observe the bottles for several minutes.

   a) Record your observations of each bottle in your notebook. What differences do you observe between the two bottles?

   b) How can you explain the difference?

Be careful picking up the beaker. The sides may be wet and slippery. Be careful with the hot water. It should not be hot enough to scald. Use a funnel to put the hot water into the bottle. Use insulated gloves or pot holders to hold the bottle.

Part C: Making Clouds

1. Fill a clean, transparent plastic 2-L bottle with very warm (but not steaming hot) tap water. Pour the water out until about two 2 cm of water remain in the bottle.

2. Light a match and immediately blow out the flame. Hold the bottle sideways and place the smoking match in the neck of the bottle to allow a little smoke to enter the bottle. Replace the cap on the bottle.

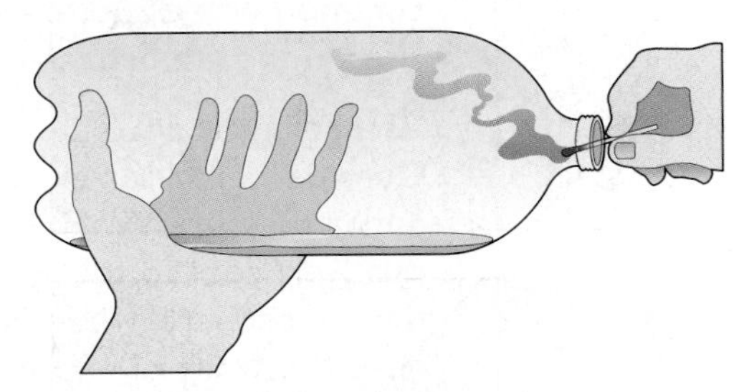

3. Crush the bottle against the edge of the table. Then make the bottle return to its original shape.

   a) Record your observations in your notebook.

   b) Why was smoke from a smoldering match introduced into the bottle?

Wear safety goggles. Place burned matches in a cup of water to soak before placing them in the trash can.

Reflecting on the Activity and the Challenge

In this activity you noted and recorded changes in the shape and movement of clouds using images and outdoor observations. You also explored the relationship between air temperature and water vapor. You explored some of the conditions under which water vapor condenses into clouds. In order to complete the **Chapter Challenge**, you will need to understand why clouds sometimes produce severe weather and create hazards. To do so, you need to understand the formation of clouds.

Digging Deeper

HOW CLOUDS FORM

All air contains water vapor, which is water that exists in the atmosphere as an invisible gas. Water vapor is not the same as clouds, fog, or steam, which are composed of tiny, visible droplets of water suspended in the atmosphere. There is an upper limit to the concentration of water vapor in air. When that upper limit is reached, the air is saturated with water vapor. The upper limit depends on temperature. Warm saturated air has more water vapor than cool saturated air. When air is saturated with water vapor, some water vapor condenses into tiny water droplets (or, if the temperature is very low, forms tiny ice crystals) that are visible in the form of clouds. The most common way clouds develop in the atmosphere is by cooling of air as it rises in the atmosphere.

Ascending air cools because of expansion. A gas cools when it expands. Consider a familiar example of expansional cooling. Air inside a bicycle tire is under pressure. When you open the tire valve, the air that escapes is relatively cool to the touch. The escaping air expands because it is entering the atmosphere, where air pressure is much less than it is inside the tire. Air does the work of expansion as it escapes the tire. Work requires energy, and that energy is drawn from the internal heat energy of the air, causing a drop in temperature.

Air pressure is the cumulative force of a multitude of air molecules colliding with a unit surface area of any object in contact with air. You can think of air pressure as the weight of a column of air acting on a unit area at the base of the column. Air pressure always decreases with increasing altitude, because the mass of air above steadily diminishes (*Figure 1*). Air that rises, as in an updraft, expands because it encounters progressively lower air pressures. As air does the work of expansion, it cools. With sufficient cooling, the air becomes saturated, and excess water vapor condenses into droplets, which form clouds. This happens in the updraft of a thunderstorm.

Conversely, when a gas (or mixture of gases) is compressed, the environment does work on the gas and its temperature rises. For example, as air is pumped into a bicycle tire it is compressed, causing the air to warm and the tire to become warm to the touch. Air descending in the atmosphere encounters steadily increasing air pressure and is compressed and warms. Compressional warming occurs in the downdraft of a thunderstorm. During the dissipating stage of a thunderstorm, the downdraft spreads through the thunderstorm cloud, and with compressional warming, the cloud vaporizes.

Geo Words

air pressure: the cumulative force of a multitude of air molecules colliding with a unit surface area of any object in contact with air.

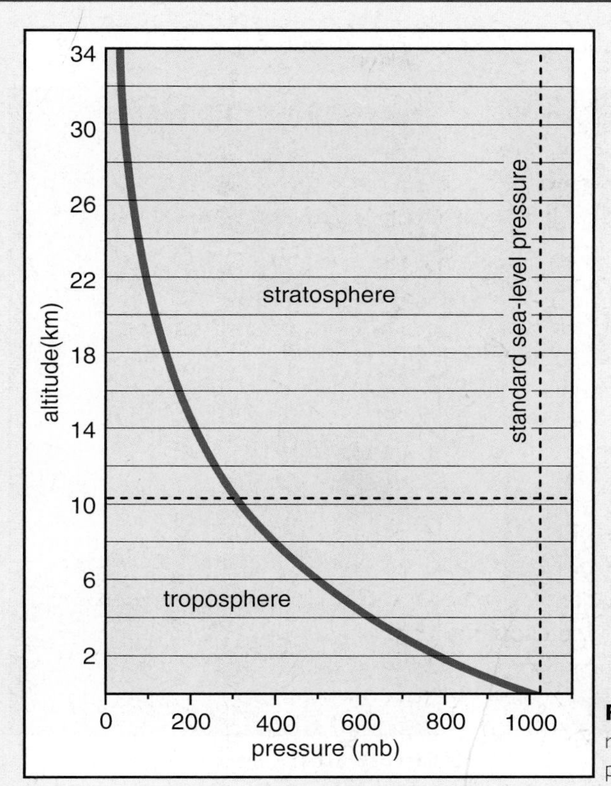

Figure 1 Graph showing the relationship between altitude and air pressure.

Geo Words

condensation nuclei: tiny solid and liquid particles that water vapor can condense on.

In addition to saturated air, cloud formation requires surfaces on which the water vapor can condense. In **Part B** of the investigation, the outside of the beaker provided a surface. In **Part C** of the investigation, smoke particles provided surfaces. Earth's atmosphere contains an abundance of tiny solid and liquid particles that water vapor can condense on. The particles are called **condensation nuclei**. Nuclei are products of many different natural and human-related activities. Forest fires, volcanic eruptions, wind erosion of soil, saltwater spray, motor vehicle exhaust, and various industrial emissions are all sources of nuclei. Some nuclei promote condensation (called condensation nuclei) and others cause formation of ice crystals (called ice-forming nuclei).

Types of Clouds

Meteorologists classify clouds into three broad categories based upon shape: cirrus, stratus, and cumulus. Cirrus clouds are wispy, stratus clouds are layered, and cumulus clouds are puffy (like cottonballs). Clouds are further classified according to their altitude (high, middle, low, or clouds with vertical development) and composition (water droplets or ice crystals), as shown in the photographs on the following pages.

High Clouds (cloud base above 6000 m, about 20,000 ft.)

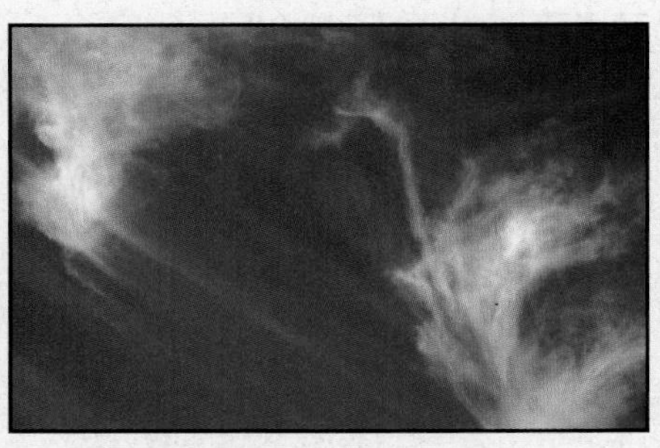

Cirrus: Thin, wispy and feathery; composed of ice crystals; also called "mares' tails."

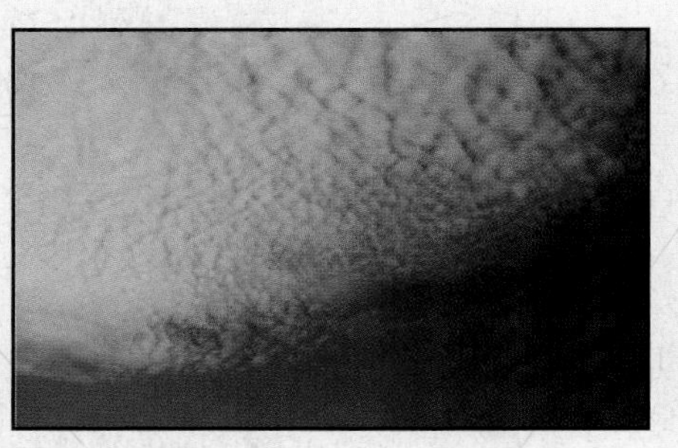

Cirrocumulus: Puffy and patchy in appearance; may form wave-like patterns, which are sometimes called "mackerel sky."

Cirrostratus: Light gray or white; thin sheets of ice crystals that usually cover much of the sky; may create a halo around the Sun or moon.

Middle Clouds (cloud base 2000–6000 m, about 6500–20,000 ft.)

Altostratus: Uniformly gray or bluish white layers composed mostly of water droplets; covers most of the sky in a thin sheet; Sun or moon may shine through as a bright spot as if viewed through frosted glass.

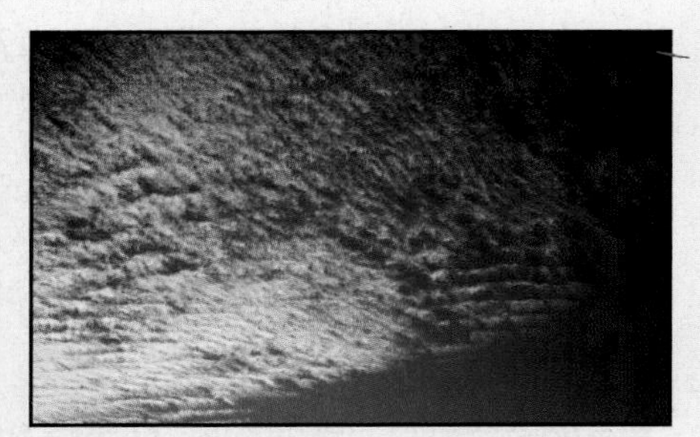

Altocumulus: Roll-like puffy, patchy clouds composed mostly of water droplets, and grouped in large sheets; also called "sheep's back" clouds.

Low Clouds (cloud base below 2000 m, about 6500 ft.)

Stratus: Light or dark gray, low cloud; covers most of the sky; may produce drizzle; fog is a stratus cloud in contact with the ground.

Nimbostratus: Thick dark gray cloud producing rain or snow; often with a ragged base.

Stratocumulus: Irregular masses of clouds, often rolling or puffy in appearance.

Vertically Developed Clouds (cloud thickness from 500 to 18,000 m, or about 1600–60,000 ft.)

Cumulus: puffy white clouds resembling cotton balls, popcorn, or cauliflower heads floating in the sky; usually have almost flat bottoms; occur individually or in groups.

Cumulonimbus: thunderstorm clouds; tall, billowing towers of puffy clouds with flat bases; can have sharp, well-defined edges or anvil shape at the top; often produces rain and sometimes hail, strong winds, or tornadoes.

Geo Words

troposphere: the portion of the atmosphere next to the Earth's surface, in which temperature generally decreases rapidly with altitude.

tropopause: the top boundary of the troposphere.

stratosphere: the outer layer of the atmosphere overlying the troposphere. The air temperature is at first constant with altitude and then increases with altitude.

The Mature and Dissipating Stages of a Thunderstorm

In the previous activity you learned that a thunderstorm begins when a cumulus cloud develops in an updraft of air. The more humid the air, the better the chance that cumulus clouds will form. If conditions in the atmosphere favor thunderstorm development, cumulus clouds will continue to billow upward and eventually the system will reach its mature stage (when precipitation begins).

In general, the more vigorous the updraft, the greater the altitude to which a thunderstorm cloud builds, and the more likely that the thunderstorm will produce severe weather. To recognize how high a thundercloud can build, consider the four different layers of the Earth's atmosphere (listed from lowest to highest): **troposphere**, **stratosphere**, mesosphere and thermosphere. The boundaries between these layers are defined by air temperature, that is, how air temperature varies with altitude in each layer, as shown in *Figure 2*.

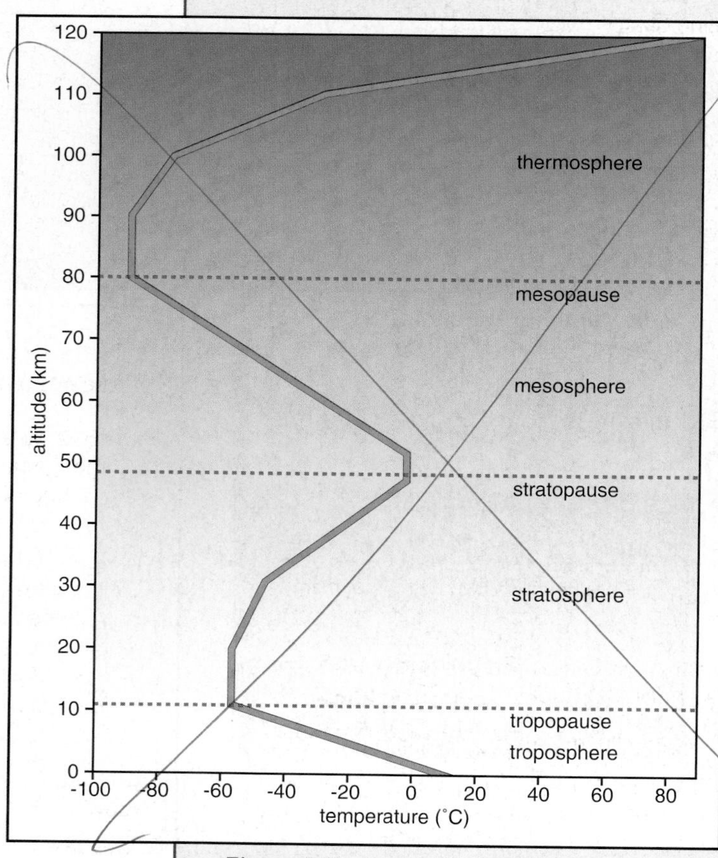

Figure 2 The layers of the atmosphere are defined on the basis of how air temperature varies with altitude.

We live in the troposphere. On average, air temperature drops with increasing altitude up to the top boundary of the troposphere, called the **tropopause**. Air temperatures are usually lower in mountainous terrain than at sea level. The next layer up is the stratosphere, in which the air temperature is at first constant with altitude and then increases with altitude. A thunderstorm cloud that pushes above the tropopause and into the lower stratosphere will be colder and denser than the surrounding air and will sink back down into the troposphere. For that reason, even very intense thunderstorms cannot build much higher than the tropopause. Usually, cumulonimbus clouds that billow up to the tropopause spread laterally and develop the characteristic flat, anvil-shaped top as shown in *Figure 3.* An anvil top is most commonly displayed by an intense thunderstorm during its mature stage.

Thunderstorm precipitation may be in the form of rain, snow, or hail. Precipitation falls through the updraft, weakening it and eventually dragging air downward, producing a downdraft alongside the updraft. The downdraft leaves the base of the cloud and flows along the ground ahead of the shaft of precipitation. The leading edge of this rain-cooled gusty air is like a miniature cold front and is known as a **gust front**. (See *Figure 3.*) This explains why a relatively cool and gusty wind often precedes an approaching thunderstorm.

Geo Words

gust front: the leading edge of the rain-cooled gusty air preceding a thunderstorm.

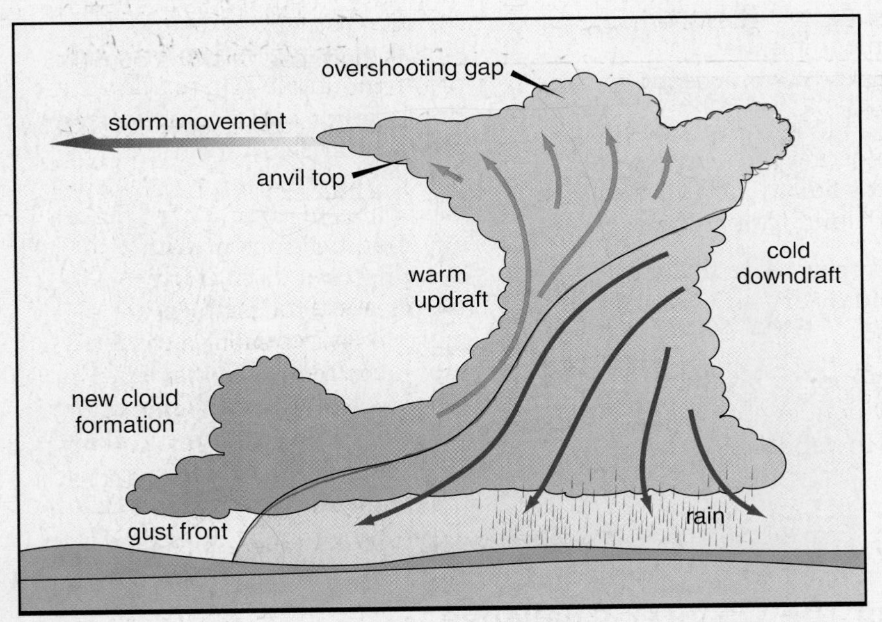

Figure 3 Profile of a thunderhead. The arrows indicate the directions of air movement, as either updrafts or downdrafts.

In an intense thunderstorm, a downdraft may strike the ground with wind speeds in excess of 100 km/h (60 mph). An intense downdraft is known as a downburst. A downburst pushes ahead of the thunderstorm and can be strong enough to uproot trees and damage buildings.

As precipitation spreads throughout the thunderstorm cloud, so does the downdraft. Eventually the downdraft overpowers the updraft. The updraft dies, precipitation comes to an end, and the cloud slowly vaporizes. This is the final, or dissipating, stage of the thunderstorm.

Check Your Understanding

1. Describe what happens to the temperature of a gas when the volume of the gas increases.
2. Describe what happens to the temperature of a gas when the gas is compressed.
3. In your own words, define what is meant by air pressure.
4. Describe what causes a downburst.
5. Describe the vertical motion of air in a thunderstorm that has reached its mature stage.

Understanding and Applying What You Have Learned

1. What hazards do cumulonimbus clouds pose to aircraft?
2. Approximately how high can a cumulonimbus cloud build? What limits a cloud's height?
3. Have you ever seen "steam" rising from a pond on a cold day? Explain why this happens.
4. In the investigation, smoke from the match acted as condensation nuclei needed to form a cloud. What are some natural and artificial sources of condensation nuclei?
5. How does lowering the temperature of humid air affect the likelihood of cloud formation?
6. Are thunderstorms more likely to form in humid or dry air? Justify your response.
7. Why does an updraft produce clouds and why does a downdraft cause clouds to vaporize?
8. On a piece of paper, draw the vertical profile of a prominent mountain range. Assume that a humid wind blowing from left to right encounters the mountain range. Because the wind cannot go through rock, it is forced to go up and over the mountain range.
   a) Using a broad smooth arrow, draw the wind flowing up the mountain on one side and down the mountain on the other. In pencil, sketch clouds on the side of the mountain where you think clouds and thunderstorms are most likely to form.
   b) Share your drawing with classmates in your group, make any modifications in your drawing, and discuss why clouds and precipitation are more likely on one side of the mountain than on the other.
   c) On which slope of the mountain range is a rain forest more likely to be found? On which slope of the mountain is a desert more likely to be found?

Preparing for the Chapter Challenge

Working with the other members of your small group, prepare a table that summarizes the characteristics of each of the three stages in a thunderstorm's life cycle. Using that table and the other components of this activity, prepare a list of observations that might warn people of the approach of a thunderstorm that has reached its mature stage. You need to add this information to your report for the entertainment company. Present your list of warning signs during a classroom discussion.

Inquiring Further

1. **Observing storm damage**

   a) Make a list of all of the types of thunderstorm-related hazards you can think of. Divide your list into three categories:

   - Hazards that may exist as a thunderstorm approaches.
   - Hazards that may exist during a thunderstorm (during the mature stage when precipitation is falling).
   - Hazards that may exist as a thunderstorm is dissipating.

   b) List all of the types of thunderstorm-related damage that you have observed in your community. If possible, identify the type of hazard that caused the damage.

   c) Could any of the damage in your community have been prevented? If so, how?

2. **Earth Systems connections**

   a) Think about what you have learned about condensation nuclei and cloud formation. Predict the type of weather changes that might occur when a volcano erupts and releases particles into the atmosphere.

   b) After you have made your prediction, conduct research on the relationship between volcanism and weather.

Activity 3 Tracking Thunderstorm Movement through Radar

Goals

In this activity you will:

- Use radar images to calculate the speed and direction of thunderstorm movement.
- Examine factors that allow thunderstorms to last for long periods of time.
- Learn how weather radar is used to track storms.
- Understand the differences in the intensities of thunderstorms.

Think about It

Although most thunderstorms last for less than an hour, some can produce severe weather and damage for several hours.

- How can you predict how long a thunderstorm will last?
- How do you know when the danger of severe weather has passed?

What do you think? Record your ideas about these questions in your *EarthComm* notebook. Be prepared to discuss your responses with your small group and the class.

Investigate

1. Look at the two images shown. They were taken from a radar screen one hour apart, at 3 P.M. and 4 P.M. The radar is located at the dot in the lower right-hand corner of the screen. The irregular blotches represent areas of precipitation detected by radar (ignore the blotch near the radar's location). The darkest areas in the center of the blotches represent the heaviest precipitation. As you go out toward the edges of the blotches, the lighter shades represent lighter precipitation.

   a) Speculate on what you think the blotches near the radar location represent. (It was suggested that you could ignore them if you were observing an approaching thunderstorm.)

2. Look closely at the 3 P.M. radar image.

   a) Which area has the most intense precipitation?

   b) How many levels of precipitation intensity does this area have?

   c) How far away is this area from the radar station?

   d) In which direction is the area moving?

3. Now look at the 4 P.M. radar image.

   a) How many levels of intensity are now indicated in the area of most intense precipitation?

   b) Describe how the precipitation intensity changed between 3 P.M. and 4 P.M.

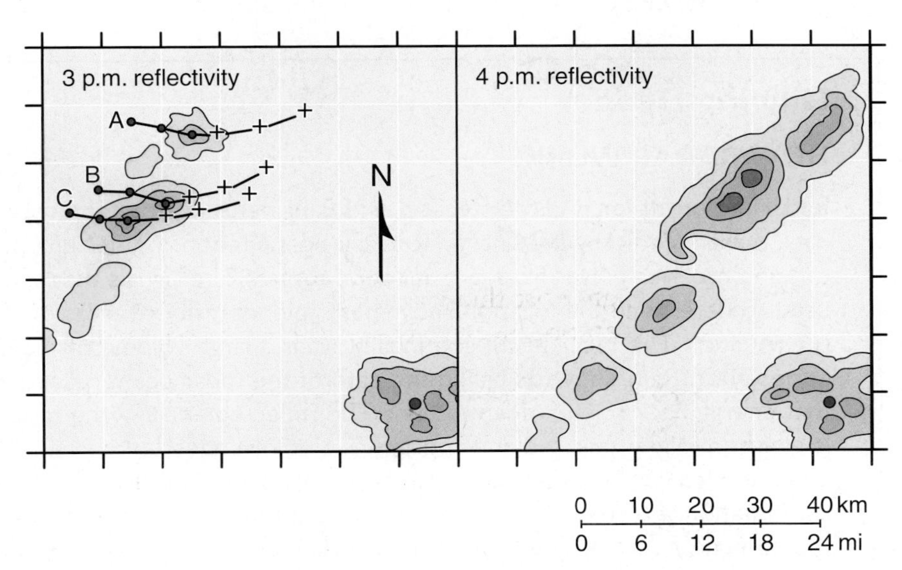

4. Compare the two images.
   a) In which general direction did the areas of precipitation move between 3 P.M. and 4 P.M.? Justify your answer.
   b) What was the speed of movement of the areas of precipitation in kilometers per hour? Include a description of the method you used to determine the speed.
   c) Where do you expect the areas to be located at 5 P.M.? Explain your answer.
   d) How do you explain the distribution and movement of the areas of precipitation?
5. Consider what you have learned in earlier activities about atmospheric conditions that produce clouds and precipitation.
   a) How do you think that radar detects precipitation?
   b) How do you think radar distinguishes heavy versus light precipitation?

Reflecting on the Activity and the Challenge

In this activity you used two radar images spaced one hour apart to determine the speed and direction of movement of thunderstorms. Think about how radar would be helpful in developing a safety plan for the concert arena proposed for your community.

Digging Deeper

HOW RADAR WORKS

Radar (acronym for *RA*dio *D*etection *A*nd *R*anging) sends out and receives back pulses of microwave energy. To track precipitation, a radar unit sends out and receives pulses from precipitation targets. The signals that return are used to determine the location, movement, and intensity of areas of precipitation. The radar beam continually scans a large volume of the lower atmosphere. Rain, snow, or hail in the path of the radar beam reflects some of that energy back to the radar antenna. The reflected energy (the radar echo) is electronically processed and appears as a color-coded blotch on a computer screen. In order to pinpoint the location and movement of areas of precipitation, a map of the region surrounding the radar is superimposed on the computer screen. The heavier the precipitation (or the larger the hail), the stronger the radar echo. Echo strength is calibrated on a color scale with light green indicating light precipitation and dark red signaling heavy precipitation.

Earth's curvature limits the range of radar. Radar signals travel along downward-bending paths somewhat less curved than Earth's surface. Radar sees well beyond the horizon but its beam gains altitude as distance from the radar increases so that radar's maximum range is about 460 km (285 mi.).

Figure 1 Radomes house weather radar antennae that receive information about developing weather.

Not all echoes on a radar screen are caused by precipitation. Nearby structures on the ground such as tall buildings or smokestacks also reflect radar signals. This reflection creates echoes known as ground clutter. In the investigation, the patterns around the radar site in the lower right-hand corner of the images you studied are ground clutter.

Successive radar images taken over time are used to track the direction and speed of areas of precipitation and associated storm centers. Each storm center is assigned an identification letter (A, B, and C in the radar images in the investigation). Past positions of the centers are denoted by dots. Lines connecting the dots show how storm centers have moved. Computer projections of future positions are indicated by crosses, which represent the positions of the centers at 15-minute intervals.

Tracking Squall Lines, Cells, and Supercells

The most common reason for a prolonged period of severe weather is that thunderstorms are often composed of many individual cells. Each thunderstorm cell is several kilometers or miles across and passes through the cumulus, mature, and dissipating stages. At a given time within a multicellular thunderstorm, some cells are reaching the mature stage while new cells are forming and old cells are dissipating.

Multicellular systems occur as **squall lines** or **mesoscale convective complexes.** An elongated band of thunderstorm cells, known as a squall line, may form in the warm humid air along or just ahead of a well-defined cold front. (See *Figures 2* and *3*.) Squall lines may bring several hours of lightning, thunder, and precipitation, and some cells in the line may produce severe weather. A mesoscale convective complex (MCC) is a nearly circular cluster of thunderstorm cells covering an area that may be a thousand times that of an individual cell—perhaps covering the entire state of Missouri. MCCs develop in warm humid air masses but are not associated with cold fronts. They are most common at night during the warm season (March through September) over the eastern two-thirds of the United States.

Geo Words

squall line: an elongated band of thunderstorm cells.

mesoscale convective complex (MCC): a nearly circular cluster of thunderstorm cells covering an area that may be a thousand times that of an individual cell.

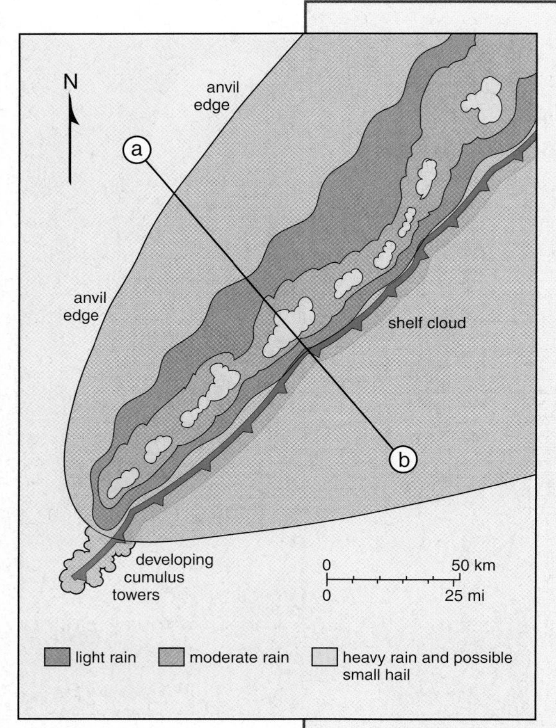

Figure 2 A squall line is an elongated band of thunderstorm cells along or just ahead of a cold front. The gust front is shown by the blue line with teeth, and the cloud anvil is the wide shaded area.

Figure 3 A line of thunderstorms in west Texas resulted in a tornado warning.

The longevity of an MCC (typically lasting 12 to 24 hours) coupled with its slow movement means that rainfall is widespread and substantial. An MCC can also produce severe weather, including flash flooding, weak tornadoes, or hail.

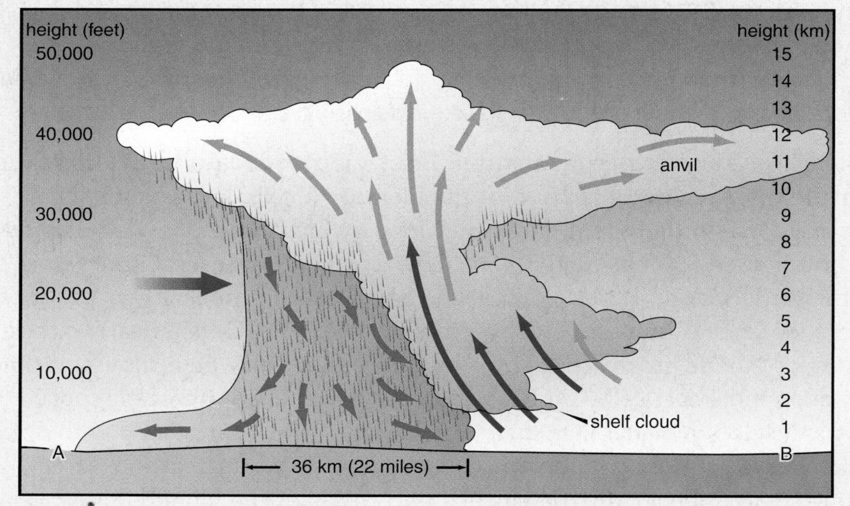

Figure 4 Large quantities of air from low levels are lifted high into the atmosphere in the leading updrafts. Air from the middle levels is brought to replace it in the trailing downdrafts.

Geo Words

supercell: a single thunderstorm cell that is much larger and longer-lasting than an ordinary cell.

The most intense type of thunderstorm is a **supercell** (see *Figure 5*), which is a single cell that is much larger and longer-lasting than an ordinary thunderstorm cell. Supercells can have exceptionally strong updrafts, estimated to be 240–280 km/h (150–175 mph) in some cases. These strong updrafts can build a cumulonimbus cloud all the way to the top of the troposphere and into the lower stratosphere (sometimes reaching an altitude of 20,000 m or 66,000 ft.). Supercells are responsible for the most powerful tornadoes and the largest, most destructive hail.

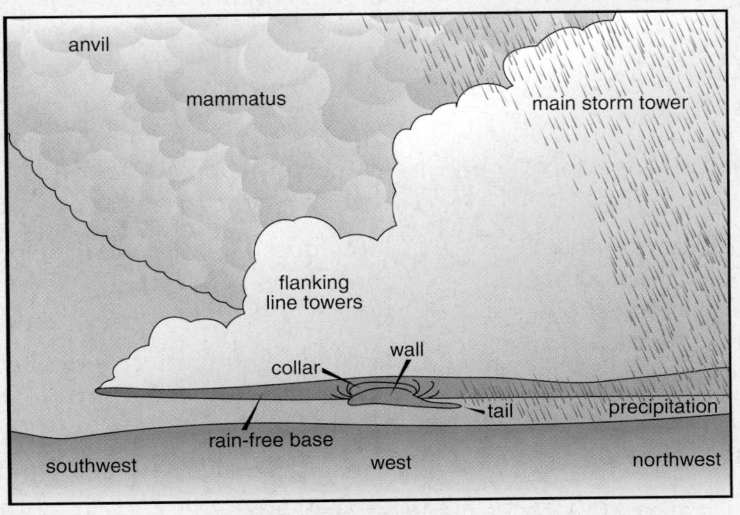

Figure 5 Diagram of a supercell thunderstorm.

Check Your Understanding

1. What man-made or natural features could interfere with weather radar?
2. How would you calculate the speed at which a thunderstorm cell is moving?
3. In your own words, describe a squall line.
4. How could you differentiate between a supercell and an ordinary thunderstorm cell?

Understanding and Applying What You Have Learned

1. Explain why thunderstorm weather may persist at a particular place for many hours.
2. In what way does a supercell thunderstorm pose a greater hazard than an ordinary thunderstorm cell?
3. Look again at the radar images from the investigation. Use a new copy of the images and cut them apart so that you can overlay one on the other for comparison.

   a) How does the predicted path of the precipitation (shown by the crosses in the 3 P.M. image) compare with the actual path (shown by the location of the precipitation in the 4 P.M. image)?

   b) How did the predicted speed of movement compare to the actual speed of movement?

   c) From what you can see in the images, are the thunderstorm cells weakening over time or becoming stronger? Explain your answer.

Preparing for the Chapter Challenge

You have learned that a thunderstorm may occur as an individual cell, a clusters of cells, and supercells. You have also learned about a valuable tool (weather radar) for tracking these cells. To prepare for the **Chapter Challenge,** do some research to locate the nearest National Weather Service radar. Keep in mind that because of Earth's curvature, the effective range of weather radar is about 460 km (285 mi.). Also consider some of the implications for your community if an approaching thunderstorm consists of many cells. How will what you've learned in this activity affect your community plan of action?

Inquiring Further

1. **Investigating advances in weather forecasting**

   Create a poster presentation that explains how Doppler radar differs from the conventional (reflection only) type of weather radar. Consider these questions:

   a) Describe in your own words the Doppler principle.

   b) What can Doppler radar do that conventional radar can't?

   c) How does Doppler radar help forecasters better predict severe weather?

   d) Is your community covered by a National Weather Service Doppler radar?

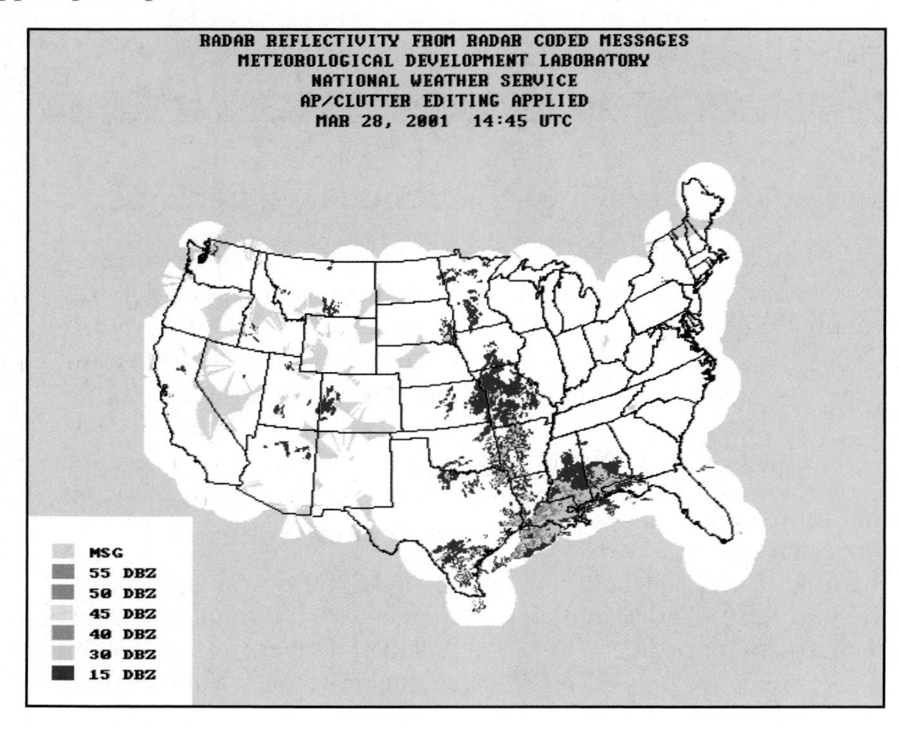

Activity 4 Severe Weather Hazards: Flash Floods

Goals

In this activity you will:

- Determine the flow of surface water using a topographic map.
- Examine factors that allow thunderstorms to persist for long periods of time.
- Explain the connection between topography and flash flooding.
- Realize the relationship between population growth and the likelihood of fatalities due to a flash flood.
- Read accounts of flash-flood disasters.
- Understand that wise land-use planning can reduce the danger from flash floods.

Think about It

Under certain conditions, thunderstorms can produce flash floods.

- What is the difference between a flood and a flash flood?
- How do flash floods impact communities?

What do you think? Record your ideas about these questions in your *EarthComm* notebook. Be prepared to discuss your responses with your small group and the class.

Investigate

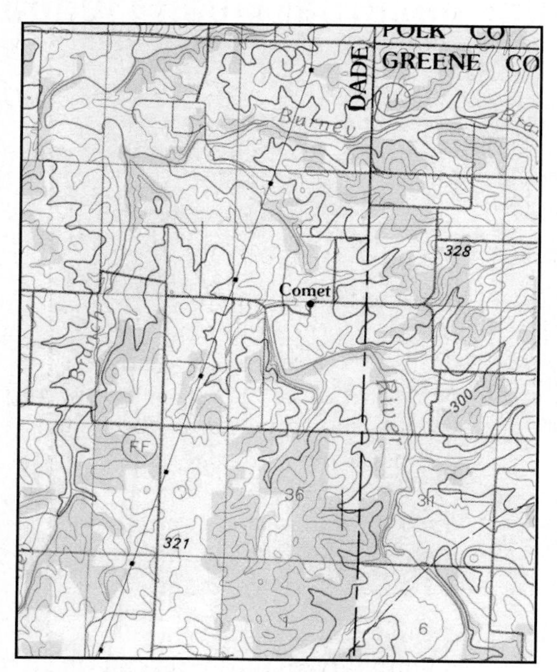

Scale 1:100,000. One centimeter represents one kilometer on the ground. Contour interval is 10 m.

1. Look at the topographic map showing the course of the Sac River, which flows past the city of Comet, Missouri.

   a) Place a clear plastic transparency sheet over the map or obtain a copy of the map. Make a dot anywhere on the map with a transparency marker. The dot represents a raindrop.

   b) Draw a line from the dot that shows how the raindrop would flow downhill after reaching the ground. In other words, with your marker, take the raindrop to contour lines of successively lower elevation. Be sure to keep the line always perpendicular to the local contours, because water flows directly downhill. If the path leads to a gully or stream, follow the gully or stream downhill until it empties into a larger river. Would the raindrop end up in the Sac River?

   c) Now look at the map again. Mark another dot on the map. This time try to pick a spot where you think the drop would flow into the Sac River. Draw a line from the spot showing the path the raindrop would take to the river. Again, make sure the path moves from higher to lower contour lines and is always perpendicular to contour lines.

   d) Using a continuous dashed line, draw the boundaries of the entire area in which all raindrops will flow into the Sac River. When you've finished, check your work by marking several raindrop dots in different parts of the area and checking that they would really flow into the river.

2. Assume that intense rains caused the Sac River to rise above its normal level.

   a) What if the Sac River rose 10 m above its normal level? Mark the area that would flood using a green transparency pen.

   b) What if the river rose 20 m? Mark the area with a blue pen.

   c) What if the river rose 30 m? Mark that area with a red pen.

   d) Under what circumstances might a thunderstorm that doesn't actually pass over Comet cause flash flooding in downtown Comet?

3. Imagine that the entire area you enclosed with a dashed line is undeveloped land that is covered with deep soil and vegetation. Now imagine that the entire area is developed and covered with asphalt and cement.
   a) Why would it make a difference whether the land was covered with vegetated soil or with pavement?
   b) In each of the two cases, describe what would happen if several inches of rain fell in the area in a couple of hours. What effect would the storm have on the river in each case?
4. Use your map and transparency as a guide.
   a) What damage would a flood cause if the Sac River rose 10 m? 20 m? 30 m?
5. Look at a topographic map of your community.
   a) Determine and record the interval (in feet) between successive contour lines.
   b) Locate the major rivers and streams that flow through your community. Pick one river or stream to study (if there are many choices, each group can pick a different one). Lay a transparency sheet on top of the map. Draw the drainage basin of your stream.
   c) What areas of the drainage basin are particularly at risk for flooding?
   d) How might flooding affect the public water supply, operations of the sewage treatment plant, automobile and public transportation, crops in the fields, homes, buildings, and industry?
   e) Assume that an intense thunderstorm stalls over the drainage basin and produces persistent rains that cause the water level to rise some 10 ft. above its normal level. By tracing the appropriate contour line, outline the region of your community that would be flooded. Do the same for a 20-foot flood of water.
   f) Would you describe your community as mostly urban or rural? How does this affect the potential for flash flooding in your community?
   g) Describe the topographic features that make your community more or less vulnerable to flash flooding.
   h) Find out if there are any flood-control structures (such as dams or levees) in your community.

Reflecting on the Activity and the Challenge

In this activity you learned how to interpret a topographic map to determine how a flash flood might affect a community. You also examined a topographic map of your area and analyzed the flood dangers in your community. Your knowledge can be applied to help planners locate a safe place for a new concert arena.

Digging Deeper

FLASH FLOODS

Thunderstorms can produce torrential rains. Usually the rain does not last very long and causes no serious problems. Sometimes heavy rains can last for hours. This can lead to flash flooding. A **flash flood** is a sudden rise in the water level of a stream, a river, or a man-made drainage channel in response to extremely heavy rains. Flash floods can also occur when a brief but heavy rain falls over the entire area of a very small watershed. In some places, water might overflow stream banks and collect in low-lying areas.

Geo Words

flash flood: a sudden rise in the water level of a stream, a river, or a man-made drainage channel in response to extremely heavy rains.

drainage basin (also watershed): the land area from which rainfall collects to reach a given point along some particular river.

The life cycle of a thunderstorm cell is typically less than an hour. How could torrential thunderstorm rains last for many hours? Flooding usually results from more than one thunderstorm cell. Flooding thunderstorms are most likely in mountainous terrain where a persistent flow of humid air up a mountain slope can cause thunderstorm cells to develop and redevelop over and over. Outside of the mountains, another possible cause of flooding rains is a succession (or "train") of slow-moving thunderstorm cells that mature over essentially the same geographical area. Thunderstorms move slowly when the steering winds in the middle and upper part of the troposphere are relatively weak.

The dotted line you drew on your map outlined the boundaries of the area where water from raindrops would drain downhill into the Sac River. A river plus all of its tributaries drain a fixed geographical area, and that area is the **drainage basin**, also called a **watershed**. Heavy rain falling on the upstream part of a drainage basin might cause flooding downstream in areas that received no rain. In other words, just because it is not raining where you are does not necessarily mean there is no flood danger.

Figure 1 Rapidly rising waters during a flash flood submerged this car.

Flash floods may be more likely in an urban area than in the surrounding countryside. The reason for this is that rain in the country can seep into the soil. In the city, rain cannot seep into asphalt or concrete parking lots, roads, and driveways. Frozen soil is also impervious to water. Instead, the water runs off these surfaces into nearby streams or other drainage ways. Storm sewers also channel water from roads into streams. Streams receiving so much water at one time can overflow their banks quickly. Also, storm sewers sometimes clog or cannot handle excessive volumes of water and back up into the streets.

The flash-flood hazard is particularly serious in mountainous terrain where river valleys are narrow and deep. Stream levels can rise very quickly because there are no broad areas next to the stream channel where the water can spread out. If the river valley also contains roads, campgrounds, or houses, a flood can be very destructive.

Figure 2 In urban areas pavement prevents rain from filtering into the soil, making cities more susceptible to the dangers of flash floods.

When a river overflows its banks in a non-mountainous area, excess water spreads over a broad flat area adjacent to the river known as a **floodplain**. Floodwaters are usually shallower than they are in mountain valleys where floodplains are very narrow or nonexistent. Also, in non-mountainous areas, roads and buildings usually are not located as close to the river as they might be in the mountains.

Geo Words

floodplain: any flat or nearly flat lowland that borders a river and which is covered by water when the river rises above flood stage.

The Big Thompson Canyon Flood

On July 31, 1976, a flash flood in the Big Thompson Canyon in Colorado claimed 139 lives (many of them campers) and caused $35.5 million in property damage. A combination of atmospheric conditions and topography contributed to the tragedy.

The Big Thompson Canyon is located in the Front Range of the Colorado Rockies about 80 km (50 mi.) northwest of Denver. During the late afternoon and evening of July 31, a persistent east-to-west flow of humid air up the mountain slopes triggered development of thunderstorm cells and heavy rainfall. Thunderstorm cells stayed in the same area because the steering winds in the upper atmosphere were weak. Runoff from the rainfall cascaded down the steep mountain slopes and into the river that winds along the narrow canyon floor. The river level rose abruptly and soon overflowed its banks as a flash flood.

Figure 3 Heavy rainfall and swiftly flowing waters filled the Big Thompson Canyon, causing extensive damage to homes and property.

The National Weather Service estimated that the upstream part of the drainage basin received 25 to 30 cm (10 to 12 in.) of rain, with perhaps 20 cm (8 in.) falling in only two hours! At one location along the river, the volume of flowing water was more than 200 times the river's normal discharge. A wall of water almost 6 m (20 ft.) high destroyed 418 houses and washed away 197 motor vehicles.

In the Big Thompson Canyon, the main road runs along the river. Many people tried desperately to escape the canyon in their cars; most of them were washed away and killed by floodwaters. These people could have survived if they had left their cars and climbed up the canyon slopes, out of reach of the water.

Fort Collins Flood

Some 21 years later, on July 28, 1997, Colorado residents were reminded of the Big Thompson Canyon tragedy when a thunderstorm system poured record rainfall on Fort Collins, Colorado. Between 5:30 and 11:00 p.m. up to 25 cm (10 in.) of rain fell on the southwest side of Fort Collins. A flash flood claimed the lives of five people along the swollen Spring Creek, just downstream from the heaviest rainfall. Floodwaters also damaged buildings, including many on the Colorado State University campus. Floodplain management, particularly along Spring Creek, likely prevented a greater loss of life and residential property damage. Management strategies enacted long before the flood included relocating buildings away from the river and constructing holding ponds near the riverbanks that could capture excess water if the river overflowed its banks.

Figure 4 Recovery efforts immediately following the Fort Collins flood.

Fort Collins is located just east of the Rocky Mountain Front Range. The atmospheric conditions that led to the Fort Collins flood were similar to those responsible for the Big Thompson Canyon flood. Unusually humid air was flowing westward over eastern Colorado toward the mountains and winds in the middle to upper troposphere were weak. Persistent upslope winds combined with weak steering winds set the stage for slow-moving thunderstorm cells that developed and redeveloped over the Front Range.

Safety Tips for Floods

Flash floods can strike any time and any place with little or no warning. In mountainous or flat terrain, distant rain may be channeled into gullies and ravines, turning a quiet streamside campsite or wash into a rampaging torrent in minutes. City streets can become rivers in seconds. Observe these flash-flood safety rules. They could save your life.

- Keep alert for signs of heavy rain (thunder and lightning), both where you are and upstream. Watch for rising water levels.
- Know where high ground is and get there quickly if you see or hear rapidly rising water.
- Don't pitch your tent in a dry streambed.
- Be especially cautious at night; the danger is harder to recognize then.
- Do not attempt to cross flowing water that may be more than knee deep. If you have doubts, do not cross.
- Do not try to drive through flooded areas.
- If your vehicle stalls, abandon it and seek higher ground immediately.
- During threatening weather, listen to commercial radio or TV, or NOAA Weather Radio for weather watch and warning bulletins.

Check Your Understanding

1. What is a flash flood?
2. Why is it dangerous to camp in a mountain valley?
3. Why are urban areas particularly vulnerable to flash flooding?
4. Why is it dangerous to drive a motor vehicle through flooded streets?
5. List five safety tips for floods that would be applicable in your community.

Understanding and Applying What You Have Learned

1. What might cause hours of thunderstorm rains along a mountain slope?
2. What might cause a prolonged period of heavy thunderstorm rains over relatively flat terrain?
3. How does knowing the drainage basin of a river help you assess the flash-flood potential of the area?
4. How is weather radar useful in determining the potential for flash flooding from thunderstorms?

Preparing for the Chapter Challenge

Use your interpretation of local topography to write a report about the potential for flash flooding in your community. You will use this report in your **Chapter Challenge.** Some of the questions below might help you prepare a report to ensure the safety of concert-goers at the proposed arena:

- Which parts of the community are particularly at risk for flooding?
- How often does this hazard affect your community?
- What weather conditions might indicate that a flash flood is possible?
- How can community members stay informed?
- What can the community as a whole do to prepare for flash floods and lessen their negative impact?
- What should citizens do if the National Weather Service issues a flash-flood watch or warning for the community?
- What can individuals do to protect themselves during a flash flood?

Inquiring Further

1. **Flash flooding in your community**

   Do some research on a flash flood that affected your community or another community in your state. Find out how weather, topography, and other factors contributed to the flood. Use the *EarthComm* web site to help you with your research.

   a) Was the community prepared for the flood? Why or why not?

   b) How did weather contribute to flooding?

   c) How much higher than usual was the stream?

   d) During the flood, was the stream water a different color than usual?

   e) Describe any evidence of flooding after the floodwaters returned to their banks.

   f) Was there any injuries, loss of life, or major property damage?

2. **City planning and severe weather**

   Conduct research in your community about the way that the city has been designed to handle the runoff from rain or snowmelt.

3. **Investigate current flash-flood events**

   Visit the *EarthComm* web site to learn more about current flash-flood warnings in the United States.

Activity 5 Lightning and Thunder

Goals

In this activity you will:

- Interpret data on casualties due to lightning.
- Describe what causes lightning.
- Explain the relationship between lightning and thunder.
- Realize that many myths and misconceptions about lightning can be dangerous.
- Know what safety precautions need to be taken during any lightning event.
- Understand that lightning is a process where electrical energy flows between the atmosphere and Earth's surface or between different clouds.

Think about It

Many people are startled or even frightened by a sudden clap or rumble of thunder. Thunder is harmless, but lightning, which causes thunder, can be deadly. In an average year, about 90 people in the United States lose their lives to lightning strikes.

- What is lightning and what causes it?
- Which state has the highest number of injuries and deaths due to lightning? Why do you think this is so?

What do you think? Record your ideas about these questions in your *EarthComm* notebook. Be prepared to discuss your responses with your small group and the class.

Investigate

1. The data table below shows the number of deaths and injuries caused by lightning in the U.S. between 1959 and 1996.

   a) Which three states had the fewest deaths per one million persons due to lightning?

   b) Which three states had the largest number of deaths per one million persons due to lightning?

   c) How many deaths and injuries per one million persons occurred in your state due to lightning during this time?

   d) Compare your state with others in the table. How does your state compare to others in terms of lightning fatalities?

   e) Give reasons to explain your state's ranking.

U.S. Lightning Deaths and Injuries 1959–96 by State

State	State Population	Deaths	Injuries	Deaths per one million persons	Injuries per one million persons
Alabama	4,351,037	89	220	20.45	50.56
Alaska	615,205	0	2	0.00	3.25
Arizona	4,667,277	60	128	12.86	27.42
Arkansas	2,538,202	110	253	43.34	99.68
California	32,682,794	21	63	0.64	1.93
Colorado	3,968,967	102	343	25.70	86.42
Connecticut	3,272,563	13	78	3.97	23.83
Delaware	744,066	15	28	20.16	37.63
D.C.	521,426	5	22	9.59	42.19
Florida	14,908,230	369	1316	24.75	88.27
Georgia	7,636,522	82	349	10.74	45.70
Hawaii	1,190,472	0	7	0.00	5.88
Idaho	1,230,923	24	81	19.50	65.80
Illinois	12,069,774	88	286	7.29	23.70
Indiana	5,907,617	81	179	13.71	30.30
Iowa	2,861,025	68	170	23.77	59.42
Kansas	2,638,667	59	182	22.36	68.97
Kentucky	3,934,310	83	206	21.10	52.36
Louisiana	4,362,758	120	239	27.51	54.78
Maine	1,247,554	22	112	17.63	89.78
Maryland	5,130,072	117	140	22.81	27.29

U.S. Lightning Deaths and Injuries 1959–96 by State (*continued*)					
State	**State Population**	**Deaths**	**Injuries**	**Deaths per one million persons**	**Injuries per one million persons**
Massachusetts	6,144,407	24	353	3.91	57.45
Michigan	9,820,231	93	654	9.47	66.60
Minnesota	4,726,411	57	131	12.06	27.72
Mississippi	2,751,335	91	215	3.07	78.14
Missouri	5,437,562	81	102	14.90	18.76
Montana	87,933	21	44	23.88	50.03
Nebraska	1,660,772	41	77	24.69	46.36
Nevada	1,743,772	6	13	3.44	7.46
New Hampshire	1,185,823	8	73	6.75	61.56
New Jersey	8,095,542	55	145	6.79	17.91
New Mexico	1,733,535	85	181	49.03	104.41
New York	18,159,175	129	482	7.10	26.54
North Carolina	7,545,828	169	543	22.40	71.96
North Dakota	637,808	11	28	17.25	43.90
Ohio	11,237,752	127	484	11.30	43.07
Oklahoma	3,339,478	91	271	27.25	81.15
Oregon	3,282,055	7	22	2.13	6.70
Pennsylvania	12,002,329	7	571	0.58	47.57
Rhode Island	987,704	4	45	4.05	45.56
South Carolina	3,839,578	82	254	21.36	66.15
South Dakota	730,789	20	61	27.37	83.47
Tennessee	5,432,679	127	360	23.38	66.27
Texas	19,712,389	175	367	8.88	18.62
Utah	2,100,562	37	87	17.61	41.42
Vermont	590,579	14	19	23.71	32.17
Virginia	6,789,225	54	222	7.95	32.70
Washington	5,687,832	54	38	9.49	6.68
West Virginia	1,811,688	5	101	2.76	55.75
Wisconsin	5,222,124	48	217	9.19	41.55
Wyoming	480,045	21	84	43.75	74.98

2. Obtain a map of the United States.
   a) On the map, shade the states in which 20 or more persons per one million were killed by lightning during the 37-year period.
   b) Why were you asked to map the number of deaths per one million persons instead of the actual number of deaths?

3. Compare your shaded map to the map of thunderstorm frequency from **Activity 1**. Focus on the region in which an average of 50 or more thunderstorm days occur annually.
   a) Describe the relationship between the map of thunderstorm frequency and your map of lightning fatalities.
   b) Which states have a high number of lightning deaths and high thunderstorm frequency? Explain why you think this is so.
   c) Which states have a low number of lightning deaths per one million persons and low thunderstorm frequency? Explain why you think this is so.
4. Look at the map on the following page showing the number of lightning strikes in a 24-hour period in the month of May 2000. How many lightning strikes occurred in the 24-hour period?

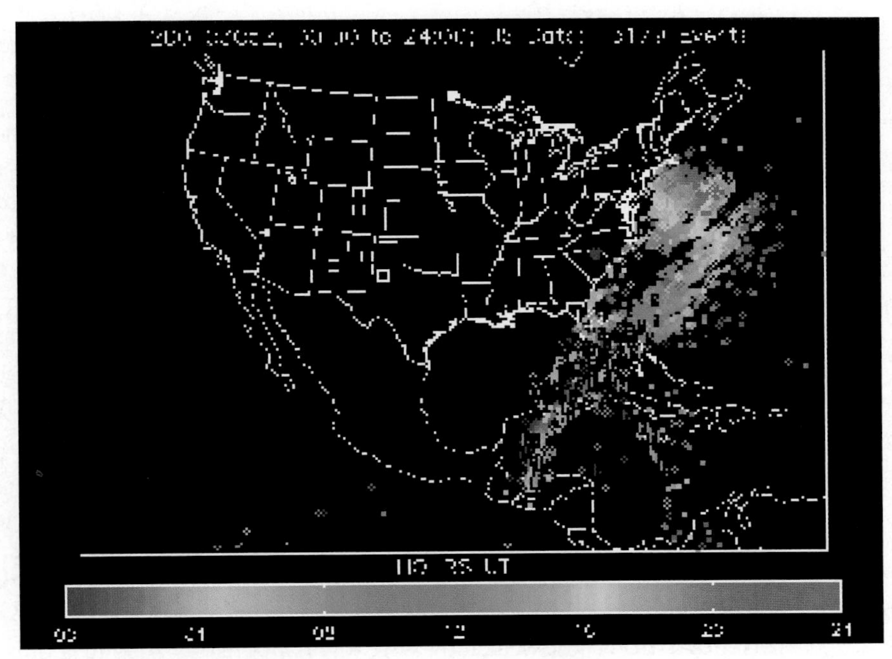

Lightning strikes in the United States over a 24-hour period.

5. Go to the *EarthComm* web site to view lightning-strike maps.

   a) Select maps from the menu and print out a map of lightning activity for the United States. Describe any patterns that you observe.

   b) Select animations from the main menu. Animations display lightning strikes every hour over the course of one month. Run one of the animations and describe any patterns and relationships that you see.

Reflecting on the Activity and the Challenge

In this activity you searched for patterns and relationships within data on the casualties associated with lightning. Plotting the data on a map of the United States allowed you to note where lightning affects people the most, and where fewer people are killed by lightning. You also compared this data to a map of thunderstorm frequency. Think about how your investigation will help you to solve the **Chapter Challenge**.

Digging Deeper

What is Lightning?

All thunderstorms, no matter how weak, produce lightning. Lightning is a powerful electrical discharge. You probably have experienced an electrical discharge when you walked across a carpet in a dry room and touched a metal doorknob. If the room was dark, you probably even saw a spark. The tips of your fingers had developed an electrical charge that was opposite to that of the knob. The electrical discharge that you felt, and might have even seen as a spark, was a flow of electrons (negatively charged subatomic particles) to the positively charged surface. An electrical discharge reestablishes electrical equilibrium (that is, evens out the difference in electric charge) between two objects.

Although lightning is also an electrical discharge, it operates on a much larger scale and involves much greater differences in electric charge than a common spark. As a cumulonimbus (thunderstorm) cloud forms, its upper regions and a much smaller region near the cloud base develop a positive charge. In between, a pancake-shaped zone of negative charge forms. Scientists don't know exactly why this happens because it is difficult and dangerous to study a thunderstorm directly. However, they think that water droplets and ice crystals within the cloud develop different charges and that updrafts and downdrafts deliver charged particles to different parts of the cloud. During this charge separation, the cumulonimbus cloud induces a positive electrical charge on the ground directly under the cloud. (Normally, Earth's surface carries a negative electrical charge.)

Figure 1 Lightning between a cumulonimbus cloud and Earth's surface.

Air is an excellent electrical insulator, so that as a cumulonimbus cloud forms and charges separate, a tremendous potential soon develops for an electrical discharge. When a thunderstorm cell enters its mature stage, the electrical resistance of air breaks down and lightning discharges, evening out the electrical charges. Lightning forges a path between the cloud and the ground or between two clouds or, less often, between two different parts of the same cloud. Of all the lightning flashes that you see when a thunderstorm is in progress, only about 20% actually strike Earth's surface.

A lightning bolt and the thunder it produces happen at the same time. Because light travels about a million times faster than sound, you see the lightning almost instantaneously but hear the thunder later. Sound travels about 330 m/s, so that in five seconds, sound travels about 1650 m, or almost one mile. Hence, every five seconds between a flash of lightning and the sound of the thunder it creates equates to about one mile from the location of the lightning bolt.

Figure 2 You see the electrical discharge of lightning before you hear the clap of thunder because light travels faster than sound.

In a 59-year period (1940–1998) of deaths due to tornadoes, hurricanes, lightning, and floods, the greatest number were caused by lightning (see *Table 1*). Over the continental 48 states, an average of 20,000,000 cloud-to-ground flashes have been detected every year since the lightning detection network covered most of the continental United States in 1989.

Table 1. Average Annual Severe-Weather Fatalities by Decade from 1940–1998.

Year	Lightning	Tornado	Flood	Hurricane
1940–1949	329.3	178.8	61.9	21.6
1950–1959	184.1	140.9	79.1	87.7
1960–1969	133.2	93.5	129.7	58.7
1970–1979	97.8	98.6	181.9	21.7
1980–1989	72.6	52.1	109.7	11.8
1990–1998	58.1	53.8	102.6	13.4
Total (59 years)	8693	6124	6547	2136

Source: National Weather Service.

What is Thunder?

Lightning heats air in the narrow path that it follows to incredibly high temperatures—estimated at 20,000° to 30,000°C! Air is heated to such high temperatures so rapidly that its density cannot respond, at least initially. The rapid rise in air temperature is, however, accompanied by a tremendous increase in air pressure locally that generates a shock wave. The shock wave then propagates outward and produces a sound wave that you hear as thunder. Sound waves are generated all along the lightning path. The first thunder heard is from the part of the lightning bolt that is closest to you (where it strikes the ground). Subsequent sound waves reach

you from segments of the bolt that are progressively farther away from your ears (at higher altitudes). This explains the persistent rumble of thunder following a lightning discharge. Thunderstorm cells more than about 20 km (12 mi.) away are too distant for thunder to be heard, although lightning can be seen. Such lightning is called heat lightning.

Lightning Safety Tips

Lightning is most frequent during the mature stage of a thunderstorm's life cycle. However, lightning can strike the ground many kilometers beyond the parent thunderstorm. That means lightning can strike areas where rain is not yet falling or has stopped falling. In other words, lightning is a hazard before, during, and after a thunderstorm hits a particular area.

Figure 3 Lightning damage can be extensive.

Lightning is most dangerous when it strikes something that readily conducts electricity. For this reason, during a thunderstorm it is wise to avoid metallic objects like lawn mowers, wire fences, and telephones. Contrary to popular belief, an automobile's tires provide occupants with little or no protection from lightning. Lightning that strikes a motor vehicle flows through the steel frame to the tires and then to the ground. Without direct contact with the metal frame, passengers usually escape injury. Lightning also tends to follow the shortest path between the cloud and the ground. Hence, when thunderstorms are threatening, stay away from high places (like hilltops) or tall objects (like isolated trees) that might attract lightning. If you are with someone who has been struck by lightning, it is important for you to know that the person does not carry an electrical charge and it is safe to touch the person and provide first aid after the lightning strike.

Check Your Understanding

1. Compare and contrast lightning and thunder.
2. Why do you hear thunder only after you see lightning?
3. Is it possible for lightning to strike without rain? Explain your answer.
4. Why is it wise to avoid metal objects during a thunderstorm?
5. Why is it wise to avoid high, open locations during a thunderstorm?

Understanding and Applying What You Have Learned

1. How is it possible that lightning ignites forest and brush fires? Wouldn't the rain that accompanies thunderstorms quickly extinguish the flames?
2. Explain why people who are near a lightning strike may suffer serious burns.
3. There is an old saying that "lightning never strikes the same place twice." From what you have learned about lightning, how realistic is this statement?
4. Why are lightning strikes particularly costly for an electrical power utility?
5. A softball game is underway at the park. The sky off toward the western horizon becomes ominously dark. The wind strengthens and becomes cool and gusty. There is no sign of lightning, you do not hear any thunder, and it is not raining yet. The people in charge decide to continue playing the game until it actually starts to rain. Is this a wise decision? Explain your response.
6. A thunderstorm is underway with lots of thunder and lightning. To protect yourself from lightning, what would you do if you were:
   a) in your house talking to a friend on the phone;
   b) in the park playing basketball;
   c) hiking in the mountains?
7. Assume that the light from a lightning bolt takes no time at all to reach your eyes.
   a) How many seconds does it take thunder to travel 1 km?
   b) Knowing these numbers, how can you figure out how far away a thunderstorm is by determining the time lapse between lightning and thunder?
   c) How can you determine whether a thunderstorm is moving toward or away from you?

Preparing for the Chapter Challenge

Write a report about the potential for lightning in your community. You will use this report in your **Chapter Challenge.** Think about how you might answer some of the following questions in your report:

- What is lightning?
- What kind of damage can it produce in your community?
- Are there any places in your community that might be particularly prone to lightning strikes (e.g., tall buildings, or hills)?
- How often does lightning affect your community?
- What weather conditions might indicate that lightning is possible?
- How can community members stay informed?
- What can the community, as a whole, do to prepare for lightning and lessen its impact?
- What can individuals do to protect themselves from lightning?

Inquiring Further

1. **Investigate your own questions about lightning**

   Write down any questions that occurred to you while studying lightning. How would you go about finding answers to your questions? Visit the *EarthComm* web site for resources to help you answer your questions.

2. **Conduct a survey**

   Interview students in your school about their experiences with lightning and find out more about the warning signs of a lightning hazard. Has anyone in the class witnessed a nearby lightning strike?

Activity 6 Severe Winds and Tornadoes

Goals

In this activity you will:

- Construct a map of tornado frequency throughout the U.S.
- Determine the likelihood of a tornado occurring where you live.
- Describe the relationship between tornado and thunderstorm distribution across the U.S.
- Explain the relationship between tornado frequency and time of year.
- Examine tornado safety procedures to follow while at school, at home, or outside.
- Understand the misconceptions related to tornado safety.

Think about It

In the United States 10,000 severe thunderstorms occur in an average year. About 10% of these produce tornadoes, and perhaps 1% of those tornadoes are violent.

- What is a tornado?
- How likely is it that your community will experience a tornado?

What do you think? Record your ideas about these questions in your *EarthComm* notebook. Be prepared to discuss your responses with your small group and the class.

Investigate

Part A: Frequency and Distribution of Tornadoes

1. Look at *Table 1*. Use the number of tornadoes per 1000 sq. mi.

   a) Which five states reported the fewest tornadoes?

   b) Which five states reported the most tornadoes?

   c) How many tornadoes occurred in your state during this period?

   d) How many tornadoes does your state average per year?

Table 1. Frequency of Tornadoes by State 1950–1994

State	Number of Tornadoes	Number of Tornadoes per 1000 sq. mi.	State	Number of Tornadoes	Number of Tornadoes per 1000 sq. mi.
Alabama	886	16.90	Montana	238	1.62
Alaska	1	0.00	Nebraska	1673	21.63
Arizona	155	1.36	Nevada	48	0.43
Arkansas	854	16.06	New Hampshire	72	7.70
California	214	1.31	New Jersey	112	12.84
Colorado	1113	10.69	New Mexico	390	3.21
Connecticut	61	11.00	New York	249	4.57
Delaware	52	20.89	North Carolina	590	10.96
Florida	2009	30.55	North Dakota	799	11.30
Georgia	888	14.94	Ohio	648	14.46
Hawaii	28	2.56	Oklahoma	2300	32.90
Idaho	115	1.38	Oregon	44	0.45
Illinois	1137	19.63	Pennsylvania	451	9.79
Indiana	886	24.33	Rhode Island	8	5.18
Iowa	1374	24.42	South Carolina	423	13.22
Kansas	2110	25.64	South Dakota	1139	14.77
Kentucky	373	9.23	Tennessee	502	11.91
Louisiana	1086	20.95	Texas	5490	20.44
Maine	145	4.10	Utah	76	0.90
Maryland	0	0.00	Vermont	32	3.33
Massachusetts	134	12.70	Virginia	279	6.52
Michigan	712	7.35	Washington	55	0.77
Minnesota	832	9.57	West Virginia	83	3.43
Mississippi	1039	21.45	Wisconsin	844	12.88
Missouri	1166	16.73	Wyoming	434	4.44

2. Using the data table and a map of the United States.

   a) Record on the map the total number of tornadoes reported in each state from 1950 to 1994.

   b) Shade in tornado frequency for each state on this map. Use the color scale shown below.

   Tornado Frequency Color Code

   - 25 or more (red)
   - 20 – 25 (orange)
   - 10 – 20 (yellow)
   - 1 – 10 (green)
   - less than 1 (blue)

   c) Look at the map of annual frequency of thunderstorm days from **Activity 1**. How do places of high tornado frequency compare to places of high thunderstorm frequency? Explain your answer in your notebook.

   d) How many thunderstorms does your state have per year?

   e) An estimated 10% of severe thunderstorms produce tornadoes. If all of the thunderstorms that occured in your state were severe, how many tornadoes would you expect to occur annually in your state?

   f) Compare your state with the rest of the United States. Calculate the percentage of all United States tornadoes that occurred in your state from 1950 to 1994. The total number of tornadoes in the United States during that period was 34,349.

3. Refer to *Table 2* showing the average number of tornadoes per month in the United States.

   a) In which four months is tornado frequency the highest?

   b) In which season is the frequency of tornadoes the highest?

   c) In which season is the frequency of tornadoes the lowest?

   d) From your experience, in which season does your state or community have the highest number of thunderstorms?

   e) How does the timing of local thunderstorm activity compare to the season of highest tornado occurrence?

Table 2. Average Number of Tornadoes per Month in the United States

Month	Average Number of Tornadoes	Month	Average Number of Tornadoes
January	17	July	85
February	20	August	65
March	55	September	40
April	105	October	25
May	155	November	20
June	140	December	15

Wear eye protection. Sweep up any sand that gets onto the floor.

Part B: Investigating Downbursts

1. Obtain a bicycle pump and a shallow pan filled with sand. Hold the nozzle of the pump about 2 cm off the sand. Have a classmate push down on the pump several times.
   a) Describe in your notebook how the stream of air disturbs the sand. Include a sketch.
   b) Imagine that the stream of air was a strong downward burst of wind, and the sand was your community. What pattern of damage do you think you would see in your community due to a strong downward burst of wind?
   c) How is this simulation different than an actual occurrence of downbursts?

Reflecting on the Activity and the Challenge

In this activity you constructed a map of tornado frequency throughout the United States. You found that thunderstorm activity and tornado occurrences are related. You also investigated some of the effects of strong winds. You will want to include the dangers associated with tornadoes and strong winds in your **Chapter Challenge**.

Geo Words

microburst: an intense downdraft impacting a relatively small area (4 km or less across).

wind shear: a sudden change in wind speed or direction with distance.

Digging Deeper

DOWNBURSTS AND TORNADOES

Downbursts that affect a relatively small area (4 km or less across) are known as **microbursts**. Microbursts cause sudden changes in wind speed and direction, called **wind shear**, which can interfere with aircraft flight (*Figure 1*). Microbursts over short distances are particularly hazardous to aircraft taking off or landing because these winds typically affect only a part of a runway and are difficult to detect. Sometimes people incorrectly attribute wind damage to an unseen tornado when in fact the damage was caused by a downburst.

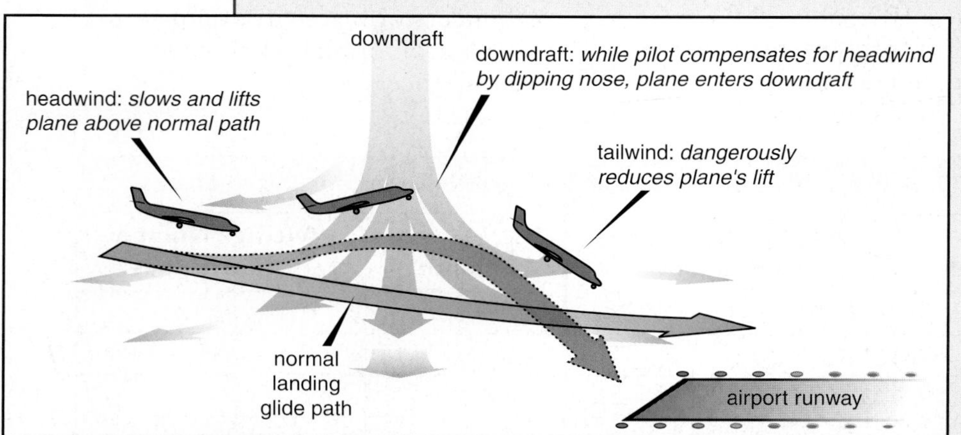

Figure 1 Wind shear can be deadly for aircraft.

Near the ground, downburst winds blow outward from a central area, whereas tornadic winds tend to swirl in circles. Microbursts can be deadly.

Tornadoes threaten people and property mainly because of their exceptionally strong winds and powerful updrafts. Winds blowing hundreds of kilometers per hour knock over trees and power poles, toss cars around as if they were toys, and destroy even well-constructed buildings. Flying glass and splintered lumber cause much of the injury and death associated with tornadoes. In violent tornadoes, the updraft near the center of the system may top 160 km/h (100 mph), strong enough to lift a house off its foundation. The most destructive tornadoes are large systems made up of two or more whirling masses of air (multi-vortex systems). Tornado intensity is rated on the Fujita scale (or F-scale) devised by the late Professor T. Theodore Fujita of the University of Chicago. (See *Table 3* below.) The F-scale is based on rotational wind speeds estimated from property damage and classifies tornadoes as weak (F-0, F-1), strong (F-2, F-3), or violent (F-4, F-5).

Tornadoes come in a variety of shapes, ranging from cylindrical cloud masses having nearly uniform lateral dimensions to long, slender rope-like pendants. A weak tornado's path on the ground is typically less than 1.5 km (1.0 mi.) long and 100 m (330 ft.) wide, with a life expectancy of only one to three minutes. Wind speeds are less than 180 km/h (110 mph). An intense tornado's path on the ground can exceed 160 km (100 miles) long and hundreds of meters wide, with a life expectancy that may exceed two hours. Estimated wind speeds in violent tornadoes range up to 500 km/h (300 mph).

Table 3. The Fujita Tornado Intensity Scale

Fujita scale	Category	Estimated wind speed	
		km/h	mph
F-0	Weak	65–118	40–73
F-1	Weak	119–181	74–112
F-2	Strong	182–253	113–157
F-3	Strong	254–332	158–206
F-4	Violent	333–419	207–260
F-5	Violent	420–513	261–318

Figure 2 Tornadoes come in a variety of shapes and sizes.

Figure 3 The funnel of a mesocyclone is made visible by water droplets.

Tornadoes usually track from southwest to northeast, but any direction is possible. Tornado paths are often erratic. Many tornadoes produce a hopscotch pattern of destruction as they alternately touch down and lift off the ground. The average forward speed of a tornado is about 48 km/h (30 mph). Winds in the vast majority of Northern Hemisphere tornadoes blow in a counterclockwise direction when viewed from above. Contrary to popular opinion, strong to violent tornadoes can occur in mountainous topography.

Each year the United States can anticipate between 700 and 1100 tornadoes, with only about 1% rated as violent (F-4 and F-5). Almost 80% of all tornadoes are relatively weak. By late February, the maximum tornado frequency is along the central Gulf states. In April, the maximum frequency shifts to the southeast Atlantic states. In May and June, the highest tornado incidence is usually over the southern Great Plains, and by early summer it has shifted to the northern Plains and the Great Lakes region.

Tornado Development

Most intense tornadoes develop in supercell thunderstorms. (Review supercells in **Activity 3**.) In a supercell thunderstorm, a powerful updraft interacts with horizontal winds to cause air to rotate about a vertical axis. The rotating air is called a **mesocyclone**. When viewed from above, the air

Geo Words

mesocyclone: a counterclockwise (viewed from above) circulation that develops in a supercell thunderstorm; may evolve into a tornado.

is rotating counterclockwise. If conditions are right, the mesocyclone circulation narrows and grows downward toward Earth's surface. As the circulation narrows, the wind speed increases. (The same thing happens when an ice skater performs a spin and brings his or her arms close to the body.) Humid air expands and cools as it is drawn inward toward the low-pressure center of the whirling system. Cooling of air causes water vapor to condense into water droplets, forming a funnel-shaped cloud extending downward from the parent cumulonimbus cloud. (See *Figure 3*.) If the funnel cloud strikes the ground, dust and debris is drawn into its circulation, and the system is described as a tornado. About 60% of mesocyclones produce tornadoes.

Most supercell thunderstorms develop as part of a squall line in the warm, humid air ahead of a well-defined cold front. A key requirement for bringing contrasting cold and warm air masses together is a strong **cyclone**. A cyclone is a stormy weather system that is plotted on a weather map as *L* or *Low* for the relatively low air pressure at its center. Here, cyclone does not refer to a hurricane or a tornado, but rather a large weather system in which surface winds blow counterclockwise and inward (viewed from above in the Northern Hemisphere). This system may be large enough to affect the weather over the eastern third of the United States at one time. As shown in *Figure 4*, surface winds in a cyclone pull together contrasting air masses to form fronts. Severe thunderstorms and tornadoes are most likely in the warm, humid air mass located southeast of the center of a cyclone and east of a cold front.

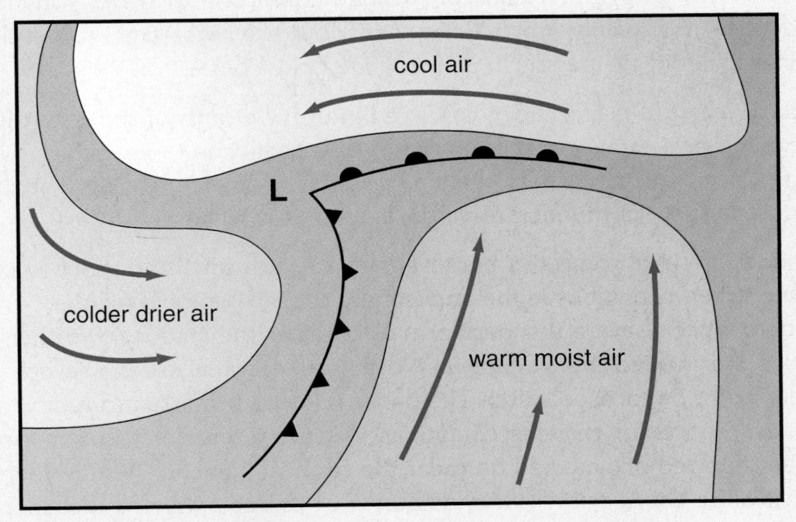

Figure 4 Circulation of air around a low-pressure system can lead to the formation of severe thunderstorms and tornadoes.

Geo Words

cyclone: a large low-pressure weather system in which surface winds blow counterclockwise and inward viewed from above in the Northern Hemisphere.

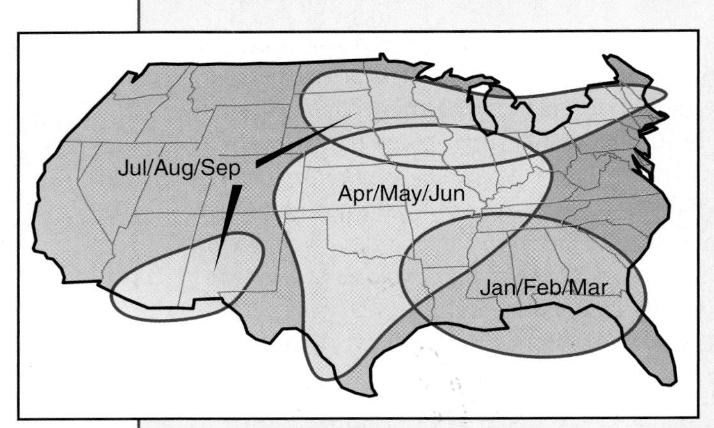

Figure 5 Although there is no "tornado season," tornadoes are more likely to occur in certain areas during certain times of the year.

Although severe thunderstorms and tornadoes have been reported in every state, they are most frequent in a north–south belt in the center of the nation, stretching from east Texas northward through Oklahoma, Kansas, Nebraska, and into southeast South Dakota. This area is called "tornado alley." The frequency of tornadoes in that region is greater than any other place in the world. Other belts of relatively high tornado frequency stretch from central Iowa eastward to central Indiana and along the Gulf Coast states. Tornadoes are infrequent in the Rockies, the Appalachians, and along the West Coast.

Why are tornadoes so frequent in the central United States? In that part of the world, all the ingredients necessary for development of severe thunderstorms come together. No mountain barriers separate the Canadian prairies and the Gulf of Mexico. That is, no barriers prevent cold air masses from surging southeastward from Canada, and warm, humid air masses from flowing northeastward from the Gulf. Where the two contrasting air masses come together, a sharply defined cold front forms. As you know, thunderstorms capable of producing tornadoes are most likely to develop in the warm, humid air mass just ahead of the cold front.

Although tornadoes have been reported in every month of the year, they are most common in spring and early summer. At that time of year, the temperature contrast in between air masses is greatest, meaning that the potential for severe thunderstorm development is highest.

Tornadoes are hard to predict because they are such small and short-lived systems. When conditions in the atmosphere appear favorable for the development of a severe thunderstorm or tornado, the National Weather Service's Storm Prediction Center in Norman, Oklahoma, issues a severe thunderstorm or tornado watch. The public is advised to listen to radio and television reports for changing conditions. If a severe thunderstorm or tornado has been spotted or detected on radar, the regional National Weather Service forecast office issues a severe thunderstorm or tornado warning that specifies the locations expected to be impacted. People living in the warning area should take all necessary precautions (see ***Tornado Safety Tips*** that follow).

The National Weather Service radar uses the Doppler effect to determine how air is moving within a thunderstorm and can provide the public with advance warning of the development of tornadoes. The Doppler effect is named for Johann Christian Doppler, the Austrian physicist who first explained the phenomenon in 1842. For this reason, weather radar with velocity-detection capability is often referred to as a Doppler radar. The Doppler effect refers to a shift in frequency of sound waves or electromagnetic waves when a source (or receiver) is moving. For example, the pitch (frequency) of an ambulance siren sounds higher as the ambulance approaches and then sounds lower as the ambulance moves away. (The Doppler effect is also used in devices that measure the speed of cars or a pitched baseball.) Doppler radar monitors the speed of precipitation (or dust) particles as they move directly toward or away from the radar antenna. With this movement, the frequency of the radar signal shifts slightly between emission and the return signal (echo). This frequency shift (the Doppler effect) is calibrated by a computer in terms of the motion of the particle. In this way, weather forecasters can detect the development of a mesocyclone before it evolves into a tornado.

Doppler radar displays are color-coded with greens and blues (cold colors) indicating motion directly toward the radar and reds and yellows (warm colors) indicating motion directly away from the radar. On a Doppler radar image, a tornado vortex signature (TVS) appears as a small region of rapidly changing wind direction within a mesocyclone. A TVS is a signal that a tornado may be forming.

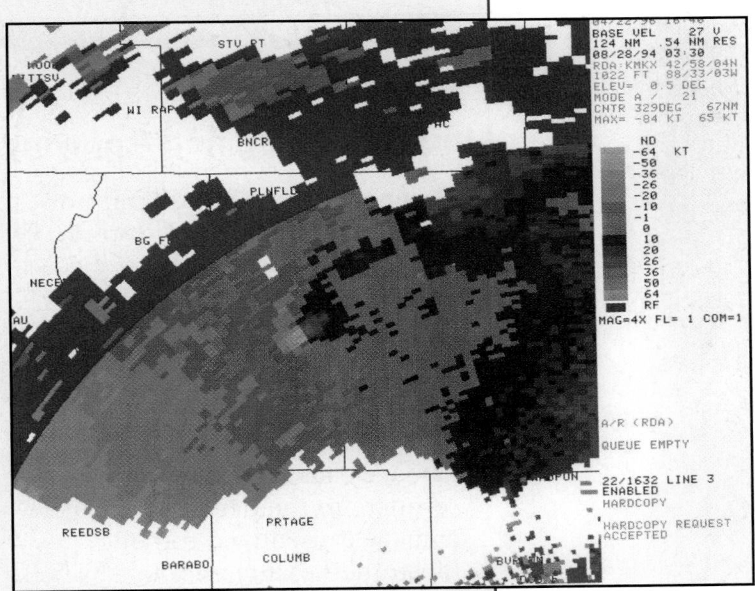

Figure 6 Doppler radar display showing a tornado vortex signature (TVS) near the center.

Tornado Safety Tips

- If you have been warned that a tornado is approaching or if you see a funnel cloud, you need to seek shelter.
- Go to a tornado shelter or steel-framed or substantially reinforced concrete building. If you're at home, go to the basement. If there is no basement, go to a small room (like a closet, bathroom, or interior hallway) in the center of the house on the lowest floor.

- Seek shelter under a mattress or a rigid piece of furniture.
- Do not go near windows and outside walls and do not open windows or exterior doors.
- If you are in an office building or school, go to an interior hallway on the lowest floor or a designated shelter area. Lie flat on the floor with your head covered. If possible, avoid auditoriums, gymnasiums, supermarkets, or other buildings that have wide, free-span roofs.
- Never try to outrun a tornado in a car. Tornadoes move quickly and erratically, so that it's hard to anticipate which direction they're going. Instead, stop and seek shelter indoors. If this is not possible, lie flat in a ravine, creek bed, or open ditch that is not flooded.
- Never seek shelter in a mobile home or car or other motor vehicle.

Check Your Understanding

1. What is the relationship between a severe thunderstorm and a tornado?
2. What is the relationship between a mesocyclone and a tornado?
3. In general terms, compare the diameter and life expectancy of a tornado with that of a supercell thunderstorm.
4. What is the difference between a weather watch and a weather warning?

Understanding and Applying What You Have Learned

1. Hail consists of jagged lumps of ice produced in a thunderstorm having an exceptionally strong updraft. Is a thunderstorm that produces hail also a good candidate to spawn a tornado? Explain your answer.
2. How does the pattern of property damage caused by a downburst compare to the pattern of property damage caused by a tornado? Illustrate the difference in a sketch.
3. Compared to other weather systems, a tornado is very small and short-lived. Why then is a tornado potentially so destructive?
4. What is the value of Doppler weather radar in safeguarding the public?
5. a) Where would you and your classmates seek shelter if a tornado warning were issued while you were in your school? Give a specific answer.

   b) Where would you and your family seek shelter if a tornado warning were issued while you were at home? Specify precisely where you would seek shelter and explain your answer.

   c) Where would the entertainment company direct their patrons to seek shelter if a tornado warning were issued?
6. During what time of the year is the threat of a tornado in your community the greatest? Explain how this might affect your advice to the entertainment company.

Preparing for the Chapter Challenge

Write an essay that explains the potential for tornadoes in your community. You will include the essay in your **Chapter Challenge**. Think about how the following questions might be useful:

- What is a tornado?
- What kind of damage can a tornado produce in your community? Are there any parts of the community that are particularly at risk for tornadoes?
- How often does this hazard affect your community?
- What weather conditions might indicate that a tornado is possible?
- How can community members stay informed before, during, and after a tornado?
- What can the community do to prepare for tornadoes ahead of time and lessen the impact?
- What should the entertainment company do if the National Weather Service issues a tornado watch for the community? A tornado warning?
- How will closing or not closing the retractable roof affect the safety of the patrons?

Inquiring Further

1. **Direction of cyclones**

   Conduct research to find out why surface winds in a cyclone blow clockwise in the Southern Hemisphere and counterclockwise in the Northern Hemisphere.

2. **Wind shear and airplanes**

   Use the Internet or the library to find newspaper articles about airplanes that crashed because of wind shear. Two examples are the Delta Airlines L-1011 in Dallas in 1985, and a USAir DC-9 in Charlotte, N.C., in 1994.

   - Choose one and write a paragraph about how the weather affected the crash.
   - Research how the Federal Aviation Administration (FAA) has reduced the chances that aircraft will encounter dangerous wind shear. What are some devices that help pilots and air traffic controllers detect wind shear?

3. **Tornadoes and Hollywood**

   Watch a non-documentary movie like *Twister*. Prepare an analysis of how the movie is or is not scientifically accurate.

Earth Science at Work

ATMOSPHERE: *Airline Pilot*
Airline pilots must evaluate how weather affects a journey from takeoff to landing.

BIOSPHERE: *Golf Pro*
Lightning and other weather-related hazards can pose danger to someone on a golf course.

CRYOSPHERE: *Bridge Inspector*
When snowfall is great in mountainous regions, spring meltwater floods often create problems for bridges and roadways.

GEOSPHERE: *Line Installer and Repairer*
Communication is essential before, during, and after a severe weather event.

HYDROSPHERE: *Police Officer*
Police officers are responsible for citizen safety when a community must be evacuated prior to a hurricane or other severe storm.

How is each person's work related to the Earth system, and to severe weather?

3

The Cryosphere ...and Your Community

3 The Cryosphere ...and Your Community

Getting Started

The origin of the term cryosphere can be traced back to the Greek word *kruos,* meaning frost. It describes the part of the Earth's surface consisting of water in its solid forms, including ice masses, ice shelves, sea ice, seasonal snow cover, lake and river ice, seasonally frozen ground, permafrost, and glaciers. There are about 30,000,000 km^3 of ice now on the Earth's surface.

- How would the melting of vast quantities of ice within the cryosphere affect other Earth systems?

What do you think? Look at the Earth systems illustration at the front of this book. In your notebook, make a table to record your ideas. In one column, write down the names of the five Earth systems. In the second column, record your ideas about how a reduction of global ice cover might affect other Earth systems. How might it affect the hydrosphere? How might it affect the atmosphere? And so on. Be prepared to discuss your ideas with your small group and the class.

Scenario

Imagine the following scenario. An asteroid is headed towards Earth. A group of astronomers believes that it will strike the Earth, raising a huge, worldwide cloud of dust. The layer of dust will block out enough solar radiation to cause the Earth's climate to become significantly cooler for many years. Another group of scientists has calculated that the asteroid would miss the Earth by a comfortable margin. This group of scientists, however, is convinced that global warming will accelerate in the next few years, and that the Earth's climate will become significantly warmer. Your community wants to be ready for either possibility and has organized a special task force to investigate global climate change. As part of this task force, your team is assigned the topic of glaciers and their effects. If the climate cools, the Earth's glaciers will begin to grow and advance. If the climate warms, the Earth's glaciers will begin to melt and retreat. Your team is charged with investigating how either of these possibilities would affect your community.

Chapter Challenge

Your challenge is to write a report for your community that would answer these questions:

- What effect would the advance of glaciers have on your community?
- What effect would the melting of glaciers have on your community?

In answering these questions, you will need to address the following items:

- How glaciers form and move in response to climate changes.
- How glacial melting and advancement affect sea level.
- How changing sea level can modify the landscape and affect your community.
- How glaciers modify the landscape.
- The influence of a cooling or warming trend on nonglacial ice on the Earth's surface.

If your community is in a part of the United States far away from any glaciers, many citizens might believe that glaciers have no impact on the community. In your report, you will need to explain why this is not so and describe the many ways in which advancing or retreating glaciers would affect your community.

You will be asked to support your conclusions in several ways. You may have to perform calculations to make your point, construct a model to demonstrate some effects of glaciers, and show what glaciers have done in the past, using evidence from the field.

Part of your task force's final report will be a public information display or video explaining the effects of glaciation. As you go through the activities in this chapter, you will be asked to think of ways to recreate the effects of glaciation realistically. Whether your display or video uses models in the laboratory, computer-generated imagery, or other methods, its realism will depend on an accurate understanding of how glaciers behave.

Assessment Criteria

Think about what you have been asked to do. Scan ahead through the chapter activities to see how they might help you to meet the challenge. Work with your classmates and your teacher to define the criteria for assessing your work. Devise a grading sheet for the assessment of the challenge. Record all this information. Make sure that you understand the criteria and the grading scheme as well as you can. Your teacher may provide you with a sample rubric to help you get started.

Activity 1 Ice Is an Unusual Material

Goals

In this activity you will:

- Calculate the time required to melt ice.
- Graph data to determine a heating curve for ice.
- Compare the relative densities of liquid and solid paraffin, and liquid water and ice.
- Investigate how pressure affects ice.
- Investigate how pressure affects snow.
- Understand that hydrogen bonds can be used to explain some of the unique properties of water.

Think about It

Ice seems to be a simple substance.

- How would you describe ice to someone who knew nothing about it and had never seen it?

What do you think? Record your ideas about this question in your *EarthComm* notebook. Be prepared to discuss your responses with your small group and the class.

Investigate

Part A: Calculating the Time Required to Melt Ice

1. Read all of the steps of this investigation. Before you do the experiment, write down your hypothesis about how the temperature of the container will change over time.

    a) How long will it take for the material to reach 10°C?

    b) How will the temperature change over time? In your notebook, sketch a graph of temperature over time.

2. From a freezer or cooler of ice, obtain a small metal container packed full of crushed ice and place it on the table beneath the ring stand. The container will have a thermometer inserted into the ice so that the bulb of the thermometer is about 2.5 cm from the bottom.

3. Work as fast as you can to arrange the container as shown in the diagram below. (You will need to tie the thermometer so that it will remain suspended while the ice melts.) Have your teacher approve your setup before you continue.

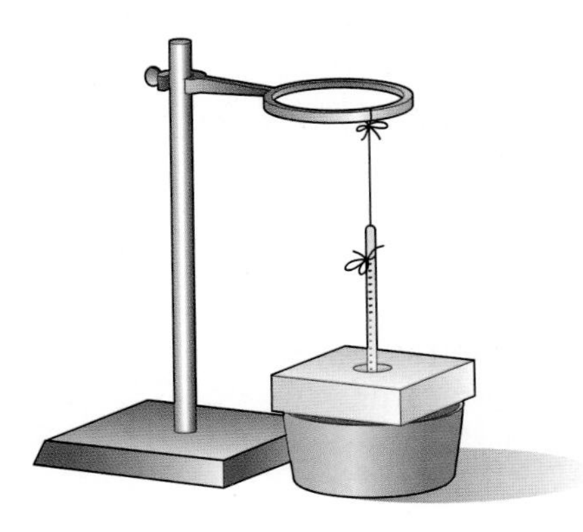

4. Take the first temperature reading as soon as possible. Continue to read the temperature every two minutes until the temperature reaches about 50°F (10°C).

    a) Record the temperatures in a data table.

    b) Plot a graph with temperature on the vertical axis and time on the horizontal axis. Choose the scales on the axis so that the curve you plot is not too steep or too gentle.

5. Compare your results to your hypothesis.

    a) How do you explain any difference?

6. From the graph, answer the following questions in your *EarthComm* notebook:

    a) At what time do you think the ice in the container began to melt?

    b) At what time do you think that all of the ice in the container had melted?

    c) Different slopes of the curve you obtained reflect different rates of increase in temperature with time. How can you explain the differing rates of increase in temperature?

    d) In most experiments, the measurement data points deviate at least slightly from a perfectly smooth curve. That effect is called "scatter in the data." How might you explain the scatter in your data?

Goggles must be worn throughout this activity. Wash your hands when finished.

Part B: Which is More Dense: the Solid Phase or Liquid Phase of a Substance?

1. Read through the steps of this part of the investigation.
   a) Predict whether paraffin will float or sink.
   b) Predict whether ice will float or sink.
2. Put on your safety goggles. Fill a small can or beaker with small chunks of paraffin wax (used for candle making and sealing preserving jars). Reserve a few chunks of wax.
3. Put the can or beaker in a pan of water and heat the water until the wax melts. **Caution!** The water and melted paraffin will be very hot. Observe the solid paraffin as the wax melts.
   a) Does the solid paraffin float or remain on the bottom of the container?

Be careful that no wax gets on the hot plate. If it does, turn off the hot plate and tell your teacher immediately.

4. When the paraffin is completely melted, carefully slide a small solid piece of wax into it. Avoid splashing any wax.
   a) Does the solid piece of wax float or sink?
5. Put an ice cube in a glass of water.
   a) Does ice float or sink in water?
   b) Which behavior do you think is more typical of materials as they melt: that of ice or that of the paraffin?
6. Compare your predictions and results.
   a) Explain any differences.

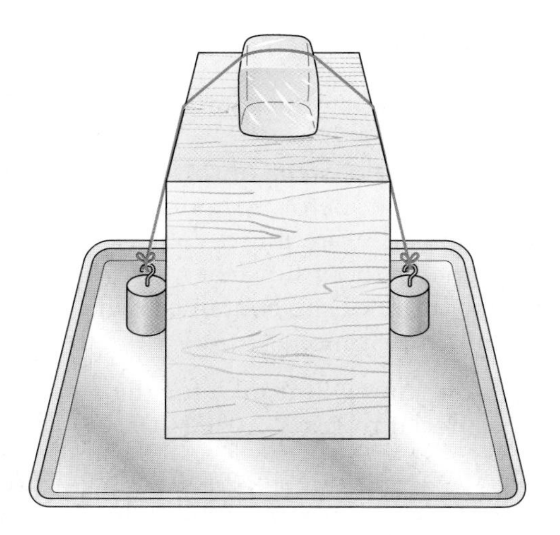

Part C: How Does Pressure Affect Ice?

1. Tie 200-g masses to each end of a very thin piece of string, wire, or fishing line as shown in the diagram.
2. Place an ice cube, fresh out of the freezer, on top of a block of wood. If the freezer is not nearby, place the ice cube in an insulated wrapper while carrying it to your station, so that it is still below the melting point at the start of the investigation.
3. Place the string over the top of the ice cube, with masses hanging down on each side of the cube.
4. Observe what happens as time goes on.
   a) Describe your observations in your notebook.
   b) How can you account for your observations?
   c) What differences do you observe depending on whether the ice cube is still below its melting temperature or at its melting temperature?

Part D: How Does Pressure Affect Snow?

1. If you have snow available, pack a snowball. Pack the snowball as hard as you possibly can. Then cut it in half using a warm table knife. Work over a pan to catch any melting water.
   a) As you pack the snowball, what happens to its volume and density?
   b) What made up most of the volume of the original snow?
   c) How has the snow in the tightly packed snowball changed?

Reflecting on the Activity and the Challenge

You have learned about several unusual properties of ice. It takes a large amount of heat to melt ice. It takes even more heat to raise the temperature of water. Unlike almost all other substances, water is less dense in the solid form than in the liquid form. The melting temperature of ice is slightly lower at high pressure than at low pressure. When pressure is applied to snow crystals, they grow together and change their shape. Understanding these properties will help you understand how glaciers form and change.

Digging Deeper

The Unusual Properties of Ice

Does it surprise you that water is an extremely unusual substance? Its unusual properties are explained by the atomic structure of the water molecule, which consists of two hydrogen atoms bonded to an oxygen atom, as shown in *Figure 1*. Because of the structure of the orbits of electrons around the three atomic nuclei, the three atoms are not arranged in a straight line; instead, the three atoms form an angle of 108°. Also, the orbiting electrons, which carry a negative electric charge, are more strongly attracted by the oxygen atom than by the hydrogen atoms. These two facts together mean that the oxygen "side" of the molecule is negatively charged and the hydrogen "side" of the molecule is positively charged, as shown in *Figure 1*. Molecules like this, with a negative charge on one side and a positive charge on the other, are called **polar molecules**.

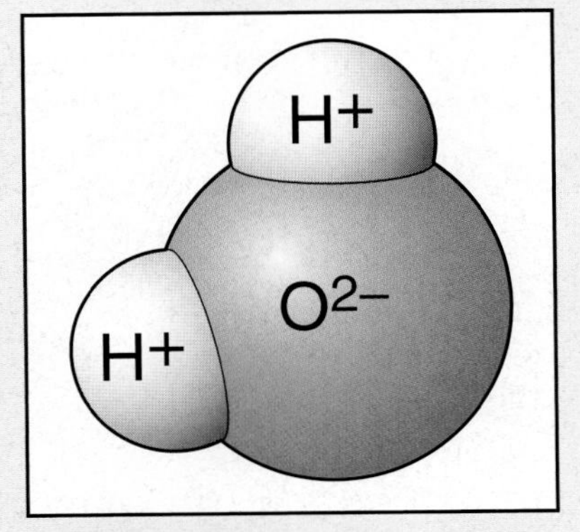

Figure 1 A water molecule is a polar molecule.

Geo Words

polar molecule: a molecule with a negative charge on one side and a positive charge on the other.

Objects with the same electric charge repel one another, and objects with different electric charges attract one another. The negative end of a water molecule repels the negative end of another water molecule, but attracts the positive end. Attraction between opposing charges in a molecule creates a **hydrogen bond**. Hydrogen bonds help to explain some of the unusual physical characteristics of water.

Geo Words

hydrogen bond: a weak chemical bond between a hydrogen atom in one polar molecule and an electronegative atom in a second polar molecule.

The temperature of any material is just a measure of the thermal vibration of its atoms and molecules. As heat energy is added to the material, the thermal vibrations become stronger, and the temperature increases. As heat is added to ice, the water molecules vibrate more and more, until eventually the vibrations break the hydrogen bonds that hold the structure together, and the ice melts to liquid water. Would you have guessed that it takes so much explanation to account for such a seemingly simple thing as melting of ice?

When liquid water freezes to form ice, the water molecules become arranged so that the negatively charged hydrogen sides of the molecules are bonded to the positively charged oxygen sides of neighboring molecules. The water molecules are arranged together in an unusually open structure, even though they are all bonded together with hydrogen bonds. When the ice melts, the water molecules are free to pack themselves more closely together. Because of the closer packing, the water molecules occupy less space; in other words, liquid water has a higher density than ice. That's why ice floats in water, as shown in *Figure 2*. Out of the zillions of substances known to science, including the paraffin wax in your investigation, only a handful have the property that the solid floats in the liquid. Have you ever thought about why plastic milk jugs have "dimples" in their sides? If the milk freezes, the dimples pop out rather than the jug splitting open as the milk (which is almost all water!) expands as it freezes.

Figure 2 Ice floats in water—an unusual but very important property of water.

As heat energy is added to liquid water, the temperature rises as the thermal vibration of the water molecules becomes stronger, just as in solid ice. As you saw in your investigation, however, it takes a lot more heat energy to raise the temperature of water than to raise the temperature of ice. The reason is that only some of the hydrogen bonds are broken when ice melts. At any given instant, a percentage of the molecules in liquid water are hydrogen-bonded to each other. As heat is added to water, a smaller and smaller percentage of the water molecules are hydrogen-bonded. Heat added to water not only increases the thermal vibration but also breaks more of the hydrogen bonds. The amount of heat needed to raise the temperature of a substance is called its **heat capacity**. Because of the heat needed to break hydrogen bonds, the heat capacity of water is far higher than any other common substance.

If you add the same amount of heat to equal masses of liquid water and dry soil, the temperature of the soil rises much faster. That's why lakes can be chilly even on sunny, warm days. Because water can absorb so much heat, the oceans are the principal heat reservoir on the Earth's surface.

For ordinary substances, higher pressure causes the melting temperature to be higher, because the high pressure tends to keep the solid from expanding to form the liquid. For water, however, it's the opposite: ice shrinks when it melts. Higher pressure helps with the shrinkage and causes the melting temperature to be slightly lower. This is why the wire melted into the ice in **Part C** of the investigation.

If you put a few ice cubes in a glass on a hot day, you will find a while later that the ice cubes have grown together into one solid mass. This phenomenon is called **regelation**. Where the ice cubes are in contact, the pressure is higher, so there's faster melting there. But melting takes heat energy, so the temperature is lowered slightly at the points of contact. That cools the adjacent ice slightly, and as the newly melted water moves to the adjacent ice, where the pressure is lower, it refreezes. Eventually the whole cube has grown together by this process of regelation. Regelation is an essential process in glaciers, as you'll see in the next activity.

Ice from the Atmosphere

Glaciers are made up of fallen snow. Each year a new layer of snow falls and buries and compresses the previous layers beneath. This compression transforms the snow into grains similar in size and shape to grains of coarse sugar. Gradually, the grains grow larger. As the air pockets between the grains get smaller, the snow slowly compacts and increases in density, much the same as the snow that you compressed into a snowball.

Geo Words

heat capacity: the amount of heat needed to raise the temperature of a substance.

regelation: a two-fold process involving the melting of ice under excess pressure and the refreezing of the derived meltwater upon the release of pressure.

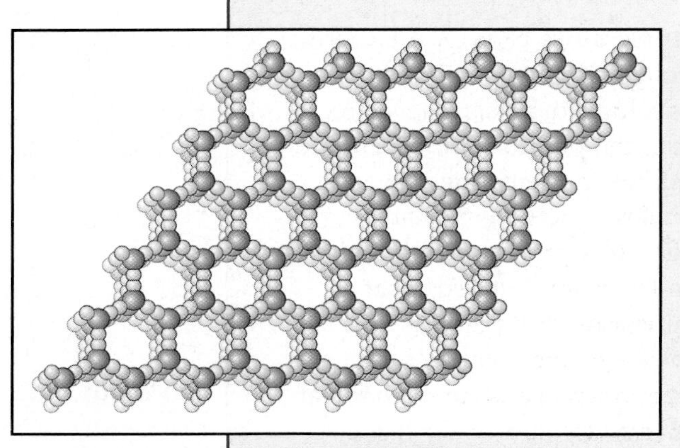

Figure 3 Expanded view of the atomic structure of ice.

Snow is not frozen rain. Snowflakes grow directly from water vapor in the atmosphere at temperatures below freezing. Snowflakes vary enormously in their shapes: no two look exactly alike. All snowflakes, however, have something in common. They all have hexagonal symmetry. A hexagon is a regular six-sided figure. Most snowflakes have six regularly spaced branches that radiate out from a central point. This hexagonal symmetry is the outward reflection of the hexagonal arrangement of water molecules in the ice structure. *Figure 3* shows an expanded view of how the water molecules are arranged together in the ice structure. From the figure you can see the hexagonal symmetry of the arrangement. *Figure 4* shows some of the many varieties of snow crystals. Flat snow crystals ("flakes") are formed when temperatures in the cloud are not far below freezing. Under some conditions, branches extend irregularly upward from the surface of the flake. Often, tiny spheres of ice are stuck to the snowflake. This happens when the falling snowflake makes contact with tiny droplets of supercooled water in the cloud, which freeze when they come in contact with the snowflake. When temperatures are much lower, however, the ice crystals grow in the form of hexagonal columns rather than flakes.

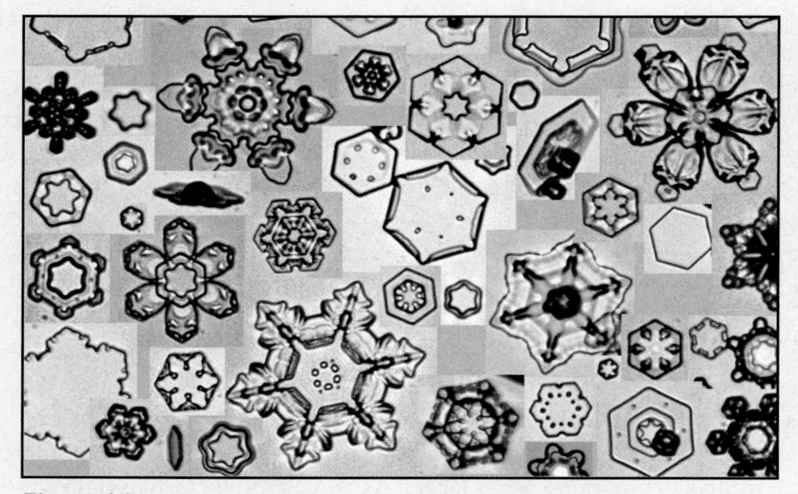

Figure 4 Snow crystals come in a variety of sizes. In the image the largest crystal measures about 140 μm, tip-to-tip diameter.

Snow is not the only form of solid precipitation. Both sleet and hail fall to the ground as solid pellets of ice, but their occurrence is very different. Sleet consists of small pellets of solid ice that form during winter storms. Sometimes in winter, cold raindrops fall through the lower layer of the atmosphere, nearest the ground, in which the temperature is well below freezing. If the raindrops freeze before they reach the ground, they fall as sleet. Sleet particles are almost always no larger than a millimeter or two in diameter. In contrast, hail forms in thunderstorms, usually in the warm season of the year. Raindrops that form in thunderclouds are sometimes caught in strong updrafts and carried up to altitudes where the air is below freezing. The raindrops freeze to form hailstones, and then fall back to lower levels. Usually, they melt again on their way down, but sometimes they reach the ground as ice particles.

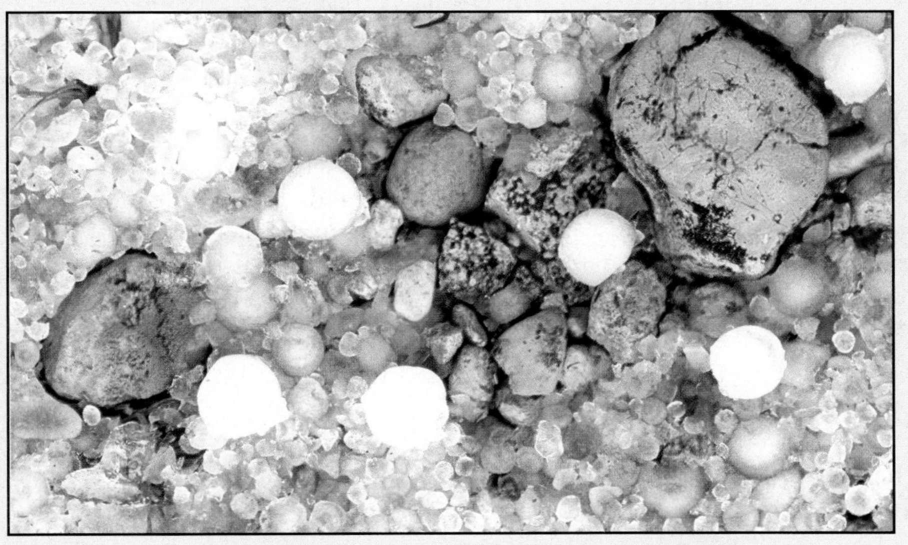

Figure 5 Hailstones. The largest hailstones in the photograph above are the diameter of a quarter.

During winter rainstorms, the ground surface and objects near the ground like trees, buildings, and telephone and power lines are sometimes well below freezing, even though the atmosphere above the ground is above freezing. This is especially common when a warm rainstorm moves into an area that had experienced very cold weather earlier. If the raindrops are even slightly below the freezing temperature, because of supercooling, they freeze almost immediately upon impact with solid surfaces on or near the ground. The result is called an ice storm.

Check Your Understanding

1. In your own words, describe the atomic structure of the water molecule. You may wish to use a diagram in your description.
2. Is it common for materials to float when they solidify? Explain your answer.
3. What are polar molecules?
4. Why does ice float in water?
5. Does ice melt more or less easily under pressure? Why?
6. How do snow crystals form?
7. What is the difference between sleet and hail?

Understanding and Applying What You Have Learned

1. Knowing what you do about the physical properties of ice, which do you think responds more slowly to seasonal changes in climate: an ice sheet or a wide ocean? Why?
2. What would happen to the Earth's oceans if ice sank instead of floated?
3. Prepare a table of the important physical properties of ice.
4. Place two ice cubes in a shallow pan. Put a heavy object on one ice cube and observe the two ice cubes melting. Use your observations and your understanding of the properties of ice to answer the following questions:
   a) Does ice with a weight on it melt faster or slower than ice without a weight on it? Why?
   b) Where does the most melting seem to take place when ice has a weight on it?
   c) How is ice with a weight on it like ice at the bottom of a glacier?
   d) Would the movement of a glacier begin at the top of the glacier or the bottom? Why?

Preparing for the Chapter Challenge

Write a few paragraphs explaining how the properties of water influence how glaciers form and how they move. Include this in your **Chapter Challenge** report. What demonstrations might you include in the public information video or display to explain some of the unusual properties of ice?

Inquiring Further

1. **Calculating the change in volume when water freezes**

   With the approval of a responsible adult, try the following investigation at home.

   - Take a plastic milk jug – one with a screw-top cap and dimples on the side (small depressions in the plastic). Fill it completely full of water. Pour the water into a large measuring cup and measure the volume of water.

   a) Record the volume, then pour the water back into the jug.

   - Cap the jug and put it in a freezer until it is frozen solid.

   b) What happens to the shape of the jug?

   - Remove the frozen jug from the freezer. Set the jug aside (perhaps until the next day) until all the ice has melted. Keep the cap on the jug to prevent evaporation.

   c) How does the water level in the jug compare with the level when you put the jug in the freezer?

   - Fill a measuring cup with water.

It is essential that the container be made of plastic. No glass should be used for this activity, regardless of its shape.

d) Record the volume of water in the cup.

- Using the measuring cup, pour water into the jug until it is brim-full. Be as careful as possible not to disturb the shape of the jug as you handle it.

e) Calculate and record the volume of water that you needed to add to fill the jug to the top. To do this you will need to subtract the final volume of water in the measuring cup from the initial volume you recorded in (d).

f) Calculate the percentage change in volume of the jug using this equation:

$$\textbf{Percentage change} = \frac{\textbf{additional volume of water added}}{\textbf{original volume of water in jug}} \times 100\%$$

g) What do you think is the purpose of the dimples in the milk jug?

h) Is your result likely to be an overestimate or an underestimate? Why?

i) What do you think might happen to soil or rock when the water trapped inside freezes?

2. **Earth Systems connections – frost wedging**

Investigate a type of physical weathering of rock called "frost wedging." How does it work? What parallels can you discover between what you find out about frost wedging and the investigations you did in this activity? How does frost wedging account for the formation of potholes in roads? Check the *EarthComm* web site for assistance with your research.

Frost wedging is a type of physical weathering.

Activity 2 How Glaciers Respond to Changes in Climate

Goals

In this activity you will:

- Make a mathematical model of an imaginary glacier.
- Calculate how the glacier would respond to hypothetical changes in climate.
- Understand the uses and limitations of this type of model in Earth science.
- Understand the mechanics of how glaciers form and move.

Think about It

Suppose the climate of the Earth got colder.

- Are there any areas in or near your state where a glacier might form? Why or why not?

What do you think? Make a list of ways that glaciers can change from year to year. For each item on your list, write down what factors might influence the rate of that change. Be prepared to discuss your responses with your small group and the class.

Investigate

Part A: Measuring the Speed of Movement of a Glacier

1. Assume the role of a glacial geologist. To measure the speed of movement of a valley glacier, you plant a row of steel stakes in the ice along a straight line across the glacier. The glacier is 1 km wide at this point, and the stakes are planted 50 m apart. You place surveying equipment on the rock slope next to the glacier and survey in the positions of the stakes. You return in half a month (fifteen days) to the surveying station, and re-survey the positions of the stakes. The table below gives the distances moved by each of the stakes, in meters.

Stake number	Distance from edge of glacier (meters)	Downglacier movement (meters)	Stake number	Distance from edge of glacier (meters)	Downglacier movement (meters)
1	50	3.6	11	550	5.2
2	100	4.2	12	600	5.0
3	150	4.5	13	650	5.0
4	200	4.8	14	700	4.9
5	250	4.9	15	750	4.7
6	300	5.0	16	800	4.4
7	350	5.2	17	850	4.0
8	400	5.4	18	900	3.7
9	450	5.5	19	950	3.3
10	500	5.3			

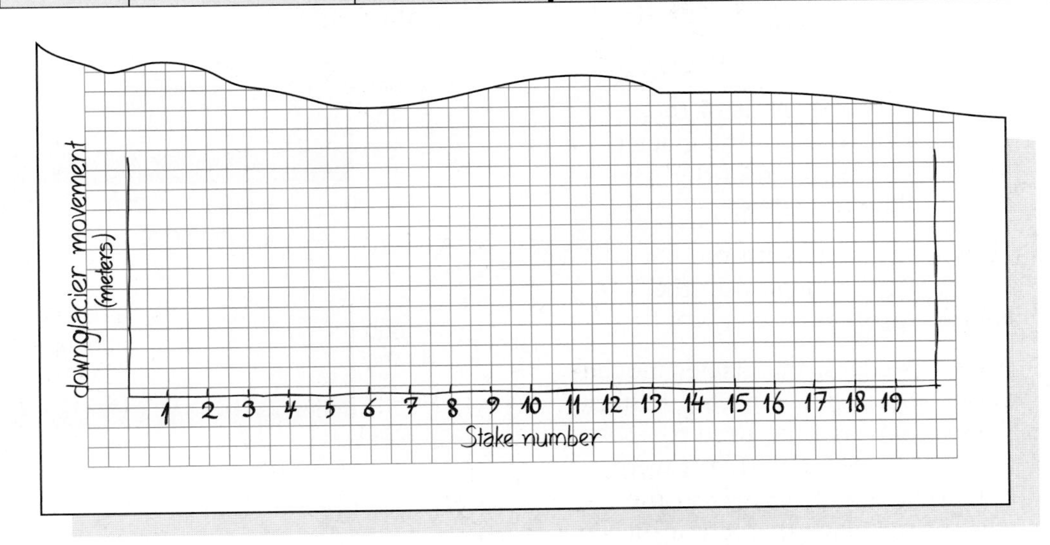

a) On a sheet of graph paper, draw a line that represents the row of stakes across the glacier.

b) Plot the positions of the nineteen stakes along the line. Add the positions of the two edges of the glacier, which are each 50 m from the outermost stakes. From these points representing the edge of the glacier, draw two heavy lines perpendicular to the row of stakes, as shown in the diagram on the previous page. Those lines represent the edge of the glacier.

c) For each point, plot the downglacier distance the stake moved.

d) Draw a smooth curve to connect all of the measurement points. Extrapolate (extend) the curve to the lines that represent the edge of the glacier. This curve is called the surface velocity profile of the glacier.

2. Use the velocity profile to answer the following questions:

a) Why is the velocity profile convex in the downglacier direction?

b) Why are the two ends of the curve, at the edge of the glacier, located downstream of the original line of stakes?

c) How can you explain the irregularities (wiggles) in the velocity profile?

d) Along its edges, the glacier moved about 3 m in half a month (fifteen days). Convert this speed to millimeters per day. If you were able to look very closely at the edge of the glacier, where it meets the bedrock base, do you think that you could detect the movement of the glacier? Explain your answer.

Part B: Modeling the Behavior of a Glacier

1. In this part of the activity, as a glacial geologist, you are monitoring a glacier in Alaska. Assume the following about your glacier:

   - It is 100 km long, 5 km wide, and 200 m thick.
   - It moves at a rate of 100 m/yr. (Note: this does not mean that the glacier gets longer by 100 m each year, but rather that any one point in the glacier moves forward 100 m in a year, as shown in the sample profiles below.)
   - It is at equilibrium. (Note: this means that it is receiving just enough snow to balance what it loses through melting. At equilibrium, the length and thickness of the glacier remain about the same.)

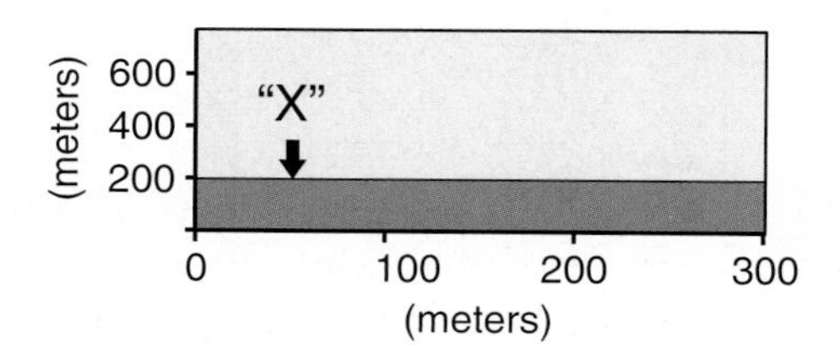

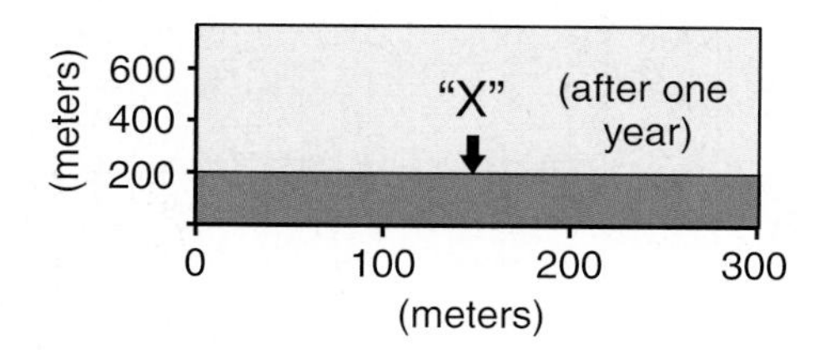

a) How long would it take a rock that falls into the ice at the head of the glacier to reach the foot? (Remember that the flow rate of the glacier is 100 m/yr and that the rock must travel the entire length of the glacier.)

2. The glacier is moving at a speed of 100 m/yr. If no ice were melting from the glacier, it would be 100 m longer after one year. However, it was assumed that the glacier maintains a constant size. Therefore, a volume of ice must be melting each year (see the sample profiles and note that the X has moved forward the same distance as was lost to the head of the glacier).

   a) How much ice is melting each year from this glacier? (If the glacier is at equilibrium, the volume of ice melted equals the distance of glacier movement per year, times the thickness of the glacier, times the width of the glacier.)

3. To be at equilibrium, the glacier must receive as much new ice each year as it loses by melting. A lot of snow that falls on a glacier simply melts and runs off without contributing anything to the glacier, especially in the warmer areas near the foot of the glacier. Assume that new ice is added only in the upper half (50 km) of the glacier.

   a) What volume of ice is needed to balance losses by melting?

   b) What thickness (depth) of ice has to be added each year to balance the melting? (Remember that the volume of ice is equal to flow per year times depth times width.)

4. On average 1 m of snow packs down into about 10 cm of ice.

   a) How much snow would have to fall on the glacier each year to create the thickness of new ice that you calculated above?

   b) Data show that 7.2 m of snowfall in the region in which the glacier is located. Is the amount of snowfall required to keep the glacier in balance realistic?

5. Imagine that the climate in the region of the glacier changes in such a way that the winter snowfall is greater by a factor of two (that is, it doubles) and the melting rate is less by a factor of two (that is, it is cut in half).

   a) How much larger will the total volume of the glacier be after 100 years?

   b) What will be the percentage increase in the size of the glacier? (To compute this, subtract the original volume of the glacier from the new volume of the glacier, divide by the length of the time interval in years, and multiply by 100 to convert to a percentage.)

6. Imagine that the climate in the region of the glacier changes in such a way that the winter snowfall is less by a factor of two and the melting rate increases by a factor of two.

   a) How many years would it take for the glacier to disappear (that is, melt completely)?

Reflecting on the Activity and the Challenge

This activity helped you gain an understanding of how glaciers move and how they respond to changes in climate. Whether the front edge of a glacier advances, retreats, or remains in the same position depends on the balance between snowfall and melting. You learned that it takes a long time for a moderate temperature change to make a big change in the quantity of glacial ice. In order to keep things simple, some assumptions were made that are not perfectly realistic. However, the activity provided you with an estimate of how much the climate had to change in order to make an increase or decrease in glacial ice. This will help you explain how glaciers move in response to climate changes when you are completing the **Chapter Challenge.**

Geo Words

glacier: a large mass of ice on the Earth's surface that flows by deforming under its own weight.

deform: to change shape.

Digging Deeper

How Glaciers Form and Move

A **glacier** is a large mass of ice on the Earth's surface that flows by deforming under its own weight. To **deform** is to change shape. A glacier is only partly a large-scale version of an ice cube sliding down a sloping tabletop; it is also like a deep puddle of honey or molasses flowing down the tabletop. Many materials act like solids on short time scales but like liquids on long time scales. Glass is a good everyday example. If you support a long, thin glass rod horizontally by its two ends in a warm room and wait patiently for weeks and months, you would find that the rod has sagged down slightly and taken on a permanent "set." Similarly, ice shatters when you hit it with a hammer, but under high pressure, deep within a glacier, it flows slowly like a liquid.

A glacier forms wherever more snow falls in winter than melts in summer, for a long period of years. As the old snow is buried by new snow, it is compressed by the weight of overlying snow. (That process is similar to the way that you compressed the snowball in Activity 1.) The crystals grow together by regelation, and the air in the spaces between the crystals is gradually forced out and upward. (Recall from Activity 1 that regelation is a twofold process involving the melting of ice under excess pressure and the refreezing of the derived meltwater upon the release of pressure.) Eventually, after several tens of meters of burial, the snow has been converted to solid glacier ice. After some further burial it begins to flow downslope as a glacier.

The largest glaciers form on broad land areas at high latitudes where summers are cool enough that not all of the previous winter's snow is melted. Large glaciers like this, called **ice sheets**, shown in *Figure 1* now cover most of Greenland and Antarctica. In the recent geological past, ice sheets also covered large areas of North America and Eurasia. Smaller glaciers, called **valley glaciers**, form in mountainous areas and flow down valleys to lower elevations, as shown in *Figure 2*. Valley glaciers are common in high-latitude mountains, but they can also form in very high mountains, even near the Equator.

Figure 1 The Antarctic ice sheet is the largest on Earth today.

Figure 2 Valley glaciers operate on much smaller scales than broad ice sheets.

Geo Words

ice sheet: a large glacier that forms on a broad land area at high latitudes where summers are cool enough so that not all of the previous winter's snow is melted.

valley glacier: a smaller glacier that forms in mountainous areas and flows down valleys to lower elevations.

Geo Words

zone of accumulation: the area in the upper part of a glacier where there is net addition of new glacier ice year after year.

zone of ablation: the area in the lower part of a glacier where there is net removal of glacier ice year after year.

snow line: the boundary between the zone of accumulation and the zone of ablation.

calving: the breaking away of a mass of ice from a glacier.

terminus: the downstream end of a glacier.

stillstand (in a glacier): the balance in accumulation and ablation when the volume of a glacier is constant and the terminus stays in the same place.

Every glacier has an area in its upper part, called the **zone of accumulation**, where there is net addition of new glacier ice year after year, and an area in its lower part, called the **zone of ablation**, where there is net removal of glacier ice year after year. The boundary between the two zones is called the **snow line**. (See *Figure 3*.) Below the snow line, all of last winter's snow is melted by the end of the summer, and old glacier ice is exposed to melting. Above the snow line, some of last winter's snow remains until the next winter. The newly formed glacier ice flows continuously downslope from the zone of accumulation to the zone of ablation. Ablation occurs mostly by melting, but where the glacier ends in the ocean, large masses of ice break away from the glacier, by a process called **calving**, and float away as icebergs.

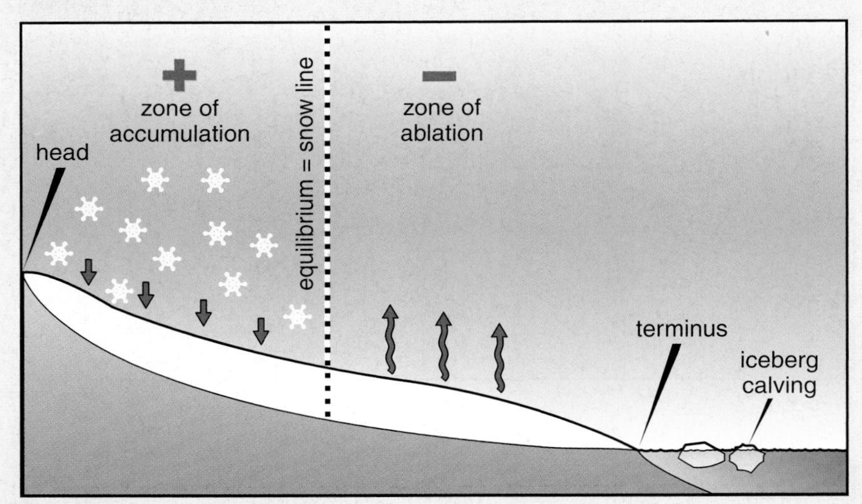

Figure 3 Diagram of a typical glacier.

Glaciologists (scientists who study glaciers) try to do "bookkeeping" of glaciers. When accumulation is greater than ablation for a period of years, the volume of the glacier increases, and the glacier lengthens. The **terminus** of the glacier (its downstream end) gradually advances downslope. When accumulation is less than ablation for a period of years, the volume of the glacier decreases, and the terminus retreats. Keep in mind, however, that even though the terminus is retreating, glacier ice is all the time being delivered to the terminus, where it finally melts. If accumulation and ablation are in balance, the volume of the glacier stays the same. The terminus then stays in the same place; that's called a **stillstand**.

It's not too difficult to measure the speed of movement of a valley glacier. During the summer melting season, you can plant a row of metal stakes across the glacier. Then set up a surveying station on the mountainside next to the glacier, and survey in the locations of the stakes. Come back next year and resurvey the stakes. The downglacier curvature of the new position of the row of stakes shows that the middle of the glacier flows faster than the edges. That's the **internal deformation** part of the movement. In internal deformation, the speed of the glacier at its edges is zero, but with **basal slip**, the speed at the edges is greater than zero.

Figure 4 Glaciologists at work.

A glacier that is below the melting temperature at its base is frozen solid to its bedrock base. Glaciers like that, with no basal slip (see **Part A** in *Figure 5*), are called cold-based glaciers. A glacier that is at its melting temperature at its base has a very thin film of water, usually no thicker than a millimeter or two, that's formed by slow flow of heat from the interior of the Earth. The lubricating film of water allows the glacier to slide on its bed (see **Part B** of *Figure 5*). Glaciers like that are called warm-based glaciers. (Does it strike you as strange that ice can be "warm" as well as "cold"?) Warm-based glaciers do a lot of geological work, because they can erode, transport, and deposit mineral and rock material as they slide over their beds.

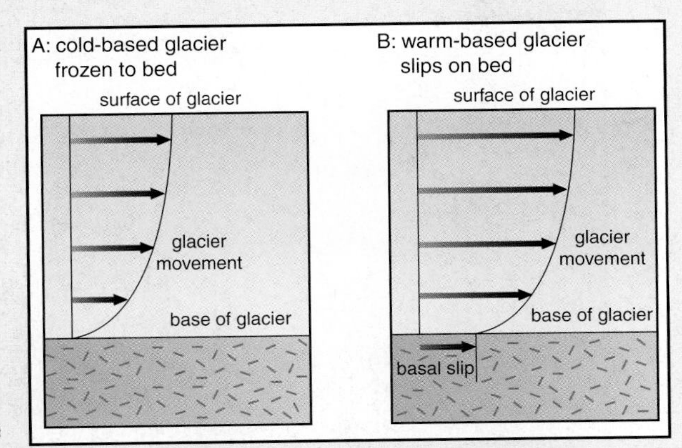

Figure 5 Two types of movement along the base of a glacier.

Geo Words

internal deformation: the part of movement of a glacier that is caused by change of shape within the glacier.

basal slip: the part of movement of a glacier that is caused by sliding of the glacier over the material beneath it, aided by a thin lubricating layer of water.

Over the past 2,000,000 years, the Earth has experienced several periods, called **ice ages**, when continental ice sheets in the Northern Hemisphere have formed and expanded to cover large areas of North America and Eurasia. Does it surprise you to learn that 20,000 years ago, an instant in terms of geologic time, the present sites of Boston, New York, and Chicago were under a mile of glacier ice? During ice ages, the Earth's climate was much colder and stormier than now. During the times in between ice ages, called **interglacials**, the Earth's climate has been much warmer. The Earth has been in an interglacial period for the past 10,000 years.

Geo Words

ice age: a time of extensive glacial activity.

interglacial: the time between ice ages.

Most geoscientists now agree that the main factor that determines the alternation of glacial periods and interglacial periods is the geometry of the Earth–Sun system. Three important things about that geometry change with time, over periods of many tens of thousands of years:

- the angle of tilt of the Earth's axis of rotation;
- the shape of the Earth's orbit of revolution around the Sun, which is not quite circular;
- the relationship between the Earth's seasons and the shape of the Earth's orbit.

Nowadays, the Earth is closest to the Sun on the 5th of January, almost in the middle of winter in the Northern Hemisphere! In another 10,000 years, however, the Earth will be farthest from the Sun during our winter, and that will make winter weather much more severe than it is today.

Figure 6 An Alaskan glacier.

Check Your Understanding

1. What is the zone of ablation?
2. What is the zone of accumulation?
3. What factors would cause a glacier to advance (grow)?
4. What factors would cause a glacier to retreat (get smaller)?

Understanding and Applying What You Have Learned

1. In **Part B** of the investigation, you modeled a hypothetical glacier.
   a) What assumptions did you make about the glacier?
   b) In what ways would real glaciers be more complicated?
   c) Are the assumptions you made realistic enough that you can draw useful conclusions from them, or are they so simplistic that they don't reflect real glaciers?
2. As the climate warmed at the end of the last ice age, it took about 5000 years for the ice to melt from the Great Lakes region. An area of the ice sheet that was 2 km thick and covered a million square kilometers melted and flowed down the Mississippi.
   a) Calculate the volume of ice that melted.
   b) Assuming that the ice sheet melted at a constant rate, calculate the rate of flow per year.
   c) What do you suppose the extra water did to the land?
3. Suppose that glaciers begin growing in the Great Lakes region at the same rate that they melted.
   a) How fast would they grow?
   b) Could they reach Indianapolis in your lifetime? (Assume that the distance from the Great Lakes region to Indianapolis is about 500 km and that you will live another 80 years.)
   c) How far would the glaciers reach in a century? In a thousand years?

Preparing for the Chapter Challenge

Calculations like those you have done in this activity will help you make a realistic evaluation of the problems that might be associated with changes in the world's glaciers. Think about how you can incorporate this in your **Chapter Challenge.**

- If you live in a high-latitude or high-elevation area, calculate how long it would take glaciers to reach your community if the climate got colder. To do this calculation, you will have to make assumptions about temperature change and the rate at which glaciers grow. Make sure to write down and explain all the assumptions you make.
- If your community is near a coastline, think about how your community might be affected if the Earth's climate warmed, causing some or all of the world's glaciers to melt. Where would all the melted ice from the glaciers go? How would that affect sea level?

Inquiring Further

1. **Monitoring present-day glacial activity**

   Do you think that most glaciers worldwide are advancing, holding steady, or retreating? Do some research to find out. Go to the *EarthComm* web site to get your research started. What are some of the theories to explain how glaciers are behaving?

2. **Investigating Precambrian glaciers**

   One of the most puzzling episodes in the Earth's history took place between 900 and 600 million years ago. There seems to have been extensive glaciation even though the continents were at low latitudes. Generate a list of hypotheses that might account for these observations. This is a real ongoing research problem, and scientists themselves do not agree on the answers. Visit the *EarthComm* web site to learn more about recent research into this question.

Activity 3 How Do Glaciers Affect Sea Level?

Goals

In this activity you will:

- Estimate, using maps, how much ice was contained in the Pleistocene ice sheets and calculate how much lower sea level was.
- Estimate the volume of ice in the Greenland and Antarctic ice sheets and calculate how much sea level would rise if they were to melt completely.
- Determine if sea-level changes would affect your community.
- Understand the mechanics of post-glacial melting and how it impacts sea level.

Think about It

Imagine that the Earth's climate suddenly became drastically warmer and all of the glaciers and ice sheets melted.

- What would happen to the sea level? Why?
- How could you predict precisely which parts of the United States would be affected by the change in sea level?

What do you think? Record your ideas about these questions in your *EarthComm* notebook. Be prepared to discuss your responses with your small group and the class.

Investigate

Part A: How Much Ice Did the Pleistocene (20 ka) Glaciers Hold?

1. Look at the images of the areas around the North and South Poles at 20,000 years ago (20 ka). Note how the squares form a grid on the screen. The diagrams are "equal-area" polar projections. In an equal-area polar projection, each grid square contains the same area. The diagrams are at slightly different scales, because the area of a grid square is different between the two projections: the area of each grid section on the north polar projection is about 800,000 km^2 (square kilometers), but the area of each grid section on the south polar projection is about 500,000 km^2 (square kilometers).

2. Use the equal-area projections of the North and South Poles at 20,000 years ago (20 ka) to answer the following questions:

   a) Add up the total area covered by Pleistocene glaciers. Express areas in square kilometers.

   b) Estimate the average thickness of the Pleistocene ice sheets by comparing them with the current Antarctic and Greenland ice sheets. The average thickness of the ice sheets today is 1.5 km.

   c) Multiply the area covered by ice by the average thickness to calculate the volume of the Pleistocene ice sheets.

3. The ice in the glaciers came from the water from the oceans, so as glaciers grew, sea level fell. As the glaciers melted, the meltwater was added to the oceans, causing sea level to rise. The change in sea level is the volume of the Pleistocene ice sheets divided by the area of the oceans.

   a) Explain why this is so.

   b) The area of the oceans is 362 million square kilometers. Using this figure, calculate the changes in sea level associated with both the growth and melting of the Pleistocene glaciers.

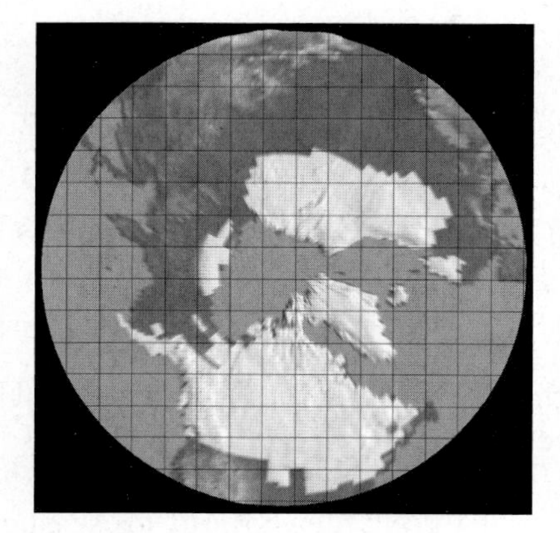

North Pole 20 ka

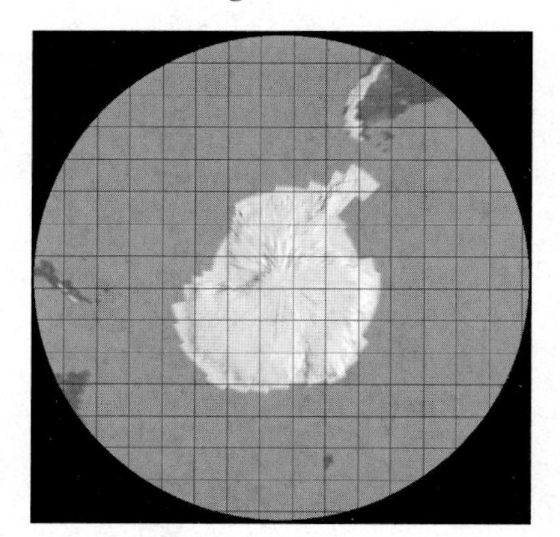

South Pole 20 ka

Part B: How Much Ice is in the Greenland and Antarctic Ice Sheets?

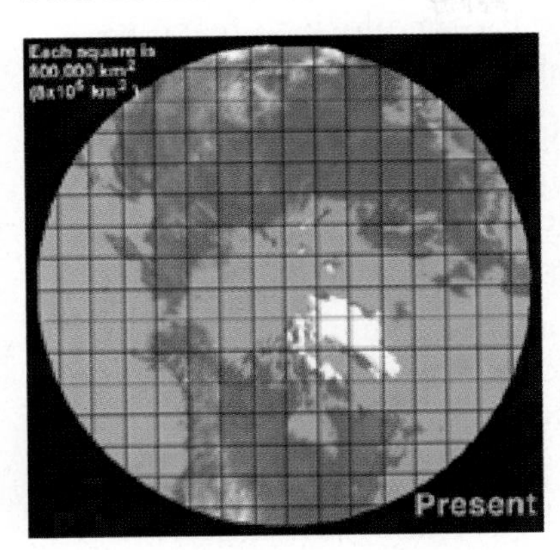

North Pole Present

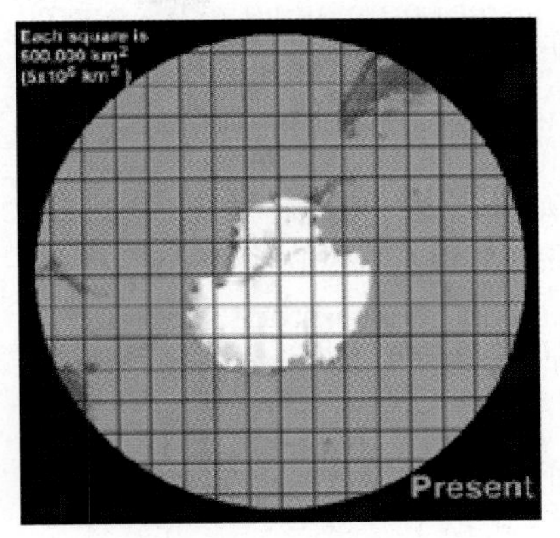

South Pole Present

1. Use the present-day equal-area projections of the areas around the North and South Poles to answer the following questions:
   a) Estimate the amount of ice currently in the Greenland and Antarctic ice sheets.
   b) Estimate the expected sea level rise if the Greenland and Antarctic ice sheets were to melt completely.

Part C: Mapping the Effect of Glaciers on Coastal Land Areas

1. You have calculated by how much the sea level would fall if the present-day glaciers expanded as much as they did during the Pleistocene.
   a) Using a relief map of the United States, outline the areas that would become dry land if the glaciers were to advance again.
2. You have also calculated how much sea level would rise if the remaining ice sheets were to melt.
   b) Using a relief map of the United States, outline the areas that would be flooded if the world's remaining glaciers were to melt.

Part D: Modeling Postglacial Rebound

1. In a mixing bowl, make a large lump of dough. To do this, put two cups of room temperature water in the mixing bowl, add two cups of sawdust and one cup of flour, and mix with a stirring spoon until uniform. Gradually add more flour, stirring continuously, until you have a stiff dough that is no longer gooey. Turn the dough out onto a sheet of floured waxed paper on a tabletop and spread it with your hands until it is about 2 cm thick. Trim the sheet of dough so that it is a square about 20 cm on a side. Trim a sheet of bubble wrap to make two squares of the same size as the dough. Place the dough between the sheets of bubble wrap, press out all the air you can, and tape all four edges watertight with the duct tape. Press the dough to mold it to the bubble wrap. This is your model of the continental crust.
2. Fill a pan or tub about half full with room temperature water and place the dough sheet on the water surface. The dough sheet should float rather than sink. Anchor the dough sheet along one side of the pan in a few places with duct tape.

3. Pick a point near the middle of the dough sheet and make a mark on the surface with a marker pen.
4. Establish two "stations" along opposite edges of the pan or tub along a line that passes directly over the mark on the dough sheet.
5. Rest a meter stick on the "stations," and tape it firmly to the edges of the pan or tub with duct tape.
6. Place a 5 cm × 5 cm × 2 cm wood block gently on the surface of the dough sheet so that one edge of the block is about one-half inch away from the mark.

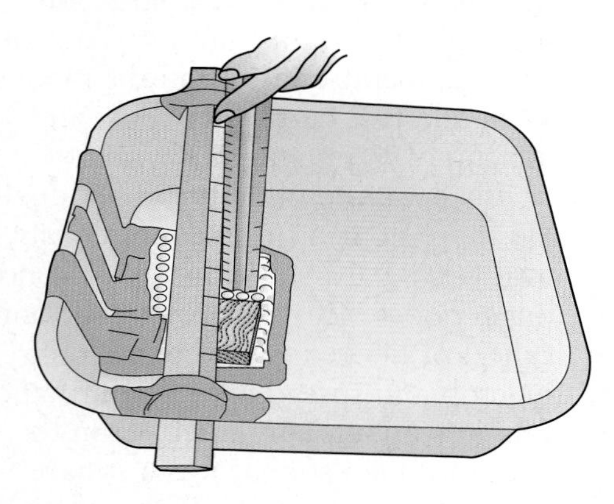

7. As soon as the wood block has sunk down to its equilibrium position (i.e., has stopped sinking), hold a centimeter ruler vertically against the yardstick and lower it down until its end touches the mark on the dough. Read the ruler to the nearest millimeter where it crosses the meter stick.
   a) Record the reading in your notebook.
8. Gently remove the block from the dough sheet. Read the elevation of the mark again.
   a) Why did it take some period of time for the block to reach its equilibrium position?
   b) What factor or factors determine the equilibrium position of the block?
   c) Did the mark on the dough end up at the same elevation when the block was removed? If so, why? If not, why not?
   d) All else being the same, would the difference in the initial and final elevations of the mark have been the same or different if you had made the dough even stiffer? thicker?
   e) All else being the same, would the difference in initial and final elevations of the mark have been the same or different if the block had been thicker? Test your prediction using a 5 cm × 5 cm × 5 cm block.
   f) If the block had been wider across, all else being the same, would the difference in initial and final elevations of the mark have been the same or different? Test your prediction using a 10 cm × 10 cm × 2 cm block.
   g) If the block had consisted of denser material, all else being the same, would the difference in initial and final elevations of the mark been the same or different?

Discard the dough as directed by your teacher. Any remaining flour that is not used for the model must be considered contaminated, and labeled as such. Goggles must be worn. Wash your hands when you are done.

Reflecting on the Activity and the Challenge

You have determined that the Pleistocene glaciers lowered sea level. You have also calculated how much the melting of the Greenland and Antarctic ice sheets would raise sea level. You investigated how the Earth's crust is depressed as a glacier develops on it and then springs back to its original position after the glacier melts. You are now prepared to determine where your community would be relative to the shoreline if sea level fell as much as it did during the Pleistocene or if sea level rose as much as it would if glaciers melted today.

Digging Deeper

How Glaciers Affect Global Sea Level

The ultimate source of the ice in glaciers is the ocean. Worldwide, there is more evaporation than precipitation over the oceans and there is more precipitation than evaporation over the continents. The snow that feeds the world's glaciers forms from water vapor that was put into the atmosphere far away by evaporation of ocean water. As the water is parked on land in the form of glacier ice, sea level around the world is lowered. As the glacier ice melts, the water flows in streams and rivers back to the oceans. This extraction of water from the oceans to form glaciers on land, and the return of the water to the oceans, is one of the important "loops" in the Earth's water cycle.

Glaciologists have a good idea of the total volume of ice in the world's glaciers. Almost all of that ice is in the Antarctic and Greenland ice sheets. To calculate the volume of ice in an ice sheet, both the area and the thickness must be known. The area of the ice sheets is easy to measure, but the thickness is more difficult. The thickness of ice sheets is measured by echo sounding. A pulse of sound is sent down through the glacier, and the echo from the bedrock at the base of the glacier returns to the surface. The thickness can be calculated from the measured two-way travel time of the sound pulse and the known speed of sound in glacier ice. Of course, for a good estimate of thickness, a very large number of soundings must be made. The ice sheets are thousands of meters thick, so the volume of ice is enormous. If all of the Antarctic ice sheet were to melt, sea level around the world would rise by about 60 m, and if all of the Greenland ice sheet were to melt, sea level would rise by about 5 m.

Figure 1 Glaciers are an important phase in the Earth's water cycle.

Geo Words

lowstand: the interval of time during a cycle of sea-level change when sea level is about at its lowest position.

As recently as 20,000 years ago (an instant, in terms of geologic history) ice sheets covered not only Greenland and Antarctica but also much of eastern and central Canada and the northern part of the United States, as well as large areas in Scandinavia and Russia. At its maximum the Laurentide ice sheet, in Canada and the United States, extended as far south as Missouri, as shown in *Figure 2*. Sea level around the world was much lower than now. Marine geologists can recognize the remnants of old shoreline features, now underwater far out to sea, that were formed during the sea-level **lowstand**, when the ice sheets were at their maximum. These shorelines are more than 100 m below present sea level. In the short space of only about 12,000 years, from 18,000 years ago to 6000 years ago, sea level rose to nearly its present position.

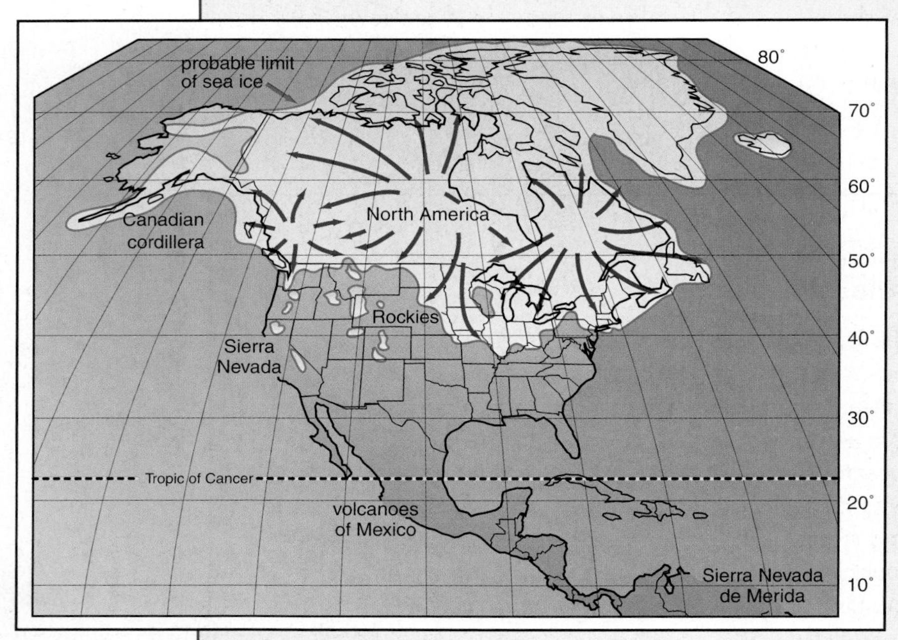

Figure 2 The approximate extent of ice sheets in North America during the last Pleistocene glaciation.

Figure 3 The mouth of the Hudson River.

During the long sea-level lowstand, the shoreline was far out near the edge of the continental shelf, 100 miles from the present shoreline. Rivers cut valleys across the shelf on their way to the ocean. Since the rise in sea level, these valleys have been flooded and partly filled with new sediment. An example on the East Coast of the United States is the submarine valley that extends over 100 miles from the present mouth of the Hudson River, in New York City, all the way to the shelf edge. On the West Coast, the Golden Gate strait, which links San Francisco Bay to the Pacific Ocean, is a flooded river valley. During the last glaciation, the Sacramento River flowed to the sea through the present Golden Gate. San Francisco Bay was a dry lowland between two mountain ranges.

Postglacial Rebound

The outermost layer of the Earth, called the **lithosphere**, which is about 100 km thick, consists of rock that is cool enough to behave like an elastic solid. Like a piece of rubber or a hacksaw blade, the lithosphere can be bent and stretched, but it returns to its original shape when the deforming force is removed. Below the lithosphere is the **asthenosphere**, which behaves like a plastic solid. When the asthenosphere is acted upon by a deforming force, it changes its shape permanently by flowing, very much the way glacier ice flows.

When an ice sheet forms, its weight bows down the lithosphere, both directly beneath the ice sheet and also for some distance next to the ice sheet. Asthenosphere material flows sideways to make room for the down-bowed lithosphere. Then when the ice sheet melts, almost instantly in terms of geologic time, the lithosphere rises up buoyantly to its original position. In **Part D** of the investigation, you simulated **subsidence** (lowering) and rebound of the lithosphere by placing a block of wood to represent a glacier on the slab of dough and then rapidly removing the load. This phenomenon is called **postglacial rebound**. Because the lithosphere is strong and the asthenosphere is very stiff, however, postglacial rebound is slow. Large areas in eastern Canada and northern New England are still rebounding after melting of the Laurentide ice sheet, at speeds of almost a meter per century! If the ice sheet was near the coastline, postglacial rebound causes the sea to retreat from the land in areas near the former position of the ice sheet. This process tends to offset the rise in sea level caused by the addition of water to the oceans as the ice sheet melts.

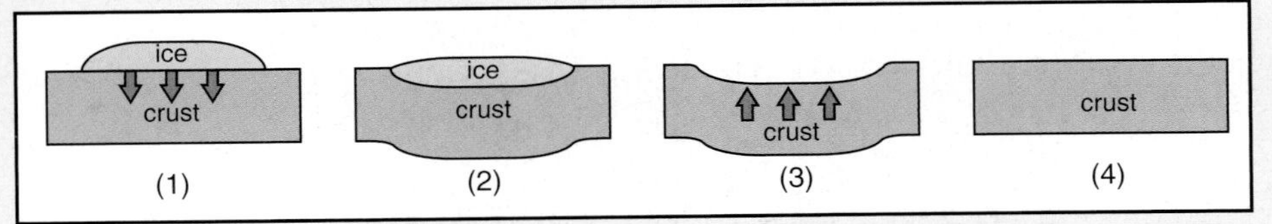

Figure 4 The addition of weight from an ice sheet (1) causes the lithosphere to bow downward (2). When the weight of the ice is removed (3), the crust rebounds to its original position.

Geo Words

lithosphere: the outermost layer of the Earth, consisting of the Earth's crust and part of the upper mantle. The lithosphere behaves like a rigid layer, in contrast to the underlying asthenosphere.

asthenosphere: the part of the mantle beneath the lithosphere. The asthenosphere undergoes slow flow, rather than behaving as a rigid block, like the overlying lithosphere.

subsidence: sinking or downward settling of the Earth's surface.

postglacial rebound: the gradual rebound (uplift) of the Earth's crust after the melting of glaciers.

Check Your Understanding

1. If the ice in glaciers is made of fresh water, why does sea level drop when the climate gets colder and the world's glaciers grow?
2. How do glaciologists measure the thickness of a glacier?
3. In very general terms describe the difference between the lithosphere and asthenosphere.
4. How does the elevation of a land area change when a glacier forms on it? When the glacier disappears again? Explain why this occurs.

Understanding and Applying What You Have Learned

1. If a glacier was 10 km from your community and was advancing at a rate of 2 m per day, how many years would it take to reach your community?
2. In what ways was the dough and plastic a good model of postglacial rebound? In what ways was it not a good model of postglacial rebound?
3. Why is it that a shoreline can shift by as much as 100 km as ice sheets change in volume, even though sea level changes by only about 100 m?
4. Compare your estimates of ice-sheet area, volume, and sea-level change to those of other students.
   a) How much variation is there in the results?
   b) How can you explain the variation?
   c) What evidence could you look for on the real Earth to check these theoretical calculations?
5. How do errors in estimating the areas of the ice sheets affect the accuracy of your estimates of how much sea level changed?
6. It is obvious that there would be problems if sea level were to rise.
   a) Would there be any serious problems if sea level were to fall?
   b) Knowing that there are possible errors in estimating sea-level change, what is the largest possible change that could reasonably be expected? the smallest?
   c) How would you explain the uncertainties to emergency planners?
   d) How would you advise planners to prepare for the different possible changes?
7. Would the melting of ice floating in the oceans have any effect on sea level? Why or why not?

Preparing for the Chapter Challenge

Write a paragraph explaining how your community would be affected if sea level rose, and a paragraph illustrating how your community would be affected if sea level fell. Include as many factors as you can, not simply whether your community would be closer or farther away from the beach. Would a change in sea level affect the economy of your community? the population? tourism? transportation? How will you use this information in your **Chapter Challenge**?

Inquiring Further

1. **The pressure under a column of ice**

   What is the pressure beneath 3 km of ice? Is that enough pressure to have any effect on rocks? (It takes 10 million kilograms per square meter to fracture rock. How much would a column of ice 1 m square and 3 km high weigh? A cubic meter of ice weighs about 900 kg.)

2. **Investigating glaciers and climate change**

   Carbon is stored in the atmosphere as carbon dioxide (CO_2). CO_2 in the atmosphere is a greenhouse gas. That means that it traps heat in the atmosphere, warming the Earth. Levels of CO_2 in the atmosphere through geologic time can be measured by examination of ice cores, because when the ice forms, CO_2 is trapped in air bubbles. Go to the *EarthComm* web site to examine ice-core data from an ice core collected at the Antarctic station Vostok and answer the following questions:

   - Do high levels of CO_2 correspond to warmer or colder temperatures?
   - In general, do glacial or interglacial periods correspond to high levels of CO_2? How would you explain this relationship?
   - If glacial and interglacial periods are identified by peaks, or maximum values, in the record, how many glacial/interglacial cycles can you identify on the Vostok record?
   - Do the cycles appear to occur with a regular frequency?
   - What implications do you think this has for current predictions regarding global warming?

Activity 4 How Rising and Falling Sea Levels Modify the Landscape

Goals

In this activity you will:

- Model sea-level changes using a stream table.
- Evaluate the uses and limitations of this type of model in Earth science.
- Describe features of a coastal landscape during a time of rising sea level.
- Describe features of a coastal landscape during a time of falling sea level.
- Determine how your community's landscape would be affected by rising and falling sea levels.
- Describe ways that sea-level changes have affected human history.

Think about It

New York City is located on several islands. The channels between these islands, like the East River, which separates Manhattan from Brooklyn and Queens, are former stream valleys flooded by rising sea level.

- Do you think the activity of glaciers in the past has affected the landscape where your community is located? What is your evidence?
- Do you think the activity of glaciers could affect your community in the future? Why or why not?

What do you think? Record your ideas about these questions in your *EarthComm* notebook. Be prepared to discuss your responses with your small group and the class.

Investigate

Part A: Modeling a Stream on the Stream Table

1. Fill a stream table with damp sand. Level the sand so that there is just enough slope to allow water to flow along it. A centimeter of rise on a stream table a meter long will work nicely. (This may seem like a very gentle slope, but bear in mind that the slope of the lower Mississippi is about a centimeter per kilometer.)
2. At the lower end, leave an embankment to simulate the edge of the continental shelf. Fill the area beyond the embankment with water to just below the lip of the embankment.
3. At the top end of the stream table, turn on a water source. Begin with a trickle, and gradually increase the flow. Let the water run long enough to cut a channel.

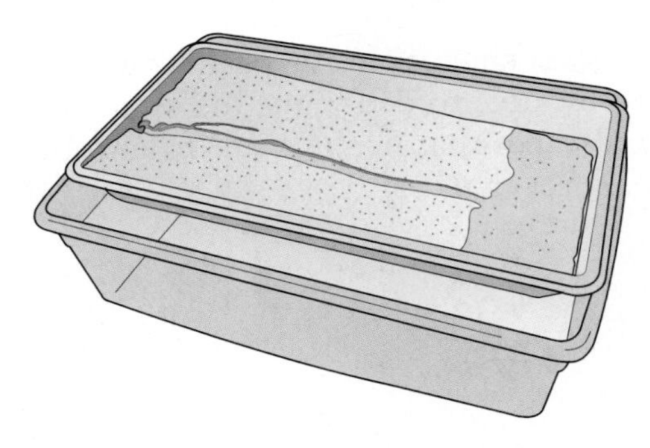

4. Watch the movement of sediment along the "river." Notice the deposition of sediment at the river's mouth (where the river empties into the "sea").

Wear goggles throughout. Clean up spills immediately. Wash your hands when done.

a) Take detailed notes and make sketches of all that you observe happening on the stream table. If you have a video camera, you might want to videotape the stream table at regular intervals to make a time-lapse video of its evolution. Some things to look for include submarine landslides, miniature canyons cut into the edge of the shelf, and submarine deltas on the edge of the shelf or at the base of the slope.

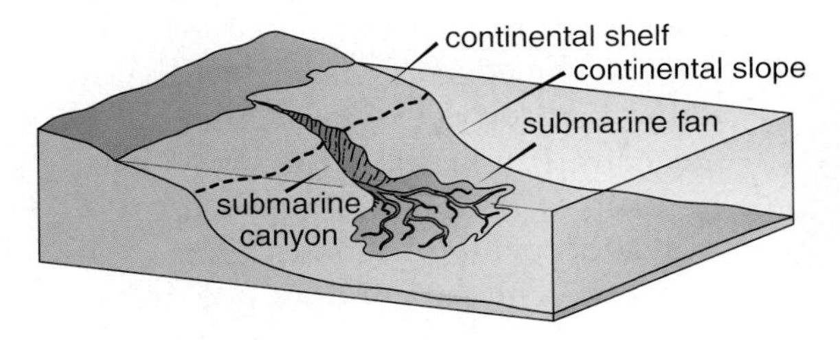

Part B: How Does a Rising Sea Level Affect the Landscape?

1. To simulate a rise in sea level, adjust the flow of water so that the water level rises and the upper surface of the sand is partially flooded. Allow the new simulation to run long enough to build a good delta. It will take only a few minutes to see changes begin to happen, but you may have to run the simulation for some time.

   a) As the water level rises, take detailed notes and make sketches of all that you observe happening. Note the retreat of the shoreline, the infilling of the now-submerged channel, and the building of a delta at the new channel mouth.

2. You learned in **Activity 3** that a sea-level rise can be caused by the melting of glaciers.
   a) What are some kinds of evidence you might look for in your community to determine whether sea level really was higher in the past?
   b) What are some of the practical problems you might encounter in finding this evidence?

Part C: How Does a Falling Sea Level Affect the Landscape?

1. To simulate a fall in sea level, adjust the drainage so that the water level falls to the edge of the shelf or below. Observe how the stream valley deepens. If you allow the water level to drop fast enough, the process of valley deepening becomes very obvious. The stream does not deepen its valley all at once. Instead, the slope near the mouth becomes steeper, and the stream begins cutting deeper at its mouth. A pronounced break in slope, called a nickpoint, migrates upstream.
   a) As the water level falls, take detailed notes and make sketches of all that you observe happening.
2. You learned in **Activity 3** that a fall in global sea level can be caused by the formation of glaciers.
   a) What are some kinds of evidence you might look for in your community to determine whether sea level really was lower in the past?
   b) What are some of the practical problems you might encounter in finding this evidence?

Reflecting on the Activity and the Challenge

Using the stream table model, you have observed landscapes associated with the rise and fall of sea level. You can use this to determine whether your community has been affected by sea-level changes in the past. You will also be able to determine the kinds of landscapes that might result from future sea-level changes.

Digging Deeper

How Rising and Falling Sea Levels Modify the Landscape

People living along or near seacoasts, and even far inland, are in areas with landscapes that were affected by sea-level rise following the retreat of the Pleistocene ice sheets. Here are some examples:

New York City is located on several islands. The channels between these islands, like the East River, which separates Manhattan from Brooklyn and Queens, are former stream valleys flooded by rising sea level.

The floor of the Hudson River for much of its length is below sea level because the river incised its valley deeply when sea level was much lower. In fact, the former channel of the Hudson River can be traced far out to sea beyond the present shoreline, as can the channels of many other rivers.

Baltimore, Maryland and Providence, Rhode Island became important seaports because of their protected harbors. The harbors of both cities are in river valleys that were flooded as sea level rose.

Figure 1 A flooded river valley forms a protected harbor in Baltimore.

The Mississippi River and its tributaries are often located on flat floodplains flanked by steep bluffs. During the Pleistocene, when sea level was much lower, the Mississippi and its tributaries cut their valleys much deeper than at present. As sea level began to rise, the valleys filled in.

One of the most important effects of the changes in sea level was the exposure and flooding of the Bering Strait land bridge, shown in *Figure 2.* The Bering Strait land bridge is simply a submerged connection between Asia and North America. When sea level is low, it is dry land. When sea level is high, it's under water. Ancestors of Native Americans may have crossed the land bridge during the last ice advance.

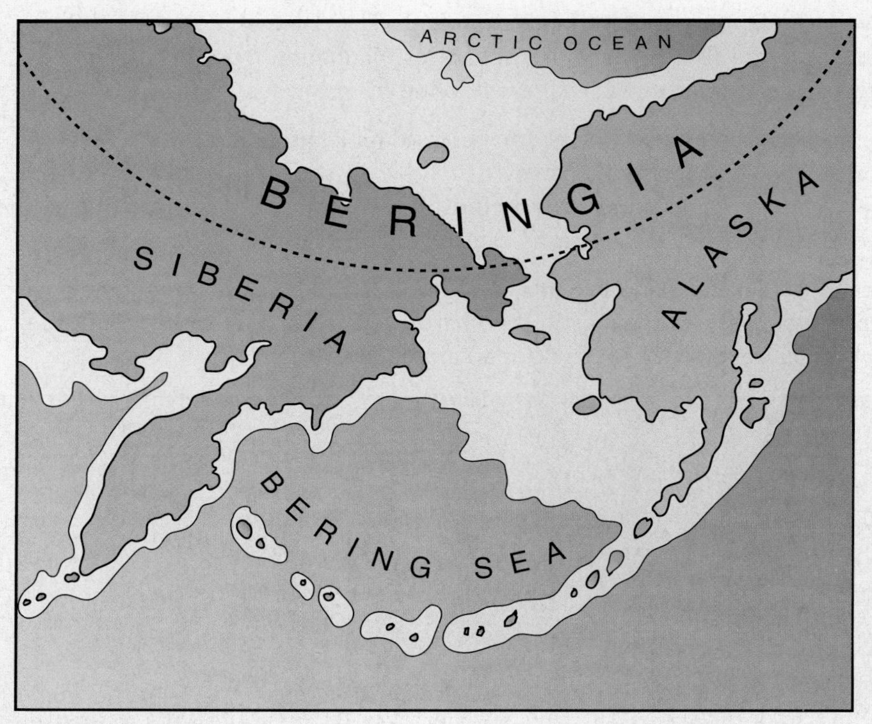

Figure 2 Shallow regions of the Bering Strait (shown in green) were exposed as land during the last ice age.

Rising sea level may have been much more important in human history than simply determining the present-day locations of cities. Around the world, settled agriculture began in many places about 7000 years ago. This was about the time sea-level rise began to slow down. Before that time, sea level was rising at a rate that would have made permanent settlement on coastlines impractical. Once sea level became stable, river deltas could form. River deltas provided fertile, well-watered sites for agriculture. Also, river valleys that had once been cut far below present sea level began to silt in. These silted-in valleys created fertile floodplains like those of the Nile.

One obvious effect of a rising sea level is the submergence of former dry land. Chesapeake Bay is a flooded river valley. Long Island and Cape Cod are ridges of glacial deposits submerged by rising sea level. The English Channel is a flooded valley. Former river systems can be traced on the sea floor of the North Sea and between Sumatra and Borneo.

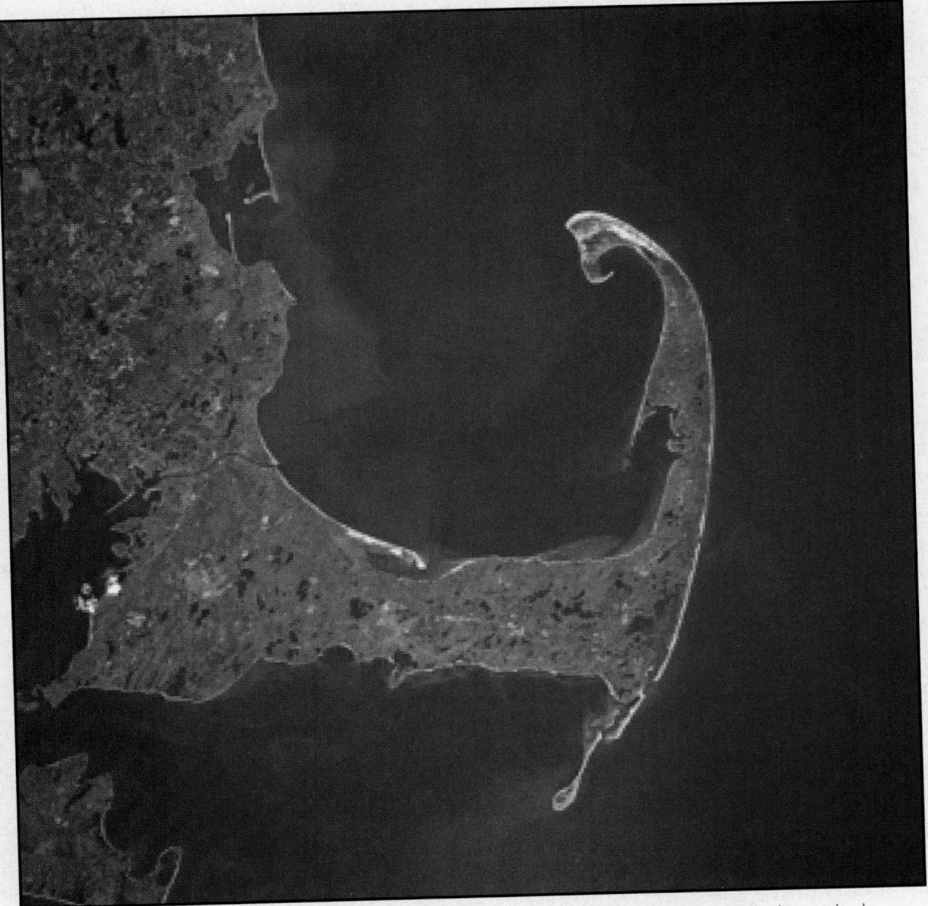

Figure 3 Aerial photo of Cape Cod. The Cape consists of glacial deposits called terminal moraines and outwash plains.

Changes in sea level due to glaciation can have effects far inland, because of the way rivers cut their valleys. In steep areas, rivers move fast and erode their beds. In areas of gentler slope, rivers slow down and deposit sediments. Rivers tend to have gentle slopes near the mouth and steep slopes near their source. The elevation of a river's mouth is called its **base level**.

Geo Words

base level: the elevation of a river's mouth.

Geo Words

nickpoint (also knickpoint): an interruption or break of slope, especially an abrupt change in the longitudinal profile of a stream or its valley.

If sea level drops, the river can now flow onto the former sea floor to a lower base level. Its slope near the mouth becomes steeper and the river begins cutting its valley deeper. The valley deepening migrates upstream until the entire valley is deepened. (Remember watching the **nickpoint** migrate upstream on your stream table?)

If sea level rises again, the lower part of the valley becomes submerged and begins to fill with sediment. Once the valley is filled in, the slope is too gentle for the stream to carry sediment. So the stream drops its sediment at the upstream end of the filled-in area. The filled-in valley bottom migrates upstream until the entire length of the valley has been filled in.

Thus, when sea level changes, rivers change too. The changes begin near the new locations of their mouths and migrate upstream. Around the world, the most important reason for sea-level change has been the melting and growing of glaciers.

Check Your Understanding

1. What is the Bering Strait land bridge? Why was it dry land during the ice ages?
2. Explain how sea-level changes after the retreat of the last continental ice sheet affected the rise in agriculture.
3. How does a change in sea level due to glaciation affect a river inland?

Understanding and Applying What You Have Learned

1. Using your work with the stream table as a guide, describe some of the features of a coastal landscape during times of rising sea level.
2. Using your work with the stream table as a guide, describe some of the features of a coastal landscape during times of falling sea level.
3. What are two limitations of using a stream table to model sea-level changes?
4. Which do you expect to be more common around the world: coastlines where former sea floor has been exposed as dry land or coastlines where former dry land has been flooded? Why? How are glaciers connected with your answer?
5. Pick a major river in or near your community. Assume that sea level rises until the coast is just downriver from your community.
   a) Draw a map showing what the landscape would look like immediately after sea level rises.
   b) Draw another map showing what the landscape might look like after the river builds a delta.
   c) Draw a final map showing what you might expect to happen if sea level drops again.
   d) Compare with classmates and discuss similarities and differences.
6. Examine the topographic map of southeast Boston. Find the topographic contour that indicates 15 m above sea level.
   a) What will happen to the area if sea level rises 15 m?
   b) Show the new coastline on the map, and shade or color the land area that will be changed.

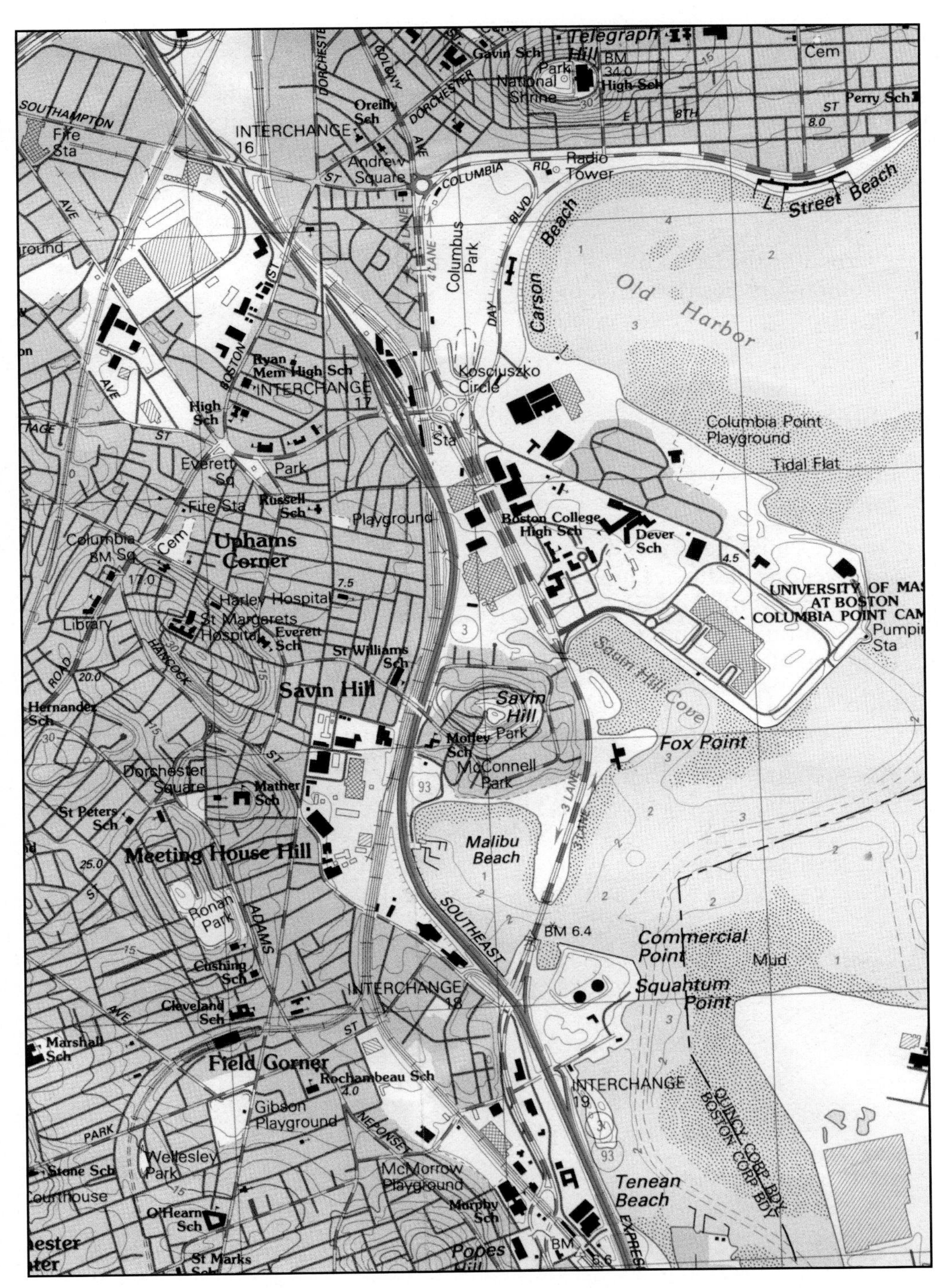

Topographic map of southeast Boston. Scale 1:25,000 (1 cm on the map represents 250 m on the ground). The contour interval is 3 m.

Preparing for the Chapter Challenge

Sea-level changes affect the landscape in ways that you have modeled using a stream table. Write a short paper in which you address the following questions:

- How has the landscape of your community been affected in the past by changes in sea level? Be as specific as you can.
- How would the landscape of your community be affected if sea level rose?
- How would the landscape of your community be affected if sea level fell?

Inquiring Further

1. **Investigating the legends of "lost continents"**

   Do you think the rise and fall of sea level during the ice ages have anything to do with legends of "lost continents" such as Atlantis? How might you test your ideas?

2. **Influence of sea-level change on coastal areas**

   Pick a coastal area of interest to you. Assume that sea level was 100 m lower 13,000 years ago and had reached its present level by 7000 years ago. Obtain a map of the area that shows the depth of water. Where was the shoreline 13,000 years ago? If sea level rose at a steady rate, how fast, in meters per year, would the shoreline have moved inland?

3. **The Beringian environment**

   Investigate paleogeographic reconstructions of Beringia. Visit the *EarthComm* web pages, where you can study animations of sea-level changes during the last 20,000 years.

Activity 5 How Glaciers Modify the Landscape

Goals

In this activity you will:

- Understand that glaciers modify the landscape by erosion and deposition.
- Model the action of glacial meltwater as it drains out of a glacier.
- Understand that the movement of glaciers can change stream drainage patterns.
- Model the effects of a glacier infringing on a stream.
- Evaluate the uses and limitations of this type of model in Earth science.
- Apply what you have observed to determine whether glaciers have affected your community in the past.

Think about It

Suppose a glacier were to cross a large river valley.

- What other materials might be in a glacier besides ice?
- How might the materials get into the glacier?

What do you think? Record your ideas about these questions in your *EarthComm* notebook. Be prepared to discuss your responses with your small group and the class.

Investigate

Part A: Modeling the Action of Glacial Meltwater

1. Spread an even layer of fine sand about 0.6 cm thick on the bottom of a baking pan. Fill the baking pan almost full of water. Put an even layer of sawdust about 1 to 2 cm thick in the bottom of the second pan. Put the pan with water inside the pan with sawdust. Put the assembly into a freezer, and wait overnight until the water is a solid block of ice. Note: it will take a long time for the water to freeze all the way to the bottom, because of the insulation of the sawdust.
2. Turn the pan upside down under warm running water until the ice block comes loose. Set the ice block aside, and cut down along two edges of the pan so that one of the narrow sides of the pan can be bent down flat, level with the bottom of the pan.
3. Replace the ice block in the pan, and wait until the block is at its melting temperature. You will know when the block has reached its melting temperature when its surface shines with a thin film of liquid water.
4. Put the pan on the floor, place a wooden block on the ice surface, and hit the board with blows from the hammer. Start very gently, and increase the force of the blows until the block shows several long cracks but has not been completely shattered.
5. Roll two long "snakes" of modeling clay in your hands, and mold them along the sides of the ice sheet.
6. Put the pan on a surface where water can drain from the pan without causing any damage. Prop up the end of the pan opposite the opening with a thin strip of wood about one inch thick, or a chalkboard eraser.

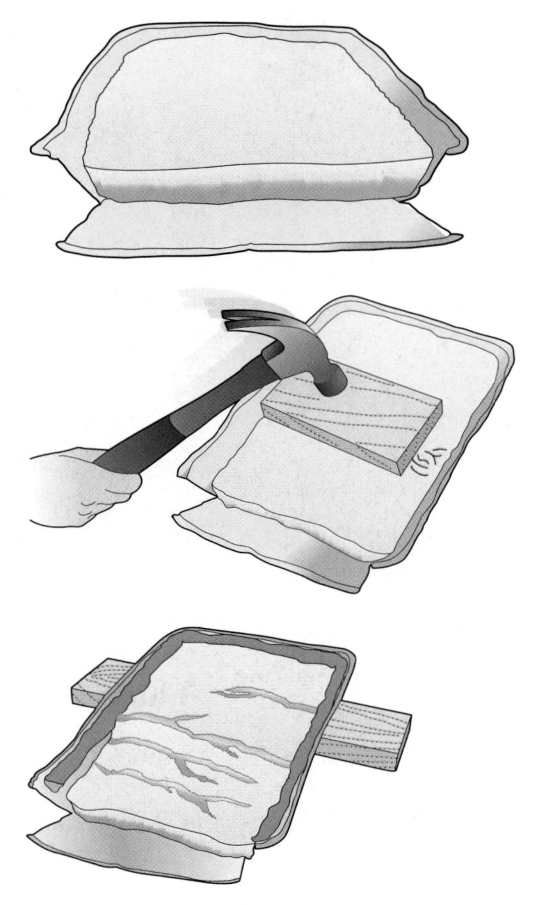

7. Spray cold water on the upper end of the ice sheet. Use just enough water so that some of the water runs down the surface of the ice sheet. Observe how the patterns of water flow and drainage change with time, as some of the ice is melted, and how the sediment at the base of the ice sheet is moved by the flowing water.

a) In your notebook, record your observations. If you have a video camera, you might want to videotape what you see for later incorporation into your **Chapter Challenge.**

b) Using your knowledge of the properties of ice and liquid water, account for the behavior you observed.

c) On a real glacier, what do you think happens to surface water (meltwater plus rainwater)?

Part B: Using a Stream Table to Model Ways that Glaciers Modify the Landscape

1. To model the ways in which glaciers modify the landscape, fill a stream table with damp sand.

2. Run water down the stream table long enough to form a well-defined channel at least a centimeter deep.

   a) Sketch the river channel in your notebook.

3. Take some finely crushed ice and mix in a small amount of colored aquarium gravel. The ice simulates a glacier. The aquarium gravel represents the sediments carried by a glacier.

4. Turn off the water in the stream table. Cover one area of the stream table with the mixture of ice and gravel. Block the channel with the mixture, except for a space around one side of the ice for a new channel to form. Pack down the ice gently (without disturbing the stream channel) to prevent seepage of water under the ice.

Wear goggles throughout this activity. Use the hammer with care. Clean up spills. Wash your hands when you are done.

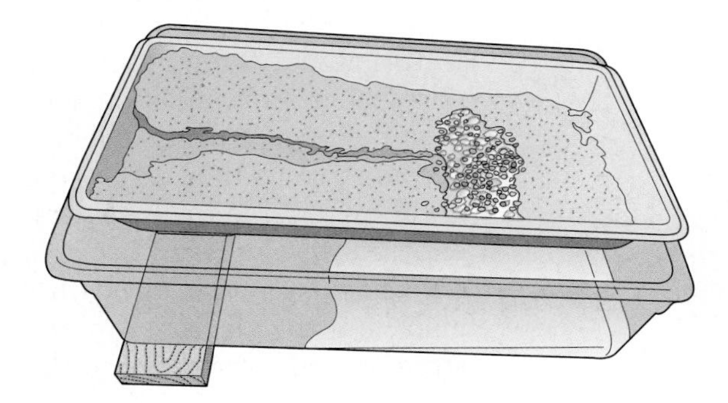

5. Turn the water back on. Let it run long enough to carve a new channel along the margins of the ice. Some water might seep under the ice along the old channel. This is acceptable because, as you saw in **Part A** of the investigation, water does flow under and within glaciers. As long as the flow under the ice does not completely prevent formation of a new channel, seepage is acceptable and even desirable.

   a) Sketch the changes that are occurring on the stream table.

6. Allow the ice to melt naturally. Observe and record the results. Complete melting will take several hours, possibly overnight.

   a) Once the ice has completely melted, sketch what you see in the stream table again.

   b) Describe any changes in the surface texture of the sand (besides the channel diversions).

   c) Describe any erosional features that are formed by the meltwater from the ice.

   d) Where did the aquarium gravel end up? Describe and sketch these changes.

Reflecting on the Activity and the Challenge

In this activity, you modeled the effects of a glacier first blocking a stream, and then melting. When glaciers move across streams they can change where and how rivers flow. Glaciers erode enormous volumes of bedrock and deposit the material beneath the glacier, at the glacier terminus, and in streams and rivers beyond the terminus. Many communities in the United States are on rivers that were modified by the Pleistocene ice sheets. This activity will help you understand and explain in the **Chapter Challenge** how the landscape in your community might have been affected in the past by glaciers, and how it might be affected in the future.

Digging Deeper

How Glaciers Erode Bedrock and Move Sediment

When a glacier forms on an area of the land, it incorporates loose soil and sediment into its base and moves it away. Once the loose material has been removed, the glacier erodes the bedrock beneath, in two ways. The rock and mineral particles that are carried at the base of the glacier, called **tools**, abrade the bedrock, in much the same way that sandpaper abrades wood. The problem is that the particles eventually wear out because the bedrock abrades them. The base of the glacier acquires new tools by plucking away blocks of the bedrock that are already cut by fractures.

Geo Words

tools: rock and mineral particles that are carried at the base of the glacier and abrade the bedrock.

Figure 1 The Finger Lakes were carved out by glaciers.

In some areas with relatively easy-to-erode bedrock, ice sheets can erode out wide and deep depressions in the bedrock. After the ice sheet retreats, such depressions are usually occupied by lakes. The Great Lakes, the Finger Lakes in central New York State (shown in *Figure 1*), and Lake Champlain are examples of large lakes that formed in this way.

Loose rock and mineral material that is carried by the glacier is called the **load** of the glacier. Much of the load is frozen into the base of the glacier. When the glacier can't hold all of its load, it deposits part of it by plastering it onto the bedrock beneath the glacier. Sediment deposited in that way is called **glacial till** (shown in *Figure* 2). Sheets of till cover large areas of North America once occupied by the Pleistocene ice sheets. Till is also deposited when the glacier ice reaches the glacier terminus. When an ice sheet is in equilibrium for a long period of time, so that its terminus stays in the same place, high ridges of sediment, called end **moraines**, are deposited. End moraines show geoscientists where the farthest advance of the ice sheet was located.

Figure 2 Moraines are composed mainly of glacial till.

Glacial Meltwater and Its Deposits

Melting of the lower areas of glaciers in summer produces enormous volumes of water. This water, together with rainwater from summer storms, flows across the glacier. The water then finds its way to the base of the glacier through fissures and holes in the ice, because water is denser than the ice. It flows at high speeds through large tunnels at the base of the ice and emerges at the terminus of the glacier. The meltwater streams that flow out from the glacier carry enormous quantities of sediment of all sizes, from clay to boulders.

As the terminus of an ice sheet retreats, much of the sand and gravel carried by meltwater streams is deposited right at the glacier terminus. It is often deposited in between large melting ice masses. After all of the ice melts from the area, these deposits are left as irregular hills and ridges. Their sizes

Geo Words

load (of a glacier): loose rock and mineral material that is carried by the glacier.

glacial till: poorly sorted, unlayered sediment carried or deposited by a glacier, usually consisting of a mixture of clay, silt, sand, gravel, and boulders ranging widely in size and shape.

moraine: a mound or ridge of chiefly glacial till deposited by the direct action of glacial ice.

and shapes vary greatly. In the northern parts of the United States these deposits are prime sources of sand and gravel for concrete. (Think about how different your life would be if there were no sand and gravel for such an ordinary but essential material like concrete.)

How Glaciers Alter River Systems

The Pleistocene ice sheet rearranged the courses of many rivers in North America. (See *Figure 3*.) Before the Pleistocene, the river drainage system of North America looked very different from today. As noted above, the Great Lakes had not yet formed. The Mississippi River was a smaller river with a much smaller drainage basin. There was no Ohio River; instead a river system extended across the middle of Indiana and Illinois to join the ancient Mississippi in the middle of Illinois.

There was no Missouri River. Rivers in the Northern Plains states that are now tributaries to the Missouri River flowed northeast into Canada. The headwaters of the Missouri, the Yellowstone, and other rivers of the Northern Plains states still flow northeast, a relic of this ancient river system. By blocking rivers that were flowing northward, the glaciers created the present Ohio and Missouri River systems.

When glaciers blocked rivers, there were many different possible outcomes. In some cases, small segments of existing river systems flooded their valleys, flowed over drainage divides, and cut new valleys, eventually connecting segments of former river systems into a new river. The Ohio River formed this way. In other cases, water flowed along the margins of the glaciers, cutting a new channel that captured all the rivers it crossed. Meanwhile, as the ice melted, it dropped debris in the former river channels, blocking them and often concealing them completely. This is how the present Missouri River formed.

As the glaciers retreated across the area now occupied by the Great Lakes, they alternately blocked and exposed outlets of the lakes. At different times, one or more of the lakes drained south through the Mississippi, southwest across Ohio and Indiana, across Ontario in a number of places, and down the Mohawk and Hudson Rivers to the Atlantic.

Check Your Understanding

1. How do glaciers erode bedrock?
2. How do glaciers deposit sediment?
3. How did the Ohio River form?
4. How did the Missouri River form?
5. How did the Great Lakes form?

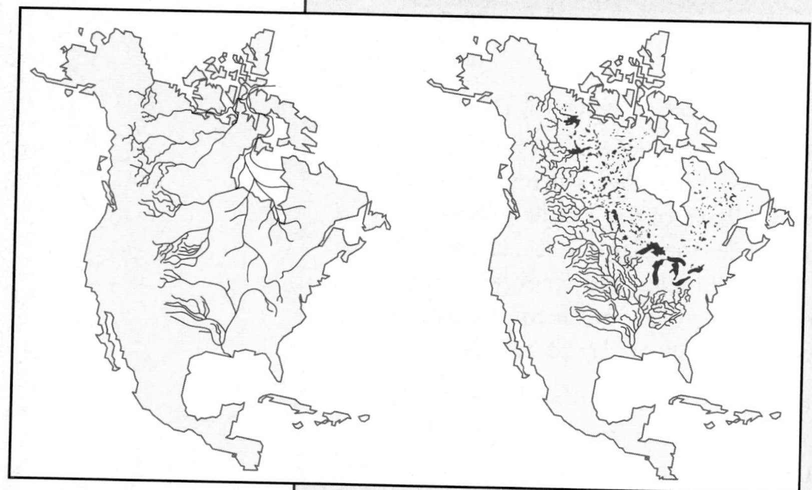

Figure 3 Drainage patterns in North America before (left) and after (right) the Pleistocene ice age.

Understanding and Applying What You Have Learned

1. How would you recognize whether a deposit of sediment on the land surface was produced by a glacier rather than by some other transporting and depositing agent, like a river? In your notebook, make a list of possible criteria.
2. When a glacier blocks a river, there are many possible outcomes.
   a) Which of the following outcomes did you observe in your investigation with the stream table? Use a sketch to illustrate each outcome.
      - The stream is diverted permanently from its old course.
      - The stream is temporarily diverted but reoccupies its old channel once the ice melts.
      - The stream changes course more than once as the ice melts.
      - Meltwater streams create channels that may or may not capture part of the drainage.

   b) Describe any other outcomes that you observed that are not listed.
3. Did the glacier on your stream table leave behind a moraine? If so, describe it.
4. If your community were located along a river that was diverted by the glaciers in the past, how would the history of your town be different if the river had not been diverted?
5. Pick the largest river nearest to your community and predict what would happen if the river becomes blocked by a glacier at various locations.
   a) Would the community be flooded?
   b) Would the river be diverted away from the town?
   c) What would you advise your community leaders to do about it?

Preparing for the Chapter Challenge

When glaciers advance and retreat, they disrupt stream patterns and sometimes change the paths of rivers. Glaciers also leave behind characteristic landforms, such as moraines. Write a short paper in which you address the following questions:

- How has the landscape of your community been affected in the past by glaciers? Be as specific as you can. Are there landforms formed by glacial erosion or deposition in or near your community?
- Did Ice Age glaciers cause stream changes in or near your community?
- Could a local river in your community be blocked or diverted by a glacier? If so, what effect would this have on your community?

Activity 6 Catastrophic Floods from Glacial Lakes

Goals

In this activity you will:

- Model the effects of ice-dam failure and catastrophic flooding.
- Evaluate the uses and limitations of this type of model in Earth science.
- Describe how catastrophic glacial floodwaters alter the landscape.
- Understand that while most geologic events are slow and gradual processes, others can be abrupt.
- Determine whether catastrophic glacial floods have affected your community in the past or could threaten your community in the future.

Think about It

A dam is a structure across a watercourse that acts to obstruct, divert, or confine the flow of water. Dams can be man-made structures built of wood, rocks, or solid masonry. They can also be natural features.

- If a glacier were blocking the flow of a river, what are some of the things that might cause the dam made of ice to break?
- What would be the result?

What do you think? Record your ideas about these questions in your *EarthComm* notebook. Be prepared to discuss your responses with your small group and the class.

Investigate

Part A: Modeling a Glacial Dam with Ice

1. On a stream table, construct a valley in the sand at least 5 cm deep and 15 cm wide.
2. Block the valley with finely crushed ice. Plug any obvious holes or weak spots.
3. Slowly fill the valley upstream of the dam with very cold water. (Let the cold tap run until the water is as cold as possible. The point is to avoid melting the ice too quickly.)
4. The dam may or may not fail during filling. If the dam does not fail, fill the valley to just below the top of the dam. Do not allow the water to run over the top of the dam.
5. Observe the ice dam periodically for signs of impending failure, like water seepage around or beneath it.
   a) Observe the dam carefully just before and during failure to determine the mechanism of failure. Record your ideas in your notebook.

Part B: Modeling a Glacial Dam with Foam Rubber

1. Again construct a valley in the sand at least 5 cm deep and 15 cm wide.
2. This time, instead of using ice as you did in Part A, use a strip of foam rubber to model the glacier. Use a strip about 5 cm by 5 cm in cross section and about 20 to 30 cm long. Pack wet sand along the base and edges of the glacial dam to create a good seal.
3. Allow the impounded lake to fill slowly. While it is filling, verify that the seal around the glacial dam is watertight.

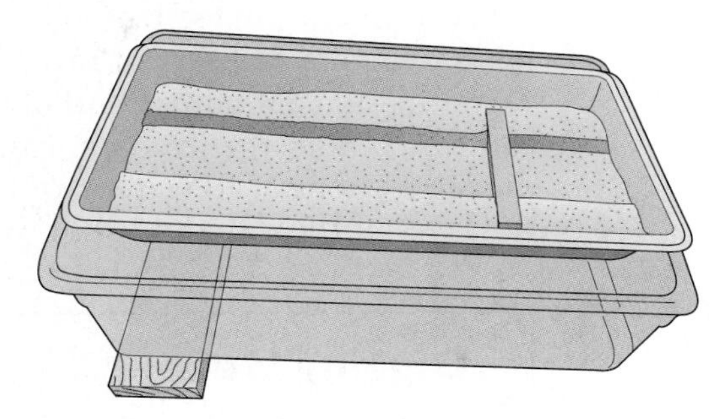

4. When the lake is about half full, seepage should become evident beneath the dam.
   a) Observe the dam carefully just before and during failure to determine what caused the failure. Record your ideas in your notebook.

Part C: Modeling the Effects of a Glacial Dam Failure

1. Again construct a valley in the sand at least 5 cm deep and 15 cm wide, and use a strip of foam rubber to model the glacier. This time also model a landscape downstream from the dam. Make a couple of tributary valleys branching off the main valley. In order to make this simulation work properly, it will probably be necessary to narrow the valley downstream from the dam.
2. Let the dam fail, and observe. Let the stream table run long enough afterward to show incision of the flood deposits, and observe.
   a) Record your observations.

Wear goggles. Clean up spills immediately. Wash your hands when you are done.

Reflecting on the Activity and the Challenge

In this activity, you have modeled some of the effects of flooding due to catastrophic drainage from glacial lakes. You have observed several ways that glacial dams can fail. Water can flow under the ice dam, causing the ice to float. The water can fill to the level of the dam and spill over the dam or the ice dam can develop weak spots that fail much like an earthen dam. If you are near a river, your **Chapter Challenge** should contain information on the possible effects of a glacier damming your local river. Glacial dams might also have affected your community in the past.

Digging Deeper

When Glacial Dams Fail

A network of enormous dry channels called the Channeled Scablands covers much of Washington State. The channels are obvious on the ground and on aerial and satellite imagery. The Scablands are often extremely rough, with jagged outcrops and thin soil, making the land unsuitable for farming.

Figure 1 The rugged terrain of the Channeled Scablands in Washington.

Between 1923 and 1932, a geologist named J. Harlan Bretz proposed that the Scablands were the result of catastrophic floods. His hypothesis remained hotly controversial until the early 1940s, when J.T. Pardee showed that a former glacial lake in Idaho and Montana, named Glacial Lake Missoula, was the source of the water. Geologists have used computers to model some of the Scablands floods. The largest floods contained more water than is flowing in all the world's rivers today. Current velocities approached 100 km/h. One of the most spectacular illustrations of the volume of flow is the fact that Wallula Gap, which is 2 km wide and 200 m deep, formed a bottleneck for floodwaters on the Columbia River. Floodwaters ponded 100 m deep in the broad valley upstream of the Wallula Gap "constriction." For a long time, geologists assumed that there was one, or at the most only a few, Scablands floods. More recently, though, evidence has shown that there were more than 50.

The Scablands floods in Washington did not, for the most part, pour directly down the gorge of the Columbia River. Instead, the Columbia was blocked in several places by the glaciers. The blockages formed two large glacial lakes. The floods poured into the lakes, overflowed, and spilled over the drainage divides. The glacial lakes provided sheltered backwaters where sediments called **backflood deposits** could be laid down. Backflood deposits enable geologists to reconstruct the floods in detail. In many cases, the flood deposits are separated by varves, which are annual layers of clay that are characteristic of glacial lakes. Varves let geologists count the actual interval in years between floods.

Geo Words

backflood deposits: sediments deposited when flooding causes water to pour over dams and into lakes.

Other Glacial Floods

Ice-dammed glacial lakes are unstable and prone to catastrophic flooding. Although few people are in danger today from such floods, failure of ice dams in the past had important effects. The Channeled Scablands provide the most spectacular evidence of what these catastrophes can do.

Figure 2 The Wisconsin Dells were formed by the catastrophic flooding of Glacial Lake Wisconsin.

Other ice-dammed lakes have failed as well. During the retreat of ice from the Midwest, a great lake called Glacial Lake Agassiz backed up against the ice sheet in Minnesota, the Dakotas, and Canada. Periodic failures of ice dams resulted in great floods of water from Lake Agassiz to the Great Lakes, then down the Mississippi River. Sudden influxes of fresh water into the Gulf of Mexico had important biological and climatic effects that scientists are still studying. In Wisconsin, a large lake called Glacial Lake Wisconsin once occupied much of the central part of the state. When its ice dam failed, the lake drained suddenly, carving a network of deep channels that are today known as Wisconsin Dells. The Dells are one of the important tourist attractions in the state. (See *Figure 2*.)

How Ice Dams Fail

Ice dams fail in a number of ways. Ice is relatively soft, so it's easy to abrade. Once water begins flowing under or on the ice, channels get bigger quickly. The water also melts the ice, further enlarging the channels. When water under high pressure is moving through the ice, frictional heating also enlarges the channels.

Catastrophic floods can also occur when meltwater accumulates beneath glaciers and finally bursts out. Such floods are believed to have occurred in several places in North America. In Iceland, meltwater accumulates under glaciers because of volcanic activity, and periodically bursts out in destructive floods.

One of the most important causes of ice-dam failure is the ability of ice to float in water. As the water begins to flow over the ice dam, the ice is actually lifted off its bed. High-pressure water begins working its way under and within the glacier, emerging with enormous force.

In a laboratory setting, the buoyancy of ice is not easy to illustrate because small amounts of ice melt quickly. In your activity, you used a strip of foam to model ice-dam failure by buoyancy.

The Role of Catastrophic Change in Geology

The emphasis on slow and gradual processes in geology leads some people to believe that catastrophic events never occurred. The idea that most geologic processes are slow and gradual is a finding based on the physical record in the rocks, not an assumption. However, that doesn't mean that catastrophes never occur. The story of the Channeled Scablands and other ice-dam collapses are excellent case studies of real, well-documented catastrophes in geology.

Check Your Understanding

1 List some of the different ways that ice dams can fail.
2. What are the consequences of a glacial flood?

Understanding and Applying What You Have Learned

1. List some of the ways in which the dams in your stream table failed.
2. How are these failures different or the same as the ways in which real ice dams can fail?
3. In the investigation, you modeled the effects of flooding due to catastrophic drainage from glacial lakes.
   a) Which aspects of your investigations were realistic?
   b) Which aspects were not realistic?
   c) Can you think of ways to improve the models?
4. Which was more realistic, the ice dam or the foam-rubber dam? Why?
5. Suppose that the nearest river to your home was blocked, and then the blockage collapsed, sending water 15 m (about 50 ft.) deeper than normal down the river.
   a) Draw a map of the areas that would be affected.
   b) Are there any places the river could cross over a ridge and affect a neighboring valley as well? Indicate these areas on your map.
6. Glacial Lake Missoula, the source of the Scablands floods, contained some 2500 km^3 of water. It spilled across Washington State and through Wallula Gap on the Columbia River. Wallula Gap is about 2 km wide and 0.2 km deep. If the water was moving at 100 km/h, how long did it take to pass through the gap? Hint: first calculate how many cubic meters passed through the gap in an hour.

Preparing for the Chapter Challenge

Many areas in the United States have been affected by glacial lakes or floods in the past. Use the questions below as a guide to prepare a short report.

- If your community was affected by glacial lakes or floods, write a paragraph or two about these lakes or floods.
- If your area was not affected by glacial lakes in the past, write a few paragraphs describing what would happen in your community if an ice dam blocked a local river, causing the formation of a glacial lake, and then failed.

Inquiring Further

1. **History of science**

   Research J. Harlan Bretz, the geologist who first proposed catastrophic flooding as a cause of the Channeled Scablands. Describe his theory and the evidence behind it. Why did other geologists originally discount his theory? Why did other geologists finally embrace his theory? Use the *EarthComm* web site to help you with your research.

Activity 7 Nonglacial Ice on the Earth's Surface

Goals

In this activity you will:

- Model and investigate the stability of structures on permafrost.
- Understand some of the problems that permafrost presents for human habitation.
- Be able to describe the interaction between sea ice and climate as an example of a positive feedback loop in Earth systems science.
- Understand some of the important implications of sea ice, icebergs, and frost within the Earth system.

Think about It

When the surface layer of deeply frozen soil thaws out, man-made structures on the land surface often suffer great damage.

- How might a man-made structure cause frozen soil to thaw?
- What are some ways of ensuring that structures built upon deeply frozen ground do not suffer damage when the ground thaws?

What do you think? Record your ideas about this question in your *EarthComm* notebook. Be prepared to discuss your responses with your small group and the class.

Investigate

Part A: Stability of Structures on Permafrost

1. Trim a sheet of Styrofoam so that it fits snugly in the bottom of a pan about 24 cm × 13 cm × 7 cm. Fill the pan almost to the brim with soil. The soil layer should be at least 2.5 cm thick. Add water very slowly at one corner of the pan, letting it soak into the soil, until the water level is just at the soil surface. Make sure there is no standing water on the surface. If there is excess water, soak it up with a paper towel. Be very careful not to slosh or jiggle the contents of the pan.
2. Put the pan in a freezer overnight, so that all of the soil is frozen solid.

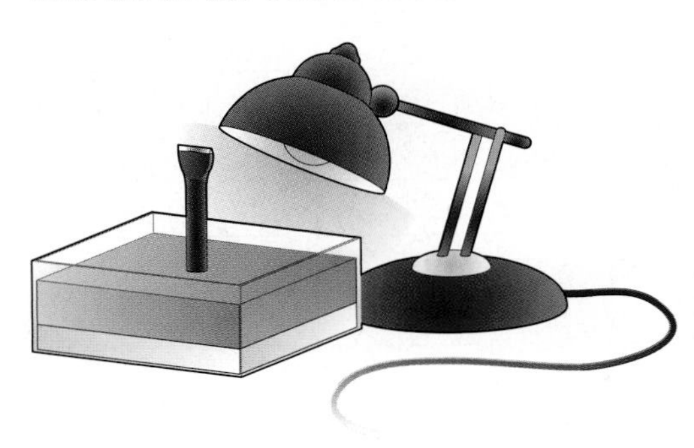

3. Remove the pan from the freezer and place it on a tabletop. Set a metal chisel upright on the frozen soil surface, flat end down, to simulate a heavy building. Set up the heat lamp to shine on the middle of the soil surface, around the base of the chisel.
   a) In your notebook, describe how the chisel behaves as the surface soil thaws.

Be careful, the heat lamp will get very hot. The metal chisel will also get hot. Handle it carefully with insulated gloves.

4. After the soil surface has thawed to a greater depth, roll a curled-up finger across the surface to simulate the passage of a motor vehicle or heavy construction equipment.
   a) In your notebook, describe the nature of the disturbance to the soil layer.
5. Use the results of your activity to answer the following:
   a) Write an explanation for the phenomena you observed.
   b) How could you have increased the stability of the chisel as the soil surface thawed?
   c) How would the behavior of the soil surface have differed if a small drain hole in the bottom of the pan had been opened after the pan was removed from the freezer?
   d) How would the behavior of the soil surface have differed if the pan had been set on a heated surface after it was removed from the freezer?

Part B: Frost-Free Days in the U.S.

1. The data in the table on the following pages lists average first and last frost dates for cities around the United States.
   a) Plot the number of frost-free days on a map of the United States.
   b) Contour the data using a contour interval of 25. Drawing contours is not always easy, especially when data points are far apart compared to how rapidly the values change from place to place. If you have trouble drawing contours, ask your teacher for help.

City	State	Last Frost Date	First Frost Date	Number of Frost-Free Days
Birmingham	Alabama	March 29	November 6	221 days
Phoenix	Arizona	February 5	December 15	308 days
Fayetteville	Arkansas	April 21	October 17	179 days
Los Angeles	California	None likely	None likely	365 days
San Francisco	California	January 8	January 5	362 days
Denver	Colorado	May 3	October 8	157 days
New Haven	Connecticut	April 15	October 27	195 days
Washington	D.C.	April 10	October 31	203 days
Jacksonville	Florida	February 14	December 14	303 days
Miami	Florida	None	None	365 days
Atlanta	Georgia	March 13	November 12	243 days
Boise	Idaho	May 8	October 9	153 days
Chicago	Illinois	April 14	November 2	201 days
Indianapolis	Indiana	April 22	October 20	180 days
Des Moines	Iowa	April 19	October 17	180 days
Topeka	Kansas	April 21	October 14	175 days
Wichita	Kansas	April 13	October 23	193 days
Louisville	Kentucky	April 1	November 7	220 days
New Orleans	Louisiana	February 20	December 5	288 days
Portland	Maine	May 10	September 30	143 days
Baltimore	Maryland	March 26	November 13	231 days
Boston	Massachusetts	April 6	November 10	217 days
Detroit	Michigan	April 24	October 22	181 days
Duluth	Minnesota	May 21	September 21	122 days
Jackson	Mississippi	March 17	November 9	236 days
St. Louis	Missouri	April 3	November 6	217 days
Helena	Montana	May 18	September 18	122 days
Lincoln	Nebraska	March 13	November 13	180 days
Las Vegas	Nevada	March 7	November 21	259 days
Albuquerque	New Mexico	April 16	October 29	196 days
Albany	New York	May 7	September 29	144 days
New York	New York	April 1	November 11	233 days

City	State	Last Frost Date	First Frost Date	Number of Frost-Free Days
Charlotte	North Carolina	March 21	November 15	239 days
Fargo	North Dakota	May 13	September 27	137 days
Columbus	Ohio	April 26	October 17	173 days
Tulsa	Oklahoma	March 30	November 4	218 days
Portland	Oregon	April 3	November 7	217 days
Pittsburgh	Pennsylvania	April 16	November 3	201 days
Charleston	South Carolina	March 11	November 20	253 days
Memphis	Tennessee	March 23	November 7	228 days
Dallas	Texas	March 18	November 12	239 days
Houston	Texas	February 4	December 10	309 days
Salt Lake City	Utah	April 12	November 1	203 days
Richmond	Virginia	April 10	October 26	198 days
Seattle	Washington	March 24	November 11	232 days
Milwaukee	Wisconsin	May 5	October 9	156 days
Cheyenne	Wyoming	May 20	September 27	130 days

2. Use your map to answer the following questions:
   a) Which areas of the map have the greatest number of frost-free days?
   b) How do you account for the pattern that you see on the map?
   c) Which areas of the country have the longest seasons for growing crops?
   d) Using your contour map, estimate the number of frost-free days in your community.
   e) How does your community compare to the rest of the United States? Would you characterize your community as having a short, medium, or long growing season?

Reflecting on the Activity and the Challenge

The cryosphere is an Earth system that extends beyond glaciers to include all of the ice and snow on the Earth's surface. In this activity, you looked at how a thaw of permafrost might affect structures on the Earth's surface, such as buildings. You also looked at frost occurrences in the United States. This will help you to think about the influence of a cooling or warming trend on nonglacial ice on the Earth's surface and your community.

Digging Deeper

Permafrost

In high-latitude land areas, where the average yearly temperature is very low, the ground is permanently frozen, often to a depth of hundreds of meters. The depth of permanent freezing depends upon two factors: the average surface temperature, and the rate of heat flow upward from the interior of the Earth. Large regions of Alaska, northern Canada, and Russia have permafrost to various depths. About 20% of the Earth's land surface has permafrost, and 85% of Alaska lies within the permafrost region. (See *Figure 1*.) In summer, a thin uppermost layer may thaw, but this layer always refreezes in the winter.

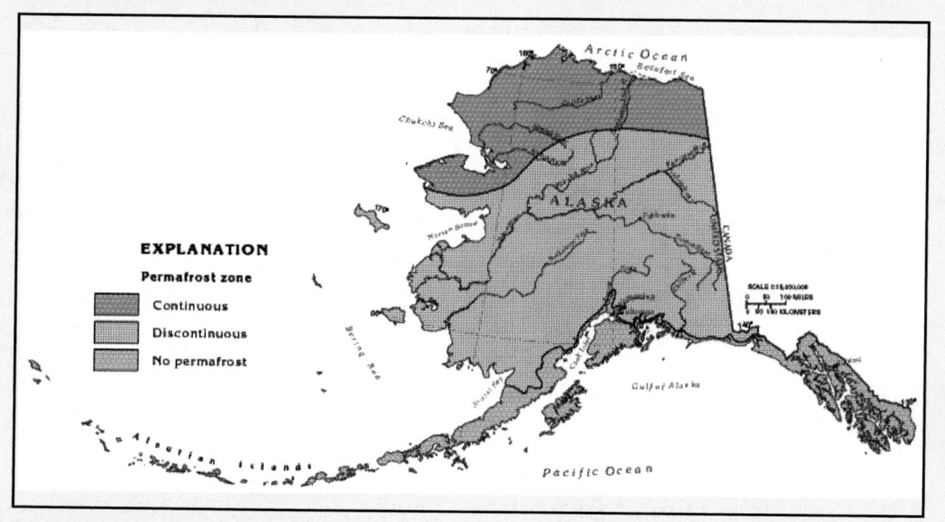

Figure 1 Permafrost distribution in Alaska.

Permafrost presents great problems for human habitation. The problems arise not because of the permafrost itself but because heated buildings, or pipelines carrying above-freezing liquids like petroleum, heat the ground and thaw the upper layer of permafrost. Because the liquid water in the thawed upper layer cannot drain downward, the layer has the consistency of thick soup. Structures sink deep into this soupy layer. To minimize these problems, buildings can be built on columns above the ground surface or on pilings that extend down below the level of summer thawing. Another technique is to place buildings on thick layers of loose gravel, which insulate the permafrost layer from the heat of the building.

Figure 2 Why is the Alaskan pipeline elevated above the ground surface?

Sea Ice

Most of the Arctic Ocean and large areas of the ocean around Antarctica are covered with a layer of ice a few meters thick for most or all of the year. This ice, called pack ice, forms on the ocean surface when the water temperature falls below about 2°C. During the summer some of the pack ice melts, leaving local wide areas of open water. During the winter the ice cover is almost continuous. The pack ice moves slowly, several kilometers per day, as it is pushed by the wind and drifts with the underlying ocean currents. This movement causes the ice to spread apart in some areas but to be pushed together in other areas to form thicker ice masses, called pressure ridges. For these reasons the pack ice does not form a perfectly smooth surface, like a frozen lake or skating rink, but instead is very rough and irregular.

The existence of the pack ice in the Arctic Ocean exerts a very important control on the Earth's climate. The layer of ice is much colder in winter than the ocean water beneath, making Arctic winters much colder than if the ice were not there. Also, ice reflects as much as 80% of the incoming solar energy back out to space, whereas the ocean surface reflects only about 20%. Many recent studies have shown that the Arctic pack ice is thinner in recent years than in the past, and the percentage of open sea is also increasing. As more ice melts, the Arctic climate warms, and that causes even more ice to melt. Earth systems scientists call such a process "positive feedback." Some geoscientists now think that the Arctic pack ice might melt entirely in the not too distant future, with great but unpredictable consequences for the Earth's climate.

Icebergs

In the Northern Hemisphere, glaciers in some places in Greenland and the Canadian arctic islands end in the ocean. Large pieces of the glacier break off, by a process called calving, and float southward, gradually melting. Because freshwater ice is only slightly less dense than sea water, most of the mass of the iceberg lies below the waterline. Because icebergs are such a hazard to shipping, there is a well-developed system for spotting and tracking North Atlantic icebergs.

Figure 3 Pieces of glaciers break off into the ocean by the process of calving.

Icebergs around Antarctica form in a different way. Along the coast of Antarctica there are two very large glaciers, called ice shelves, that float in the ocean but are anchored to the land on one side. Occasionally, large areas of these ice shelves become separated and drift away as ice islands. These ice islands are truly gigantic: they can be many tens of kilometers across and several hundred meters thick. Some have been reported to be as large as the entire state of Rhode Island! It takes years, not just months, for them to melt completely. There have even been proposals to tow these large icebergs thousands of miles to coastal cities in need of fresh water.

Frost

Frost is a deposit of ice that forms on the ground surface during clear and cold nights with no wind. All materials radiate heat energy. The ground surface, and low vegetation growing on the ground, radiate heat upward on clear nights. Some of this radiated heat is absorbed by gases in the atmosphere and radiated back down to the ground surface, but much of it escapes into space. You know that cold air is denser than warm air, so the chilled layer of air near the ground remains there, provided that there is no wind to mix it with the warmer air overhead. If the temperature of the chilled layer of air near the ground falls to the point where the air cannot hold all of the water vapor it contains, dew is formed (if the air temperature is above freezing) or frost is formed (if the air temperature is below freezing). Valleys and depressions are especially susceptible to frost. The reason is that chilled air that forms on surrounding slopes flows slowly downslope into the valley, because it is slightly denser than the overlying air. The temperature near the ground can differ by several degrees over short distances because of this effect.

Figure 4 Farmers will often spray citrus crops with water to protect them from frost.

Many kinds of plants are damaged or killed by frost. Vegetable crops like tomatoes, peppers, beans, and squash are killed by early frosts. Many kinds of flowers last no longer than the first frost. Citrus crops are often heavily damaged by frost, even though the citrus trees themselves survive. The price of orange juice depends very strongly on how much of the year's orange crop is damaged by frosts. The growing season in a given area is usually considered to be the number of days between the average date of the last spring frost and the average date of the first fall frost. The growing season varies from much less than 100 days in many areas of the northern United States to all year-round in certain areas of the southern United States.

Check Your Understanding

1. Why does permafrost cause damage to man-made structures?
2. How does the pack ice in the Arctic influence the Earth's climate?
3. How does frost form?

Understanding and Applying What You Have Learned

1. If you were given the responsibility of designing a building in an area of deep permafrost, what construction techniques might you try to implement to minimize permafrost damage?
2. If you were given the responsibility of designing a major research project to monitor long-term changes in pack ice in the Arctic Ocean, what research techniques would you try to implement? For each, what are the advantages and disadvantages? Assume that the level of funding is unlimited! (That's every scientist's daydream.)
3. If you were given the responsibility of choosing the best site for an apple orchard, what criteria would you try to use?

Preparing for the Chapter Challenge

Write a short paper in which you address the following:

- How would a global cooling trend affect nonglacial ice on the Earth's surface?
- How would a global warming trend affect nonglacial ice on the Earth's surface?
- How would the change in the distribution of nonglacial ice on the Earth's surface impact your community?

This paper can be included in your **Chapter Challenge** to help convince citizens that a change in global climate affect the community in some way. You may want to use this essay as an introduction, to help grab the attention of your audience. This may be particularly valuable if you live in an area where glaciers seem to pose little threat.

Inquiring Further

1. **Permafrost and the biosphere**

   How do animals adapt to permafrost environments?

2. **Technological design in permafrost**

   As a structure warms the ground (or as permafrost thaws because of natural changes in the climate), the ground subsides. Design a building that would move and adjust to the subsidence of the ground because of thawing of permafrost. The *EarthComm* web site will help you with this.

3. **Investigate types of permafrost**

   You have probably heard the saying that Native Americans in Alaska have many different words for snow. Likewise, there is no single type of permafrost. Investigate one of the various kinds of permafrost listed below and the implications for building structures on that particular type of permafrost.

   Cold permafrost — Remains below 30° F, and which may be as low as 10° F as on the North Slope; tolerates introduction of considerable heat without thawing.

   Ice-rich — 20% to 50% visible ice.

   Thaw-stable — Permafrost in bedrock, in well-drained, coarse-grained sediments such as glacial outwash gravel, and in many sand and gravel mixtures. Subsidence or settlement when thawed is minor, foundation remains essentially sound.

   Thaw-unstable — Poorly drained, fine-grained soils, especially silts and clays. Such soils generally contain large amounts of ice. The result of thawing can be loss of strength, excessive settlement and soil containing so much moisture that it flows.

   Warm permafrost — Remains just below 32° F. The addition of very little additional heat may induce thawing.

Earth Science at Work

ATMOSPHERE: *Aviation Safety Inspector*
Ice and snow on the wings of planes during takeoff present safety concerns.

BIOSPHERE: *Homeowner/Gardener*
The length of the growing season is important to gardeners.

CRYOSPHERE: *Mountain Rescuer*
Monitoring the amount of snowfall is important to the safe operation of a ski resort.

GEOSPHERE: *Pipeline Worker*
Oil companies must deal with the effects of permafrost along Alaskan pipeline routes.

HYDROSPHERE: *Ship Captain*
Ship captains are responsible for safe passage of their vessels through high latitude oceans (iceberg threat).

How is each person's work related to the Earth system, and to the cryosphere?

Unit IV

EarthComm®

Earth System Science in the Community

Earth's Natural Resources

1

Energy Resources ...and Your Community

1 Energy Resources ...and Your Community

Getting Started

As the population of a community increases, so does the demand for energy resources for transportation, electricity, and heating fuels. Ensuring a supply of energy to meet the growing needs of a community requires careful analysis and planning.

Think about all of the ways that you have used energy resources in your day so far, starting with when you got up in the morning until you arrived at this class.

- What was the source of energy for each of the activities?

What do you think? Sketch out some of your ideas about energy resources on paper. First, make a list of all the ways that energy resources have been used in your day so far, starting with the moment you woke up (the alarm clock, for instance) and ending with your present classroom. There should be at least five items on your list. Next, make a column for the energy resource responsible for the activity (coal burned to produce electricity for the alarm clock, for instance). Make another column for the source of this energy resource (coal deposit in New Mexico, for example). Finally, make a column for energy alternatives for each activity. For instance, one energy alternative to taking the bus to school is to ride your bike to school. An energy alternative to taking a ten-minute hot shower is to take a five-minute hot shower. The alternatives do not necessarily have to reduce the energy required. They could have other benefits, such as reducing an environmental hazard. Be prepared to discuss your table with your small group and the class.

Scenario

Community leaders often depend upon experts to outline a 10-year plan that addresses the impact of an increase in population on energy consumption and supply. Your community has called upon your *EarthComm* classroom to evaluate energy consumption and use in your community. Community leaders need realistic solutions or alternatives for energy use in your community, to help prevent possible energy shortages while maintaining the quality of the environment.

Chapter Challenge

Your challenge is to produce a report that is written for a general audience. In the report, you must critically analyze energy use in your community on the assumption of a population growth of 20%, and to provide realistic solutions to avoid an energy-supply shortage. Your report needs to help community members understand the origin, production, and consumption (use) of energy resources from an Earth System perspective. You need to address the following points:

- What are the current energy uses and consumption rates in your community?
- How will energy needs change if the population of your community increases by 20%?
- Will your community be able to meet these needs, and at what costs?
- What solutions or alternatives exist that may help to avoid a potential energy crisis?

Your report should explain at least one example for each of the following questions:

- What is the difference between renewable and nonrenewable energy resources?
- How are energy resources used to do work?
- How are energy resources formed, discovered, and produced?
- How much energy conservation is possible in your community? What steps can be taken?
- Does the production and consumption of energy affect the environment?

Assessment Criteria

Think about what you have been asked to do. Scan ahead through the chapter activities to see how they might help you to meet the challenge. Work with your classmates and your teacher to define the criteria for assessing your work. Record all of this information. Make sure that you understand the criteria as well as you can before you begin. Your teacher may provide you with a sample rubric to help you get started.

Activity 1 Exploring Energy Resource Concepts

Goals

In this activity you will:

- Investigate heat transfer by the processes of conduction, convection, and radiation.
- Investigate the conversion of mechanical energy into heat.
- Learn about the Second Law of Thermodynamics and how it relates to the generation of electricity.

Think about It

A car moving along a mountain road has energy. It has energy due to its motion (kinetic energy), energy due to its position in a gravity field (potential energy), and energy stored as fuel in its gas tank (chemical energy).

- Classify each item below as having kinetic energy, potential energy, or chemical energy:
  a) a rock balanced at the edge of a cliff;
  b) a piece of coal;
  c) a landslide;
  d) a roller-coaster car;
  e) a diver on a 10-m platform;
  f) a car battery;
  g) tides.

What do you think? Record your ideas in your *EarthComm* notebook. Be prepared to discuss your responses with your small group and the class.

Investigate

Energy can neither be created nor destroyed (except in nuclear reactions), but it can be changed from one form into another. The following activities will help you to explore basic concepts that govern the use of energy.

Part A: Heat Transfer

Station 1

1. Put your hand close to a 100-W light bulb and notice the heating that occurs in your hand. This is similar to the heat generated from direct sunlight.
   a) Describe what happens to the temperature of your hand as you move it slowly toward and away from the bulb.
   b) Hold a piece of paper between your hand and the light bulb. Describe and explain the change in temperature of your hand.
   c) Compare and explain the temperature difference of your hand when you hold it above the light bulb versus holding it near the side of the bulb.

Be careful not to touch the hot bulb.

Station 2

1. Which cup will keep the water hot for a longer amount of time, a metal cup or a Styrofoam® cup? Why?
   a) Write down your hypothesis.
2. Design an experiment to test your hypothesis.
   a) Record your experimental design in your notebook.
   b) With the approval of your teacher, carry out your experiment, and record your observations.
3. Five minutes after you fill the cup, place your hand around each of the cups.
   a) Which one feels hotter? Why?

Be sure your teacher approves your design before you begin. The water should not be hot enough to scald. Wipe up spills immediately. Use alcohol thermometers only.

Station 3

1. Set up two solar cookers as shown in the diagrams below. One is a standard solar cooker and the other is an identical solar cooker inside an insulated box.
   a) What differences do you expect in the temperature inside the two solar cookers over time? Write down your hypothesis in your notebook.

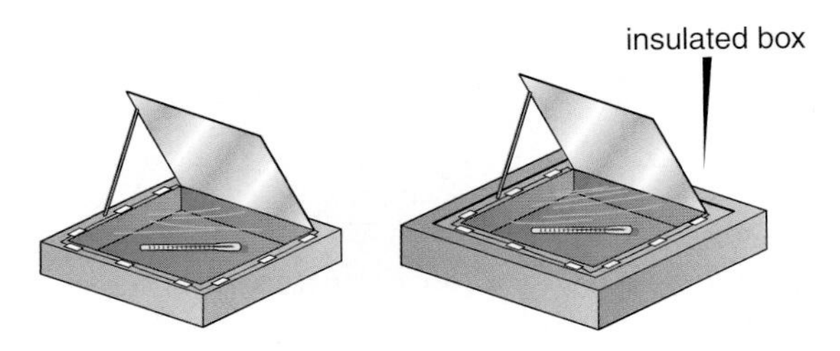

2. Design an investigation to test your hypothesis. Your design should include a plan to measure the temperature in each solar cooker and to record data every minute for at least 25 minutes.
   a) Set up a table to record your data.

3. Place a thermometer in each solar cooker and close the lids. You will want to be able to read the thermometer without blocking the path of solar energy and without opening the boxes.
   a) Record and graph the data.
4. Use the evidence that you have collected to answer the following questions:
   a) How did your results compare with your hypothesis?
   b) What heating mechanism causes the cookers to heat up in the first place?
   c) What are the different heat transfer mechanisms that are taking place in the cookers? Use diagrams to record your ideas in your notebook.
   d) What mechanism keeps the heat from escaping?
   e) What improvements could be made to the cooker if you had to do it over again?

Be careful when you touch items after they have been in the solar cooker. They will be hot.

Part B: Kinetic Energy, Potential Energy, and Heat

1. The following is a thought experiment. The graph shows the path of a small lump of modeling clay that is thrown into the air.
   a) Copy the graph onto a sheet of graph paper.
2. Imagine that you had thrown the clay into the air so that it landed on a tabletop. In your group, discuss and record your ideas about the following:
   a) How does the kinetic energy of the piece of clay change over time? When is it highest? When is it lowest?
   b) How does the potential energy of the lump of clay change over time? When is it highest? When is it lowest?
   c) How was kinetic energy transformed into potential energy? When did this happen?
   d) How was kinetic energy transformed into heat? When did this happen?
   e) Find a way to represent the changes in these three forms of energy over time. Record your ideas on the sheet of graph paper that shows the path of the modeling clay.

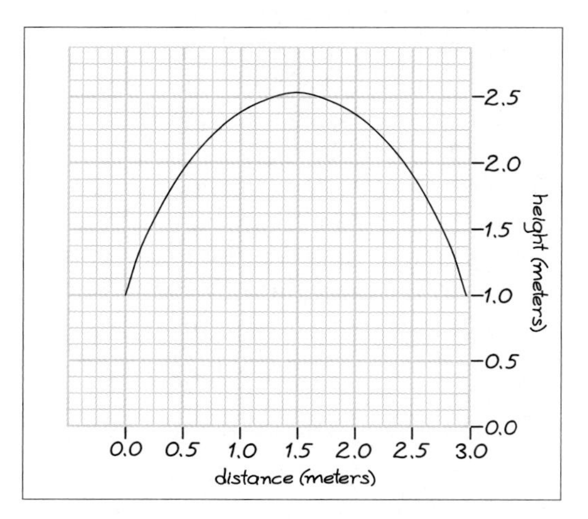

Part C: Energy Units and Conversions

1. Look at the conversion table. (In this activity you will record all your data in metric units. The table gives both metric and English equivalents to all the units that you will be using in this activity. Refer to this table whenever necessary.)

   a) Begin a concept map to show how the units are interconnected. Complete the concept map as you work through this part of the activity.

Energy Conversion Table
Heat
1 kcal (kilocalorie) = the heat needed to raise the temperature of one kilogram of water from 14.5°C to 15.5°C 1 Btu (British thermal unit) = the heat needed to raise the temperature of one pound of water from 60°F to 61°F 1 kcal = 1000 cal = 3.968 Btu
Force, mass, and velocity
1 kg = 0.069 slug acceleration of gravity (g) = 9.8 m/s^2 = 32 ft/s^2 1 N (newton) = 1 J/m (joule per meter) = 0.225 pounds 1 m/s = 3.28 fps (feet per second) = 2.24 mph (miles per hour)
Energy and work (the mechanical equivalent of heat)
1 kcal = 1000 cal = 4184 J (joules) 1 Btu = 252 cal = 777.9 ft-lb (foot-pounds) = 1055 J 1 kWh = 3,600,000 J = 3413 Btu 1 quad (Q) = 10^{15} Btu
Power (the rate at which work is done)
1 W (watt) = 1 J/s (joules per second) 1 hp (horsepower) = 550 ft-lb/s = 746 W

2. Do you think that you can produce power equal to that of a 100-W light bulb? Obtain and weigh a steel ball.

   a) Record the weight of the ball in newtons. As shown by the conversion tables, a newton is a unit of force. The weight of the ball is the same as the force exerted on the ball by the pull of gravity. Show your work in your *EarthComm* notebook.

3. Work is defined as the product of a force times the distance through which the force acts. The work needed to lift the steel ball a certain vertical distance is the force (weight of the ball, in newtons) times the vertical distance, or

$$W = F \cdot d,$$

where W is work in joules (J), F is force in newtons (N), and d is the height it is raised in meters (m).

a) In order for an object to obtain kinetic energy, work must be done on it. Calculate the work necessary to lift the steel ball to a height of 2 m.

4. Power is the rate at which work is done. The power you produce when you lift the ball is equal to the work divided by the time it took to lift the ball. If you lift the ball a number of times in a certain time period, the average power you produce is equal to the work of each lift, times the number of lifts, divided by the total time it took to do all of the lifting. Remember that the work is measured in joules and the time is measured in seconds.

$$P = W/t$$

where P is power in watts (W), W is work in joules (J), and t is time in seconds (s).

a) Calculate the power produced by lifting the ball 10 times in one minute. Note from the table that the unit for power is the watt. One watt = one joule per second.

5. In your group, discuss what a person would have to do to produce as much power as a 100-W bulb. Examples include running (how fast?) or climbing stairs (how fast?). Do this as a "thought experiment", one that you will describe (with calculations) but not conduct.

a) Record your thought experiment. Show your calculations.

6. The energy it took to produce the power to the ball came from chemical energy. In this case, the chemical energy was energy stored in the food you ate for breakfast. Assume that your body was 100% efficient (all of the stored energy is converted into kinetic energy).

a) Calculate the number of times you could lift the ball to equal a 200 Calorie candy bar (use the table and remember that one food calorie = 1 kilocalorie or 1000 calories).

b) In nature, no energy change is 100% efficient. Some energy is lost to the environment. In the case of lifting the ball, what form does the lost energy take?

7. As a class discuss the question of whether a person can produce as much power as a 100-W light bulb.

a) Record the results of your discussion.

Reflecting on the Activity and the Challenge

In Part A of this activity you looked at different ways that heat transfer occurs. Part B helped you to understand the concepts of potential and kinetic energy. In Part C you explored concepts of work, power, and units of energy. You also completed calculations to determine whether or not the exertion required to lift a ball can be equivalent to the power produced by a 100-W light bulb. These activities will help you think about how energy is transformed into a form that you can use. It will also help you think about ways to conserve energy resources so that your community can meet its growing energy needs.

Digging Deeper

HEAT AND ENERGY CONVERSIONS

Heat Transfer

Heat is really the **kinetic energy** of moving molecules. **Temperature** is a measure of this motion. The term **heat transfer** refers to the tendency for heat to move from hotter places to colder places. Many of the important aspects of heat transfer (see *Figure 1*) that you observed with the solar cooker had to do with heat conduction, which is one of the processes of heat transfer. All matter consists of atoms. At temperatures above **absolute zero** (about –273°C, the coldest anything can be!), the atoms vibrate. You sense those vibrations as the temperature of the material. The stronger the vibration, the hotter the material. When a hotter material is in contact with a colder material, collisions between adjacent vibrating atoms in the two materials cause the energy of the vibrations to even out, cooling the hot material and warming the cold material.

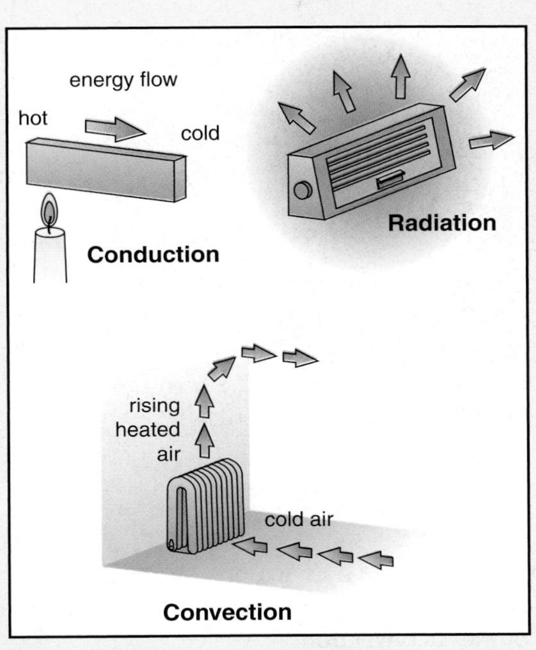

Figure 1 Three types of heat transfer.

Conduction is the type of heat transfer you experience when you take a hot bath, when you heat a piece of metal, or when the air cools a cup of hot coffee left on top of a table. For instance, when you put a metal pot on the stove, only the bottom of the pot is in contact with the burner, yet the heat flows through the entire pot all the way to the handle. Materials differ greatly in how well they conduct heat. In **thermal insulators**, like Styrofoam, crumpled paper, or a down jacket, the heat flows slowly. Thermal insulators like these contain a large amount of trapped air. Air is a poor conductor because the air molecules are not in constant contact. Metals, on the other hand, are very good conductors of heat. Heat conduction is very important in your community. Keeping your home warm in the winter

Geo Words

heat: kinetic energy of atoms or molecules associated with the temperature of a body of material.

kinetic energy: a form of energy associated with motion of a body of matter.

temperature: a measure of the energy of vibrations of the atoms or molecules of a body of matter.

heat transfer: the movement of heat from one region to another.

absolute zero: the temperature at which all vibrations of the atoms and molecules of matter cease; the lowest possible temperature.

conduction: a process of heat transfer by which the more vigorous vibrations of relatively hot matter are transferred to adjacent relatively cold matter, thus tending to even out the difference in temperature between the two regions of matter.

thermal insulator: a material that impedes or slows heat transfer.

Geo Words

convection: motion of a fluid caused by density differences from place to place in the fluid.

convection cell: a pattern of motion in a fluid in which the fluid moves in a pattern of a closed circulation.

electromagnetic radiation: the movement of energy, at the speed of light, in the form of electromagnetic waves.

would be very difficult (and expensive) without the insulating properties of the walls and the roof. Improving the insulation of your home by using insulating materials like those shown in *Figure 2* can greatly reduce the amount of energy needed to heat or cool your home.

Figure 2 Thermal insulation helps to keep your home warm. It conserves energy needed for space heating.

Another form of heat transfer is **convection**, which is important in liquids and gases. When a liquid or a gas is heated, its density decreases. That causes it to rise above its denser surroundings. In a room heated with a wood stove or a steam or hot water radiator, for instance, a natural circulation pattern is developed. The hot air from the stove rises towards the ceiling and cooler air travels down the walls and across the floor towards the stove. That kind of circulation is called a **convection cell**. Heat convection is also very important to your community. Many of the features of weather, such as sea breezes and thunderstorms, are caused by convection. Also, the way that you heat or cool your home depends strongly on heat convection.

Figure 3 The Sun emits electromagnetic radiation that warms the surface of the Earth.

A third form of heat transfer is **electromagnetic radiation**. Everything emits electromagnetic radiation. Examples of electromagnetic radiation are radio and television waves, visible light, ultraviolet light, and X-rays. Hotter materials emit more energy of electromagnetic radiation than colder materials. The warmth you feel from a hot fire, the Sun, or a light bulb is due to electromagnetic radiation traveling (at the speed of light!) from the hot object to you.

Radiation is important to the community for many reasons. Solar radiation causes things in the community to be heated. Solar radiation heated the solar cooker. It also heats someone standing in the sunshine on a cold winter day, or a parked car in the Sun in the summer with all its windows closed. If a building is designed appropriately, the heat from the Sun can substitute for heat from other energy resources for space heating and hot water. Using insulation or light-colored reflective materials reduces solar heating in warmer months when heat is not desired.

Energy, Work, and Power

In the investigation, you dealt with four forms of energy: energy of motion, called kinetic energy; energy of position, called **potential energy**; energy stored in the chemical bonds of a substance, called **chemical energy**, and heat. Kinetic energy and potential energy together are called **mechanical energy**. You know that objects in motion have energy, because of what they can do to you when they hit you. The energy of motion is called kinetic energy. The more mass the body has, and the faster it is moving, the more kinetic energy it has. When you threw (or imagined throwing) the lump of modeling clay up in the air, you gave it kinetic energy. The kinetic energy was gradually converted to potential energy. When the lump reached its highest point, its kinetic energy was at a minimum. On the way down, the lump regained its kinetic energy. When it hit the table, all of its kinetic energy was changed to heat. The change in temperature was so small that you would need a very sensitive thermometer to measure it. That's an example of how kinetic energy is changed to heat energy by **friction**. When you rub your hands together to keep them warm, you are converting kinetic energy to heat by friction. Of course, you are always resupplying your hands with kinetic energy by the action of your arm muscles.

Figure 4 A coal-powered train is an example of how chemical energy stored in coal is converted into heat energy that in turn is converted to mechanical energy.

Geo Words

potential energy: mechanical energy associated with position in a gravity field; matter farther away from the center of the Earth has higher potential energy.

chemical energy: energy stored in a chemical compound, which can be released during chemical reactions, including combustion.

mechanical energy: the sum of the kinetic energy and the potential energy of a body of matter.

friction: the force exerted by a body of matter when it slides past another body of matter.

Geo Words

work: the product of the force exerted on a body and the distance the body moves in the direction of that force; work is equivalent to a change in the mechanical energy of the body.

force: a push or pull exerted on a body of matter.

power: that time rate at which work is done on a body or at which energy is produced or consumed.

watt: a unit of power.

horsepower: a unit of power.

biomass: the total mass of living matter in the form of one or more kinds of organisms present in a particular habitat.

In physics, the term **work** has a very specific meaning. Work is equal to the **force** you exert on some object multiplied by the distance you move the object in the direction of the force. The importance of work is that it causes a change in the mechanical energy (kinetic and/or potential) of the object. When you threw the lump of modeling clay up in the air, your hand did the work. It exerted an upward force on the clay for a certain distance to give it its kinetic energy.

Power is the term used for the rate at which work is done or at which energy is produced or used. Think once more about the now-famous lump of modeling clay. You could have given it its upward kinetic energy by swinging your arm upward slowly for a long distance, generating low power but for a long time. Or, you could have swung your arm upward fast over only a short distance, generating high power but for only a short time. Whenever your muscles move your own body or some other object, you are generating power. The **watt** is the unit of power that is commonly used to describe the power of electrical devices. **Horsepower** is the unit of power that is often used to describe the power of other mechanical devices.

Converting Heat into Mechanical Energy

Figure 5 Coal is fed by a conveyor into a combustion chamber, where it is burned.

You have explored the idea that mechanical energy always tends to be converted into heat by friction. Nothing on Earth is completely frictionless, although some things, like air-hockey pucks, involve very little friction. Only in the emptiness of outer space can bodies move without friction. But how about energy conversion in the opposite direction: from heat to mechanical energy?

The conversion of heat into mechanical energy is central to most of the processes for producing electricity from energy resources. These resources include coal, natural gas, petroleum, sunlight, **biomass**, and nuclear energy. In these processes, water is heated to produce steam. When water boils (at atmospheric pressure) it undergoes about a thousand-fold increase in volume. The pressure of the steam exerts a force that does work to increase the kinetic energy of a turbine. The steam pressure is used to turn a turbine that generates electricity.

The **Second Law of Thermodynamics** states that you can never completely convert heat into mechanical energy. In fact, in converting any form of energy into another, there is always a decrease in the amount of "useful" energy. Stated in general terms, the **efficiency** of a machine or process is the ratio of the desired output (work or energy) to the input:

$$\% \text{ Efficiency} = \frac{\text{useful energy or work out}}{\text{energy or work in}} \times 100$$

Electrical power plants have efficiencies of about 30%. An efficiency of 33% means that for every three trainloads of coal that are burned to produce electricity, the chemical heat energy from only one of those trainloads is converted to electricity.

Some methods for generating electricity are not based on the conversion of heat to mechanical energy. Hydropower and wind power are examples. In hydropower, the mechanical energy of the falling water is converted directly to the mechanical energy of the rotating turbine. The efficiency of hydropower is only about 80% rather than 100%, however, because of friction and the incomplete use of available mechanical energy. Similarly, the wind already has mechanical energy. The efficiency of wind power is no greater than about 60%, mainly because some of the wind goes around the turbine without adding to its rotation. Actual efficiencies of most wind turbines range from 30% to 40% (The windmills shown in *Figure 6* have an efficiency of only 16%.) By comparison, the efficiency of a normal automobile engine is about 22%.

Figure 6 The efficiency of these windmills is only about 16%. Modern wind turbines have efficiencies between 30% and 40%.

Geo Words

Second Law of Thermodynamics: the law that heat cannot be completely converted into a more useful form of energy.

thermodynamics: a branch of physics that deals with the relationships and transformations of energy.

efficiency: the ratio of the useful energy obtained from a machine or device to the energy supplied to it during the same time period.

Check Your Understanding

1. What are some of the methods for generating energy that are based on the conversion of heat to mechanical energy?
2. Describe the three processes of heat transfer.
3. In your own words define mechanical energy.
4. Why can't the efficiency of a device be more than 100%?
5. Why is the efficiency of a device always less than 100%?

Understanding and Applying What You Have Learned

1. a) Explain how all of the different parts of the solar cooker work in terms of different heat transfer processes.
   b) How would you adapt your solar cooker to make it more effective and efficient?
2. Describe how a one-liter and two-liter container of water in the same oven differ in their heat and their temperature.
3. Describe how you think heat is transferred in the following situations:
   a) A cold room becomes warm after turning on a hot-water radiator.
   b) Your hand is heated as you grasp the handle of a heated pan on the stove.
   c) The bottom of a pan is heated when placed on an electric burner.
   d) A cold room becomes warm after window drapes are opened on a sunny day.
4. If the energy input of a system is 2500 cal and the energy output is 500 cal, what is the efficiency of the system?
5. A 300 hp engine is equivalent to how many foot pounds per second? In your own words state what this means.
6. When you drive in a car, energy is not lost, even though gasoline is being used up. Use what you have learned in this activity to explain what happens to this energy.

Preparing for the Chapter Challenge

Your **Chapter Challenge** is to help community members think about how they will meet their growing energy needs. Draft an introduction to your report. Use what you have learned in this activity to explain how energy resources are used to do work. Help people to understand how mechanical energy is converted to heat in the devices they use in their everyday lives. You might also begin to think about steps that community members might take to improve their energy efficiency.

Inquiring Further

1. **Perpetual-motion machines**

   The United States Patent Office receives many applications for perpetual-motion machines. All the applications are turned down.

   What is a perpetual-motion machine, and why can no one get a patent for one?

2. **Improving efficiencies of electricity generation**

   Innovative methods for power generation are now being developed to improve the efficiency of generating electricity from energy resources. What are some new methods for generating electricity from coal, natural gas, or oil that have improved efficiencies? Visit the *EarthComm* web site to help you find this information.

3. **History of science**

   Research the work of James Prescott Joule. A Scottish physicist, Joule conducted a famous experiment to observe the conversion of mechanical energy to heat energy. How did the experiments help Joule to conclude that heat is a form of energy?

4. **Solar cooking applications**

   In your investigation, you explored a model of one kind of solar food cooker. Research:

   - How people are using solar cookers and reducing the consumption of wood and fossil fuels for cooking food.
   - Where are solar cookers most commonly used?
   - Are they a suitable energy alternative for your community?
   - How does the use of a solar cooker reduce the effect on the biosphere?

Activity 2 Electricity and Your Community

Goals

In this activity you will:

- Compare energy resources used to generate electricity in the United States to other countries.
- Identify the major energy sources used to produce electricity in the United States and your state.
- Identify trends and patterns in electricity generation.
- Understand the difference between electric power and electric energy.
- Be able to describe commonly used methods of generating electric power.

Think about It

Electricity is a key part of life in the United States. Factories, stores, schools, homes, and most recreational facilities depend upon a supply of electricity. The unavailability of electricity almost always makes the news!

- What is electric energy?
- What are some consequences of not having electricity when it is needed?

What do you think? Record your ideas in your *EarthComm* notebook. Be prepared to discuss your responses with your small group and the class.

Investigate

Part A: Global and United States Electricity Generation

1. Use the data table showing world net electricity generation by type to answer the questions on the following page.

World Net Electricity Generation by Type, 1998 (in billion-kilowatt hours)					
Region Country	**Fossil Fuels**	**Hydro**	**Nuclear**	**Geothermal and Other**[1]	**Total**
North America					
Canada	148.7	328.6	67.7	6.1	551.1
Mexico	134.2	24.4	8.8	5.4	172.8
United States	2550.0	318.9	673.7	75.3	3617.9
The Caribbean & South America					
Bolivia	2.0	1.4	0.0	0.1	3.5
Brazil	15.6	288.5	3.1	9.7	316.9
Puerto Rico	17.0	0.3	0.0	0.0	17.3
Venezuela	21.6	52.5	0.0	0.0	74.0
Western Europe					
France	52.9	61.4	368.6	2.8	485.7
Spain	93.2	33.7	56.0	3.5	186.4
Switzerland	2.3	33.1	24.5	1.1	61.1
United Kingdom	235.3	5.2	95.1	6.2	341.9
Eastern Europe & Former U.S.S.R.					
Bulgaria	20.2	3.3	15.5	0.0	39.0
Romania	27.5	18.7	4.9	0.0	51.1
Russia	530.1	157.9	98.3	0.0	786.3
Middle East					
Cyprus	2.8	0.0	0.0	0.0	2.8
Saudi Arabia	116.5	0.0	0.0	0.0	116.5
Africa					
Angola	0.5	1.4	0.0	0.0	1.9
Egypt	47.1	12.1	0.0	0.0	59.2
Ethiopia	0.0	1.6	0.0	0.0	1.6
Morocco	11.6	1.7	0.0	0.0	13.4
Far East & Oceania					
China	880.2	202.9	13.5	0.0	1096.5
Japan	571.3	91.6	315.7	23.8	1002.4
Malaysia	52.5	4.8	0.0	0.0	57.3
Mongolia	2.5	0.0	0.0	0.0	2.5
Nepal	0.1	1.1	0.0	0.0	1.2
Total	**5535.7**	**1645.1**	**1745.4**	**134**	**9060.3**

[1]Geothermal and Other consists of geothermal, solar, wind, wood, and waste generation.
Source: Adapted from the Energy Information Administration/International Energy Annual 1999 report.

a) List the three countries that generate the most electricity.

b) What type of electricity generation is used most by these areas? Which is used least?

c) List the three countries that generate the least electricity.

d) What type of electricity generation is used most by these areas? Which is used least?

e) How do the resources for electricity generation differ between the top and bottom regions? How do you account for the differences?

2. Consider electricity generation by fuel type.

a) Without including the United States, rank global electricity generation by fuel type from highest to lowest.

b) Rank United States electricity generation by fuel type from highest to lowest.

c) How does the electricity generation in the United States match global electricity generation? Why do you think this is so?

3. Devise a way to represent the data for the United States graphically (either a pie chart or a histogram plot).

a) Complete your graph in your *EarthComm* notebook.

Part B: Investigating Electricity Generation in your State

1. Go to the *EarthComm* web site to visit the Department of Energy's Energy Information Administration web page. Click on your state to obtain your "state electricity profile." Use the information at the site to answer the following questions:

a) Is your state a net importer or net exporter of electricity? What does this mean?

b) What energy resources are used to generate electricity in your state?

c) Which energy resource does your state depend upon the most to generate electricity?

d) What are the trends in energy resource use for electric power generation in your state over time?

e) How has the generation of electricity changed in your state over time? How much has it grown in the most recent 10-year period? What is the annual rate of growth?

f) How does the growth in electricity generation in your state compare to the rate of population growth in your state? How might you explain the relationship?

g) How do the types of resources used by your state for electricity generation compare with the averages for the world and the United States, which you looked at in Part A of the investigation?

h) What sector (residential, commercial, industrial, or other) consumes the most energy for electricity in your state? What sector consumes the least?

i) What resources does your state use most for residential and commercial purposes? for transportation? for industry?

2. Review the data and figures for the electricity profile for your state.

   a) How do they help you to meet your **Chapter Challenge**?

Reflecting on the Activity and the Challenge

In this activity, you looked at the different types of resources that are used to generate electricity. You also investigated the trends in energy resource use and electricity generation in your state. Knowing where and how your state uses energy resources to generate electricity will help you to think about how to meet the energy needs of an expanding population in the **Chapter Challenge**.

Digging Deeper

MEETING ELECTRICITY NEEDS

Energy Resources

As the world's population increases and there is continued comparison to the current western European, Japanese, and North American living standards, there is likely to be demand for more energy. Several different energy resources can be used either for heating or cooling, electricity generation, industrial processes, and fuels for transportation. Energy resources available in the world include coal, **nuclear fission**, hydroelectric, natural gas, petroleum, wood, wind, solar, refuse-based, biomass, and oceanic (tides, waves, vertical temperature differences). In addition, nuclear fusion has been proposed as the long-term source, although research progress toward making it a reality has been very slow.

Figure 1 Nuclear power can come from the fission of uranium, plutonium, or thorium or the fusion of hydrogen into helium.

Geo Words

nuclear fission: the process by which large atoms are split into two parts, with conversion of a small part of the matter into energy.

Geo Words

electric power: power associated with the generation and transmission of electricity.

electric energy: energy associated with the generation and transmission of electricity.

fossil fuel: fuel derived from materials (mainly coal, petroleum, and natural gas) that were generated from fossil organic matter and stored deep in the Earth for geologically long times.

geothermal energy: energy derived from hot rocks and/or fluids beneath the Earth's surface.

photovoltaic energy: energy associated with the direct conversion of solar radiation to electricity.

turbine: a rotating machine or device that converts the mechanical energy of fluid flow into mechanical energy of rotation of a shaft.

Generating Electric Energy

Energy resources are used to generate electricity, an energy source with which you are very familiar. **Electric power** is the rate at which electricity does work—measured at a point in time. The unit of measure for electric power is a watt. The maximum amount of electric power that a piece of electrical equipment can accommodate is the capacity or capability of that equipment. You can check the tags or labels on electrical appliances for this information, such as a "1200-W hair dryer" or "40-W stereo receiver". **Electric energy** is the amount of work that can be done by electricity. The unit of measure for electric energy is the watt-hour. A 1200-W hair dryer used for 15 min would require (theoretically) 300 Wh of electric energy.

Fossil fuels supply about 70% of the energy sources for electricity in the United States. Coal, petroleum, and natural gas are currently the dominant fossil fuels used by the electrical power industry. When fossil fuels are burned to generate electricity, a variety of gases and particulates are formed. If these gases and particulates are not captured by some pollution control equipment, they are released into the atmosphere. Other sources of energy can also be converted into electricity, including **geothermal energy**, solar thermal energy, **photovoltaic energy**, and biomass. These alternative sources of energy have many advantages over fossil fuels, as you will explore in Activity 5.

Fossil Fuels and Nuclear Energy

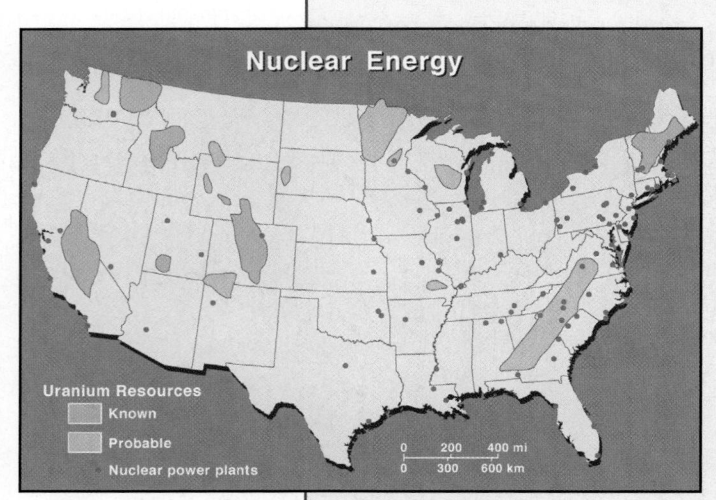

Figure 2 The location of nuclear power plants and uranium resources in the continental United States.

Most of the electricity in the United States is produced in steam **turbines**. A turbine converts the kinetic energy of a moving fluid (liquid or gas) to mechanical energy. In a fossil-fueled steam turbine, the fuel is burned in a boiler to produce steam. The resulting steam then turns the turbine blades that turn the shaft of the generator to produce electricity. In a nuclear-powered steam turbine, the boiler is replaced by a reactor containing a core of nuclear fuel (primarily enriched uranium). Heat produced in the reactor by fission of the uranium is used to make steam. The steam is then passed through the turbine generator to produce electricity, as in the fossil-fueled steam turbine. *Figure 2* shows the nuclear-power plant locations and uranium resources available in the United States.

Hydroelectric Power

Water is the leading **renewable energy source** used by electric utilities to generate electric power. **Hydroelectric power** is the result of a process in which flowing water is used to spin a turbine connected to a generator. Hydroelectric plants operate where suitable waterways are available. Many of the best of these sites have already been developed. (*Figure 3* shows the hydroelectric plants developed in the United States.) Seventy percent of the hydroelectric power in the United States is generated in the Pacific and Rocky Mountain States. The two basic types of hydroelectric systems are those based on falling water and natural river current. In a falling-water system, water accumulates in reservoirs created by dams. This water then falls through conduits and applies pressure against the turbine blades to drive the generator to produce electricity. In the second system, called a run-of-the-river system, the force of the river current (rather than falling water) applies pressure to the turbine blades to produce electricity. Because they do not store water, these systems depend upon seasonal changes and stream flow.

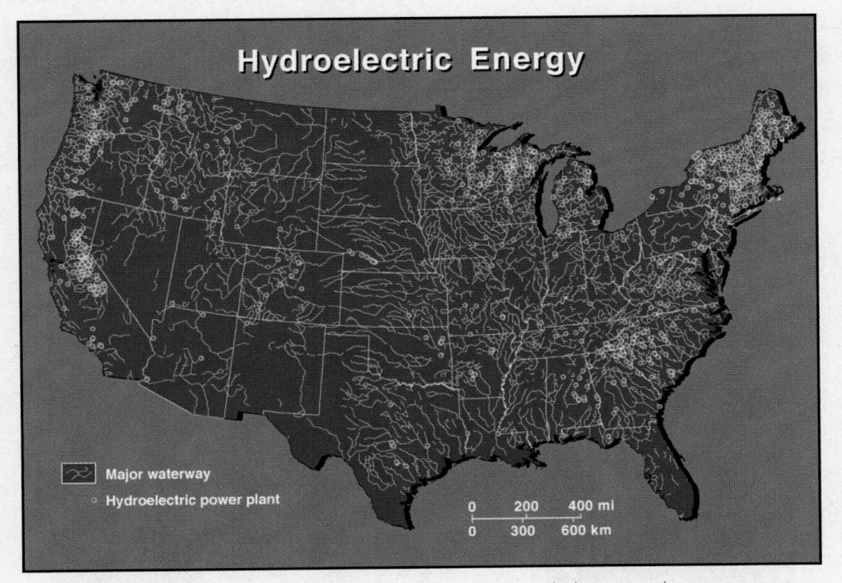

Figure 3 The location of hydroelectric power plants relative to major waterways in the continental United States.

Generating electricity using water has several advantages. The major advantage is that water, a renewable resource, is a source of cheap power. In addition, because there is no fuel combustion, there is little air pollution

Geo Words

renewable energy source: an energy source that is powered by solar radiation at the present time rather than by fuels stored in the Earth.

hydroelectric power: electrical power derived from the flow of water on the Earth's surface.

in comparison with fossil fuel plants and limited thermal pollution compared with nuclear plants. They can start quickly because they do not need to wait for water to be heated into steam. Also, the flow of water can be adjusted to make quick changes in power output during peak demands for electricity. Like other energy sources, the use of water for generation has limitations, including environmental impacts caused by damming rivers and streams, which affects the habitats of the local plants and animals. Another disadvantage to some hydroelectric power plants is that they depend upon the flow of water which varies with seasons and during droughts.

Other Resources Used to Generate Electricity

Currently, renewable resources (other than water) and geothermal energy supply less than 1% of the electricity generated by electric utilities. They include solar, wind, and biomass (wood, municipal solid waste, agricultural waste, etc.). Although geothermal energy is a nonrenewable resource, we will not run out of it. Geothermal power comes from heat energy buried beneath the surface of the Earth. Most of this heat is at depths beyond current drilling methods. In some areas of the country, magma—the molten matter under the Earth's crust from which igneous rock is formed by cooling—flows close enough to the surface of the Earth to produce steam. That steam can then be harnessed for use in conventional steam-turbine plants. *Figure 4* shows the locations of geothermal plants within the United States. Most are found in the western United States where magma is close enough to the surface to supply steam.

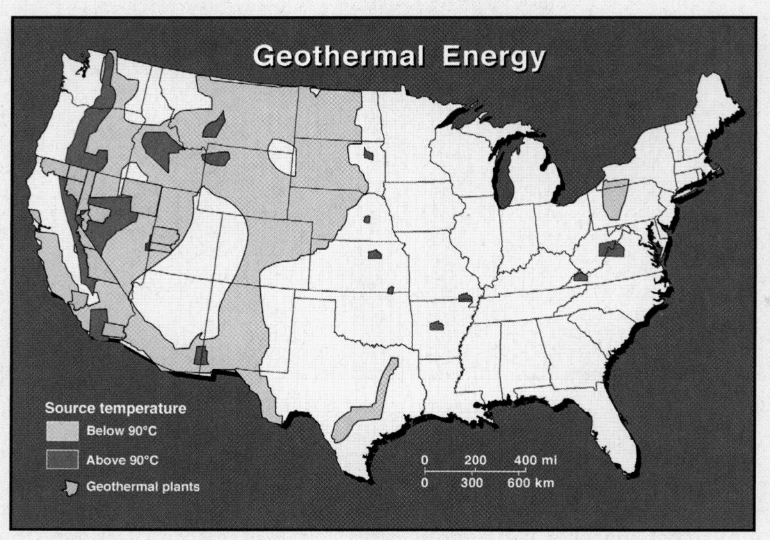

Figure 4 The location of geothermal power plants and source temperatures in continental United States.

Check Your Understanding

1. What is the difference between electric energy and electric power?
2. Compare and contrast steam turbine versus hydroelectric power generation. How are they similar? How are they different?
3. From an Earth systems perspective, what are the advantages and disadvantages of hydroelectric power?
4. In your own words, explain why biomass and wind are called renewable energy sources.

Understanding and Applying What You Have Learned

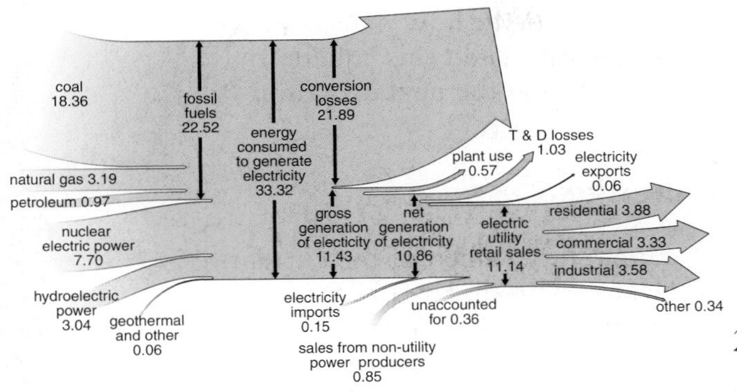

Electricity flow - electric utilities 1999
(quadrillion Btus)

1. Use the diagram above to answer the following questions about the generation of electricity in the United States:

   a) Which of the energy sources listed above are nonrenewable? What percentage of energy consumed to produce electricity in 1999 did they represent?
   b) Which of the energy sources listed above are renewable? What percentage of energy consumed in 1999 did they represent?
   c) In Activity 1 you learned about energy efficiency. Compare the value for energy consumed to produce electricity versus net generation of electricity in 1999. What is the overall efficiency of electricity generation systems in the United States?
   d) Conversion losses accounted for 21.89 quads of the energy consumed to generate electricity. Explain what this means in relation to your answer to question 3(c) above.

2. Converting one form of energy into another comes with a loss in usable energy (waste heat). The "net efficiency" of a multi-step process is the product of the efficiencies of each step. For example, if each step in a two-step process was 50% efficient, the net efficiency would be $0.50 \times 0.50 = 0.25 = 25\%$.

   a) Use the information below to calculate the net efficiency of a home water heater.
   b) The lowest efficiency of this process is the generation of electricity. Why do you think this is so?

3. Each year, the United States uses more than 71 trillion Btus of solar energy. One million Btu equals 90 lb. of coal or eight gallons of gasoline.

Step	Efficiency (in percent)
Production of coal	96
Transporting coal to power plant	97
Generating electricity	33
Transmission of electricity (from plants to homes)	85
Heating efficiency of the hot water heater	70
Net Efficiency	?

a) How many tons of coal per year does this equate to?
b) How many gallons of gasoline does this equate to?
c) If the United States increased its use of solar energy 5% annually, how many tons of coal would this equate to during the next 10 years?

Preparing for the Chapter Challenge

Write a short paper in which you review the energy resources used in your community and how their use affects your community. Be sure to include information on how energy resources are used to generate electricity and fuel in your community, which of the current means of producing power your community relies on most, and which seems the least important. This paper will help you to think about what resources your community is using before the population increase.

Inquiring Further

1. **Storage of solar energy**

   Electricity from solar energy offers a clean, reliable, and renewable source of energy for home use. Batteries are used to store electrical energy for use at night or when bright sunlight is not available. Research recent advances in technologies used to store solar energy. How do car batteries perform in comparison to solar batteries in solar systems? Present your report to the class.

2. **Energy from the oceans**

   Three novel ways have been proposed to use the energy of the oceans for electrical generation: tides, waves, and vertical temperature differences. Choose one of these, and research the theory and techniques involved in its use for generating electricity.

3. **Other methods of generating electricity**

   Research one of the following methods of generating electricity. How do they work? Where are they being used? What are the ideal conditions for using these alternatives? What are the advantages of these methods over fossil fuels? What are the disadvantages or limitations? Visit the *EarthComm* web site for sources of information.

   - photovoltaics
   - geothermal
   - solar thermal

Activity 3 Energy from Coal

Goals

In this activity you will:

- Classify the rank of coal using physical properties.
- Interpret a map of coal distribution in the United States.
- Understand that fossils fuels represent solar energy stored as chemical energy.
- Understand what coal is made of and how coal forms.
- Be able to explain in your own words why coal is a nonrenewable resource.

Think about It

Coal is the largest energy source for electricity in the United States. Known coal reserves are spread over almost 100 countries. At current production levels, proven coal reserves are estimated to last over 200 years.

- How does coal form?
- Why is coal referred to as "stored solar energy"?

What do you think? Record your ideas in your *EarthComm* notebook. Be prepared to discuss your responses with your small group and the class.

Investigate

Part A: Types of Coal

1. Obtain a set of samples of four different types of coal. Examine the samples. Look for evidence of plant origin, hardness, luster (the way light is reflected off the surface of the sample), cleavage (tendency to split into layers), and any other characteristics that you think distinguish the four types of coal. Discuss similarities and differences with members of your group.

   a) In your *EarthComm* notebook, create a data table and record your observations.

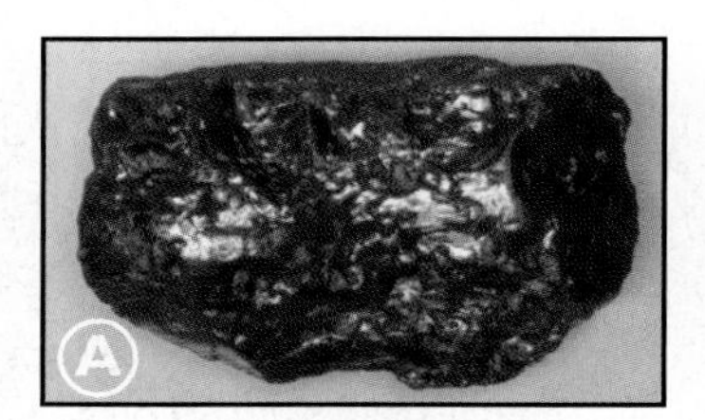

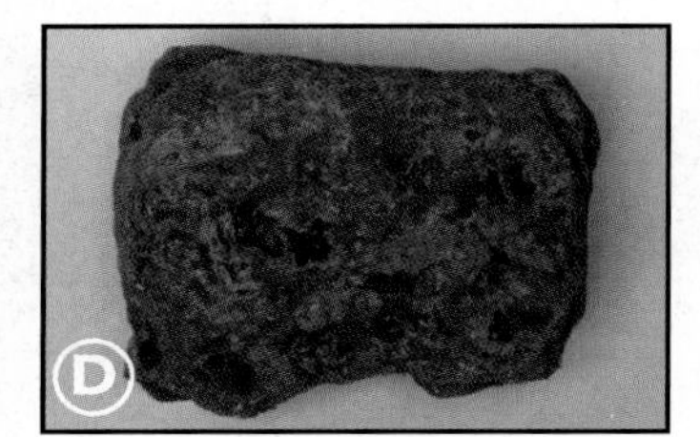

2. Use your completed data table, to answer the following questions:

   a) How might the samples be related?

   b) Put the samples in order from least compacted to most compacted.

   c) Which sample do you think contains the most stored energy? Why do you think so?

3. The next step of this activity will be a demonstration. You will observe your teacher igniting a small piece of coal held over a Bunsen burner. The sample will be removed from the flame and you will observe how the sample burns. Four types of coal will be tested.

   a) Note the speed of ignition, color of the flame, speed of burning, and odor. Note any other characteristic that you think distinguishes the samples. Summarize your observations in a data table or chart.

   b) In your *EarthComm* notebook, summarize the major differences you observed.

   c) Review your answer to Step 2(c). Have your ideas changed? Explain.

Wear goggles. Clean up all loose pieces of coal. Wash your hands after you are done.

Part B: Coal Resources

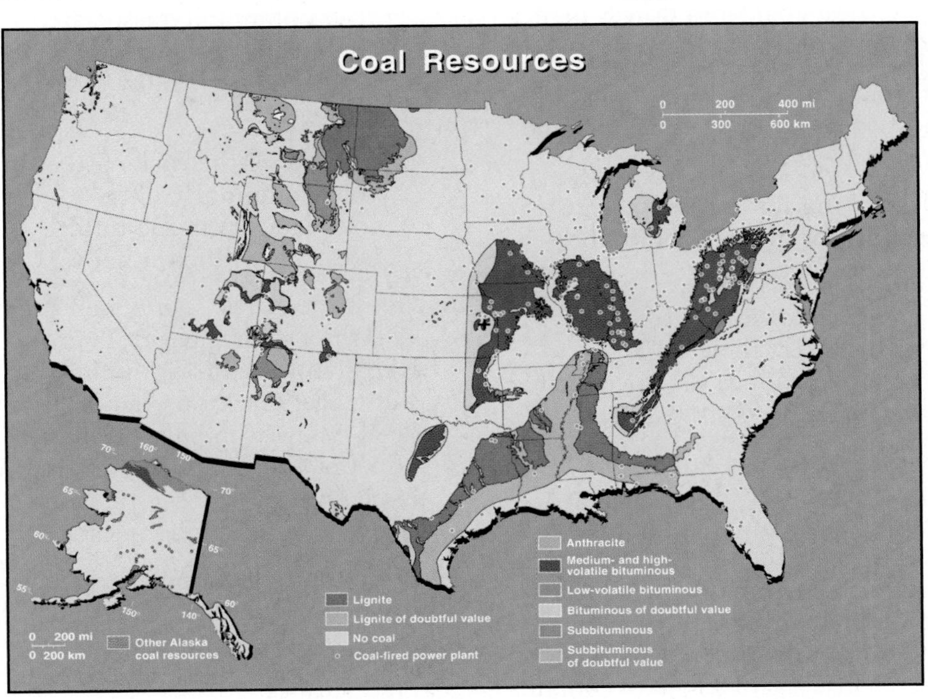

1. Examine the map of coal resources in the United States.
   a) Summarize the major trends, patterns, or relationships in the distribution of coal and the locations of coal-fired power plants.
2. Refer to the map to answer the following questions about coal resources in the United States. Record your responses in your *EarthComm* notebook.
   a) How many states contain coal deposits? Which states are they?
   b) What are the main types of coal present in the following regions: east-central, southeast, and west-central?
   c) Can you infer from the map which state has or produces the most coal? Why or why not?
   d) Measure how far your community is from a source of coal.
   e) Look at the distribution of coal-fired power plants. Why do you think coal is used in these plants rather than petroleum or natural gas?
   f) What type of coal is closest to your community?
3. Refer to the map to answer the following questions about coal-fired power plants in the United States. Record your responses in your *EarthComm* notebook.
   a) Is there a coal-fired power plant in your state? If so, where is it located?

b) How far from your community is the closest coal-fired power plant? Where is the plant located?

c) In what part of the country are the most coal-fired power plants found? Why do you think this is so?

d) Identify the states that have coal-fired power plants but do not have coal deposits.

e) How might distance from coal deposits affect the cost of electricity produced by a coal-fired power plant?

4. Examine a geologic map of your community or state. A geologic map shows the distribution of bedrock at the Earth's surface. The bedrock shown on the map might be exposed at the surface, or it might be covered by a thin layer of soil or very recent sediment. Every geologic map contains a legend that shows the kinds of rocks that are present in the area.

   a) Write down or note the names of the different rock types present in your community, the names of the rock units, their ages, and their locations.

5. Compare the geologic map to the map that shows the distribution of coal deposits in the United States.

   a) Are any of the rocks present in your community associated with coal deposits? What kinds of rocks are they? How old are the rocks?

   b) Are any of the rocks present in your community likely to yield coal deposits in the future? Which ones? How do you know?

6. Go to the *EarthComm* web site to visit the Department of Energy's Energy Information Administration web page. Click on your state to open a new web page. Use the "Energy Consumption, Prices, and Expenditures" profiles page to answer the following questions:

   a) In your state, approximately what percentage of primary energy consumption for residential purposes comes from coal?

   b) In your state, what percentages of primary energy consumption for commercial, industrial, and transportation purposes come from coal?

   c) How would an increase in the population of your community affect this usage?

Reflecting on the Activity and the Challenge

In this activity you described four types of coal and observed how the different types of coal react when burned. You then examined maps to determine the distribution of coal deposits in the United States and also to determine if coal deposits are found near your community. You looked at the location of coal-fired power plants relative to coal deposits and thought about how the distance between the two can affect the cost of electricity production. These activities will help you to determine how coal is used as an energy resource in your community and the impacts of its use.

Digging Deeper

COAL AS A FOSSIL FUEL

There are only four primary sources of energy available for use by humankind. They are solar radiation, the Earth's interior heat, energy from decay of radioactive material in the Earth, and the tides. You can assume that starlight, moonlight, and the kinetic energy of meteorites hitting the Earth are so small that they can be neglected. The energy in coal is energy from solar radiation that is stored as chemical energy in rock.

The energy in coal originates as solar energy. Plants in the biosphere store this solar energy by a process called **photosynthesis**. During photosynthesis, green plants convert solar energy into chemical energy in the form of **organic (carbon-based) molecules**. Only 0.06% of the solar energy that reaches the Earth is stored through photosynthesis, but the amount of energy that is stored in the Earth's vegetation is enormous. Photosynthesis yields the carbohydrate glucose (a sugar) and water, as shown in the equation below:

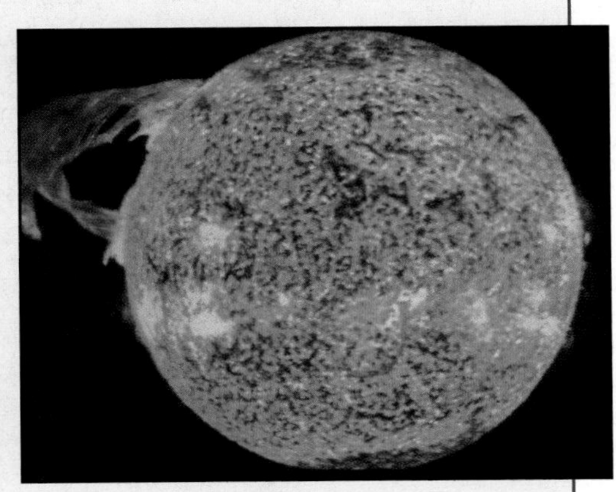

Figure 1 Solar radiation is one of the primary sources of energy on Earth.

$6CO_2$	+	$6H_2O$	Sunlight used →	$C_6H_{12}O_6$	+	$6O_2$
Carbon dioxide removed from atmosphere and biosphere		Water removed from hydrosphere	Solar energy stored as chemical energy	Glucose (sugar) for biosphere		Oxygen released to atmosphere

The energy stored in glucose (a sugar) and other organic molecules can be released when broken in a reaction with oxygen (oxidized). **Oxidation** occurs through **respiration, decomposition**, or combustion (burning).

$6O_2$	+	$C_6H_{12}O_6$	Chemical energy released →	$6H_2O$	+	$6CO_2$
Oxygen removed from atmosphere and biosphere		Glucose from biosphere	Heat released to organism by respiration or to environment by decomposition or combustion	Water returned to hydrosphere		Carbon dioxide returned to atmosphere and biosphere

Geo Words

photosynthesis: the process by which plants use solar energy, together with carbon dioxide and nutrients, to synthesize plant tissues.

organic (carbon-based) molecules: molecules with the chemical element carbon as a base.

oxidation: the chemical process by which certain kinds of matter are combined with oxygen.

respiration: physical and chemical processes by which an organism supplies its cells and tissues with oxygen needed for metabolism.

decomposition: the chemical process of separation of matter into simpler chemical compounds.

Geo Words

sediments: loose particulate materials that are derived from breakdown of rocks or precipitation of solids in water.

lithosphere: the rigid outermost shell of the Earth, consisting of the crust and the uppermost mantle.

nonrenewable resource: an energy source that is powered by materials that exist in the Earth and are not replaced nearly as fast as they are consumed.

You can see from the reactants (on the left sides of the two equations) and the products (on the right sides of the two equations) that photosynthesis and oxidation are the reverse of each other.

The energy that enters the biosphere by photosynthesis is nearly equal to the energy lost from the biosphere by oxidation, as shown in *Figure 2*. Most of the carbon in the biosphere is soon returned to the atmosphere (or to the ocean) as carbon dioxide, but a very small percentage is buried in **sediments**. It is protected from oxidation and becomes part of the **lithosphere** (in the form of peat, coal, and petroleum) for future use as fossil fuels. The rate of storage of energy used by organisms for photosynthesis is very slow. However, these processes have been active throughout much of Earth's history. Therefore, the amount of energy stored as fossil fuels is enormous (32×10^{20} Btus). Fossil fuels are consumed far faster than they form, and therefore they are classified as **nonrenewable resources**.

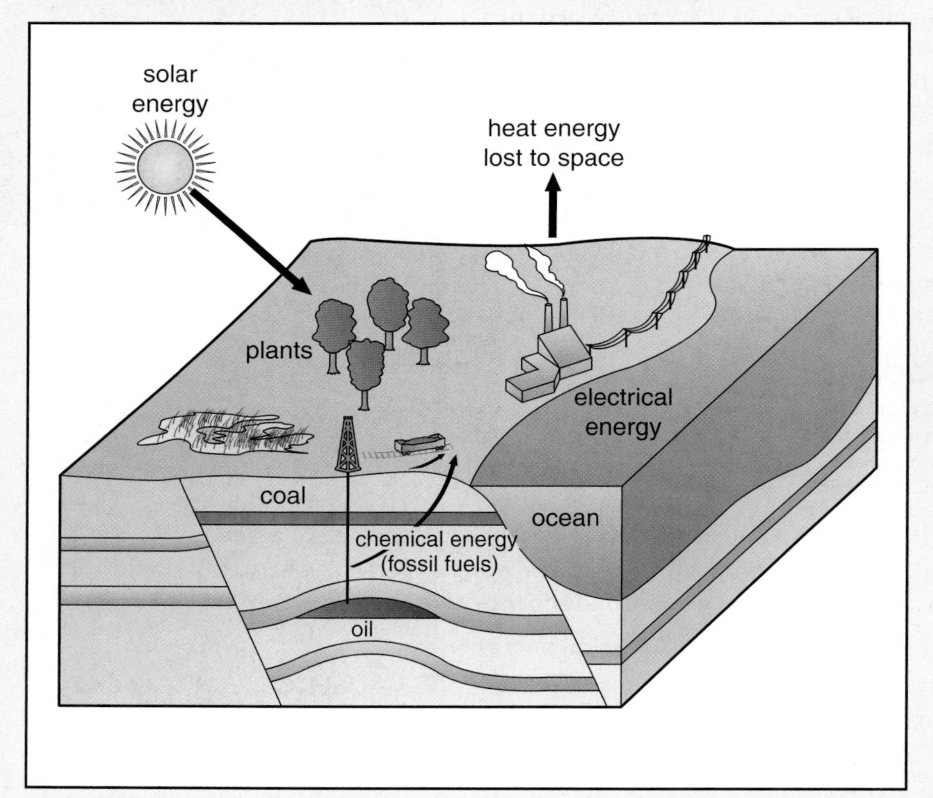

Figure 2 The flow of energy from the Sun, to plants, to storage in fossil fuels, and loss back into space.

The Formation of Coal

Most coal starts out as **peat**. Peat is an unconsolidated and porous deposit of plant remains from a bog or swamp. Structures of the plant matter, like stems, leaves, and bark can be seen in peat. When dried, peat burns freely, and in some parts of the world it is used for fuel. Today, most peat comes from peat bogs that formed during the retreat of the last ice sheets, between about ten thousand and twenty thousand years ago.

Coal, by definition, is a combustible rock with more than 50% by weight of carbonaceous material. Coal is formed by compaction and hardening of plant remains similar to those in peat. The plant remains are altered physically and chemically through a combination of bacterial decay, compaction, and heat. Most coal has formed by lush growth of plants in coastal fresh-water swamps, called coal swamps, in low-lying areas that are separated from any sources of mud and sand. Plants put their roots down into earlier deposits of plant remains, and they in turn die and serve as the medium for the roots of even later plants. In such an environment, the accumulation of plant debris exceeds the rate of bacterial decay of the debris. The bacterial decay rate is reduced because the available oxygen in organic-rich water is completely used up by the decay process. Thick deposits of almost pure plant remains build up in this way. Coal swamps are rare in today's world, but at various times in the geologic past they were widespread.

For the plant material to become coal, it must be buried by later sediment. Burial causes compaction, because of the great weight of overlying sediment. During compaction, much of the original water that was in the pore spaces of the plant material is squeezed out. Gaseous products (methane is one) of alteration are expelled from the deposit. The percentage of the deposit that consists of carbon becomes greater and greater. As the coal becomes enriched carbon, the coal is said to increase in rank. (See *Figure 3*.) The stages in the rank of coal are in the following order: peat, lignite, sub-bituminous coal, bituminous coal, anthracite coal, and finally graphite (a pure carbon mineral). It is estimated that it takes many meters of original peat material to produce a thickness of one meter of bituminous coal.

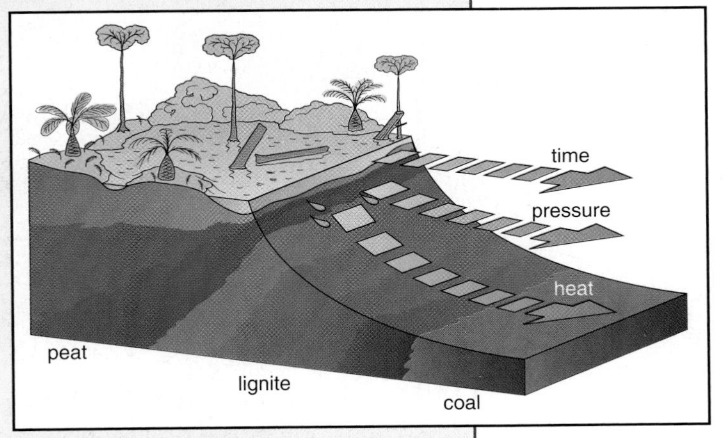

Figure 3 Increasing time, pressure, and heat result in the formation of progressively higher ranking coal.

Geo Words

peat: a porous deposit of partly decomposed plant material at or not far below the Earth's surface.

coal: a combustible rock that had its origin in the deposition and burial of plant material.

Geo Words

sedimentary rock: a rock, usually layered, that results from the consolidation or lithification of sediment.

As shown in *Figure 4*, coal is always interbedded with other **sedimentary rocks**, mainly sandstones and shales. Environments of sediment deposition change with time. An area that at one time was a coal swamp might later have become buried by sand or mud from some nearby river system. Eventually, the coal swamp might have become reestablished. Upon burial, the plant material is converted to coal, and the sand and mud form sandstone and shale beds. In this way, there is an alternation of other sedimentary rock types with the coal beds. Some coal beds are only centimeters thick, and are not economically important. Coal beds that are mined can be up to several meters thick.

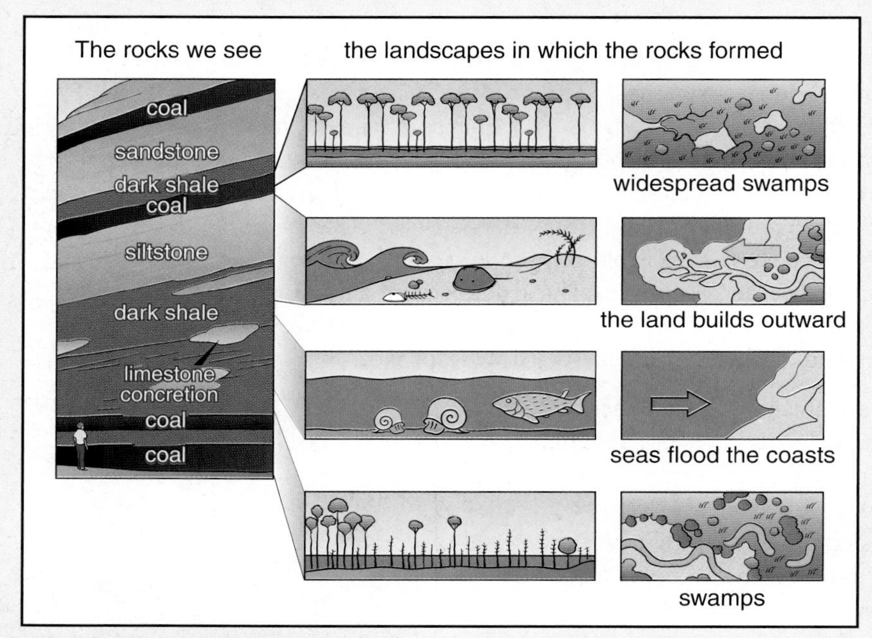

Figure 4 Rocks reflect the environments of the past.

Types of Coal

The type of coal that is found in a given region depends partly on the composition of the original plant material (together with any impurities that were deposited in small quantities and at the same time). But mainly the type of coal depends on the depth and temperature of later burial. Coal that is buried very deeply attains high rank, because of the high pressures and temperatures associated with deep burial. Low-rank coals (lignite and sub-bituminous) have not been buried as deeply.

Coal varies greatly in composition. One of the most important features of coal composition is sulfur content. Sulfur is important because it is released into the atmosphere as sulfur dioxide when the coal is burned. The sulfur dioxide then combines with water in the atmosphere to form sulfuric acid, causing acid rain. The sulfur content of coal can range from a small fraction of 1% to as much as 5%, depending mainly on the sulfur content of the original plant material. The carbon content of coal increases with increasing rank of coal, from lignite to anthracite, as illustrated in *Table 1*. The heat content is also an important feature of coal. The greater the heat content, the smaller the mass of coal that needs to be burned to produce the needed heat. The heat content of coal increases with the rank of the coal, because it depends mainly on the carbon content.

Table 1 Percentages of Carbon and Volatile Matter in Coal

Coal Rank	Carbon Content (%)	Volatile matter (%)	Btus per pound
Lignite	25–35	up to ~50%	4000–8300
Sub-bituminous	35–45	up to ~30%	8300–13,000
Bituminous	45–86	less than 20%	10,500–15,000
Anthracite	86–98	less than 15%	15,000

The carbon content of coal supplies most of its heat energy per unit weight. The amount of energy in coal is expressed in British thermal units (Btu) per pound. A Btu is the amount of heat needed to raise the temperature of one pound of water one degree Fahrenheit. Peat can be used as a source of fuel, but it has a very low heat content per pound of the fuel burned. Lignite (also called brown coal), the least buried and usually the youngest type of coal, is used mainly for electric power generation. Sub-bituminous coal is a desirable heat source because of its often low sulfur content. Sub-bituminous coal is found in the western United States and Alaska. Bituminous coal is the most abundant coal in the United States, with a large deposit in the Appalachian Province of the East. Bituminous coal is used mainly for generating electricity and making coke for the steel industry. Anthracite, found in a very small supply in the eastern United States, has been used mainly for home heating.

Some of the constituents of coal are not combustible. The part of coal that is not consumed by burning is called ash. Most of the ash consists of sand, silt, and clay that was deposited in the coal swamp along with the vegetation. The purest coal has only a small fraction of 1% ash. The ash content of usable coal can be much higher. Some of the ash remains in the combustion chamber, and some goes up the flue. Ash in coal is undesirable because it reduces the heat content slightly. It also must be removed from the combustion chamber and discarded.

Check Your Understanding

1. Explain how solar energy is stored and released by the processes of photosynthesis and oxidation.
2. Describe, in your own words, the formation of coal, from plant matter to anthracite.
3. Why is the heat content of coal (in Btu/pound) important in determining how the coal is used?
4. What is the origin of the sulfur content of coal, and why is sulfur an undesirable constituent of coal?

Understanding and Applying What You Have Learned

1. How much impact do you think the proximity of your community or state to a coal deposit has on the use of coal for energy generation? Support your response with the information you collected in the investigation.
2. Discuss what happens to the carbon content of peat if it is allowed to decay in the presence of oxygen.
3. Use what you have learned about coal in this activity to provide two reasons why it would be a better idea to burn anthracite in a home fireplace than lignite.
4. What are the advantages to using coal to meet energy needs? What are the disadvantages?

Preparing for the Chapter Challenge

You have learned how coal deposits form, and you looked at the distribution of coal deposits in the United States and your community. You also considered how reliant your community is on coal as an energy resource. Write a paper in which you consider how the number of coal deposits in the United States, along with the time required to produce a coal deposit, shape the way that your community uses coal today and how it will use coal in the future, in light of projected population growth.

Inquiring Further

1. **Model the formation of coal**

   Line a plastic shoebox (or two-liter bottle with the top cut off) with plastic wrap. Pour water into the container to a depth of four inches. Spread about two inches of sand on the bottom. Drop small leaves, sticks, and pieces of fern on the sand. Let it stand for two weeks. Record what you observe as changes in color and decomposition occur. Gently sift fine sand or mud on top of the plant layer to a depth of two inches. Wait two weeks and drain any remaining water. Let it sit and dry for another two weeks. Remove the "formation" out of the container. Slice it open to see how you have simulated coal where the plants were, and fossil imprints from the plant leaves.

   Wash your hands after handling any material in this activity. Wear goggles. Complete the activity under adult supervision.

2. **Plants associated with coal**

   Research the ancient plants whose remains formed the United States coal deposits.

Activity 4 Coal and Your Community

Goals

In this activity you will:

- Investigate the production and consumption of coal in the United States.
- Investigate how coal sources are explored.
- Understand methods of coal mining.
- Determine how coal is used in your state.
- Evaluate possible practices to conserve coal resources.

Think about It

Coal production in the United States decreased in 1999 and 2000, but coal consumption continues to increase.

- Why do you think that the production of coal has decreased?
- Why do you think that the consumption of coal has increased?
- Given that coal is a finite, nonrenewable source of energy, what are some ways to extend the supply of coal?

What do you think? Record your ideas in your *EarthComm* notebook. Be prepared to discuss your responses with your small group and the class.

Investigate

Part A: Trends in Coal Production and Consumption

Table 1 United States Coal Production, by Region 1991–2000 (in million short tons[1])

Year	Appalachian	Interior	Western	Total
1991	457.8	195.4	342.8	996.0
1992	456.6	195.7	345.3	997.5
1993	409.7	167.2	368.5	945.4
1994	445.4	179.9	408.3	1033.5
1995	434.9	168.5	429.6	1033.0
1996	451.9	172.8	439.1	1063.0
1997	467.8	170.9	451.3	1089.9
1998	460.4	168.4	488.8	1117.5
1999	425.6	162.5	512.3	1100.4
2000	420.9	144.7	509.9	1075.5

Source: United States Energy Information Administration, 1996, 1998, 2000
[1]One short ton = 2000 pounds.

1. *Table 1* gives recent trends in coal production in the United States during the last 10 years. Using these values, summarize the trends in coal production for the three major coal-producing regions, and the trend in total coal production.
   a) Appalachian
   b) Interior
   c) Western
   d) Total
2. Make one graph showing coal production in the three regions. Leave room at the end of the graph to project coal production for the next 50 years.
   a) Include your graph in your *EarthComm* notebook.
3. Extrapolate the trends in the data to the year 2050. To do this, you will have to produce a best-fit line through the data points and estimate what you think to be the trend in coal production. There is a little guesswork involved in determining how much of the data you think represents the trend. Is it the last four years or the last eight years?
   a) On the basis of your projections only (you are ignoring all other factors that might influence future production), which coal-producing region do you predict will be the first to exhaust its supply of coal? In what year will this happen?
   b) In your group, identify at least three factors that might affect actual coal production. Record your ideas in your notebook.

4. Examine the data in *Table 2*. Summarize the major trends in coal consumption for each sector from 1991 to 2000, and the total coal consumption.
   a) Electric power
   b) Coke plants (steel manufacturing)
   c) Other industrial plants
   d) Residential and commercial users
   e) Total
   f) What percentage of total coal consumption did electrical power make in 1991? In 2000?

Table 2 United States Coal Consumption by Sector, 1991–2000 (in million short tons[1])

	Consumption by Sector (million short tons)				
Year	**Electric Power**	**Coke Plants**	**Other Industrial Plants**	**Residential and Commercial Users**	**Total**
1991	777.2	33.9	75.4	6.1	892.5
1992	786	32.4	74.0	6.2	898.5
1993	820.8	31.3	74.9	6.2	933.9
1994	826.7	31.7	75.2	6.0	939.6
1995	849.8	33.0	73.1	5.8	961.7
1996	896.9	31.7	70.9	6.0	1005.6
1997	922.0	30.2	71.5	6.5	1030.1
1998	937.8	28.2	67.4	4.9	1038.3
1999	946.8	28.1	65.5	4.9	1045.3
2000	970.7	29.3	65.5	4.9	1070.5

Source: United States Energy Information Administration, 1996, 1998, 2000
[1]One short ton = 2000 pounds.

5. Using the data from *Table 2*, make a graph that shows coal consumption for electrical power.
   a) Include the graph in your *EarthComm* notebook.
6. Extrapolate the trend in the data to the year 2050, as in Step 3.
   a) On the basis of recent trends, how much coal will be needed for electrical power generation in the year 2020? 2050?
7. Coal consumption for electrical power increased at an average rate of 1.65% per year between 1996 and 2000.
   a) On the basis of this average, predict coal consumption for electrical power for the years 2010 and 2020 (Hint: Begin by multiplying the value for the year 2000 by 1.0165. This gives you a prediction for the year 2001.)

8. Assume that by conserving electricity the average rate of growth in consumption of coal is cut in half, to 0.825% per year.

   a) Predict the amount of coal consumed in 2010 and 2020.

9. Draw a new graph that shows your prediction of coal production and coal consumption for the period 2000 to 2050. Do this by superimposing the curves you fitted in Steps 3 and 6. Also, in a third curve, take into account the prediction you made in Step 8, on the assumption of savings by conservation.

   a) Include your graph in your *EarthComm* notebook.

   b) Do you predict that consumption will exceed production? If so, how do you think that the shortfall in production will be made up? Or do you predict that production will exceed consumption? If so, do you think that that is a reasonable or likely scenario?

   c) In what ways do you think that production and consumption are related? Does production drive consumption? If so, how and to what extent? Or does consumption drive production? If so, how, and to what extent?

Table 3 Logs for Core Holes (elevations are in feet above sea level; ms = mudstone, ss = sandstone).

Location 1		Location 2		Location 3		Location 4		Location 5	
3930–4000	ms	3990–4000	ms	3890–4020	ss	3960–3990	ss	3980–4030	ss
3920–3930	coal	3970–3990	ss	3750–3890	ms	3900–3960	ms	3710–3980	ms
3870–3920	ms	3890–3970	ms	3660–3750	ss	3895–3900	ss	3675–3710	coal
3835–3870	coal	3885–3890	coal	3620–3660	ms	3720–3895	ms	3665–3675	ms
3660–3835	ms	3855–3885	ms	3570–3620	coal	3715–3720	ss	3610–3665	coal
3590–3660	coal	3850–3855	coal	3500–3570	ms	3690–3715	ms	3500–3610	ms
3500–3590	ms	3760–3850	ms		coal	3650–3690	coal		
	ms	3710–3760	coal		coal	3645–3650	ms		
		3700–3710	ms			3590–3645	coal		
		3630–3700	coal			3500–3590	ms		
		3600–3630	ms						

Part B: Coal Exploration

1. In 1998, the Western Region overtook the Appalachian Region as the largest coal-producing region in the United States with 488.8 million short tons produced, up by 8.3% over 1997. The low-sulfur Powder River Basin coal fields in Wyoming dominated growth in coal production in the Western Region. The diagram on the following page shows a cross section of five core holes drilled along an east–west line. *Table 3* provides drilling results for these wells. The results include the elevations and types of rock units in the core holes.

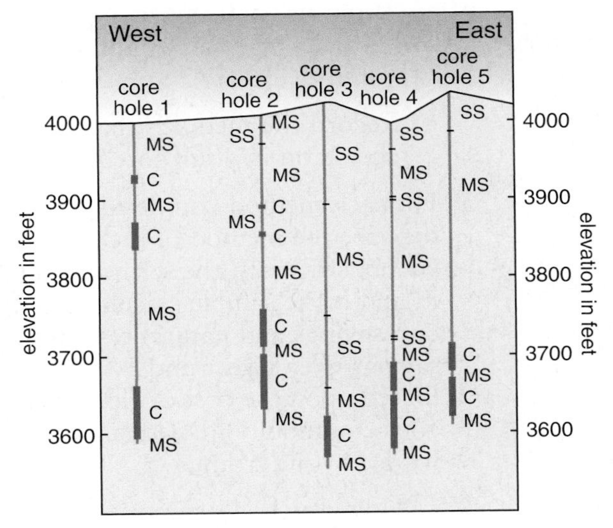

a) Complete the cross section in the diagram. Match up the rock units as you imagine them to be connected in the subsurface. (Hint: The mudstone at the bottom of all five wells is the same unit.)

b) Compare your results with those from other members of your group. How do they compare? How do you explain the differences?

c) Where would you drill your next well in order to determine whether or not your interpretation of the cross section is correct? Explain.

d) Is the information in the cross section sufficient to determine the coal seam with the greatest volume of coal? Why or why not?

e) What is the average thickness of coal in the lower coal seam?

f) Assume that the lower coal seam covers an area of 300,000 acres and that each acre-foot yields 1770 short tons of coal. (An acre-foot is the volume of a 1-foot thick layer that covers an area of 1 acre.) How many short tons of coal does the lower coal seam contain?

Part C: Conserving Coal Resources

1. A 100-W bulb burning for 10 h uses 1 kWh (kilowatt-hour) of electricity (the same as ten 100-W bulbs burning for one hour).

   a) Calculate the kilowatt-hours of electricity used in one year for a 100-W bulb running continuously.

   b) Assume that electricity costs $0.06/kWh. (Your teacher may give you a more accurate figure for your community.) Calculate the yearly cost of running the 100-W bulb continuously.

   c) Assume that one pound of good-quality coal can produce about 1 kWh of electricity. Calculate the amount of coal required per year to keep the 100-W bulb running continuously.

   d) How much coal is required each year to keep a 100-W bulb burning continuously in 20 million households?

   e) What do you think are the environmental consequences of burning this much coal?

   f) The average electricity bill for a family of four in the United States is about $50 per month (this is for homes where cooking and heating are by natural gas or oil). Estimate the yearly amount of electricity (in kilowatt-hours) that this is equivalent to. How many tons of coal are needed?

2. A variety of researchers are currently seeking methods to reduce our consumption of electricity, much of which is produced from coal. In your group decide on five ways to make

homes (or offices) in your community more energy efficient. Go to the *EarthComm* web site for useful links that will help you to explore ways to conserve energy resources. Suggestions are provided below.

- Water heaters — How do you ensure the best energy efficiency of an electric home water heater and reduce energy costs?
- Major home appliances — What are the most efficient ways to use major home appliances and how can you improve their efficiency?
- Home tightening — How can you slow the escape of heat energy from buildings, saving money and making them more comfortable?
- Insulation — How can insulation be used to reduce energy consumption?
- Home cooling — What strategies can you suggest for keeping a building cool in summer and improving the efficiency of air conditioning units?

a) Record the methods that you decide on in your notebook.

3. For each method you decided on, describe the method of improving energy efficiency, the science behind it, and make sample calculations of cost savings and natural resource savings over 1 year and 10 years (based upon the cost of electricity in your community). Divide up the work in your group.

   a) How much electricity would be saved?

   b) How much coal would be conserved over the course of a year? Over 10 years?

4. Conservation (reduction in use) is another way to reduce energy usage.

   a) In your group, decide on 10 ways to conserve energy.

   b) Calculate the savings.

Digging Deeper

COAL EXPLORATION AND MINING

In 2000, over 52% of electricity in the United States was produced from coal (see *Figure 1*). There was a fall in coal production around the middle of the last century. This was in part due to the decline of the steel industry, which uses coal in steel production. In addition, oil and gas have largely replaced coal for transportation and home heating. Since the early 1960s, however, there has been a steady increase in the production of coal. This is mainly because of the increasing demand for electricity. There has also been a development of major coalfields in the western United States and Canada.

Source: Energy Information Administration, Electric Power Monthly March 2001.

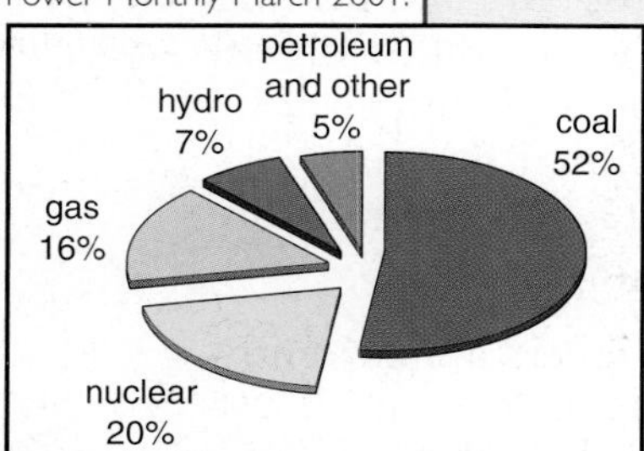

Figure 1 Percentage share of United States electric power industry net generation.

Of all fossil fuels, the largest reserves are contained in coal deposits (92%). Three major factors determine which coals are currently economical for mining. One factor is the cost of transportation to areas where the coal is utilized. Another factor is the environmental concern associated with the mining and use of coal. From a geologic perspective, the quality, thickness, volume, and depth of coal are important in determining whether or not a coal is mined.

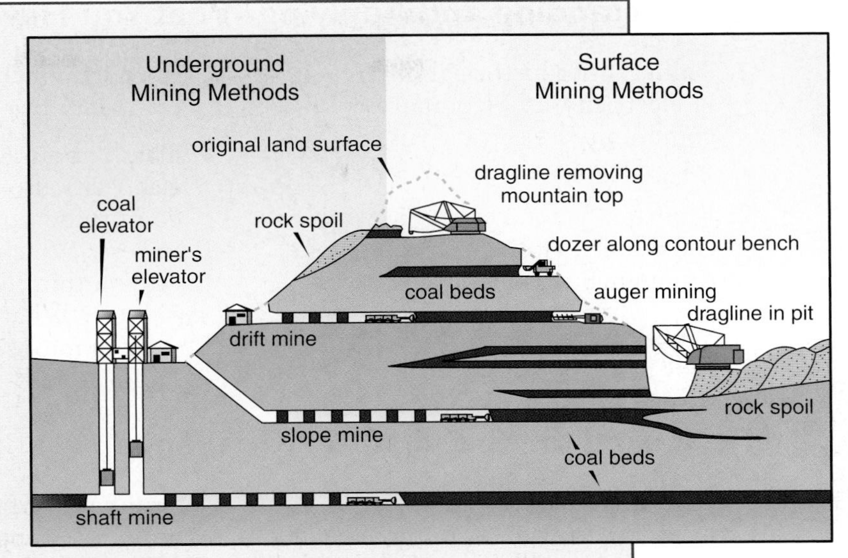

Figure 2 Different methods used to mine coal from the ground.

Figure 2 illustrates underground and surface mining methods (Note: You would not typically see all these mining methods used in one location.) Underground mining methods include drift, slope, and shaft mining.

Surface mining methods include area, contour, mountaintop removal, and auger mining. (An auger is a tool used for boring a hole.) *Figure 3* shows a surface coal mine.

Figure 3 A mine high wall showing layered sandstone and mudstone in a Gulf Coast coal mine. The haul truck provides a sense of scale.

Check Your Understanding

1. What are the major factors that determine whether a coal seam is economically and geologically suitable for mining?
2. What are the major factors that have determined the trends in coal consumption during the 20th century?

Understanding and Applying What You Have Learned

1. Think about historical differences in electricity use. Compare your community's electricity use 10 years ago and today.
2. Estimate the total energy use of your community. How much could your community reasonably conserve?
3. Compare electricity usage of a single college student vs. a family of four. Which group uses more electricity per person?
4. Think about regional differences in electricity use. Which part of the country do you think would be the lowest? Which part of the country do you think would be the highest? What are the reasons for your opinions?

Preparing for the Chapter Challenge

Assume that your community is expected to grow 10% in the next 10 years. If electricity usage remains the same, this means that the electrical capacity required to meet this increase must also increase. This may be undesirable and involve the cost of building a new power plant or an increased environmental hazard.

Determine specific steps that could be taken by the community to conserve electricity. How would the reduced demand for electricity offset the growth in the community? What incentives would be given to encourage the community to implement these proposed changes?

Inquiring Further

1. **Investigating a coal mine**

   Investigate the coal mine closest to your community.

   a) What mining method do they use?
   b) What is their annual production?
   c) How has the mine influenced the economy of the community?
   d) Has the mine impacted the environment?
   e) What steps has the mining company taken to reduce the environmental impact of the mine?

   Have your teacher arrange a field trip to this mine and discuss the trip with your class.

2. **Electricity usage in other countries**

   Find out information about electricity usage in other countries. What can you learn from other countries about energy conservation? (For instance, Japan has a high gross national product (GNP) yet very low energy use. How do they manage this?)

3. **Personal electricity usage**

   Learn how to read your electricity meter, and with your family conduct some conservation experiments for a few days (or weeks) and report the results of these experiments to the class.

Activity 5 Environmental Impacts and Energy Consumption

Goals

In this activity you will:

- Examine one of the environmental impacts of using coal.
- Understand how the use of fossil fuels relates to one of Earth's major geochemical cycles—the carbon cycle.
- Understand the meaning of pH.
- Understand how weather systems and the nature of bedrock geology and soil affect the impact of acid rain.
- Determine ways that energy resources affect your community.
- Analyze the positive and negative effects of energy use on communities.

Think about It

Sulfur dioxide and nitrogen oxides are the primary causes of acid rain. In the United States, about two-thirds of all sulfur dioxide and one-quarter of all nitrogen oxides comes from electric power generation that relies on burning fossil fuels like coal.

- How does acid rain affect the environment?
- What can be done to reduce the amount of sulfur dioxide and nitrogen oxides released into the atmosphere?

What do you think? Record your ideas in your *EarthComm* notebook. Be prepared to discuss your responses with your small group and the class.

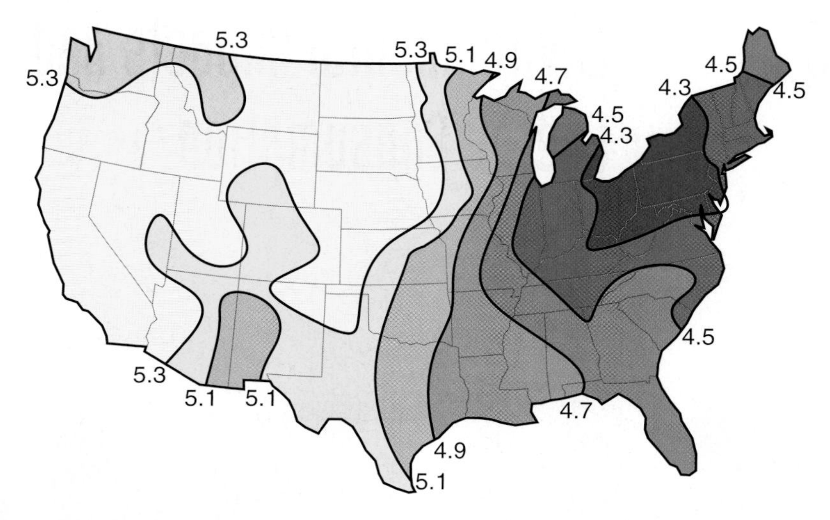

Investigate

1. The map shows the acidity of rainfall across the United States. Normal rainfall is slightly acidic (pH 5.6, where 7.0 is neutral). Carbon dioxide in the atmosphere reacts with water to form a weak acid called carbonic acid. The lower the pH, the more acidic is the rainwater.
   a) What part of the United States has the most acidic rainwater?
   b) How does the pattern of rainwater pH correlate with the locations of coal-producing regions and coal-fired power plants? (See Activity 3.)
   c) What parts of the United States have the least acidic rainwater? Why?
   d) What can you infer about the direction in which wind and weather move across the United States?

2. Your teacher will pour an "acid rain" solution (pH of about 4.5) through samples of crushed limestone and granite. Containers below the crushed rock contain pH indicator solution. A change in the color of the pH indicator reveals a change in the pH.

3. Note your answers to the following questions in your *EarthComm* notebook:
   a) Describe any color changes in the collection containers.
   b) What was the pH of the liquids after passing through the two types of crushed rock?
   c) Which type of rock—limestone or granite—neutralized more of the acid?
   d) Where would you expect a greater environmental impact from acid rain—in a region with granite bedrock or one with limestone bedrock? Explain.

e) What does this investigation suggest about how areas may differ in their sensitivity to acid rainwater?

4. Your teacher will repeat the investigation using crushed samples of rock from your community.

   a) Record all your observations and explain your results.

Reflecting on the Activity and the Challenge

In this activity you examined one of the major environmental impacts of using coal to produce energy. You explored how the extent to which acid rain affects a region depends upon several factors, like the nature of the rock and soil, weather patterns, and the location of coal-fired power plants. This will help you to understand the benefits of understanding how the growing consumption of energy resources impacts Earth systems.

Digging Deeper

ENERGY RESOURCES AND THEIR IMPACT ON THE ENVIRONMENT

Fossil Fuels and the Carbon Cycle

When a fossil fuel is burned, its carbon combines with oxygen to yield carbon dioxide gas (CO_2). Fossil fuels supply 84% of the primary energy consumed in the United States. They are responsible for 98% of United States emissions of carbon dioxide. The amount of carbon dioxide produced depends on the carbon content of the fuel. For each unit of energy produced, natural gas emits about half, and petroleum fuels about three-quarters, of the carbon dioxide produced by coal. Important questions that scientists are trying to answer are what is happening to all of this carbon dioxide, and how does it affect the Earth system?

According to the **First Law of Thermodynamics**, the amount of energy always remains constant. The amount of chemical energy consumed when a fossil fuel is burned equals the amount of heat energy released. Scientists think of matter in a similar way. Matter cannot be created or destroyed. The **Law of Conservation of Matter** helps you to understand what happens when one type of matter is changed into another type. For example, fossil fuels consist mainly of hydrocarbons, which are made up of hydrogen

Geo Words

First Law of Thermodynamics: the law that energy can be converted from one form to another but be neither created nor destroyed.

Law of Conservation of Matter: that law that in chemical reactions, the quantity of matter does not change.

and carbon. When a fossil fuel is burned (to heat homes or power cars), the carbon-containing compounds are changed chemically into carbon dioxide. The amount of carbon consumed equals the amount of carbon produced. The carbon never "goes away", but moves from one reservoir within the Earth system to another.

Geo Words

carbon cycle: the global cycle of movement of carbon, in all of its forms, from one reservoir to another.

anthropogenic: generated or produced by human activities.

The idea that many processes work together in a global movement of carbon from one reservoir to another is known as the **carbon cycle** (see *Figure 1*). You have already learned about several important reservoirs of carbon. The atmospheric reservoir of carbon is carbon dioxide gas. Biomass contains carbon (mostly carbon in plants and soil). In the geosphere, carbon is found in solids (coal, limestone), liquids (petroleum hydrocarbons), and gas (methane). The ocean is the largest reservoir of carbon (exclusive of carbon-bearing bedrock), found in the form of bicarbonate ions.

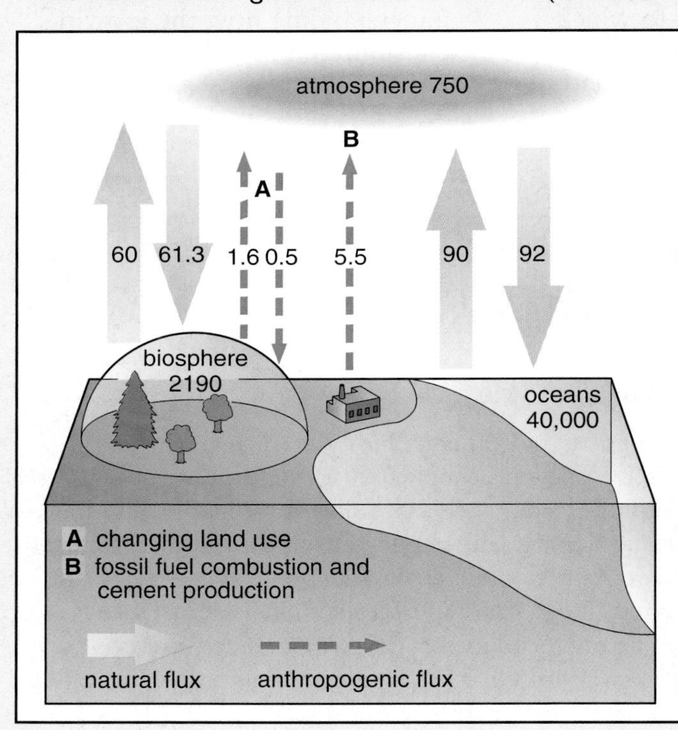

Figure 1 Global carbon cycle (billion metric tons).

Photosynthesis and respiration dominate the movement (flux) of carbon dioxide between the atmosphere, land, and oceans (the wide blue arrows show this flux in *Figure 1*). The smaller red arrows indicate the flux of carbon related to human (**anthropogenic**) activities. Natural processes like photosynthesis can remove some of the net 6.6 billion metric tons of anthropogenic carbon dioxide emissions produced each year. This means that an estimated 3.3 billion metric tons of this carbon is added to the atmosphere annually in the form of carbon dioxide.

By measuring concentrations of carbon dioxide in the atmosphere over time, scientists have learned that levels of carbon dioxide in the atmosphere have increased about 25% in the last 150 years (see *Figure* 2). Scientists believe that human activity has caused this growth. This coincides with the beginning of the industrial age. The increase is due largely to the burning of fossil fuels and to deforestation (*Figure 1*). How does this affect the Earth system? Carbon dioxide is one of several **greenhouse gases** (gases that slow the escape of heat energy from the Earth to space). Some scientists believe that the rapid addition of carbon dioxide to the atmosphere is changing the energy budget of the Earth, causing global climate to warm.

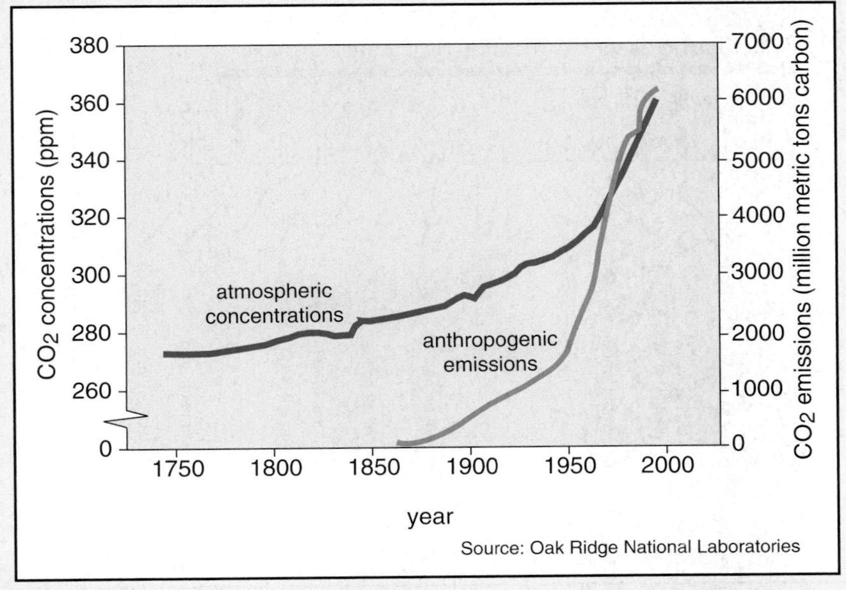

Figure 2 Atmospheric CO_2 concentrations and anthropogenic CO_2 emissions over time.

Geo Words

greenhouse gases: gases in the Earth's atmosphere that absorb certain wavelengths of the long-wavelength radiation emitted to outer space by the Earth's surface.

acid: a compound or solution with a concentration of hydrogen ions greater than the neutral value (corresponding to a pH value of less than 7).

base: a compound or solution with a concentration of hydrogen ions less than the neutral value (corresponding to a pH value of greater than 7).

neutral: having a concentration of hydrogen ions that corresponds to a value of pH of 7.

Coal and Acid Rain

Acidity is a measure of the concentration of hydrogen ions (H^+) in an aqueous (water) solution. A solution with a high concentration of hydrogen ions is acidic. A solution with a low concentration of hydrogen ions is basic. The concentration of hydrogen ions is important, because hydrogen ions are very reactive with other substances. Mixing **acids** and **bases** can cancel out their effects, much as mixing hot and cold water evens out the water temperature. That's because the concentration of hydrogen ions in the mixture lies between that of the two original solutions. A substance that is neither acidic nor basic is said to be **neutral**.

The pH scale shown in *Figure 3* measures how acidic or basic a substance is. It ranges from 0 to 14. A pH of 7 is neutral, a pH less than 7 is acidic, and a pH greater than 7 is basic. Each whole pH value below 7 is 10 times more acidic than the next higher value, meaning that the concentration of hydrogen ions is 10 times as great. For example, a pH of 4 is 10 times more acidic than a pH of 5 and 100 times (10 times 10) more acidic than a pH of 6. The same holds true for pH values above 7, each of which is 10 times more alkaline (another way to say basic) than the next lower whole value. For example, a pH of 10 is 10 times more alkaline than a pH of 9.

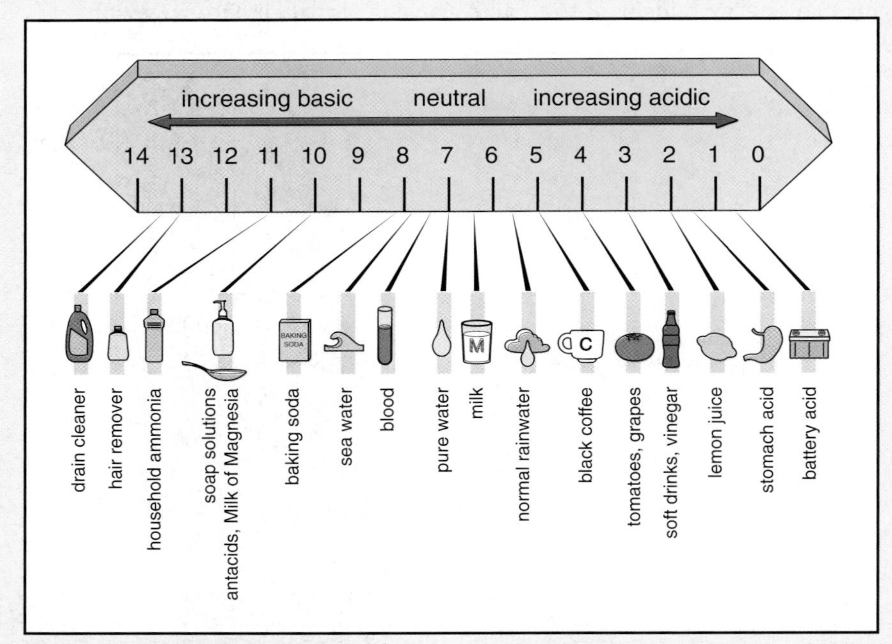

Figure 3 The pH scale.

Pure water is neutral, with a pH of 7.0. Natural rainwater, however, is mildly acidic, with a pH of about 5.7. The reason is that carbon dioxide in the atmosphere dissolves in rainwater, and some of the dissolved carbon dioxide reacts with the water to form a weak acid, called carbonic acid, H_2CO_3. As the carbon dioxide content of the atmosphere has gradually increased in recent decades, because of the burning of fossil fuels, natural rainwater has become slightly more acidic, even aside from the serious problem of acid rain caused by sulfur dioxide.

By measuring concentrations of carbon dioxide in the atmosphere over time, scientists have learned that levels of carbon dioxide in the atmosphere have increased about 25% in the last 150 years (see *Figure 2*). Scientists believe that human activity has caused this growth. This coincides with the beginning of the industrial age. The increase is due largely to the burning of fossil fuels and to deforestation (*Figure 1*). How does this affect the Earth system? Carbon dioxide is one of several **greenhouse gases** (gases that slow the escape of heat energy from the Earth to space). Some scientists believe that the rapid addition of carbon dioxide to the atmosphere is changing the energy budget of the Earth, causing global climate to warm.

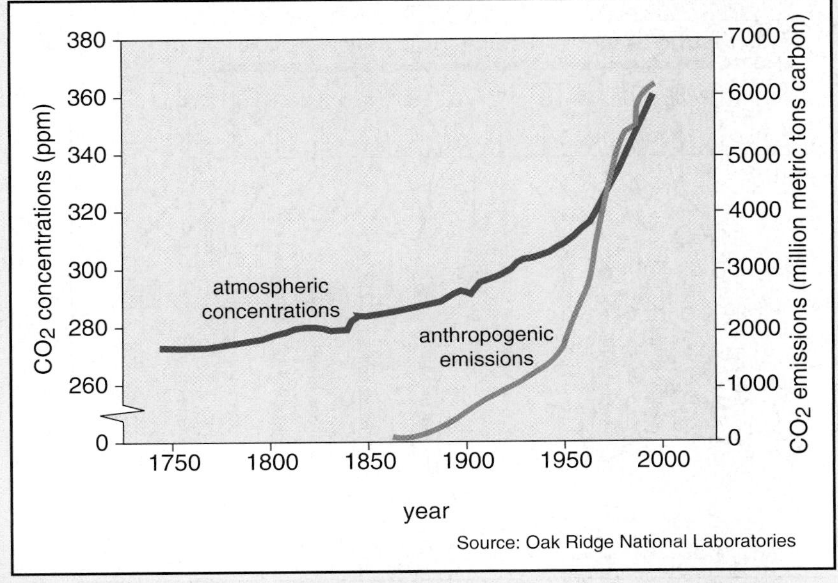

Figure 2 Atmospheric CO_2 concentrations and anthropogenic CO_2 emissions over time.

Geo Words

greenhouse gases: gases in the Earth's atmosphere that absorb certain wavelengths of the long-wavelength radiation emitted to outer space by the Earth's surface.

acid: a compound or solution with a concentration of hydrogen ions greater than the neutral value (corresponding to a pH value of less than 7).

base: a compound or solution with a concentration of hydrogen ions less than the neutral value (corresponding to a pH value of greater than 7).

neutral: having a concentration of hydrogen ions that corresponds to a value of pH of 7.

Coal and Acid Rain

Acidity is a measure of the concentration of hydrogen ions (H^+) in an aqueous (water) solution. A solution with a high concentration of hydrogen ions is acidic. A solution with a low concentration of hydrogen ions is basic. The concentration of hydrogen ions is important, because hydrogen ions are very reactive with other substances. Mixing **acids** and **bases** can cancel out their effects, much as mixing hot and cold water evens out the water temperature. That's because the concentration of hydrogen ions in the mixture lies between that of the two original solutions. A substance that is neither acidic nor basic is said to be **neutral**.

The pH scale shown in *Figure 3* measures how acidic or basic a substance is. It ranges from 0 to 14. A pH of 7 is neutral, a pH less than 7 is acidic, and a pH greater than 7 is basic. Each whole pH value below 7 is 10 times more acidic than the next higher value, meaning that the concentration of hydrogen ions is 10 times as great. For example, a pH of 4 is 10 times more acidic than a pH of 5 and 100 times (10 times 10) more acidic than a pH of 6. The same holds true for pH values above 7, each of which is 10 times more alkaline (another way to say basic) than the next lower whole value. For example, a pH of 10 is 10 times more alkaline than a pH of 9.

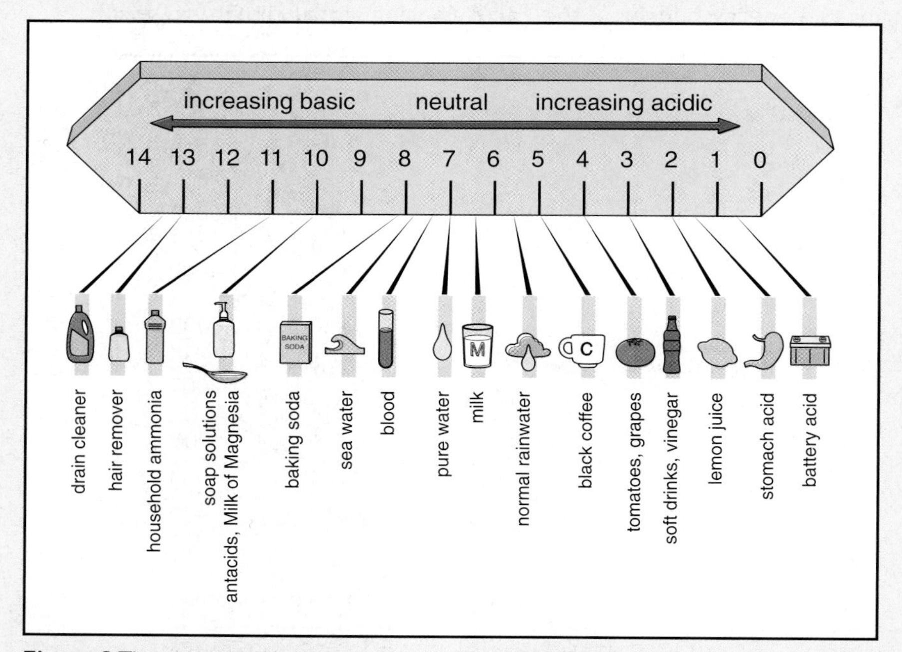

Figure 3 The pH scale.

Pure water is neutral, with a pH of 7.0. Natural rainwater, however, is mildly acidic, with a pH of about 5.7. The reason is that carbon dioxide in the atmosphere dissolves in rainwater, and some of the dissolved carbon dioxide reacts with the water to form a weak acid, called carbonic acid, H_2CO_3. As the carbon dioxide content of the atmosphere has gradually increased in recent decades, because of the burning of fossil fuels, natural rainwater has become slightly more acidic, even aside from the serious problem of acid rain caused by sulfur dioxide.

You learned in a preceding activity that coal contains as much as 5% sulfur. When the coal is burned, the sulfur is emitted as sulfur dioxide gas, SO_2. The sulfur dioxide then reacts with water in the atmosphere to form sulfuric acid, a strong acid. The last reaction in this process is written as follows:

$$SO_3 + H_2O \rightarrow H_2SO_4$$

Some of the sulfuric acid then dissociates into hydrogen ions and sulfate ions in solution in the water. This reaction can be broken down into two parts, as follows:

$$H_2SO_4 \rightarrow H^+ + HSO_3^- \rightarrow 2H^+ + SO_4^{2-}$$

The H^+ ions reach the Earth's surface dissolved in raindrops, either over land or over the ocean. Burning of fuels also produces nitrogen oxide gases, which react with water to form nitric acid, another strong acid. Some of the nitric acid breaks down to release hydrogen ions, by a reaction similar to that of sulfuric acid. Burning of coal is not the only source of sulfuric acid and nitric acid: petroleum-fueled power plants, smelters, mills, refineries, and motor vehicles produce these acids as well.

Acid rain is especially damaging to lakes. If the pH of lake water becomes too acidic from acid rain, it kills fish, insects, aquatic plants, and plankton. Environmental scientists describe lakes that have suffered heavily from acid rain as "dead," because the entire food web of the lake has been disrupted to the point where little is left alive in the lake.

Acid rain is a serious problem locally in many areas of the United States, but the biggest problem is in the Northeast. The reason is that there are a large number of coal-fueled electric power plants in the Midwest, especially in the heavily populated areas of Ohio, Michigan, and Illinois, and these plants burn mostly the relatively high-sulfur coal that is mined in the Interior coal province (see Activity 3). Winds blow mostly from west to east in the mid-latitudes of North America, and that tends to bring the acids to the Northeast, where large amounts fall with rain.

Figure 4 This marble gravestone from 1921 is being slowly dissolved by acid rain.

Geo Words

scrubbing: removal of sulfur dioxide, ash, and other harmful byproducts of the burning of fossil fuels as the combustion products pass upward through a stack or flue.

Another reason for the impact of acid rain in the northeastern United States has to do with the types of rock and soil in that region. In the activity, you learned that limestone has a greater ability to neutralize acid than granite. If the soil upon which acid rain falls is rich in limestone or if the underlying bedrock is either composed of limestone or marble, then the acid rain may be neutralized. This is because limestone and marble are more alkaline (basic) and produce a higher pH when dissolved in water. The higher pH of these materials dissolved in water offsets or buffers the acidity of the rainwater, producing a more neutral pH.

If the soil is not rich in limestone or if the bedrock is not composed of limestone or marble, then no neutralizing effect takes place, and the acid rainwater accumulates in the bodies of water in the area. This applies to much of the northeastern United States where the bedrock is typically composed of granite and similar rocks. Granite has no neutralizing effect on acid rainwater. Therefore, over time more and more acid precipitation accumulates in lakes and ponds.

Coal supplies about half of all electricity generated in the United States. In recent years, as the problem of acid rain has grown, various technologies have been developed to reduce the emission of sulfur dioxide from coal-fueled electric power plants. The general term for these processes is flue-gas desulfurization, also called **scrubbing**.

Several kinds of scrubbers are in use. The most common is the wet scrubber. In wet scrubbing, the flue gas from the power plant is sprayed with a calcium carbonate solution in the form of a slurry. The SO_2 is oxidized to form calcium sulfate. Scrubbers of this kind can remove up to 95% of the SO_2 that is emitted from the power plant. Scrubbers can have a useful byproduct. Certain kinds of scrubbers are now in operation that produce pure calcium sulfate (gypsum) as a byproduct. This gypsum can be used industrially, to make plaster and wallboard, so that gypsum does not have to be mined and processed.

Advantages and Disadvantages of Energy Resources

Energy resources affect the community in positive and negative ways. Energy resources allow you to maintain a very high quality of life. Energy resources also indirectly provide employment for a large sector of the community. Use of energy resources, however, can have obvious and not-so-obvious negative impacts on the community. One of the most obvious negative impacts is a deterioration of air quality due to the burning of fossil fuels. Not-so-obvious impacts include acid rain, possible global warming, ground-water contamination, and the financial burden on households. It is important to realize that you, both personally and as a member of the community, can control how you use energy resources to potentially reduce the negative impacts.

Check Your Understanding

1. Describe the carbon cycle in relation to the consumption of fossil fuels.
2. Explain what is meant by the phrase "The Earth is a closed system with respect to matter." How does this statement relate to the Law of Conservation of Matter?
3. In your own words, explain the concept of pH.
4. Why is acid rain a greater problem in the Northeast than elsewhere in the United States?
5. How is sulfur dioxide removed from the gases emitted from coal-fueled electric power plants?

Understanding and Applying What You Have Learned

1. What are some of the positive impacts associated with the use of energy resources?
2. What are some of the negative impacts associated with the use of energy resources?
3. The unit used to measure the amount of SO_2 gas emitted by a coal-fired power plant is pounds of SO_2 per million Btu (lb. SO_2/MBtu).
   a) Assume a 1000-megawatt power plant emits 2.5 lb. SO_2/MBtu. Calculate the amount of SO_2 gas emitted in one year, assuming that the plant operates on average at 75% capacity.
   b) Assume that the coal used has an energy output of 10,000 Btu/pound and use a heat-to-electricity efficiency of 30%. How many tons of coal does the plant use each year?
4. At the present time the United States has about 470 billion tons of coal reserves and the average annual production is about 1.05 billion tons.
   a) Calculate how many years the reserves will last at the present production rate.
   b) What would cause the reserves to run out sooner than this number?
   c) What would cause the reserves to run out later than this number?
5. Assume that you live in a small community in the midwest. The electricity for the community is provided by a 1000-megawatt power plant, and the electricity is produced entirely by the burning of coal. The power plant is located in the community and provides employment to about 75 employees. The coal comes from an underground coal mine about 50 miles away. The coal mine produces about one million tons of coal a year and employs about 150 people. Many of the people that work in the mine also live in your community. The sulfur content of the coal is about 4.5% and the ash content is about 11%. The amendments to the Clean Air Act require that by the year 2010 the power plant must reduce SO_2 emissions by 60%. Also, it is possible that by the year 2010 a "carbon tax" will be levied on electricity produced from coal to address global warming issues. There are a number of ways that this can be done, including retrofitting the power plant to burn natural gas (natural gas emits half the CO_2 gas and almost no SO_2 gas), install a wet scrubber, conserve electricity, etc.

   Answer the following questions:

   a) How will the implementation of environmental laws affect your community?
   b) What alternatives does your community have to address changes that could be made to minimize the impact?
   c) How will the community gain from the environmental laws?

Preparing for the Chapter Challenge

Your challenge is to prepare a plan that will help your community to meet its growing energy needs. Part of your plan must address how the use of energy resources impacts the environment. Based upon what you have learned in this activity, write an essay about ways to reduce the impact of acid rain on communities.

Inquiring Further

1. **The Clean Air Act and environmental regulation**

   Look up information on the amendments to the Clean Air Act having to do with SO_2 emissions and report your findings to the class.

2. **Acid rain and other Earth systems**

   How has acid rain affected your community? How do the processes that lead to acid rain also work to lower the pH of ground water? The latter is a challenge that community officials must face in areas that are mined for coal and some kinds of minerals. Extend your investigation into acid rain by examining acid mine drainage. The *EarthComm* web site will help you to extend your investigations.

3. **Coal mine**

   Investigate the coal mine closest to your community. What are the percent sulfur, weight of SO_2 per million Btu, and percent ash for the coal? What reclamation steps are taken at the mine?

4. **Power plant**

   Investigate the power plant closest to your community. If possible, have your teacher arrange a field trip to this plant.

   a) Which fuels are used at the power plant?
   b) What are the sources of these fuels?
   c) What is the power plant's capacity, in megawatts?
   d) How many people are employed at the power plant?
   e) What is done with the ash from the power plant (i.e., the solid materials that are caught as they pass up through the flue of the plant)?

Activity 6 Petroleum and Your Community

Goals

In this activity you will:

- Recognize the overwhelming dependence of today's society on petroleum.
- Graph changes in domestic oil production and foreign imports to predict future needs and trends.
- Understand why oil and gas production have changed in the United States over time.
- Describe the distribution of oil and gas fields across the United States.
- Understand the origin of petroleum and natural gas.
- Investigate the consumption of oil and natural gas in your community.

Think about It

Oil accounts for nearly all transportation fuel. It is also the raw material for numerous products that you use.

- What percentage of oil used every day in the United States is produced in the United States?
- Where are oil and natural gas found in the United States?

What do you think? Record your ideas in your *EarthComm* notebook. Be prepared to discuss your responses with your small group and the class.

Investigate

Part A: Trends in Oil Production

Table 1 United States Petroleum Production, Imports, and Total Consumption from 1954 to 1999

Year	Total U.S. Wells (thousands)	Total U.S. Production (thousand barrels/day)	Total Foreign Imports (thousand barrels/day)	Total U.S. Total Consumption (thousand barrels/day)
1954	511	7030	1052	8082
1957	569	7980	1574	9559
1960	591	7960	1815	9775
1963	589	8640	2123	10,763
1966	583	9580	2573	12,153
1969	542	10,830	3166	13,996
1972	508	11,180	4741	15,921
1975	500	10,010	6056	16,066
1978	517	10,270	8363	18,633
1981	557	10,180	5996	16,176
1984	621	10,510	5437	15,947
1987	620	9940	6678	16,618
1990	602	8910	8018	16,928
1993	584	8580	8620	17,200
1996	574	8290	9478	17,768
1999	554	7760	10,551	18,311

Source: U.S. Energy Information Agency

1. *Table 1* shows statistics for petroleum every three years from 1954 to 1999. Copy the data table into your notebook. If you have Internet access, you can download the complete set of data (annual data since 1954) at the *EarthComm* web site as a spreadsheet file.
   a) Construct a graph of United States (domestic) petroleum production, foreign petroleum imports, and petroleum consumption. Leave room at the right side of your graph to extrapolate the data for 20 years.

2. What are the major trends in petroleum needs during the last 45 years? Use your graph and the data table to answer the following questions:
   a) Describe how domestic production has changed during the 45-year period. About when did it peak?
   b) Describe how petroleum imports have changed during that period.
   c) Describe how total petroleum consumption has changed over the period.

d) In what year did the United States begin to import more petroleum than it produced?

e) What percentage of the total petroleum consumption was met by domestic production in 1954? In 1999? How does this compare to your answer to the first **Think about It** question at the start of this activity?

f) In 1954, the estimated population of the United States was 163 million. In 1999, the estimated population was 273 million. How many barrels of petroleum per person per day were needed in the United States in 1954? In 1999?

g) A barrel of oil is equal to 42 gallons. Convert your answers in Step 2(f) to gallons per person per year needed in the United States in 1954 versus 1999.

h) Has the consumption of petroleum changed at the same rate as the growth in population? What might account for the change?

i) Use the data in the table to calculate the average well yield for oil production in the United States.

3. What do you predict about future needs for and sources of petroleum in the United States?

a) Extrapolate the three curves to the year 2020. Explain your reasoning behind your extrapolations for each of the three curves.

b) On the basis of your extrapolations, how much petroleum will be produced domestically, imported into the United States, and consumed in the year 2010? In 2020?

c) Identify several factors that might make the actual curves for production, import, and consumption different from what you have predicted. Record your ideas in your *EarthComm* notebook.

4. Share your results from Step 3.

a) Set up a data table to record the predictions from each person in the class for the years 2010 and 2020. An example is provided below.

b) Calculate the minimum, maximum, and average for the class.

c) How do you explain the differences in predictions?

Sample Data Table. Results of Class Predictions of United States Oil Production, Imports, and Total Production in 2010 and 2020.

	United States Production		Imports		Total	
	2010	2020	2010	2020	2010	2020
Student 1						
Student 2						
Student...						
Class Average						

Part B: United States Oil and Gas Resources

In Part A you learned that petroleum production has declined in the United States. The questions below will help you to investigate where oil and gas are found in the United States, and where these natural resources are refined into products that communities depend upon.

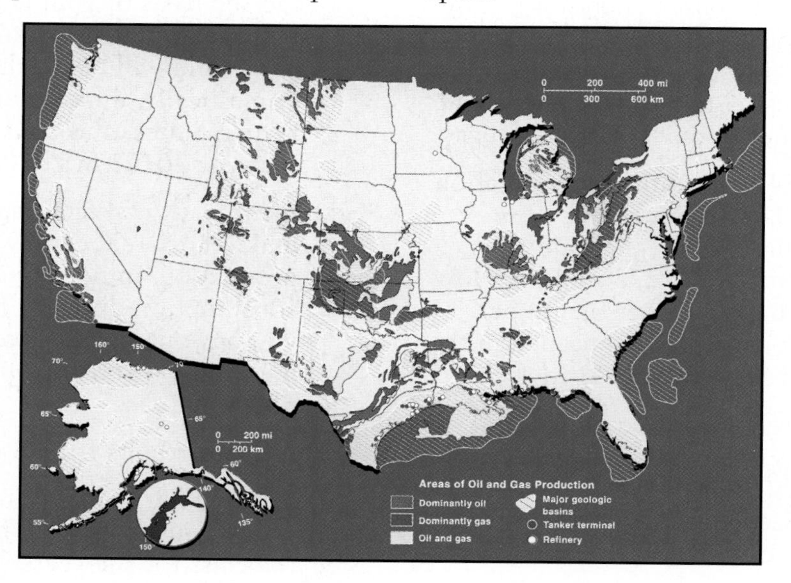

1. Use the map of oil and gas production to explore general trends and patterns in petroleum and gas fields.

   a) Are oil and gas deposits distributed evenly across the country?

   b) Which states have no oil or gas production?

   c) What patterns or trends do you see in oil-producing and gas-producing regions?

   d) Are oil and gas always found together? Give at least two examples.

   e) The map also shows the locations of major sedimentary basins. Sedimentary basins are depressions of the Earth's crust where thick sediments have accumulated. What is the relationship between oil and gas production and sedimentary basins?

2. Refer to the map to investigate where oil and gas are found, refined, and distributed in relation to your community.

   a) How far is your community from the nearest oil field?

   b) How far is your community from the nearest gas field?

   c) How far from your community is the nearest refinery located?

   d) Do you think the factors above might affect the price of petroleum products (gasoline, heating oil, or propane) in your community? Explain your answer.

Part C: Oil and Gas Resource Use in Your Community

1. Investigate trends in petroleum and natural gas production, consumption, and distribution in your state. Go to the *EarthComm* web site to obtain links to the data sets for your state available at the Energy Information Agency. This will allow you to examine several sources of data that will help you with the **Chapter Challenge.**
2. Develop your own questions as you examine the data and review the **Chapter Challenge.**
   a) How have oil and gas consumption in your state changed during the last 40 years? How can you use the data to predict future needs?
   b) How much petroleum is consumed per capita in your state? How much natural gas is consumed? How can you use this information to predict how much oil and gas your community will need 20 years from now if the population grows by 20%?
   c) How do oil and gas reach your community? If switching to another type of fuel is part of your plan for your community's energy future, how will these resources get to your community if they are not local?

Reflecting on the Activity and the Challenge

In this activity you graphed the recent history of oil consumption in the United States. You used these trends to predict what the future needs might be. You examined a map of oil and gas fields in the United States and learned that these resources are not distributed evenly throughout the country. In Part C, you investigated trends in oil and gas resource use in your community. Think about how your work in this activity will help you to solve the **Chapter Challenge** on the energy future of your community.

Digging Deeper

PETROLEUM AND NATURAL GAS AS ENERGY RESOURCES

The Nature and Origin of Petroleum and Natural Gas

Petroleum, also called **crude oil**, is a liquid consisting mainly of organic compounds that range from fairly simple to highly complex. **Natural gas** consists mainly of a single organic compound, methane (with the chemical formula CH_4). Oil and natural gas reside in the pore spaces of some sedimentary rocks.

The raw material for the generation of oil and gas is organic matter deposited along with the sediments. The organic matter consists of the remains of tiny plants and animals that live in oceans and lakes and settle to the muddy bottom when they die. Much of the organic matter is oxidized before it is permanently buried. But, enormous quantities are preserved and buried along with the sediments. As later sediments cover the sediments more and more deeply, the temperature and pressure increase. These higher temperatures and pressures cause some of the organic matter to be transformed into oil and gas. Petroleum geologists use the term "maturation" for this process of change in the organic matter. They call the range in burial depth that is appropriate for generation of oil and gas the "oil window" or "gas window" and the regions rich in organic matter subjected to these depths "the kitchen." Depending upon the details of the maturation process, sometimes mostly oil is formed, and sometimes mostly gas.

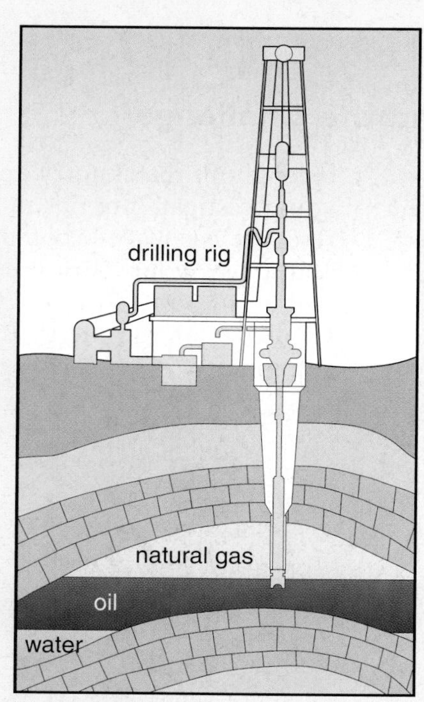

Figure 1 Schematic of a petroleum trap.

The mudstone and shale that form the **source rocks** for oil and gas are impermeable. Fluids can flow through them very slowly. The oil and gas rise very slowly and percolate through the source rocks, because they are less dense than water, which is the main filler of the pore spaces. Once they reach fractures and much more permeable rocks like some kinds of sandstone and limestone, however, they can migrate much more rapidly. Much of the oil and gas rise all the

Geo Words

petroleum: an oily, flammable liquid, consisting of a variety of organic compounds, that is produced in sediments and sedimentary rocks during burial of organic matter; also called crude.

crude oil: (see petroleum).

natural gas: a gas, consisting mainly of methane, that is produced in sediments and sedimentary rocks during burial of organic matter.

source rocks: sedimentary rocks, containing significant concentrations of organic matter, in which petroleum and natural gas are generated during burial of the deposits.

seal: an impermeable layer or mass of sedimentary rock that forms the convex-upward top or roof of a petroleum reservoir.

reservoir: a large body of porous and permeable sedimentary rock that contains economically valuable petroleum and/or natural gas.

way to the Earth's surface, to form oil and gas seeps, and escape to the atmosphere or produce tar mats. If the rocks are capped deep in the Earth by an impermeable layer called a **seal**, the oil and gas are detained in their upward travel. A large volume of porous rock containing oil and gas with a seal above is called a **reservoir**. The oil and gas can be brought to the surface by drilling deep wells into the reservoir, as shown in *Figure 1*. Petroleum reservoirs around the world, in certain favorable geological settings, range in age from more than a billion years to geologically very recent, just a few million years old. Oil and gas are nonrenewable resources, however, because the generation of petroleum operates on time scales far longer than human lifetimes.

Oil and natural gas are useful fuels for several reasons. They have a very high heat content per unit weight. They cost less than coal to transport and are fairly easy to transport. Oil and gas can also be refined easily to form many kinds of useful materials. Petroleum is used not only for fuel. It is used as the raw material (called **feedstock**) for making plastics and many other synthetic compounds like paints, medicines, insecticides, and fertilizer.

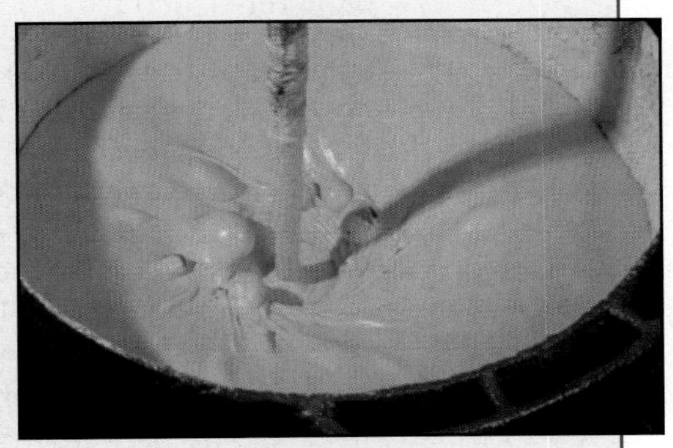

Figure 2 Petroleum is found in many common products, including paint.

Geo Words

feedstocks: raw materials (for example, petroleum) that are supplied to a machine or processing plant that produces manufactured material (for example, plastics).

The History of Oil Production in the United States

The United States has had several basic shifts in its energy history (see *Figure 3*). The overall pattern of energy consumption in the United States has been that of exponential growth. The burning of wood to heat homes and provide energy for industry was the primary energy source in about 1885. Coal surpassed wood in the late 1800s, only to be surpassed by petroleum and natural gas in the 1950s. Hydroelectric power appeared in 1890

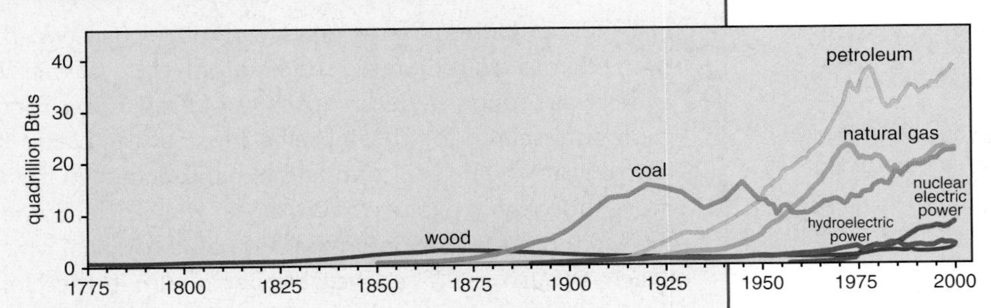

Figure 3 Energy consumption, by source, in the United States since 1775.

Geo Words

sedimentary basin: an area of the Earth's crust where sediments accumulate to great thicknesses.

and nuclear electric power appeared in about 1957. Solar photovoltaic, advanced solar thermal, and geothermal technologies also represent recent developments in energy sources. Of all the recently developed energy sources, petroleum and natural gas have had the greatest impact.

Figure 4 This early 1900's drilling rig was used to prospect for oil. Drilling rigs have advanced greatly since then.

The first successful oil-drilling venture was Colonel Titus Drake's drilling rig in 1859 in northwestern Pennsylvania. Drake reached oil at a depth of 70 ft. Petroleum got major boosts with the discovery of Texas's vast Spindletop Oil Field in 1901. With the advent of mass-produced automobiles, consumption of oil began to grow more rapidly in the 1920s. In the years after World War II, coal lost its position as the premier fuel in the United States. Trucks that ran on petroleum took business once dominated by railroads, which began to switch to diesel locomotives themselves. Labor troubles and safety standards drove up the cost of coal production. Natural gas, which began as a source for lighting (gas lanterns and streetlights), replaced coal in many household ranges and furnaces. By 1947 consumption of petroleum and natural gas exceeded that of coal and then quadrupled in a single generation. No source of energy has ever become so dominant so quickly.

United States Petroleum Production

In 1920, the United States produced more than two-thirds of the world's oil. By 1998, the United States supplied only 12% of the world's oil needs. Oil and gas are found in **sedimentary basins** in the Earth's crust. To find oil, geologists search for these basins. Most of the basins in the United States have already been explored and exploited. In fact, the United States is the most thoroughly explored country in the world. Some people estimate that the United States has less than 35% of its original oil remaining. There is not nearly enough easily recoverable petroleum in the United States to meet the demand, although domestic production will continue for years to come.

Oil and gas resources are not evenly distributed throughout the country. The top ten oil fields account for 33% of all remaining oil in the United States (including a field in California that was discovered in 1899!). Texas has been the largest producing state since the late 1920s, when it surpassed California. In the late 1980s, Alaska rivaled Texas.

Check Your Understanding

1. What is the origin of oil and natural gas?
2. Why do some people refer to the present times as the "petroleum age"?
3. What are the advantages of petroleum and natural gas as fuels?
4. Why are oil and gas considered nonrenewable resources?

Understanding and Applying What You Have Learned

1. What is the world's future supply of petroleum relative to world demand?
2. Is the future decrease in petroleum production likely to be abrupt or gradual? Why?
3. Look at *Table 1* from the investigation.
   a) Calculate the average yield per well in barrels of petroleum per day for each year in the table.
   b) Graph the total number of wells producing and the average well yield over time.
   c) Describe any relationships that you see. How is well yield changing over time?
4. Think back to your projections of future oil production, foreign imports, and consumption.
   a) How might tax incentives for switching to renewable energy sources affect oil and gas consumption rates in the future?
   b) How might a change in the cost of fuels (gasoline, jet fuel, etc.) affect consumption rates in the future?
   c) How might new discoveries of oil affect domestic production?
   d) Identify two other factors that you think might affect one of your three projections.

Inquiring Further

1. **United States oil and gas fields**
   Research one of the top ten United States oil and gas fields. Go the *EarthComm* web site for links that will help you to research the geology and production history of the top ten United States oil fields. Why are these fields so large? What is the geologic setting? When was oil or gas discovered? What is the production history of the field? How much oil and gas remains?

   Oil Fields
   Alaska (Prudhoe Bay; Kuparuk River), California (Midway-Sunset; Belridge South; Kern River; Elk Hills), Texas (Wasson; Yates; Slaughter), Gulf of Mexico (Mississippi Canyon Block 807)

   Gas Fields
   New Mexico (Blanco/Ignacio-Blanco; Basin), Alaska (Prudhoe Bay), Texas (Carthage; Hugoton Gas Area [also in Kansas and Oklahoma); Wyoming (Madden), Alabama (Mobile Bay), Colorado (Wattenberg); Utah (Natural Buttes); Virginia (Oakwood).

Activity 7 Oil and Gas Production

Goals

In this activity you will:

- Design investigations into the porosity and permeability of an oil reservoir.
- Understand factors that control the volume and rate of production in oil and gas fields.
- Explore the physical relationships between natural gas, oil, and water in a reservoir.
- Understand why significant volumes of oil and gas cannot be recovered and are left in the ground.
- Use this knowledge to understand the variability in estimates of remaining oil and gas resources.
- Appreciate the importance of technological advances in maximizing energy resources.

Think about It

When oil and gas are discovered, petroleum geologists calculate the volume of the oil or gas field. They also need to estimate how much they can recover (remove). This helps them to determine the potential value of the discovery.

- When oil is discovered, what percentage of it can actually be recovered and brought to the surface?
- When will global oil production begin to decline? When will global gas production begin to decline?

What do you think? Record your ideas in your *EarthComm* notebook. Be prepared to discuss your responses with your small group and the class.

Investigate

Part A: Reservoir Volume

1. In your group, discuss the questions below. Record your ideas in your notebook.

   a) Which material can hold more oil: gravel or sand? Why do you think this is so?

   b) How might this knowledge be useful in oil and gas exploration?

2. Develop a hypothesis related to the question in Step 1. a).

   a) Record your hypothesis in your notebook.

3. Using the following materials, design an investigation to test your hypothesis:
   two 500-mL (16-oz.) clear plastic soda bottles (with bottoms removed)
   fine cheesecloth
   electrical tape
   stands with clamps to hold the plastic bottles
   sand
   aquarium gravel
   50-mL graduated cylinder
   200 mL of vegetable oil
   water

   a) Record your procedure. Note any safety factors.

4. When your teacher has approved your design, conduct your investigation. (Note: At the end of your investigation, do not empty out the sediments and oil. You will need them for Part B.)

   a) Record your results.

5. Use your results to answer the following questions:

   a) What is the porosity (percentage of open space) of your gravel sample?

   Porosity is the ratio of pore space to total volume of solid material and pore space, usually expressed as a percentage by multiplying the ratio by 100. 1 mL = 1 cm^3 (cubic centimeter). For example, suppose you poured 25 mL of sediment into a container and found that 10 mL of water completely filled the spaces between the grains. One milliliter is equal to one cubic centimeter. Expressed as a percentage, the porosity of the sediment would be:

$$(10\ cm^3/25\ cm^3) \times 100 = 40\%\ \text{porosity}$$

   b) What is the porosity of your sand sample?

   c) Imagine that a geologist discovers an oil field in a sandy reservoir (in petroleum geology, a reservoir is a porous material that contains oil or gas). The total volume of the reservoir (sand plus oil) is equivalent to 60 million barrels of oil. How many barrels of oil are in the reservoir?

   d) If the reservoir described above were gravel, how many barrels of oil would it contain?

Wear goggles. Clean up spills. Wash your hands when you are done.

Part B: Recovery Volume

1. Suppose you were to remove the cap at the bottom of the plastic bottle of gravel and oil.

   a) What volume of oil do you think would drip out of the container if you left it out overnight? Explain your answer.

2. Make sure that the cheesecloth is secured to the bottle with electrical tape. Unscrew the cap from the bottom of the container of gravel and oil. Allow the oil to seep through the container and drain into a collection device overnight.
3. Record your answers to the following questions:
   a) What is the volume of oil recovered?
   b) Calculate the percentage of oil recovered.
   c) How do you explain your results?
   d) If your model represented a discovery of 100 million barrels of oil, how many barrels of oil would actually be removed from the reservoir?
4. Brainstorm about ways to improve your results.
   a) How can you recover a greater percentage of oil? If your situation permits, test your method of removing more oil.

Part C: Factors that Affect Oil and Gas Production Rates

1. In your group, discuss your ideas about the following question: How does the grain size of sediment affect the ease with which fluids flow through them?
2. Develop a hypothesis about the question above. You will be testing the rate of flow of water through gravel, coarse sand, fine sand, and silt/clay. Your hypothesis should include a prediction (what you expect) and a reason for your prediction.
   a) Record your hypothesis.
3. Write down your plan for conducting the investigation. An idea about how to set up the equipment is shown in the diagram below.
   a) What is your independent variable?
   b) What is your dependent variable?
   c) How will you control other variables that might affect the validity of your results?
   d) How will you record and keep track of your results?

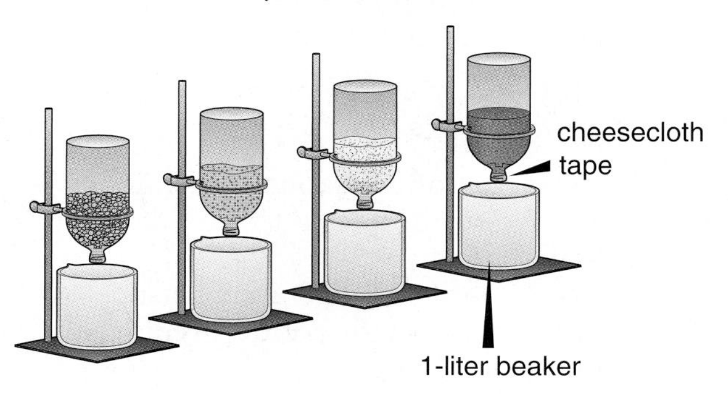

4. When you are ready, conduct your investigation.
   a) Describe your results. Was your hypothesis correct?
   b) Describe the relationship between grain size and permeability.
   c) How do your results relate to the production of oil?
5. If time permits, develop and test a hypothesis about the relationship between the viscosity of a fluid and its rate of flow through material. Viscosity is a term used to describe a fluid's internal friction or resistance to flow. In this case, you might wish to compare the speed at which oil, water, and a third fluid flow through a particular type of sediment.

Part D: Properties of Fluids in Reservoirs

1. Pour water into a 500 mL clear plastic soda bottle until the bottle is about half full. Add vegetable oil until the bottle is three-fourths full. Screw the cap onto the bottle. The water, vegetable oil, and air in this model represent water, crude oil, and natural gas. The bottle represents an oil and gas reservoir.
2. Slowly turn the bottle upside down (if you shake the bottle or turn it quickly, you will create bubbles in the oil, which will interfere with your observations).
   a) In your notebook, sketch a diagram to represent the relationship between the oil, water, and air.
   b) Why do you think that the material stacks in the order that it does?
   c) If a well is drilled into an oil and gas reservoir, what material would it encounter first? second? third?
3. Tilt the bottle at a 45° angle and hold it there. Repeat this with a 10° angle. Look at the bottle from the side, and then from the top.
   a) Draw a diagram of the side view of the bottle when it is tilted at a 45° angle versus a 10° angle.
   b) How does the surface area covered by the oil change with the angle of the oil reservoir (bottle)?
   c) Imagine drilling a vertical well through the bottle. This represents a well being drilled into reservoirs that are sloping at different angles in the subsurface. Would it penetrate a greater thickness of oil when the reservoir is at a 10° angle or when the reservoir is at a 45° angle? Why?
4. Look at the data table below. It shows the results of drilling of four wells. The elevation of the top of each well was 5000 ft. above sea level.

Feature	Well A	Well B	Well C	Well D
Elevation of top of VanSant Sandstone	3650 ft.	3850 ft.	3950 ft.	3900 ft.
Elevation of base of VanSant Sandstone	3450 ft.	3700 ft.	3825 ft.	3800 ft.
Result	Water (dry hole)	Oil at 3850 ft.; water at 3800 ft.	Gas at 3950 ft.; oil at 3875 ft.	Gas at 3900 ft.; oil at 3875 ft.

The locations of the wells are shown on the map below. The map shows the elevation of top of the VanSant Sandstone. Symbols on the map show the results of drilling.

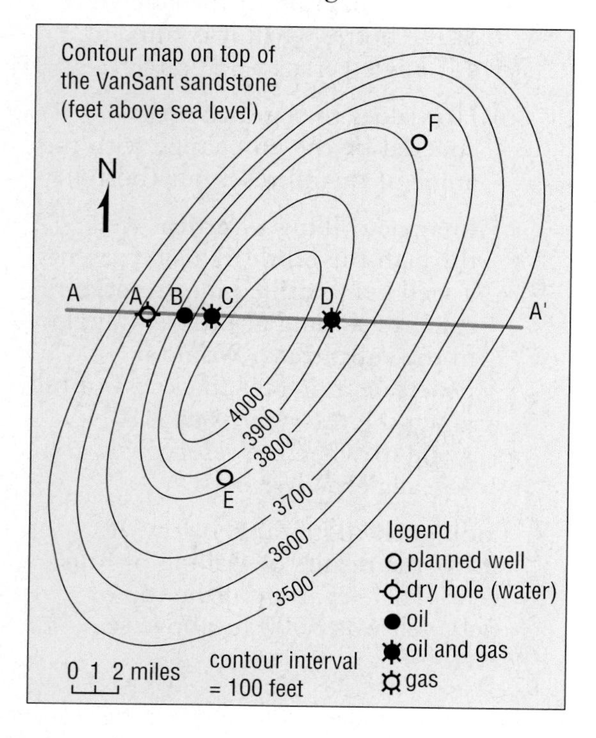

a) Draw a cross section across the oil field along the east–west line labeled A–A′. Plot the top of the VanSant Sandstone and the base of the VanSant Sandstone.

b) Use the results of drilling to show the level of gas, oil, and water in the cross section.

c) Use the results and your cross section to map the areal extent of the oil and gas. Color the gas red and the oil green on the map.

d) An example from another oil and gas field is provided below.

A geologist proposes to drill two more wells at locations E and F. Use your map and cross section to decide whether or not you would support the plan.

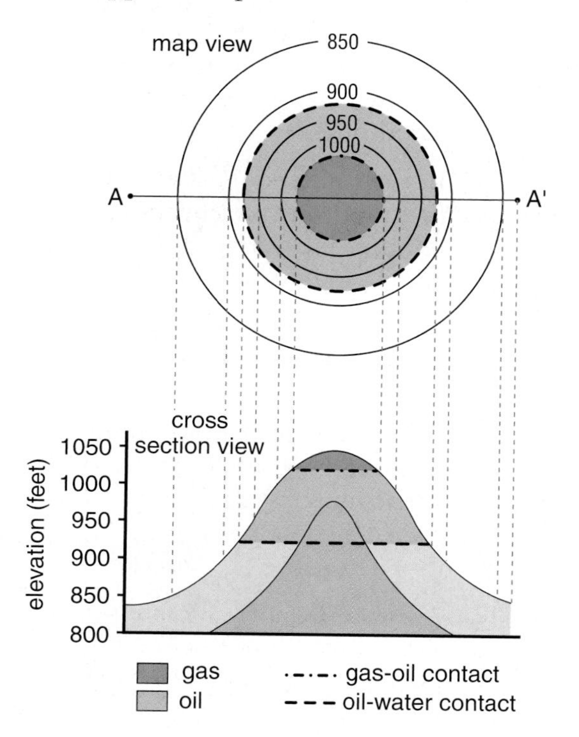

Reflecting on the Activity and the Challenge

In this activity, you explored the percentage of open space in various kinds of sediment. You saw that materials hold different amount of oil and that it is not easy to get all of the oil out of the material. You also explored the rate at which fluids flow through different materials. In the last part of the activity, you investigated the relationship between gas, oil, and water, and used the relationships to make predictions about where oil and gas would be in an oil and gas field. As you read further, you will see how the factors you have explored relate to the ability to predict how much oil and gas can be produced to meet future energy needs.

Digging Deeper

PETROLEUM RECOVERY

Porosity and Permeability

Most sedimentary rocks have open spaces in addition to the solid materials. These open spaces are called pores. Sedimentary rocks are the most porous rocks. Think about what a sandstone looks like, on the inside, when it is first deposited by flowing water. The sand particles are in contact with one another in the form of a framework. This framework is much like the way oranges are stacked in the supermarket, except that the sand particles usually have a much less regular shape than oranges. As you learned in the investigation, the **porosity** is defined as the volume of pore spaces divided by the bulk volume of the material, multiplied by 100 to be expressed as a percentage.

The porosity of loose granular material like sand can be as much as 30%–40%. If the sand particles have a wide range of sizes, however, the porosity is less, because the smaller particles occupy the spaces between the larger particles. When the sand slowly becomes buried in sedimentary basins to depths of thousands of meters, where temperatures and pressures are much higher, new mineral material, called **cement**, is deposited around the sand grains. The cement serves to make the sediment into a rock. The cement in sedimentary rocks is not the same as the cement that's used to make concrete, although it serves the same purpose. The addition of the cement reduces the porosity of the rock, sometimes to just a few percent.

Geo Words

porosity: the ratio of pore space to total volume of a rock or sediment, multiplied by 100 to be expressed as a percentage.

cement: new mineral material precipitated around the particles of sediment when it is buried below the Earth's surface.

The most productive petroleum reservoir rocks are those in which the pore spaces are mostly connected rather than isolated. When the pore spaces are all connected to one another, fluid can flow slowly through the rock. Fluid flows through a porous rock when the fluid pressure differs from one place to another. Here's a simple "thought experiment" to show the concept. Attach a long horizontal pipe to the base of a large barrel. Stuff the pipe with sand, and put a screen over the downstream end of the pipe. When you fill the barrel with water, the water flows through the sand in the pipe, because the pressure at the bottom of the barrel is high but the pressure at the open end of the pipe is low. The water finds its way down the pipe through the connected pore spaces in the sand. The rate of flow through the porous material in response to a pressure difference is related to the **permeability** of the material. For a rock to have a high permeability, it generally has to have high porosity, but the pores must be fairly large and also well connected with one another. Good petroleum reservoir rocks must have high porosity, to hold the oil or gas, and also high permeability, so that the oil or gas can flow through the rock to the bottom of a well. Some reservoir rocks are highly fractured, and the oil and gas can move along open fractures as well as through pore spaces.

Geo Words

permeability: the ease with which a fluid can be forced to flow through a porous material by a difference in fluid pressure from place to place in the material.

Fluids also flow in porous rocks because of differences in density. Most sedimentary rocks at depth in the Earth have their pore spaces filled with water. When droplets of oil or bubbles of gas are formed in the rock during maturation of the organic matter in the rock, they tend to rise upward through the pore water. They do this because both oil and gas are less dense than water. Only when they are trapped and accumulate beneath a seal do they form a petroleum reservoir.

For a mass of rock to form a seal, it has to be impermeable. In its simplest form, it also has to have a convex-upward shape, like an umbrella. Of course, it is usually much more complicated in its geometry than an umbrella. Just imagine raindrops falling upward into the inside of your umbrella, rather than downward to be shed off the umbrella! The vertical distance from the crest or top of the seal to its lower edge, where the petroleum can spill out, is called the closure of the reservoir. Giant petroleum reservoirs can have closures of many tens of meters, as well as horizontal dimensions of thousands of meters. Reservoirs that large can hold hundreds of millions of barrels of oil.

Recovery of Petroleum

Oil and gas are removed from reservoirs by drilling deep wells into the reservoir. At greater depths, the wells are only a few inches in diameter, but they can extend down for as much as several thousand meters. The technology of drilling has become very advanced. Wells can be drilled vertically downward for some distance and then diverted and angled off to the side at precise orientations. Sometimes the oil is under enough pressure to flow out of the well to the ground surface without being pumped, but usually it has to be pumped out of the well. The pumping lowers the pressure in the well, and the oil flows slowly into the well from the surrounding reservoir.

Figure 1 Pumps are used to extract oil.

Much of the oil in a reservoir remains trapped in place after years of production. The main reason is that oil adheres to the walls of the pores in the form of coatings. The coatings are especially thick around the points of contact of sediment particles, where the pore spaces narrow down to small "throats." Several techniques have been developed to recover some of the remaining petroleum. Use of such techniques is called **secondary recovery**. One of the most common methods of secondary recovery is to inject large quantities of very hot steam down into a well, which mobilizes some of the remaining petroleum by causing it to flow more easily.

Geo Words

secondary recovery: the use of techniques to recover oil still trapped among sediment particles after years of production.

Estimates of Petroleum Reserves

Petroleum geologists attempt to estimate the future recoverable reserves of petroleum in the world. Estimates of this kind are very important, because they are needed to predict when world petroleum production will start to level off and then decline. This is a very difficult task, because it has to include regions of the world which have not yet been explored in detail but which seem to be similar, geologically, to areas already explored. Estimates of recoverable reserves have increased steadily with time, as petroleum has been discovered in regions that were not believed to contain petroleum. That can't go on forever, of course, because there is only a finite volume of petroleum in the Earth. The most recent estimate from the United States Geological Survey gives a good idea of recoverable reserves. Rather than being a single-value estimate, it is based on probability. The 95% probable value is about 2250 billion barrels of oil. That means that the probability that recoverable reserves are greater than 2250 billion barrels is 95%. The 50% estimate is about 3000 billion barrels, and the 5% estimate is about 3900 billion barrels (a one-in-twenty chance). You can see from this great range in values how indefinite such estimates are.

Figure 2 A drilling platform offshore.

The world's supply of petroleum is finite. Most of the land areas of the world have been intensively explored already, and the major new areas for petroleum exploration and production (called "plays," in the language of the oil business) lie in deep offshore areas of the oceans. It's generally agreed among petroleum geologists that such areas are the last major exploration frontier.

Gradually, over the next few decades, petroleum production will level off and start to decline, although only slowly. Because demand is likely to keep on rising, there will be a growing shortfall in petroleum. That shortfall will have to be made up by use of various alternative energy sources. Solid organic matter contained in certain sandstones, called tar sands, and in certain shales, called oil shales, are already becoming economically favorable to produce, and they will certainly be a major factor in future energy production. Canada is especially rich in tar sands, and there are abundant oil shales in the western United States. Developing fuel-efficient vehicles and giving greater attention to conservation of fuel will also help communities to address the shortfall of petroleum.

Figure 3 Heavy oil found in tar sands at Vernal, Utah.

Check Your Understanding

1. In your own words, describe the difference between porosity and permeability.
2. What causes the permeability of a sediment deposit to decrease as it is buried deeply and turned into a sedimentary rock?
3. Why is it not possible to extract all of the oil from a reservoir?
4. What are the benefits of predicting oil and gas reserves?

Understanding and Applying What You Have Learned

1. a) Why does the permeability of a fine-grained sedimentary rock tend to be less than that of a coarse-grained rock?
   b) Why does the permeability of a rock with particles of approximately equal size tend to be greater than a rock with particles of a wide range of sizes?
2. Why does a porous and permeable sedimentary rock need to have an upper seal to have the potential to be a petroleum reservoir?
3. What do you think replaces the oil in the pore spaces of a reservoir rock when the oil is pumped out? Why?
4. It is very expensive to drill wells and recover oil and gas. Use what you have learned about porosity and permeability to explain why geologists evaluate these features of reservoirs.

Preparing for the Chapter Challenge

Communities depend upon a fairly constant supply of the fuels and other products that are developed from oil and gas. Use what you have learned about oil and gas exploration and production to write a brief essay. The essay should outline the basics of oil and gas exploration and production. It should also help community members to understand the scientific basis for differences in estimates about the future supply of oil and gas.

Inquiring Further

1. **Secondary recovery methods**

   Research modern methods of secondary recovery:

   - carbon dioxide flooding,
   - nitrogen injection,
   - horizontal drilling,
   - hydraulic fracturing.

2. **Protection of sensitive environments**

   Investigate how oil and gas exploration and production companies are using advanced technologies to obtain resources while minimizing environmental impacts. Examples include:

   - The use of slimhole rigs (75% smaller than conventional rigs) to reduce the effect on the surface environment.
   - Using directional drilling technology to offset a drilling rig from a sensitive environment.
   - Using downhole separation technology to reduce the volume of water produced from a well.

3. **Careers in the petroleum industry**

   Investigate careers in the oil and gas industry. No single industry employs as many people as the petroleum industry.

4. **Conserving transportation fuels**

   Investigate methods of reducing consumption of transportation fuels.

Activity 8 Renewable Energy Sources—Solar and Wind

Goals

In this activity you will:

- Construct a solar water heater and determine its maximum energy output.
- Construct a simple anemometer to measure wind speeds in your community.
- Evaluate the use of the Sun and the wind to reduce the use of nonrenewable energy resources.
- Understand how systems based upon renewable energy resources reduce consumption of nonrenewable resources.

Think about It

The solar energy received by the Earth in one day would take care of the world's energy requirements for more than two decades, at the present level of energy consumption.

- How can your community take advantage of solar energy to meet its energy needs?
- Is your community windy enough to make wind power feasible for electricity generation?

What do you think? Record your ideas in your *EarthComm* notebook. Be prepared to discuss your responses with your small group and the class.

Investigate

Part A: Solar Water Heating

1. Set up a solar water heater system as shown in the diagram. Put the source container above the level of the heater so that water will flow through the tubing by gravity. Do the experiment on a sunny day and face the surface of the solar water heater directly toward the Sun. This will enable you to calculate maximum output.

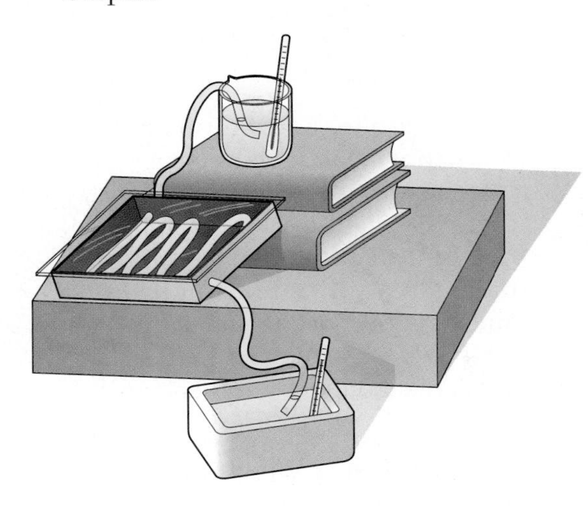

2. In its simplest form, a solar water heater uses the energy of the Sun to increase the temperature of water as the water flows through a coil of tubing.

   A Btu (British thermal unit) is the amount of heat energy required to raise the temperature of one pound of water one degree Fahrenheit. You need to determine the number of Btus of heat generated. If you divide this value by the surface area of the heater and by the time it takes to heat the water, you will have the Btu/ft^2/min output of your solar water heater. It will help to know that 1000 mL of water weighs about 1 kg, or 2.205 pounds.

   a) Decide how you will do the investigation and record the data.

   b) Prepare a data table to record relevant data.

   c) Think about the conversions that you will need to make in order to end up with Btu/ft^2/min.

3. Run about one liter of cold water through the heater. Refill the source container with two liters of cold tap water and run the experiment.

4. Answer the following questions:

   a) What is the output of your solar water heater in Btu/ft^2/min?

   b) Convert the output of your solar water heater from Btu/ft^2/min to Btu/ft^2/hour.

   c) Assume that you could track the Sun (automatically rotate the device so that it faces the Sun all day long). Assuming an average of 12 hours of sunlight per day, what would its output be per day?

   d) Convert this value to kWh/m^2/day by dividing by 317.2.

5. Examine the maps on the following page of average daily solar radiation for the United States. The first map shows the values for January, and the second one shows the values for July. The maps represent 30-year averages and assume 12 hours of sunlight per day and 30 days in a month.

a) How does the average daily solar radiation per month for January compare with that of July? How might you explain the difference?

b) Find your community on the map. Record the values for average daily solar radiation for January and July.

c) How does the daily output of your solar water heater compare to the average daily solar radiation for the two months shown?

d) Percent energy efficiency is the energy output divided by the energy input, multiplied by 100. Calculate the estimated efficiency of your solar water heater. If you have Internet access, you can obtain the average values for the month in which you do this activity. Otherwise, calculate the efficiency for whatever month is closest—January or July.

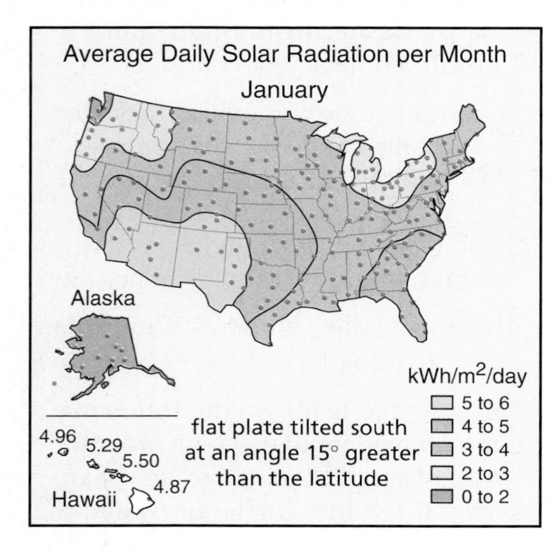

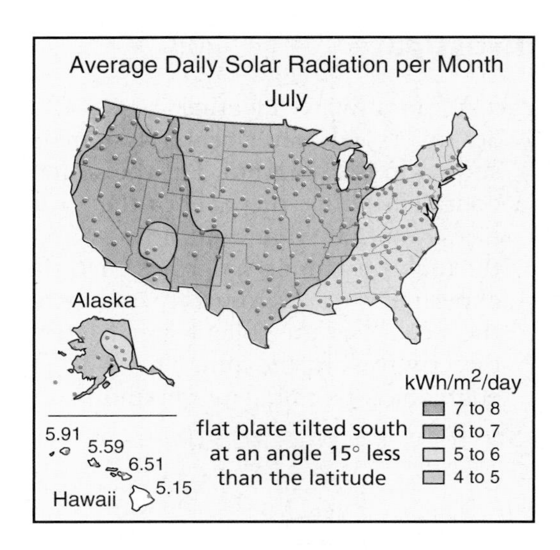

6. In your group, brainstorm about ways to increase the efficiency of your solar water heater. Variables that you might explore include the role of insulation, diameter of the tubing, type of material in the tubing, and flow rate of the water through the system. If time and materials are available, conduct another test.

Part B: Harnessing Wind Energy

1. Set up an anemometer as shown in the diagram below.

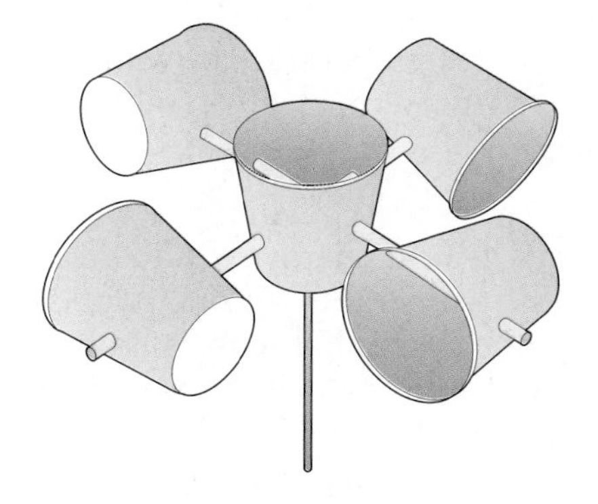

2. To find the wind speed, count the number of revolutions per minute. Next, calculate the circumference of the circle (in feet) made by the rotating paper cups. Multiply the revolutions per minute by the circumference of the circle (in feet per revolution), and you will have the velocity of the wind in feet per minute. (This will be an underestimate, because the cups do not move quite as fast as the wind.)
3. Take your anemometer home during a weekend. This will give you a chance to make measurements at different times of the day and over two days.
4. Make a map of your house and the yard around the house. Plan to place the anemometer at different locations and make velocity measurements at different times of the day.
5. Using a watch, count the number of times the colored cups spins around in one minute. Use your calibration to convert this to wind velocity.
   a) Keep a record of the wind speeds you are measuring for the next few days.
6. Measure the wind speed at different times of the day.
   a) Is the wind speed the same in the morning; the afternoon; the evening?
   b) Move your anemometer to another location. Is it windier in other places?
   c) Do trees or buildings block the wind?
7. Back in class, determine the best location for a wind turbine at your home.
   a) On the basis of the data that you have taken, make a plot of wind velocity throughout the day. You will assume that the wind will blow the same way every day. (Is this a good assumption?)
8. Following the example given below, calculate the electricity that can be produced at your home for a wind turbine with a 2-m blade (the cost of this wind turbine is about $3000).
   a) What percentage of your household electricity needs does this represent?
   b) How much wind power could be produced with a 3-m blade? (Cost about $5000.)

Example calculation of wind power: Wind power per square meter of turbine area is equal to 0.65 times the cube of the wind velocity. For instance, if a wind turbine has a blade 2 m long, then the area that the turbine sweeps is $\pi(1 \text{ m})^2 = 3.14 \text{ m}^2$.

If a constant wind of 10 m/s (22 mph) is blowing, the power is $(10^3)(0.65)(3.14) = 2000$ W. The efficiency of wind turbines is about 40%, so the actual power that can be produced is about 800 W. If the wind blew a constant 10 m/s every day of the year, this wind turbine could produce 7000 kWh of electricity.

9. Get together as a class and discuss this activity.

   a) How do your calculations compare with those of other students?

   b) What are the characteristics of the locations of homes that have the highest wind velocities?

   c) What are the characteristics of those with the lowest wind velocities?

   d) Overall, how well suited is your community for wind power?

   e) Which locations in your community are optimally suited for wind turbines?

   f) How realistic is the assumption that two days' worth of measurements can be extrapolated to yearly averages?

   g) For how long should measurements of wind velocity be taken to get an accurate measure of wind velocities in your community?

Reflecting on the Activity and the Challenge

Water heating can consume as much as 40% of total energy consumption in a residence. In this investigation, you determined the energy output of a flat-plate collector, a device used to heat water using solar energy. You also built a simple anemometer in the classroom and used it to measure wind speed around your home at different times of the day. Finally, you calculated whether it would be feasible and cost-effective for you to generate part of your family's electricity needs using wind power. You may wish to consider the use of different forms of solar energy as an alternative energy source for your community.

Digging Deeper

SOLAR ENERGY

Forms of Solar Energy

Forms of solar energy include direct and indirect solar radiation, wind, photovoltaic, biomass, and others. Tidal energy is produced mainly at the expense of the Earth–Moon system, but the Sun contributes to the tides as well. The three forms of solar energy with the most potential are solar-thermal (direct solar radiation to generate electricity), wind power, and photovoltaic cells.

Solar energy is produced in the extremely hot core of the Sun. In a process called nuclear fusion, hydrogen atoms are fused together to form helium atoms. A very small quantity of mass is converted to energy in this process. The ongoing process of nuclear fusion in the Sun's interior is what keeps the Sun hot. The surface of the Sun, although much cooler than the interior, is still very hot. Like all matter in the universe, the surface radiates electromagnetic energy in all directions. This electromagnetic radiation is mostly light, in the visible range of the spectrum, but much of the energy is also in the ultraviolet range (shorter wavelengths) and in the infrared range (longer wavelengths). The radiation travels at the speed of light and reaches the Earth in about eight minutes. Only an extremely small fraction of this energy reaches the Earth, because the Earth is small and very far away from the Sun. It is far more than enough, however, to provide all of the Earth's energy needs, if it could be harnessed in a practical way. The Sun has been producing energy in this way for almost five billion years, and it will continue to do so for a few billion years more.

Capturing the Sun's energy is not easy, because solar energy is spread out over such a large area. In any small area it is not very intense. The rate at which a given area of land receives solar energy is called **insolation**. (That's not the same as insulation, which is material used to keep things hot or cold.) Insolation depends on several factors: latitude, season of the year, time of day, cloudiness of the sky, clearness of the air, and slope of the land surface. If the sun is directly overhead and the sky is clear, the rate of solar radiation on a horizontal surface at sea level is about 1000 W/m^2 (watts per square meter). This is the highest value insolation can have on the Earth's surface except by concentrating sunlight with devices like mirrors or lenses. If the Sun is not directly overhead, the solar radiation received on the surface is less because there is more atmosphere between the Sun and the surface to absorb some of the radiation. Note also that insolation decreases when a surface is not oriented perpendicular to the Sun's rays. This is because the surface presents a smaller cross-sectional area to the Sun.

Geo Words

insolation: the rate at which a given area of land receives solar energy.

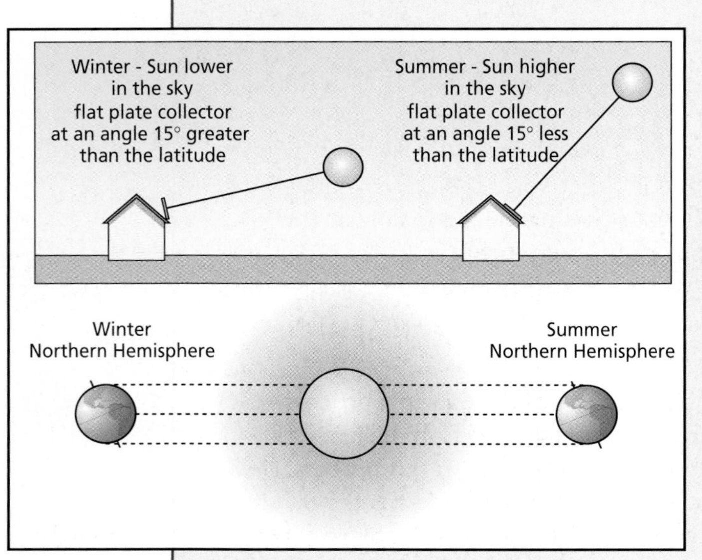

Figure 1 The tilt of the Earth on its axis during winter and during summer.

The Earth rotates on its axis once a day, and it revolves in an almost circular but slightly elliptical orbit around the Sun once a year. The Earth's rotation axis is tilted about 23° from the plane of the Earth's orbit around the Sun, as shown in *Figure 1*. It points toward the Sun at one end of its orbit (in summer) and away from the Sun at the other end (in winter). In summer, when the Earth's axis points toward the Sun, the Sun traces a path high in the sky, and the days are long. In winter, the Sun's path across the sky is much closer to the horizon, and the days are short. This has important effects on seasonal insolation and on building designs.

Solar Heating

Home heating is one of the main uses of solar energy. There are two basic kinds of solar heating systems: active and passive. In an active system, special equipment, in the form of a solar collector, is used to collect and distribute the solar energy. In a passive system, the home is designed to let in large amounts of sunlight. The heat produced from the light is trapped inside. Passive systems do not rely on special mechanical equipment, but they are not as effective as active systems.

Figure 2 Flat plate collector.

Water heating is another major use of solar energy. Solar water heating systems for buildings have two main parts: a solar collector and a storage tank. The solar energy is collected with a thin, flat, rectangular box with a transparent cover. As shown in *Figure 2*, the bottom of the collector box is a plate that is coated black on the upper surface and insulated on the lower surface. The solar energy that strikes the black surface is converted to heat. Cool water is circulated through pipes from the hot collector box to a storage tank. The water is warmed as it passes through the collector box. These collectors are not very expensive, and in sunny regions they can provide most or all of the need for hot water in homes or swimming pools.

Many large commercial buildings such as the one shown in *Figure 3*, can use solar collectors to provide more than just hot water. A solar ventilation system can be used in cold climates to preheat air as it enters a building. The heat from a solar collector can even be used to provide energy for cooling a building, although the efficiency is less than for heating. A solar collector is not always needed when using sunlight to heat a building. Buildings can be designed for passive solar heating. These buildings usually have large, south-facing windows with overhangs. In winter, the Sun shines directly through the large windows, heating the interior of the building. In summer, the high Sun is blocked by the overhang from shining directly into the building. Materials that absorb and store the Sun's heat can be built into the sunlit floors and walls. The floors and walls then heat up during the day and release heat slowly at night. Many designs for passive solar heating also provide day lighting. Day lighting is simply the use of natural sunlight to brighten up the interior of a building.

By tapping available renewable energy, solar water heating reduces consumption of conventional energy that would otherwise be used. Each unit of energy delivered to heat water with a solar heating system yields an even greater reduction in use of fossil fuels. Water heating by natural gas, propane, or fuel oil is only about 60% efficient, and although electric water heating is about 90% efficient, the production of electricity from fossil fuels is generally only 30% to 40% efficient. Reducing use of fossil fuel for water heating not only saves stocks of the fossil fuels but also eliminates the air pollution and the emission of greenhouse gases associated with burning those fuels.

Figure 3 Solar collectors on the roofs of these buildings can be used to provide energy for heating water and regulating indoor climates.

Photovoltaics

Generation of electricity is another major use of solar energy. The most familiar way is to use photovoltaic (PV) cells, which are used to power toys, calculators, and roadside telephone call boxes. PV systems convert light energy directly into electricity. Commonly known as solar cells, these systems are already an important part of your lives. The simplest systems power many of the small calculators and wristwatches. More complicated systems provide electricity for pumping water, powering communications equipment, and even lighting homes and running appliances. In a surprising number of cases, PV power is the cheapest form of electricity for performing these tasks. The efficiency of PV systems is not high, but it is increasing as research develops new materials for conversion of sunlight to electricity. The use of PV systems is growing rapidly.

Wind Power

People have been using wind power for hundreds of years to pump water from wells. Only in the past 25 years, however, have communities started to use wind power to generate electricity. The photograph in *Figure 4* is taken from an electricity-generating wind farm near Palm Springs, California. The triple-blade propeller is one of the most popular designs used in wind turbines today. Wind power can be used on a large scale to produce electricity for communities (wind farms), or it can be used on a smaller scale to meet part or all of the electricity needs of a household.

Figure 4 Wind turbines on a wind farm in California.

Figure 5 shows typical wind velocities in various parts of the country. California leads the country in the generation of electricity from wind turbines (in 2000, California had a wind-power capacity of 1700 MW, with 20,000 wind turbines), Many other areas in the country have a high potential for wind power as well. These areas include the Rocky Mountains, the flat Midwest states, Alaska, and many other areas. Commercial wind turbines can have blades with a diameter as large as 60 m.

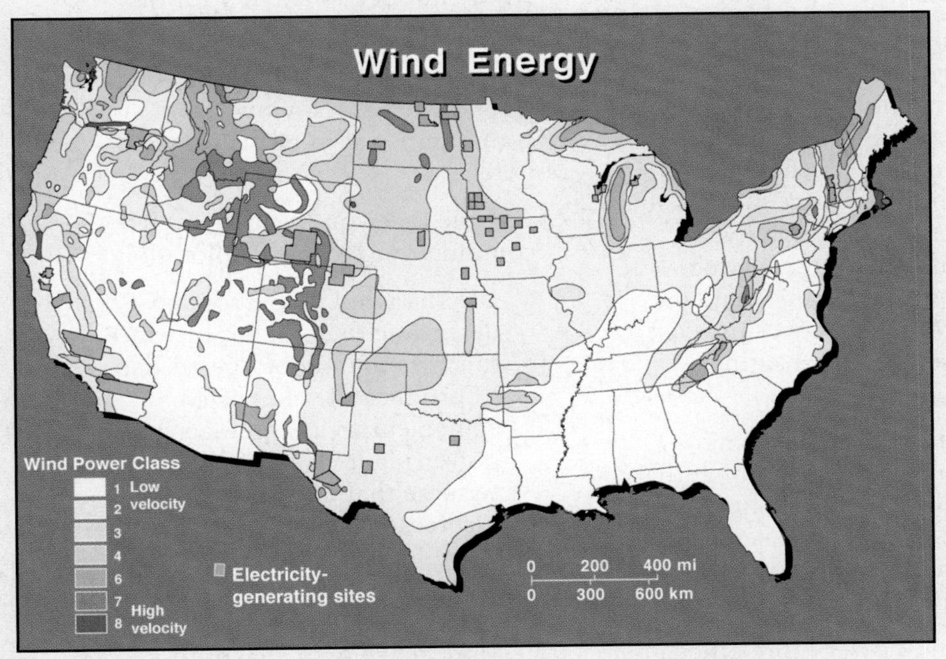

Figure 5 Location of electricity-generating wind-power plants relative to wind-power classes in the United States.

The cost of producing electricity from wind has dropped significantly since the early 1990s. This is due mainly to design innovations. In 2000, the cost of wind power was less than six cents per kilowatt-hour, which makes it competitive with electricity produced by coal-fired plants. The price is expected to decrease below five cents per kilowatt-hour by 2005.

Check Your Understanding

1. How does the Sun produce solar energy?
2. What factors determine the rate at which solar energy is received by a given area of the Earth's land surface?
3. What are the differences between passive solar heating systems and active solar heating systems?
4. Why must solar energy systems occupy a much larger area, per unit of energy produced, than conventional systems that burn fossil fuels?

Understanding and Applying What You Have Learned

1. Suppose that you live in a region that requires 1 million Btus (or 293 kWh) per day to heat a home. An average solar water heating system with an output of 720 Btu/ft^2/day costs $50 per square foot to install.

   a) What size of system would be required to meet these heating requirements?
   b) What would the system cost?
   c) Find out how much it costs to heat a home for five months in the winter where you live (November through March). Calculate how many years it would take to recover the cost of the solar water-heating system.

2. Why is metal used for tubing (rather than the plastic tubing that you used in the activity) in flat plate solar collectors? Use what you have learned about heat conduction to explain.

3. Why is wind power more efficient than generating electricity from all the methods that involve heating water to make steam?

4. Why is wind power less efficient than hydropower?

5. A wind turbine is placed where it is exposed to a steady wind of eight mph for the entire day except for two hours when the wind speed is 20 mph. Calculate the wind energy generated in the two windy hours compared with the 22 less windy hours.

6. Looking at the United States map of wind velocities, where does your community fall in terms of wind potential?

7. Assume that you live in a community in eastern Colorado.

   a) Use the wind map above to determine the number of wind turbines that would be needed to produce the electricity equivalent to a large electric power plant (1000 MW). Assume that each of the wind turbines has a blade diameter of 50 feet.
   b) If these wind turbines are replacing a coal-fired power plant that burns coal with a ratio of pounds of SO_2 per million Btu of 2.5, calculate the reduction in the amount of SO_2 emitted per year.
   c) What are some problems that might be encountered with the wind farm?

Preparing for the Chapter Challenge

Your challenge is to find a way to meet the growing energy needs of your community. You have also been asked to think about how to reduce the environmental impacts of energy resource use. Using what you have learned, describe how some of the energy needs of your community could be accommodated by wind or solar power.

Inquiring Further

1. **Solar-thermal electricity generation**

   Research how a solar thermal power plant, like the LUZ plant in the Mojave Desert in California, produces electricity. Diagram how such a system works. What are the kilowatt-hour costs of producing electricity using this method? What does the future hold for power plants of this kind?

2. **Photovoltaic electricity**

   Investigate photovoltaic electricity generation and discuss the results of your investigation with the class.

3. **History of wind energy or solar energy**

   Investigate how people in earlier times and in different cultures have harnessed the Sun or wind energy. What developments have taken place in the past hundred years? How is wind energy being used today? Include diagrams and pictures in your report.

4. **Wind farms**

   Prepare a report on how electricity is generated on wind farms. Describe types of wind generators, types and sizes of wind farms, the economics of electricity production on wind farms, and the locations of currently operating wind farms in the United States. Include diagrams.

Earth Science at Work

ATMOSPHERE: *Air-Monitoring Technician*
Sophisticated equipment is used by government agencies to monitor air quality. Reports are issued that inform the public as to the type and amount of pollutants present on any given day.

BIOSPHERE: *Environmental Scientist*
Companies are concerned about the impact they may have on the environment. Environmental impact studies are completed before new projects are undertaken.

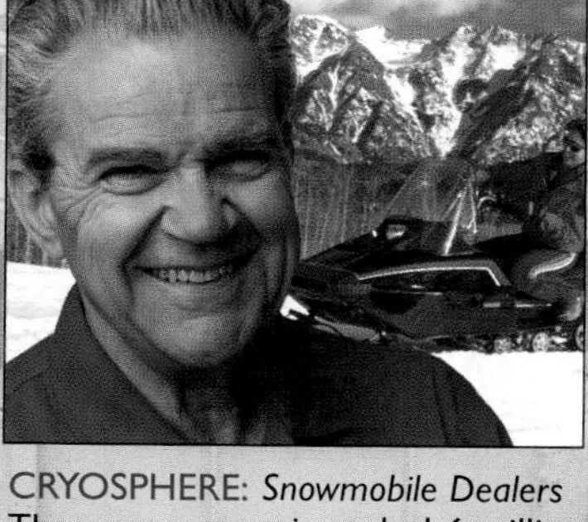

CRYOSPHERE: *Snowmobile Dealers*
There are approximately 1.6 million registered snowmobiles in the United States. They are often used for recreation, but in some areas, they are the preferred forms of transportation during winter months.

GEOSPHERE: *Exploration Geologist*
Before a company initiates any drilling operations, extensive surveys are completed. Geologists construct field maps, examine rock samples, and study seismic data to search for oil and gas deposits.

HYDROSPHERE: *Turbine Manufacturer*
In the generation of hydroelectricity, falling water strikes a series of blades attached around a shaft, causing a turbine to rotate.

How is each person's work related to the Earth system, and to energy resources?

2

Mineral Resources ...and Your Community

2 Mineral Resources ...and Your Community

Getting Started

The Earth contains a vast array of useful minerals, including metals like gold, silver, and copper, and nonmetallic minerals like quartz and salt. Minerals are a natural resource. Their formation, extraction, and use occur within a complex Earth system. Mineral resources are nonrenewable. Some kinds of mineral deposits can no longer be formed on Earth, even given enough time, because conditions on and within the Earth have changed through geologic time.

- What are minerals?
- What Earth systems are involved in forming a mineral?
- How might the use of minerals change one of the Earth's systems?

What do you think? Look at the diagram of the Earth Systems at the front of this book. In your notebook, draw a picture to show what you know about the Earth systems that are involved in forming a mineral. Draw a second diagram to show how the use of a mineral might change the Earth system. Be sure to label the diagram so that others can understand what you have drawn. Be prepared to discuss your pictures with your small group and the class.

Scenario

A company in your community has developed a new line of beverages, which they claim are the most environmentally friendly drinks on the market. In order to maintain this promise, the company is seeking to design beverage containers that are produced with minimal environmental impacts. Additionally, this relatively new company does not have a large amount of money, and therefore would like the container to be both inexpensive and practical to manufacture.

Chapter Challenge

Your challenge is to use mineral resources, preferably from your community, to design a beverage container that has a minimal environmental impact at all stages, from developing the raw materials to recycling the empty containers. It should also be cost-effective and practical.

You must also produce a report that substantiates your design and addresses these questions:

- What are mineral resources?
- What mineral resources do you plan to use to make your beverage container?
- Where were the mineral resources you plan to use to make your container formed?

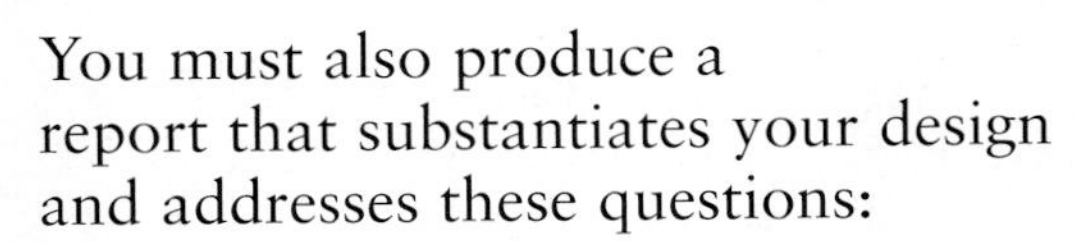

- How are the mineral resources that are being used to make the container found?
- How are the mineral resources extracted?
- How can you use these minerals to make a practical container?
- What impact does the use of these mineral resources have on the environment?
- What name will you give the beverage that conveys how unusual and environmentally friendly it is? What is your rationale for the choice of name?

Assessment Criteria

Think about what you have been asked to do. Scan ahead through the chapter activities to see how they might help you to meet the challenge. Work with your classmates and your teacher to define the criteria for assessing your work. Devise a grading sheet for the assessment of the challenge. Your teacher may provide you with a sample rubric to help you get started. Record all this information. Make sure that you understand the criteria and the grading scheme as well as you can before you begin.

Activity 1 Materials Used for Beverage Containers in Your Community

Goals

In this activity you will:

- Investigate the kinds of beverage containers found in your community.
- Determine what materials are commonly used to make beverage containers, specifically those found in your community.
- Determine why certain materials are used to make the beverage containers.

Think about It

Different materials are sometimes used to perform the same function but with different levels of performance and cost. For example, softball bats can be made of aluminum or wood. The use of materials for a specific function can also vary by geographic location.

- When your community was first founded, what did the settlers use for beverage containers?
- Are those containers commercially practical today?

What do you think? Record your ideas in your *EarthComm* notebook. Be prepared to discuss your responses with your small group and the class.

Investigate

1. Break into small groups and discuss some of the variables that are important in designing a container to hold liquids. Your teacher will provide your group with some empty containers collected from the school or home to analyze.
   a) What are the pros and cons of using each of these containers?
   b) Produce a list of variables that you think are important.
2. As a class, decide which variables are most significant and how they can be measured.
   a) Record the list in your *EarthComm* notebook.
3. On the basis of the variables you selected, design a data table to help in the collection of information on materials used for beverage containers in your school cafeteria. What additional important information should be included on the data sheet (name, location where data are collected, date, etc.)?
   a) Keep a record of your data table in your notebook.

Obtain permission before entering the school cafeteria. Follow guidelines provided to help maintain sanitary conditions (wash hands, do not touch utensils or cooking surfaces).

4. Use your table to analyze the beverage containers used in your school cafeteria.
   a) Collect and interpret the data.
   b) What hypotheses can you make about the various materials used to make containers?
5. After you have collected information from your school cafeteria, come back to the classroom and have a spokesperson from your group explain what you have learned so far.
6. As a class, discuss what your data means.
   a) Is your cafeteria representative of your whole community when it comes to use of beverage containers? Explain.
   b) How would your choice of location to sample influence your results?
7. Complete a survey of the beverage containers used in your community. As a class, decide where in your community data should be collected. Every student should have a different location. You might consider locations like supermarkets, neighborhood markets, convenience stores, vending machines, discount stores, gas stations that sell snacks, fast-food restaurants, or sports stadiums. Half of the class should collect the information from disposed containers at home or school. You will be using the form you previously developed, or a modified version of it, to collect information on all the kinds of beverages, kinds of containers for each, and frequency of use of each container in your community.
   a) Visit the location decided on in class and fill out your data table.

Obtain permission from store managers before conducting surveys in the store. Have an adult with you.
When collecting data about disposed containers, do not handle the containers or other trash with bare hands.

8. After you have collected your data, share it with the rest of the class and create a large database. On the basis of the entire data set, break into small groups and discuss and answer the following questions:
   a) What types of materials are used in your community to hold liquids?
   b) From the data you collected, why do you think certain materials are used for certain containers?
   c) In your community, what is the most popular material for a new beverage container?
   d) In your community, what is the most common material for a recyclable beverage container?
   e) What are the raw ingredients necessary to make these containers both directly and indirectly?
   f) From where do the raw ingredients come to make the containers?
   g) Do you think your community has all the resources necessary to make these containers?

Reflecting on the Activity and the Challenge

In this activity, you discussed the variables that are important to consider when designing a beverage container. You then developed a method to survey the kinds of beverage containers found at your school cafeteria and throughout your community. This will help you to understand what materials are commonly used to make beverage containers and get you thinking about what factors you need to consider in designing your own container.

Digging Deeper

MATERIALS USED FOR BEVERAGE CONTAINERS

Have you ever been on an impossible scavenger hunt? The items on your list just cannot seem to be found anywhere. Well, here is a sample list of items that you probably could not buy or find in your community today: tortoise-shell comb, plywood car, glass insulators, and a hand mirror made of gutta-percha (an early plastic material used in the mid-1800s). These are all real items, but today they can be found in a museum, not a store. What has changed? Although you can buy products that perform the same function, they are made from very different materials.

Figure 1 Glass insulators were first produced in the 1850s for use with telegraph lines.

Cost was formerly one of the main deciding factors in choosing a raw material for the manufacture of a product. Today, many other factors are involved, like manufacturing technology, product performance, and environmental issues. The choice of materials that products are made of affects everyone. The selection of materials influences the choice of resources used, technology development, environmental impacts, and worldwide competitiveness. Everyone has some influence on the choice of material. The manufacturer chooses a material that results in an acceptable product with a low price. The consumer influences a material choice by deciding how much he or she is willing to spend for a product. Society influences material decisions through concern about product standards, safety and environmental regulations, and bans on some material uses.

To a manufacturer the most important performance measure for a soft-drink container is the ability of the container to hold the beverage under pressure while minimizing the amount of escaping carbonation. If too much carbonation is lost, the beverage tastes flat.

Aluminum, steel, glass, and plastic are the main materials for beverage containers. The 12-oz aluminum can dominates the market for reasons of performance and cost. At the 12-oz size plastic (PET — polyethylene terephthalate) bottles cannot compete with aluminum because too much carbonation leaks through the walls of the container, shortening the product's shelf life. About 33 aluminum cans can be made from one pound of aluminum. It is possible to make only 20 steel cans and only 12 glass bottles from one pound of each material. Glass is heavy and thus it costs about 10 times as much to handle and distribute than steel or aluminum.

Figure 2 One pound of aluminum yields about 33 aluminum cans.

Aluminum is preferred over steel largely because the prolific recycling of aluminum makes it much cheaper to produce aluminum cans. Over 65% of all aluminum beverage cans are recycled. The manufacture of aluminum from bauxite ore is extremely energy intensive, but the cost of making aluminum from recycled material is inexpensive. The manufacture of steel from iron ore uses much less energy than aluminum, so the cost savings in recycling steel are not as great.

Why do you not see one-liter soda cans made from aluminum? Compared to the total amount of carbonation present in a one-liter volume, the amount of leakage is acceptable for a normal shelf life. An aluminum container would need a much thicker wall to withstand the carbonation pressure and still be stackable, and this would make the product too expensive. The raw material cost of plastic is about one-fifth that of aluminum and steel, and half that of glass. Thus, plastic is the material of choice for the larger containers.

The impact on the environment is now a major factor in designing new products and choosing materials for making those products. "Design for the Environment" (DFE) is now an important part of engineering a new product. Government estimates indicate that manufacturing operations in the United States account for 50% of all waste, over 11 billion tons annually, whereas household garbage accounts for only 2% of the waste generated each year. DFE includes keeping track of the environmental cost of components in the product. For example, if a computer has a circuit board that requires use of a chlorinated solvent to manufacture, that circuit board should carry a higher cost, because many chlorinated solvents pollute ground water and can cause cancer if not handled properly. Another aspect of DFE is using recycled materials to manufacture a product, hence reducing use of raw materials. Still another aspect is making a product easier to recycle when its life is over by making the product as simple as possible, using fewer different materials, and making the product easy to disassemble for recycling.

Figure 3 Aluminum cans can be recycled and the aluminum can be reused.

Check Your Understanding

1. What are the main materials used for producing beverage containers?
2. Compare and contrast the use of aluminum and glass for beverage containers.
3. Describe two factors that influence the selection of materials for making products.

Understanding and Applying What You Have Learned

1. The glass industry produced 10 million tons of glass containers in 1997. The population of the United States is 281 million people. Assume that none of the containers were exported. Answer the following questions. Be sure to show your work and to explain any other assumptions that you make.
   a) How many pounds of glass containers were produced per capita in the United States in 1997?
   b) What is the population of your state? Your community?
   c) How many pounds of glass containers were produced for your state in 1997?
   d) How many pounds of containers were produced for your community in 1997?
2. In 1999, 1.93 billion pounds of aluminum scrap were collected. Assume that one pound of aluminum yields 33.1 beverage containers.
   a) How many aluminum cans were recycled in 1999?
   b) 102.2 billion aluminum cans were produced in 1999. What percentage of cans were recycled?
3. Most glass beverage containers are soda-lime silicate glass, which is made from soda, lime, and silica. A typical soda-lime glass is 15% soda (Na_2O), 10% lime (CaO), and 75% silica (SiO_2). Assume that soda, lime, and silica sand cost \$0.171, \$0.269, and \$0.05, per pound, respectively.
   a) What is the per-pound cost of soda-lime glass?
   b) If 12 glass bottles can be produced per pound of glass, what is the cost per bottle?
   c) What other expenses factor into using glass bottles for beverage containers?

Preparing for the Chapter Challenge

Write a short paper that reviews the common materials used for beverage containers in your community. Which of these materials do you think would be useful in designing your beverage container? Why? Do you think that your community has all of the resources necessary to make your container? You can include this paper as an introduction to your **Chapter Challenge**, to help justify your choice of materials.

Inquiring Further

1. **Beverage containers through time**

   How do you think beverage containers have changed over time? How do you think beverage containers have changed from when your parents were your age? From when your grandparents were your age? Visit the *EarthComm* web site to help you get started with your research on how beverage containers have changed through time.

2. **Beverages containing carbon dioxide**

   Strength and flexibility are two important mechanical properties of beverage containers. There are two main reasons for this:

   - Many beverages, like soda, beer, and champagne, are under pressure from dissolved carbon dioxide gas.
   - Many beverage containers need to withstand the force exerted from inside when the liquid freezes to ice.

   Using your school library, community library, or the Internet (visit the *EarthComm* web site to get started), research the following topics, which are essential to a clear understanding of the two points listed above:

   a) What is the maximum concentration of carbon dioxide that can be dissolved in water, and how does that concentration vary with temperature?

   b) How much does water expand when it freezes, and how much force can be generated by the expansion?

Activity 2 What Are Minerals?

Goals

In this activity you will:

- Define the term "mineral" in your own words.
- Evaluate the usefulness of various physical properties for describing and identifying different minerals.
- Explore how mineral crystals are constructed and how the external form of a crystal reflects its ionic structure.
- Identify a variety of mineral specimens according to their physical properties.

Think about It

There are an estimated 9000 different species of birds around the world.

- What are some ways to identify birds?

What do you think? Even if you are not a birdwatcher, try to list at least four or five different features that could help you identify different kinds of birds. Make sure you consider other features besides purely visual ones. Record your ideas in your *EarthComm* notebook. Be prepared to discuss your responses with your small group and the class.

Investigate

Check your list of properties with your teacher before testing the minerals. Some properties, while useful, should only be tested with teacher direction.

Part A: Properties of Minerals

1. In a small group, study a set of mineral samples.
   a) Make a list of properties you can use to describe the minerals. For example, color may be a property that you would use to describe a particular mineral.
   b) Write a brief description of each mineral sample using the properties that you listed.
   c) Which properties are most useful in describing an individual mineral sample?
   d) Which properties are the least useful?

2. Make a class list of all the properties the different groups came up with to describe the minerals. Discuss the usefulness of the properties.
   a) In your notebook keep a record of the class list of the most useful properties in describing minerals.

Part B: Breaking Minerals

1. Your teacher will supply your small group with a crystal of halite (NaCl) and a crystal of calcite ($CaCO_3$). Put on your safety goggles. Place the halite crystal on a large sheet of paper. Hit it with the hammer, starting very gently and increasing the force of the impact until the crystal breaks.

Wear goggles.

2. Examine the broken pieces with a magnifying glass under a strong light.
   a) Describe the shape of the fragments, and sketch one or more of them in your *EarthComm* notebook.
3. Select one or a few of the fragments, and break them again with the hammer. Again, examine them with the magnifying glass.
   a) Describe and sketch the fragments in your *EarthComm* notebook.
4. Repeat Steps 1–3 with the calcite crystal.
5. From the results of your investigation, answer the following questions:
   a) How does the characteristic shape of the halite fragments differ from the characteristic shape of the calcite fragments?
   b) Why do you think that the fragments of the crystals have regular and distinctive shapes?
   c) Why do you think that the halite fragments and the calcite fragments have different shapes?

Part C: Stacking Spheres

1. Cut four squares of corrugated cardboard. Make the edges of each piece just long enough so that five marbles can line up along the edges. Tape the pieces together to form a square enclosure with vertical walls, as shown in the diagram below.

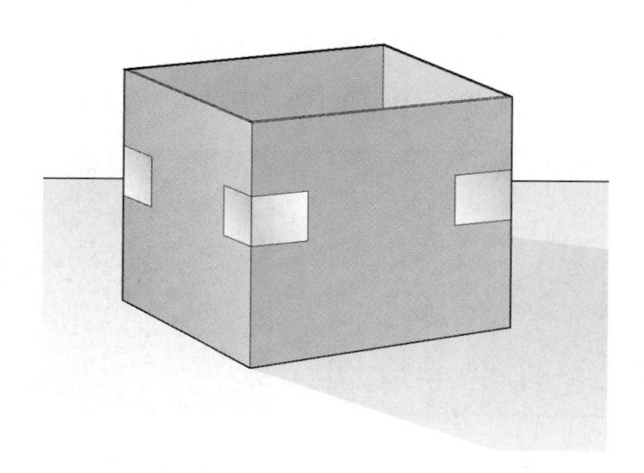

2. Line the bottom of the cardboard enclosure with a single layer of marbles. Next, carefully add a second layer of marbles, with each marble in the second layer directly above a marble of the lowest layer. Continue adding layers until you have filled the enclosure. You have created a regular three-dimensional array of marbles.
   a) In your *EarthComm* notebook, describe the geometry of the array of marbles.
   b) How does the geometry of this array compare to the geometry of the crystal fragments you produced in Part B of the activity?

3. There is another way to pack spheres in a regular three-dimensional array. Build another cardboard enclosure, adjusting its size so that you can arrange the first layer of marbles as shown in the diagram below. Lay in the marbles of the second layer in the "pockets" formed by three adjacent marbles of the first layer. One marble of the second layer is shown in the sketch below. Add layers until you have filled the enclosure.
   a) In your *EarthComm* notebook, describe the geometry of this array of marbles.
   b) How does the geometry of this array compare to the geometry of the crystal fragments you produced in Part B of the investigation?

   c) Can you think of a third way of building a regular three-dimensional array of marbles in a cardboard enclosure? If so, sketch and describe the array in your *EarthComm* notebook.

Be sure to immediately pick up any marbles that drop on the floor. They can cause accidents.

Part D: Mineral Identification

1. Work in small groups to identify the mineral samples you examined in Part A of this activity. Your teacher will give you a test kit containing:
   - a piece of unglazed porcelain tile (called a streak plate),
   - a magnet or paper clip, and
   - some of the minerals of the Mohs scale to test hardness.

   Decide how you could use each piece of equipment to help you identify the minerals. Read the appropriate sections of **Digging Deeper** to help you determine what tests would be appropriate.

   Test as many of the properties as you can to describe each mineral as completely as possible.
   a) Record your findings in a data table. Be sure to note any special properties exhibited by your samples.
   b) Use the Mineral Identification Key on the following pages to assign mineral names to your samples. Compare your list of mineral properties and names to those of other groups. Your teacher will lead a discussion on the correct mineral names and why some teams might have gotten different names for the same mineral.

Check your plan with your teacher before proceeding.

Mineral Identification Key

Mineral Name	Hardness	Streak	Specific Gravity	Cleavage	Crystal Shape	Color	Other Properties
Corundum	9	White	4 (Med. – High)	None	Commonly six-sided crystals	Gray, red, brown, blue	
Topaz	8	White	3.5 (Med. – High)	One perfect	Orthorhombic or massive	Colorless, yellow, blue, or brown	
Quartz	7	White	2.7 (Medium)	None	Hexagonal or massive	Any color to colorless, greasy luster	Conchoidal fracture
Potassium feldspar	6	White	2.6 (Medium)	Two at 90°	Monoclinic or triclinic	White, pink, or brown	
Plagioclase feldspar	6	White	2.6 (Medium)	Two at 90°	Triclinic (rare)	Blue-gray, black, or white	Striations on some cleavage planes
Magnetite	6	Dark gray	5.2 (High)	None	Massive	Dark gray to black	Magnetic, metallic luster
Pyrite	6	Dark gray	5.0 (High)	None	Cubic crystals common	Brass yellow, may tarnish brown	Metallic luster, brittle
Apatite	5	White	3.1 (Medium)	One poor	Common as six-sided crystals	Brown, green, blue, yellow, or black	
Hematite	5	Red to red-brown	5.0 (High)	None	Hexagonal	Red or steel gray	Red form – earthy luster, gray form – metallic luster
Fluorite	4.5	White	3.0 (Medium)	Octahedral	Crystals usually cubic	Colorless purple, blue, yellow, green	

Mineral Identification Key (continued)

Mineral Name	Hardness	Streak	Specific Gravity	Cleavage	Crystal Shape	Color	Other Properties
Calcite	3	White	2.8 (Medium)	Three perfect	Hexagonal	Colorless, white, yellow, gray	Transparent to translucent, reacts with HCl
Muscovite mica	2.5	White	2.7 (Medium)	One perfect	Monoclinic	Colorless, yellow, light brown	Elastic, flexible sheets
Biotite mica	2.5	Gray-brown	2.7 (Medium)	One perfect	Monoclinic	Very dark brown to black	Elastic, flexible sheets
Galena	2.5	Gray	7.5 (Very high)	Cubic	Cubic crystals common	Silvery gray	Metallic luster
Halite	2.5	White	2.5 (Medium)	Cubic	Cubic crystals	Colorless, white	Salty taste
Talc	1	White	2.7 (Medium)	One perfect	Monoclinic (rare)	White, gray, yellow	Soapy feel, pearly or greasy luster

Reflecting on the Activity and the Challenge

In this activity, you saw that there are many different ways to describe minerals. Some properties are more useful than other properties when describing and comparing minerals. Being able to accurately describe minerals helped to identify them. You also learned about how atoms are arranged in mineral crystals, and how the geometry of the arrangement affects the physical properties of the mineral. You must be able to identify minerals and understand their properties to be able to select the best mineral to use for your beverage container.

Digging Deeper

MINERALS

Types of Minerals

Minerals have been important to humans for a long time. Early humans used red hematite and black manganese oxide to make cave paintings. People in the Stone Age made tools out of hard, fine-grained rocks. In the Bronze Age, people discovered how to combine copper and tin from minerals into an alloy called bronze. Later, in the Iron Age, people made tools of iron, which is contained in minerals like hematite (Fe_2O_3) and magnetite (Fe_3O_4).

Figure 1 The minerals red hematite and black manganese oxide were used by early humans to draw their cave painting.

Today, minerals are used in thousands of ways. Feldspar is used to make porcelain. Calcite is used to make cement. Iron and manganese, together with small amounts of several other metals, make steel, which is used to make buildings, trains, cars, and many other things. Gypsum is used to make plaster and wallboard. These are just a few examples of how minerals are used in your daily lives.

Geo Words

mineral: a naturally occurring inorganic, solid material that consists of atoms that are arranged in a regular pattern and has characteristic chemical composition, crystal form, and physical properties.

To be considered a mineral, a material must meet five criteria:

- Minerals are solid, not gas or liquid.
- Minerals are inorganic, which means they are not alive and never have been.
- Minerals are naturally occurring, not manufactured.
- Minerals have definite chemical compositions, which can be expressed as a formula of elemental symbols (such as SiO_2, Ag, or Fe_2O_3).
- Minerals have a regular three-dimensional arrangement of atoms (called a crystal structure).

Some minerals, called **native-element minerals** consist of only one element (*Figure 2*). A good example is gold (Au), which is often found as nuggets or pieces of pure gold not combined with any other element. Copper (Cu), iron (Fe), and silver (Ag) are other native elements. Most minerals, however, are combinations of elements. Quartz (*Figure 3*), for example, with the formula SiO_2, is made of atoms of silicon and oxygen. Calcite ($CaCO_3$) is made of calcium, carbon, and oxygen.

Geo Words

native-element mineral: a mineral consisting of only one element.

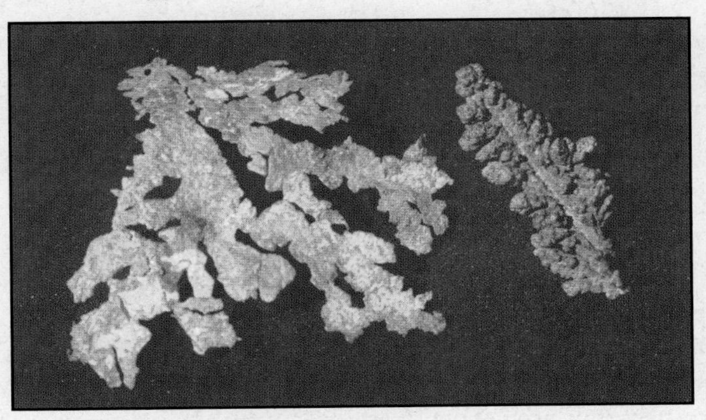

Figure 2 Copper is a native-element mineral.

Figure 3 Quartz is a mineral composed of the elements silicon and oxygen.

Geo Words

rocks: naturally occurring aggregates of mineral grains.

ores: rocks that contain valuable minerals.

ions: atoms that have an electric charge because one or more electrons have been added to the atom or removed from the atom.

electrons: particles with a negative electric charge, which orbit around the nucleus of the atom.

Rocks are naturally occurring aggregates of mineral grains. Some rocks consist of only one mineral, but most contain several different kinds of minerals. Sometimes you can see the mineral grains, and sometimes they are too small to see without magnification. Granite consists mostly of large crystals of feldspar, quartz, and mica, as shown in *Figure 4*. Basalt consists mostly of tiny crystals of feldspar and pyroxene.

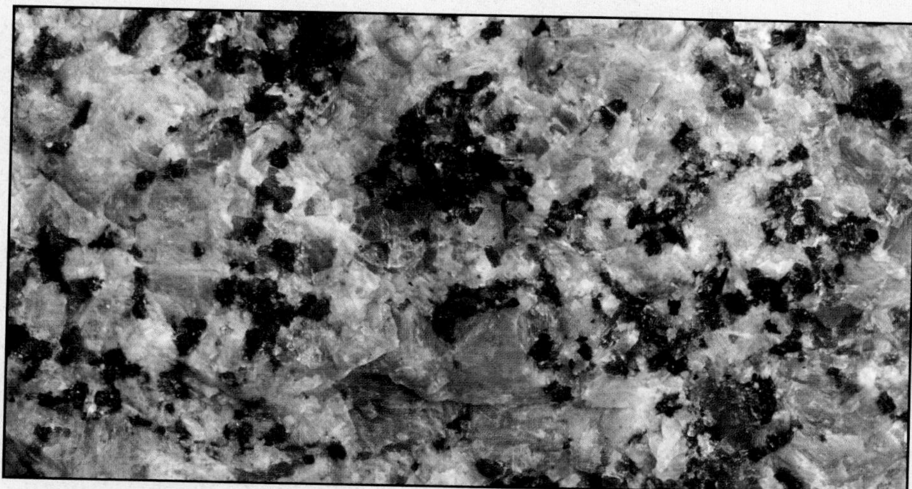

Figure 4 Granite is a rock composed of several different types of minerals.

Rocks that contain valuable minerals are called **ores**. To remove the valuable minerals, the ore first has to be mined, by digging or blasting. Then the desired mineral is separated from the rest of the ore by processes like crushing, sieving, melting, or settling through a liquid. Most metals and many important nonmetals are refined from ores. Valuable minerals are distributed very unevenly in the Earth's crust. Finding new deposits to meet the needs of industry and technology depends upon understanding the characteristics of the minerals and the ores that contain them.

The Chemistry and Structure of Minerals

As you saw in the activity, minerals are very different from one another. That is because all minerals have a particular chemical composition. Minerals consist of atoms of one or more chemical elements. Each chemical element has different chemical and physical behaviors.

The atoms in minerals are arranged in a regular three-dimensional array (*Figure 5*). The atoms in almost all minerals are in the form of **ions**. Ions are atoms that have an electric charge because one or more **electrons** (particles with a negative electric charge, which orbit around the nucleus of the atom) have been added to the atom or removed from the atom. The ions in a mineral are packed together in an arrangement that brings the ions as close together as possible. The packing arrangement puts positively charged ions in close contact with negatively charged ions. Objects with unlike electric charges are attracted to each other, and it is these attractive forces that hold the mineral together as a solid.

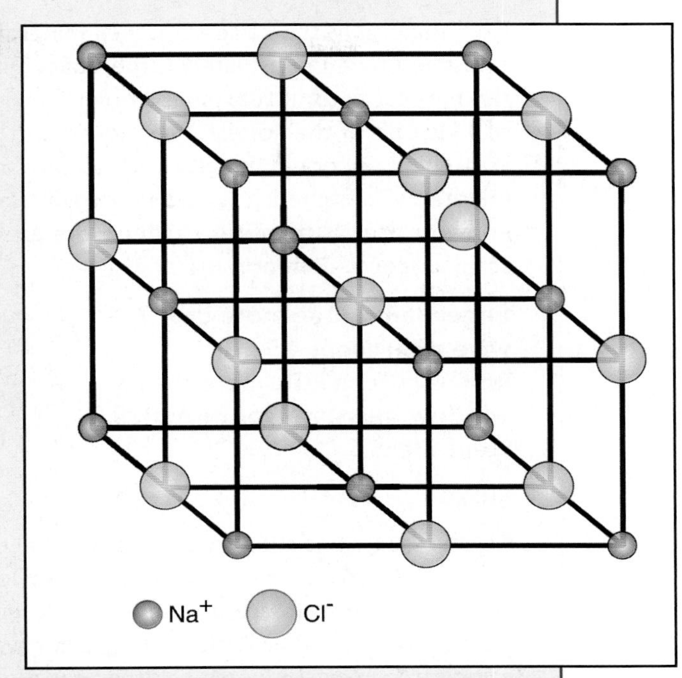

Figure 5 This expanded view of halite shows the orderly three-dimensional arrangement of sodium and chlorine atoms.

Identifying Minerals

The properties of the atoms in a mineral, and also their geometrical arrangement, affect the color, shape, hardness, and other properties of the mineral. Geologists use a variety of tests to describe, compare, and identify minerals. Some of these tests are simple, and can be done with simple equipment. Other tests require special (and expensive!) equipment.

Hardness

Hardness is the resistance of a mineral to scratching. Mineralogists use a relative scale of hardness, called the **Mohs scale**, given below, to test the hardness of a mineral:

1. Talc
2. Gypsum
3. Calcite
4. Fluorite
5. Apatite
6. Orthoclase
7. Quartz
8. Topaz
9. Corundum
10. Diamond

Geo Words

hardness: the resistance of a mineral to scratching.

Mohs scale: a standard of 10 minerals by which the hardness of a mineral may be rated.

Each mineral in the scale scratches minerals earlier in the scale and is scratched by minerals later in the scale. Diamond (with a hardness of 10) is the hardest natural substance known, and the mineral talc (with a hardness of 1) is one of the softest. The way to test the hardness of a mineral is to scratch an unknown mineral with a material of known hardness. If the mineral is scratched, it is not as hard. If the unknown mineral scratches the known material, then it is harder. Here are the hardnesses, on the Mohs scale, of some common materials:

fingernail: a little more than 2
wire nail: about 4.5
knife blade: a little more than 5
window glass, masonry nail: 5.5
steel file: 6.5

Geo Words

luster: the reflection of light from the surface of a mineral, described by its quality and intensity.

Luster

Luster describes the way a mineral reflects light. Luster is either metallic or nonmetallic. Minerals with metallic luster look like polished metal. Nonmetallic lusters are often further described as glassy (or vitreous), waxy, pearly, earthy, or dull. Pyrite and galena have metallic luster. Quartz and calcite have a glassy luster. Feldspar has a pearly luster. (See *Figure 6A–D*).

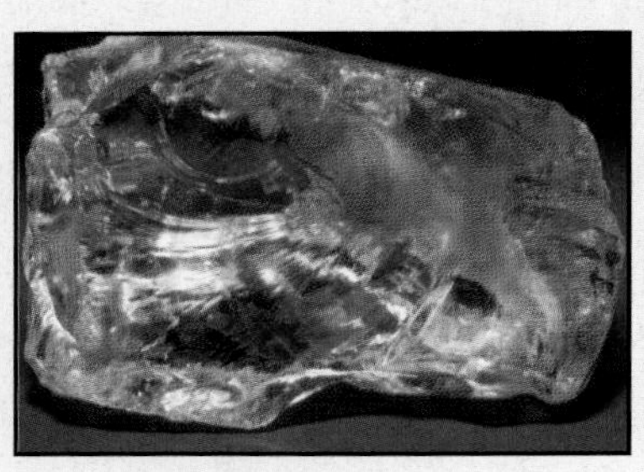

Figure 6A Quartz has a vitreous luster.

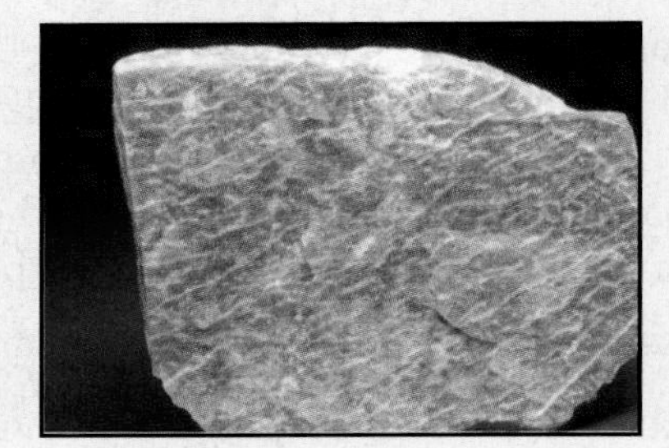

Figure 6B Feldspar has a pearly luster.

Figure 6C Galena shows a metallic luster.

Figure 6D Pyrite has a metallic luster.

Streak

Streak is the color of the powdered mineral. To determine the streak of a mineral, scratch the mineral across a piece of unglazed porcelain tile, called a streak plate. Many minerals have a distinctive streak color. Streak color may be different from the color of the mineral sample. For example, hematite is often dark gray in color, but it always has a red streak.

Specific Gravity

Specific gravity is a ratio of the weight of the mineral to the weight of an equal volume of water, which has a specific gravity of 1. Galena, a mineral containing lead, has a specific gravity of approximately 7. That means it is about 7 times as dense as water. Nonmetallic minerals like quartz, feldspar, and calcite mostly have specific gravities less than 3.

Cleavage

Many minerals have **cleavage**, which means that when they break, they tend to break along regularly oriented planes of weakness, as shown in *Figure 7A–D*. Cleavage planes form along planes of weak atomic bonds in the mineral. For example, mica splits easily into sheets, because there are very weak atomic bonds between the layers of atoms in mica. Galena and halite break in cubes; they have cleavage in three directions, all at right angles to one another. Feldspar has cleavage in two directions, at nearly right angles, but it breaks irregularly in other directions. Some minerals, like quartz, have no cleavage. Quartz breaks into irregular shapes and often shows a curved surface called **conchoidal fracture**.

Figure 7A Muscovite has one direction of cleavage.

Figure 7B Halite has three directions of cleavage. They are at 90° to each other.

Figure 7C Calcite has three directions of cleavage. They are not at 90° to each other.

Figure 7D Feldspar has two directions of cleavage.

Geo Words

streak: the color of a mineral in its powdered form, usually obtained by scratching the mineral on a streak plate and observing the mark it leaves.

specific gravity: the ratio of the weight of a given volume of a substance to the weight of an equal volume of water.

cleavage: the breaking of a mineral along regularly oriented planes of weakness, thus reflecting crystal structure.

conchoidal fracture: a type of mineral fracture that gives a smoothly curved surface.

Crystal Shape

When minerals grow in unconfined spaces, they usually develop a regular crystal shape. Quartz crystals grow as six-sided (hexagonal) columns with pointed tops (*Figure 8*). Garnets often grow in regular twelve-sided shapes called dodecahedra (*Figure 9*).

Figure 8 Quartz crystals grow in hexagonal columns.

Figure 9 Garnets often grow in dodecahedral shapes.

Color

Color is usually the first thing you notice about a mineral, but it is the least reliable property in mineral identification. Many minerals have different colors depending on what impurities are present. Corundum (Al_2O_3) is sometimes tinted red by small amounts of chromium; these crystals are known as rubies. The same mineral tinted blue by small amounts of titanium is called a sapphire. Quartz is usually transparent, but it also occurs in a great many colors, depending on what impurities are present. Some minerals tarnish or change color when surfaces are exposed to air. Many minerals have the same color as other minerals. Many prospectors in the gold-rush days were fooled by pyrite (fool's gold), which has a metallic luster and a color similar to gold. Pyrite has a lower specific gravity than gold, is brittle (gold is malleable), and leaves a black streak on a white porcelain tile (gold has a gold-colored streak).

Other Properties

Some minerals have special properties that make them easy to identify. Some of these properties also make these minerals useful for specific purposes.

- Metals tend to be good conductors of electricity. This makes them useful in the production and distribution of power, and in machinery. Most metals are also malleable (meaning that they can be changed in shape under pressure without breaking) and ductile (meaning that they can be stretched into wire).
- Some carbonate minerals fizz when a drop of weak hydrochloric acid is applied on the mineral. Acid breaks down the chemical bonds in the carbonate and releases CO_2 gas. Acid is a good test to identify the calcium carbonate mineral calcite.
- A few minerals are radioactive, releasing subatomic particles and radiation as the unstable atoms within them decay. For example, uranium minerals can be detected with a Geiger counter.
- Some minerals are magnetic. Magnetite, an important ore of iron, is naturally magnetic.
- Minerals like fluorite that change ultraviolet light to other wavelengths are called fluorescent. A few minerals that store light energy and release it gradually, are called phosphorescence.

Check Your Understanding

1. What is a mineral?
2. Why do different minerals have different properties?
3. Is color a good identifying property of a mineral? Why or why not?
4. What is the difference between cleavage and crystal shape of a mineral?

Understanding and Applying What You Have Learned

1. Refer back to the list of materials used for beverage containers that you compiled in Activity 1.
   a) Which of these materials are made from minerals?
   b) What other materials might be made from minerals and used as beverage containers?
2. Give at least one advantage and one disadvantage to using a native element to produce a product (for example, a beverage container).
3. How might the physical appearance or properties of a mineral be misleading when evaluating its potential for use as a beverage container?
4. Describe at least one possible disadvantage to producing beverage containers from a mineral that has high specific gravity.
5. A student claims that diamond is the hardest mineral because carbon, from which diamond is made, is a very hard element. Use what you have learned about minerals to provide a different explanation as to why diamond is the hardest mineral.
6. Correct the following misconception: "Quartz is always clear or transparent."

7. Make a concept map that demonstrates your understanding of minerals. Include the following terms: mineral, element, rock, ore, compound.

8. When you broke halite with a hammer, you observed smaller cubic-shaped crystals. What would you observe if you shattered a quartz crystal into many smaller pieces? Why?

Preparing for the Chapter Challenge

Write a paper that will serve to help the president of the beverage company understand what mineral resources are. Be sure to define "mineral" and explain how minerals are identified. Also, explain how the arrangement of atoms in different minerals affects their physical properties and therefore their potential for use as materials to produce beverage containers.

Inquiring Further

1. **Mineral make-up of the Earth**

   Investigate the proportions of various materials in the crust of the Earth. Which minerals are most common? Which elements make up most of the minerals?

2. **Metallic and nonmetallic resources from minerals**

   Which minerals are the source of metals like iron, silver, lead, and copper? What are some nonmetallic resources? Which minerals are the sources of these resources?

Activity 3 Where Are Mineral Resources Found?

Goals

In this activity you will:

- Identify the mineral resources and commodities of the United States.
- Identify the mineral resources and commodities within your community and state.
- Understand how different minerals are formed and which minerals are best suited for particular tasks.
- Describe the uses of your state's major mineral resources.

Think about It

Suppose that your community was located on a small island in the middle of the ocean.

- Would you find mineral resources on the island?
- Is it possible to find a single place on Earth that has no resources?

What do you think? Record your ideas in your *EarthComm* notebook. Be prepared to discuss your responses with your small group and the class.

Investigate

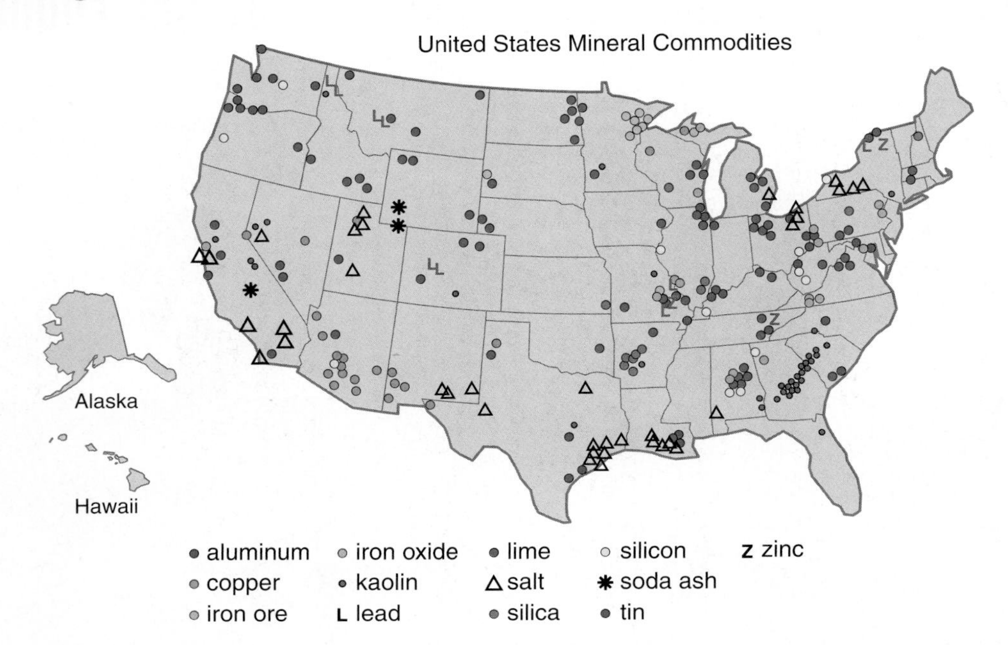

1. State and federal governments compile maps of the locations of mineral deposits. The deposits are usually designated by the mineral commodity name. For example, chalcopyrite is an important copper mineral; the mineral commodity is copper. Examine the map of the United States that shows the distribution of several mineral resource commodities and processing plants.

   a) What mineral commodities are shown on the map?

   b) What minerals would you look for to find those commodities?

   c) What trends do you see in the distribution of the resources? Discuss the possible reasons for this distribution.

   d) If the whole United States were your community, could you find the mineral resources necessary to make all of the beverage containers you use in your community? Explain.

2. Use a mineral resource map of your state (which usually shows resources by commodity, not by mineral name) to answer the following questions:

   a) Are the resources evenly distributed around the state or are they concentrated in some areas?

   b) Which mineral commodities are found in your state?

   c) How do you think geologists decided whether a mineral deposit should be included on the map?

   d) Does a deposit have to be mined to be on the map?

3. Depending on where you live, you may need to obtain mineral-resource maps from a nearby state to create your map. Create a new map of the area within a 100-km radius of your community.

   a) Make your map fit on one piece of graph paper. Include your map in your *EarthComm* notebook.

   b) If there are no commodities within 100 km of your community, how far might you have to go to locate some within your state?

Reflecting on the Activity and the Challenge

In this activity, you identified the mineral resources of the United States and those within your state. This will allow you to determine where the materials needed to make your beverage container will be obtained. If your community does not have all the mineral resources necessary to make all the products you use in your community, you must begin to think about relying on an exchange of resources and materials between other communities in order to make all of the products you use.

Digging Deeper

MINERAL RESOURCES ON EARTH

Mineral Resources in the Ocean

Almost all of the Earth's mineral resources are located on the continents. The waters of the ocean contain staggering quantities of many chemical elements, but the concentrations are for the most part extremely small. For most of these chemical elements the costs of extraction make it impractical to use them as mineral resources. Gold is a good example. Its concentration in the ocean is only 0.011 μg (millionths of a gram) per liter, but that adds up to more than ten billion (10^{10}) total kilograms of gold in the ocean! That's far greater than the known reserves of gold in continental ore deposits. The technology for extracting gold from seawater is so difficult and expensive, however, that it is nowhere close to being economically feasible.

Figure 1 Due to the low concentration of most chemicals in ocean water, it is not practical to "mine" minerals from the waters of the ocean.

Figure 2 Iron-manganese nodules form slowly on the ocean floor.

Great expanses of the deep-ocean floor, especially in the Pacific Ocean, are covered with dark-colored, rounded masses, called iron–manganese nodules, as shown in *Figure 2*. They range in size from golf balls to large fists. They consist of very fine-grained minerals of iron and manganese, with many other chemical elements in smaller concentrations. Techniques for mining them from the ocean floor have been developed, but there are two problems. The sediment stirred up and suspended in the water during mining would have a harmful effect on the deep ocean environment. Also, the open oceans belong to no one country but to humankind as a whole.

Mineral Resources of the Continents

One of the most basic facts of geology is that the Earth's continents are geologically old. The oldest continental bedrock is known to be four billion years old (only half a billion years younger than the Earth itself), and large areas of the Earth have bedrock that is older than a billion years. On the other hand, geological processes that generate the bedrock record on the continents operate on fairly short geological time scales. That means that there has been plenty of time for the geological record of the continents to become extremely varied and complicated. Just a casual glance at a geologic map of United States like the one in *Figure 3* shows a jumble of irregularly shaped areas, colored in with a great variety of colors and patterns corresponding to rocks of different types and different ages. That should leave you with the impression that a large part of the United States has complex geology.

Figure 3 Geologic map of the continental United States.

Only a very tiny fraction of the rocks of a continent like North America are ores. Most rocks, like sandstones, shales, limestones, granites, basalts, schists, and gneisses, do not contain economically valuable minerals (although many are used as building stones, and limestones are used in making cement and glass). Ores are formed only in certain very specific and unusual conditions. Metal mining operations in the United States occupy much less than one-tenth of 1% of the total land area. If they blindfolded you, put a parachute on you, put you aboard an airplane, and arranged for you to make your skydive over a random point in the United States, the chance that you would land on or near an ore deposit is extremely small!

Look again at the geologic map of the United States in *Figure 3*. One very striking thing about it is that large areas are occupied by only one or a few colors representing sedimentary rocks. In these areas, the deeper rocks of the Earth's crust are covered by a blanket of nearly horizontally layered sedimentary rocks that in most places have never been buried very deeply or subjected to ore-forming processes. In contrast, most areas of the western United States and some areas of the eastern United States show a complex pattern of rock types. These areas are called **orogenic belts**. The word *orogenic* means "mountain-building." These are areas where collisions between the Earth's **lithospheric plates** have resulted in great uplift of the land surface to make mountains, which in many places have later been worn down from their originally great heights. In those areas, igneous activity (the movement and crystallization of previously melted rock, called **magma**) has led to deposition of various kinds of ores in certain places.

One of the most important ore-forming processes is called **hydrothermal activity**. Magmas contain many economically valuable chemical elements in very small concentrations. They become concentrated in the water-rich "juices" that are left over after most of the magma has crystallized to form ordinary igneous rocks, because these elements tend not to be included in the main minerals that crystallize from the magma. These juices work their way upward toward the surface. As they move upward they cool, causing a great variety of unusual minerals to crystallize. Many deposits such as the one shown in *Figure 4*, called hydrothermal

Figure 4 The backhoe removes silver ore from a hydrothermal deposit in an underground mine.

Geo Words

orogenic belt: a region that has been subjected to folding and deformation during the process of formation of mountains (orogeny).

lithospheric plate: a rigid, thin segment of the outermost layer of the Earth, consisting of the Earth's crust and part of the upper mantle. The plate can be assumed to move horizontally and adjoins other plates.

magma: naturally occurring molten rock material, generated within the Earth, from which igneous rocks are derived through solidification and related processes.

hydrothermal activity: pertaining to hot water, the action of hot water, or to the products of this action, such as a mineral deposit precipitated from a hot aqueous solution.

deposits, are valuable ores. Much of the copper, zinc, tin, lead, mercury, gold, silver, and platinum, among others, come from hydrothermal ore deposits.

Ores of iron and aluminum, the two most-used metals, are a different story. Almost all iron ore comes from special sedimentary rocks, very rich in iron minerals, that were deposited in the oceans far back in geologic time, about two billion years ago. In the United States, these rocks are mined for iron ore in northern Minnesota and northern Michigan. In contrast, aluminum ore, called bauxite, consists of aluminum oxides that are formed when rocks containing aluminum are weathered at the Earth's surface in warm and humid climates. Some bauxite is mined in the United States, but most comes from other countries.

Mineral Resources around the World

Some countries of the world are richer in mineral deposits than others. Countries that include large areas of very old continental rocks are especially rich in mineral resources. Canada, Russia, Congo, South Africa, Brazil, and Australia are especially rich in mineral deposits.

The United States has abundant energy reserves (coal, oil, natural gas) but is not as rich in most mineral deposits as many other large countries. The United States has fairly abundant iron, copper, and tin deposits, but almost all of the aluminum, as well as most of the ores of several special metals that are important in making steel, like nickel, cobalt, or chromium, must be imported. The United States government maintains stockpiles of important metals, in case supplies from other countries become reduced or cut off in the future.

Check Your Understanding

1. What mineral resources are retrieved from the oceans?
2. What mineral resources are produced through hydrothermal activity?
3. What are the two most commonly used metals and where do they come from?

Understanding and Applying What You Have Learned

1. Look up additional information about the mineral resources found in your state.
   a) How are your state's mineral resources used?
   b) What products are manufactured from them?
2. Refer back to the mineral resources map of your community that you produced. Could any of the minerals or commodities near your community be used to make a beverage container? If yes, which ones? If no, how far would you need to go to locate minerals for making beverage containers?
3. Why is such a small proportion of the area of the United States underlain by ore deposits?

4. Why are many ores associated with orogenic belts? Describe an example of an important ore in the United States that is not associated with mountain building.

5. If producing a beverage container depended primarily upon how close your community was to the source of the primary mineral resource, would your community produce containers made of glass (silica sand), steel (iron ore), or aluminum? Explain.

Preparing for the Chapter Challenge

Prepare a short paper in which you address the following questions:

- Which of the minerals in or near your community could be used to make any container that could hold a liquid?
- Which minerals in or near your community could contribute to the manufacture of a beverage container? What could those minerals contribute?
- If your community were not allowed to import a mineral commodity like aluminum, what impact would the lack of aluminum have? What minerals could be used instead?

Inquiring Further

1. **Worldwide distribution of mineral resources**

   On a map showing worldwide distribution of mineral deposits, find the possible sources of minerals to make beverage containers.

2. **Communities and mineral resources**

   Describe the locations of communities in relation to the mineral resources of your state. What relationships seem to exist? What industries grew because of the raw materials available locally?

3. **The Hall process**

   Research the Hall process used to produce aluminum from aluminum ore. How does the process work? Why did this process impact the aluminum industry? Who was Hall, and why did he become famous and wealthy?

Activity 4 How Are Minerals Found?

Goals

In this activity you will:

- Construct a model of a mineral deposit and map the relative positions of the deposits.
- Understand how geologists explore for mineral resources by conducting a survey to locate ore deposits in another group's model.
- Use your survey results to drill for the ore.
- Understand the necessity and benefits of exploratory surveys in locating minerals.

Think about It

Suppose that your neighbor's dog buried your family car keys somewhere in your backyard.

- What tools or information would you want to help you to locate your keys?

What do you think? Record your ideas in your *EarthComm* notebook. Be prepared to discuss your suggestions with your small group and the class.

Investigate

Part A: Creating a Model of a Mineral Deposit

1. Draw a grid of 1-cm squares on one side of a 2-L (half-gallon) paper orange juice or milk carton. On graph paper draw the same number of squares in the same configuration. Do the same for the bottom of the carton.
   Label the drawings and the carton with corresponding letters.
   Also draw a graph of the top of the carton (what will be the exposed sand surface). Label the drawing.
   a) Write your group's name on the graph paper and on the carton.
2. Use scissors to remove one side of the carton. Place the carton so that the open side is up, as shown.

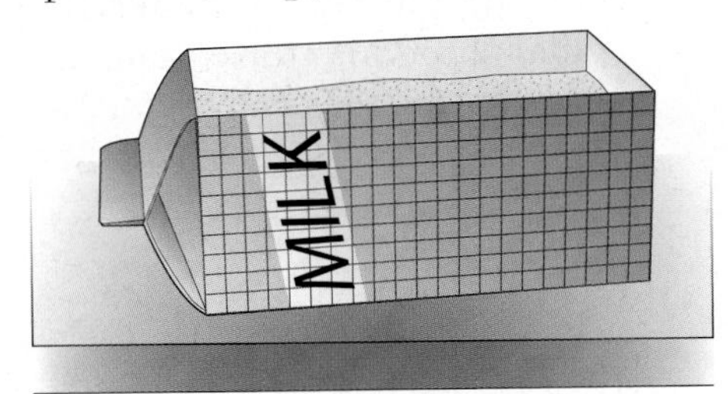

3. Next, fill up the carton to within 2 cm of the top using dampened sand to represent sediment and black sand (magnetite) or iron filings to represent ore bodies. You will need to distribute the "ore" throughout the sand in lines (called veins or ore shoots), in concentrated layers, or in pods. Do not bury the filings too deeply, and make sure that the deposits you produce are large enough to deflect a compass needle.

   Wear goggles.

   a) As you fill the carton, sketch ("map") the position of the "ore" onto the graph paper representations of the side, the end, and the top of the carton.
4. When your model is complete, give it your teacher. Do not let other groups see your model or your sketches, because later each group will try to find the ore in another group's model. Keep your map for later use.

Part B: Exploring a Model of a Mineral Deposit

1. Obtain another group's model to explore. You will have a limited amount of money (an imaginary $100,000) with which to conduct your exploration. You will use a magnetic compass to survey the model. The buried magnetite or iron filings will cause the compass needle to be deflected from north. Each measurement made with the compass will cost you $500.
2. First, run your compass quickly over the model to see the maximum amount of deflection. This will help you scale your measurements. Design symbols to represent no needle deflection, maximum needle deflection, and one or two magnitudes in between. Then plan where you intend to make measurements.
   a) Draft a legend to explain the symbols.
   b) Using a piece of graph paper, make a map showing all the locations on the model where you will measure the deflection of the compass needle.

3. Now it's time to complete the magnetic survey. Pick one person to operate the compass, one to record the measurements for each location in a data table, and a third person to plot the data on your map.
   a) Measure at each location marked on your map. On the map and in the data table, write the symbol for the strength of the measurement based on the movement of the compass needle.
   b) After a few measurements, you may decide to alter your plan or make more or fewer measurements. Record any changes you make in your survey plan, along with the reasons for them.
   c) When you are satisfied that you have as many measurements as you want (or can afford), draw a cross section of what you think your deposit looks like.

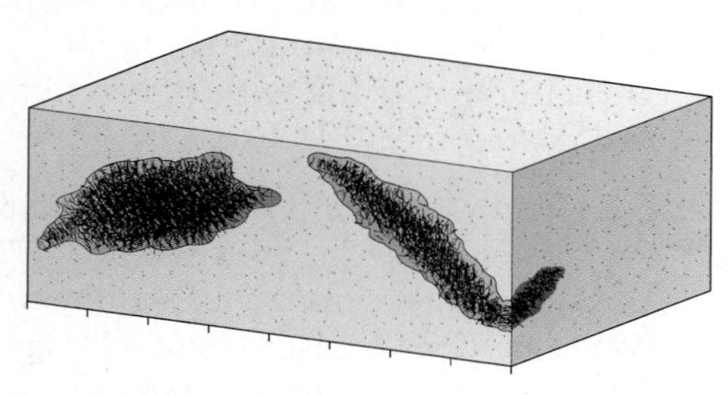

4. To confirm your hypothesis of the location of the deposit, select sites to "drill" holes with a soda straw. Each drill hole will cost your group $30,000.
   a) Mark the locations of your planned holes on your map. Be sure to total your survey account, because you cannot drill more holes than your company can pay for.
5. Mark the locations of the drill holes on the surface of the sand with "survey flags" (toothpicks and masking tape). Carefully push a straw all the way down into a location marked by a survey flag. (It may be necessary to dampen the sand again if enough time has passed for the model to dry out. The sand should stick together as a column inside the straw.)
6. Carefully pull out the straw and inspect your core through the clear plastic, or insert a wooden dowel and push the core out onto scrap paper for viewing.
   a) Estimate the percent recovery of the "rocks" in the drill core.
   b) Create a vertical description (called a log) of each hole by recording the type of material at regular intervals (for example, every centimeter). Supplement your descriptions with sketches.

"Hole A"

Depth	Description
0-1 cm	100% light tan sand
1-2 cm	mostly sand (75%), small amount of black iron deposit
2-3 cm	100% black iron deposit
3-4 cm	50-50 light tan sand and black iron deposit
4-5 cm	mostly sand (80%), small amount of black iron deposit
5-6 cm	100% sand
6-7 cm	100% light tan sand
7-8 cm	100% light tan sand

7. Complete "drilling" at all locations.
   a) When all the holes have been drilled and logged, draw a cross section of the ore deposit. On a piece of graph paper plot the drill holes and the drilling results. Connect the plotted points to show where the ore is located based on your observations.
   b) Record how much money your group spent on your project, and return the models to your teacher for use in the next activity.
8. Report your exploration venture to the class, and discuss the various groups' outcomes and experiences. Obtain a copy of the cross section produced by the group that explored your model. However, do not yet reveal the original sketch (map) of your model to the other group.
   a) In your *EarthComm* notebook summarize your class discussions.

Dispose of the straws after use or mark them as lab equipment. Wash your hands after you have completed the activity.

Part C: Mineral Commodities in Your Community

1. In the previous activity you researched the mineral commodities located near your community. Use the information you collected to answer the following questions:
   a) What types of deposits do the minerals found in your community typically form?
   b) Which of the minerals have been found by geochemical or geophysical techniques?
   c) Which of the mineral resources are located by geologic mapping?
   d) Are techniques of seismic wave propagation and gravity surveys used in your state? For which mineral deposits are these methods used?

Reflecting on the Activity and the Challenge

Mineral-bearing ores are difficult to locate from the surface. This activity illustrates that exploration is expensive, and not always accurate. However, the costs of your project would be much higher if the drilling had been done by random sample without any survey. Although the ore body shown on your cross-section and exploration map probably differs in size or shape from the actual ore body, you would not have found it at all without the survey to guide your test drilling. Think about how you can use what you have learned to address an important aspect of your **Chapter Challenge**—explaining how mineral resources used to make beverage containers are found.

Geo Words

geologic map: a map on which the distribution, nature, and age relationships of rock units are recorded.

geochemist: a geologist who studies the distribution and amounts of chemical elements in minerals, ores, rocks, soils, water, and the atmosphere, and their circulation in nature.

geophysicist: a geologist who studies the physical properties of the Earth, or applies physical measurement to geological problems.

Digging Deeper

RESOURCE EXPLORATION

Finding Minerals

Exploration for mineral resources relies heavily on experience, observation, data, analysis, inference, and a good understanding of geology, geochemistry, and geophysics. Even with all that, it still involves quite a lot of uncertainty. There is never a guarantee that minerals will be in the place prospectors expect them to be. Also, even if the minerals are located, there is no guarantee that the concentration will be high enough for the mineral to be extracted profitably.

Exploration is very expensive. Drilling is the most expensive way to explore, but it is the only sure way to confirm the type and amount of minerals present. In mineral exploration, drill holes are often at least 300 m deep, and they may cost $150 per meter. In oil exploration, a single drill hole may cost millions of dollars. To avoid spending money on "dry holes," geologists use other techniques to find out more about what is below the surface and to eliminate areas with low exploration potential.

Mapping

Geologists map the surface by examining rock types in the field. Geologists know that all mineral deposits are associated with specific kinds of rocks. Therefore, they can look for those specific rocks as "guides to ore," before doing any other kind of work. They then look for folds, faults, and fractures in the rocks, any unusual colors, and rock formations that seem inconsistent with the surroundings. They also may take rock samples to analyze in the laboratory. Then they record all the information on a map. **Geologic maps** like the one shown in *Figure 1* help geologists infer what lies below the surface.

Figure 1 Geologic map of Southern Colorado.

Geochemistry

There is no guarantee that what is on the surface matches what is deeper underground. **Geochemists** are geologists who specialize in analyzing the chemistry of rocks and minerals. Samples taken at the surface, from pits dug in the ground, or from a few drilled holes, can be analyzed to determine all the chemical elements present. These elements may be present in very small concentrations, often just a few parts per million (ppm), as halos above or around valuable mineral deposits. In addition to sampling rocks and soil, geochemists may also sample vegetation and water, because elements from the deposit can be absorbed by plant roots or dissolved in water. Their presence can help geochemists get an idea about whether a mineral deposit might exist below the surface. However, the geochemist cannot tell how far below the surface a mineral deposit might exist.

Geo Words

seismic wave: a general term for all elastic waves in the Earth, produced by earthquakes or generated artificially by explosions.

geophone: a seismic detector placed on or in the ground that responds to ground motion at its point of location.

Geophysics

The **geophysicist** is another important part of the exploration team. Oil companies and mineral companies employ thousands of geophysicists worldwide. Geophysics involves measuring properties of the Earth and of specific rocks and minerals. In oil exploration, the most common geophysical technique is **seismic-wave** reflection. A vibration is created by a small explosion or by striking the ground with a steel plate. The vibrations travel through the crust until they are reflected back to the surface by a rock layer. The returning vibration is recorded on a sensitive instrument called a **geophone**, as shown in *Figure 3*. The time it takes for the vibrations to return to the surface tells the geophysicist about the depth to the reflecting layer. Using this information, the geophysicist can more accurately determine if the subsurface is folded, faulted, or has other features where oil, gas, or mineral deposits may accumulate.

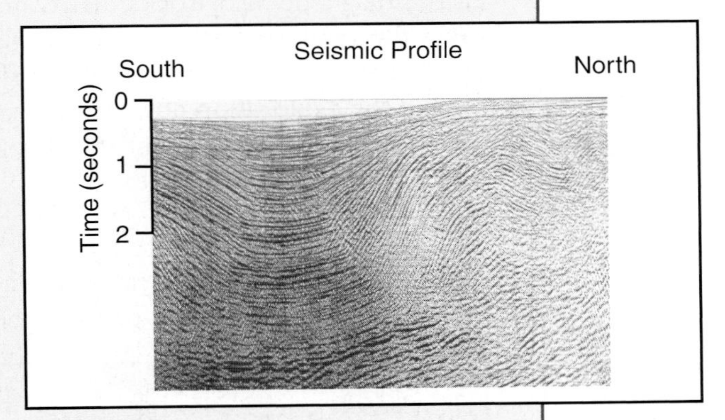

Figure 2 Seismic section.

Figure 3 A geophone stuck into the Earth and connected to the geophone cable with two clips.

Geophysicists also measure the local strength of the Earth's gravity field, which is affected by the density of the rocks below the Earth's surface. Dense rocks cause a slightly greater gravitational attraction than the less dense rocks around them, so gravity at the surface is slightly greater. Most ores are denser than the common rocks in the Earth's crust, like granite, basalt, sandstone, or limestone. The gravity field over most ore bodies is therefore slightly stronger than over nearby areas. An instrument called a **gravimeter** as shown in *Figure 4*, which measures the acceleration due to gravity, can detect these small differences in the Earth's gravity field. You would weigh slightly more over such an ore body than elsewhere, although not enough to feel it.

Other Techniques

An instrument called a **magnetometer** is used to detect changes in the Earth's magnetic field. Rocks that contain abundant iron-bearing minerals affect the local magnetic field. Geologists can also measure how well rocks conduct electricity. Most of the electrical current that geologists put in the ground for exploration flows through water in the pore spaces of rock. If a rock conducts electricity very well, it is likely to contain either a lot of water or a lot of metallic minerals.

Geo Words

gravimeter: an instrument for measuring variations in Earth's gravitational field.

magnetometer: an instrument for measuring variations in Earth's magnetic field.

The Bottom Line

No exploration technique can give a complete picture of the subsurface. The exploration geologist must be very good at reading the story of the rocks and discovering all the clues to successfully find minerals. Millions of dollars are spent on a typical exploration venture, but only a very small number of such ventures become economically profitable mines.

Figure 4 Geologists gather data using gravimeters. The silver plate at the lower left is a gravimeter.

Check Your Understanding

1. What are the benefits of mineral exploration?
2. What geophysical technique did you use to survey the model in class?
3. How do maps help geologists be more successful in exploration?

Understanding and Applying What You Have Learned

1. Compare your original diagram of the model your group built to the map made by the group that surveyed it.
   a) What differences exist between the two maps? Why do they not match?
   b) What other exploration techniques might you have used to gather more information about your model? What instruments would have been needed?
2. a) How could you have reduced the cost of your exploration?
   b) How do you think mineral explorers decide how much money and time to spend on exploration?
3. How do exploratory surveys help to reduce the cost of obtaining mineral resources?
4. The map on the right shows the thickness of a rock formation made of limestone and dolostone. The data for the map came from drilling into the Earth, taking core samples, and measuring the thickness of the formation in each core. The map also shows the percentage of dolostone in the formation. The dolostone contains the highest percentage of lead-zinc ore. Use a copy of the map to complete the following questions:
   a) Areas where dolostone exceeds 50% have profitable ore. Color these areas green.
   b) What is the thickness range of dolostone in the profitable area?
   c) What is the surface area (in square feet) of dolostone in the profitable area?
   d) If the average profits on the dolostone is $0.30 per cubic yard, what is the approximate company gain?
   e) Assume that the top of the formation is level. Draw a cross section between core holes 1, 5, 7 and 8. Indicate the profitable area on the cross section.

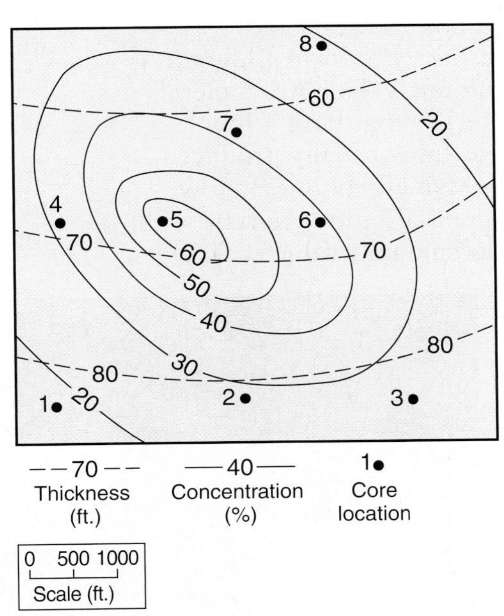

Preparing for the Chapter Challenge

On the basis of the mineral commodities located near your community and the types of deposits such minerals might form, discuss how you could explore for the type of deposit most economically feasible for making a beverage container. Answer the following questions:

- How big do you expect the deposit to be?
- Is the deposit visible from the surface?
- How deep might the deposit be?
- Does a model exist for the deposit? How does the model affect the way you would explore for it?
- What laws apply in your community to mineral exploration and staking claims?

Inquiring Further

1. **Mineral supply and demand**

   Are common minerals and materials likely to have a high or low economic value? Are minerals and materials that have a low economic value usually produced in large or small volumes? How far are minerals and materials with low economic unit value typically transported? What about minerals and materials with high economic unit value? Explain why this is so.

2. **Maps and mineral deposits**

   How do geologists decide which mineral deposits should be included on survey maps?

Activity 5

What Are the Costs and Benefits of Mining Minerals?

Goals

In this activity you will:

- Design a procedure for mining the model used in Activity 4 that optimizes time and cost, and reduces environmental impact.
- Analyze the outcomes of mining the model to understand demand and market value, and the costs of labor, refining, transportation, and environmental reclamation.
- Understand the benefits and drawbacks to mineral exploration and mining.
- Analyze the economic importance of mining in your state.

Think about It

Minerals are excavated from the Earth through mining.

- What is a mine, and what does it look like?
- What does the process of mining involve?

What do you think? Have you ever seen a working or abandoned mine? In your *EarthComm* notebook, sketch a picture of a mine that you have seen or one that you imagine. Describe what processes take place in the mine. Be prepared to discuss your ideas with your small group and the class.

Investigate

1. Using the same model that you surveyed in Activity 4, make a plan for excavating the ore. Here are the rules for excavation:

   - Your mining company has \$200,000 to spend on this project.
   - You have 15 minutes to complete mining.
   - It costs \$1000 to remove each gram of sand from the model.
   - It costs \$1000 for every minute spent mining.
   - It costs \$500 to replace each gram of sand for environmental restoration after the mine is closed.
   - If your company disrupts more than 25% of the surface of the model, you must pay a \$5000 restoration fee.
   - Your company will receive \$4000 for each gram of ore recovered.

   Using the cross section you drew in the previous activity, formulate a plan for removing the ore. Consider the factors of ore concentration, depth, and surrounding material in your plan. Discuss the ways you will keep your excavation localized to avoid the \$5000 restoration fee. Be sure every member of your group understands his or her role in the mining operation before you begin, because time is worth money.

   a) Outline your plan in your *EarthComm* notebook.

2. You will use plastic spoons to remove the sand and ore. A timekeeper will start the clock on your mining. As you remove material from the model, separate the sand from the ore by placing a sheet of scrap paper over the ore and touching the "unrefined" ore with a magnet through the paper. Transfer the magnet, the paper, and whatever ore sticks to it to another container. When you lift the magnet away from the paper, the released ore will be caught in the container. Do this several times to remove all the ore from the sand. Save all materials to take to the "refinery."

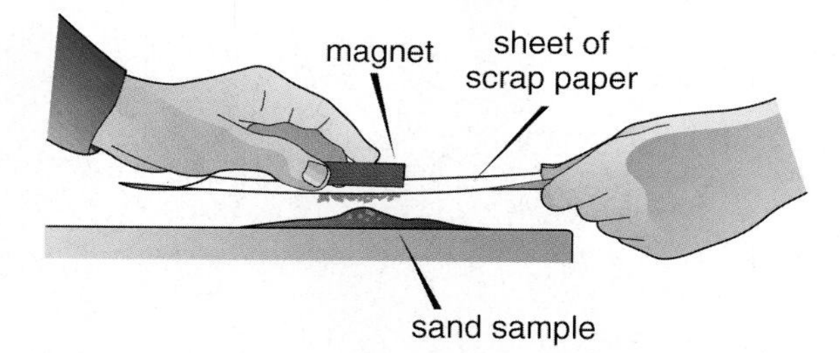

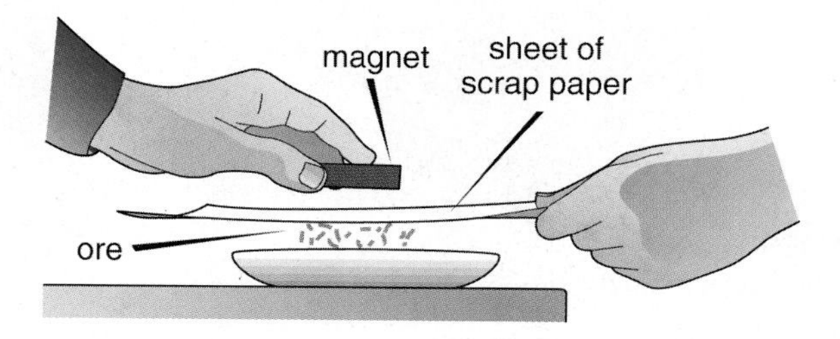

3. When you bring the containers of sand and ore and the remains of the model to the refinery (weigh-in station), the clock stops for your group.

   a) Write down the number of minutes you used in mining.

Wear goggles throughout the activity. Wash your hands when you are done.

4. Calculate the costs for environmental reclamation.
   a) Find the mass of the sand and record it in your notebook.
   b) Record in your notebook the total cost for returning the sand to the model.
   c) Measure the surface area of your mine and compare that to the total surface area of the model. Did you disturb more than 25% of the surface? If so, charge your company $5000 for environmental reclamation.
5. Calculate your income.
   a) Find the mass of the ore and calculate your company's income determined by the current market value of the ore ($4000 per gram).
   b) Write the information in your *EarthComm* notebook.

Reflecting on the Activity and the Challenge

Your model mine illustrated some of the factors involved in recovering mineral deposits from the Earth. The depth of the deposit, the type of material around it, and the concentration of the ore are all geological factors that influence mining. These factors influence the decision to mine and the costs of the mineral resources used to produce goods, such as containers.

Digging Deeper

REMOVING MINERAL RESOURCES FROM THE GROUND

Types of Mining

There are two main ways to get minerals out of the ground: surface mining and underground mining. The method used depends on the location, size, depth, and grade or quality of the ore. Both surface mining and underground mining require planning and money to ensure that the ore is removed in the most cost-effective, safe, and environmentally sound way possible.

Surface Mining

Geo Words

spoil: overburden and other waste material removed in mining, quarrying, dredging, or excavating.

In the United States, most minerals are removed from the ground by a kind of surface mining called open-pit mining. This method of mining has advantages when the ore body is very large and close to the surface. Large earth-moving equipment removes the soil and rocks above the ore. The original topsoil is stored in large piles for later use in reclamation, and the underlying rock material that covers the ore body, called **spoil**, is stored elsewhere. The operation continues from a series of large benches and roads that wind their way down into a large, deep pit. The benches serve as both working areas and as roads to haul the ore to the surface. Ore and waste in the pit is drilled, blasted, and loaded into huge trucks that carry it to crushers or waste piles. The crushed ore is then transported to facilities for storage and refining. The Bingham Canyon copper mine, shown in *Figure 1*, near Salt Lake City, Utah, is the largest open-pit in the world. It is approximately two miles in diameter and more than a half mile deep.

Figure 1 The Bingham Canyon copper mine in Utah is the largest open-pit mine in the world.

Resources like sand, gravel, coal, phosphate, as well as iron ore and copper ore, are extracted from surface mines. Much of the coal in the western United States comes from open-pit mines called strip mines. Strip mining is used where the deposit is in the form of thin but widespread sheet-like layers near the surface.

As world reserves of petroleum become scarcer in the coming decades, the great reserves of oil shale in Colorado and Wyoming are likely to be mined from strip mines, for conversion to petroleum. In Canada, deposits called tar sands, which contain enormous reserves of petroleum, are already being mined from strip mines, and are likely to be a major source of petroleum for both Canada and the United States in the future.

Figure 2 Strip-mined land not yet reclaimed.

Dimension stone quarries are special types of open-pit mines. Large blocks of granite, marble, limestone, or sandstone are removed intact (not crushed). Quarrying is used when the desired final product is large blocks of the rock itself.

Underground Mining

Underground mining is used when the mineral resources lie deep beneath the surface, or when an ore body has irregular geometry. Shafts and

tunnels provide access and ventilation. Depending upon the orientation of the ore body, the mineshaft may be vertical, horizontal, inclined, or shaped like a corkscrew. Tunnels, on the other hand, are close to being horizontal. After the ore body is reached, additional tunnels are excavated through the body and the ore is hauled to the surface by trains, loaders, trucks, or elevators. Many mining operations use trackless semi-automated equipment to speed extraction and reduce the number of people exposed to the hazards of working underground. In most underground mines, large "rooms" are formed when the ore is removed. Pillars of ore and/or waste rock material are left to support the mine roof. The grade of the ore and the strength of the overlying rock help determine the size of the rooms and the support pillars. Modern techniques include back-filling tunnels to recover ore pillars and prevent cave-ins and environmental problems like acid mine drainage.

Figure 3 Silver ore is retrieved by blasting tunnels in underground mines.

Environmental and Safety Concerns

A surface mine disturbs the land in many ways. Removing **overburden** (rock material that overlies a mineral deposit) and vegetation destroys established ecosystems. Patterns of surface water runoff are changed. Dissolved chemicals from some of the rocks and minerals exposed at the surface during mining can contaminate surface water and ground water.

Geo Words

overburden: rock material that overlies a mineral deposit and must be removed prior to mining.

All states require mining operations to file environmental reclamation plans before any ground is disturbed. During reclamation, the mining company attempts to stabilize the ecosystem, returning the land to productivity unrelated to mining. The first step is to spread the spoil back over the strip-mined area to restore the original topography of the land. Then any available topsoil from the original stripping is spread over the spoil, and native trees and shrubs are replanted.

Figure 4 After mining operations finish at a site, native vegetation is replanted.

Mining involves heavy equipment in both surface mines and underground mines, and worker safety is an ongoing concern. Underground mine disasters involving the collapse of tunnels are much less frequent in the United States than in many other countries, but disasters do occasionally happen. Most mining deaths in the United States are related to improperly maintained equipment, failure to wear or use safety equipment, or working in unsafe locations within the mine.

Many mining operations can also expose workers to hazardous materials. Long-term exposure to certain minerals causes health problems for miners and people who process mined ore. Silica dust, heavy metals, asbestos fibers, and other materials can cause serious illnesses. The presence of these materials and the levels of exposure must be constantly monitored to protect mine workers. This raises the cost of mineral excavation.

Check Your Understanding

1. What are the two main kinds of mining? Which kind did your model simulate?
2. What types of mineral resources are extracted from surface mines?
3. What are some of the hazards associated with underground mining?

Understanding and Applying What You Have Learned

1. In your simulation, what real-life mining cost did the fee of $1000 per minute represent? What real-life mining cost was represented by the fee of $500 per gram of sand to be replaced?
2. Analyze your group's mining venture.
   a) Did your company make a profit or a loss?
   b) What factors affected your profitability?
   c) What was the largest expense associated with your mining venture?
3. Imagine your model enlarged to a depth of one meter.
   a) What changes would you incorporate in the plan to recover ore?
   b) What risks would be involved?
4. How might delays like equipment repair, environmental claims, labor strikes, etc., affect your profit?
5. A company has approached your community leaders, seeking approval to open a new mine nearby. Both the mine and the associated refining plant will employ people in your community. In class, designate some members as government representatives, some as company representatives, and the rest as members of the community.
   - If you are designated a government representative, come up with at least three questions to ask the company representatives about the mine.
   - If you are a company representative, decide on the commodity to be mined and the method by which it will be extracted. Prepare a brief presentation on the benefits of the mine and be prepared to answer questions.
   - If you are a member of the community, come up with at least three questions for the government council and company representatives.

   a) Write your questions or your presentation points in your notebook.
   b) With your class members role-playing the parts outlined, conduct a town hall meeting to discuss the pros and cons of opening the mine. Write the questions and answers given in your notebook. After the council, discuss as a class how the meeting went, and summarize the outcome in writing.

Preparing for the Chapter Challenge

Choose a mineral commodity in your community that could be used to make a beverage container and research how it is mined. You may want to take a class field trip to the mine. Use the questions below to conduct an interview with an engineer or geologist at the mine. Write an article about the mine. Your article should include the following information plus other information you have learned in previous activities:

- Is the deposit mined from the surface or underground?
- How large an area does the mine affect on the surface?
- How deep is the mine?
- How many people work at the mine?
- What happens to the minerals that are removed from the ground?
- What effect, if any, does the mine have on the environment?
- What laws apply to mining in your community?

Inquiring Further

1. **Mineral resources and state revenue**

   Visit the *EarthComm* web site, or use encyclopedias, almanacs, or tourist information, to investigate the contribution of mining to your state's economic structure. How much of the state's revenue is generated by mining-related operations?

2. **Environmental impacts of mining**

   What kinds of environmental impact result from surface mining? From underground mining? What kinds of environmental impact regulations for mining are in place for your state? Visit the *EarthComm* web site to get you started with your research.

3. **Sources of minerals**

   Investigate the sources of minerals used in local manufacturing. Why are many minerals found in the United States imported from other countries? Why are most construction materials obtained from local sources?

Activity 6 How Are Minerals Turned into Usable Materials?

Goals

In this activity you will:

- Model the separation of a mineral from a rock.
- Model the separation of an element from a mineral.
- Understand some of the techniques involved in mineral processing and the problems associated with them.

Think about It

When making your favorite cookies, you mix the ingredients in the right proportions and in the right order. You heat the cookie dough to create a chemical reaction that will result in cookies. Now imagine taking one of the cookies and trying to get the original ingredients back out of it. Imagine if the only way to get flour was to extract it from cookies.

- How would that change the cost of flour?
- How would it change how careful you are about the way you use flour?

What do you think? Record your ideas in your *EarthComm* notebook and be prepared to discuss them with the class.

Investigate

Part A: Separating a Mineral from a Rock

1. Fill a clear-plastic jar one-third full of crushed sulfide ore and steel shot. Keep a few pieces of ore and shot aside for comparison purposes during the investigation.
2. Add enough water to cover the ore and shot with 1 cm of water. Cover the jar securely with a lid.
3. Wrap the jar in a towel and shake the jar for two minutes.
4. Screen the mixture by pouring it over a wire screen into a plastic cup to collect the slurry.
5. Observe the mixture you poured off (the slurry) and the ore and shot.
   a) Record your observations in your *EarthComm* notebook.

6. Return the oversize ore pieces and steel shot to the jar and recycle the water by pouring the captured water back into the container. Repeat the process three to five times.
7. After the last repetition of the process, do not recycle the materials. Add about a tablespoon of water to the jar to retrieve the settled ore by sloshing it around and pouring it into the plastic cup.
8. Save the slurry (the screened liquid and ore) by pouring it into a plastic cup.
9. Separate the large pieces of ore and steel shot and put them on a paper towel to dry.
10. Add 4-6 teaspoons of liquid bubble-bath solution to the plastic cup of slurry and stir thoroughly. Place a long straw with a squeeze bulb in the mixture and blow gently.
    a) Make observations about what happens and record this in your notebook.
11. Scrape off the bubbles with an index card and place them on a paper towel to dry. Repeat the process three times.
    a) Record your observations each time.

Wear goggles and lab apron. Dispose of the straw after the activity.

Part B: Separating an Element from a Mineral

1. Fold a piece of circular paper as shown in the diagram. Open the paper so that it forms a cone. Place the paper cone in a funnel.

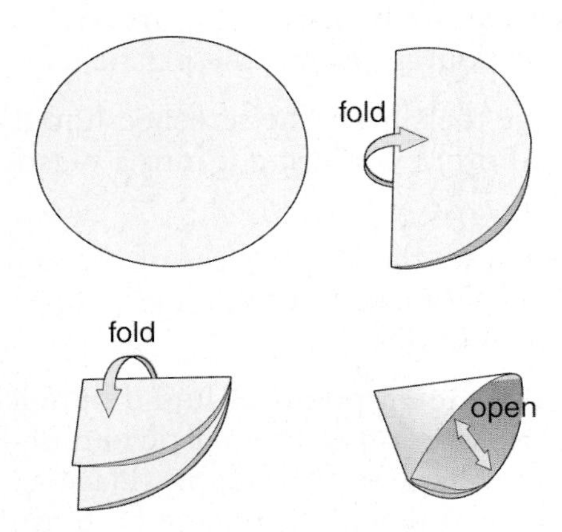

2. Fill the cone three-quarters full with crushed copper oxide ore.
3. Hold the funnel over a collection bottle. Gently pour vinegar through the funnel. The material you collect is the "leaching solution."
4. Put the funnel over a second bottle. Pour the leaching solution from the first bottle back through the funnel.
5. Repeat Step 4 several times to make the leaching solution as strong as possible. A greener solution indicates increasing concentration of copper.
   a) Record your observations after each pour.
6. Clean the battery contacts of a dry-cell battery by rubbing them with steel wool. Use alligator clips to clip insulated wire to connect a spoon to the negative terminal of the battery. Use the insulated wire and clips to connect a lead fishing weight to the positive terminal of the battery.
7. Suspend the lead weight and the spoon in the leached solution so they do not touch each other, as shown in the diagram. Let them sit until you see a change on the spoon.
   a) Keep track of the time. Record the changes on the spoon.

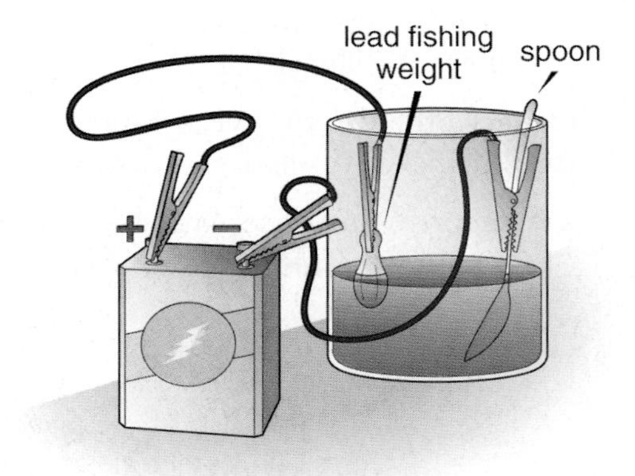

8. Switch the clips to the battery so that the spoon is attached to the positive terminal and the lead weight is attached to the negative terminal. Let them sit until you see a change.
   a) Record your observations.

Wear goggles, gloves, and a lab apron. Do not leave the battery unattended while it is connected to the objects in the solution. Mark the spoon as lab equipment.

Reflecting on the Activity and the Challenge

In the first part of the investigation, you modeled the separation of minerals from rocks. In the second part of the investigation, you examined how elements are removed from minerals. These investigations should help you to think about the processes needed to prepare a mineral resource for actual use.

Digging Deeper

PROCESSING MINERAL ORES

Gold is one of the few elements that can be found in its native state as a mineral. If gold occurs as placer deposits (nuggets in stream gravel) or in a vein with quartz, it does not take much processing other than to separate it from the host rock and then collect enough to melt into a bullion bar. Most of the gold produced in the United States, however, does not come from placer or vein deposits; most comes from ore deposits in the western United States that have such fine flakes of gold disseminated throughout the rock that the gold cannot be seen with the unaided eye. Deposits like these represent a special challenge in separating the gold from the rest of the rock.

Geo Words

smelting: melting ores to separate impurities from pure metal.

The rock containing the gold is first crushed to a very small size. The rock is mixed with a cyanide solution, and the cyanide dissolves the gold out of the rock. The solution containing the gold is passed through a charcoal filter and the gold is precipitated onto the charcoal. The cyanide can be reused. The gold is then separated from the charcoal.

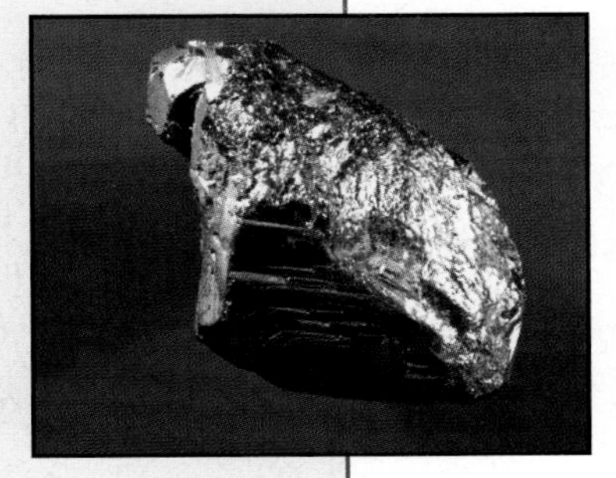

Figure 1 Gold ore.

Scientists have found that some naturally occurring bacteria can help to separate minerals from rocks, and elements from minerals, without using so many potentially dangerous chemicals. One such bacterium is called *Thiobacillus fero-oxidans*. This bacterium was discovered in coal-mine drainage waters in 1947. It prefers a warm environment with a pH of 2 to 3.5, lots of sulfur, and a little nitrogen and carbon dioxide. *Thiobacillus* generates its energy by catalyzing the oxidation of iron. Because many copper minerals contain iron and sulfur, *Thiobacillus* can be used to help separate copper from some minerals. Other bacteria are being experimented with to help process gold without the use of cyanide.

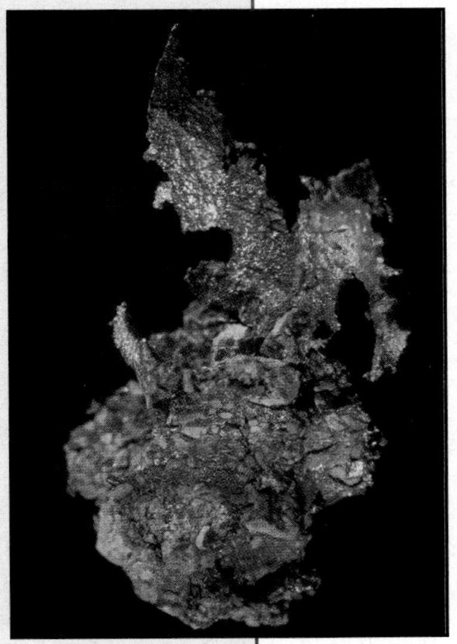

Figure 2 Native copper.

Some minerals cannot easily be leached with a chemical or eaten by bacteria. Iron and some copper minerals have to be **smelted** in order to separate the useful elements from the less useful elements. In the case of copper, the copper is wanted, and the iron and sulfur are waste. In the case of iron, the iron is wanted, but not the oxygen or sulfur. Smelting involves the use of high temperatures in order to break the chemical bonds holding the atoms together. In the past, smelters emitted a lot of pollution, especially sulfuric acid, which tends to cause acid rain. Modern smelters use filters to

Figure 3 Agricultural lime makes use of waste from zinc mines.

collect the harmful gases before they escape into the atmosphere. Copper smelters produce liquid sulfuric acid, which is sold to companies that use it to produce fertilizer. Both copper and iron smelters have a solid waste product called **slag**. Slag is recycled as a construction material and is used as ballast on railroad tracks.

A consequence of all mineral processing is the production of enormous quantities of waste. In a typical copper mine, for every 1 kg of copper that is produced there is 200 kg of waste rock to be disposed of. For a typical gold mine, for every 1 kg of gold produced, there is approximately 1600 kg of waste. For a typical bauxite mine, for every 1 kg of aluminum that is produced, there is 3 kg of waste. This waste is one of the biggest environmental problems faced by the mining industry, because it must be stored on the mine property or at the processing site. Although uses have been found for some waste products, many more are waiting for a clever person to come along and find a use for them.

Geo Words

slag: solid waste product that comes from the smelting of metal ore or when making steel.

Check Your Understanding

1. Explain how gold is separated from rock.
2. What is smelting?
3. Identify a problem associated with mineral processing.

Understanding and Applying What You Have Learned

1. Answer the following questions about Part A of the investigation:
   a) What is the purpose of the jar, and what does it represent in a mineral processing plant?
   b) What is the purpose of the steel shot, and what does it represent in a mineral processing plant?
   c) What is the purpose of the screen?
   d) Why is this process necessary?
   e) Froth flotation is a method of separating the metal-bearing mineral from unwanted material in an ore. Ore is ground to a powder and mixed with water and frothing agents. Air is blown through the mixture. The mineral particles cling to the bubbles, which rise to form a froth on the surface. The waste material settles to the bottom. The froth is skimmed off and the water and chemicals are removed, leaving concentrated metal. Explain how you modeled froth flotation in the investigation.
2. Answer the following questions about Part B of the investigation:
   a) Is all of the copper mineral collected after a few trials?
   b) What happened to the vinegar solution during the leaching experiment?

c) Were copper ions attracted to the anode or cathode during the first observation? How about the second time?
d) Why didn't the copper ore stay at the bottom?
e) What happens to the waste?
f) How could the waste be used?
g) Where does the copper mineral go from here?
h) Why did the changes take place in the vinegar solution?
i) How long did you leave the spoon in the solution?
j) Would a longer time make a thicker coating?
k) What would happen if you added more solution?
l) What happened when the wires were switched?

3. Explain how the properties of your state's minerals allow them to be used in their present states for various applications. What processes must your state's mineral resources go through to be refined prior to use?

Preparing for the Chapter Challenge

Pick a mineral deposit in your area that could contribute to the manufacture of a beverage container (the same deposit you have been studying) and research how the mineral from that deposit is processed to produce a final material. Create a flow chart that illustrates the process from mining to finished material. Write a report in your notebook based on your research.

Consider what impact mineral processing has on the environment and how scientists and engineers minimize that impact. Examine the mineral deposit you researched in the context of how the deposit and its use (past or present) has affected the environment. Consider all aspects of the environment, including air, water, land, plants, animals, and the related impact of transporting, using, and disposing of the related products.

Inquiring Further

1. **Minerals used to make common materials**

   Using data sources from reference books and the Internet, make a table that gives information on several common materials used in daily life that are derived from mineral resources.

   Materials to investigate: aluminum, copper, gold, gypsum, magnesium, silver, steel, titanium, zinc.

   Information: major ores, percentage produced in the United States, where produced in the United States, methods of production, major uses, annual consumption in United States, percent recycled, known world reserves, estimate of how long reserves will last, given annual world consumption.

Earth Science at Work

ATMOSPHERE: *Automotive Engineer*
New design techniques and the use of new materials can make cars lighter and more aerodynamic. This translates into a car that is more energy efficient.

BIOSPHERE: *Bioengineer*
The human body is a complex machine that can break down. Medical technology has developed a variety of artificial body parts to help people continue to enjoy an active lifestyle.

CRYOSPHERE: *Groundskeeper*
The removal of ice from sidewalks and roads often involves the use of chemicals. The sources of these chemicals and their effect on the environment are important considerations for a groundskeeper.

GEOSPHERE: *Commodities Broker*
The value of shares in a mining company is very dependent on the quantity and types of minerals that have been discovered, as well as the potential outcome of future explorations.

HYDROSPHERE: *Plumber*
The mineral content in water can cause damage to home plumbing. Water softeners can be installed to avoid some of the problems associated with "hard" water.

How is each person's work related to the Earth system, and to mineral resources?

3

Water Resources ...and Your Community

Water Resources ...and Your Community

Getting Started

A person requires less than one gallon of fresh water a day to survive. That's what you need to replace the fluids you lose daily. Yet the per capita use of fresh water in the United States is nearly 2000 gallons per day.

- Why is so much water used each day?
- How does your use of water change Earth systems?
- How do changes in Earth systems affect your use of water?

What do you think? Look at the diagram of the Earth systems at the front of this book. Think about where water is stored in the various spheres within the Earth system. Think about how water moves from one sphere to another. Use your reflections about water in the Earth system to answer the three questions above. Be prepared to discuss your answers with your small group and the class.

Scenario

Developers would like to build an industrial park, a mini-mall, and a planned residential area in your community. They estimate that the industrial park will use 100,000 gallons of water per day and the mini-mall will use 20,000 gallons per day. The planned residential area will include two city parks, one community hospital, and one golf course. As a member of a citizens' planning group, you have been assigned the task of providing the city or county planners with information related to these increased needs.

Chapter Challenge

Your report will need to include enough information to answer the following questions:

- How much water does the residential development require?
- Does your town have enough extra water to supply the four new developments?
- If your town does not have enough extra water, what are ways that the town could increase its supply?
- Will the construction and operation of these developments pollute your town's drinking water? What factors would determine that?
- Will the addition of these developments become a strain on the water-treatment capabilities of your town? Will they add a lot of expense?

Assessment Criteria

Think about what you have been asked to do. Scan ahead through the chapter activities to see how they might help you to meet the challenge. Work with your classmates and your teachers to define the criteria for assessing your work. Devise a grading sheet for the assessment of the challenge. Your teacher may provide you with a sample rubric to help you get started. Record all this information. Make sure that you understand the criteria and the grading scheme as well as you can before you begin.

Activity 1 Sources of Water in the World and in Your Community

Goals

In this activity you will:

- Identify and analyze the various sources and distribution of salt water and fresh water on Earth.
- Interpret data and a topographic map to determine the water sources that your community uses for drinking water.
- Generate a graphical model of the transport of water between reservoirs within the water cycle.
- Develop a method of determining the amount of fresh water that could be collected from your school roof and on the entire area of your community in one year.

Think about It

Imagine you are watching the news on television, and the meteorologist says, "We received one inch of rain yesterday."

- How much water was that?
- Where is all that water today?

What do you think? Record your ideas about these questions in your *EarthComm* notebook. Be prepared to discuss your responses with your small group and the class.

Investigate

Part A: Water in the Hydrosphere

1. Fill five 4-L milk jugs with water. These five jugs of water represent all the water on the Earth.

   a) Calculate how many milliliters are in the 20 L (five milk jugs). Record this value. Note: the actual amount of water may not be exactly 20 L, but for the purpose of this model it will be satisfactory.

2. Ice (mostly in the form of glaciers) holds 1.81% of all the water on Earth (see the data table below).

   a) If 20 L represents all the water on Earth, calculate the number of milliliters that represents the water found in glaciers.

3. Remove this amount from the milk jugs and pour it into a separate container labeled "glaciers."

4. Repeat Steps 2 and 3 for the water found in ground water, saltwater and freshwater lakes and streams, and the atmosphere.

   a) Record your calculations.

   b) Calculate the number of milliliters of water in the oceans, but leave the water in the milk jugs.

   c) Find the sum of the six values. How do you account for the "missing" water?

Clean up spills. Dispose of the water.

Distribution of Water in the Hydrosphere

Reservoir	Percentage of total water	Percentage of fresh water (ice and liquid)	Percentage of fresh water (liquid only)
Oceans	97.54	—	—
Ice (mostly glaciers)	1.81	73.9	—
Ground water	0.63	25.7	98.4
Saltwater lakes and streams	0.007	—	—
Freshwater lakes and streams	0.009	0.36	1.4
Atmosphere	0.001	0.04	0.2

These figures account for 99.9% of all water. They do not add up to 100%, because some water is tied up in the biosphere and as soil moisture.

5. Develop your own method for making a model of the percentages of each category shown in the table for fresh water (ice and liquid) and liquid fresh water. Use something other than water as the physical material for your model.

   a) From your work with the models, write down several observations or discoveries that you found most surprising or striking. Explain your observations.

Have your teacher check the plan for your model.

Part B: Local Water Use

1. Obtain a set of data on water use in your county. Use the data and a topographic map of your community to answer these questions:
   a) How much water does your city or county use each year?
   b) What and where are the sources of your community's drinking water?
   c) What are other water sources your community might use in the future? Record your ideas.
   d) Where are the water treatment facilities located?

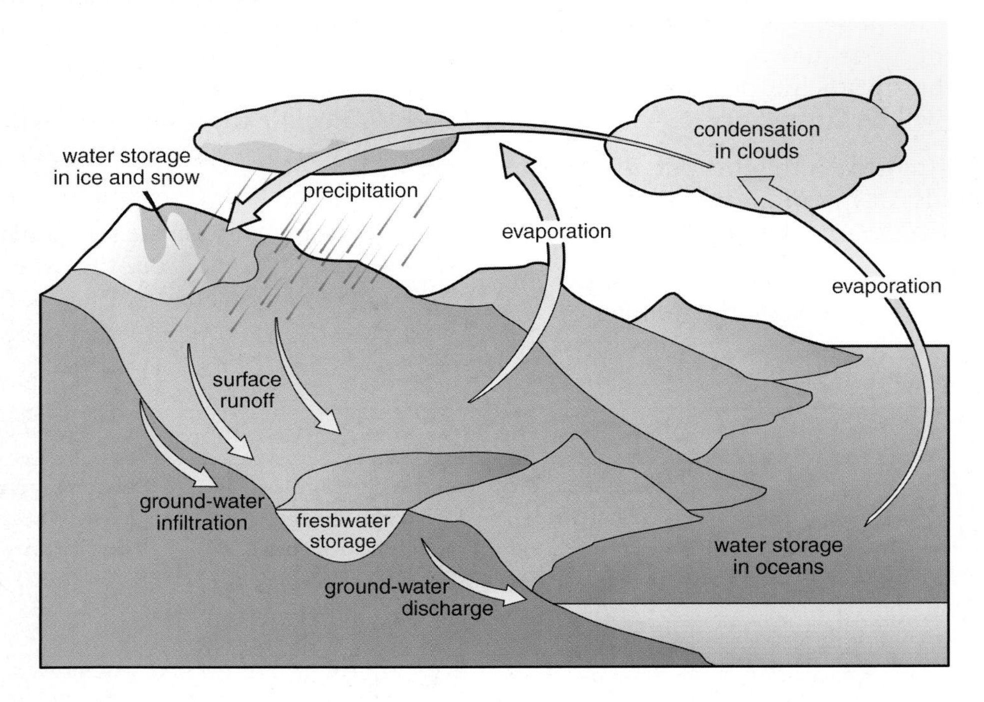

Part C: Modeling the Water Cycle

1. The total volume of water near the Earth's surface is almost constant, but the water is in constant motion. The water cycle describes how the Earth's surface water moves from place to place in an endless cycle. Study the diagram above that shows a simplified version of the water cycle.
2. On the following page is a more complete list of the components of the water cycle. There are also definitions of some terms with which you may not be familiar. The list is divided into two parts: *reservoirs* (places where water is stored) and *processes* (ways that water is moved from place to place).
   a) Using blank sheets of paper, draw a rectangular box for each reservoir item. Try to keep the dimensions of the boxes less than about 2.5 cm. Write the name of each reservoir in a box. You will have to write small.
   b) Draw a circle for each process item. Make the diameter of each circle less than 2.5 cm. Write the name of each process in a circle.

Reservoirs:

oceans
atmosphere
clouds
glaciers
soil moisture
ground water
lakes
rivers
vegetation

Processes:

evaporation from the ocean surface
precipitation onto the ocean surface
evaporation from the land surface
precipitation onto the land surface
precipitation onto glaciers
condensation to form clouds
melting of glaciers
calving of glaciers
surface runoff into rivers
surface runoff into lakes
infiltration of surface water
ground-water flow
river flow
transpiration from plants
uptake of water by plant roots

Definitions:

calving: Some glaciers end in the ocean. As the glacier ice moves forward into the ocean water, it breaks away from the glacier in huge masses, to float away as icebergs, which gradually melt.

ground water: Some of the liquid water at the Earth's surface moves downward through porous Earth materials until it reaches a zone where the material is saturated with water. This water flows slowly beneath the Earth's surface until it reaches rivers, lakes, or the ocean.

infiltration: Some of the rain that falls on the Earth's surface sinks directly into the soil.

soil moisture: Water, in the form of liquid, vapor and/or ice, resides in the Earth's soil layer. It is the water that remains in the soil after rainfall moves downward toward the ground water zone. Soil moisture is available for plants. What is not used by plants gradually moves back up to the soil surface, where it evaporates into the atmosphere.

surface runoff: Some of the rain that falls on the Earth's surface flows across the land surface, eventually reaching a stream, a river, a lake, or the ocean.

transpiration: Water taken up by the roots of plants is delivered to the leaves. Some of this water is used to make new plant tissue, and some is emitted from the leaves in the form of water vapor, by a process called transpiration.

3. Cut out all of the boxes and circles with a pair of scissors.
4. On a piece of poster board, draw a horizontal line lengthwise across the middle of the sheet. This represents the Earth's surface in a vertical cross-section view.
   a) On the left half of the poster board, draw some mountains to represent a continent.
   b) On the right half of the poster board, draw a small island or a sailboat to represent a large ocean.
   c) Using the simplified water-cycle diagram as a model, place the boxes and circles that you have created where you think they belong. Tape them to the poster board with small pieces of removable tape. Using removable tape allows you to adjust the positions of the boxes and circles as needed.

d) With colored pencils, draw arrows between the various boxes and circles to show the movement or transport of water from place to place on or near the Earth's surface. Remember that a circle (representing a process) will be located in the middle of an arrow between two different boxes (representing two different storage places). Think about whether the movement or transport is in the form of liquid water, water vapor, or ice (or two or three of these at the same time). Use blue for liquid water, red for water vapor, and green for ice.

5. Once everyone in your small group has agreed upon the best version of the water cycle, compare your results with those of the other groups. Answer the following questions:

   a) Is there net movement of water vapor from the oceans to the continents, or from the continents to the oceans? Explain your answer.

   b) Is there net movement of liquid water from the oceans to the continents, or from the continents to the oceans? Explain your answer.

   c) How does the nature of the water cycle vary with the seasons?

Part D: Collecting Rainwater in Your Community

1. Suppose your community needed to consider ways of adding to their supply of fresh water. One alternative is to collect and store rain and melted snow that runs off from the roofs of buildings.

   a) In your notebook, make a list of what you would need to know to calculate an estimate of the total amount of annual runoff from your school roof.

   b) Develop a plan for making this estimate. Compare your plan with the plans of others. Refine it on the basis of your discussions.

   c) Calculate the total runoff for one year's precipitation.

   d) How does this amount compare to the amount of water that your city or county uses each year? What percentage is it?

   e) What modifications to school grounds and buildings might need to be made to accommodate a runoff collection plan?

   f) How could a plan for runoff collection benefit communities seeking to conserve available water?

   g) How much money would this water save your school? Find out how much your community charges for each 1000 gallons of water. Assume that untreated roof water is worth half as much as treated city water. Calculate the value of the water that could be collected from the roof of your school for one year.

   h) What could be some concerns or problems with using roof runoff as a source of water?

   i) Calculate how much water falls as precipitation on the total area of your community in an average year, and compare that volume with your community's annual water use.

Reflecting on the Activity and the Challenge

You have learned that of all the water on Earth, less than 1% is available for drinking water. You also explored the nature of the water cycle, and how Earth system science can be used to characterize it. You have also learned where your community's drinking water comes from. This will help you determine whether or not your community has enough water for the additional development. It also provides you with some ideas about where extra water could be obtained.

Digging Deeper

THE WATER CYCLE

Water is the only common substance that exists at the Earth's surface as a solid, a liquid, and a gas. Water is present at or near the surface everywhere on Earth. In many places, the presence of water is obvious, in the form of lakes, rivers, glaciers, and the ocean. Even in the driest of deserts, however, it rains now and then, and although the humidity there is usually very low, there is at least some water vapor in the air.

Figure 1 Nearly two percent of Earth's water exists as ice.

Geo Words

water cycle (or hydrologic cycle): the constant circulation of water from the sea, through the atmosphere, to the land, and its eventual return to the atmosphere by way of transpiration and evaporation from the land and evaporation from the sea.

closed system: a system in which material moves from place to place but is not gained or lost from the system.

evaporation: the change of state of matter from a liquid to a gas. Heat is absorbed.

precipitation: water that falls to the surface from the atmosphere as rain, snow, hail, or sleet.

Water, in the form of liquid, solid, or vapor, is in a continuous state of change and movement. Water resides in many different kinds of places, and it takes many different kinds of pathways in its movement. The combination of all of these different movements is called the **water cycle**, or the **hydrologic cycle**.

The water cycle is called a cycle because the Earth's surface water forms a **closed system**. In a closed system, material moves from place to place within the system but is not gained or lost from the system. The Earth's surface water is actually not exactly a closed system, because relatively small amounts are gained or lost from the system. Some water is buried with sediments and becomes locked away deep in the Earth for geologically long times. Volcanoes release water vapor contained in the molten rock that feeds the volcanoes. Nonetheless, these gains and losses are very small compared to the volume of water in the Earth's surface water cycle.

Evaporation and **precipitation** are the major processes in the water cycle. The balance between evaporation and precipitation varies from place to place and from time to time. It's known, however, that there is more evaporation than precipitation over the surface of the Earth's oceans, and there is more precipitation than evaporation over the surface of the Earth's continents. That fact has a very important implication. There is net movement of water vapor from the oceans to the continents, and net movement of liquid (and solid) water from the continents to the oceans.

Figure 2 What part of the water cycle does the fog over the San Francisco Bay illustrate?

The oceans cover about three-quarters of the Earth. Ocean water is constantly evaporating into the atmosphere. If enough water vapor is present in the air, and if the air is cooled sufficiently, the water vapor condenses to form tiny droplets of liquid water. If these droplets are close to the ground, they form fog. (See *Figure 2*.) If they form at higher altitudes, by rising air currents, they form clouds.

All of the solid or liquid water that falls to Earth from clouds is called precipitation. Snow, sleet, and hail are solid forms of precipitation. Rain and drizzle are liquid forms of precipitation.

When rain falls on the Earth's surface, or snow melts, several things can happen to the water. Some evaporates back into the atmosphere. Some water flows downhill on the surface, under the pull of gravity, and collects in streams and rivers. This flowing water is called **surface runoff**. Most rivers empty their water into the oceans. Some rivers, however, end in closed basins on land. Death Valley and the Great Salt Lake are examples of such closed basins.

Figure 3 Some of the water that falls to the Earth's surface collects in streams.

Some precipitation soaks into the ground rather than evaporating or running off. The water moves slowly downward, percolating through the open pore spaces of porous soil and rock material. Eventually the water reaches a zone where all of the pore spaces are filled with water. This water is called **ground water**. Some water, called soil moisture, remains behind in the surface layer of soil. (See *Figure 4*.)

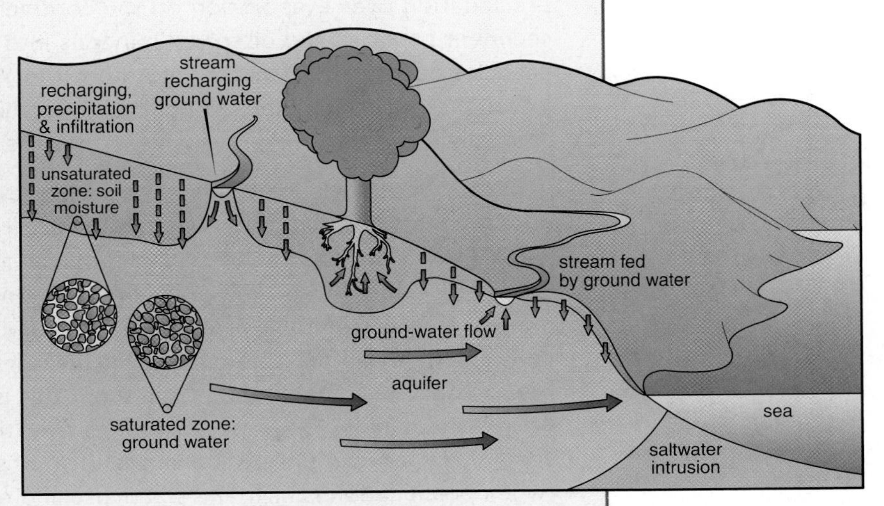

Figure 4 Schematic diagram of ground-water flow.

Geo Words

surface runoff: the part of the water that travels over the ground surface without passing beneath the surface.

ground water: the part of the subsurface water that is in the zone of saturation, including underground streams.

The roots of plants absorb some of the water that soaks into the soil. This water travels upward through the stem and branches of the plant into the leaves and is released into the atmosphere as a vapor in a process called **transpiration**.

Figure 5 Plants like these broadleaf trees play an important part in the water cycle.

It has been estimated that each year about 36,000 km^3 of water flows from the surface of the continents into the oceans. That represents the excess of precipitation over evaporation on the continents. This water carries sediment particles and dissolved minerals into the ocean. The sediment particles come to rest on the ocean bottom. When seawater evaporates, it leaves the dissolved materials behind. Over geologic time, this process has gradually made the oceans as salty as they are now.

In Earth systems science, the water cycle is viewed as a flow of matter and energy. Each place that holds water is called a **reservoir**. The rate at which water flows from one reservoir to another in a given time is called a **flux**. Energy is required to make water flow from one reservoir to another. On average, the total amount of water in all reservoirs combined is nearly constant. Although the data table in the investigation suggests that reservoirs have a constant amount of water in them, this is not the case. The amount of water stored in any one reservoir varies over time. For example, in many areas there may be more water in the form of ground water during the spring (when precipitation is high, and water use and evaporation is low) than in the summer (when precipitation is low, and evaporation and water use are high).

Geo Words

transpiration: the process by which water absorbed by plants, usually through the roots, is emitted into the atmosphere from the plant surface in the form of water vapor.

reservoir: a place in the Earth system that holds water.

flux: the movement of water from one reservoir to another.

Check Your Understanding

1. In your own words, describe the water cycle.
2. Explain why the water cycle can be viewed as a closed system.
3. Describe three "paths" of the water cycle that precipitation can follow once it reaches the surface of the Earth.

Understanding and Applying What You Have Learned

1. Describe the different conditions on Earth under which water is a solid, a liquid, or a gas.
2. If 36,000 km^3 of water flow from the surface of the Earth into the oceans each year, how many cubic kilometers of water evaporate from the oceans each year?
3. The data table in the investigation defines the hydrosphere somewhat differently than the image shown in the front of the book. Explain any differences you note between the data table and the image.
4. Using the same techniques you used in Part C of the investigation, construct a model, with boxes and arrows, that shows sources, volumes, and transport paths and processes of water in your community. Use the water data that you collected in the investigation for your county or municipality. You will need to know the sources and volumes of water, and the uses.
5. How can the destruction of large areas of rain forest affect climate?

Preparing for the Chapter Challenge

Using what you have learned about water in your community and the water cycle, write a short essay on water resources in your community that can be included in your final **Chapter Challenge.** Include the following information in your essay:

- A complete description of the sources of drinking water for your community.
- An analysis of the stage or component of the water cycle that is most critical to your community.
- At least two new water sources that you have identified to provide for increased demands.
- A description of how this information can be applied to the questions raised by the **Chapter Challenge.**

Inquiring Further

1. **Volcanic eruptions and the water cycle**

   Volcanic eruptions release large amounts of water vapor. After you have done some research, construct a water cycle diagram that shows the pathways and flow of water and water vapor in a volcanic region.

2. **Drinking water from the ocean**

   How can the ocean be used as a source of drinking water? Visit the *EarthComm* web site to do some research on the technology currently available.

3. **Dating water**

   Investigate how chlorofluorocarbons (CFCs) released from aerosols and tritium (hydrogen 3) released during global nuclear testing in the 1950s and early 1960s are used to determine the age of ground water.

Activity 2 How Does Your Community Maintain Its Water Supply?

Goals

In this activity you will:

- Create and manipulate physical models of surface-water and ground-water supply systems.
- Explain how a change in one part of the water-supply system creates changes in other parts of the system.
- Understand the main ways that a community can increase its water supply.
- Compare and contrast surface-water systems and ground-water systems.
- Analyze the water-supply system in your community.

Think about It

You and others in your community expect to always have the water you need when you turn on the faucet.

- How can any community guarantee that there will be enough fresh water available to meet the needs for personal, recreational, business, industrial, and agricultural use?
- Suppose your region were experiencing a severe and prolonged drought. Would ground water or surface water be a more reliable water supply? Explain your response.

What do you think? Record your ideas in your *EarthComm* notebook. Be prepared to discuss your responses with your small group and the class.

Investigate

Part A: Modeling a Surface-Water Supply System

1. Set up a coffee can, a clear, 750 mL soda bottle (top cut off), and a shallow container as shown in the diagram.
2. Line the inside of the bottle with coffee filter paper. (This keeps sand from entering the plastic tubing.)
3. Pour sand into the filter until the bottle is one-third full of sand.
4. One kind of surface-water reservoir is one in which a river is dammed to make a lake that a community can use for its water supply. Assume that the ideal situation is a balance between a flow of water into the reservoir and the flow out of the reservoir. Water flows out of the reservoir not only for community use but also to return to the river downstream of the community. Water flow for both of these uses can be controlled by a system of valves.

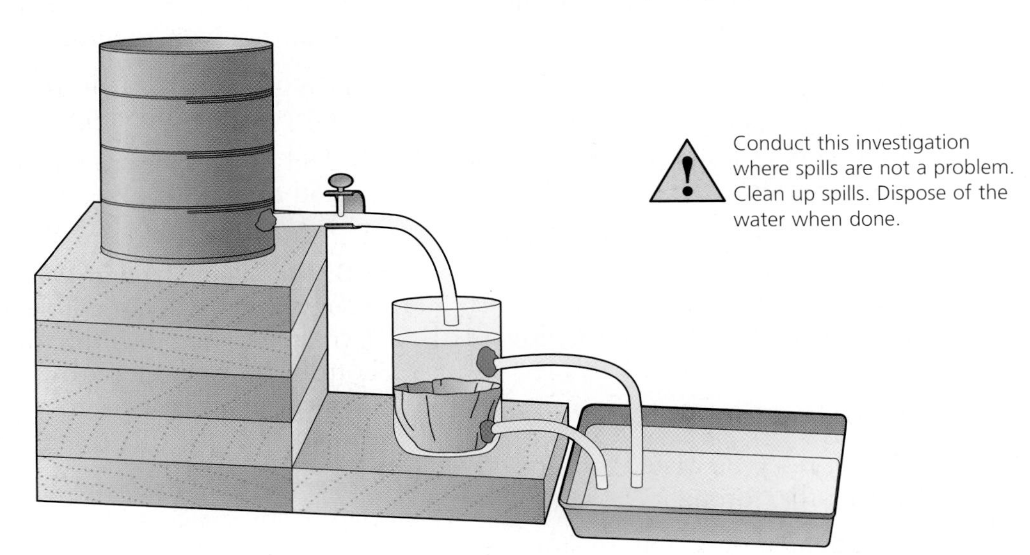

Conduct this investigation where spills are not a problem. Clean up spills. Dispose of the water when done.

Identify the following parts of the water supply system on the model that you set up:

- precipitation, river flow, reservoir inflow;
- surface-water reservoir;
- ground-water reservoir;
- ground-water withdrawal/outflow (for the community);
- surface-water withdrawal/outflow (for the community);
- total community consumption.

a) Draw a box model (systems diagram) of the physical model, with reservoirs and flows. Label the reservoirs and flows.

5. Release the clamp slightly to allow water to flow into the bottle. Observe what happens with the flow of water out of the bottle. Adjust the flow until you have a steady-state system (in which reservoir inflow equals reservoir outflow).

   a) Record what you see.

   b) How did the two water-supply reservoirs respond to the increase in precipitation?

6. Release the clamp completely to allow a full flow of water from the can. Observe what this does to the flow of water out of the bottle.

   a) Record what you see.

   b) How did the water-supply reservoirs respond to a decrease in precipitation?

7. Tighten the clamp so that the flow comes to a trickle. Again, allow the system to reach a steady state. Observe what this does to the flow of water out of the bottle.

   a) Record what you see.

   b) Which reservoir lasted longer during the "drought"?

8. Finally, tighten the clamp to stop the flow of water.

   a) Observe and record the results.

9. In your model, you did not control community consumption (water flowed out of the reservoirs despite a decrease in recharge into the reservoirs).

   a) Where would you place additional clamps on the model to keep water in both reservoirs during a drought?

   b) What would these clamps represent in the real world?

10. What would you add to the model to show the return of water from the community to the system shown in your model and to other parts of the system not shown (for example, communities downstream)?

    a) Make a diagram to show your ideas.

11. How could you use a graduated cylinder and a stopwatch so that your model allows you to measure the rate of flow of water from one system reservoir to another (flux)?

    a) Record your ideas in your *EarthComm* notebook.

    b) Add flux arrows to the box model you drew.

Part B: Modeling Ground-Water Supply

1. Your teacher will cut a hole about 1 cm across in a dish tub, on the bottom and very close to where the surface curves upward to one of the end walls of the tub.

2. Cover the hole from the inside with two or three pieces of duct tape.

   With a sharp pencil, punch a circular hole in the duct tape, from the inside of the tub, to be about 3 mm in diameter.

3. Pour sand evenly into the dish tub, making a small depression (a low spot) near the center of the sand surface. Make the depression about 5 cm deep and about 10 cm across, with gently sloping sides. The depression represents a shallow lake

Wear safety goggles. Clean up spills immediately. Dispose of water when you are done.

or wetland. The diagram illustrates the arrangement in the tub.

4. Place the tub where water can drain harmlessly from the hole in the bottom.

5. Cover one end of a piece of the rigid tubing with a layer of cheesecloth and tape it to the sides of the tubing with duct tape.

   Bury the tubing to stand vertically in the sand, about 10 cm away from the center of the depression. The lower end of the tubing should be about 5 cm from the bottom of the tub. This represents a town water supply well.

6. Cover one end of another piece of rigid tubing with a layer of cheesecloth and tape the cheesecloth to the sides of the tubing.

   Bury the tubing to stand vertically in the sand, about 10 cm away from the center of the depression. The lower end of the tubing should be only about 0.5 cm from the bottom of the tub. This represents a monitoring well.

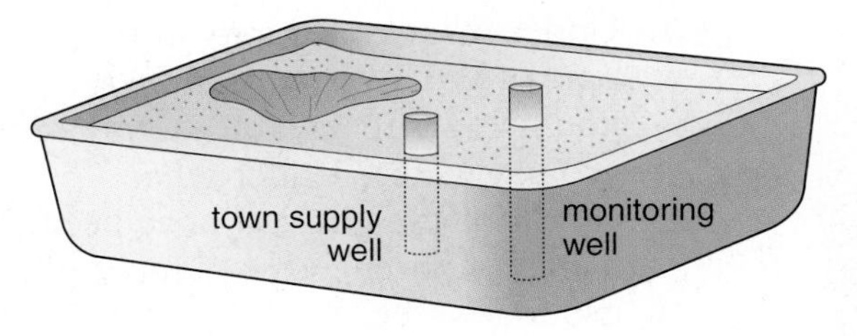

7. Each member of your group needs to assume one of the following roles: a water supplier, a consumer, a well monitor, and a data recorder.

8. Establish an equilibrium ground-water system with a water table high enough for there to be water in the lake. Follow these steps:

   - The water supplier starts supplying water at a rate of 1000 mL/min. The water supplier pours a steady supply of water onto the sand at the "upstream" end of the tub (the end opposite the hole in the bottom of the tub). The best way to do this is to measure out a constant volume of water in a measuring cup and pour that water into the tub every 15 or 30 seconds.
   - The data recorder can then record the rate of supply. This represents ground-water recharge from infiltration of precipitation into the ground surface.
   - The well monitor will monitor the test well by dipping a chopstick ruler gently into the well until it rests on the bottom and then reading the position of the water surface in the well.

9. Continue supplying water and monitoring the well until the level of the water table stops changing.

   If there is no surface water in the depression, increase the rate of water supply and monitor the test well until the height of the ground-water table stops changing.

   Repeat the process until there is water in the depression.

10. Now it's time to start extracting water from the supply well.

    To do this, tape a drinking straw to the end of the turkey baster and seal the joint with a wrap of duct tape.

11. The consumer should then squeeze the bulb of the baster, insert the straw into the bottom of the well, and **slowly** release the bulb.

12. Keep track of the volume of water extracted per minute. Try to keep the rate of extraction constant, minute after minute.

    The data recorder should record the rate of withdrawal.

13. Meanwhile, the well monitor should monitor the test well and record the height of the water table every minute.

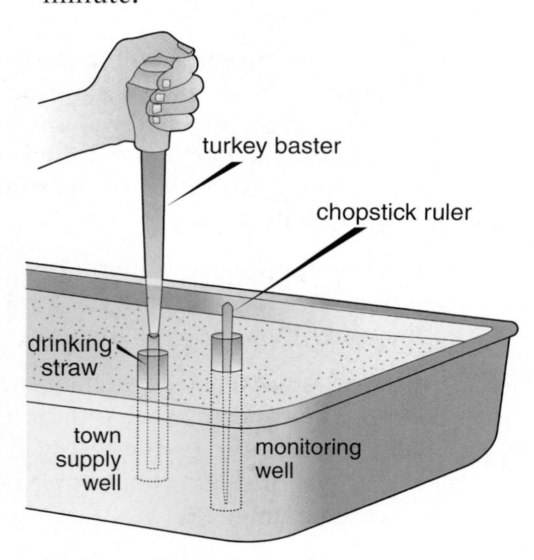

14. Depending on the rate of extraction relative to the rate of recharge, three outcomes of this part of the investigation are possible:

Mark the baster and chopstick ruler as laboratory equipment. Dispose of the drinking straw.

- A new equilibrium height of the water table in the test monitoring well is established, and some water remains in the depression.
- A new equilibrium height of the water table in the test monitoring well is established, but the depression has lost its water.
- The supply well runs dry; that is, the rate of withdrawal cannot be maintained.

    Try to reproduce each of these outcomes, by adjusting the rate of withdrawal from relatively low to relatively high.

15. Use the results of this part of the investigation to answer the following questions:

    a) What is the difference between a renewable ground-water supply and a nonrenewable ground-water supply? What do you think it means to "mine" ground water?

    b) Under what conditions in this part of the investigation were you using a renewable supply of ground water? Under what conditions might you have been using a nonrenewable supply of ground water?

    c) Do you think it would have been less likely to lower the water table if you had obtained a given rate of supply by pumping water from two nearby supply wells instead of only one? Explain your answer.

Part C: Water Supply in Your Community

1. You will be able to find the answers to some of these questions in your water supplier's consumer confidence report, also called a water quality report. For some of the other questions, appoint one student in your group or class to call your community's water supplier to ask. The *EarthComm* web site also provides links.
   a) What kind of municipal water system does your community use? What factors do you think affected their selection of the system?
   b) How does your community evaluate potential development projects in light of water supply?
   c) What contingency plans does your community have in order to deal with potential droughts?
   d) Every freshwater system requires maintenance and upkeep. What does this mean for your community, and what does it cost?
   e) Determine the average cost per gallon of water for your community and three others in or near your county. List the factors that account for the differences in the cost per gallon. Does the cost vary? Why?
   f) Determine the gallons-per-person use for your own community and two neighboring communities. List the factors that could account for the differences.

Reflecting on the Activity and the Challenge

You learned that many factors affect the operation of a community's water supply system. Some of these factors are drought, increased demand, and the needs of other communities. You also learned that a system consists of many parts and that a change in one part of the system affects other parts of the system. You also took a closer look at the impact of consumption on groundwater supplies. You then analyzed the water-supply system in your community. Understanding these factors will help you evaluate whether your community's water-supply system is capable of handling the extra demands of the three new developments proposed in the **Chapter Challenge**.

Digging Deeper

WATER SUPPLIES

Sources of Water Supplies

In colonial times in the United States, most people took water from rivers or dug their own wells. Today, cities and towns need reliable and safe supplies of water for their citizens. Water must be collected, stored, and treated. There must be enough water to see a community through times of drought and times of increased water use. The two main water sources are surface waters, from rivers and lakes, and ground water.

Figure 1 Surface water from freshwater lakes is a valuable source for human use.

Figure 2 Large pumps are used to supply water to a community.

There are six ways to increase the supply of water to a community:

- withdrawing water from ground-water aquifers;
- withdrawing water directly from nearby rivers or lakes;
- building dams to create reservoirs to store runoff;
- improving the efficiency of water use through water conservation;
- transporting water from a distant area by means of aqueducts;
- converting salt water to fresh water.

Many of these choices affect other communities. For example, if one town takes water from a river, it decreases the amount of water available to towns downstream.

Surface Water

In most rivers, water flow is too variable from season to season to be a reliable direct source of water supply. Most surface-water supplies are from large lakes of artificial reservoirs, which fluctuate less from season to season or from year to year.

Dams, such as the one shown in *Figure 3*, are beneficial in providing a water source and controlling floods, but they have disadvantages as well. Reservoirs behind dams displace wildlife and people, and they cover cropland. Dams can disrupt the natural migration of fish. The sediment carried by a river is deposited in the reservoir behind the dam. Over time, the reservoir slowly fills up with sediment, leaving less room for water.

Ground Water

Ground water for water supply is pumped from porous material below the surface. Three concepts are important in understanding ground-water supplies. **Porosity** is a measure of the percentage of pores (open spaces) in a material. **Permeability** is a measure of how easy it is to force water to flow through a porous material. An **aquifer** is any body of sediment or rock that has sufficient size and sufficiently high porosity and permeability to provide an adequate supply of water from wells. The best aquifers consist of loose and porous sand and gravel, although fractured bedrock can also form good aquifers.

Figure 3 A large reservoir was created from the Colorado River at the Glen Canyon Dam, Arizona.

Geo Words

porosity: a measure of the percentage of pores (open spaces) in a material.

permeability: a measure of how easy it is to force water to flow through a porous material.

aquifer: any body of sediment or rock that has sufficient size and sufficiently high porosity and permeability to provide an adequate supply of water from wells.

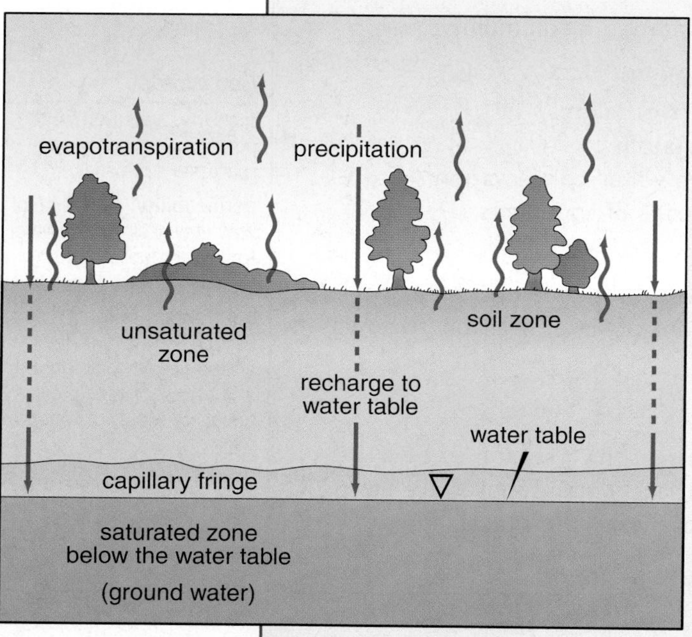

Figure 4 The main two zones of the ground-water system—saturated and unsaturated. The water table marks the upper surface of the saturated zone.

Down to a certain depth below the surface, the pores in the sediment and rock are mostly filled with air, except when water is percolating downward after a heavy rain. This is called the unsaturated zone. Eventually the downward-moving water reaches a zone called the **saturated zone**, where all of the pores are filled with water. The top of the saturated zone is called the **water table**. These zones are illustrated in *Figure 4*. The water table can be located at the surface in places next to rivers and lakes, and also in wetlands. In some areas it can lie many tens of meters below the surface.

Because ground water must move through small pores it flows very slowly. Ground water speeds of a meter per day are considered high. Speeds as low as a meter per year are common. In general, the smaller the pore spaces between the grains, the slower the ground water flows. Ground water moves from areas where the water table is relatively high to areas where it is relative low. *Figure 5* shows the flow of ground water in a typical landscape.

Geo Words

saturated zone: the zone, beneath the water table where all of the pores are filled with water.

water table: the surface between the saturated zone and the unsaturated zone (zone of aeration).

evapotranspiration: loss of water from a land area through transpiration of plants and evaporation from the soil and surface water.

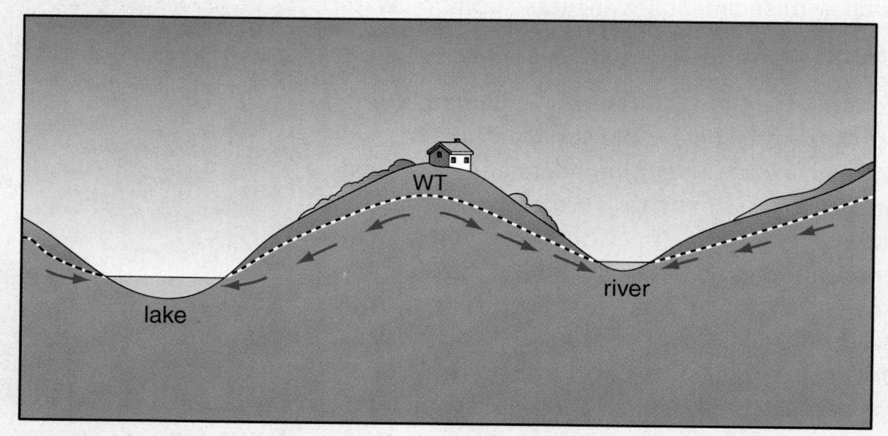

Figure 5 The water table (WT) is shown as a dashed line. The blue arrows show the direction of ground-water flow.

In rural areas, most homes have their own wells to tap ground water. Many towns, and even some small cities, obtain part or all of their water supply from several large wells that pump ground water from large aquifers.

Most aquifers, called **unconfined aquifers**, have a free connection upward to the surface. The water in these aquifers is replenished by downward percolation of surface water from directly above. The sand in your model was an unconfined aquifer. Addition of new water to an aquifer by downward flow of surface water is called **recharge**.

Figure 6 The signs show the approximate position of the land surface in 1925, 1955, and 1977 in the San Joaquin Valley, California. Switching to surface water slowed the rate of land subsidence, but new ground-water pumping during a drought from 1987 to 1992 caused further subsidence.

Some aquifers are isolated from the surface by an overlying layer of very impermeable material, called an **aquiclude**. Layers of fine clay are especially effective aquicludes. Confined aquifers cannot be recharged from directly above. The recharge area for a confined aquifer may be located far away, tens or even hundreds of kilometers.

In many areas where ground water has been used for a long time, the land surface has subsided (sunk down) because so much water has been withdrawn. For example, at Edwards Air Force Base in California, the land has subsided more than two meters, damaging some of the runways once used by the Space Shuttle.

Geo Words

unconfined aquifer: an aquifer that has a free connection upward to the surface.

recharge: addition of new water to an aquifer by downward flow of surface water.

aquiclude: a body of rock that will absorb water slowly, but will not transmit it fast enough to supply a well.

Aqueducts

In areas where water use is greater than local supplies, as in southern California, water must be brought in from distant areas where water is abundant. About two-thirds of California's precipitation falls in the north, but about two-thirds of the population lives in the south. A system of reservoirs in northern California supplies southern California with water that is transported through a system of large surface pipes and channels, called **aqueducts**, like the one shown in *Figure 7*. Over long distances the water in the aqueducts flows downhill under gravity, but in some places enormous pumping plants must raise the water up over hills and mountains.

Figure 7 This aqueduct in southern California carries much needed water over long distances.

Desalination

Converting sea water to fresh water, a process called **desalination**, is still too expensive to be widely used. It is used in some arid countries that are located near the ocean. Israel and Saudi Arabia obtain much of their water from desalination. As new and less expensive techniques for desalination become developed, the process will be a more and more important source of fresh water in many countries.

Water Conservation

Conservation is a great way for a community to stretch its water supplies further without having to develop new supplies. Although the water supply stays the same, if the community uses less, then there will be more water for new development. You will learn more about conservation in later activities.

Geo Words

aqueduct: a system of large surface pipes and channels used to transport water.

desalination: the process of removing dissolved salts from sea water in order to make it potable.

Check Your Understanding

1. Describe six possible ways to increase the supply of water to a community.
2. What are the advantages of building a dam to provide a source of surface water? What are the disadvantages?
3. Explain how porosity and permeability of Earth materials are important when considering ground water as a water source.
4. Why is desalination of ocean water not a practical source of water supply?

Understanding and Applying What You Have Learned

1. Look at the diagram of the surface-water reservoir.
   a) Write down the major parts of the water-supply system. Also, think about the parts that may not be shown.
   b) Which part of the system do people have the least control over? Why?
   c) Which part of the system do people have the greatest control over? Why?
   d) How might the volume of water entering the reservoir from the river vary from season to season?
   e) How might the amount of fresh water that a community needs vary from season to season?
   f) Assume there is a severe drought. How might the system respond in order to guarantee the amount of fresh water needed by the community?
   g) What other factors can be manipulated in times of drought to make the system operate as efficiently as possible?

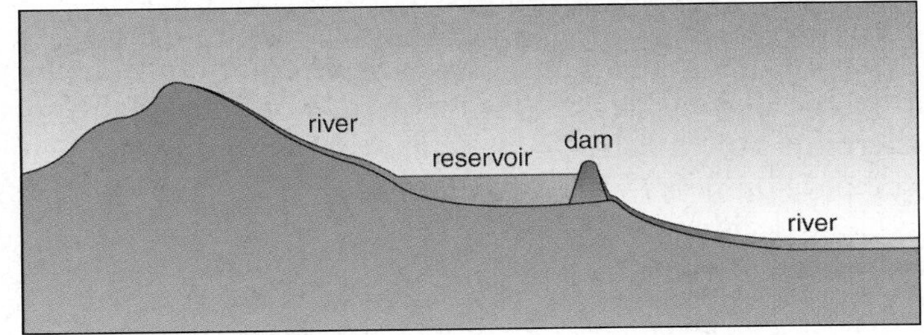

A surface-water reservoir.

2. Look at the diagram of the confined aquifer.
   a) What factors would affect the level of fresh water in this aquifer?
   b) What will happen if the volume of water entering at the recharge area decreases and the demand for fresh water from the wells remains the same? When would this situation be likely to occur? Which wells would be affected first?
   c) Assume that the ground water enters the aquifer at a constant rate. What would the community need to know before agreeing to a development project that would result in a significant increase in water use?

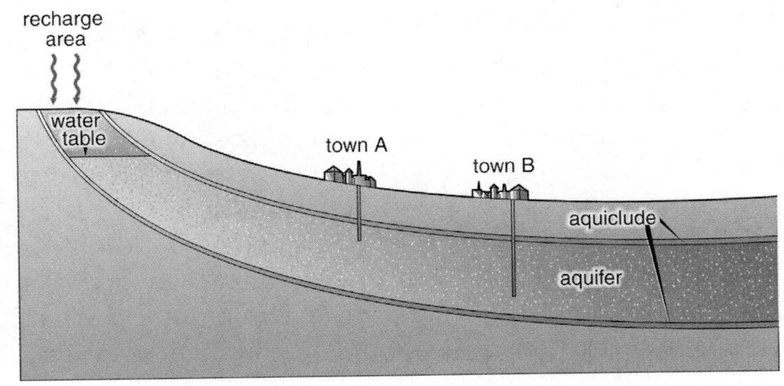

A confined aquifer.

3. a) Suppose that the capacity of the ground-water supply is sufficient to meet the increased needs of a new development in your community. The development is approved. How will this affect other communities using the same ground water in the future?

   b) Should this have any effect on your community's decision to approve the development? Explain why or why not.

4. Compare and contrast the advantages and disadvantages of using ground water versus surface water.

5. Suppose your community's water supplier proposed building a dam on a nearby river to increase water supply. Make a list of the pros and cons of damming a river in your community.

Preparing for the Chapter Challenge

Using what you have learned in this activity and your community's water quality report, determine how much water the proposed developments will use per year. As a class, decide on the number of houses that will be included in the area.

Does your community have an adequate water supply to meet this demand now and over the next 25 years?

Inquiring Further

1. **First American reservoirs**

   Visit the *EarthComm* web site to do some research on the first reservoirs constructed in America. When and where were they built? How did they work?

2. **First American ground-water systems**

   Visit the *EarthComm* web site to conduct some research on the first ground-water systems developed in America. When and where were they built? How did they work?

3. **Water supplies in desert cities**

   Pick a large desert city, such as Las Vegas or Phoenix. Conduct some research to find out how these cities maintain a water supply.

4. **Water needs of a golf course**

   How much water does it take to run a golf course? You have already established an estimate for your community's water use. Pick several communities in parts of the United States with a very different climate from yours. Find out how much water an average golf course uses in those communities. What accounts for the differences in water use?

Activity 3 Using and Conserving Water

Goals

In this activity you will:

- Design a method for determining how much fresh water your school uses every day.
- Analyze statistics on water use for your county and an adjacent county and explain any differences in per-person water use between your county and an adjacent county.
- Differentiate between uses of fresh water in the United States and identify these uses as consumptive or nonconsumptive.
- Explore methods of water conservation and make suggestions about which methods would benefit your school and home.

Think about It

You use water for everyday activities, probably without considering how much water you are actually using.

- How much water does it take to brush your teeth?
- Devise a plan to arrive at an accurate estimate of how much water you typically use to do this.

What do you think? Record your ideas in your *EarthComm* notebook. Be prepared to discuss your responses with your small group and the class.

Investigate

Part A: Water Use in Your School

1. Working in small groups, brainstorm and develop a list of the various ways that water is used at your school each day.
   a) Record your list in your notebook.
2. Share your list with the class and arrive at a combined list.
3. As a class, rank all the items on the board from the highest estimated water use to the lowest.
   a) Record this list and ranking.
4. Each group should take one category of water use and devise a plan for determining the volume of water used daily for this activity.
   a) Design your plan and write it out, detailing specifically how you will calculate a fairly accurate estimate.
5. Share your plan with the class. Be prepared to make changes on the basis of comments from the class.
6. With the approval of your teacher, carry out your plan.
   a) Write down all of the steps that you took to determine the average daily water use for your category.
   b) Write down your group's estimate of water use.
   c) Write down ways that your school could reduce this amount.

Any water collected should be disposed of. Wipe up spills immediately.

   d) Estimate how much daily water use could be reduced if these water conservation methods were adopted in your school.
7. Combine the estimates from each group to come up with one estimate for the average daily water use at your school.
8. Find out how much water the school really does use every day. This is something you might be able to find out from the office, or the school system headquarters, or the school's water supplier. You might need to average a few monthly water bills and then divide it by the number of days in a month to come up with the daily water use.
   a) How close was your group's estimate to the actual number?
   b) How close was your class's estimate to the actual number?
   c) What might have caused the difference?
9. Working in small groups, brainstorm and develop a list of the various ways that water use could be conserved at your school.
   a) Record your list in your notebook.
10. As a class, discuss how much water could be saved at your school each day by using the methods each group came up with for saving water.
    a) How much water would these measures save for a whole school year?
    b) Calculate the cost savings that your plan would have.

Part B: Water Use in Your Community

1. Gather data on the water use in your county and one additional nearby county. If this information is not available for your county, use the two counties closest to your community.
2. Construct a copy of the data table shown below for each county.
   a) Enter the data value for each category. You may wish to change or add categories that are characteristic of the counties you are studying.
3. Construct a bar graph showing the volume of use for surface water and ground water for each of the following categories: domestic, commercial, industrial, and agricultural.
   a) Use different-colored bars for each category. Have separate bars for each category's sub-use area. The vertical axis of the bar graph should be used to show the volume of water used per year.

Water Use in Two Local Counties				
County name:				
Total population of county:				
	Ground water (units)	**Surface water (units)**	**Ground water (units)**	**Surface water (units)**
Total use				
Domestic (residential)				
Commercial				
Industrial: mining				
Industrial: power plants				
Agricultural: farming				
Agricultural: livestock				
Agricultural: irrigation				

4. Calculate the per capita water use of ground water and surface water for both your county and your neighboring county. Use the equation below.

$$\frac{\text{Volume used per year}}{\text{Population of county}} = \text{per capita water use per year}$$

   a) Show your calculations and the result in your notebook.
5. Use the results of this part of the investigation to answer the following questions:
   a) Which county has the higher per capita water use?
   b) Consider the uses of water in both counties. Why is the per capita water use of each so high? What could account for a difference in water use by communities in the same area?

c) Consider the uses of water in your county and the other county. What water conservation methods would you suggest for them?

d) Share your ideas on water conservation with the class. Add one or two methods to your list besides those your group has already recorded.

e) If your community conserved 20 gallons per person per year, how many more people could live there and not exceed your current yearly water use?

Reflecting on the Activity and the Challenge

You learned approximately how much water your school uses every day and came up with ways to reduce this amount. Reducing water use day to day is a good way of conserving water for future use. You also analyzed your county's water consumption and compared it to a neighboring county. This provides you with knowledge of areas that you may consider when developing a conservation plan. Together, this information could help your community accommodate the new developments proposed in the **Chapter Challenge.**

Digging Deeper

FRESH-WATER USE

Types of Water Use

Figure 1 shows the source, use, and disposition of fresh water in the United States in 1995. The amounts shown are in millions of gallons per day (Mgal/day). Irrigation is the largest single use of fresh water in the United States. An estimated 134 billion gallons were used to irrigate crops every day in 1995 (that's equal to ten stacks of one-gallon milk jugs that reach to the moon!). The domestic use of water (26,100 Mgal/day) accounts for only about 7.7% of daily fresh water use. Irrigation and thermoelectric use make up 78% of daily demand for fresh water. That is why it is important to develop methods of irrigation that minimize consumption (reduce water loss).

- ***Public supply*** implies water used for street cleaning, fire fighting, public facilities like swimming pools or fountains, and supply to public buildings like a city hall or municipal museum.
- ***Domestic*** implies water used in homes.
- ***Commercial*** implies water used in businesses, large and small, but exclusive of manufacturing processes.
- ***Industrial*** implies water used in manufacturing processes.
- ***Thermoelectric*** implies generation of electricity by oil-fired, coal-fired, gas-fired, or nuclear-powered power plants. Almost all of this water is used for cooling the steam that is used to produce the rotational motion that drives the electrical generators. Most of the water is said to be self-supplied; that is, it is taken from special reservoirs or wells connected to the power plant rather than from municipal supplies. Also, most of the water is returned to rivers, streams, or the ocean, warmer than before it was used for cooling.

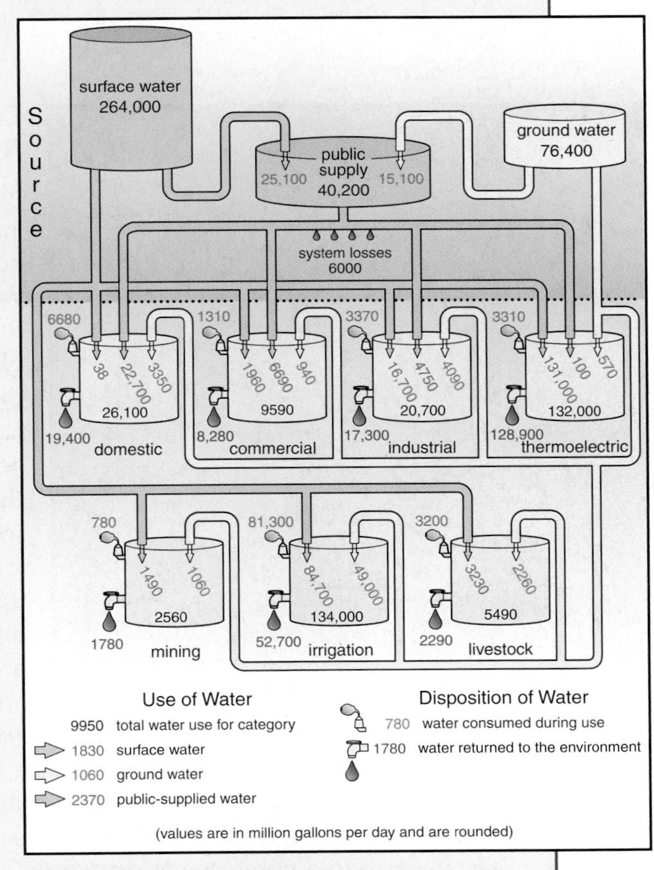

Figure 1 Water use and disposition in the United States in 1995.

Figure 2 Cooling tower of a nuclear power plant.

Figure 3 About 12% of the demand for fresh water is for public supply.

Figure 4 Is this an example of consumptive or nonconsumptive use?

Consumptive Use versus Nonconsumptive Use

Most of the total water use shown in *Figure 1* is said to be nonconsumptive. **Consumptive water** is water that is returned to the atmosphere as water vapor. **Nonconsumptive water** is water that is returned, in liquid form, to the natural environment after use. Some nonconsumptive water goes into rivers, lakes, or the ocean, and some soaks into the ground. In your home, for example, almost all the water you use, except for what you drink and then return to the atmosphere by breathing and perspiring, is nonconsumptive. The portion of water used in lawn and garden sprinkling that evaporates before soaking into the ground is also consumptive water. Think about what happens to water used in bathing, clothes washing, and toilet flushing: it goes either into a municipal sewer system or into a home septic system.

Geo Words

consumptive water: water that is returned to the atmosphere as water vapor.

nonconsumptive water: water that is returned, in liquid form, to the natural environment after use.

Water Conservation

As you learned in the last activity, common ways of increasing water supply are construction of dams and reservoirs, transportation of water from one place to another through pipelines or aqueducts, and development of groundwater sources. In many parts of the country, however, population growth is outstripping the ability to meet the growing demand for water. In communities where new water sources are limited, water conservation programs are widely encouraged. Such programs will become more and more important as the years go by, even in the United States, which is blessed with a water supply that is more abundant than in most countries of the world.

Some examples of ways citizens can save water at home are low-flush toilets, low-flow showerheads, and faucet aerators. In public bathrooms, faucets with motion sensors ensure that the water is turned off when no one is using it.

Figure 5 In a drip irrigation system water flows into main tubing and then into smaller tubes that branch out along rows of crops.

In agriculture, farmers can conserve water by using **drip irrigation** for crops, as shown in *Figure 5*. Traditional methods of irrigation are very inefficient: much of the water evaporates in the air or soaks into the soil between the crop rows rather than being delivered to plant roots. In drip irrigation, water is supplied slowly to the soil through special piping and tubing that is punctured with many small holes. The water drips or oozes out around the plant roots. Xeriscaping is an excellent way of saving water in cities and towns in arid regions. **Xeriscaping** means planning residential or commercial landscaping to use as little water as possible. It involves using plants that are naturally adapted to the climate of the area rather than plants like grasses, which need frequent watering during hot, dry weather. *Figure 6* shows an example of xeriscaping. Other ways of conserving water in landscaping is to use mulch around the bases of plants, as well as drip irrigation.

Figure 6 Landscaping with plants that require little water.

Geo Words

drip irrigation: a form of irrigation in which water is supplied slowly to the soil through special piping and tubing that is punctured with many small holes that allow the water to drip or ooze out.

xeriscaping: residential or commercial landscaping planned to use as little water as possible.

Check Your Understanding

1. What categories account for the major water use in the United States?
2. How can water conservation effectively increase the water supply of a community?
3. Why does drip irrigation save water?
4. What is the difference between consumptive and nonconsumptive water use?

Understanding and Applying What You Have Learned

1. Examine *Figure 1* on page R173. Each of the seven cylinders has two small faucets. The top faucet shows the amount of water that is "used up," or consumed. The bottom faucet shows the amount of water returned back to the environment (nonconsumptive use). Use the diagram to answer the following questions:
   a) For each of the seven major uses of water, calculate the percentage of water consumed during use and returned to the environment.
   b) Which use of water has the highest percentage of nonconsumptive use? (highest percentage returned to the environment).
2. Think about the water cycle. Some of the water used by plants is returned to the atmosphere by transpiration. Describe two other ways that irrigation water reenters the water cycle.
3. Suppose that irrigation water is withdrawn from a confined aquifer. Why might this cause a problem in the water supply?

Preparing for the Chapter Challenge

1. Would you expect the proposed developments in your community to increase demand, decrease supply, or a combination of both? Explain your answer. You will use this material in your **Chapter Challenge.**
2. Using what you've learned about water conservation and water use in your community, write a short essay about three water conservation methods that you would suggest for the proposed developments. Defend your choices. Explain why your methods might be useful if water demand increases or if water supply decreases.
3. If your town does not have enough extra water for the new developments, what are some ways the town could get more water? Include water conservation ideas as well as ideas for building more capacity. You can incorporate these ideas into your final report.

Inquiring Further

1. **Water use in your home**

   Design a plan for conducting a water audit in your home. Calculate a reasonable estimate for the total amount of water used in your home daily, monthly, and over a period of one year.

Be sure to obtain parental permission before carrying out your plan.

2. **Water use in a different community**

   Find the gallons-per-day-per-person water use for a town in a different part of the United States. Is it very different from your town? What could account for the difference?

Activity 4 Water Supply and Demand: Water Budgets

Goals

In this activity you will:

- Construct a water budget of your community from data sets.
- Explain the influence of local climate on the water budget.
- Identify times of year when supply and demand of water are greatest and lowest.
- Describe key controls on the quantity and availability of surface and ground water in your community.
- Construct and analyze a box model of an irrigation water budget.
- Explain how ground-water development affects the ground-water system.

Think about It

Take a moment to consider the source and amount of money you have during a month's time and how you spend it.

- Are there times of the year when you have more money and times of the year when you have less money?
- What are the advantages of planning or budgeting for lean times? How would you go about doing this?

What do you think? Record your ideas in your *EarthComm* notebook. Be prepared to discuss your responses with your small group and the class.

Investigate

Part A: Your Local Water Budget

1. Work with members of your group to construct and interpret your local water budget. Find data for the average temperature and amount of precipitation in or near your community on a monthly basis. Visit the *EarthComm* web site or your school library to help you to find this information.
2. Make a climatograph to show the monthly average temperature and precipitation for your community. Use the sample shown to help you construct your graph.
   a) Keep a copy of your graph in your *EarthComm* notebook. Be sure that the axes are correctly labeled and that the graph has a title.

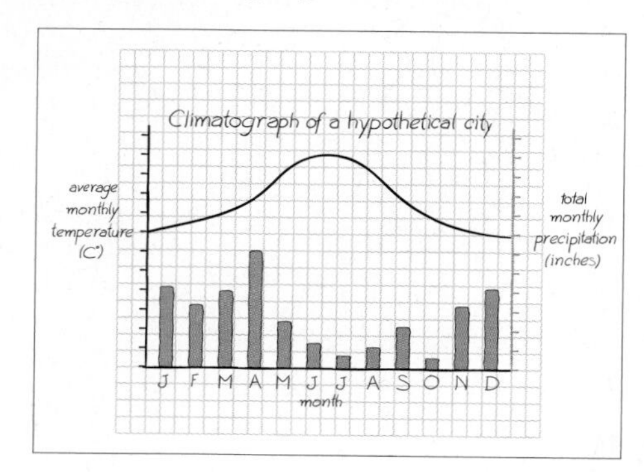

3. Using your graph as a resource, answer the following questions:
   a) During which month or months does your community receive the least amount of precipitation?
   b) During which month or months does your community receive most of its precipitation, or is precipitation spread evenly through the year?
   c) Does your community receive a great deal of snowfall during the winter? If yes, in which month, or months, is snowfall heaviest? In which month or months is snowmelt likely to be greatest?
   d) From your work in previous activities, does your community rely mostly on surface water or on ground water?
   e) In what months would you expect the surface water to be the lowest?
   f) Suggest ways the community would have to adjust if these levels became extremely low.
   g) In what months would you expect ground water to be the lowest?
   h) Suggest ways the community would have to adjust if these levels became extremely low.
   i) Predict the month or months for which there is the greatest demand for water. Explain why. Do these times of increased demand coincide with high water supply or low water supply?

Part B: Constructing the Irrigation Water Budget for the United States

1. To irrigate crops, farmers divert water from existing supplies. Crops use some of the water, some is lost in transit, some evaporates, and some returns to the sources of supply in liquid form. All of the water can be accounted for, in one way or another, in terms of a budget. As with any budget, all of the amounts have to add up in a consistent way (that is, "be accounted for").

Below, in alphabetical order, are several components of the water budget for irrigation. Together, these components account for all of the water involved in irrigation. The significance of the percentages is described following the list.

- consumption by crops: 41%
- delivery to irrigated crops: 78%
- diversions of irrigation water from supply sources: 100%
- losses during delivery: 22%
- losses of water during irrigation use: 37%
- losses that are returned to supply sources: 46%
- net depletion of supply sources: 54%
- permanent losses by evaporation: 13%
- total losses of irrigation water during delivery and use: 59%

a) On blank sheets of paper, draw a box for each of the items above and write in the name of the item and the percentage.

b) Cut out the boxes and arrange them on a large sheet of paper in a way you think shows the natural "flow" of the irrigation water.

c) Draw arrows from one box to another box to show the pathways followed by the irrigation water. The starting box is the one titled "diversion of irrigation water from supply sources."

To understand the meaning of the percentages, read the two notes below:

- If two arrows come into a given box, the percentages in the boxes where the arrows come from have to add up to the percentage in the given box.
- If two arrows leave from a given box, the percentages in the boxes where the arrows go have to add up to the percentage in the given box.

2. Within your small group, discuss the nature of the water budget and try to agree upon a "flow diagram."

   a) Compare your result with those of the other student groups.

3. Use the class results to answer the following questions:

   a) What do you think are the important kinds of sources for irrigation water?

   b) What processes do you think are involved in the various water losses listed above?

   c) In what ways can farmers minimize these water losses?

   d) How is the ground-water system affected by irrigation?

Reflecting on the Activity and the Challenge

You learned how much precipitation falls in your community during the different months of the year. You also speculated about which months your community uses more water than usual. You also examined an irrigation water budget, which should make you think about the water budget in your own community. This will help you anticipate the possible impact of the new developments on your community's water supply.

Geo Words

water budget: an accounting of the sources of water supply and water demand, and of how the supply is divided among the various uses that make up the demand.

discharge: the volume of water that flows past a point on the river per unit of time.

Digging Deeper

NATURAL FLUCTUATIONS IN WATER RESOURCES

Water Budgets

In the financial world, a budget is a plan for the finances of a government or an organization or a person for a period of time, like a year, giving expected expenses and the income to cover those expenses. The term is used somewhat differently in science. A **water budget**, for example, is an accounting of the sources of water supply and water demand, and of how the supply is divided among the various uses that make up the demand. In this activity you dealt with your community's water budget, and also the irrigation water budget for the entire United States. Studying a water budget is useful because it organizes your thinking about water use and can help to identify aspects of water use that need the attention of natural scientists, engineers, planners, or government officials.

In most regions of the United States, rainfall varies considerably from season to season. For example, in some locations, riverbeds are dry during much of the year and flow only during storms. In other places, rivers flow all year long, even during months with little rain. Can you explain these differences? One advantage of using ground water, the other major source of water supply in the United States, is that it is not as vulnerable to short-term changes in precipitation as surface water. As you will see below, however, the ground water table fluctuates up and down as well.

Rivers

The volume of water that flows past a point on the river per unit of time is called the **discharge** of the river. It is measured in cubic feet per second or cubic meters per second. Large rivers, like the Missouri–Mississippi river system can have discharges as great as 16,800 m^3/s. In most rivers, discharge varies greatly between times of drought and times of floods.

Figure 1 A river with a large discharge.

Figure 2 River bed during a drought.

Most cities and towns that use rivers for water supply take out only a small fraction of the river discharge. In some places, however, the demand for river water is so great that the natural discharge of the river is greatly decreased. For example, so much of the discharge of the Colorado River in the Southwest is used for water supply that by the time the river reaches its mouth in the Gulf of California, the flow is only a trickle!

Not all of the water in rivers comes from surface runoff. Rain also infiltrates into the Earth to become ground water. Remember from previous activities that after a certain depth below the land surface, the rock and/or sediment is saturated with ground water. Flowing rivers, as well as lakes and springs, are places where the water table comes out to the land surface (*Figure 3*). At times of high river flow, some of the river water feeds the ground water as shown in *Figure 4A*. At times of low river flow, however, the ground water feeds the river as in *Figure 4B*. This is why most large rivers flow even during long droughts. On average, ground water supplies as much as 40% of the water that flows in streams and rivers. During times of drought the percentage is much greater, and during times of flood the percentage is much smaller.

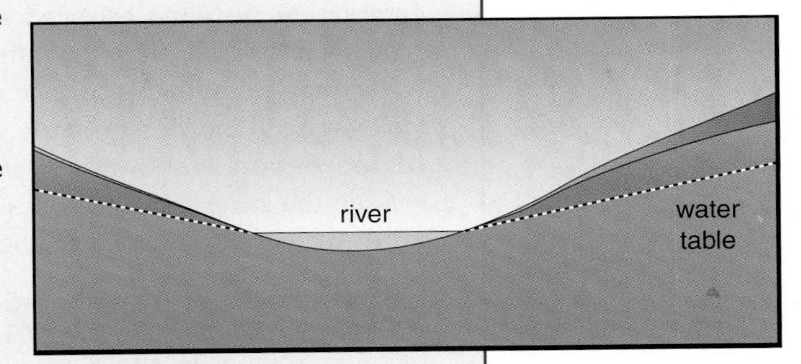

Figure 3 The water table comes to the land surface at flowing rivers.

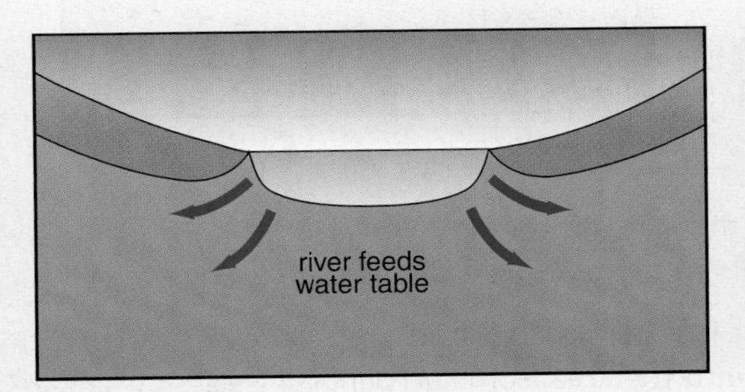

Figure 4A High river flow.

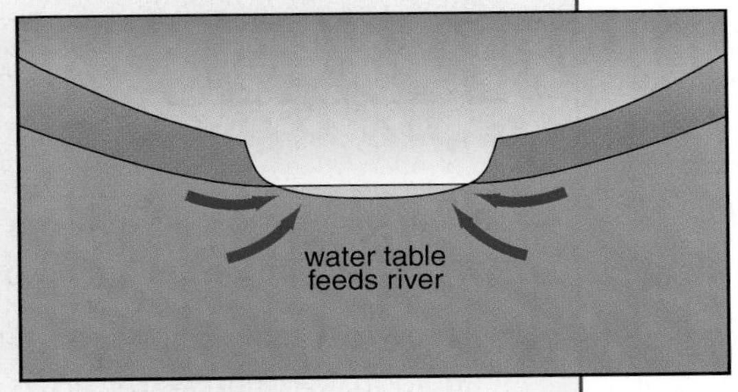

Figure 4B Low river flow.

Ground Water

At some depth below the Earth's surface, the porous rock and sediment are saturated with ground water. The upper limit of saturation is called the water table. Above the water table, in the unsaturated zone, most of the pore spaces are occupied by air. Below the water table, in the saturated zone, all of the pore spaces are filled with water, which flows slowly toward areas where the water table is lower.

Geo Words

cone of depression: the conical shape of the water table near a well, caused by the drawing down of the water table as water is pumped from the well.

Users of ground water would find life easier if the height of the water table in the aquifer that supplied their well always stayed the same. Variations in rainfall affect the height of the water table. During dry spells, ground water keeps flowing toward lakes, rivers, and (in coastal areas) the ocean, but there is no recharge to keep the water table from falling. During rainy spells, on the other hand, recharge more than makes up for ground water drainage, and the water table rises. In many areas the difference in the height of the water table between dry spells and rainy spells can be many meters.

Before wells are drilled into an aquifer, discharge and recharge of the ground-water system is in a condition of long-term balance: the amount of water entering the system is balanced with the amount leaving, and the height of the ground-water table stays the same, on average. Pumping of ground water for water supply upsets this balance. As water is pumped from a well, the water table drops in the vicinity of the well and takes on a conical shape called a **cone of depression**, as shown in *Figure 5A*. If water is being pumped from several wells in the same area, the cones of depression merge with one another, and the water table is depressed over the entire area, as shown in *Figure 5B*.

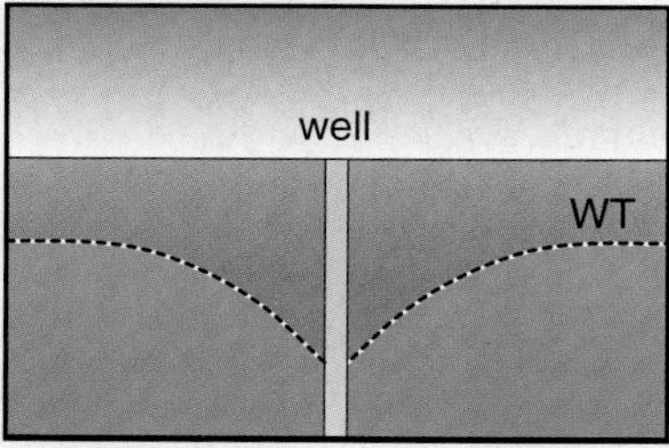

Figure 5A

Figure 5B

As you may have realized from the investigation in Activity 2, when ground water is pumped from a field of wells, one of two things can result:

- The water table is lowered, but it becomes stabilized at some lower height as additional ground water flows into the area from surrounding areas.
- Inflow of ground water from surrounding areas is not great enough to stabilize the water table, and the level keeps falling indefinitely.
- In the latter case, the ground water can no longer be thought of as a renewable resource, because it is being consumed faster than it can be replenished. In a real sense, the ground water is then being mined. In some areas of the United States the imbalance between recharge and discharge has lowered the water table by as much as a 100 m over the last 50 years. Eventually the expense of pumping from great depths becomes prohibitive (or, if the wells are not deep enough, they run dry!), and new sources must be developed.

Check Your Understanding

1. How can a river still have water in it during a drought?
2. What is the difference between the time for surface water and ground water to respond to precipitation?
3. What happens to the water table when ground water is pumped from a well?

Understanding and Applying What You Have Learned

1. Visit the *EarthComm* web site or the school library to find out the number of acres in your county. One inch of rain falling on one acre is equal to 27,154 gallons (102.9 cubic meters) of water. This is called one acre-foot of water.
   a) Calculate the number of gallons of water that reach the Earth in your county in a year's time.
   b) Is that more or less than the amount of water that your county uses in a year?
   c) Assume for a moment that your county's demand for water exceeded the amount of precipitation that fell. How is it possible that streams are still flowing and that water is still in the ground?
2. How would the "income" and "expenditures" in the water budgets of developing countries that lack the advances in technology that have occurred in the Unites States differ from our country?
3. How much does the distribution of annual rainfall affect your community's water supply?

Preparing for the Chapter Challenge

In your community, what times of the year do high use and low availability limit your water resources? How would the addition of the three proposed developments affect this situation? Write answers to these questions in a short essay to include with your **Chapter Challenge.**

Inquiring Further

1. **Water use in other countries**

   Calculate and compare the total volume of rainfall, per unit of land area (for example, per square kilometer) for two other countries: one that has an arid climate, and one that has a rainy climate. Determine the types of water resources these countries have available. Compare their use and rainfall patterns with those of your community. Explain any differences.

2. **High Plains Aquifer**

   The High Plains Aquifer (formerly the Ogallala Aquifer), an enormous confined aquifer that stretches from Kansas to Texas, is one of the largest sources of ground water for irrigation in the United States. Go to the *EarthComm* web site to research the magnitude of lowering of the water table in the aquifer from 1950 to 2000. What do you think might be potential alternative sources of irrigation water, if and when it becomes too costly to pump water from the High Plains Aquifer to the surface?

Activity 5 Water Pollution

Goals

In this activity you will:

- Construct and analyze a physical model of the movement of pollutants in ground water and determine how the pumping of water from wells influences this movement.
- Measure the level of nitrates in a stream within your community.
- Conduct a mathematical analysis (case study) of pollution of surface waters by road salt.
- Identify and describe ways that human activity affects surface water and ground water quality.

Think about It

We rely on clean sources of water to sustain ourselves.

- How do pollutants get into surface water?
- How do pollutants get into ground water?

What do you think? Record your ideas in your *EarthComm* notebook. Be prepared to discuss your responses with your small group and the class.

Investigate

Part A: Ground-Water Pollution

1. Prepare the dish tub the same way you did in the investigation in Activity 2. Fill the tub with about 10 cm of sand. You are constructing a model that will enable you to infer how pollutants reach the water table and then move with the ground-water flow.

2. Wrap a small piece of cheesecloth around the end of two pieces of rigid tubing, and tape it in place with duct tape. Bury the pieces of tubing vertically in the sand. The lower end of the tubing should be about 5 cm from the bottom of the tub. Locate the wells as shown in the diagram below.

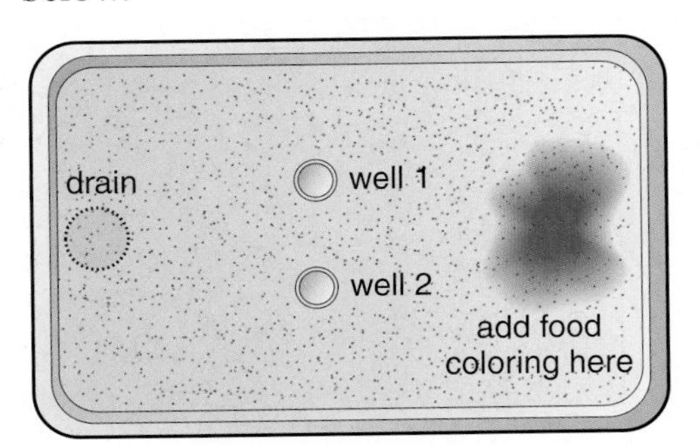

3. As you did in the investigation in Activity 2, attach a drinking straw to the end of the turkey baster.

4. With a measuring cup, pour water onto the "upstream" end of the dish tub (the end opposite the drain hole) at a constant rate of about 1000 mL/min. Wait until the outflow from the drain hole is constant.

5. With an eyedropper, squirt food coloring onto the surface of the sand at the place where you add water. Use as much food coloring as the eyedropper can hold. Start the stopwatch as soon as the food coloring begins to be washed into the sand by the water supply.

6. With the baster, extract small volumes of water, about 1 mL each time, from Well 1 every five seconds and squirt the water into a clean plastic cup.

   a) When you first detect the food coloring in the water samples, note the time elapsed. Keep taking samples until there is no noticeable food coloring in them, and note the time elapsed.

7. With the eyedropper, squirt several more drops of food coloring into Well 1. Begin taking small water samples from Well 2 with the baster. Continue to extract water from the well until the water draining from the tub shows some food coloring.

   a) Did you detect any of the food coloring in Well 2?

8. With the eyedropper, squirt several more drops of food coloring into Well 1. With the baster, extract about 100 mL/min from Well 2 for a time long enough that the water draining from the tub shows some food coloring.

   a) Again, determine whether you can detect any food coloring in Well 2.

9. Use the results of your investigation to answer the following:

   a) What factors do you think determine how long it took for the food coloring to be detected in Well 1 in Step 6?

Wear goggles throughout. Clean up spills. Mark baster as lab equipment.

b) Why do you think that the time during which some food coloring could be detected in Well 1 in Step 6 was much longer than the time it took to squirt the food coloring on the sand surface originally?

c) Why do you think that no food coloring could be detected in Well 2 in Step 7?

d) Why do you think that some food coloring could be detected in Well 2 in Step 8?

Dispose of water and straw after use.

Part B: Testing Water for Nitrate

How does the quality of water from local streams, ponds, or untreated wells compare to water that has been processed by a local water treatment plant? To find out, you will test for nitrate chemical levels in two types of water. One source will be an untreated water sample from one of your community's streams. The second will be tap water that was treated by your community's water treatment facility.

1. If you have a stream near your school, collect a water sample and label its location. Your teacher will provide you with a safety protocol for collecting the sample. If such water sources are not available nearby, your teacher will provide you with a sample.
2. Read over the instructions for the nitrate test kits. Be sure you are familiar with the testing procedures as outlined.
3. Test your untreated water sample for nitrate.

   a) Record your results.

Follow the test-kit instructions carefully. Dispose of chemicals immediately and in the manner described in the directions.

4. Test the treated tap water for nitrate.

   a) Record your results.

5. Use the results to answer the following question:

   a) How does the quality of water from local streams, ponds, or untreated wells compare to water that has been processed by a local water treatment plant?

Part C: Modeling Water Pollution from Road Salt

1. The sketch map shows a reservoir that is located next to an interstate highway. It's winter. The highway crew salt the highway three times in one day. Two days later a storm dumps 2.5 cm of rain on the area.

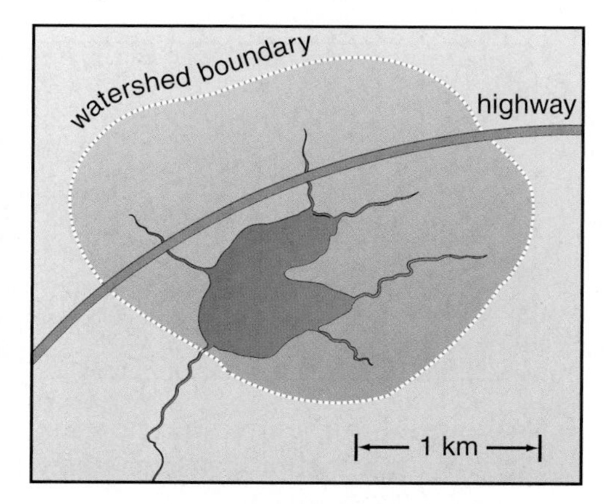

You have the following information:

- The trucks spread 70 kg of salt per kilometer on each traffic lane of the highway.
- The highway has three traffic lanes in each direction.
- The length of the highway that runs through the watershed area of the reservoir is 5 km.

- The watershed area of the reservoir is 10 km^2.
- All of the rain that falls on the watershed area of the reservoir runs off into the reservoir. (The ground is frozen solid, so the part of the rain that infiltrates into the ground to become ground water is negligible.)
- The average depth of the reservoir is 5 m.
- The surface area of the reservoir is one square kilometer.

Possible assumptions:

- The reservoir water that is displaced by the new runoff from the rainstorm contains none of the salt.
- The mixing time of the reservoir is shorter than the time elapsed before the next major rainstorm.

Additional helpful facts:

- There are 1000 L in a cubic meter.
- There are 1000 mL in a liter.

2. Using this information, answer the following questions:
   a) What is the average concentration of salt in the runoff that enters the reservoir from its watershed? Express your answers in grams of salt per liter of water.
   b) What is the concentration of salt in the reservoir after the rainstorm and before the next rainstorm? Express your answers in milligrams of salt per liter of water.
   c) The United States government has issued what is called a health advisory about sodium in public water supplies. A health advisory provides guidance about levels of pollutants, without specifying mandatory maximum concentrations. According to the health advisory, levels of up to 20 mg of sodium per liter of water should not result in adverse health effects over short time periods for most people. How does your

calculated concentration of salt in the reservoir compare with this figure for sodium? Keep in mind that the composition of the salt is sodium chloride (NaCl) and that the sodium accounts for about 40% of the weight of salt.

d) Can you think of any problems with the assumptions you were given above? If so, how do you think they would affect the results from your model? In other words, how realistic is your model?

3. Fill a one-gallon plastic milk jug with water. The jug holds about 3.8 L, but assume that it holds a full 4 L. A teaspoon of salt weighs about 0.5 g. Put a teaspoon of salt in the jug, shake it well, and pour small amounts into plastic drinking cups, one for each member of your group. Taste the water, and decide whether you can taste any of the salt. If not, add another teaspoon of salt and try again. Repeat the process until everyone can taste the salt. Compare the salt concentrations (in grams per liter) at which each member of your group was first able to taste the salt.

   a) What is the minimum concentration of salt in water that causes the water to taste salty to members of your group?

   b) How do these values compare with the salt concentration you calculated for the reservoir?

   c) How do they compare with the United States government health advisory on sodium given earlier?

Reflecting on the Activity and the Challenge

You modeled the pollution of a ground water aquifer, modeled pollution of a reservoir by road salt, and tested the quality of water in your community. Understanding ways that water can become polluted will help you understand how the three developments proposed in the **Chapter Challenge** might affect the quality of your community's water supply.

Digging Deeper

POLLUTION IN SURFACE WATER AND GROUND WATER

Computational Models

Scientists often try to develop computational models to represent natural processes. Such models serve to make predictions, as well as to help organize thinking about a problem. Later, when there is an opportunity to make actual measurements in the field, the predictions of the model are either verified or turn out to be wrong. If they are verified, that tells the scientists that the hypotheses that went into their model are probably right. If the predictions are wrong, however, then the scientists know that they need to rethink the hypotheses, because they are either wrong or incomplete (or both!). In Part C of the investigation, you also developed a computational model in which you assumed some data and made some hypotheses.

Kinds of Pollutants

Both surface water and ground water can become polluted. Surface water generally contains a greater variety of pollutants, from a greater variety of sources. That is because there are so many ways a pollutant can get into a river or lake. Polluted ground water may contain fewer kinds of pollutants, but there may be a larger amount of each pollutant than in a typical surface water supply. It is usually much more difficult to solve problems of ground-water pollution than problems of surface-water pollution. That is because ground water moves far more slowly than surface water. It might take many human lifetimes for pollution to clear from a large aquifer by natural ground water flow.

Figure 1 Livestock waste can pollute water supplies.

Figure 2 Nutrients in fertilizers applied to fields can pollute water supplies.

Many kinds of substances can pollute water. Below is a list of the most common pollutants. Most of these fall into one of two categories: organic, or non-organic.

Organic:

sewage
livestock wastes
pathogenic (disease-causing) microorganisms

Non-organic:

nutrients in fertilizers (mainly nitrates and phosphates)
industrial and commercial chemicals
road salts
agricultural pesticides
acidic mining wastes
waste heat
radioactive waste

Domestic Sewage

About three-quarters of homes in the U.S. are served by municipal sewage systems. The remaining one-quarter discharge their sewage into home septic systems (or even directly into the ground!). The human wastes in sewage are not themselves especially harmful, but disease-causing microorganisms, which are common in sewage, are hazardous to health. Many illnesses, like cholera and typhoid, are caused by contact with sewage. Such illnesses are much less common in the United States than in areas of the world where people are more likely to be exposed to untreated sewage. Coliform bacteria which live in the intestinal tracts of all warm-blooded animals (including humans), are a common type of bacteria that are generally not harmful. However, their presence in sewage is commonly used as a signal of sewage contamination.

Figure 3 Water testing at Lake Cachuma, the water storage reservoir above Santa Barbara, California.

Nutrients from Fertilizers

Plants need many inorganic chemical nutrients for good growth. The most important of these are the elements nitrogen and phosphorus. Nitrogen is the most abundant gas in the Earth's atmosphere, but it cannot be used directly by plants. Certain soil bacteria convert nitrogen gas into soluble forms of nitrogen, mainly nitrate ions, which can be used by plants. Lightning bolts do the same thing, and of course man-made chemical fertilizers contain soluble nitrogen. The problem with nitrogen-containing fertilizers is that they can be "too much of a good thing": excess nitrogen promotes the growth of algae in natural surface waters, and when the algae die and decay they rob the waters of the oxygen needed by aquatic animals. Runoff from croplands is not the only culprit; fertilizer on lawns, golf courses, and home gardens often carries even more excess nitrogen than croplands.

Human wastes contain water-soluble nitrogen compounds. In areas where untreated or inadequately treated sewage finds its way into ground water supplies, nitrate levels in drinking water can be excessive. Human infants can be harmed by drinking water that has high nitrate levels because nitrate reduces the amount of oxygen carried by red blood cells. Too much nitrate can cause "blue-baby syndrome," which can lead to suffocation.

Phosphorus is used by plants in the form of phosphate ions. The most common form of phosphorus in fertilizers is ground-up phosphate rock, a type of sedimentary rock. Deposits of a material called guano, the excrement of bats and seabirds, are also mined for phosphorus. The problem with soluble phosphorus is similar to that with nitrogen: it causes algal blooms, such as the one shown in *Figure 4*, in natural waters. Another problem with phosphorus is that the phosphates contained in household soaps and detergents can pollute ground water supplies. In many heavily populated communities that rely on ground water from water supplies, restrictions have been placed on the phosphate content of detergents.

Figure 4 Soluble phosphorus can cause algal blooms such as the one shown on the left.

Toxic Chemicals

In earlier times, the number of artificial chemicals and materials used by people in their daily lives was small. Today, however, such substances number in the tens of thousands, and the list grows rapidly. A large percentage of these are potentially or definitely a health hazard. Many of these were found to be toxic only after long use. Some of the major chemical pollutants are gasoline, chemical solvents, agricultural pesticides and herbicides, and compounds of heavy-metal elements like copper, zinc, lead, mercury, and cadmium used in industrial processes.

Unfortunately, the rate of development of new substances is far greater than the ability of government and private testing laboratories to investigate their toxicity. Inevitably, many of these toxic substances find their way into surface water supplies and ground water supplies. Partly this is from everyday use. In many places, however, toxic chemicals have been dumped, illegally, on the land

Figure 5 Photograph of a toxic-waste site.

surface or stored in decaying drums at dumpsites. Cleaning up sites such as the one shown in *Figure 5*, all around the United States, is a multi-billion-dollar task.

Waste Heat

Most of the electricity in the United States is generated in power plants where burning of fossil fuels produces steam to drive turbines, which in turn drive electrical generators. Enormous quantities of water are used to cool and condense the steam back into liquid water, to be recycled in the power plant. The cooling water is put into rivers, lakes, or the ocean. It is typically 5°–10°C warmer than the water bodies that it enters. This causes three major problems. Warm water cannot hold as much dissolved oxygen as cool water, therefore less oxygen is available for aquatic animals. The higher temperatures also tend to lead to the disappearance of some species of organisms and the replacement of these species with warmer-water organisms. Sudden temperature changes can kill organisms outright, by what is called thermal shock.

Figure 6 Diversion dam for the Pit River Power Plant in central California.

Check Your Understanding

1. What are some of the health hazards of domestic sewage?
2. If coliform bacteria are generally not harmful, why is there concern about them?
3. How do chemical fertilizers cause water pollution?
4. What risks do power plants pose to the environment?

Understanding and Applying What You Have Learned

1. In some northern states, harmful quantities of salt are appearing in streams, ponds, and even shallow ground water aquifers. What might be the source of the salt and how is it getting into these water sources? How might this salty water affect fresh water ecosystems?
2. The High Plains Aquifer is one of the largest ground water aquifers in the United States. It is located under Colorado, Nebraska, and parts of Texas, Oklahoma, Kansas, South Dakota, and Iowa. It is a nonrenewable water resource and has shown an increase in levels of nitrate. List the potential sources of these increases. Describe how the use of the topsoil above this aquifer may be related to its pollution.
3. Make a list of ways that citizens in your community can reduce water pollution.
4. What is the relevance of the experiment you completed in Part A of the investigation with respect to water pollution? Is your community affected by this problem? Could it be? What steps could be taken to reduce the effect?

Preparing for the Chapter Challenge

How will the construction or operation of the proposed developments pollute your town's drinking water? What factors will influence the type or amount of water pollution? What can be done to minimize water pollution from these developments? Your group should prepare a poster and present a report to your citizens' planning group addressing these questions. Prepare this poster for an audience of an intelligent group of concerned citizens who are not knowledgeable about water pollution.

Inquiring Further

1. **Local water quality**

   Research how well your local water provider has maintained the water quality in your community.

   a) Look in the provider's water quality report.

   b) Also try looking for past violations that have been reported to the United States Environmental Protection Agency (EPA). Go to the *EarthComm* web site to help you find this information. Write down the nature of any violations and the dates in which they occurred.

c) Also try searching EPA Envirofacts Warehouse—facilities information (visit the *EarthComm* web site to find this information) to research the status of industries in your county. This resource indicates if a particular industry is in compliance with federal pollution laws.

2. **Pollution from pets**

   Feces and urine from cats and dogs is a major source of water pollution in some urban and suburban areas (especially those without "pooper-scooper" ordinances).

   a) Try to determine the size of the cat and dog population from your community's authorities, and then, by making assumptions about the average daily output of cats and dogs, arrive at an estimate of the total weight of cat and dog feces and the volume of cat and dog urine deposited on the yards, sidewalks, streets, and other public areas of your community per year.
   b) What do you think are the effects of this material on water quality in your community or in other areas surrounding your community?

3. **Pollution in films**

   Water pollution has been the topic of motion pictures. As a class, view a film about water pollution. After watching, have a class discussion on the problems associated with the following issues:

   a) detecting and cleaning up past industrial pollution of water sources;
   b) determining whether small concentrations of industrial pollutants are harmful to human health or not;
   c) whether industries that in the past used substances that are now known to be hazardous but were not thought to be hazardous at the time should be held legally liable for cleanup and for damage awards.

4. **Thermal pollution**

   Locate the power plant nearest to your school, or, if there is none in your community, the power plant nearest to your community. Contact the operators of the power plant to find out their method of cooling. Some power plants use air cooling, but most use water cooling. If the plant uses water for cooling, find out the source of the water, the volume used, and the method of disposal. If the cooling water is discharged into a river or lake, try to obtain data on the temperature rise in the water body.

Activity 6 Water Treatment

Goals

In this activity you will:

- Model and explain key processes and stages in water treatment and filtration.
- Understand how and why water is treated prior to human consumption.
- Research and describe the water treatment process used by your community.

Think about It

On a backpack trip high in the mountains you become separated from your group. You realize that it may be several days before you are found.

- What concerns would you have about drinking water?
- What contaminants might be found in the water?
- What precautions should you take?

What do you think? Record your ideas in your *EarthComm* notebook. Be prepared to discuss your responses with your small group and the class.

Investigate

Part A: Modeling Water Treatment

1. Place 1 L of untreated water into a plastic bottle or beaker. Label this container "no alum" and set it aside.
2. Place 1 L of untreated water into a second plastic bottle or beaker. Label this beaker "alum."
   a) In your notebook, note any odors associated with the samples.
3. Pour the untreated water in the "alum" beaker back and forth using a third empty container. Do this 15 times.
   a) Note your observations about the appearance and odor of the water.
   b) If there is an increase in odor, explain why.
4. Add 50 g of silt or clay to each water sample. Add 20 g of alum to the "alum" beaker.
5. Stir the water in each container slowly for five minutes.
   a) What effect did the alum have on the water?
6. Allow each of the containers to sit for 15 minutes.
   a) Record your observations every five minutes. Make your observations while holding a blank sheet of white paper behind each container.

   Note: Keep the water in the "alum" beaker, undisturbed for use in Part B. While waiting you may construct the filtering device for Part B.

Use only water samples provided by your teacher. Do not taste the water. Smell the water carefully by waving fumes toward your nose. No chemicals should be consumed. Wash hands well after the activity.

Part B: Modeling Water Filtration

1. Using a rubber band, securely attach a nylon screen or piece of stocking across the mouth of a 2-L plastic soda bottle with the bottom 5 cm cut off. Turn the bottle upside down and set it in a 500-mL beaker. Using another beaker, measure 800 mL of sand and pour it gently into the upside down bottle. Add 400 mL of gravel on top of the sand, as shown in the diagram. The sand and gravel will serve as your filter. Clean the filter by pouring 4-L of tap water slowly into it. Pour off the filtered water.

2. When the water with alum from Part A of the investigation has been undisturbed for 15 minutes, carefully pour the water into the filtering device. Try not to disturb the sediment at the bottom of the beaker.

3. After the water has gone through the filter, place two drops of non-chlorine bleach into the clean filtered water.
4. Compare the "no alum" sample of water with your filtered results. (Do not drink the water!)
   a) Record your observations in your notebook.
5. Dispose of the water as instructed.

Part C: Water Treatment in Your Community

1. Obtain a copy of your community's water-quality report. If your community has a population of more than 100,000, this report is required to be available to the public. Contact your local municipality or check their web page. If the information you require to complete the following steps is not in the water-quality report, have one student call the water provider and ask the question(s) for the class.
2. Examine the water-treatment process that your community uses.
   a) Outline your community's water-treatment process. Use diagrams if necessary.
   b) How much does it cost to treat water in your community?
   c) How much water is treated daily?
   d) What is the maximum capacity of the system?
3. Investigate the water supply for your community's water-treatment facility.
   a) What would happen to the water-treatment process if a great deal of soil were added to the system at once? Is there such a thing as sediment pollution?
   b) List some activities that destroy vegetation that anchors the soil.
   c) Would any of these activities occur during the construction of the proposed development?

Reflecting on the Activity and the Challenge

You have just modeled two important steps in treating water to make it safe for drinking. These two steps are probably used at your local water-treatment plant, which you investigated. Cleaning up water is not free. The more pollutants there are in the water, the more complicated and costly it is to treat it. You will need to consider the pollution the new developments will add to your community's water system.

Digging Deeper

WATER TREATMENT

Water purification occurs in nature. Evaporation and condensation separates the water from substances that are dissolved in it. Bacteria convert dissolved organic contaminants into simple and harmless compounds. Sand and gravel filter out suspended material that makes water cloudy. Rainwater and ground water that is purified by bacteria and filtered by sediment provide clean drinking water. Unfortunately, the demand for water is much greater than nature's supply, so communities rely on municipal water-treatment plants.

No matter where a community's water comes from, it must be treated in a facility similar to the one shown in *Figure 1*, because even where water is used carefully, pollutants are bound to be introduced. You cannot wash clothes, take a shower, or flush the toilet without adding pollutants to the water! The goal of water treatment is to produce water that is safe for drinking and does not have an objectionable taste, smell, or appearance. Treatment of wastewater from municipal sewage systems involves a series of steps: screening, flocculation, filtering, and disinfecting. Depending on your local water sources and treatment needs, your community's water-treatment process may be slightly different than what is presented here. Keep in mind also that the wastewater that passes through home septic systems is not treated at all: it goes directly into the ground and tends to add to the pollutant levels in local ground water supplies.

Figure 1 Water is made safe for human use at a water-treatment plant.

Screening, Flocculation, and Settling

Sewage is first passed through screens to remove large pieces of debris and other solids. Then it flows slowly through sedimentation tanks, where gritty mineral material settles to the bottom and lightweight sludge floats to the top. To aid in the settling process, alum (aluminum sulfate, $Al_2(SO_4)_3$) and slaked lime (calcium hydroxide, $Ca(OH)_2$) are sometimes added to the sewage to cause particles of sediment to coagulate (stick together) to form larger clumps called flocs, which settle relatively fast. The sludge is sent to a separate container where the organic matter in it is digested by microorganisms.

Filtration

In the next treatment step, called filtration, the water passes through sand, gravel, and activated carbon (charcoal). These filters help remove minute particles and even odors that give water a bad smell or taste. If the water contains inorganic (non-living) components that cannot be adequately removed by filtering, a chemical exchange process–called ion exchange–can be used. This chemical process is effective in treating hard water, which is water that does not easily produce suds in the presence of soap. It can also be used to treat arsenic, chromium, excess fluoride, nitrates, radium, and uranium. As part of this stage, the wastewater is aerated (exposed to the oxygen of the atmosphere), which greatly speeds up the natural oxidation and decomposition of organic matter in the water by the activity of microorganisms.

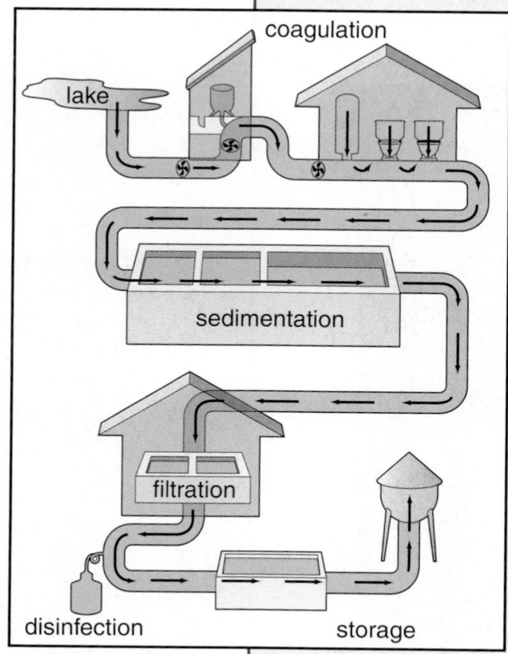

Figure 2 A schematic of a water-treatment plant.

Disinfection

Water is disinfected before it enters the distribution system. Disinfectants kill remaining dangerous microbes. Chlorine is a very effective disinfectant, and residual concentrations of chlorine guard against biological contamination in the water distribution system. Ozone is a powerful disinfectant, but it is not effective in controlling biological contaminants in the distribution pipes. Sometimes the disinfecting process produces byproducts. Disinfecting byproducts (DBPs) are contaminants that form when chlorine reacts with organic matter that is in treated drinking water. Long-term exposure to some DBPs may increase the risk of cancer or have other adverse health effects. The United States Environmental Protection Agency regulates the presence of four DBPs in drinking water, and scientists are continuing to study the issue.

Levels of Treatment

Primary treatment involves only settling and perhaps chlorination. Such treatment is better than nothing, but the water is unsatisfactory for human use. Secondary treatment involves filtering and oxidation by microorganisms, and results in wastewater with much better quality. In some places, tertiary treatment is used. Tertiary treatment, involves flocculation, disinfecting, and use of specific additives to remove undesirable excess compounds of iron, manganese, etc., in addition to the processes of primary and secondary treatment. Tertiary treatment is designed to produce high-quality domestic water that can be added to public water supplies—although the public is still not very willing knowingly to drink treated sewage!

Treatment Costs

On a per-gallon basis, water is cheap. On average, water costs are slightly more than $2 per 1000 gallons, although the costs tend to be lower for large water systems. Treatment accounts for about 15% of that cost. Other costs are for equipment (the treatment plant and distribution system) and labor for operation and maintenance of the system.

Water Hardness

As water moves through soil, sediment, or bedrock, it dissolves some of the minerals. Water hardness reflects the concentration of dissolved solids in the water, mainly calcium and magnesium. Ground water tends to be harder than surface water, because it has had more time to be in contact with the solid materials. Hard water is also more common in areas where the bedrock is limestone, which consists of the calcium carbonate mineral calcite, which is slightly soluble in water. Some degree of hardness in drinking water is considered healthy, because calcium and magnesium are essential nutrients. Hardness also lowers the concentration of lead dissolved from the lead-based solder that is often used in copper plumbing systems. A disadvantage of hard water is that it reduces soapsuds. It also can leave hard, white deposits in teapots, on shower walls, and in water heaters and boilers. Water softeners reduce hardness by replacing the calcium and magnesium ions with sodium ions. A problem with water softeners is that high sodium levels in drinking water are undesirable for persons with high blood pressure.

Figure 3 Damaged pipes.

Check Your Understanding

1. Describe how water is purified in nature.
2. Draw a diagram of a typical water-treatment process. Label the processes and show the flow of water from source to community water mains.
3. What are the advantages and disadvantages of softening water during treatment?

Understanding and Applying What You Have Learned

1. Assume for a moment that the new development will require that your community waterworks be expanded. Create a list of three or four concessions that developers could make that would lessen the economic impact of the proposed expansion.
2. In the investigation, you treated a water sample. What steps in the complete water treatment process were you modeling? What additional steps would be required to make your sample safe for drinking?
3. How do concerns regarding the pollution of ground water differ from those surrounding the pollution of surface water? Does your community treat these water sources differently? Why or why not?

Preparing for the Chapter Challenge

Write a short essay for inclusion in your chapter report in which you address the following questions:

1. The construction of these new developments can add sediment to local streams. How might this adversely affect your water-treatment plant? How could this negative effect be decreased?
2. How will your community's water-treatment plants handle the extra water required by these new developments? How will the additional developments affect the cost of water in your community?

Inquiring Further

1. **Softening water**

   Line three funnels with filter paper. Place sand in one, Calgon in one, and leave the third empty. Use this to design an experiment to find out whether Calgon™ or natural filtration by sand is a better method of softening water.

   Check your plan with your teacher before carrying it out. Do not drink the water. Wash hands after the activity.

2. **Recycling water in space**

   Investigate and report on current techniques involved with recycling water on long-term space ventures.

3. **Purified, distilled, spring, and mineral water**

   Investigate the differences between purified water, distilled water, spring water, and mineral water.

4. **Water analyses in your community**

   Investigate the following analyses conducted at water-treatment facilities: bacteriological analyses, general chemical and physical analyses, metal analysis, organic-compound analyses, radionuclides.

Earth Science at Work

ATMOSPHERE: *Maintenance Engineer*
Airlines employ engineers to maintain, inspect, and repair aircraft. They are responsible for maintaining wheels and brakes, flight control systems, and ice and rain protection systems.

BIOSPHERE: *Medical Technologist*
Some bacteria, viruses, and fungi are hazardous to human health. Medical-laboratory technologists are trained to detect and identify any microorganism.

CRYOSPHERE: *Figure Skater*
An adequate supply of water to "flood" an ice-skating surface is essential in maintaining good ice conditions. Figure skaters depend on a smooth ice surface to safely perform their jumps, spins, and intricate footwork.

GEOSPHERE: *Well Driller*
Many areas rely on ground water as their municipal water supply. Wells must be drilled to reach these water reservoirs.

HYDROSPHERE: *Public Works Employee*
The planning, administration, operation, and maintenance of the wastewater collection of a city require an understanding of wastewater treatment options available. This is the responsibility of a public works department.

How is each person's work related to the Earth system, and to water resources?

Unit V

EarthComm®

Earth System Science
in the Community

Earth System Evolution

1

Astronomy ...and Your Community

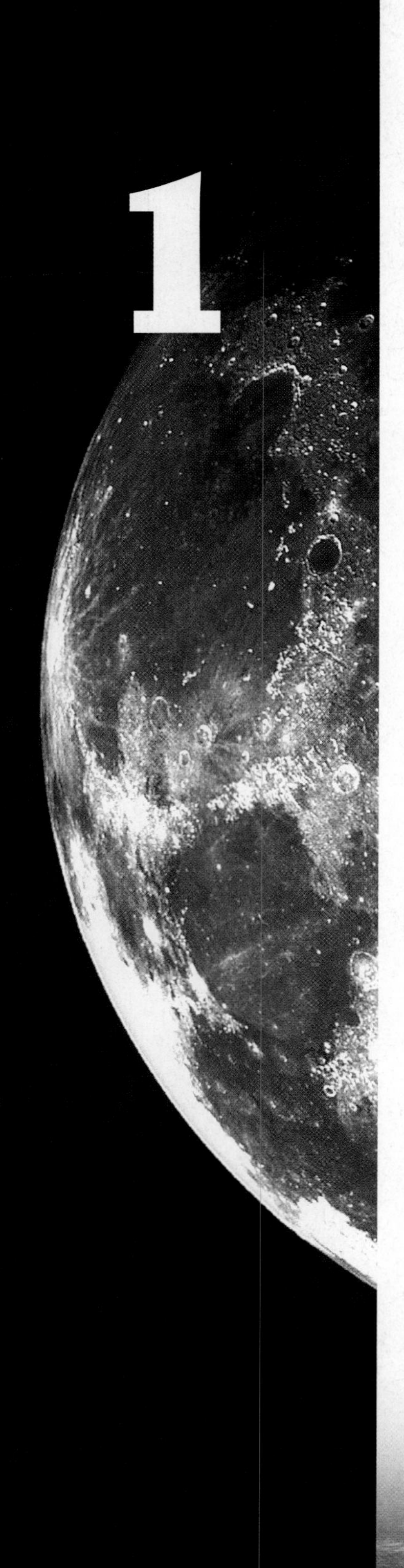

1 Astronomy ...and Your Community

Getting Started

Throughout time, all systems in the universe are affected by processes and outside influences that change them in some way. This includes Earth and the solar system in which it exists. You have years of experience with life on the third planet from the Sun, and you know a lot about your tiny corner of the universe. Think about the Earth in relation to its neighbors in the solar system.

- What objects make up the solar system?
- How far is the Earth from other objects in the solar system?
- Which objects in the solar system can influence the Earth?
- Can you think of any objects or processes outside the solar system that might affect the Earth?

Write a paragraph about Earth and its place in this solar system. After that, write a second paragraph about processes or events in the solar system that could change Earth. Describe what they do and how Earth is, or might be, affected. Try to include answers to the questions above.

Scenario

Scientists recently announced that an asteroid 2-km wide, asteroid 1997XF11, would pass within 50,000 km of Earth (about one-eighth the distance between the Earth and the Moon) in October 2028. A day later, NASA scientists revised the estimate to 800,000 km. News reports described how an iron meteorite blasted a hole more than 1 km wide and 200 m deep, and probably killed every living thing within 50 km of impact. That collision formed Arizona's Meteor Crater some 50,000 years ago. Such a collision would wipe out a major city today. These reports have raised concern in your community about the possibility of a comet or asteroid hitting the Earth. Your class will be studying outer space and the effects that the Sun and other objects in the solar system can have on the Earth. Can you share your knowledge with fellow citizens and publish a booklet that will discuss some of the possible hazards from outer space?

Chapter Challenge

In your publication, you will need to do the following:

- Describe Earth and its place in the universe. Include information about the formation and evolution of the solar system, and about the Earth's distance from and orbit around the Sun. Be sure to mention Earth's place in the galaxy, and the galaxy's place in the universe.
- Describe the kinds of solar activities that influence the Earth. Explain the hazardous and beneficial effects that solar activity (sunspots and radiation, for example) have on the planet. Discuss briefly the Sun's composition and structure, and that of other stars.
- Discuss the Earth's orbital and gravitational relationships with the Sun and the Moon.
- Explain what comets and asteroids are, how they behave, how likely it is that one will collide with Earth in your lifetime, and what would happen if one did.
- Explain why extraterrestrial influences on your community are a natural part of Earth system evolution.

The booklet should have a model of the solar system that will help citizens understand the relative sizes of and distances between solar-system bodies.

Assessment Criteria

Think about what you have been asked to do. Scan ahead through the chapter activities to see how they might help you to meet the challenge. Work with your classmates and your teachers to define the criteria for assessing your work. Record all of this information. Make sure that you understand the criteria as well as you can before you begin. Your teacher may provide you with a sample rubric to help you get started.

Activity 1 The History and Scale of the Solar System

Goals

In this activity you will:

- Produce a scale model of the solar system.
- Identify some strengths and limitations of scale models.
- Calculate distances to objects in the solar system in astronomical units (AU), light- years, and parsecs.
- Explain, in your own words, the nebular theory of the formation of the solar system.
- Explain the formation of the universe.

Think about It

Earth is part of a large number of objects that orbit around a star called the Sun.

- What objects make up the solar system? Where are they located in relation to Earth?

What do you think? Record your ideas in the form of a diagram of the solar system in your *EarthComm* notebook. Without looking ahead in this book, draw the Sun and the planets, and the distances from the Sun to the planets, as nearly to scale as you can. Be prepared to discuss your diagram with your small group and the class.

Investigate

1. Use the data in *Table 1* to make a scale model of the solar system. Try using the scale 1 m = 150,000,000 km.

   a) Divide all the distances in the first column by 150,000,000 (one hundred and fifty million). Write your scaled-down distances in your notebook, in meters.

   b) Divide all the diameters in the second column by 150,000,000. Write your scaled-down diameters in your notebook, in meters.

   c) Looking at your numbers, what major drawback is there to using the scale 1 m = 150,000,000 km?

Table 1 Diameters of the Sun and Planets, and Distances from the Sun

Object	Distance from Sun (km)	Diameter (km)
Sun	0	1,391,400
Mercury	57,900,000	4878
Venus	108,209,000	12,104
Earth	149,598,770	12,756
Mars	227,900,000	6794
Jupiter	778,200,000	142,984
Saturn	1,429,200,000	120,536
Uranus	2,875,000,000	51,118
Neptune	4,504,400,000	49,528
Pluto	5,915,800,000	2302

2. Now try another scale: 1 m = 3,000,000 km (three million kilometers).

   a) Divide all the distances in the first column by 3,000,000. Write your scaled-down distances in your notebook in meters.

   b) Divide all the diameters in the second column by 3,000,000. Write your scaled-down diameters in your notebook in meters.

   c) Looking at your numbers, what major drawback is there to using the scale 1 m = 3,000,000 km?

3. Using what you have learned about scaling distances and diameters in the solar system, make models of the Sun and the planets. Each of the planets can be drawn on a different sheet of paper using a ruler to lay out the correct sizes for the different planets and the Sun.

4. To represent the distances from the Sun to the planets you will need to use a tape measure. You may want to measure the size of your stride and use this as a simple measuring tool.

To do this, stand behind a line and take five steps in as normal a way as possible and note where your last step ended. Now measure the distance from where you started to the end. Divide by five to determine how far you walk with each step. Knowing the length of your stride is an easy way to determine distances.

a) Explain the scale(s) you decided to use and your reasons for your choices.

b) Is it possible to make a model of the solar system on your school campus in which both the distances between bodies and the diameters of the bodies are to the same scale? Why or why not?

Reflecting on the Activity and the Challenge

In this activity you used ratios to make a scale model of the solar system. You found out that scale models help you appreciate the vastness of distances in the solar system. You also found out that there are some drawbacks to the use of scale models. Think about how you might use the model you made as part of your **Chapter Challenge**.

Geo Words

astronomical unit: a unit of measurement equal to the average distance between the Sun and Earth, i.e., about 149,600,000 (1.496×10^8) km.

light-year: a unit of measurement equal to the distance light travels in one year, i.e., 9.46×10^{12} km.

Digging Deeper

OUR PLACE IN THE UNIVERSE

Distances in the Universe

Astronomers often study objects far from Earth. It is cumbersome to use units like kilometers (or even a million kilometers) to describe the distances to the stars and planets. For example, the star nearest to the Sun is called Proxima Centauri. It is 39,826,600,000,000 km away. (How would you say this distance?)

Astronomers get around the problem by using larger units to measure distances. When discussing distances inside the solar system, they often use the **astronomical unit** (abbreviated as AU). One AU is the average distance of the Earth from the Sun. It is equal to 149,598,770 km (about 93 million miles).

Stars are so far away that using astronomical units quickly becomes difficult, too! For example, Proxima Centauri is 266,221 AU away. This number is easier to use than kilometers, but it is still too cumbersome for most purposes. For distances to stars and galaxies, astronomers use a unit called a **light-year**. A light-year sounds as though it is a unit of time, because a year is a unit of time, but it is really the distance that light travels in a year. Because light travels extremely fast, a light-year is a very large distance. For example, the Sun is only 8 light *minutes* away from Earth, and the nearest stars are several light-years away. Light travels at a speed of 300,000 km/s. This makes a light-year 9.46×10^{12} (9,460,000,000,000) km. Light from Proxima Centauri takes 4.21 years to reach Earth, so this star is 4.21 light-years from Earth.

Astronomers also use a unit called the **parsec** (symbol pc) to describe large distances. One parsec equals 3.26 light-years. Thus, Proxima Centauri is 1.29 pc away. The kiloparsec (1000 pc) and megaparsec (1,000,000 pc) are used for objects that are extremely far away. The nearest spiral galaxy to the Milky Way galaxy is the Andromeda galaxy. It is about 2.5 million light-years, or about 767 kpc (kiloparsecs), away.

The Nebular Theory

As you created a scale model of the solar system, you probably noticed how large the Sun is in comparison to most of the planets. In fact, the Sun contains over 99% of all of the mass of the solar system. Where did all this mass come from? According to current thinking, the birthplace of our solar system was a **nebula**. A nebula is a cloud of gas and dust probably cast off from other stars that used to live in this region of our galaxy. More than 4.5 billion years ago this nebula started down the long road to the formation of a star and planets. The idea that the solar system evolved from such a swirling cloud of dust is called the nebular theory.

You can see one such nebula in the winter **constellation** Orion (see *Figure 1*), just below the three stars that make up the Belt of Orion. Through a pair of binoculars or a small, backyard-type telescope, the Orion Nebula looks like a faint green, hazy patch of light. If you were able to view this starbirth region through a much higher-power telescope, you would be able to see amazing details in the gas and dust clouds. The Orion Nebula is very much like the one that formed our star, the Sun. There are many star nurseries like this one scattered around our galaxy. On a dark night, with binoculars or a small telescope, you can see many gas clouds that are forming stars.

Figure 1 Orion is a prominent constellation in the night sky.

Geo Words

parsec: a unit used in astronomy to describe large distances. One parsec equals 3.26 light-years.

nebula: general term used for any "fuzzy" patch on the sky, either light or dark; a cloud of interstellar gas and dust.

constellation: a grouping of stars in the night sky into a recognizable pattern. Most of the constellations get their name from the Latin translation of one of the ancient Greek star patterns that lies within it. In more recent times, more modern astronomers introduced a number of additional groups, and there are now 88 standard configurations recognized.

Figure 2 The Keyhole Nebula. Imaged by the Hubble Space Telescope.

In the nebula that gave birth to our solar system, gravity caused the gases and dust to be drawn together into a denser cloud. At the same time, the rate of rotation (swirling) of the entire nebula gradually increased. The effect is the same as when a rotating ice skater draws his or her arms in, causing the rate of rotation to speed up. As the nebular cloud began to collapse and spin faster, it flattened out to resemble a disk, with most of the mass collapsing into the center. Matter in the rest of the disk clumped together into small masses called **planetesimals**, which then gradually collided together to form larger bodies called **protoplanetary bodies**.

At the center of the developing solar system, material kept collapsing under gravitational force. As the moving gases became more concentrated, the temperature and pressure of the center of the cloud started to rise. The same kind of thing happens when you pump up a bicycle tire with a tire pump: the pump gets warmer as the air is compressed. When you let the air out of a tire, the opposite occurs and the air gets colder as it expands rapidly. When the temperature in the center of the gas cloud reached about 15 million degrees Celsius, hydrogen atoms in the gas combined or fused to create helium atoms. This process, called **nuclear fusion**, is the source of the energy from the Sun. A star—the Sun—was born!

Fusion reactions inside the Sun create very high pressure, and like a bomb, threaten to blow the Sun apart. The Sun doesn't fly apart under all this outward pressure, however. The Sun is in a state of equilibrium. The gravity of the Sun is pulling on each part of it and keeps the Sun together as it radiates energy out in all directions, providing solar energy to the Earth community.

The Birth of the Planets

The rest of the solar system formed in the swirling disk of material surrounding the newborn Sun. Nine planets, 67 satellites (with new ones still being discovered!), and a large number of comets and asteroids formed. The larger objects were formed mostly in the flat disk surrounding where the Sun was forming.

Geo Words

planetesimal: one of the small bodies (usually micrometers to kilometers in diameter) that formed from the solar nebula and eventually grew into protoplanets.

protoplanetary body: a clump of material, formed in the early stages of solar system formation, which was the forerunner of the planets we see today.

nuclear fusion: a nuclear process that releases energy when lightweight nuclei combine to form heavier nuclei.

Four of these planets, shown in *Figure 3*—Mercury, Venus, Earth, and Mars—are called the **terrestrial** ("Earth-like") **planets**. They formed in the inner part of our solar system, where temperatures in the original nebula were high. They are relatively small, rocky bodies. Some have molten centers, with a layer of rock called a mantle outside their centers, and a surface called a crust. The Earth's crust is its outer layer. Even the deepest oil wells do not penetrate the crust.

The larger planets shown in *Figure 3*—Jupiter, Saturn, Uranus, and Neptune—consist mostly of dense fluids like liquid hydrogen. These **gas giants** formed in the colder, outer parts of the early solar nebula. They have solid rocky cores about the size of Earth, covered with layers of hydrogen in both gas and liquid form. They lie far from the Sun and their surfaces are extremely cold.

Pluto is the most distant planet from the Sun. Some astronomers do not even classify it as a planet, and there is controversy about whether Pluto should be included among the "official" planets. Pluto is very different from the terrestrial or the gaseous planets. Some scientists think that it may not have been part of the original Solar System but instead was captured later by the Sun's gravity. Others think that it was once a moon of an outer planet because it resembles the icy moons of the gas giants.

Figure 3 Composite image of the planets in the solar system, plus the Moon.

All life, as we know it on Earth, is based on carbon and water. Although there is evidence that water may once have been plentiful on Mars, liquid water has not been found on any of the planets in the Solar System. Without water, these planets cannot support life as we know it. There is evidence, however, that Jupiter's moon Europa may be covered with ice, and that an ocean may lie beneath the ice. Scientists have found that life on Earth can thrive in even some of the most extreme environments. Powerful telescopes have discovered organic molecules in interstellar gas and have discovered planets around many nearby stars. Although life may not exist on the other planets in the solar system, the possibility of life on Europa and in other places in the universe exists!

Geo Words

terrestrial planets: any of the planets Mercury, Venus, Earth, or Mars, or a planet similar in size, composition, and density to the Earth. A planet that consists mainly of rocky material.

gas giant planets: the outer solar system planets: Jupiter, Saturn, Uranus, and Neptune, composed mostly of hydrogen, helium, and methane, and having a density of less than 2 gm/cm^2.

Geo Words

comet: a chunk of frozen gases, ice, and rocky debris that orbits the Sun.

asteroid: a small planetary body in orbit around the Sun, larger than a meteoroid (a particle in space less than a few meters in diameter) but smaller than a planet. Many asteroids can be found in a belt between the orbits of Mars and Jupiter.

There are trillions of **comets** and **asteroids** scattered throughout the solar system. Earth and other solar-system bodies are scarred by impact craters formed when comets and asteroids collided with them. On Earth, erosion has removed obvious signs of many of these craters. Astronomers see these comets and asteroids as the leftovers from the formation of the solar system. Asteroids are dark, rocky bodies that orbit the Sun at different distances. Many are found between the orbits of Mars and Jupiter, making up what is called the asteroid belt. Many others have orbits outside of the asteroid belt. Comets are mixtures of ice and dust grains. They exist mainly in the outer solar system, but when their looping orbits bring them close to the Sun, their ices begin to melt. That is when you can see tails streaming out from them in the direction away from the Sun. Some comets come unexpectedly into the inner solar system. Others have orbits that bring them close to the Sun at regular intervals. For example, the orbit of Halley's comet brings it into the inner solar system every 76 years.

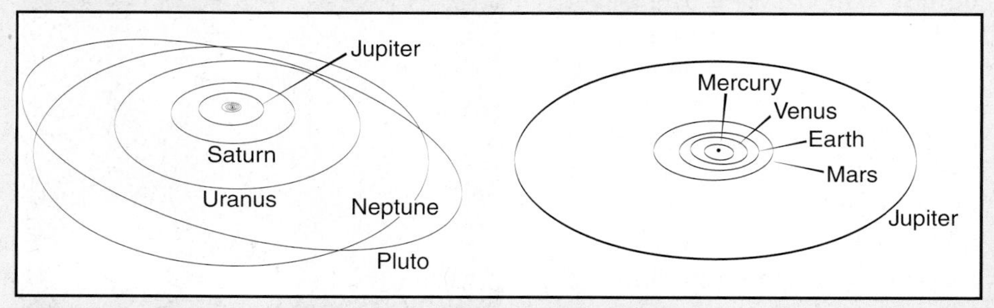

Figure 4 Two diagrams are required to show the orbits of the planets to scale.

Where is the Solar System in Our Galaxy?

Have you ever seen the Milky Way? It is a swath of light, formed by the glow of billions of stars, which stretches across the dark night sky. From Earth, this band of celestial light is best seen from dark-sky viewing sites. Binoculars and backyard-type telescopes magnify the view and reveal individual stars and nebulae. Unfortunately, for those who like to view the night sky, light pollution in densely populated areas makes it impossible to see the Milky Way even on nights when the atmosphere is clear and cloudless.

Galaxies are classified according to their shape: elliptical, spiral, or irregular. Our home galaxy is a flat spiral, a pinwheel-shaped collection of stars held together by their mutual gravitational attraction. Our galaxy shown in *Figure 5* is called the Milky Way Galaxy, or just the galaxy. Our solar system is located in one of the spiral arms about two-thirds of the way out from the center of the galaxy. What is called the Milky Way is the view along the flat part of our galaxy. When you look at the Milky Way, you are looking out through the galaxy parallel to the

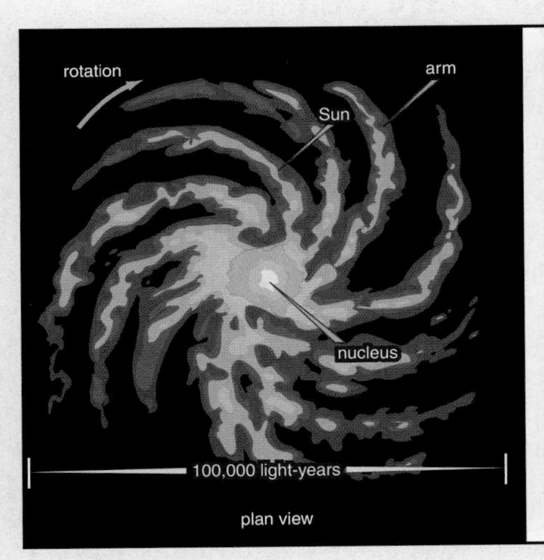

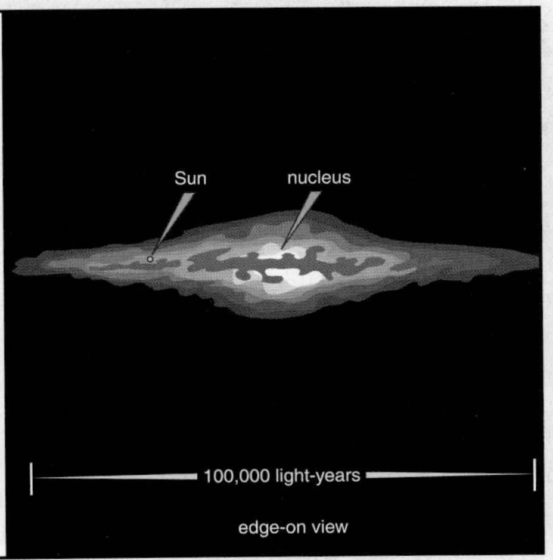

Figure 5 The Milky Way Galaxy. Our solar system is located in a spiral band about two-thirds of the way from the nucleus of the galaxy.

plane of its disk. The individual stars you see dotting the night sky are just the ones nearest to Earth in the galaxy. When you view the Milky Way, you are "looking through" those nearest stars to see the more distant parts of the galaxy. In a sense, you are looking at our galaxy from the inside. In other directions, you look through the nearest stars to see out into intergalactic space!

Our Milky Way Galaxy formed about 10 billion years ago and is one of billions of galaxies in the universe. The universe itself formed somewhere between 12 to 14 billion years ago in an event called the Big Bang. This sounds like the universe began in an explosion, but it did not. In the beginning, at what a scientist would call "time zero," the universe consisted almost entirely of energy, concentrated into a volume smaller than a grain of sand. The temperatures were unimaginably high. Then the universe expanded, extremely rapidly, and as it expanded the temperature dropped and matter was formed from some of the original energy. **Cosmologists** (scientists who study the origin and dynamics of the universe) think that most of the matter in the universe was formed within minutes of time zero! The expansion and cooling that started with the Big Bang continues to this day.

The galaxies and stars are the visible evidence of the Big Bang, but there is other, unseen evidence that it happened. It's called the cosmic background radiation, which is radiation that is left over from the initial moments of the Big Bang. Astronomers using special instruments sensitive to low-energy radio waves have detected it coming in from all directions from the universe. The existence of the cosmic background radiation is generally considered to be solid evidence of how the Big Bang happened.

Geo Words

cosmologist: a scientist who studies the origin and dynamics of the universe.

Check Your Understanding

1. What are the distances represented by a light-year, an astronomical unit, and a parsec?
2. Which of the units in Question 1 would you use to describe each of the following? Justify your choice.
   a) Distances to various stars (but not our Sun)?
   b) Distances to various planets within our solar system?
   c) Widths of galaxies?
3. In your own words, explain the nebular theory for the beginning of our solar system.
4. Briefly describe the origin of the universe.

Understanding and Applying What You Have Learned

1. Using the second scale (1 m = 3,000,000 km) you used for distance in your model of the solar system:
   a) How far away would Proxima Centauri be from Earth?
   b) How far away would the Andromeda galaxy be on your scale, given that Andromeda is 767 kiloparsecs or 2.5 million light-years away?
2. The Moon is 384,000 km from Earth and has a diameter of 3476 km. Calculate the diameter of the Moon and its distance from the Earth using the scale of the model you developed in the **Investigate** section.
3. Refer again to *Table 1*. If the Space Shuttle could travel at 100,000 km/hr, how long would it take to go from Earth to each of the following objects? Assume that each object is as close to the Earth as it can be in its orbit.
   a) The Moon?
   b) Mars?
   c) Pluto?
4. What is the largest distance possible between any two planets in the solar system?
5. Use your understanding of a light-year and the distances from the Sun shown in *Table 1*. Calculate how many minutes it takes for sunlight to reach each of the nine planets in the solar system. Then use the unit "light-minutes" (how far light travels in one minute) to describe the *distances* to each object.
6. Write down your school address in the following ways:
   a) As you would normally address an envelope.
   b) To receive a letter from another country.
   c) To receive a letter from a friend who lives at the center of our galaxy.
   d) To receive a letter from a friend who lives in a distant galaxy.

Preparing for the Chapter Challenge

Begin to develop your brochure for the **Chapter Challenge**. In your own words, explain your community's position relative to the Earth, Sun, the other planets in our solar system, and the entire universe. Include a few paragraphs explaining what your scale model represents and how you chose the scale or scales you used.

Inquiring Further

1. **Solar-system walk**

   Create a "solar-system walk" on your school grounds or your neighborhood. Draw the Sun and the planets to scale on the sidewalk in chalk. Pace off the distances between the Sun and the nine planets at a scale that is appropriate for the site.

2. **Scaling the nearest stars**

   Look up the distances to the five stars nearest to the Sun. Where would they be in your scale model? To show their location, would you need a map of your state? Country? Continent? The world?

3. **Nuclear fusion**

   Find out more about the process of nuclear fusion. Explain how and why energy is released in the process by which hydrogen atoms are converted into helium atoms within the Sun. Be sure to include Albert Einstein's famous equation, $E = mc^2$, in your explanation, and explain what it means.

4. **Star formation**

   Write a newspaper story about star formation. Visit the *EarthComm* web site to find information available on the web sites of the Hubble Space Telescope and the European Southern Observatory to find examples of star-forming nebulae in the galaxy. How are they similar? How are they different? What instruments do astronomers use to study these nebulae?

Activity 2 The Earth–Moon System

Goals

In this activity you will:

- Investigate lunar phases using a model and observations in your community.
- Investigate the general idea of tidal forces.
- Understand the role of the Earth, the Moon, and the Sun in creating tides on Earth.
- Understand the Earth–Moon system and the Moon's likely origin.
- Compare the appearance of the Moon to other solar-system bodies.

Think about It

Think about the last time that you gazed at a full Moon.

- What happened to make the Moon look the way it does?
- What is the origin of the Moon?
- How does the Moon affect the Earth?

What do you think? Record your ideas about these questions in your *EarthComm* notebook. Be prepared to discuss your responses with your small group and the class.

Investigate

Part A: Lunar Phases

1. Attach a pencil to a white Styrofoam® ball (at least 5 cm in diameter) by pushing the pencil into the foam. Set up a light source on one side of the room. Use a lamp with a bright bulb (150-W) without a lampshade or have a partner hold a flashlight pointed in your direction. Close the shades and turn off the overhead lights.
2. Stand approximately 2 m in front of the light source. Hold the pencil and ball at arm's length away with your arm extended towards the light source. The ball represents the Moon. The light source is the Sun. You are standing in the place of Earth.
   a) How much of the illuminated Moon surface is visible from Earth? Draw a sketch of you, the light source, and the foam ball to explain this.
3. Keeping the ball straight in front of you, turn 45° to your left but stay standing in one place.
   a) How much of the illuminated Moon surface is visible from Earth?
   b) Has the amount of light illuminating the Moon changed?
   c) Which side of the Moon is illuminated? Which side of the Moon is still dark? Draw another diagram in your notebook of the foam ball, you, and the light source in order to explain what you see.
4. Continue rotating counterclockwise away from the light source while holding the ball directly in front of you. Observe how the illuminated portion of the Moon changes shape as you turn 45° each time.
   a) After you pass the full Moon phase, which side of the Moon is illuminated? Which side of the Moon is dark?
   b) How would the Moon phases appear from Earth if the Moon rotated in the opposite direction?

Be careful not to poke the sharp end of the pencil into your skin while pushing the pencil into the foam. Use caution around the light source. It is hot. Do not touch the Styrofoam to the light.

Part B: Observing the Moon

1. Observe the Moon for a period of at least four weeks. During this time you will notice that the apparent shape of the Moon changes.
   a) Construct a calendar chart to record what you see and when you see it. Sketch the Moon, along with any obvious surface features that you can see with the naked eye or binoculars.
   b) Do you always see the Moon in the night sky?
   c) How many days does it take to go through a cycle of changes?
   d) What kinds of surface features do you see on the Moon?
   e) Label each phase of the Moon correctly and explain briefly the positions of the Sun, the Earth, and the Moon during each phase.

Tell an adult before you go outside to observe the Moon.

Part C: Tides and Lunar Phases

1. Investigate the relationship between tides and phases of the Moon.
   a) On a sheet of graph paper, plot the high tides for each city and each day in January shown in *Table 1*. To prepare the graph, look at the data to find the range of values. This will help you plan the scales for the vertical axis (tide height) and horizontal axis (date).
   b) On the same graph, plot the Moon phase using a bold line. Moon phases were assigned values that range from zero (new Moon) to four (full Moon).
2. Repeat this process for low tides.
3. Answer the following questions in your *EarthComm* notebook:
   a) What relationships exist between high tides and phases of the Moon?
   b) What relationships exist between low tides and phases of the Moon?
   c) Summarize your ideas about how the Moon affects the tides. Record your ideas in your *EarthComm* notebook.

Table 1 Heights of High and Low Tides in Five Coastal Locations during January 2001
(All heights are in feet.)

			Breakwater, Delaware		Savannah, Georgia		Portland, Maine		Cape Hatteras, North Carolina		New London, Connecticut	
Date	**Moon Phase**	**Moon Phase**	**High**	**Low**	**High**	**Low**	**High**	**Low**	**High**	**Low**	**High**	**Low**
1/3/01	First Quarter	2	3.6	0.2	7.3	0.5	8.5	1	2.6	0.2	2.4	0.3
1/6/01	Waxing Gibbous	3	4.5	0	8.2	0.5	9.7	0.1	3.4	−0.4	3	−0.2
1/10/01	Full Moon	4	5.6	−0.9	9.4	−1.5	11.6	−1.9	4.2	−0.8	3.5	−0.7
1/13/01	Waning Gibbous	3	5.1	−0.7	8.8	−0.9	11	−1.4	3.7	−0.6	3	−0.5
1/16/01	Last Quarter	2	4.1	−0.1	7.9	−0.2	9.7	0.1	3	−0.2	2.7	0
1/20/01	Waning Crescent	1	4.3	0.1	7.3	0.2	9.4	0.2	3.2	0	2.8	0
1/24/01	New Moon	0	4.6	0	8.1	−0.1	9.7	−0.1	3.3	−0.1	2.8	−0.1
1/30/01	Waxing Crescent	1	3.7	0.1	7.4	0.1	8.7	0.6	2.6	0	2.4	0.2

4. *Table 1* shows data from the month of January 2001. At the *EarthComm* web site, you can obtain tidal data during the same period that you are doing your Moon observations. Select several cities nearest your community.
   a) Record the highest high tide and the lowest low tide data for each city. Choose at least eight different days to compare. Correlate these records to the appearance of the Moon during your observation period. Make a table like *Table 1* showing high and low tides for each location.
   b) What do you notice about the correlation between high and low tides and the appearance of the Moon?

Part D: Tidal Forces and the Earth System

1. Use the data in *Table 2*.
   a) Plot this data on graph paper. Label the vertical axis "Number of Days in a Year" and the horizontal axis "Years before Present." Give your graph a title.
   b) Calculate the rate of decrease in the number of days per 100 million years (that is, calculate the slope of the line).
2. Answer the following questions:
   a) How many fewer days are there every 10 million years? every million years?
   b) Calculate the rate of decrease per year.
   c) Do you think that changes in the number of days in a year reflect changes in the time it takes the Earth to orbit the Sun, or changes in the time it takes the Earth to rotate on its axis? In other words, is a year getting shorter, or are days getting longer? How would you test your idea?

Table 2 Change in Rotation of Earth Due to Tidal Forces

Period	Date (millions of years ago)	Length of Year (days)
Precambrian	600	424
Cambrian	500	412
Ordovician	425	404
Silurian	405	402
Devonian	345	396
Mississippian	310	393
Pennsylvanian	280	390
Permian	230	385
Triassic	180	381
Jurassic	135	377
Cretaceous	65	371
Present	0	365.25

Reflecting on the Activity and the Challenge

In this activity you used a simple model and observations of the Moon to explore lunar phases and surface characteristics of the Moon. You also explored the relationship between tides and the phases of the Moon. The tides also have an effect that decreases the number of days in a year over time. That's because tides slow the rotation of the Earth, making each day longer. You now understand that tides slow the rotation of the Earth, and how this has affected the Earth. This will be useful when describing the Earth's gravitational relationships with the Moon for the **Chapter Challenge**.

Geo Words

accretion: the process whereby dust and gas accumulated into larger bodies like stars and planets.

Digging Deeper

THE EVOLUTION OF THE EARTH–MOON SYSTEM

The Formation of the Earth and Moon

Figure 1 The Moon is the only natural satellite of Earth.

You learned in the previous activity that during the formation of the solar system, small fragments of rocky material called planetesimals stuck together in a process called **accretion**. Larger and larger pieces then came together to form the terrestrial planets. The leftovers became the raw materials for the asteroids and comets. Eventually, much of this material was "swept up" by the newborn inner planets. Collisions between the planets and the leftover planetesimals were common. This was how the Earth was born and lived its early life, but how was the Moon formed?

Scientists theorize that an object the size of Mars collided with and probably shattered the early Earth. The remnants of this titanic collision formed a ring of debris around what was left of our planet. Eventually this material accreted into a giant satellite, which became the Moon. Creating an Earth–Moon system from such a collision is not easy. In computer simulations, the Moon sometimes gets thrown off as a separate planet or collides with the Earth and is destroyed. However, scientists have created accurate models that predict the orbit and composition of both the Earth and Moon from a collision with a Mars-sized object. The Moon's orbit (its distance from the Earth, and its speed of movement) became adjusted so that the gravitational pull of the Earth is just offset by the centrifugal force that tends to make the Moon move off in a straight line rather than circle the Earth. After the Earth–Moon system became stabilized, incoming planetesimals continued to bombard the two bodies, causing impact craters. The Earth's surface has evolved since then. Because the Earth is geologically an active place, very few craters remain. The Moon, however, is geologically inactive. *Figure 2* shows the Moon's pockmarked face that has preserved its early history of collisions.

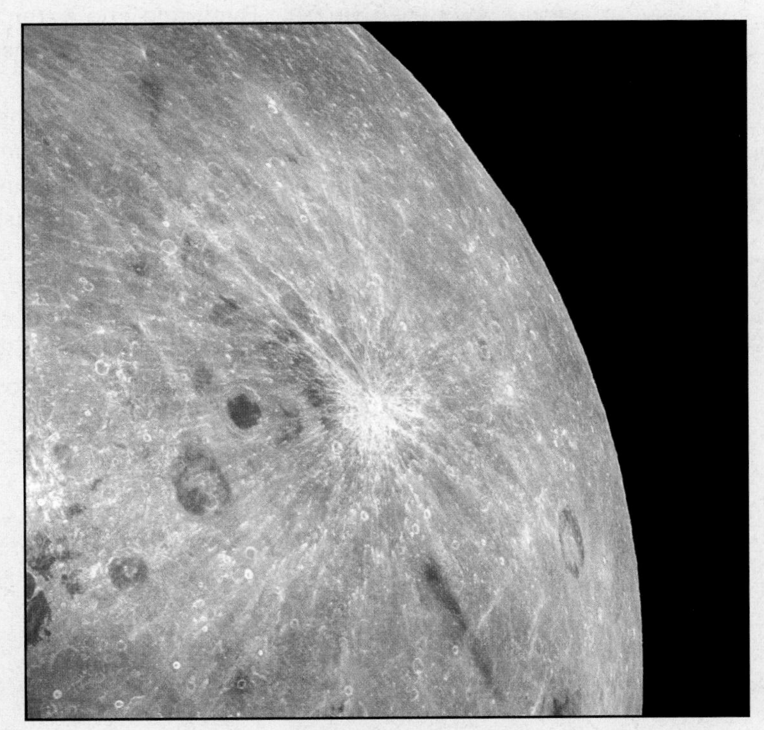

Figure 2 Impact craters on the Moon.

When the Earth was first formed, its day probably lasted only about six hours. Over time, Earth days have been getting longer and longer. In other words, the Earth takes longer to make one full rotation on its axis. On the other hand, scientists have no reason to think that the time it takes for the Earth to make one complete revolution around the Sun has changed through geologic time. The result is that there are fewer and fewer days in a year, as you saw in the **Investigate** section. Why is the Earth's rotation slowing down? It has to do with the gravitational forces between the Earth, the Moon, and the Sun, which create ocean tides.

Tides

The gravitational pull between the Earth and the Moon is strong. This force actually stretches the solid Earth about 20 cm along the Earth–Moon line. This stretching is called the Earth tide. The water in the oceans is stretched in the same way. The stretching effect in the oceans is greater than in the solid Earth, because water flows more easily than the rock in the Earth's interior. These bulges in the oceans, called the ocean tide, are what create the high and low tides (see *Figure 3*). It probably will seem strange

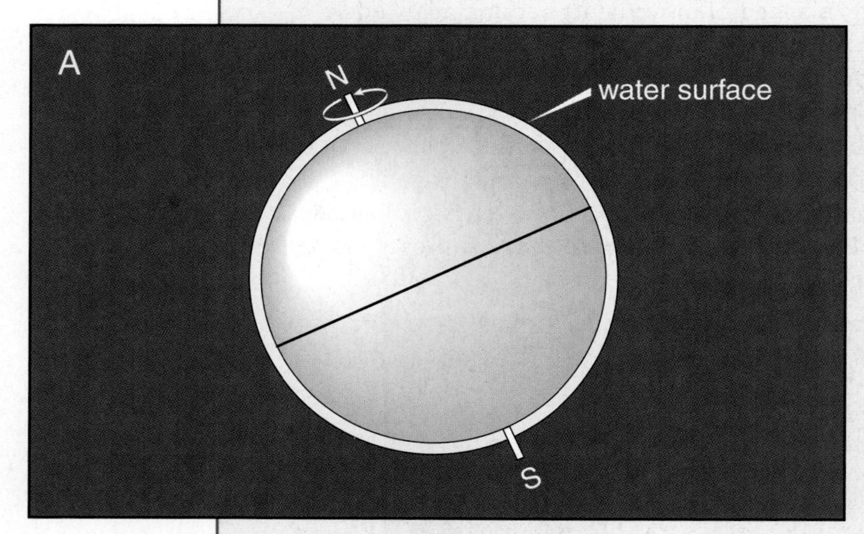

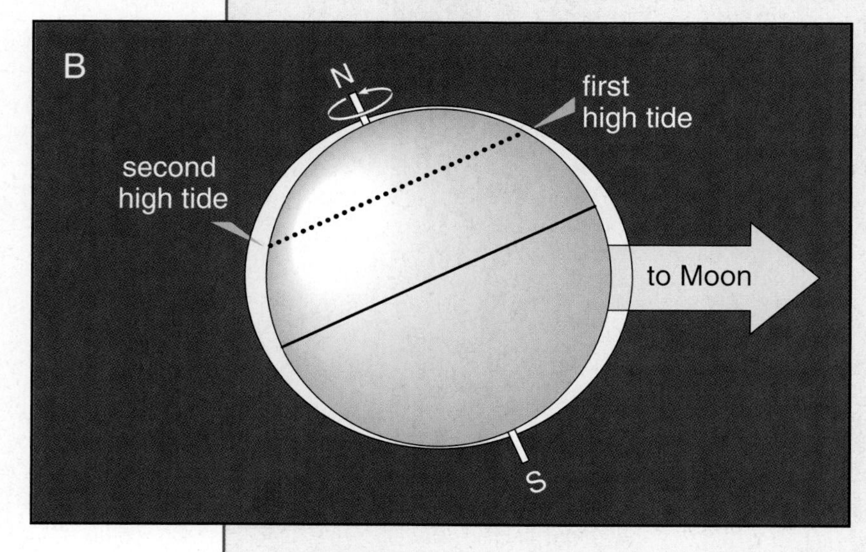

Figure 3 Schematic diagram of tides. Diagram A illustrates how the ocean surface would behave without the Moon and the Sun (no tides). Diagram B illustrates, that in the presence of the Moon and the Sun, shorelines away from the poles experience two high tides and two low tides per day.

to you that there are two bulges, one pointing toward the Moon and the other away from the Moon. If the tides are caused by the pull of the Moon, why is there not just one bulge pointing toward the Moon? The explanation is not simple. If you are curious, you can pursue it further in the **Inquiring Further** section of this activity.

As the Earth rotates through a 24-h day, shorelines experience two high tides—one when the tidal bulge that points toward the Moon passes by, and once when the tidal bulge that points away from the Moon passes by. The tidal cycle is not exactly 24 h. By the time the Earth has completed one rotation (in 24 h), the Moon is in a slightly different place because it has traveled along about 1/30 of the way in its orbit around the Earth in that 24-h period. That's why the Moon rises and sets about 50 min. later each day, and why high and low tides are about 50 min. later each day. Because there are two high tides each day, each high tide is about 25 min. later than the previous one.

The gravitational pull of the Sun also affects tides. Even though it has much greater mass than the Moon, its tidal effect is not as great, because it is so much farther away from the Earth. The Moon is only 386,400 km away from the Earth, whereas the Sun is nearly 150,000,000 km away. The Moon exerts 2.4 times more tide-producing force on the Earth than the Sun does. The changing relative positions of the Sun, the Moon, and the Earth cause variations in high and low tides.

The lunar phase that occurs when the Sun and the Moon are both on the same side of the Earth is called the new Moon. At a new Moon, the Moon is in the same direction as the Sun and the Sun and Moon rise together in the sky. The tidal pull of the Sun and the Moon are adding together, and high tides are even higher than usual, and low tides are even lower than usual. These tides are called **spring tides** (see *Figure 4*). Don't be confused by this use of the word "spring." The spring tides have nothing to do with the spring season of the year! Spring tides happen when the Sun and Moon are in general alignment and raising larger tides. This also happens at another lunar phase: the full Moon. At full Moon, the Moon and Sun are on opposite sides of the Earth. When the Sun is setting, the full Moon is rising. When the Sun is rising, the full Moon is setting. Spring tides also occur during a full Moon, when the Sun and the Moon are on opposite sides of the Earth. Therefore, spring tides occur twice a month at both the new-Moon and full-Moon phases.

Geo Words

spring tide: the tides of increased range occurring semimonthly near the times of full Moon and new Moon.

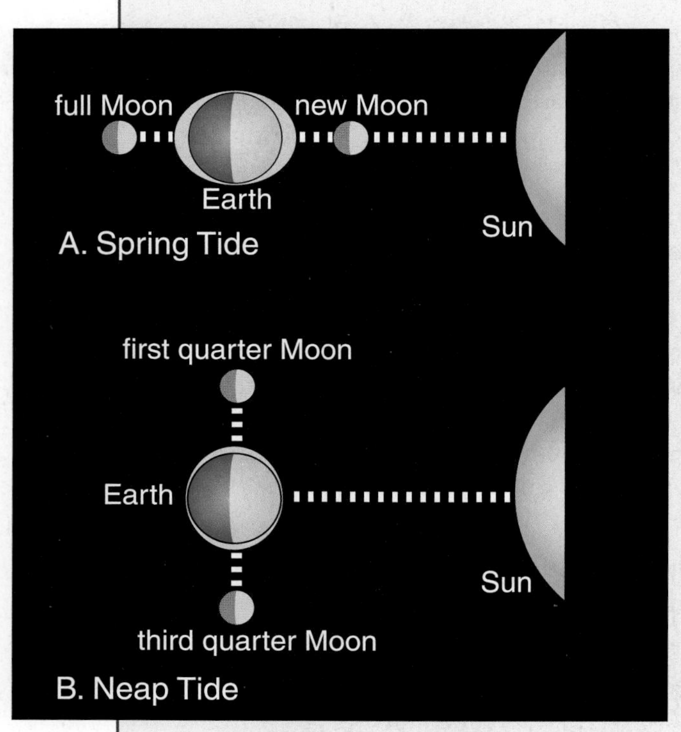

Figure 4 Schematic diagrams illustrating spring and neap tides.

When the line between the Earth and Sun makes a right angle with the line from the Earth to the Moon, as shown in *Figure 4*, their tidal effects tend to counteract one another. At those times, high tides are lower than usual and low tides are higher than usual. These tides are called **neap tides**. They occur during first quarter and third-quarter Moons. As with spring tides, neap tides occur twice a month.

The tide is like a kind of ocean wave. The high and low tides travel around the Earth once every tidal cycle, that is, twice per day. This wave lags behind the Earth's rotation, because it is forced by the Moon to travel faster than it would if it were free to move on its own. That's why the time of high tide generally does not coincide with the time that the Moon is directly overhead. The friction of this lag gradually slows the Earth's rotation.

Geo Words

neap tide: the tides of decreased range occurring semimonthly near the times of the first and last quarter of the Moon.

Another way to look at tides is that the tidal bulges are always located on the sides of the Earth that point toward and away from the Moon, while the Earth with its landmasses is rotating below the bulges. Each time land on the spinning Earth encounters a tidal bulge there is a high tide at that location. The mass of water in the tidal bulge acts a little like a giant brake shoe encircling the Earth. Each time the bulge of water hits a landmass, energy is lost by friction. The water heats up slightly. (This is in addition to the energy lost by waves hitting the shore, which also heats the water by a small fraction of a degree.) Over long periods of time, the tidal bulge has the effect of slowing down the rotation of the Earth, and actually causing the Moon to move away from the Earth. The current rate of motion of the Moon away from the Earth is a few centimeters a year. This has been established by bouncing laser beams off of reflectors on the Moon to measure its distance. Although the Moon's orbit is not circular and is complex in its shape, measurements over many years have established that the

Moon is indeed moving away from the Earth. Special super-accurate clocks have also established that the day is gradually becoming slightly longer as well, because of this same phenomenon, called "tidal friction." The day (one rotation of the Earth on its axis) has gradually become longer over geologic time. As the Earth system evolves, cycles change as well.

In this activity you limited the factors that cause and control the tides to the astronomical forces. These factors play only one part. The continents and their different shapes and ocean basins also play a large role in shaping the nature of the tides. Although many places on Earth have two high tides and two low tides every day (a semidiurnal tide), some places experience only one high tide and one low tide every day (a diurnal tide). There are still other places that have some combination of diurnal and semidiurnal tides (mixed tides). In these places (like along the west coast of the United States) there are two high tides and two low tides per day, but the heights of the successive highs and lows are considerably different from one another.

Figure 5 How do tides affect coastal communities?

Check Your Understanding

1. How did the Moon likely form?
2. Describe the relative positions of the Earth, the Moon, and the Sun for a spring tide and for a neap tide.
3. What effect have tides had on the length of a day? Explain.

Understanding and Applying What You Have Learned

1. Refer back to the graph of the changing length of the day that you produced in the investigation. Think about the causes of tidal friction and the eventual outcome of tidal friction. Predict how long you think the day will eventually be. Explain the reasoning for your prediction.

2. Think about the roles that the Sun and Moon play in causing the ocean tides.

   a) If the Earth had no Moon, how would ocean tides be different? Explain your answer.
   b) How would the ocean tides be different if the Moon were twice as close to the Earth as it is now?
   c) What differences would there be in the ocean tides if the Moon orbited the Earth half as fast as it does now?

3. Look at *Figure 3B* in in the **Digging Deeper** section. Pretend that you are standing on a shoreline at the position of the dotted line. You stand there for 24 h and 50 min, observing the tides as they go up and down.

   a) What differences would do you notice, if any, between the two high tides that day?
   b) Redraw the diagram from *Figure 3B*, only this time, make the arrow to the Moon parallel to the Equator. Make sure you adjust the tidal bulge to reflect this new position of the Moon relative to the Earth. What differences would you now see between the two high tides that day (assuming that you are still at the same place)?
   c) Every month, the Moon goes through a cycle in which its orbit migrates from being directly overhead south of the Equator to being directly overhead north of the Equator and back again. To complicate things, the maximum latitude at which the Moon is directly overhead varies between about 28.5° north and south, to about 18.5° north and south (this variation is on a 16.8 year cycle). How do you think the monthly cycle relates to the relative heights of successive high tides (or successive low tides)?

4. Return to the tide tables for the ocean shoreline that is nearest to your community (your teacher may provide a copy of these to you).

   a) When is the next high tide going to occur? Find a calendar to determine the phase of the Moon. Figure out how to combine these two pieces of information to determine whether this next high tide is the bulge toward the Moon or away from the Moon.
   b) The tide tables also provide the predicted height of the tides. Look down the table to see how much variation there is in the tide heights. Recalling that the Sun also exerts tidal force on the ocean water, try to draw a picture of the positions of the Earth, Moon, and Sun for:
      i) The highest high tide you see on the tidal chart.
      ii) The lowest high tide you see on the tidal chart.
      iii) The lowest low tide you see on the tidal chart.

5. The questions below refer to your investigation of lunar phases.

   a) Explain why the Moon looks different in the sky during different times of the month.

b) What advantage is there to knowing the phases of the Moon? Who benefits from this knowledge?

c) It takes 27.32166 days for the Moon to complete one orbit around the Earth. The Moon also takes 27.32166 days to complete the rotation about its axis. How does this explain why we see the same face of the Moon all the time?

Preparing for the Chapter Challenge

Write several paragraphs explaining the evolution of the Earth–Moon system, how mutual gravitational attraction can affect a community through the tides, and how the changing length of the day could someday affect the Earth system. Be sure to support your positions with evidence.

Inquiring Further

1. **Tidal bulge**

   Use your school library or the library of a nearby college or university, or the Internet, to investigate the reason why the tidal bulge extends in the direction away from the Moon as well as in the direction toward the Moon. Why does the Earth have two tidal bulges instead of just one, on the side closest to the Moon?

2. **Tidal forces throughout the solar system**

   Tidal forces are at work throughout the solar system. Investigate how Jupiter's tidal forces affect Jupiter's Moons Europa, Io, Ganymede, and Callisto. Are tidal forces involved with Saturn's rings? Write a short report explaining how tidal friction is affecting these solar-system bodies.

3. **Impact craters**

   Search for examples of impact craters throughout the solar system. Do all the objects in the solar system show evidence of impacts: the planets, the moons, and the asteroids? Are there any impact craters on the Earth, besides the Meteor Crater in Arizona?

Activity 3 Orbits and Effects

Goals

In this activity you will:

- Measure the major axis and distance between the foci of an ellipse.
- Understand the relationship between the distance between the foci and eccentricity of an ellipse.
- Calculate the eccentricity of the Earth's orbit.
- Draw the Earth's changing orbit in relation to the Sun.
- Explain how the Earth's changing orbit and its rotation rate could affect its climate.
- Draw the orbits of a comet and an asteroid in relation to the Earth and the Sun.

Think about It

The Earth rotates on its axis as it revolves around the Sun, some 150,000,000 km away. The axis of rotation is now tilted about 23.5°.

- What is the shape of the Earth's orbit around the Sun?
- How might a change in the shape of the Earth's orbit or its axis of rotation affect weather and climate?

What do you think? In your *EarthComm* notebook, draw a picture of the Earth's orbit around the Sun, as seen from above the solar system. Record your ideas about how this shape affects weather and climate. Be prepared to discuss your responses with your small group and the class.

Investigate

1. Draw an ellipse using the following steps:
   - Fold a piece of paper in half.
   - Use a straightedge to draw a horizontal line across the width of the paper along the fold.
   - Put two dots 10 cm apart on the line toward the center of the line. Label the left dot "A" and the right dot "B."
   - Tape the sheet of paper to a piece of thick cardboard, and put two pushpins into points A and B. The positions of the pushpins will be the foci of the ellipse.
   - Tie two ends of a piece of strong string together to make a loop. Make the knot so that when you stretch out the loop with your fingers into a line, it is 12 cm long.
   - Put the string over the two pins and pull the loop tight using a pencil point, as shown in the diagram.
   - Draw an ellipse with the pencil. Do this by putting the pencil point inside the loop and then moving the pencil while keeping the string pulled tight with the pencil point.

     a) Draw a small circle around either point A or point B and label it "Sun."

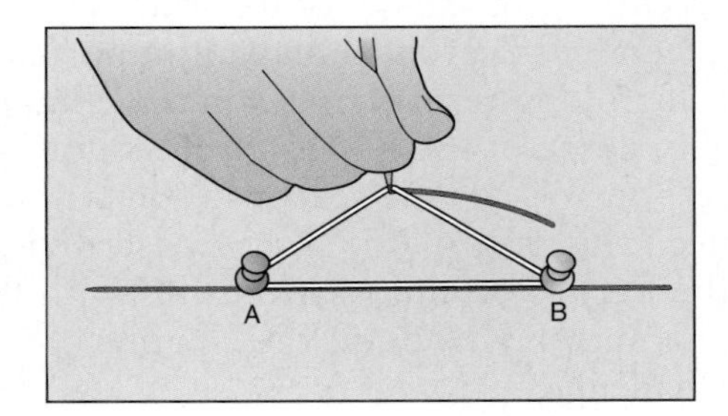

2. Repeat the process using the following measurements and labels:
   - Two points 8 cm apart labeled C and D (1 cm inside of points A and B).
   - Two points 6 cm apart, labeled E and F (2 cm inside of points A and B).
   - Two points 4 cm apart, labeled G and H (3 cm inside points A and B).
   - Two points 2 cm apart, labeled I and J (4 cm inside points A and B).
3. Copy the data table on the next page into your notebook.

   a) Measure the width (in centimeters) of ellipse "AB" at its widest point. This is the major axis, L (see diagram on the next page). Record this in your data table.

   b) Record the length of the major axis for each ellipse in your data table.

   c) Record the distance between the two foci, d (the distance between the two pushpins) for each ellipse in your data table (see diagram).

   d) The eccentricity E of an ellipse is equal to the distance between the two foci divided by the length of the major axis. Calculate the eccentricity of each of your ellipses using the equation $E = d/L$, where d is the distance between the foci and L is the length of the major axis. Record the eccentricity of each ellipse.

Be sure the cardboard is thicker than the points of the pins. If not, use two or more pieces of cardboard.

Ellipse	Major Axis (*L*) (cm)	Distance between the Foci (*d*) (cm)	Eccentricity *E = d/L*
AB		10	
CD		8	
EF		6	
GH		4	
IJ		2	

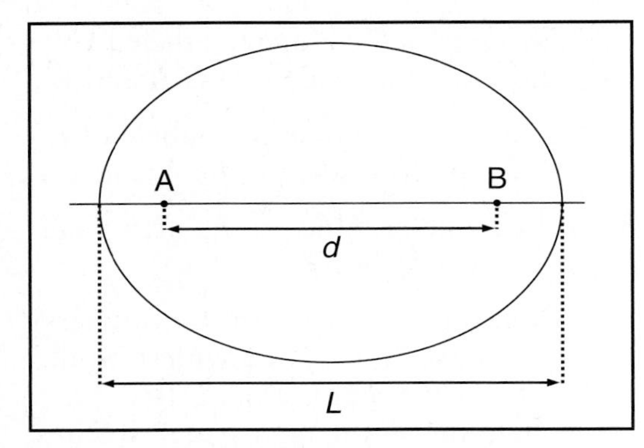

Ellipse with foci AB showing major axis length *L* and distance between the foci *d*.

4. Study your data table to find a relationship between the distance between the foci and the eccentricity of an ellipse.

   a) Record the relationship between the distance between the foci and the eccentricity in your notebook.

   b) Think of your ellipses as the orbits of planets around the Sun. Does the distance to the center of the Sun stay the same in any orbit?

   c) Which orbit has the least variation in distance from the Sun throughout its orbit? Which has the most?

5. Earth's orbit has an eccentricity of about 0.017. Compare this value to the ellipse with the lowest eccentricity of those you drew.

   a) Why does it make sense to describe Earth's orbit as "nearly circular"?

Reflecting on the Activity and the Challenge

In this activity, you explored a geometric figure called an ellipse. You also learned how to characterize ellipses by their eccentricity. The orbits of all nine planets in our solar system are ellipses, with the Sun at one focus of the ellipse representing the orbit for each planet. As you will see, although Earth's orbit is very nearly circular (only slightly eccentric), the shape of its orbit is generally believed to play an important role in long-term changes in the climate. The shape of the Earth's orbit is not responsible for the seasons. You will need this information when describing the Earth's orbital relationships with the Sun and the Moon in the **Chapter Challenge.**

Digging Deeper

ECCENTRICITY, AXIAL TILT, PRECESSION, AND INCLINATION

Eccentricity

After many years of analyzing observational data on the motions of the planets, the astronomer Johannes Kepler (1571–1630) developed three laws of planetary motion that govern orbits. The first law states that the orbit of each planet around the Sun is an ellipse with the Sun at one focus. The second law, as shown in *Figure 1*, explains that as a planet moves around the Sun in its orbit, it covers equal areas in equal times. The third law states that the time a planet takes to complete one orbit is related to its average distance from the Sun.

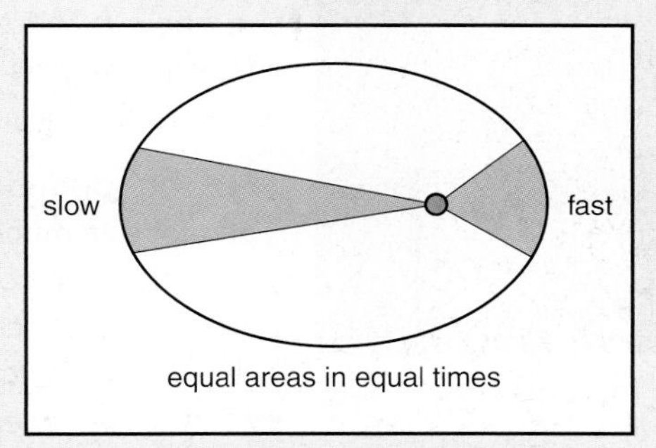

Figure 1 Kepler's Second Law states that a line joining a planet and the Sun sweeps out equal areas in equal intervals of time.

As you saw in the **Investigate** section, the shape of an ellipse can vary from a circle to a very highly elongated shape, and even to a straight line. The more flattened the ellipse is, the greater its eccentricity. Values of **eccentricity** range from zero for a circle, to one, for a straight line. (A mathematician would say that the circle and the line are "special cases" of an ellipse.) The two planets in the solar system with the most elliptical orbits are Mercury, the closest planet, and Pluto, the farthest one. Both have eccentricities greater than 0.2. The orbit of Mars is also fairly elliptical, with an eccentricity of 0.09. In comparison, the Earth's orbit is an ellipse with an eccentricity of 0.017. (This is a much lower value than even the eccentricity of ellipse IJ in the investigation, which looked much like a circle.) If you were to draw an ellipse with an eccentricity of 0.017 on a large sheet of paper, most people would call it a circle. It's eccentric enough, however, to make the Earth's distance from the Sun vary between 153,000,000 km and 147,000,000 km. To make things more complicated, the eccentricity of the Earth's orbit changes over time, because of complicated effects having to do with the weak gravitational pull of other planets in the solar system. Over the course of about 100,000 years, the Earth's orbit ranges from nearly circular (very close to zero eccentricity) to more elliptical (with an eccentricity of about 0.05).

Planetary scientists have found that some solar-system objects have highly elliptical orbits. Comets are a well-known example. As they move closer

Geo Words

eccentricity: the ratio of the distance between the foci and the length of the major axis of an ellipse.

Figure 2 The comet's head or coma is the fuzzy haze that surrounds the comet's true nucleus.

to the Sun, the icy mix that makes up a comet's nucleus begins to turn into gas and stream away. The result is a ghostly looking tail and a fuzzy "shroud," that you can see in *Figure 2*. It is called a **coma**, and it forms around the nucleus. When a comet gets far enough away from the Sun, the ices are no longer turned to gas and the icy nucleus continues on its way.

Another good example is the distant, icy world Pluto. Throughout much of its year (which lasts 248 Earth years) this little outpost of the solar system has no measurable atmosphere. It does have a highly eccentric orbit, and its distance from the Sun varies from 29.5 to 49.5 AU. The strength of solar heating varies by a factor of almost four during Pluto's orbit around the Sun. As Pluto gets closest to the Sun, Pluto receives just enough solar heating to vaporize some of its ices. This creates a thin, measurable atmosphere. Then, as the planet moves farther out in its orbit, this atmosphere freezes out and falls to the surface as a frosty covering. Some scientists predict that when Pluto is only 20 years past its point of being closest to the Sun, its atmosphere will collapse as the temperature decreases.

Axial Tilt (Obliquity)

The Earth's axis of rotation is now tilted at an angle of 23.5° to the plane of the Earth's orbit around the Sun, as seen in *Figure 3*. Over a cycle lasting about 41,000 years, the axial tilt varies from 22.1° to 24.5°. The greater the angle of tilt, the greater the difference in solar energy, and therefore temperature, between summer and winter. This small change, combined with other long-term changes in the Earth's orbit, is thought to be responsible for the Earth's ice ages.

Precession

The Earth also has a slight wobble, the same as the slow wobble of a spinning top. This wobble is called the **precession** of the Earth's axis. It is caused by differences in the gravitational pull of the Moon and the Sun on the Earth. It takes about 25,725 years for this wobble to complete a cycle. As the axis wobbles, the timing of the seasons changes. Winter occurs when a hemisphere, northern or southern, is tilted away from the Sun. Nowadays, the Earth is slightly closer to the Sun during winter (January 5) in the Northern Hemisphere. Don't let anybody tell you that winter happens because the Earth is farthest from the Sun! The Earth's orbit is nearly circular! Also, even if a particular hemisphere of the Earth is tilted towards the Sun, it is not significantly closer to the Sun than the other hemisphere, which is tilted away.

Geo Words

coma: a spherical cloud of material surrounding the head of a comet. This material is mostly gas that the Sun has caused to boil off the comet's icy nucleus. A cometary coma can extend up to a million miles from the nucleus.

precession: slow motion of the axis of the Earth around a cone, one cycle in about 26,000 years, due to gravitational tugs by the Sun, Moon, and major planets.

The precession of the Earth's axis is one part of the precession cycle. Another part is the precession of the Earth's orbit. As the Earth moves around the Sun in its elliptical orbit, the major axis of the Earth's orbital ellipse is rotating about the Sun. In other words, the orbit itself rotates around the Sun! These two precessions (the axial and orbital precessions) combine to affect how far the Earth is from the Sun during the different seasons. This combined effect is called precession of the equinoxes, and this change goes through one complete cycle about every 22,000 years. Ten thousand years from now, about halfway through the precession cycle, winter will be from June to September, when the Earth will be farthest from the Sun during the Northern Hemisphere winter. That will make winters there even colder, on average.

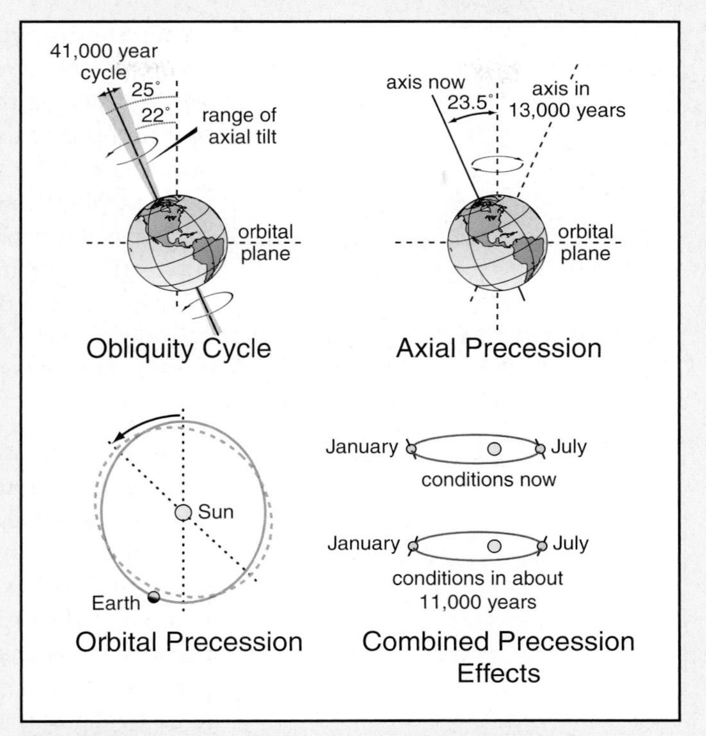

Figure 3 The tilt of the Earth's axis and its orbital path about the Sun go through several cycles of change.

Inclination

When you study a diagram of the solar system, you notice that the orbits of all the planets except Pluto stay within a narrow range called the **orbital plane** of the solar system. If you were making a model of the orbits of the planets in the solar system you could put many of the orbits on a tabletop. However, Pluto's orbit, as shown in *Figure 4*, could not be drawn or placed on a tabletop (a plane, as it is called in geometry). Pluto's path around the Sun is inclined 17.1° from the plane described by the Earth's motion around the Sun. This 17.1° tilt is called its orbital **inclination**.

Geo Words

orbital plane: (also called the ecliptic or plane of the ecliptic). A plane formed by the path of the Earth around the Sun.

inclination: the angle between the orbital plane of the solar system and the actual orbit of an object around the Sun.

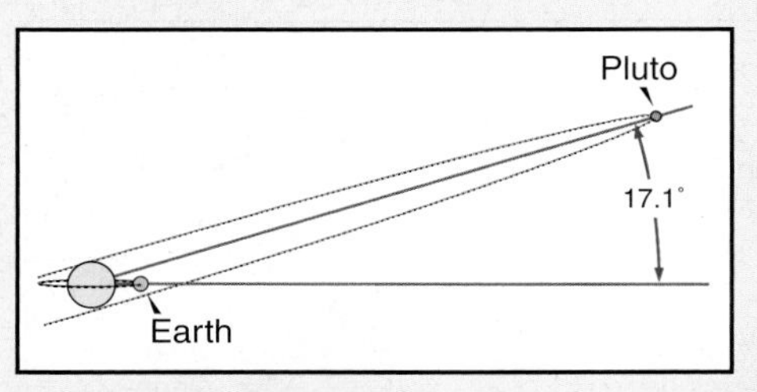

Figure 4 Pluto's orbit is inclined 17.1° to the orbital plane of the rest of the solar system.

What are the orbital planes of asteroids and comets? Both are found mainly in the part of the solar system beyond the Earth. Although some asteroids can be found in the inner solar system, many are found between the orbits of Mars and Jupiter. The common movie portrayal of the "asteroid belt" as a densely populated part of space through which one must dodge asteroids is wrong. The asteroids occupy very little space. Another misconception is that asteroids are the remains of a planet that exploded.

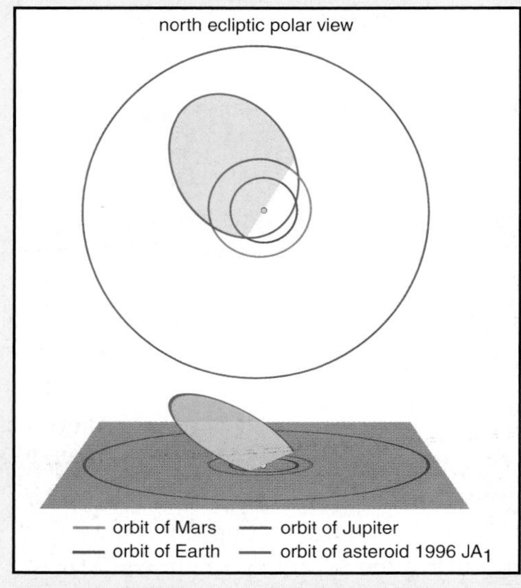

Figure 5 The orbit of the Earth-approaching asteroid 1996 JA, in relation to the Earth.

The orbits of asteroids are more eccentric than the orbits of the planets, and they are often slightly inclined from the orbital plane. As the Earth orbits the Sun, it can cross the orbital paths of objects called Earth-approaching asteroids. There is a great deal of interest in finding Earth-approaching or Earth-crossing asteroids. A collision with an object a few miles across could be devastating, because of its very high velocity relative to the Earth. Astronomers search the skies for asteroids and map their orbits. In this way they hope to learn what's coming toward Earth long before it poses a danger to your community.

Comets are "loners" that periodically visit the inner solar system. They usually originate in the outer solar system. As shown in *Figure 5*, they have very high-inclination orbits—some as much as 30° from the plane of the solar system. In addition, their orbits are often highly eccentric. Astronomers also search the skies for comets. Once a comet is discovered, its orbit is calculated and the comet is observed as it moves closer to the Sun and changes. A collision of a comet's nucleus with the Earth would be serious, but a collision with a comet's tail is much more likely. A collision with the tail would have little, if any, effect on the Earth, because the tail consists mainly of glowing gas with very little mass.

Check Your Understanding

1. In your own words, explain what is meant by the eccentricity of an ellipse.
2. For an ellipse with a major axis of 25 cm, which one is more eccentric; the one with a distance between the foci of 15 cm or with a distance between the foci of 20 cm? Explain.
3. How does the precession of the Earth's axis of rotation affect the seasons? Justify your answer.
4. Why is there a danger that a large asteroid might strike the Earth at some time in the future?

Understanding and Applying What You Have Learned

1. The major axis of the Earth's orbit is 299,200,000 km, and the distance between the foci is 4,999,632 km. Calculate the eccentricity of the Earth's orbit. How does this value compare to the value noted in the **Digging Deeper** reading section?
2. On the GH line on the ellipse that you created for your **Investigate** activity, draw the Earth at its closest position to the Sun and the farthest position away from the Sun.
3. Refer to the table that shows the eccentricities of the planets to answer the following questions:
   a) Which planet would show the greatest percentage variation in its average distance from the Sun during its year? Explain.
   b) Which planet would show the least percentage variation in its average distance from the Sun throughout its year? Explain.
   c) Is there any relationship between the distance from the Sun and the eccentricity of a planet's orbit? Refer to *Table 1* in the **Investigate** section of **Activity 1.**
   d) Why might Neptune be farther away from the Sun at times than Pluto is?
   e) Look up the orbital inclinations of the planets and add them to a copy of the table.

Eccentricities of the Planets	
Planet	**Eccentricity**
Mercury	0.206
Venus	0.007
Earth	0.017
Mars	0.093
Jupiter	0.048
Saturn	0.054
Uranus	0.047
Neptune	0.009
Pluto	0.249

4. Draw a scale model to show changes in the Earth's orbit of about the same magnitude as in nature (a cycle of 100,000 years).
   a) Draw the orbit of the Earth with a perfectly circular orbit at 150,000,000 km from the Sun. Use a scale of 1 cm = 20,000,000 km. Make sure that your pencil is sharp, and draw the thinnest line possible.
   b) Make another drawing of the actual shape of the Earth orbit—an ellipse. This ellipse has 153,000,000 km as the farthest distance and 147,000,000 km as the closest point to the Sun.
   c) Does the difference in distance from the Sun look significant enough to cause much difference in temperature? Explain.
5. Draw the solar system as viewed from the plane of the ecliptic (orbital plane).

a) How will Pluto's orbit look with its 17° inclination?
b) How will the orbits of the other planets look?
c) Draw in the orbits of Earth-crossing asteroids with inclinations of 20° and 30° to the orbital plane.
d) Draw in the orbits of several comets with high inclinations. Some typical high-inclination comets are Comet Halley (162.22°) and Ikeya-Seki (141.86°).

6. Now that you know that the Earth's orbit is elliptical (and the Moon's is too), you can think about a third astronomical factor that controls the nature of tides. Tidal forces are stronger when the Moon is closer to the Earth, and when the Earth is closer to the Sun.
   a) Draw a diagram to show the positions of the Moon, Earth, and Sun that would generate the highest tidal ranges (difference in height between high and low tides) of the year.
   b) Draw a diagram to show the positions of the Moon, Earth, and Sun that would generate the lowest tidal ranges (difference in height between high and low tides) of the year.

Preparing for the Chapter Challenge

1. In your own words, explain the changes in the Earth's orbital eccentricity, and how it might have affected your community in the past. What effect might it have in the future?
2. Describe the orbits of comets and asteroids and how they are different from those of the planets. What effect could comets and asteroids have on your community if their orbits intersected the Earth's orbit and the Earth and the comet or asteroid were both at that same place in their orbits?

Inquiring Further

1. **The gravitational "slingshot" effect on spacecraft**

   The gravitational tug of the Sun and the planets plays a role in shaping the orbits of solar-system bodies. NASA has used the gravitational pull of Jupiter and Saturn to influence the paths of spacecraft like Pioneer and Voyager. Investigate how this gravitational "slingshot" effect works and its role in moving small bodies from one orbit to another.

2. **Investigate the orbits of comets and asteroids**

   Look up the orbital information for some typical comets and asteroids. Try to include some with high inclinations and orbital eccentricities.

Activity 4 Impact Events and the Earth System

Goals

In this activity you will:

- Investigate the mechanics of an impact event and make scale drawings of an impact crater.
- Calculate the energy (in joules) released when an asteroid collides with Earth.
- Compare natural and man-made disasters to the impact of an asteroid.
- Understand the consequences to your community should an impact event occur.
- Investigate the chances for an asteroid or comet collision.

Think about It

Meteor Crater in Arizona is one of the best-preserved meteor craters on Earth. It is 1.25 km across and about 4 km in circumference. Twenty football games could be played simultaneously on its floor, while more than two million spectators observed from its sloping sides.

- How large (in diameter) do you think the meteor was that formed Meteor Crater?
- How would the impact of the meteor have affected living things near the crater?

What do you think? Record your ideas about these questions in your *EarthComm* notebook. Be prepared to discuss your responses with your small group and the class.

Investigate

1. Given the following information, calculate the energy released when an asteroid collides with Earth:

   - The spherical, iron–nickel asteroid has a density of 7800 kg/m^3.
   - It is 40 m in diameter.
   - It has a velocity of 20,000 m/s relative to the Earth.

   Note: It is very important to keep track of your units during these calculations. You will be expressing energy with a unit called a "joule." A joule is 1 kg m^2/s^2.

   a) Find the volume of the asteroid in cubic meters. The equation for the volume of a sphere is as follows:

   $$V = \frac{4}{3}\pi r^3$$

   where V is volume of the sphere, and r is the radius of the sphere.

   b) Multiply the volume by the density to find the total mass of the asteroid.

   c) Calculate the energy of the asteroid. Because the asteroid is moving, you will use the equation for kinetic energy, as follows:

   $$KE = \frac{1}{2}mv^2$$

   where KE is kinetic energy,
   m is the mass, and
   v is the velocity.

   Express your answer in joules. To do this express mass in kilograms, and velocity in meters per second.

   For some perspective, a teenager uses over 10,000 kJ (kilojoules) of energy each day. (There are 1000 J (joules) in a kilojoule.)

2. The combination of calculations that you just performed can be summarized as:

   $$\textbf{Energy} = \frac{2}{3}\pi\rho r^3 v^2$$

   where r is the radius,
   ρ is the density, and
   v is the velocity of the object.

   a) Suppose an object makes an impact with the Earth at 10 times the velocity of another identical object. By what factor will the energy of the object increase?

   b) Suppose an object makes an impact with the Earth, and it has 10 times the radius of another object traveling at the same speed. By what factor will the energy of the object increase?

   c) How do these relationships help to explain how small, fast-moving objects can release a tremendous amount of energy as well as larger yet slower-moving objects?

3. The asteroid described in **Step 1** above was the one responsible for Meteor Crater in Arizona.

   a) Copy the following table into your notebook. Enter your calculation for Meteor Crater.

   b) Calculate the energy released by the impacts shown in the table.

Object	Radius (m)	Density (kg/m^3)	Impact Velocity (m/s)	Energy (joules)	Richter Scale Magnitude Equivalent
Asclepius	100	3000	30,000		
Comet Swift-Tuttle	1000	5000	60,000		
Chicxulub impactor	5000	3000	32,000		
SL9 Fragment Q	2150	1000	60,000		
Meteor Crater	20	7800	20,000		

Note:

- Asclepius is an asteroid that passed within 690,000 km of Earth in 1989.
- Comet Swift-Tuttle is a future threat to the Earth–Moon system, having passed Earth in 1992 and being scheduled for return in 2126.
- SL9 Fragment Q is a fragment of Comet Shoemaker-Levy that impacted Jupiter in 1994.
- Chicxulub impactor is the name of the asteroid that triggered the extinction of the dinosaurs 65 million years ago.

4. Use the table below to compare the energy from all these events to known phenomena.
   a) In your notebook, explain how the energies of these four impact events compare to some other known phenomena.

Phenomena	Kinetic Energy (joules)
Annual output of the Sun	10^{34}
Severe earthquake	10^{18}
100-megaton hydrogen bomb	10^{17}
Atomic bomb	10^{13}

5. Make a scale drawing of the Chicxulub impactor compared with Earth. The diameter of the Earth is 12,756 km.
   a) If you made the diameter of the Chicxulub impactor 1 mm, what would the diameter of the Earth be?
   b) If you made the diameter of the Chicxulub impactor 0.5 mm, which is probably about as small as you can draw, what would the diameter of the Earth be?

6. How do these impact events compare with the energy released in an earthquake? If you have a calculator capable of handling logarithms, answer the following questions:

   a) Calculate the Richter scale equivalent of the energy released by the four impact events. Use the following equation:

   $$M = 0.67 \log_{10}E - 5.87$$

   where M is the equivalent magnitude on the Richter scale, and E is the energy of the impact, in joules.

   b) How do your results compare with the table below, which shows the five greatest earthquakes in the world between 1900 and 1998? Which impacts exceed the world's greatest earthquakes?

Location	Year	Magnitude
Chile	1960	9.5
Prince William Sound, Alaska	1964	9.2
Andreanof Islands, Aleutian Islands	1957	9.1
Kamchatka	1952	9.0
Off the Coast of Ecuador	1906	8.8

Reflecting on the Activity and the Challenge

You have calculated the energy released when asteroids of different sizes hit the Earth's surface, and you have compared these to other energy-releasing events. This comparison will be helpful as you explain the hazards associated with an impact in your **Chapter Challenge** brochure.

Digging Deeper

ASTEROIDS AND COMETS

Asteroids

Asteroids are rocky bodies smaller than planets. They are leftovers from the formation of the solar system. In fact, the early history of the solar system was a period of frequent impacts. The many scars (impact craters) seen on the Moon, Mercury, Mars, and the moons of the outer planets are the evidence for this bombardment. Asteroids orbit the Sun in very elliptical orbits with inclinations up to 30°. Most asteroids are in the region between Jupiter and Mars called the asteroid belt. There are probably at least 100,000 asteroids 1 km in diameter and larger. The largest, called Ceres, is about 1000 km across. Some of the asteroids have very eccentric orbits that cross Earth's orbit. Of these, perhaps a few dozen are larger than one kilometer in diameter. As you learned in the activity, the energy of an asteroid impact event increases with the cube of the radius. Thus, the larger asteroids are the ones astronomers worry about when they consider the danger of collision with Earth.

The closest recent approach of an asteroid to Earth was Asteroid 1994 XM 11. On December 9, 1994, the asteroid approached within 115,000 km of Earth. On March 22, 1989 the asteroid 4581 Asclepius came within 1.8 lunar distances, which is close to 690,000 km. Astronomers think that asteroids at least 1 km in diameter hit Earth every few hundred million years. They base this upon the number of impact craters that have been found and dated on Earth. A list of asteroids that have approached within two lunar distances of Earth (the average distance between Earth and the Moon) is provided in *Table 1* on the following page. Only close-approach distances less than 0.01 AU for asteroids are included in this table.

Figure 1 Image of the asteroid Ida, which is 58 km long and 23 km wide.

Table 1 Asteroids with Close-Approach Distances to Earth

Name or Designation	Date of Close Earth Approach	Distance	
		(AU)	(LD)
1994 XM1	1994–Dec–09	0.0007	0.3
1993 KA2	1993–May–20	0.001	0.4
1994 ES1	1994–Mar–15	0.0011	0.4
1991 BA	1991–Jan–18	0.0011	0.4
1996 JA1	1996–May–19	0.003	1.2
1991 VG	1991–Dec–05	0.0031	1.2
1999 VP11	1982–Oct–21	0.0039	1.5
1995 FF	1995–Apr–03	0.0045	1.8
1998 DV9	1975–Jan–31	0.0045	1.8
4581 Asclepius	1989–Mar–22	0.0046	1.8
1994 WR12	1994–Nov–24	0.0048	1.9
1991 TU	1991–Oct–08	0.0048	1.9
1995 UB	1995–Oct–17	0.005	1.9
1937 UB (Hermes)	1937–Oct–30	0.005	1.9
1998 KY26	1998–Jun–08	0.0054	2.1

(AU) – Astronomical distance Unit: 1.0 AU is roughly the average distance between the Earth and the Sun.
(LD) – Lunar Distance unit: 1.0 LD is the average distance from the Earth to the Moon (about 0.00257 AU).

Most (but not all) scientists believe that the extinction of the dinosaurs 65 million years ago was caused by the impact of an asteroid or comet 10 km in diameter. Such a large impact would have sent up enough dust to cloud the entire Earth's atmosphere for many months. This would have blocked out sunlight and killed off many plants, and eventually, the animals that fed on those plants. Not only the dinosaurs died out. About 75% of all plants and animals became extinct. One of the strong pieces of evidence supporting this hypothesis is a 1-cm-thick layer of iridium-rich sediment about 65 million years old that has been found worldwide. Iridium is rare on Earth but is common in asteroids.

Our planet has undergone at least a dozen mass-extinction events during its history, during which a large percentage of all plant and animal species became extinct in an extremely short interval of geologic time. It is likely that at least some of these were related to impacts. It is also likely that Earth will suffer another collision sometime in the future. NASA is currently forming plans to discover and monitor asteroids that are at least 1 km in size and with orbits that cross the Earth's orbit. Asteroid experts take the threat from asteroids very seriously, and they strongly suggest that a program of systematic observation be put into operation to predict and, hopefully avoid an impact.

Comets

Comets are masses of frozen gases (ices) and rocky dust particles. Like asteroids, they are leftovers from the formation of the solar system. There are many comets in orbit around the Sun. Their orbits are usually very eccentric with large inclinations. The orbits of many comets are very large, with distances from the Sun of greater than 20,000 astronomical units (AU). The icy head of a comet (the nucleus) is usually a few kilometers in diameter, but it appears much larger as it gets closer to the Sun. That's because the Sun's heat vaporizes the ice, forming a cloud called a coma. Radiation pressure and the action of the **solar wind** (the stream of fast-moving charged particles coming from the Sun) blow the gases and dust in the coma in a direction away from the Sun. This produces a tail that points away from the Sun even as the comet moves around the Sun. Halley's Comet, shown in *Figure 2*, is the best known of these icy visitors. It rounds the Sun about every 76 years, and it last passed by Earth in 1986.

Figure 2 Halley's Comet last appeared in the night sky in 1986.

Comets have collided with the Earth since its earliest formation. It is thought that the ices from comet impacts melted to help form Earth's oceans. In 1908 something hit the Earth at Tunguska, in Siberian Russia. It flattened trees for hundreds of miles, and researchers believe that the object might have been a comet. Had such an event occurred in more recent history in a more populated area, the damage and loss of life would have been enormous. A list of comets that have approached within less than 0.11 AU of Earth is provided in *Table 2* on the following page.

Geo Words

solar wind: a flow of hot charged particles leaving the Sun.

Geo Words

meteoroid: a small rock in space.

meteor: the luminous phenomenon seen when a meteoroid enters the atmosphere (commonly known as a shooting star).

meteorite: a part of a meteoroid that survives through the Earth's atmosphere.

Table 2 Close Approaches of Comets

Name	Designation	Date of Close Earth Approach	Distance (AU)	Distance (LD)
Comet of 1491	C/1491 B1	1491–Feb–20	0.0094	3.7*
Lexell	D/1770 L1	1770–Jul–01	0.0151	5.9
Tempel-Tuttle	55P/1366 U1	1366–Oct–26	0.0229	8.9
IRAS-Araki-Alcock	C/1983 H1	1983–May–11	0.0313	12.2
Halley	1P/ 837 F1	837–Apr–10	0.0334	13
Biela	3D/1805 V1	1805–Dec–09	0.0366	14.2
Comet of 1743	C/1743 C1	1743–Feb–08	0.039	15.2
Pons-Winnecke	7P/	1927–Jun–26	0.0394	15.3
Comet of 1014	C/1014 C1	1014–Feb–24	0.0407	15.8*
Comet of 1702	C/1702 H1	1702–Apr–20	0.0437	17
Comet of 1132	C/1132 T1	1132–Oct–07	0.0447	17.4*
Comet of 1351	C/1351 W1	1351–Nov–29	0.0479	18.6*
Comet of 1345	C/1345 O1	1345–Jul–31	0.0485	18.9*
Comet of 1499	C/1499 Q1	1499–Aug–17	0.0588	22.9*
Schwassmann-Wachmann 3	73P/1930 J1	1930–May–31	0.0617	24

* Distance uncertain because comet's orbit is relatively poorly determined.
(AU) – Astronomical distance Unit: 1.0 AU is roughly the average distance between the Earth and the Sun.
(LD) – Lunar Distance unit: 1.0 LD is the average distance from the Earth to the Moon (about 0.00257 AU).

Meteoroids, Meteors, and Meteorites

Meteoroids are tiny particles in space, like leftover dust from a comet's tail or fragments of asteroids. Meteoroids are called **meteors** when they enter Earth's atmosphere, and **meteorites** when they reach the Earth's surface. About 1000 tons of material is added to the Earth each year by meteorites, much of it through dust-sized particles that settle slowly through the atmosphere. There are several types of meteorites. About 80% that hit Earth are stony in nature and are difficult to tell apart from Earth rocks. About 15% of meteorites consist of the metals iron and nickel and are very dense. The rest are a mixture of iron–nickel and stony material. Most of the stony meteorites are called chondrites. Chondrites may represent material that was never part of a larger body like a moon, a planet, or an asteroid, but instead are probably original solar-system materials.

Figure 3 Lunar meteorite.

Check Your Understanding

1. Where are asteroids most abundant in the solar system?
2. How might a major asteroid impact have caused a mass extinction of the Earth's plant and animal species at certain times in the geologic past?
3. Why do comets have tails? Why do the tails point away from the Sun?
4. What are the compositions of the major kinds of meteorites?

Understanding and Applying What You Have Learned

1. Look at the table of impact events shown in the **Investigate** section. Compare the densities of the object that formed Meteor Crater and SL9 Fragment Q from the Shoemaker-Levy Comet. Use what you have learned in this activity to explain the large difference in densities between the two objects.
2. If an asteroid or comet were on a collision course for Earth, what factors would determine how dangerous the collision might be for your community?
3. How would an asteroid on a collision course endanger our Earth community?
4. Comets are composed largely of ice and mineral grains. Assume a density of 1.1 g/cm^3:
   a) How would the energy released in a comet impact compare to the asteroid impact you calculated in the **Investigate** section? (Assume that the comet has the same diameter and velocity as the asteroid.)
   b) Based upon your calculation, are comets dangerous if they make impact with the Earth? Explain your response.
5. From the information in the **Digging Deeper** reading section, and what you know about the eccentricities and inclinations of asteroid orbits, how likely do you think it is that an asteroid with a diameter of 1 km or greater will hit the Earth in your lifetime? Explain your reasoning. Can you apply the same reasoning to comets?
6. Add the asteroid belt to the model of the solar system you made in the first activity. You will need to think about how best to represent the vast number of asteroids and their wide range of sizes. Don't forget to add in some samples of Earth-approaching asteroids and the orbit of one or two comets.

Preparing for the Chapter Challenge

Assume that scientists learn several months before impact that a large asteroid will hit near your community. Assume that you live 300 km from the impact site. What plans can your family make to survive this disaster? What are some of the potential larger-scale effects of an asteroid impact? Work with your group to make a survival plan. Present your group's plan to the entire class. Be sure to record suggestions made by other groups. This information will prove useful in completing the **Chapter Challenge.**

Inquiring Further

1. **Impact craters on objects other than the Earth**

   In an earlier activity you studied impact sites on the Moon. Look at Mercury, Mars, and the moons of Saturn, Uranus, and Neptune to see other examples of impact craters in the solar system. How are these craters similar to Meteor Crater? How are they different?

2. **Modeling impact craters**

   Simulate an asteroid or comet hitting the Earth. Fill a shoebox partway with plaster of Paris. When the plaster is almost dry, drop two rocks of different sizes into it from the same height. Carefully retrieve the rocks and drop them again in a different place, this time from higher distance. Let the plaster fully harden, then examine, and measure the craters. Measure the depth and diameter and calculate the diameter-to-depth ratio. Which is largest? Which is deepest? Did the results surprise you?

3. **Earth-approaching asteroids**

   Do some research into current efforts by scientists to map the orbits of Earth-approaching asteroids? Visit the *EarthComm* web site to help you get started with your research. How are orbits determined? What is the current thinking among scientists about how to prevent impacts from large comets or asteroids?

4. **Barringer Crater**

   Research the Barringer Crater (Meteor Crater). The crater has been named for Daniel Moreau Barringer, who owned the property that contains the crater. Explain how scientists used Barringer Crater to understand how craters form. Study the work of Dr. Eugene Shoemaker, who was one of the foremost experts on the mechanics of impact cratering.

Wear goggles while modeling impact craters. Work with adult supervision to complete the activity.

Activity 5 The Sun and Its Effects on Your Community

Goals

In this activity you will:

- Explore the structure of the Sun and describe the flow of solar energy in terms of reflection, absorption, and scattering.
- Understand that the Sun emits charged particles called the solar wind, and how this wind affects "space weather."
- Explain the effect of solar wind on people and communities.
- Understand sunspots, solar flares, and other kinds of solar activities and their effects on Earth.
- Learn to estimate the chances for solar activity to affect your community.

Think about It

Every day of your life you are subjected to radiation from the Sun. Fortunately, the Earth's atmosphere and magnetic field provides protection against many of the Sun's strong outbursts.

- In what ways does solar radiation benefit you?
- In what ways can solar radiation be harmful or disruptive?

What do you think? Record your ideas about these questions in your *EarthComm* notebook. Be prepared to discuss your response with your small group and the class.

Investigate

1. Use the data in *Table 1* to construct a graph of sunspot activity by year.
   a) Plot time on the horizontal axis and number of sunspots on the vertical axis.
   b) Connect the points you have plotted.
   c) Look at your graph. Describe any pattern you find in the sunspot activity.

Table 1 Sunspot Activity 1899 to 1998

Year	Number of Sunspots	Year	Number of Sunspots	Year	Number of Sunspots	Year	Number of Sunspots
1899	12.1	1924	16.7	1949	134.7	1974	34.5
1900	9.5	1925	44.3	1950	83.9	1975	15.5
1901	2.7	1926	63.9	1951	69.4	1976	12.6
1902	5.0	1927	69.0	1952	31.5	1977	27.5
1903	24.4	1928	77.8	1953	13.9	1978	92.5
1904	42.0	1929	64.9	1954	4.4	1979	155.4
1905	63.5	1930	35.7	1955	38.0	1980	154.6
1906	53.8	1931	21.2	1956	141.7	1981	140.4
1907	62.0	1932	11.1	1957	190.2	1982	115.9
1908	48.5	1933	5.7	1958	184.8	1983	66.6
1909	43.9	1934	8.7	1959	159.0	1984	45.9
1910	18.6	1935	36.1	1960	112.3	1985	17.9
1911	5.7	1936	79.7	1961	53.9	1986	13.4
1912	3.6	1937	114.4	1962	37.6	1987	29.4
1913	1.4	1938	109.6	1963	27.9	1988	100.2
1914	9.6	1939	88.8	1964	10.2	1989	157.6
1915	47.4	1940	67.8	1965	15.1	1990	142.6
1916	57.1	1941	47.5	1966	47.0	1991	145.7
1917	103.9	1942	30.6	1967	93.8	1992	94.3
1918	80.6	1943	16.3	1968	105.9	1993	54.6
1919	63.6	1944	9.6	1969	105.5	1994	29.9
1920	37.6	1945	33.2	1970	104.5	1995	17.5
1921	26.1	1946	92.6	1971	66.6	1996	8.6
1922	14.2	1947	151.6	1972	68.9	1997	21.5
1923	5.8	1948	136.3	1973	38.0	1998	64.3

The number of sunspots on the visible solar surface is counted by many solar observatories and is averaged into a single standardized quantity called the sunspot number. This explains the fractional values in the table.

2. *Table 2* contains a list of solar flares that were strong enough to disrupt terrestrial communications and power systems.
   a) Plot the data from *Table 2* onto a histogram.
   b) What pattern do you see in the activity of solar flares?

3. Compare the two graphs you have produced.
   a) What pattern do you see that connects the two?
   b) How would you explain the pattern?

Table 2 Strongest Solar Flare Events 1978–2001

Date of Activity Onset	Strength	Date of Activity Onset	Strength
August 16, 1989	X20.0	December 17, 1982	X10.1
March 06, 1989	X15.0	May 20, 1984	X10.1
July 07, 1978	X15.0	January 25, 1991	X10.0
April 24, 1984	X13.0	June 09, 1991	X10.0
October 19, 1989	X13.0	July 09, 1982	X 9.8
December 12, 1982	X12.9	September 29, 1989	X9.8
June 06, 1982	X12.0	March 22, 1991	X9.4
June 01, 1991	X12.0	November 6, 1997	X9.4
June 04, 1991	X12.0	May 24, 1990	X9.3
June 06, 1991	X12.0	November 6, 1980	X9.0
June 11, 1991	X12.0	November 2, 1992	X9.0
June 15, 1991	X12.0		

The X before the number is a designation of the strongest flares.
Source: IPS Solar Flares & Space Service in Australia.

Reflecting on the Activity and the Challenge

In this activity you used data tables to plot the number of sunspots in a given year and to correlate strong solar-flare activity with larger numbers of sunspots. You found out that the number of sunspots varies from year to year in a regular cycle and that strong solar flares occur in greater numbers during high-sunspot years. In your **Chapter Challenge,** you will need to explain sunspots and solar flares, their cycles, and the effects of these cycles on your community.

Geo Words

photosphere: the visible surface of the Sun, lying just above the uppermost layer of the Sun's interior, and just below the chromosphere.

chromosphere: a layer in the Sun's atmosphere, the transition between the outermost layer of the Sun's atmosphere, or corona.

corona: the outermost atmosphere of a star (including the Sun), millions of kilometers in extent, and consisting of highly rarefied gas heated to temperatures of millions of degrees.

Digging Deeper

THE SUN AND ITS EFFECTS

Structure of the Sun

From the Earth's surface the Sun appears as a white, glowing ball of light. Like the Earth, the Sun has a layered structure, as shown in *Figure 1*. Its central region (the core) is where nuclear fusion occurs. The core is the source of all the energy the Sun emits. That energy travels out from the core, through a radiative layer and a convection zone above that. Finally, it reaches the outer layers: the **photosphere**, which is the Sun's visible surface, the **chromosphere**, which produces much of the Sun's ultraviolet radiation, and the superheated uppermost layer of the Sun's atmosphere, called the **corona**.

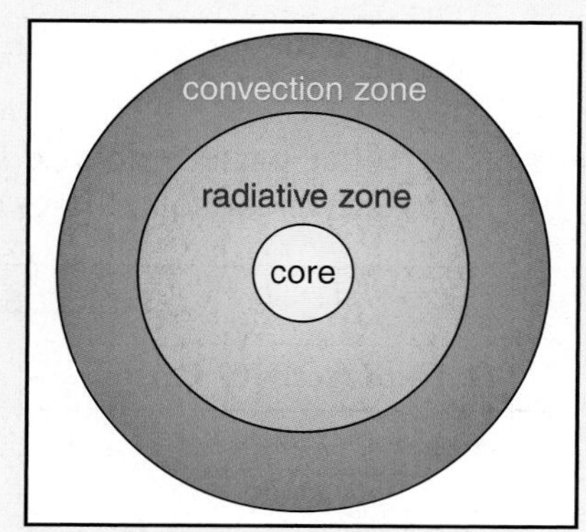

Figure 1 The layered structure of the Sun.

The Sun is the Earth's main external energy source. Of all the incoming energy from the Sun, about half is absorbed by the Earth's surface (see *Figure 2*). The rest is either:

- absorbed by the atmosphere, or
- reflected or scattered back into space by the Earth or clouds.

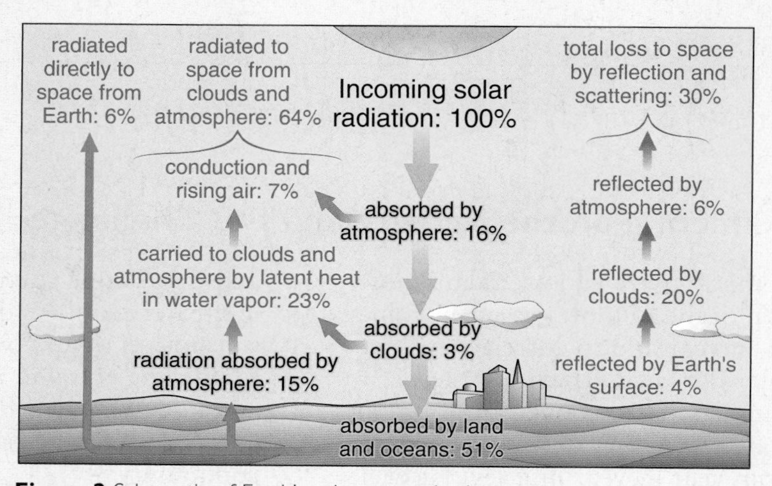

Figure 2 Schematic of Earth's solar energy budget.

Molecules of dust and gas in the atmosphere interfere with some of the incoming solar radiation by changing its direction. This is called scattering, and it explains the blue color of the sky. The atmosphere scatters shorter visible wavelengths of visible light, in the blue range, more strongly than longer visible wavelengths, in the red and orange range. The blue sky you see on a clear day is the blue light that has been scattered from atmospheric particles that are located away from the line of sight to the Sun. When the Sun is low on the horizon, its light has to travel through a much greater thickness of atmosphere, and even more of the blue part of the spectrum of sunlight is scattered out of your line of sight. The red and orange part of the spectrum remains, so the light you see coming directly from the Sun is of that color. The effect is greatest when there is dust and smoke in the atmosphere, because that increases the scattering. The scattered light that makes the sky appear blue is what makes it possible for you to see in a shaded area.

Figure 3 Dust and smoke in the atmosphere enhance the beauty of sunsets.

Most of the sunlight that passes through the atmosphere reaches the Earth's surface without being absorbed. The Sun heats the atmosphere not directly, but rather by warming the Earth's surface. The Earth's surface in turn warms the air near the ground. As the Earth's surface absorbs solar radiation, it re-radiates the heat energy back out to space as infrared radiation. The wavelength of this infrared radiation is much longer than that of visible light, so you can't see the energy that's re-radiated. You can feel it, however, by standing next to a rock surface or the wall of a building that has been heated by the Sun.

Geo Words

albedo: the reflective property of a non-luminous object. A perfect mirror would have an albedo of 100% while a black hole would have an albedo of 0%.

The reflectivity of a surface is referred to as its **albedo.** Albedo is expressed as a percentage of radiation that is reflected. The average albedo of the Earth, including its atmosphere, as would be seen from space, is about 0.3. That means that 30% of the light is reflected. Most of this 30% is due to the high reflectivity of clouds, although the air itself scatters about 6% and the Earth's surface (mainly deserts and oceans) reflects another 4%. (See *Figure 2* in the **Digging Deeper** section.) The albedo of particular surfaces on Earth varies. Thick clouds have albedo of about 0.8, and freshly fallen snow has an even higher albedo. The albedo of a dark soil, on the other hand, is as low as 0.1, meaning that only 10% of the light is reflected. You know from your own experience that light-colored clothing stays much cooler in the Sun than dark-colored clothing. You can think of your clothing as having an albedo, too!

The Earth's Energy Budget

The amount of energy received by the Earth and delivered back into space is the Earth's energy budget. Like a monetary budget, the energy resides in various kinds of places, and moves from place to place in various ways and by various amounts. The energy budget for a given location changes from day to day and from season to season. It can even change on geologic time scales. Daily changes in solar energy are the most familiar. It is usually cooler in the morning, warmer at midday, and cooler again at night. Visible light follows the same cycle, as day moves from dawn to dusk and back to dawn again. But overall, the system is in balance. The Earth gains energy from the Sun and loses energy to space, but the amount of energy entering the Earth system is equal to the amount of energy flowing out, on a long-term average. This flow of energy is the source of energy for almost all forms of life on Earth. Plants capture solar energy by photosynthesis, to build plant tissue. Animals feed on the plants (or on one another). Solar energy creates the weather, drives the movement of the oceans, and powers the water cycle. All of Earth's systems depend on the input of energy from the Sun. The Sun also supplies most of the energy for human civilization, either directly, as with solar power and wind power, or indirectly, in the form of fossil fuels.

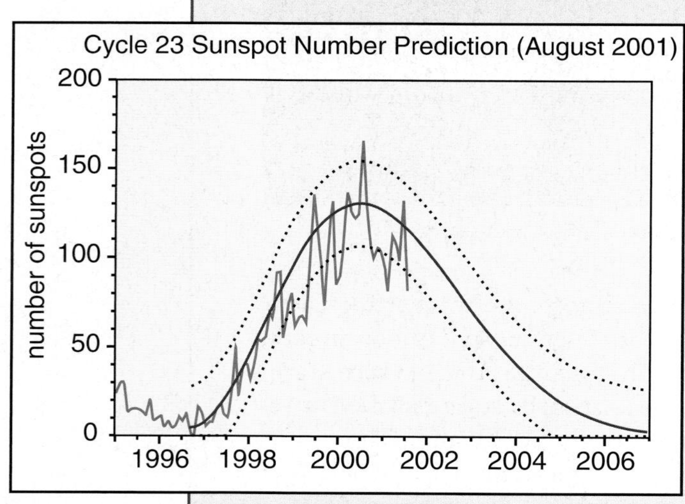

Figure 4 The jagged line represents the actual number of sunspots; the smooth dark line is the predicted number of sunspots.

Harmful Solar Radiation

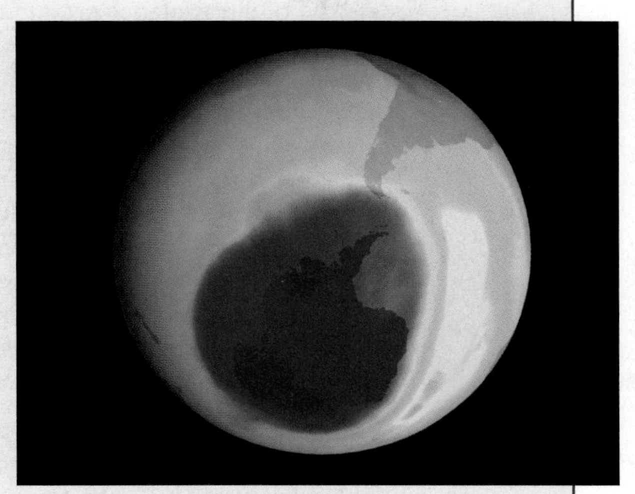

Figure 5 Depletion in the ozone layer over Antarctica. Rather than actually being a hole, the ozone hole is a large area of the stratosphere with extremely low concentrations of ozone.

Just as there are benefits to receiving energy from the Sun, there are dangers. The ill effects of sunlight are caused by ultraviolet (UV) radiation, which causes skin damage. The gas called ozone (a molecule made up of three oxygen atoms) found in the upper atmosphere shields the Earth from much of the Sun's harmful UV rays. The source of the ozone in the upper atmosphere is different from the ozone that is produced (often by cars) in polluted cities. The latter is a health hazard and in no way protects you. Scientists have recently noted decreasing levels of ozone in the upper atmosphere. Less ozone means that more UV radiation reaches Earth, increasing the danger of Sun damage. There is general agreement about the cause of the ozone depletion. Scientists agree that future levels of ozone will depend upon a combination of natural and man-made factors, including the phase-out, now under way, of chlorofluorocarbons and other ozone-depleting chemicals.

Sunspots and Solar Flares

Sunspots are small dark areas on the Sun's visible surface. They can be as small as the Earth or as large as Uranus or Neptune. They are formed when magnetic field lines just below the Sun's surface are twisted and poke through the solar photosphere. They look dark because they are about 1500 K cooler than the surrounding surface of the Sun. Sunspots are highly magnetic. This magnetism may cause the cooler temperatures by suppressing the circulation of heat in the region of the sunspot.

Sunspots last for a few hours to a few months. They appear to move across the surface of the Sun over a period of days. Actually, the sunspots move because the Sun is rotating. The number of sunspots varies from year to year and tends to peak in 11-year cycles along with the number of dangerously strong solar flares. Both can affect systems here on Earth. During a solar

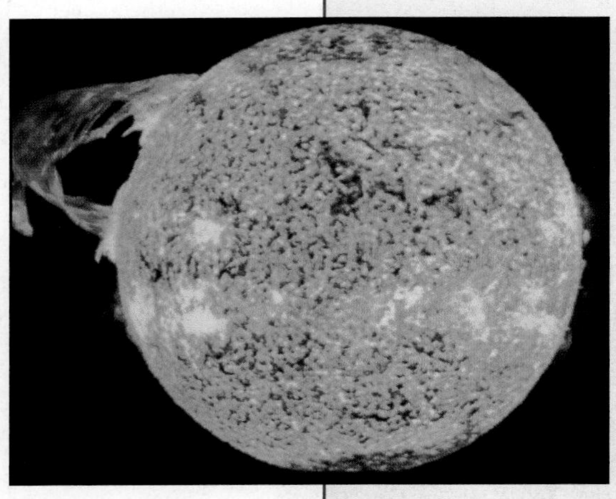

Figure 6 A solar flare.

flare like the one shown in *Figure 6*, enormous quantities of ultraviolet, x-ray, and radio waves blast out from the Sun. In addition, protons and electrons stream from flares at 800 km/hr. These high-radiation events can be devastating to Earth-orbiting satellites and astronauts, as well as systems on the ground. In 1989 a major solar flare created electric currents that caused a surge of power that knocked out a power grid in Canada, leaving hundreds of thousands of people without power. More recently, in 1997, radiation from a flare affected an Earth-orbiting satellite that carried telecommunications traffic. For at least a day people whose beeper messages went through that satellite had no service.

Geo Words

plasma: a state of matter wherein all atoms are ionized; a mixture of free electrons and free atomic nuclei.

ionosphere: the part of the Earth's atmosphere above about 50 km where the atoms are significantly ionized and affect the propagation of radio waves.

ion: an atom with one or more electrons removed (or added), giving it a positive (or negative) charge.

The flow of charged particles (also called a **plasma**) from the Sun is called the solar wind. It flows out from the solar corona in all directions and is responsible for "space weather"—the environment outside our planet. Like severe storms in our atmosphere, space weather can cause problems for Earth systems. Strong outbursts in this ongoing stream of charged particles can disrupt radio signals by disturbing the upper layers of the atmosphere. The sounds of your favorite short-wave radio station or the signals sent by a ham radio operator travel as radio waves (a form of electromagnetic radiation). These signals travel around the Earth by bouncing off the **ionosphere**, a layer of the atmosphere 80 to 400 km above the Earth's surface. The ionosphere forms when incoming solar radiation blasts electrons out of the upper-atmosphere gases, leaving a layer of electrons and of charged atoms, called **ions**. The ionosphere acts like a mirror, reflecting a part of the radio waves (AM radio waves in the 1000 kHz range) back to Earth.

Solar flares intensify the solar wind, which makes the ionosphere thicken and strengthen. When this happens, radio signals from Earth are trapped inside the ionosphere. This causes a lot of interference. As discussed above, solar activity can also be a problem for satellite operations. Astronauts orbiting the Earth and people aboard high-flying aircraft—particularly those who fly polar routes, where exposure to radiation may be greatest—also have cause to worry about space weather. To provide up-to-date information about current solar activity, the United States government operates a Space Environment Center Web site called "Space Weather Now."

At least one effect of space weather is quite wonderful. When the solar wind encounters the Earth's magnetic field, it excites gases in the Earth's atmosphere, causing them to glow. The charged particles from the solar wind end up in an oval-shaped area around the Earth's magnetic poles. The result is a beautiful display called an **aurora**, seen in *Figure 7*. People who live in northern areas see auroras more often than those who live near the Equator do. During periods of heavy solar activity, however, an aurora can be seen as far south as Texas and New Mexico. Auroras are often called the northern lights (aurora borealis) or southern lights (aurora australis). From the ground, they often appear as green or red glows, or shimmering curtains of white, red, and green lights in the sky.

Figure 7 The aurora borealis or northern lights light up the sky in the Northern Hemisphere.

Collecting Data about the Sun

How do astronomers collect data about the Sun? From the ground, they use solar telescopes—instruments outfitted with special sensors to detect the different kinds of solar activity. There are dozens of solar telescope sites around the world. They include the Sacramento Peak Solar Observatory in New Mexico, the McMath Solar telescope in Arizona, and the Mount Wilson solar observatory in California. From space, they study the Sun using orbiting spacecraft like the Yohkoh satellite (*Yohkoh* is the Japanese word for "sunbeam"). Other missions include the Transition Region and Coronal Explorer (TRACE), the Ulysses Solar-Polar mission, the Solar and Heliospheric Observatory, the GOES satellites, and many others. These spacecraft are equipped with detectors sensitive to x-rays, radio waves, and other wavelengths of radiation coming from the Sun. In this way, scientists keep very close track of solar activity and use that information to keep the public informed of any upcoming dangers.

Some scientists theorize that sunspot cycles affect weather on Earth. They think that during times of high sunspot activity, the climate is warmer. During times of no or low sunspot activity, the climate is colder. A sharp decrease in sunspots occurred from 1645 to 1715. This period of lower solar activity, first noted by G. Sporer and later studied by E.W. Maunder, is called the Maunder Minimum. It coincided with cooler temperatures on Earth, part of a period now known as the "Little Ice Age." Similar solar minimums occurred between 1420–1530, 1280–1340, and 1010-1050 (the Oort minimum). These periods preceded the discovery of sunspots, so no correlation between sunspots and temperature is available. Solar astronomers number the solar cycles from one minimum to the next starting with number one, the 1755–1766 cycle. Cycle 23 peaked (was at a maximum) in the year 2000. (See *Figure 4*.) There is still much debate about the connection between sunspot cycles and climate.

Geo Words

aurora: the bright emission of atoms and molecules near the Earth's poles caused by charged particles entering the upper atmosphere.

Check Your Understanding

1. How do solar flares interfere with communication and power systems?
2. In your own words, explain what is meant by the term "solar wind." How does the Sun contribute to "space weather?"
3. Describe the Earth's energy budget.

Understanding and Applying What You Have Learned

1. Study the graph that you made showing sunspot activity. You have already determined that sunspot activity occurs in cycles. Using graph paper, construct a new graph that predicts a continuation of the cycle from 2001 to 2015. Indicate which years you think would see increased solar-flare activity and more dangerous "space weather."
2. The latest sunspot maximum occurred in 2001. Using the data from your sunspot-activity data table, predict the next sunspot minimum.
3. Make lists of the possible consequences of solar flares to the following members of your community: an air traffic controller, a radio station manager, and the captain of a ship at sea. Can you think of other members of your community who would be affected by solar activity?
4. You have read that Earth's albedo is about 0.30.
   a) In your own words, describe what this means.
   b) Is the Earth's albedo constant? Why or why not?
   c) How does changing a planet's albedo change a planet's temperature? Why does this occur?
   d) If Earth's albedo was higher, but Earth was farther from the Sun, could the Earth have the same temperature? Why or why not?

Preparing for the Chapter Challenge

You have been asked to help people in your community to understand how events from outside the Earth affect their daily lives. Write a short paper in which you address the following questions:

1. How has the Sun affected your community in the past?
2. How has the Sun affected you personally?
3. How might the Sun affect your community in the future?
4. What are some of the benefits attained from a study of the Sun?
5. What are some of the problems caused by sunspots and solar flares?
6. Explain how auroras are caused. Explain also why they can or cannot be viewed in your community.
7. Compare the chances of dangerous effects from the Sun with the chances of an impact event affecting the Earth.

Inquiring Further

1. **Viewing sunspots**

   If you have a telescope, you can view sunspots by projecting an image of the Sun onto white cardboard. Never look directly at the Sun, with or without a telescope. Stand with your back to the Sun, and set up a telescope so that the large (front) end is pointing toward the Sun and the other end is pointing toward a piece of white cardboard. You should see a projection of the Sun on the cardboard, including sunspots. If you map the positions of the sunspots daily, you should be able to observe the rotation of the Sun over a couple of weeks. Use the *EarthComm* web site to locate good science sites on the Internet that show daily images of solar activity. Search them out and compare your observations of sunspots to what you see from the large observatories.

   Work with an adult during this activity. Do not look at the bright image for long periods of time.

2. **Aurorae**

   Have people in your community ever seen the northern lights? Even if your community is not very far north, do some research to see if the auroras have ever been spotted from your community.

3. **Solar radiation and airplanes**

   Periods of sunspot maximum increase the dosage of radiation that astronauts and people traveling in airplanes receive. Do some research on how much radiation astronauts receive during sunspot minima and maxima. How much radiation do airplane passengers receive? How do the amounts compare to the solar radiation you receive at the Earth's surface? How do scientists balance safety with the issue of the extra weight that would be added to aircraft, spacecraft, or space suits to provide protection?

4. **The hole in the ozone layer**

   People who live near the South Pole of the Earth are at risk for increased ultraviolet exposure from the Sun. This is due to a thinning in the atmosphere called the ozone hole. Research this ozone hole. Is there a northern ozone hole? Could these ozone holes grow? If so, could your community be endangered in the future?

5. **History of science**

   Research the life of British physicist Edward Victor Appleton, who was awarded the Nobel Prize in physics in 1947 for his work on the ionosphere. Other important figures in the discovery of the properties of the upper atmosphere include Oliver Heaviside, Arthur Edwin Kennelly, F. Sherwood Rowland, Paul Crutzen, and Mario Molina.

Activity 6 The Electromagnetic Spectrum and Your Community

Goals

In this activity you will:

- Explain electromagnetic radiation and the electromagnetic spectrum in terms of wavelength, speed, and energy.
- Investigate the different instruments astronomers use to detect different wavelengths in the electromagnetic spectrum.
- Understand that the atoms of each of the chemical elements have a unique spectral fingerprint.
- Explain how electromagnetic radiation reveals the temperature and chemical makeup of objects like stars.
- Understand that some forms of electromagnetic radiation are essential and beneficial to us on Earth, and others are harmful.

Think about It

Look at the spectrum as your teacher displays it on the overhead projector. Record in your *EarthComm* notebook the colors in the order in which they appear. Draw a picture to accompany your notes.

- What does a prism reveal about visible light?
- The Sun produces light energy that allows you to see. What other kinds of energy come from the Sun? Can you see them? Why or why not?

What do you think? Record your ideas in your *EarthComm* notebook. Be prepared to discuss your responses with your small group and the class.

Investigate

Part A: Observing Part of the Electromagnetic Spectrum

1. Obtain a spectroscope, similar to the one shown in the illustration. Hold the end with the diffraction grating to your eye. Direct it toward a part of the sky away from the Sun. (**CAUTION**: never look directly at the Sun; doing so even briefly can damage your eyes permanently.) Look for a spectrum along the side of the spectroscope. Rotate the spectroscope until you see the colors going from left to right rather than up and down.
   a) In your notebook, write down the order of the colors you observed.
   b) Move the spectroscope to the right and left. Record your observations.
2. Look through the spectroscope at a fluorescent light.
   a) In your notebook, write down the order of the colors you observed.
3. Look through the spectroscope at an incandescent bulb.
   a) In your notebook, write down the order of the colors you observed.

Do not look directly at a light with the unaided eye. Use the spectroscope as instructed.

4. Use your observations to answer the following questions:
   a) How did the colors and the order of the colors differ between reflected sunlight, fluorescent light, and the incandescent light? Describe any differences that you noticed.
   b) What if you could use your spectroscope to look at the light from other stars? What do you think it would look like?

Part B: Scaling the Electromagnetic Spectrum

1. Tape four sheets of photocopy paper end to end to make one sheet 112 cm long. Turn the taped sheets over so that the tape is on the bottom.
2. Draw a vertical line 2 cm from the left edge of the paper. Draw two horizontal lines from that line, one about 8 cm from the top of the page, and one about 10 cm below the first line.
3. On the top line, plot the frequencies of the electromagnetic spectrum on a logarithmic scale. To do this, mark off 24 1-cm intervals starting at the left vertical line. Label the marks from 1 to 24 (each number represents increasing powers of 10, from 10^1 to 10^{24}).
4. Use the information from the table of frequency ranges ($\log_{10}$) to divide your scale into the individual bands of electromagnetic radiation. For the visible band, use the entire band, not the individual colors.

Frequency Range Table			
EMR Bands	**Frequency Range (hertz)**	**Log_{10} Frequency Range (hertz)**	**10^{14} Conversions**
Radio and Microwave	Near 0 to 3.0×10^{12}	0 to 12.47	.
Infrared	3.0×10^{12} to 4.6×10^{14}	12.47 to 14.66	.
Visible	4.6×10^{14} to 7.5×10^{14}	14.66 to 14.88	4.6×10^{14} to 7.5×10^{14}
Red	4.6×10^{14} to 5.1×10^{14}	14.66 to 14.71	4.6×10^{14} to 5.1×10^{14}
Orange	5.1×10^{14} to 5.6×10^{14}	14.71 to 14.75	5.1×10^{14} to 5.6×10^{14}
Yellow	5.6×10^{14} to 6.1×10^{14}	14.75 to 14.79	5.6×10^{14} to 6.1×10^{14}
Green	6.1×10^{14} to 6.5×10^{14}	14.79 to 14.81	6.1×10^{14} to 6.5×10^{14}
Blue	6.5×10^{14} to 7.0×10^{14}	14.81 to 14.85	6.5×10^{14} to 7.0×10^{14}
Violet	7.0×10^{14} to 7.5×10^{14}	14.85 to 14.88	7.0×10^{14} to 7.5×10^{14}
Ultraviolet	7.5×10^{14} to 6.0×10^{16}	14.88 to 16.78	.
X-ray	6.0×10^{16} to 1.0×10^{20}	16.78 to 20	.
Gamma Ray	1.0×10^{20} to...	20 to ...	.

5. To construct a linear scale, you will need to convert the range of frequencies that each band of radiation covers for the logarithmic scale. This will allow you to compare the width of the bands of radiation relative to each other. Convert the frequency numbers for all bands (except visible) to 10^{14} and record them in the table.

   Example: 10^{17} is 1000 times greater than 10^{14}, so $2.5 \times 10^{17} = 2500 \times 10^{14}$.

6. On the lower horizontal line, mark off ten 10-cm intervals from the vertical line. Starting with the first interval, label each mark with a whole number times 10^{14}, from 1×10^{14} to 10×10^{14}. Label the bottom of your model "Frequency in hertz." Plot some of the 10^{14} frequencies you calculated on the bottom line of your constructed model. Plot the individual colors of the visible spectrum and color them.

   a) Compare the logarithmic and linear scales. Describe the differences.

7. Look at the range of ultraviolet radiation.

   a) How high do the ultraviolet frequencies extend (in hertz)?

   b) Using the same linear scale that you constructed in **Step 6** (10 cm = 1×10^{14} Hz), calculate the width (in centimeters) of the ultraviolet electromagnetic radiation band.

   c) Using this same scale what do you think you would need to measure the distance from the beginning of the ultraviolet band of the electromagnetic radiation to the end of the ultraviolet band of the electromagnetic radiation?

   d) Using your calculations above and the linear scale you created in **Steps 5** and **6,** how much wider is the ultraviolet band than the entire visible band? How does this compare to the relative widths of these two bands on the log scale you created in Steps 1-4?

8. X-rays are the next band of radiation.

   a) Using the same linear scale (10 cm $= 1 \times 10^{14}$ Hz) calculate the distance from the end of the ultraviolet band to the end of the x-ray band. Obtain a map from the Internet or use a local or state highway map to plot the distance.

   b) Based on your results for the width of the x-ray band, what would be your estimate for the width of the gamma-ray band of radiation? What would you need to measure the distance?

Part C: Using Electromagnetic Radiation in Astronomy

1. Astronomers use electromagnetic radiation to study objects and events within our solar system and beyond to distant galaxies. In this part of the activity, you will be asked to research a space science mission and find out how astronomers are using the electromagnetic spectrum in the mission and then report to the rest of the class. The *EarthComm* web site will direct you to links for missions that are either in development, currently operating, or operated in the past. A small sampling is provided in the table on the next page. Many missions contain multiple instruments (it is very expensive to send instruments into space, so scientists combine several or more studies into one mission), so you should focus upon one instrument and aspect of the mission and get to know it well. The **Digging Deeper** reading section of this activity might help you begin your work.

   Questions that you should try to answer in your research include:

   - What is the purpose or key question of the mission?
   - How does the mission contribute to our understanding of the origin and evolution of the universe or the nature of planets within our solar system?
   - Who and/or how many scientists and countries are involved in the mission?
   - What instrument within the mission have you selected?
   - What wavelength range of electromagnetic radiation does the instrument work at?
   - What is the detector and how does it work?
   - What does the instrument look like?
   - How are the data processed and rendered? Images? Graphs?
   - Any other questions that you and your teacher agree upon.

2. When you have completed your research, provide a brief report to the class.

Descriptions of Selected Missions	
Mission/Instrument	**Description**
Hubble – NICMOS Instrument	Hubble's Near Infrared Camera and Multi-Object Spectrometer (NICMOS) can see objects in deepest space—objects whose light takes billions of years to reach Earth. Many secrets about the birth of stars, solar systems, and galaxies are revealed in infrared light, which can penetrate the interstellar gas and dust that block visible light.
Cassini Huygens Mission to Saturn and Titan	The Ultraviolet Imaging Spectrograph (UVIS) is a set of detectors designed to measure ultraviolet light reflected or emitted from atmospheres, rings, and surfaces over wavelengths from 55.8 to 190 nm (nanometers) to determine their compositions, distribution, aerosol content, and temperatures.
SIRTF	The Space InfraRed Telescope Facility (SIRTF) is a space-borne, cryogenically cooled infrared observatory capable of studying objects ranging from our solar system to the distant reaches of the Universe. SIRTF is the final element in NASA's Great Observatories Program, and an important scientific and technical cornerstone of the new Astronomical Search for Origins Program.
HETE-2 High Energy Transient Explorer	The High Energy Transient Explorer is a small scientific satellite designed to detect and localize gamma-ray bursts (GRB's). The primary goals of the HETE mission are the multi-wavelength observation of gamma-ray bursts and the prompt distribution of precise GRB coordinates to the astronomical community for immediate follow-up observations. The HETE science payload consists of one gamma-ray and two x-ray detectors.
Chandra X-Ray Observatory	NASA's Chandra X-ray Observatory, which was launched and deployed by Space Shuttle Columbia in July of 1999, is the most sophisticated x-ray observatory built to date. Chandra is designed to observe x-rays from high-energy regions of the universe, such as the remnants of exploded stars.

Reflecting on the Activity and the Challenge

The spectroscope helped you to see that visible light is made up of different color components. Visible light is only one of the components of radiation you receive from the Sun. In the second part of the activity, you explored models for describing the range of frequencies of energy within electromagnetic radiation.

Finally, you researched a space mission to learn how astronomers are using electromagnetic radiation to understand the evolution of the Earth system. Radiation from the Sun and other objects in the universe is something you will need to explain to your fellow citizens in your **Chapter Challenge** brochure.

Digging Deeper

ELECTROMAGNETIC RADIATION

The Nature of Electromagnetic Radiation

In 1666, Isaac Newton found that he could split light into a spectrum of colors. As he passed a beam of sunlight through a glass prism, a spectrum of colors appeared from red to violet. Newton deduced that visible light was in fact a mixture of different kinds of light. About 10 years later, Christiaan Huygens proposed the idea that light travels in the form of tiny waves. It's known that light with shorter wavelengths is bent (refracted) more than light with longer wavelengths when it passes through a boundary between two different substances. Violet light is refracted the most, because it has the shortest wavelength of the entire range of visible light. This work marked the beginning of **spectroscopy**—the science of studying the properties of light. As you will learn, many years of research in spectroscopy has answered many questions about matter, energy, time, and space.

In your study of the Sun you learned that the Sun radiates energy over a very wide range of wavelengths. Earth's atmosphere shields you from some of the most dangerous forms of electromagnetic radiation. You are familiar with the wavelengths of light that do get through and harm you—mostly in the form of sunburn-causing ultraviolet radiation. Now you can take what you learned and apply the principles of spectroscopy to other objects in the universe.

In the **Investigate** section, you used a **spectroscope** to study the Sun's light by separating it into its various colors. Each color has a characteristic wavelength. This range of colors, from red to violet, is called the **visible spectrum**. The visible spectrum is a small part of the entire spectrum of **electromagnetic radiation** given off by the Sun, other stars, and galaxies.

Electromagnetic radiation is in the form of electromagnetic waves that transfer energy as they travel through space. Electromagnetic waves (like ripples that expand after you toss a stone into a pond) travel at the speed of light (300,000 m/s). That's eight laps around the Earth in one second. Although it's not easy to appreciate from everyday life, it turns out that electromagnetic radiation has properties of both particles and waves. The colors of the visible spectrum are best described as waves, but the same energy that produces an electric current in a solar cell is best described as a particle.

Geo Words

spectroscopy: the science that studies the way light interacts with matter.

spectroscope: an instrument consisting of, at a minimum, a slit and grating (or prism) which produces a spectrum for visual observation.

visible spectrum: part of the electromagnetic spectrum that is detectable by human eyes. The wavelengths range from 350 to 780 nm (a nanometer is a billionth of a meter).

electromagnetic radiation: the energy propagated through space by oscillating electric and magnetic fields. It travels at 3×10^8 m/s in a vacuum and includes (in order of increasing energy) radio, infrared, visible light (optical), ultraviolet, x-rays, and gamma rays.

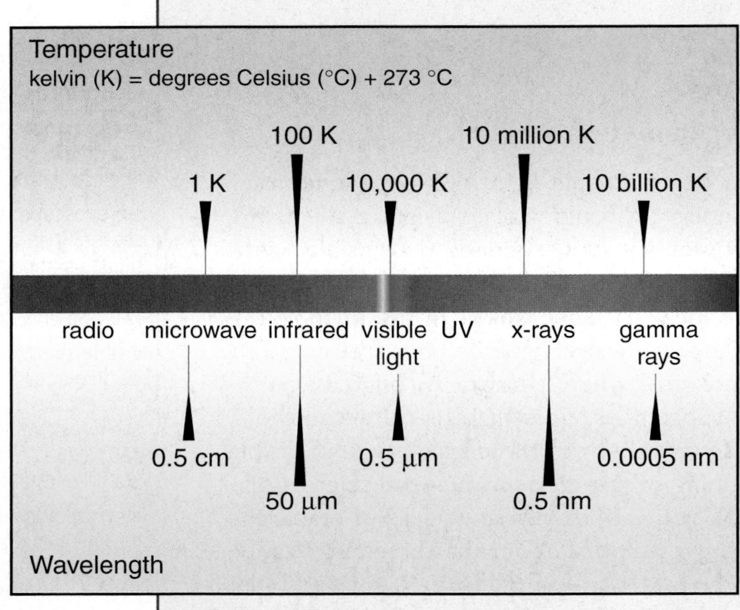

Figure 1 The electromagnetic spectrum. Wavelengths decrease from left to right, and energy increases from left to right. The diagram shows that a relationship exists between the temperature of an object and the peak wavelength of electromagnetic radiation it emits.

Figure 1 summarizes the spectrum of energy that travels throughout the universe. Scientists divide the spectrum into regions by the wavelength of the waves. Long radio waves have wavelengths from several centimeters to thousands of kilometers, whereas gamma rays are shorter than the width of an atom.

Humans can see only wavelengths between 0.4 and 0.7 µm, which is where the visible spectrum falls. A micrometer (µm) is a millionth of a meter. This means that much of the electromagnetic radiation emitted by the Sun is invisible to human eyes. You are probably familiar, however, with some of the kinds of radiation besides visible light. For example, **ultraviolet** radiation gives you sunburn. **Infrared** radiation you detect as heat. Doctors use x-rays to help diagnose broken bones or other physical problems. Law-enforcement officers use radar to measure the speed of a motor vehicle, and at home you may use microwaves to cook food.

Geo Words

ultraviolet: electromagnetic radiation at wavelengths shorter than the violet end of visible light; with wavelengths ranging from 5 to 400 nm.

infrared: electromagnetic radiation with wavelengths between about 0.7 to 1000 µm. Infrared waves are not visible to the human eye.

radio telescope: an instrument used to observe longer wavelengths of radiation (radio waves), with large dishes to collect and concentrate the radiation onto antennae.

Astronomy and the Electromagnetic Spectrum

Humans have traveled to the Moon and sent probes deeper into our solar system, but how do they learn about distant objects in the universe? They use a variety of instruments to collect electromagnetic radiation from these distant objects. Each tool is designed for a specific part of the spectrum. Visible light reveals the temperature of stars. Visible light is what you see when you look at the stars through telescopes, binoculars, or your unaided eyes. All other forms of light are invisible to the human eye, but they can be detected.

Radio telescopes like the Very Large Array (VLA) and Very Large Baseline Array (VLBA) in New Mexico are sensitive to wavelengths in the radio range. Radio telescopes produce images of celestial bodies by recording the different amounts of radio emission coming from an area of the sky observed. Astronomers process the information with computers to produce an image. The VLBA has 27 large dish antennas that work together as a single instrument. By using recorders and precise atomic clocks installed at each antenna, the signals from all antennas are combined after the observation is completed.

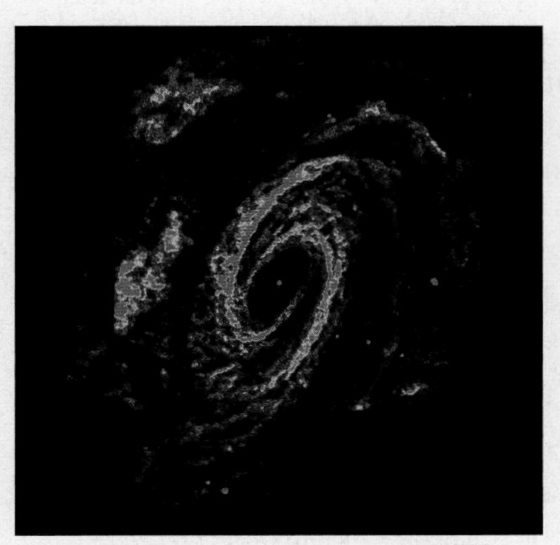

Figure 2 The galaxy M81.

The galaxy M81 is a spiral galaxy about 11 million light-years from Earth and is about 50,000 light-years across. The spiral structure is clearly shown in *Figure 2*, which shows the relative intensity of emission from neutral atomic hydrogen gas. In this pseudocolor image, red indicates strong radio emission and blue weaker emission.

Geo Words

x-ray telescope: an instrument used to detect stellar and interstellar x-ray emission. Because the Earth's atmosphere absorbs x-rays, x-ray telescopes are placed high above the Earth's surface.

The orbiting Chandra **x-ray telescope** routinely detects the highly energetic radiation streaming from objects like supernova explosions, active galaxies, and black holes. The Hubble Space Telescope is outfitted with a special infrared instrument sensitive to radiation being produced by star-forming nebulae and cool stars. It also has detectors sensitive to ultraviolet light being emitted by hot young stars and supernova explosions.

A wide array of solar telescopes both on Earth and in space study every wavelength of radiation from our nearest star in minute detail. The tools of astronomy expand scientists' vision into realms that human eyes can never see, to help them understand the ongoing processes and evolution of the universe.

Figure 3 Astronauts working on the Hubble Space Telescope high above the Earth's atmosphere.

Geo Words

peak wavelength: the wavelength of electromagnetic radiation with the most electromagnetic energy emitted by any object.

Using Electromagnetic Radiation to Understand Celestial Objects

The wavelength of light with the most energy produced by any object, including the Sun, is called its **peak wavelength**. Objects that are hot and are radiating visible light usually look the color of their peak wavelength. People are not hot enough to emit visible light, but they do emit infrared radiation that can be detected with infrared cameras. The Sun has its peak wavelength in the yellow region of the visible spectrum. Hotter objects produce their peaks toward the blue direction. Very hot objects can have their peaks in the ultraviolet, x-ray, or even gamma-ray range of wavelength. A gas under high pressure radiates as well as a hot solid object. Star colors thus reflect temperature. Reddish stars are a "cool" 3000 to 4000 K (kelvins are celsius degrees above absolute zero, which is at minus 273°C). Bluish stars are hot (over 20,000 K).

One of the most important tools in astronomy is the spectrum—a chart of the entire range of wavelengths of light from an object. Astronomers often refer to this chart as the spectrum of the star. These spectra come in two forms: one resembles a bar code with bright and dark lines (see *Figure 4*), and the other is a graph with horizontal and vertical axes (see *Figure 5*). Think of these spectra as "fingerprints" that reveal many kinds of things about an object: its chemical composition, its temperature and pressure, and its motion toward or away from us.

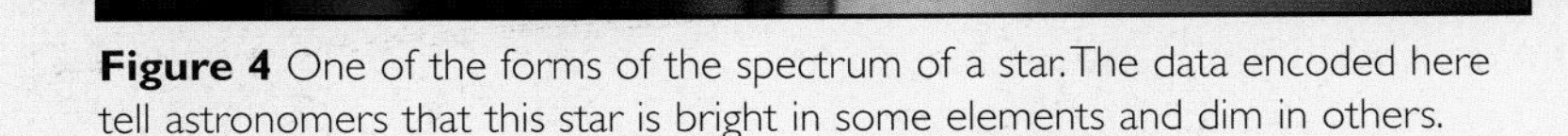

Figure 4 One of the forms of the spectrum of a star. The data encoded here tell astronomers that this star is bright in some elements and dim in others.

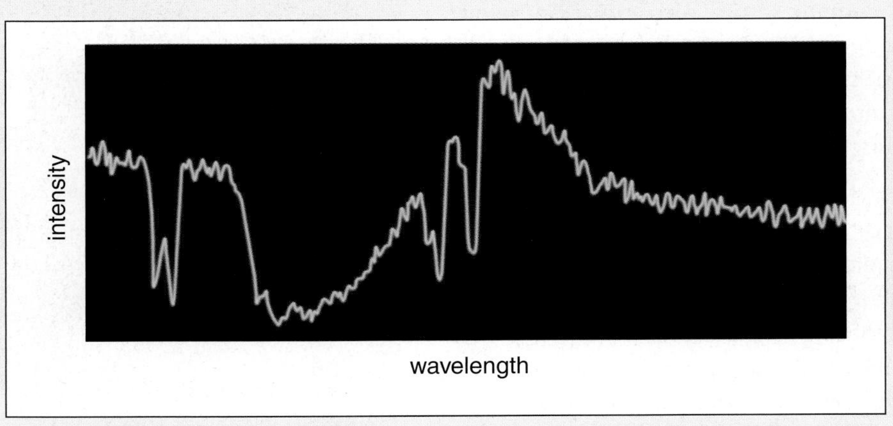

Figure 5 The spectrum of a star can also be represented as a graph with horizontal and vertical axes.

Each chemical element in the universe has its own unique spectrum. If you know what the spectrum of hydrogen is, you can look for its fingerprint in a star. If you suspect that a star may have a lot of the elements helium or calcium, for example, you can compare the spectrum of the star with the known spectra of helium or calcium. If you see bright lines in the stellar spectrum that match the patterns of bright lines in the helium or calcium spectra, then you have identified those elements in the star. This kind of spectrum is known as an **emission spectrum**. If you look at the star and see dark lines where you would expect to see an element—especially hydrogen—it is likely that something between you and the star is absorbing the element. This kind of spectrum is known as an **absorption spectrum**.

The positions of lines in a spectrum reveal the motion of the star toward or away from Earth, as well as the speed of that motion. You have experienced the effect yourself, when a car, a truck, or a train passes by you with its horn blowing. The pitch of the sound increases as the object approaches you, and decreases as the object passes by and moves away from you. That is because when the object is coming toward you, its speed adds to the speed of the sound, making the wavelength of the sound seem higher to your ear. The reverse happens when the object is moving away from you. The same principle applies to the spectrum from a distant object in space, which might be moving either toward Earth or away from Earth.

Geo Words

emission spectrum: a spectrum containing bright lines or a set of discrete wavelengths produced by an element. Each element has its own unique emission spectrum.

absorption spectrum: a continuous spectrum interrupted by absorption lines or a continuous spectrum having a number of discrete wavelengths missing or reduced in intensity.

Check Your Understanding

1. What are the colors of the spectrum of visible sunlight, from longest wavelength to shortest? Are there breaks between these colors, or do they grade continuously from one to the next? Why?
2. Which wavelengths of light can be more harmful to you than others? Why?
3. What tools do astronomers use to detect different wavelengths of light?
4. How can the speed of a distant object in space be measured?

Understanding and Applying What You Have Learned

1. Imagine that you are on a distant planet. Name two parts of the electromagnetic spectrum that you would use to investigate Earth. Explain the reasoning for your choices.
2. Refer to *Figure 1* to answer the questions below.
   a) Describe the relationship between wavelength and energy in the electromagnetic spectrum.
   b) Based upon this relationship, why do astronomers use x-ray telescopes to study supernova explosions and black holes?
3. The Sun looks yellow, can warm the surface of your skin, and can also give you a bad sunburn. Explain these three everyday phenomena in terms of the electromagnetic spectrum and peak wavelength.

Preparing for the Chapter Challenge

Recall that your challenge is to educate people about the hazards from outer space and to explain some of the benefits from living in our solar system. Electromagnetic radiation has both beneficial and harmful effects on life on Earth. Use what you have learned in this activity to develop your brochure.

1. Make a list of some of the positive effects of electromagnetic radiation on your community. Explain each item on the list.
2. Make a list of some of the negative effects of electromagnetic radiation on your community. Explain each item on the list.
3. Make a list of celestial radiation sources and any effects they have on Earth systems. What are the chances of a stellar radiation source affecting Earth?

Inquiring Further

1. **Using radio waves to study distant objects**

   Radio waves from the Sun penetrate the Earth's atmosphere. Scientists detect these waves and study their strength and frequency to understand the processes inside the Sun that generate them. Do some research on how these waves are studied.

2. **Detecting electromagnetic radiation**

   Investigate some of the instruments that astronomers use to detect electromagnetic radiation besides light. Where are you likely to find ultraviolet detectors? Describe radio telescope arrays.

3. **Technologies and the electromagnetic spectrum**

   Research some of the technologies that depend on the use of electromagnetic radiation. These might include microwave ovens, x-ray machines, televisions, and radios. How do they work? How is electromagnetic radiation essential to their operation? What interferes with their operation?

Activity 7 Our Community's Place Among the Stars

Goals

In this activity you will:

- Understand the place of our solar system in the Milky Way galaxy.
- Study stellar structure and the stellar evolution (the life histories of stars).
- Understand the relationship between the brightness of an object (its luminosity) and its magnitude.
- Estimate the chances of another star affecting the Earth in some way.

Think about It

When you look at the nighttime sky, you are looking across vast distances of space.

- As you stargaze, what do you notice about the stars?
- Do some stars appear brighter than others? Larger or smaller? What about their colors?

What do you think? Record your impressions and sketch some of the stars in your *EarthComm* notebook. Be prepared to discuss your thoughts with your small group and the class.

Investigate

Part A: Brightness versus Distance from the Source

1. Set a series of lamps with 40-, 60-, and 100-W bulbs (of the same size and all with frosted glass envelopes) up at one end of a room (at least 10 m away). Use the other end of the room for your observing site. Turn all the lamps on. Close all of the shades in the room.
   a) Can you tell the differences in brightness between the lamps?
2. Move the lamp with the 40-W bulb forward 5 m toward you.
   a) Does the light look brighter than the 60-W lamp?
   b) Does it look brighter than the 100-W lamp?
3. Shift the positions of the lamps so that the 40-W lamp and the 100-W lamp are in the back of the room and the 60-W lamp is halfway between you and the other lamps.
   a) How do the brightnesses compare?
4. Using a light meter, test one bulb at a time. If you do not have a light meter, you will have to construct a qualitative scale for brightness.
   a) Record the brightness of each bulb at different distances.
5. Graph the brightness versus the distance from the source for each bulb (wattage).
   a) Plot distance on the horizontal axis of the graph and brightness on the vertical axis. Leave room on the graph so that you can extrapolate the graph beyond the data you have collected. Plot the data for each bulb and connect the points with lines.
   b) Extrapolate the data by extending the lines on the graph using dashes.
6. Use your graph to answer the following questions:
   a) Explain the general relationship between wattage and brightness (as measured by your light meter).
   b) What is the general relationship between distance and brightness?
   c) Do all bulbs follow the same pattern? Why or why not?
   d) Draw a light horizontal line across your graph so that it crosses several of the lines you have graphed.
   e) Does a low-wattage bulb ever have the same brightness as a high-wattage bulb? Describe one or two such cases in your data.
   f) The easiest way to determine the absolute brightness of objects of different brightness and distance is to move all objects to the same distance. How do you think astronomers handle this problem when trying to determine the brightness and distances to stars?
7. When you have completed this activity, spend some time outside stargazing. Think about the relationship between brightness and distance as it applies to stars.
   a) Write your thoughts down in your *EarthComm* notebook.

Do not stare at the light bulbs for extended periods of time.

Part B: Luminosity and Temperature of Stars

1. An important synthesis of understanding in the study of stars is the Hertzsprung-Russell (HR) diagram. Obtain a copy of the figure below. Examine the figure and answer the following questions:

   a) What does the vertical axis represent?

   b) What does the horizontal axis represent?

   c) The yellow dot on the figure is the Sun. What is its temperature and luminosity?

   d) Put four more dots on the diagram labeled A through D to show the locations of stars that are:

      A. hot and bright
      B. hot and dim
      C. cool and dim
      D. cool and bright

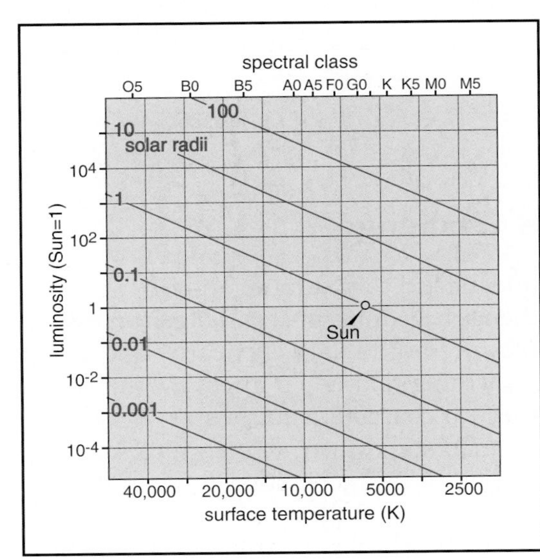

2. Obtain a copy of the *Table 1* and the HR diagram that shows the locations of main sequence stars, supergiants, red giants, and white dwarfs.

   a) Using the luminosity of the stars, and their surface temperatures, plot the locations of stars shown in *Table 1* on a second HR diagram.

3. Classify each of the stars into one of the following four categories, and record the name in your copy of the table:

   — Main sequence
   — Red giants
   — Supergiants
   — White dwarfs

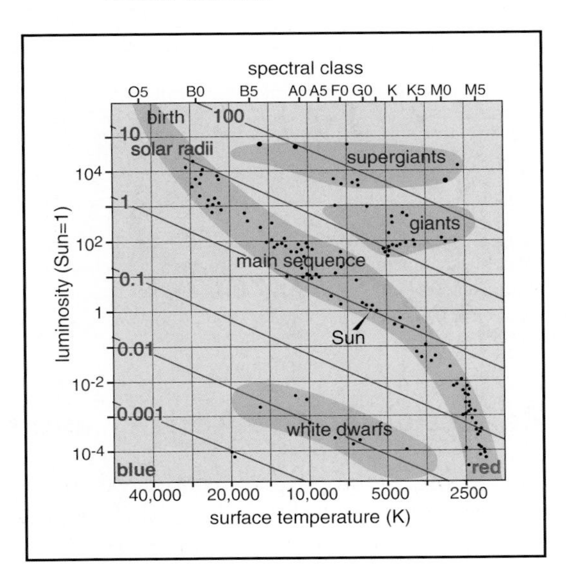

Table 1 Selected Properties of Fourteen Stars							
Star	**Surface Temperature (K)**	**Luminosity (Relative to Sun)**	**Distance (Light-Years)**	**Mass (Solar Masses)**	**Diameter (Solar Diameters)**	**Color**	**Type of Star**
Sirius A	9100	22.6	8.6	2.3	2.03	Blue	
Arcturus	4300	115	36.7	4.5	31.5	Red	
Vega	10300	50.8	25.3	3.07	3.1	Blue	
Capella	5300	75.8	42.2	3	10.8	Red	
Rigel	11000	38,679	733	20	62	Blue	
Procyon A	6500	7.5	11.4	1.78	1.4	Yellow	
Betelgeuse	2300	4520–14,968 (variable)	427	20	662	Red	
Altair	7800	11.3	65.1	2	1.6	Yellow	
Aldebaran	4300	156–171 (variable)	65	25	51.5	Red	
Spica	25300	2121	262	10.9	7.3	Blue	
Pollux	4500	31	33	4	8	Red	
Deneb	10500	66,500	1600	25	116	Yellow	
Procyon B	8700	0.0006	11.2	0.65	0.02	White	
Sirius B	24000	0.00255	13.2	0.98	0.008	Blue-white	

Note: Mass, diameter, and luminosity are given in solar units. For example, Sirius A has 2.3 solar masses, has a diameter 2.03 that of the Sun, and has luminosity 22.6 times brighter than the Sun.
1 solar mass = 2×10^{30} kg = 330,000 Earth masses; 1 solar diameter = 700,000 km = 110 Earth diameters.

Reflecting on the Activity and the Challenge

Measuring the apparent differences in brightness of the light bulbs at different distances helps you to see that distance and brightness are important factors in helping you understand the objects in our universe. When you look at the stars at night, you are seeing stars at different distances and brightnesses. In your **Chapter Challenge** you will be telling people about the effects of distant objects on the Earth. When you assess danger from space, it is important to understand that stars in and of themselves don't pose a danger unless they are both relatively nearby and doing something that could affect Earth. The spectral characteristics of stars help you to understand their temperature, size, and other characteristics. In turn, that helps you to understand if a given star is or could be a threat to Earth. The light from distant stars can also be used to understand our own star, and our own solar system's makeup and evolution.

Digging Deeper

EARTH'S STELLAR NEIGHBORS

Classifying Stars

You already know that our solar system is part of the Milky Way galaxy. Our stellar neighborhood is about two-thirds of the way out on a spiral arm that stretches from the core of the galaxy. The galaxy contains hundreds of billions of stars. Astronomers use a magnitude scale to describe the brightness of objects they see in the sky. A star's brightness decreases with the square of the distance. Thus, a star twice as far from the Earth as an identical star would be one-fourth as bright as the closer star. The first magnitude scales were quite simple—the brightest stars were described as first magnitude, the next brightest stars were second magnitude, and so on down to magnitude 6, which described stars barely visible to the naked eye. The smaller the number, the brighter the star; the larger the number, the dimmer the star.

Today, scientists use a more precise system of magnitudes to describe brightness. The brightest star in the sky is called Sirius A, and its magnitude is –1.4. Of course, the Sun is brighter at –27 and the Moon is –12.6! The dimmest naked-eye stars are still sixth magnitude. To see anything dimmer than that, you have to magnify your view with binoculars or telescopes. The best ground-based telescopes can detect objects as faint as 25th magnitude. To get a better view of very faint, very distant objects, you have to get above the Earth's atmosphere. The Hubble Space Telescope, for example can detect things as dim as 30th magnitude!

Figure 1 This NASA Hubble Space Telescope near-infrared image of newborn binary stars reveals a long thin nebula pointing toward a faint companion object which could be the first extrasolar planet to be imaged directly.

Perhaps you have seen a star described as a G-type star or an O-type star. These are stellar classifications that depend on the color and temperature of the stars. They also help astronomers understand where a given star is in its evolutionary history. To get such information, astronomers study stars with spectrographs to determine their temperature and chemical makeup. As you can see in the table below, there are seven main categories of stars:

Stellar Classification	Temperature (kelvins)
O	25,000 K and higher
B	11,000–25,000 K
A	7500–11,000 K
F	6000–7500 K
G	5000–6000 K
K	3500–5000 K
M	less than 3500 K

Geo Words

luminosity: the total amount of energy radiated by an object every second.

molecular cloud: a large, cold cloud made up mostly of molecular hydrogen and helium, but with some other gases, too, like carbon monoxide. It is in these clouds that new stars are born.

The Lives of Stars

Astronomers use the term **luminosity** for the total rate at which a star emits radiation energy. Unlike apparent brightness (how bright the star appears to be) luminosity is an intrinsic property. It doesn't depend on how far away the star is. In the early 1900s Ejnar Hertzsprung and Henry Norris Russell independently made the discovery that the luminosity of a star was related to its surface temperature. In the second part of this activity, you worked with a graph that shows this relationship. It is called the Hertzsprung-Russell (HR) diagram in honor of the astronomers who discovered this relationship. The HR diagram alone does not tell you how stars change. By analogy, if you were to plot the IQ versus the weight of everyone in your school, you would probably find a very poor relationship between these two variables. Your graph would resemble a scatter plot more than it would a line. However, if you plotted the height versus weight for the same people, you are more likely to find a strong relationship (data would be distributed along a trend or line). The graph doesn't tell you why this relationship exists — that's up to you to determine. Similarly, the HR diagram shows that stars don't just appear randomly on a plot of luminosity versus temperature, but fall into classes of luminosity (red giants, white dwarfs, and so on).

The life cycle of a star begins with its formation in a cloud of gas and dust called a **molecular cloud**. The material in the cloud begins to clump together, mixing and swirling. Eventually the core begins to heat as more material is drawn in by gravitational attraction. When the temperature in the center of the cloud reaches 15 million kelvins, the stellar fusion reaction starts up and a star is born. Such stars are called main-sequence stars. Many stars spend 90% of their lifetimes on the main sequence.

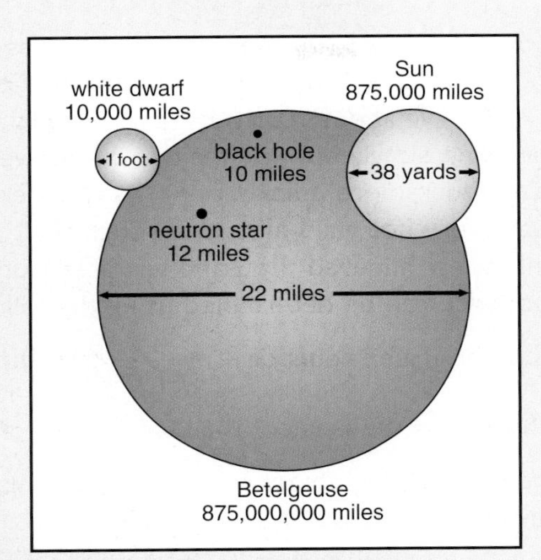

Figure 2 Scaling stars to 10,000 miles to one foot reveals the relative sizes of various stars.

Newborn stars are like baby chickens pecking their way out of a shell. As these infant stars grow, they bathe the cloud surrounding them in strong ultraviolet radiation. This vaporizes the cloud, creating beautiful sculpted shapes in the cloud. In the photograph in *Figure 3*, the Hubble Space Telescope studied a region of starbirth called NGC 604. Notice the cluster of bright white stars in the center "cavern" of the cloud of gas and dust. Their ultraviolet light has carved out a shell of gas and dust around the stellar newborns.

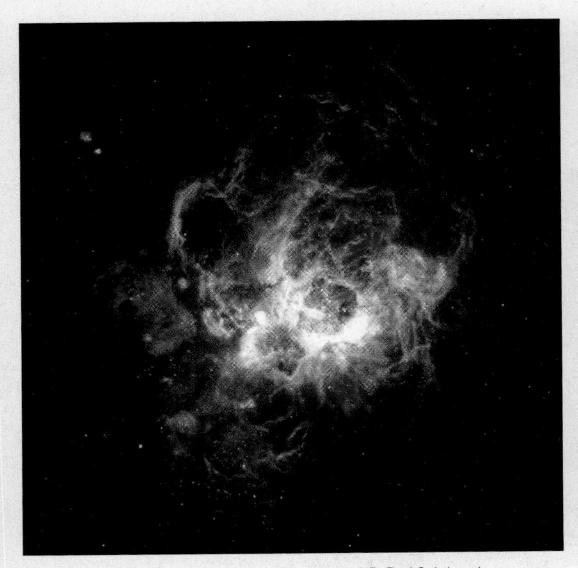

Figure 3 The starforming region NGC 604 in the galaxy M33.

Figure 4 The Orion Nebula is an example of a molecular cloud, from which new stars are born.

How long a star lives depends on its mass (masses of selected stars are shown in *Table 1* in the **Investigate** section of the activity). Stars like our Sun will live about 10 billion years. Smaller, cooler stars might go on twice that long, slowly burning their fuel. Massive supergiant stars consume their mass much more quickly, living a star's life only a few tens of millions of years. Very hot stars also go through their fuel very quickly, existing perhaps only a few hundred thousand years. The time a star spends on the main sequence can be determined using the following formula:

$$\text{Time on main sequence} = \frac{1}{M^{2.5}} \times \mathbf{10\ billion\ years}$$

where M is the mass of the star in units of solar masses.

Even though high-mass stars have more mass, they burn it much more quickly and end up having very short lives.

In the end, however, stars of all types must die. Throughout its life a star loses mass in the form of a stellar wind. In the case of the Sun this is called the solar wind. As a star ages, it loses more and more mass. Stars about the size of the Sun and smaller end their days as tiny, shrunken remnants of their former selves, surrounded by beautiful shells of gas and dust. These are called planetary nebulae. In about five billion years the Sun will start to resemble one of these ghostly nebulae, ending its days surrounded by the shell of its former self.

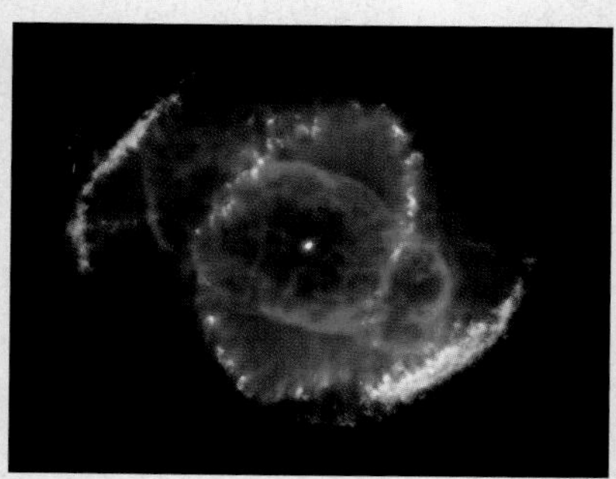

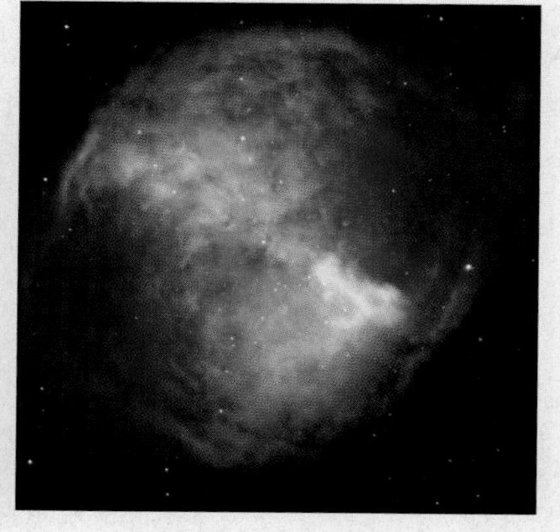

Figures 5A, 5B, 5C Three examples of the deaths of stars about the size of the Sun. **A:** The Butterfly Nebula. **B:** The Cat's-Eye Nebula. In both cases at least the dying star lies embedded in a cloud of material exhaled by the star, as it grew older. **C:** The Dumbbell Nebula. European Southern Observatory.

Massive stars (supergiants tens of times more massive than the Sun) also lose mass as they age, but at some point their cores collapse catastrophically. The end of a supergiant's life is a cataclysmic explosion called a **supernova**. In an instant of time, most of the star's mass is hurled out into space, leaving behind a tiny remnant called a **neutron star**. If the star is massive enough, the force of the explosion can be so strong that the remnant is imploded into a **stellar black hole**—a place where the gravity is so strong that not even light can escape.

The material that is shed from dying stars (whether they end their days as slowly fading dwarf stars, or planetary nebulae, or supernovae) makes its way into the space between the stars. There it mixes and waits for a slow gravitational contraction down to a new episode of starbirth and ultimately star death. Because humans evolved on a planet that was born from a recycled cloud of stellar mass, they are very much star "stuff"—part of a long cycle of life, death, and rebirth.

Astronomers search the universe to study the mechanics of star formation. Star nurseries and star graveyards are scattered through all the galaxies. In some cases, starbirth is triggered when one galaxy collides with (actually passes through) another. The clouds of gas and dust get the push they need to start the process.

Scientists also search for examples of planetary nebulae. They want to understand when and how these events occur. Not only are these nebulae interesting, but also they show scientists what the fate of our solar system will be billions of years from now.

What would happen if there were a supernova explosion in our stellar neighborhood sometime in the future? Depending on how close it was, you could be bombarded with strong radiation and shock waves from the explosion. The chances of this happening are extremely small—although some astronomers think that a supernova some five billion years ago may have provided the gravitational kick that started our own proto–solar nebula on the road to stardom and planetary formation.

Figure 6 The Crab Nebula is the remnant of a supernova explosion first observed in the year AD 1054.

Geo Words

supernova: the death explosion of a massive star whose core has completely burned out. Supernova explosions can temporarily outshine a galaxy.

neutron star: the imploded core of a massive star produced by a supernova explosion.

stellar black hole: the leftover core of a massive single star after a supernova. Black holes exert such large gravitational pull that not even light can escape.

Check Your Understanding

1. How do astronomers classify stars?
2. Write a brief outline of how stars are born.
3. What determines the way a star dies?

Understanding and Applying What You Have Learned

1. Using an astronomy computer program or a guidebook to the stars, make a list of the 10 nearest stars, and their distances, magnitudes, and spectral classes. What do their classes tell you about them?
2. What is mass loss and how does it figure in the death of a star? Is the Sun undergoing mass loss?
3. What happens to the material left over from the death of a star?
4. Two identical stars have different apparent brightnesses. One star is 10 light-years away, and the other is 30 light-years away from us. Which star is brighter, and by how much?
5. Refer to *Table 1* to answer the questions below:
   a) Calculate how long the Sun will spend on the main sequence.
   b) Calculate how long Spica will spend on the main sequence.
   c) Relate your results to the statement that the more massive the star, the shorter they live.
6. Explain the relationships between temperature, luminosity, mass, and lifetime of stars.

Preparing for the Chapter Challenge

You are about to complete your **Chapter Challenge**. In the beginning you were directed to learn as much as you could about how extraterrestrial objects and events could affect the Earth and your community. In order to do this you have explored the stars and planets, looking at all the possibilities. By now you have a good idea about how frequently certain kinds of events occur that affect Earth. The Sun is a constant source of energy and radiation.

In this final activity you learned our solar system's place in the galaxy, and you read about how stars are born and die. Because the birth of our solar system led directly to our planet, and the evolution of life here, it's important to know something about stars and how they come into existence.

You now know that the solar system is populated with comets and asteroids, some of which pose a threat to Earth over long periods of time.

The evolution of the Earth's orbit and its gravitational relationship with the Moon make changes to the Earth's climate, length of year, and length of day. The solar system is part of a galaxy of other stars, with the nearest star being only 4.21 light-years away. The Sun itself is going through a ten-billion-year-long period of evolution and will end as a planetary nebula some five billion years in the future. Finally, our Milky Way galaxy is wheeling toward a meeting with another galaxy in the very, very distant future. Your challenge now that you know and understand these things is to explain them to your fellow citizens and help them understand the risks and benefits of life on this planet, in this solar system, and in this galaxy.

Inquiring Further

1. **Evolution of the Milky Way galaxy**

   The Milky Way galaxy formed some 10 billion years ago, when the universe itself was only a fraction of its current age. Research the formation of our galaxy and find out how its ongoing evolution influenced the formation of our solar system.

2. **Starburst knots in other galaxies**

   Other galaxies show signs of star birth and star death. You read about a starbirth region called NGC 604 in the **Digging Deeper** reading section of this activity. Astronomers have found evidence of colliding galaxies elsewhere in the universe. In nearly every case, such collisions have spurred the formation of new stars. In the very distant future the Milky Way will collide with another galaxy, and it's likely that starburst knots will be formed. Look for examples of starbirth nurseries and starburst knots in other galaxies and write a short report on your findings. How do you think such a collision would affect Earth (assuming that anyone is around to experience it)?

Earth Science at Work

ATMOSPHERE: *Astronaut*
There is no atmosphere in space; therefore, astronauts must have pressurized atmosphere in their spacecraft cabins. Protective suits protect them when they perform extra-vehicular activities.

BIOSPHERE: *Exobiologist*
"Did life ever get started on Mars?" By learning more about the ancient biosphere and environments of the early Earth, exobiologists hope that they may be able to answer such questions when space missions return with rocks gathered on Mars.

CRYOSPHERE: *Glaciologist*
Ice is abundant on the Earth's surface, in the planetary system, and in interstellar space. Glaciologists study processes at or near the base of glaciers and ice sheets on Earth and other planets.

GEOSPHERE: *Planetary geologist*
By researching Martian volcanism and tectonism, or the geology of the icy satellites of Jupiter, Saturn, and Uranus, planetary geologists hope to develop a better understanding of our place in the universe.

HYDROSPHERE: *Lifeguard*
Surfers and other water-sport enthusiasts rely on lifeguards to inform them of the time of high and low tides. Low tides or high tides can create dangerous situations.

How is each person's work related to the Earth system, and to Astronomy?

2

Climate Change ...and Your Community

CLIMATE CHANGE ...and Your Community

Getting Started

The Earth's climate has changed many times over geologic history.

- What kinds of processes or events might cause the Earth's climate to change?

What do you think? Write down your ideas about these questions in your *EarthComm* notebook. Be prepared to discuss your ideas with your small group and the class.

Scenario

Your local newspaper would like to run a series of articles about global warming. However, the newspaper's science reporter is unavailable. The newspaper has come to your class to ask you and your classmates to write the articles. These feature articles and an editorial will be run in the Science and Environment section of the newspaper. The newspaper editor wants to give the readers of the paper a thorough scientific background to understand the idea of global climate change.

Chapter Challenge

Article 1: How Has Global Climate Changed Over Time?

Many people are not aware that the Earth's climate has changed continually over geologic time. This article should contain information about:

- the meaning of "climate," both regional and global;
- examples of different global climates in the geologic past;
- how geologists find out about past climates, and
- a description of your community's present climate and examples of past climates in your part of the country.

Article 2:
Causes of Global Climate Change
Some people might not be aware that human production of greenhouse gases is not the only thing that can cause the Earth's climate to change. There are many different factors that may affect how and when the Earth's climate changes. This article should include information about:

- Milankovitch cycles;
- plate tectonics;
- ocean currents, and
- carbon dioxide levels.

Article 3:
What is "Global Warming" and How Might It Affect Our Community?

Although almost everyone has heard the terms "greenhouse gases" and "global warming," there is a lot of confusion about what these terms actually mean. This article should contain information about:

- greenhouse gases;
- how humans have increased the levels of carbon dioxide in the atmosphere;
- why scientists think increased carbon dioxide might lead to global warming;
- possible effects of global warming, focusing on those that would have the greatest impact on your community, and
- why it is difficult to predict climate change.

Editorial
The final piece is not an article but rather an editorial in which the newspaper expresses its opinion about a particular topic. In the editorial, you should state:

- whether your community should be concerned about global warming and why, and
- what steps, if any, your community should take in response to the possibility of global warming.

Assessment Criteria

Think about what you have been asked to do. Scan ahead through the chapter activities to see how they might help you to meet the challenge. Work with your classmates and your teachers to define the criteria for assessing your work. Record all of this information. Make sure that you understand the criteria as well as you can before you begin. Your teacher may provide you with a sample rubric to help you get started.

Activity 1 Present-Day Climate in Your Community

Goals

In this activity you will:

- Identify factors of the physical environment.
- Use a topographic map to gather data about elevation and latitude, and physical features.
- Interpret data from a climate data table.
- Compare and contrast climate information from two different parts of the United States.
- Understand how physical features can influence the climate of an area.

Think about It

A friend e-mails you from Italy to ask what your environment and climate are like. You plan to e-mail her a reply.

- How would you describe the physical environment of your community?
- How would you describe the climate of your community?

What do you think? Record your ideas about these questions in your *EarthComm* notebook. Be prepared to discuss your responses with your small group and the class.

Investigate

Part A: Physical Features and Climate in Your Community

1. Depending on where your community is located and how large it is, you might wish to expand your definition of "community" to include a larger area, like your county or state. For example, your town does not have to be right on the ocean to have its climate influenced by the ocean.

   a) Write a "definition" of the area that you will examine as your "community."

2. Use the climate data tables provided on the following pages and topographic maps of your "community" to describe the climate in your community.

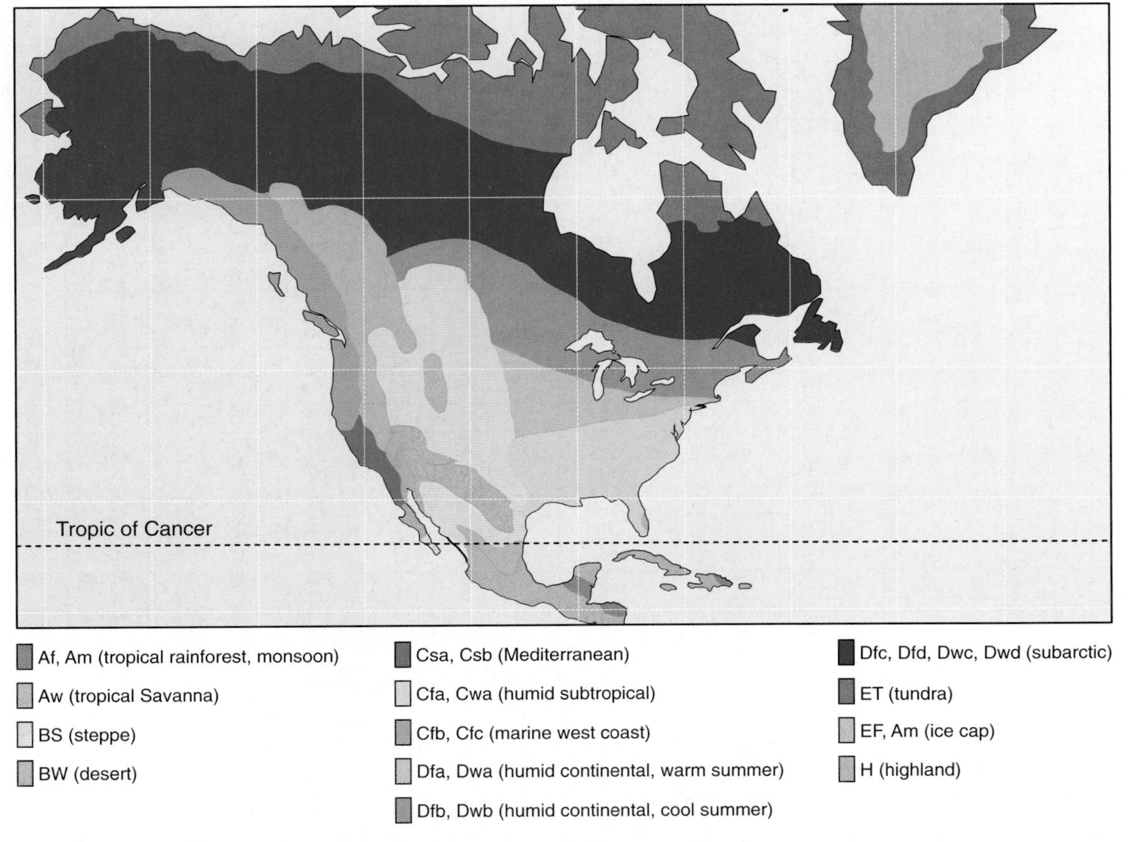

Abbreviations for climate zones are based upon the Koppen classification system.

Temperatures in the United States														
Country and Station	Latitude	Longitude	Elevation	Temperature										
				Record Length	Average Daily								Extreme	
					January		April		July		October			
					Max.	Min.	Max.	Min.	Max.	Min.	Max.	Min.	Max.	Min.
United States (Conterminous):	′ ″	′ ″	Feet	Yrs.	°F	°F	°F	°F	°F	°F	°F	°F	°F	°F
Albuquerque, NM	35 03N	106 37W	5311	30	46	24	69	42	91	66	71	45	105	–17
Asheville, NC	35 26N	82 32W	2140	30	48	28	67	42	84	61	68	45	100	–16
Atlanta, GA	33 39N	84 26W	1010	30	52	37	70	50	87	71	72	52	105	– 8
Austin, TX	30 18N	97 42W	597	30	60	41	78	57	95	74	82	60	109	– 2
Birmingham, AL	33 34N	86 45W	620	30	57	36	76	50	93	71	79	52	106	–10
Bismarck, ND	46 46N	100 45W	1647	30	20	0	55	32	86	58	59	34	114	–45
Boise, ID	43 34N	116 13W	2838	30	36	22	63	37	91	59	65	38	112	–28
Brownsville, TX	25 54N	97 26W	16	30	71	52	82	66	93	76	85	67	106	12
Buffalo, NY	42 56N	78 44W	705	30	31	18	53	34	80	59	60	41	99	–21
Cheyenne, WY	41 09N	104 49W	6126	30	37	14	56	30	85	55	63	32	100	–38
Chicago, IL	41 47N	87 45W	607	30	33	19	57	41	84	67	63	47	105	–27
Des Moines, IA	41 32N	93 39W	938	30	29	11	59	38	87	65	66	43	110	–30
Dodge City, KS	37 46N	99 58W	2582	30	42	20	66	41	93	68	71	46	109	–26
El Paso, TX	31 48N	106 24W	3918	30	56	30	78	49	95	69	79	50	112	– 8
Indianapolis, IN	39 44N	86 17W	792	30	37	21	61	40	86	64	67	44	107	–25
Jacksonville, FL	30 25N	81 39W	20	30	67	45	80	58	92	73	80	62	105	7
Kansas City, MO	39 07N	94 36W	742	30	40	23	66	46	92	71	72	49	113	–23
Las Vegas, NV	36 05N	115 10W	2162	30	54	32	78	51	104	76	80	53	117	8
Los Angeles, CA	33 56N	118 23W	97	30	64	45	67	52	76	62	73	57	110	23
Louisville, KY	38 11N	85 44W	477	30	44	27	66	43	89	67	70	46	107	–20
Miami, FL	25 48N	80 16W	7	30	76	58	83	66	89	75	85	71	100	28
Minneapolis, MN	44 53N	93 13W	834	30	22	2	56	33	84	61	61	37	108	–34
Missoula, MT	46 55N	114 05W	3190	30	28	10	57	31	85	49	58	30	105	–33
Nashville, TN	36 07N	86 41W	590	30	49	31	71	48	91	70	76	49	107	–17
New Orleans, LA	29 59N	90 15W	3	30	64	45	78	58	91	73	80	61	102	7
New York, NY	40 47N	73 58W	132	30	40	27	60	43	85	68	66	50	106	–15
Oklahoma City, OK	35 24N	97 36W	1285	30	46	28	71	49	93	72	74	52	113	–17
Phoenix, AZ	33 26N	112 01W	1117	30	64	35	84	50	105	75	87	55	122	16
Pittsburgh, PA	40 27N	80 00W	747	30	40	25	63	42	85	65	65	45	103	–20
Portland, ME	43 39N	70 19W	47	30	32	12	53	32	80	57	60	37	103	–39
Portland, OR	45 36N	122 36W	21	30	44	33	62	42	79	56	63	45	107	– 3
Reno, NV	39 30N	119 47W	4404	30	45	16	65	31	89	46	69	29	106	–19
Salt Lake City, UT	40 46N	111 58W	4220	30	37	18	63	36	94	60	65	38	107	–30
San Francisco, CA	37 37N	122 23W	8	30	55	42	64	47	72	54	71	51	109	20
Sault Ste. Marie, MI	46 28N	84 22W	721	30	23	8	46	30	76	54	55	38	98	–37
Seattle, WA	47 27N	122 18W	400	30	44	33	58	40	76	54	60	44	100	0
Sheridan, WY	44 46N	106 58W	3964	30	34	9	56	31	87	56	62	33	106	–41
Spokane, WA	47 38N	117 32W	2356	30	31	19	59	36	86	55	60	38	108	–30
Washington, D.C.	38 51N	77 03W	14	30	44	30	66	46	87	69	68	50	104	–18
Wilmington, NC	34 16N	77 55W	28	30	58	37	74	51	89	71	76	55	104	0

Source: NOAA

Average Precipitation in the United States													
Country and Station	**Average Precipitation**												
United States (Conterminous):	**Record Length Years**	**Jan. In.**	**Feb. In.**	**Mar. In.**	**Apr. In.**	**May In.**	**Jun. In.**	**Jul. In.**	**Aug. In.**	**Sep. In.**	**Oct. In.**	**Nov. In.**	**Dec. In.**	**Year In.**
Albuquerque, NM	30	0.4	0.4	0.5	0.5	0.8	0.6	1.2	1.3	1.0	0.8	0.4	0.5	8.4
Asheville, NC	30	4.2	4.0	4.8	4.0	3.7	3.5	5.9	4.9	3.6	3.1	2.8	3.6	48.1
Atlanta, GA	30	4.4	4.5	5.4	4.5	3.2	3.8	4.7	3.6	3.3	2.4	3.0	4.4	47.2
Austin, TX	30	2.4	2.6	2.1	3.6	3.7	3.2	2.2	1.9	3.4	2.8	2.1	2.5	32.5
Birmingham, AL	30	5.0	5.3	6.0	4.5	3.4	4.0	5.2	4.9	3.3	3.0	3.5	5.0	53.1
Bismarck, ND	30	0.4	0.4	0.8	1.2	2.0	3.4	2.2	1.7	1.2	0.9	0.6	0.4	15.2
Boise, ID	30	1.3	1.3	1.3	1.2	1.3	0.9	0.2	0.2	0.4	0.8	1.2	1.3	11.4
Brownsville, TX	30	1.4	1.5	1.0	1.6	2.4	3.0	1.7	2.8	5.0	3.5	1.3	1.7	26.9
Buffalo, NY	30	2.8	2.7	3.2	3.0	3.0	2.5	2.6	3.1	3.1	3.0	3.6	3.0	35.6
Cheyenne, WY	30	0.5	0.6	1.2	1.9	2.5	2.1	1.8	1.4	1.1	0.8	0.6	0.5	15.0
Chicago, IL	30	1.9	1.6	2.7	3.0	3.7	4.1	3.4	3.2	2.7	2.8	2.2	1.9	33.2
Des Moines, IA	30	1.3	1.1	2.1	2.5	4.1	4.7	3.1	3.7	2.9	2.1	1.8	1.1	30.5
Dodge City, KS	30	0.6	0.7	1.2	1.8	3.2	3.0	2.3	2.4	1.5	1.4	0.6	0.5	19.2
El Paso, TX	30	0.5	0.4	0.4	0.3	0.4	0.7	1.3	1.2	1.1	0.9	0.3	0.5	8.0
Indianapolis, IN	30	3.1	2.3	3.4	3.7	4.0	4.6	3.5	3.0	3.2	2.6	3.1	2.7	39.2
Jacksonville, FL	30	2.5	2.9	3.5	3.6	3.5	6.3	7.7	6.9	7.6	5.2	1.7	2.2	53.6
Kansas City, MO	30	1.4	1.2	2.5	3.6	4.4	4.6	3.2	3.8	3.3	2.9	1.8	1.5	34.2
Las Vegas, NV	30	0.5	0.4	0.4	0.2	0.1	*	0.5	0.5	0.3	0.2	0.3	0.4	3.8
Los Angeles, CA	30	2.7	2.9	1.8	1.1	0.1	0.1	*	*	0.2	0.4	1.1	2.4	12.8
Louisville, KY	30	4.1	3.3	4.6	3.8	3.9	4.0	3.4	3.0	2.6	2.3	3.2	3.2	41.4
Miami, FL	30	2.0	1.9	2.3	3.9	6.4	7.4	6.8	7.0	9.5	8.2	2.8	1.7	59.9
Minneapolis, MN	30	0.7	0.8	1.5	1.9	3.2	4.0	3.3	3.2	2.4	1.6	1.4	0.9	24.9
Missoula, MT	30	0.9	0.9	0.7	1.0	1.9	1.9	0.9	0.7	1.0	1.0	0.9	1.1	12.9
Nashville, TN	30	5.5	4.5	5.2	3.7	3.7	3.3	3.7	2.9	2.9	2.3	3.3	4.2	45.2
New Orleans, LA	30	3.8	4.0	5.3	4.6	4.4	4.4	6.7	5.3	5.0	2.8	3.3	4.1	53.7
New York, NY	30	3.3	2.8	4.0	3.4	3.7	3.3	3.7	4.4	3.9	3.1	3.4	3.3	42.3
Oklahoma City, OK	30	1.3	1.4	2.0	3.1	5.2	4.5	2.4	2.5	3.0	2.5	1.6	1.4	30.9
Phoenix, AZ	30	0.7	0.9	0.7	0.3	0.1	0.1	0.8	1.1	0.7	0.5	0.5	0.9	7.3
Pittsburgh, PA	30	2.8	2.3	3.5	3.4	3.8	4.0	3.6	3.5	2.7	2.5	2.3	2.5	36.9
Portland, ME	30	4.4	3.8	4.3	3.7	3.4	3.2	2.9	2.4	3.5	3.2	4.2	3.9	42.9
Portland, OR	30	5.4	4.2	3.8	2.1	2.0	1.7	0.4	0.7	1.6	3.6	5.3	6.4	37.2
Reno, NV	30	1.2	1.0	0.7	0.5	0.5	0.4	0.3	0.2	0.2	0.5	0.6	1.1	7.2
Salt Lake City, UT	30	1.4	1.2	1.6	1.8	1.4	1.0	0.6	0.9	0.5	1.2	1.3	1.2	14.1
San Francisco, CA	30	4.0	3.5	2.7	1.3	0.5	0.1	*	*	0.2	0.7	1.6	4.1	18.7
Sault Ste. Marie, MI	30	2.1	1.5	1.8	2.2	2.8	3.3	2.5	2.9	3.8	2.8	3.3	2.3	31.3
Seattle, WA	30	5.7	4.2	3.8	2.4	1.7	1.6	0.8	1.0	2.1	4.0	5.4	6.3	39.0
Sheridan, WY	30	0.6	0.7	1.4	2.2	2.6	2.6	1.2	0.9	1.2	1.1	0.8	0.6	15.9
Spokane, WA	30	2.4	1.9	1.5	0.9	1.2	1.5	0.4	0.4	0.8	1.6	2.2	2.4	17.2
Washington, D.C.	30	3.0	2.5	3.2	3.2	4.1	3.2	4.2	4.9	3.8	3.1	2.8	2.8	40.8
Wilmington, NC	30	2.9	3.4	4.0	2.9	3.5	4.3	7.7	6.9	6.3	3.0	3.1	3.4	51.4

Source: NOAA

Average Snowfall in the United States														
Country and Station United States (Conterminous): Data through 1998	**Average Snowfall (includes ice pellets)**													
	Record Length	**Jan.**	**Feb.**	**Mar.**	**Apr.**	**May**	**Jun.**	**Jul.**	**Aug.**	**Sep.**	**Oct.**	**Nov.**	**Dec.**	**Year**
	Years	**In.**	**In.**	**In.**	**In.**	**In.**	**In.**	**In.**	**In.**	**In.**	**In.**	**In.**	**In.**	**In.**
Albuquerque, NM	59	2.5	2.1	1.8	0.6	0	T	T	T	T	0.1	1.2	2.6	10.9
Asheville, NC	34	5	4.3	2.8	0.6	T	T	T	T	0	T	0.7	2	15.4
Atlanta, GA	62	0.9	0.5	0.4	T	0	0	0	0	0	T	0	0.2	2
Austin, TX	57	0.5	0.3	T	T	T	0	0	0	0	0	0.1	T	0.9
Birmingham, AL	55	0.6	0.2	0.3	0.1	T	T	T	0	T	T	T	0.3	1.5
Bismarck, ND	59	7.6	7	8.6	4	0.9	T	T	T	0.2	1.8	7	7	44.1
Boise, ID	59	6.5	3.7	1.6	0.6	0.1	T	T	T	T	0.1	2.3	5.9	20.8
Brownsville, TX	59	T	T	T	0	0	0	0	T	0	0	T	T	T
Buffalo, NY	55	23.7	18	11.9	3.2	0.2	T	T	T	T	0.3	11.2	22.8	91.3
Cheyenne, WY	63	6.6	6.3	11.9	9.2	3.2	0.2	T	T	0.9	3.7	7.1	6.3	55.4
Chicago, IL	39	10.7	8.2	6.6	1.6	0.1	T	T	T	T	0.4	1.9	8.1	37.6
Des Moines, IA	57	8.3	7.2	6	1.8	0	T	T	0	T	0.3	3.1	6.7	33.4
Dodge City, KS	56	4.3	3.9	5	0.8	T	T	T	T	0	0.3	2.1	3.6	20
El Paso, TX	57	1.3	0.8	0.4	0.3	T	T	T	0	T	0	0.9	1.6	5.3
Indianapolis, IN	67	6.6	5.6	3.4	0.5	0	T	0	T	0	0.2	1.9	5.1	23.3
Jacksonville, FL	57	T	0	0	T	0	T	T	0	0	0	0	0	T
Kansas City, MO	64	5.7	4.4	3.4	0.8	T	T	T	0	T	0.1	1.2	4.4	20
Las Vegas, NV	48	0.9	0.1	0	T	0	0	0	T	0	T	0.1	0.1	1.2
Los Angeles, CA	62	T	T	T	0	0	0	0	0	0	0	0	T	T
Louisville, KY	51	5.4	4.6	3.3	0.1	T	T	T	0	0	0.1	1	2.1	16.6
Miami, FL	56	0	0	0	0	T	0	0	0	0	0	0	0	T
Minneapolis, MN	60	10.2	8.2	10.6	2.8	0.1	T	T	T	T	0.5	7.9	9.4	49.7
Missoula, MT	54	12.3	7.3	6	2.1	0.7	T	T	T	T	0.8	6.2	11.3	46.7
Nashville, TN	56	3.7	3	1.5	0	0	T	0	T	0	0	0.4	1.4	10
New Orleans, LA	50	0	0.1	T	T	T	0	0	0	0	0	T	0.1	0.2
New York, NY	130	7.5	8.6	5.1	0.9	T	0	T	0	0	0	0.9	5.4	28.4
Oklahoma City, OK	59	3.1	2.4	1.5	T	T	T	T	T	T	T	0.5	1.8	9.3
Phoenix, AZ	61	T	0	T	T	T	0	0	0	0	T	0	T	T
Pittsburgh, PA	46	11.7	9.2	8.7	1.7	0.1	T	T	T	T	0.4	3.5	8.2	43.5
Portland, ME	58	19.6	16.9	12.9	3	0.2	0	0	0	T	0.2	3.3	14.6	70.7
Portland, OR	55	3.2	1.1	0.4	T	0	T	0	T	T	0	0.4	1.4	6.5
Reno, NV	54	5.8	5.2	4.3	1.2	0.8	0	0	0	0	0.3	2.4	4.3	24.3
Salt Lake City, UT	70	13.8	10	9.4	4.9	0.6	T	T	T	0.1	1.3	6.8	11.7	58.6
San Francisco, CA	69	0	T	T	0	0	0	0	0	0	0	0	0	T
Sault Ste. Marie, MI	55	29	18.4	14.7	5.8	0.5	T	T	T	0.1	2.4	15.8	31.1	117.8
Seattle, WA	48	2.9	0.9	0.6	0	T	0	0	0	0	T	0.7	2.2	7.3
Sheridan, WY	58	11	10.4	12.6	9.9	2	0.1	T	0	1.3	4.6	9.2	10.9	72
Spokane, WA	51	15.6	7.5	3.9	0.6	0.1	T	0	0	T	0.4	6.3	14.6	49
Washington, D.C.	55	5.5	5.4	2.2	T	T	T	T	T	0	0	0.8	2.8	16.7
Wilmington, NC	47	0.4	0.5	0.4	T	T	T	T	0	0	0	T	0.6	1.9

Trace (T) is recorded for less than 0.05 inch of snowfall.

Last updated on 5/25/2000 by NRCC.

Source: The National Climatic Data Center/NOAA

Frost-Free Days in the United States			
City/State	**Last Frost Date**	**First Frost Date**	**No. of Frost-Free Days per Year**
Albany, NY	May 7	September 29	144 days
Albuquerque, NM	April 16	October 29	196 days
Atlanta, GA	March 13	November 12	243 days
Baltimore, MD	March 26	November 13	231 days
Birmingham, AL	March 29	November 6	221 days
Boise, ID	May 8	October 9	153 days
Boston, MA	April 6	November 10	217 days
Charleston, SC	March 11	November 20	253 days
Charlotte, NC	March 21	November 15	239 days
Cheyenne, WY	May 20	September 27	130 days
Chicago, IL	April 14	November 2	201 days
Columbus, OH	April 26	October 17	173 days
Dallas, TX	March 18	November 12	239 days
Denver, CO	May 3	October 8	157 days
Des Moines, IA	April 19	October 17	180 days
Detroit, MI	April 24	October 22	181 days
Duluth, MN	May 21	September 21	122 days
Fargo, ND	May 13	September 27	137 days
Fayetteville, AR	April 21	October 17	179 days
Helena, MT	May 18	September 18	122 days
Houston, TX	February 4	December 10	309 days
Indianapolis, IN	April 22	October 20	180 days
Jackson, MS	March 17	November 9	236 days
Jacksonville, FL	February 14	December 14	303 days
Las Vegas, NV	March 7	November 21	259 days
Lincoln, NB	March 13	November 13	180 days
Los Angeles, CA	None likely	None likely	365 days
Louisville, KY	April 1	November 7	220 days
Memphis, TN	March 23	November 7	228 days
Miami, FL	None	None	365 days
Milwaukee, WI	May 5	October 9	156 days
New Haven, CT	April 15	October 27	195 days
New Orleans, LA	February 20	December 5	288 days
New York, NY	April 1	November 11	233 days
Phoenix, AZ	February 5	December 15	308 days
Pittsburgh, PA	April 16	November 3	201 days
Portland, ME	May 10	September 30	143 days
Portland, OR	April 3	November 7	217 days
Richmond, VA	April 10	October 26	198 days
Salt Lake City, UT	April 12	November 1	203 days
San Francisco, CA	January 8	January 5	362 days
Seattle, WA	March 24	November 11	232 days
St. Louis, MO	April 3	November 6	217 days
Topeka, KS	April 21	October 14	175 days
Tulsa, OK	March 30	November 4	218 days
Washington, D.C.	April 10	October 31	203 days
Wichita, KS	April 13	October 23	193 days

Include the following important climatic factors in your description of the climate in your community:

- Average daily temperatures in the winter and summer.
- Record high and low temperatures in the winter and summer.
- Average monthly precipitation in the winter and summer.
- Average winter snowfall.
- Growing season (number of days between last spring frost and first autumn frost).

3. Inspect a topographic map of your town.
   a) What is the latitude of your town?
   b) What is the elevation of your school?
   c) What is the highest elevation in your town?
   d) What is the lowest elevation in your town?
   e) Which of the following physical features can be found in or fairly near your community: mountains, rivers, valleys, coasts, lakes, hills, plains, or deserts? Specify where they are in relation to your community.
   f) Describe some of the ways that the physical features of your community might influence the climate.

Part B: Physical Features and Climate in a Different Community

1. Select a community that is in a part of the United States that is very different from where you live. For example, if you live in the mountains, pick a community on the plains. If you live near an ocean, pick a community far from a large body of water.
   a) Record the community and the reason you chose that community in your *EarthComm* notebook.
2. Describe the climate in this community.
   a) Include information for the same climatic factors that you used to describe your own community.
3. Inspect a topographic map of this community.
   a) Describe the same physical features that you did for your community.
4. Compare the physical features and climates of the two communities.
   a) In what ways might the physical features influence climate in the two places?

Part C: Heating and Cooling of Land versus Water

1. How do the rates at which rock and soil heat and cool compare to the rate at which water heats and cools? How might this affect climate in your community?
   a) Write down your ideas about these two questions.
   b) Develop a hypothesis about the rate at which rock or soil heat and cool compared to the rate at which water heats and cools.
2. Using materials provided by your teacher, design an experiment to investigate the rates of cooling and heating of soil or rock and water. Note the variable that you are

manipulating (the independent variable), the variable that you are measuring (the dependent variable), and the controls within your experiment.

a) Record your design, variables, controls, and any safety concerns in your notebook.

3. When your teacher has approved your design, conduct your experiment.

4. In your *EarthComm* notebook, record your answers to the following questions:

a) Which material heated up faster?

b) Which material cooled more quickly?

c) How did your results compare with your hypothesis?

d) How does this investigation relate to differences in climate between places near a body of water, versus places far from water?

Have your teacher approve your design before you begin your experiment. Do not touch any heat source. Report any broken thermometers to your teacher. Clean up any spills immediately.

Reflecting on the Activity and the Challenge

In this activity, you learned about the physical features and climate of your community. You also compared the physical features and climate of your community to that of another community in the United States. This helped you begin to see ways in which physical features influence climate. This will help you explain the meaning of the term "climate" and describe the climate of your community in your newspaper article.

Digging Deeper

WEATHER AND CLIMATE

Factors Affecting Climate

Weather refers to the state of the atmosphere at a place, from day to day and from week to week. The weather on a particular day might be cold or hot, clear or rainy. **Climate** refers to the typical or average weather at a place, on a long-term average. For example, Alaska has a cold climate, but southern Florida has a tropical climate. Minnesota's climate is hot in the summer and cold in the winter. Western Oregon has very rainy winters. Each of these regions has a definite climate, but weather that varies from day to day, often unpredictably.

The climate of a particular place on Earth is influenced by several important factors: latitude, elevation, and nearby geographic features.

Geo Words

weather: the state of the atmosphere at a specific time and place.

climate: the general pattern of weather conditions for a region over a long period of time (at least 30 years).

Geo Words

latitude: a north-south measurement of position on the Earth. It is defined by the angle measured from the Earth's equatorial plane.

elevation: the height of the land surface relative to sea level.

glacier: a large long-lasting accumulation of snow and ice that develops on land and flows under its own weight.

windward: the upwind side or side directly influenced to the direction that the wind blows from; opposite of leeward.

leeward: the downwind side of an elevated area like a mountain; opposite of windward.

rain shadow: the reduction of precipitation commonly found on the leeward side of a mountain.

Latitude

Latitude is a measure of the distance of a point on the Earth from the Equator. It is expressed in degrees, from zero degrees at the Equator to 90° at the poles. The amount of solar energy an area receives depends upon its latitude. At low latitudes, near the Equator, the Sun is always nearly overhead in the middle of the day, all year round. Near the poles, the Sun is low in the sky even in summer, and in the winter it is nighttime 24 hours of the day. As a result, regions near the Equator are much warmer than regions near the poles. Assuming a constant elevation, temperatures decrease by an average of about one degree Fahrenheit for every three degrees latitude away from the Equator.

Elevation

Elevation, the height of a point on the Earth's surface above sea level, also affects the physical environment. Places at high elevations are generally cooler than places at low elevations in a given region. On average, temperatures decrease by about 3.6°F for every 1000 ft. (300 m) gain in elevation. In many places at high elevation, **glaciers** form because the summers are not warm enough to melt all of the snow each year. The mountains in Glacier National Park in Montana, which are located at high elevation as well as fairly high latitude, contain glaciers, as shown in *Figure 1*.

Figure 1 Mountains at high elevation and high latitudes often contain glaciers.

Geographic Features

Geographic features, like mountain ranges, lakes, and oceans, affect the climate of a region. As shown in *Figure 2*, mountains can have a dramatic effect on precipitation in nearby areas. The **windward** side of a mountain chain often receives much more rainfall than the leeward side. As wind approaches the mountains, it is forced upwards. When the air rises, it cools, and water vapor condenses into clouds, which produce precipitation. Conversely, the **leeward** side of a mountain range is in what is called a **rain shadow**. It often receives very little rain. That is because the air has already lost much of its moisture on the windward side. When the air descends the leeward slope of the mountain, it warms up as the greater air pressure compresses it. That causes clouds to evaporate and the humidity of the air to decrease. The deserts of the southwestern United States have low rainfall because they are in the rain shadow of the Sierra Nevada and other mountain ranges along the Pacific coast.

Figure 2 The rain shadow effect. Most North American deserts are influenced by this effect.

Large bodies of water can also affect climate dramatically. The ocean has a moderating effect on nearby communities. Temperatures in coastal communities vary less than inland communities at similar latitude. This is true both on a daily basis and seasonally. The effect is especially strong where the coast faces into the prevailing winds, as on the West Coast of the United States. Kansas City's average temperature is 79°F in July and 26°F in January. San Francisco's average temperature is 64°F in July and 49°F in January. On average, New York City's January temperatures vary only 11°F during a day, whereas Omaha's January temperatures vary 20°F during a day. In each of these two cases the difference between the two cities' climates is too great to be related to their latitude, which is only different by less than about 1.5°. Instead, the differences in climate in these two places are because water has a much higher **heat capacity** than soil and rock. That means that much more heat is needed to raise the temperature of water than to raise the

Geo Words

heat capacity: the quantity of heat energy required to increase the temperature of a material or system; typically referenced as the amount of heat energy required to generate a 1°C rise in the temperature of 1 g of a given material that is at atmospheric pressure and 20°C.

Geo Words

lake-effect snow: the snow that is precipitated when an air mass that has gained moisture by moving over a relatively warm water body is cooled as it passes over relatively cold land. This cooling triggers condensation of clouds and precipitation.

global climate: the mean climatic conditions over the surface of the Earth as determined by the averaging of a large number of observations spatially distributed throughout the entire region of the globe.

Little Ice Age: the time period from mid-1300s to the mid-1800s AD. During this period, global temperatures were at their coldest since the beginning of the Holocene.

temperature of soil or rock. In the same way, water cools much more slowly than soil and rock. Land areas warm up quickly during a sunny day and cool down quickly during clear nights. The ocean and large lakes, on the other hand, change their temperature very little from day to day. Because the ocean absorbs a lot of heat during the day and releases it at night, it prevents daytime temperatures in seaside communities from climbing very high and prevents nighttime temperatures from falling very low (unless the wind is blowing from the land to the ocean!). By the same token, oceans store heat during the summer and release it during the winter, keeping summers cooler and winters warmer than they would be otherwise.

Lake-effect snow is common in late autumn and early winter downwind of the Great Lakes in north–central United States. Cold winds blow across the still-warm lake water, accumulating moisture from the lake as they go. When they reach the cold land, the air is cooled, and the water precipitates out of the clouds as snow. The warm oceans also supply the moisture that feeds major rainstorms, not just along the coast but even far inland in the eastern and central United States.

Global Climate

Climates differ from one region to another, depending on latitude, elevation, and geographical features. However, the entire Earth has a climate, too. This is called **global climate**. It is usually expressed as the year-round average temperature of the entire surface of the Earth, although average rainfall is also an important part of global climate. Today, the average temperature on the surface of the Earth is about 60°F. But the Earth's climate has changed continually over geologic time. During the Mesozoic Era (245–65 million years ago), when the dinosaurs roamed the Earth, global climate was warmer than today. During the Pleistocene Epoch (1.6 million–10,000 years ago), when mastodons and cave people lived, global climate seesawed back and forth between cold glacial intervals and warmer interglacial intervals. During glacial intervals, huge sheets of glacier ice covered much of northern North America. Just a few hundred years ago, the climate was about 3°F cooler than now. The time period from about the mid-1300s to the mid-1800s is called the **Little Ice Age**, because temperatures were generally much colder than today, and glaciers in many parts of the world expanded. Global temperatures have gradually increased since then, as the Earth has been coming out of the Little Ice Age.

Over the next several activities, you will be looking at what causes these changes in global climate, and how human activity may be causing global climate change.

Check Your Understanding

1. What is the difference between weather and climate?
2. What is the difference between regional climate and global climate?
3. Compare the climate of a city along the Pacific coast with that of a city with a similar latitude but located inland.
4. Explain how there can be snow on the top of a mountain near the Equator.
5. What is the "Little Ice Age"?

Understanding and Applying What You Have Learned

1. What is the nearest body of water to your community? In what ways does it affect the physical environment of your community?
2. Identify one physical feature in or near your community and explain how it affects the climate. You may need to think regionally. Is there a mountain range, a large lake, or an ocean in your state?
3. How would changing one feature of the physical environment near your community affect the climate? Again, you may need to think on a regional scale. Name a feature and tell how the climate would be different if that feature changed. How would this change life in your community?
4. Can you think of any additional reasons to explain the differences in climate between your community and the other community you looked at in the **Investigate** section?

Preparing for the Chapter Challenge

1. Clip and read several newspaper articles about scientific topics.
2. Using a style of writing appropriate for a newspaper, write a few paragraphs in which you:
   - explain the term climate;
   - describe the climate of your community (including statistics about seasonal temperatures, rainfall, and snowfall);
   - explain what physical factors in your community or state combine to produce this climate;
   - explain the difference between regional climate and global climate, and
   - describe the Earth's global climate.

Inquiring Further

1. **Weather systems and the climate of your community**

   Investigate how weather systems crossing the United States affect the climate in your community. In many places, weather systems travel in fairly regular paths, leading to a somewhat predictable series of weather events.

2. **Jet stream**

   Do some research on the jet stream. How does its position affect the climate in your community?

Activity 2 Paleoclimates

Goals

In this activity you will:

- Understand the significance of growth rings in trees as indicators of environmental change.
- Understand the significance of ice cores from glaciers as indicators of environmental change.
- Investigate and understand the significance of geologic sediments as indicators of environmental change.
- Examine the significance of glacial sediments and landforms as evidence for climate change.
- Investigate and understand the significance of fossil pollen as evidence for climate change.

Think about It

The cross section of a tree trunk shows numerous rings.

- What do you think the light and dark rings represent?
- What might be the significance of the varying thicknesses of the rings?

What do you think? Record your ideas about these questions in your *EarthComm* notebook. Be prepared to discuss your responses with your small group and the class.

Investigate

Part A: Tree Rings

1. Examine the photo that shows tree growth rings from a Douglas fir. Notice the arrow that marks the growth ring that formed 550 years before the tree was cut.

   a) Where are the youngest and the oldest growth rings located?

   b) Not all the growth rings look identical. How are the rings on the outer part of the tree different from those closer to the center?

   c) Mark the place on a copy of the picture where the change in the tree rings occurs.

2. Using the 550-year arrow as a starting point, count the number of rings to the center of the tree. Now count the number of rings from the arrow to where you marked the change in the way the rings look.

   a) Record the numbers.

   b) Assuming that each ring represents one year, how old is the tree?

   c) Assume that the tree was cut down in the year 2000. What year did the tree rings begin to look different from the rings near the center (the rings older than about 550-years old)?

3. Compare the date you calculated in **Step 2 (c)** to the graph that shows change in temperature for the last 1000 years.

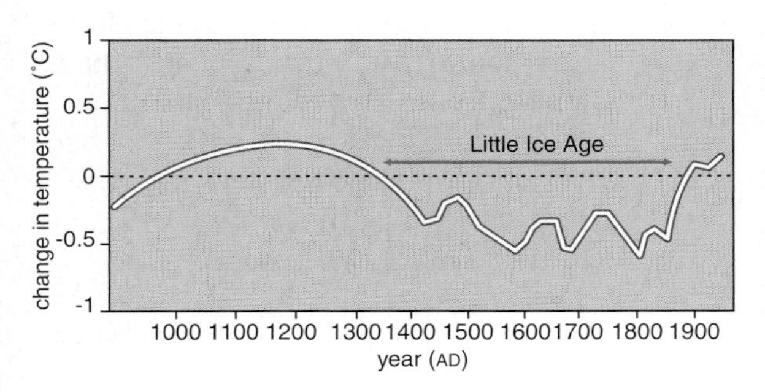

   a) What was happening to the climate during that time?

   b) From these observations, what would you hypothesize is the correlation between the thickness of tree rings and climate?

4. Examine the diagram that shows temperature change for the last 150,000 years.

a) How does the duration of the Little Ice Age compare to the duration of the ice age (the time between the two interglacial periods) shown in this figure?

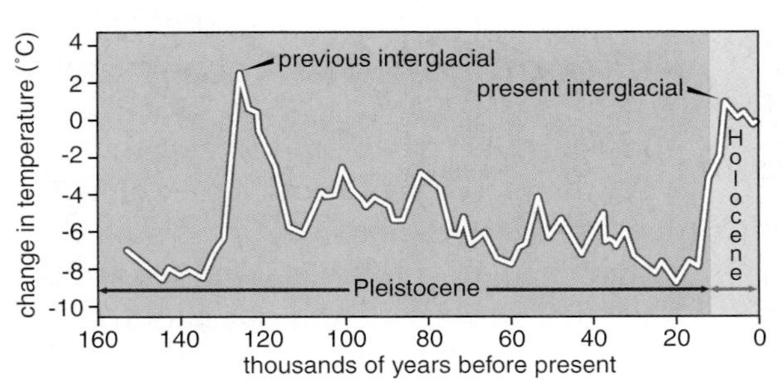

Part B: Fossil Pollen

1. Using blue, red, green, and yellow modeling clay, put down layers in a small container. You may put them down in any order and thickness. The container represents a lake or pond, and the clay represents sediment that has settled out over a long time.
2. When you have finished laying down your "sediments," use a small pipe to take a core. Push the pipe straight down through all of the layers. Then carefully pull the pipe back up. Use a thin stick to push the core of sediments out of the pipe.

Place folded paper towels or other padding between your hand and the upper end of the pipe before pushing it into the clay.

   a) Draw a picture of the core in your notebook. Note which end is the top.

   b) Measure and record the thickness of each layer of sediment to the nearest tenth of a centimeter.

3. The different colors of clay represent sediments that have settled out of the lake water at different times. For this exercise, imagine that each centimeter of clay represents the passage of 1000 years.

   a) How many years does your core represent from top to bottom?

   b) How many years does each layer represent?

4. Imagine that the different colors of clay represent the following:
   - Blue: sediments containing pollen from cold-climate plants like spruce and alder trees.
   - Red: sediments containing pollen from warm-climate plants, like oak trees and grasses.
   - Green: sediments containing mostly spruce and alder pollen, with a little oak and grass pollen.
   - Yellow: sediments containing mostly oak and grass pollen, with a little spruce and alder pollen.

   a) How do you think the pollen gets into the lake sediments?

   b) Describe what the climate around the lake was like when each layer of sediment was deposited.

   c) Write a paragraph describing the climate changes over the period of time represented by your core. Make sure you say at how many years before the present each climate change occurred. Note whether transitions from one type of climate to another appear to have happened slowly or quickly.

Reflecting on the Activity and the Challenge

You will need to explain some of the ways that geologists know about climates that existed in the geologic past. The activity helped you understand two of the ways that geologists find out about ancient climates. Geologists study tree growth rings. They relate the thickness of the tree rings to the climate. Geologists also collect cores from layers of sediments and study the kinds of pollen contained in the sediments. The pollen shows what kinds of plants lived there in the past, and that shows something about what the climate was like.

Digging Deeper

HOW GEOLOGISTS FIND OUT ABOUT PALEOCLIMATES

Direct Records and Proxies

A **paleoclimate** is a climate that existed sometime in the past: as recently as just a few centuries ago, or as long as billions of years ago. For example, in the previous activity you learned that the world was warmer in the Mesozoic Era (245 – 65 million years ago) and experienced periods of glaciation that affected large areas of the Northern Hemisphere continents during the Pleistocene Epoch (1.6 million–10,000 years ago). At present the Earth is experiencing an interglacial interval—a period of warmer climate following a colder, glacial period. The Earth today has only two continental ice sheets, one covering most of Greenland and one covering most of Antarctica.

The last retreat of continental glaciers occurred between about 20,000 years ago and 8000 years ago. That was before the invention of writing, so there is no direct record of this change. Systematic records of local weather, made with the help of accurate weather instruments, go back only about 200 years. A global network of weather stations has existed for an even shorter time. Historical accounts exist for individual places, most notably in China. For certain places in China records extend back 2000 years. They are useful, but more extensive information is required to understand the full range of climate variability. **Paleoclimatologists** use a variety of methods to infer past climate. Taken together, the evidence gives a picture of the Earth's climatic history.

Unfortunately, nothing gives a direct reading of past temperature. Many kinds of evidence, however, give an indirect record of past temperature. These are called **climate proxies**. Something that represents something else indirectly is called a proxy. In some elections, a voter can choose another person to cast the vote, and that vote is called a proxy. There are many proxies for past climate, although none is perfect.

Fossil Pollen

Pollen consists of tiny particles that are produced in flowers to make seeds. Pollen is often preserved in the sediments of lakes or bogs, where it is blown in by the wind. For example, a layer of sediment may contain a lot of pollen from spruce trees, which grow in cold climates. From that you can infer that the climate around the lake was cold when that layer of sediment was being deposited. Geologists collect sediment from a succession of

Geo Words

paleoclimate: the climatic conditions in the geological past reconstructed from a direct or indirect data source.

paleoclimatologist: a scientist who studies the Earth's past climate.

climate proxy: any feature or set of data that has a predictable relationship to climatic factors and can therefore be used to indirectly measure those factors.

sediment layers. They count the number of pollen grains from different plants in each layer. Then they make charts that can give an idea of the climate changes that have taken place. (See *Figure 1*.) Pollen is easy to study because there is so much of it. Geologists also study fossil plants and insects to reconstruct past climates.

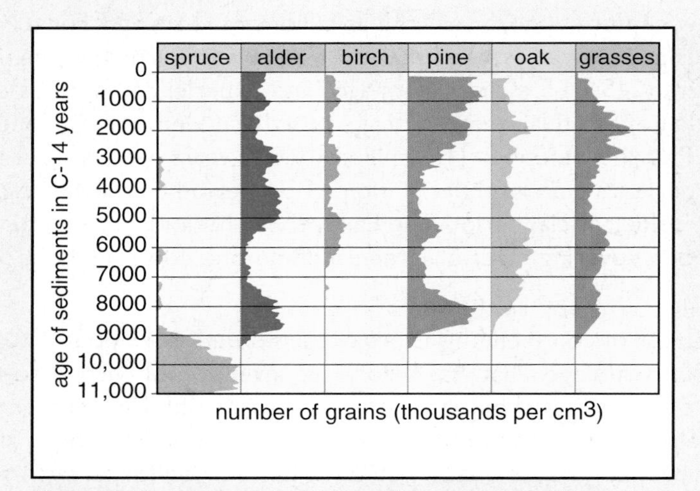

Figure 1 Changes over time in the relative amounts of different types of pollen from various trees and grasses give clues as to how climate has changed in the past.

Ice Cores

Figure 2 Scientists are able to obtain clues into past climatic conditions from air bubbles trapped in ice cores.

In recent years, study of cores drilled deep down into glaciers, like the one shown in *Figure 2*, has become a very powerful technique for studying paleoclimate. Very long cores, about 10 cm (4 in.) in diameter, have been obtained from both the Greenland and Antarctic ice sheets. The longest, from Antarctica, is almost 3400 m (about 1.8 mi.) long. Ice cores have been retrieved from high mountain glaciers in South America and Asia.

Glaciers consist of snow that accumulates each winter and does not melt entirely during the following summer. The snow is gradually compressed into ice as it is buried by later snow. The annual layers can be detected by slight changes in dust content. The long core from Antarctica provides a record of climate that goes back for more than 400,000 years. See *Figure 3*.

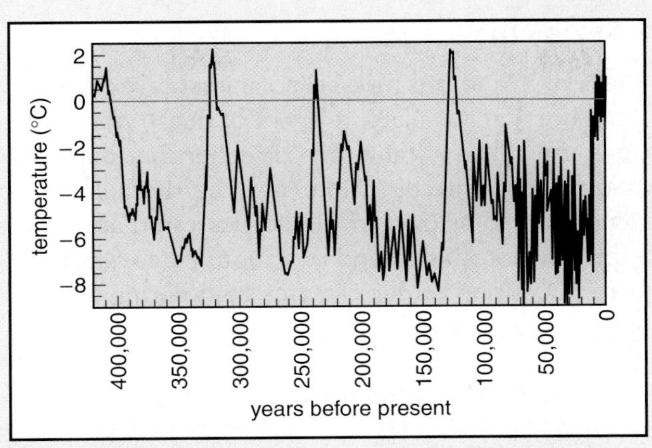

Figure 3 Temperature variation over the past 420,000 years, relative to the modern surface temperature at Vostok (−55.5°C).

Geo Words

isotope: one of two or more kinds of atoms of a given chemical element that differ in mass because of different numbers of neutrons in the nucleus of the atoms.

Bubbles of air trapped in the ice contain samples of the atmosphere from the time when the snow fell. Paleoclimatologists study the oxygen in the water molecules in the ice. Almost all of the oxygen atoms in the atmosphere are in two forms, called **isotopes**. The two isotopes are oxygen-16 (written ^{16}O) and oxygen-18 (written ^{18}O). They are the same chemically, but they have slightly different weights. ^{18}O is slightly heavier than ^{16}O. The proportion of these two isotopes in snow depends on average global temperatures. Snow that falls during periods of warmer global climate contains a greater proportion of ^{18}O, and snow that falls during periods of colder global climate contains a smaller proportion. The ratio of ^{18}O to ^{16}O can be measured very accurately with special instruments. Another important way of using the glacier ice to estimate global temperature is to measure the proportions of the two naturally occurring isotopes of hydrogen: ^{1}H, and ^{2}H (which is called deuterium).

The air bubbles in the ice contain carbon dioxide. The amount of carbon dioxide in the glacier ice air bubbles depends on the amount of carbon dioxide in the air at that time. The amount of carbon dioxide in the atmosphere can be correlated to global temperatures. During times when the paleoclimate is thought to have been warm, the ice core record shows relatively higher levels of atmospheric carbon dioxide compared to times of interpreted colder climate. Measurements of carbon dioxide taken from the cores give a global picture, because carbon dioxide is uniformly distributed in the global atmosphere.

A third component of the ice that yields clues to paleoclimates is dust. During colder climates, winds tend to be stronger. The stronger winds erode more dust, and the dust is deposited in small quantities over large areas of the Earth.

Geo Words

foraminifera: an order of single-celled organisms (protozoans) that live in marine (usually) and freshwater (rarely) environments. Forams typically have a shell of one or more chambers that is typically made of calcium carbonate.

loess: the deposits of wind-blown silt laid down over vast areas of the mid-latitudes during glacial and postglacial times.

Deep-Sea Sediments

Sand-size shells of a kind of single-celled animal called **foraminifera** ("forams," for short) accumulate in layers of ocean-bottom sediment. During warm climates, the shells spiral in one direction, but during cold climates, the shells spiral in the opposite direction. Also, the shells consist of calcium carbonate, which contains oxygen. Geologists can measure the proportions of the two oxygen isotopes to find out about paleoclimates. The shells contain more ^{18}O during colder climates than during warmer climates.

Glacial Landforms and Sediments

Glaciers leave recognizable evidence in the geologic record. Glacial landforms are common in northern North America. Glaciers erode the rock beneath, and then carry the sediment and deposit it to form distinctive landforms. Cape Cod, in Massachusetts, and Long Island, in New York, are examples of long ridges of sediment deposited by glaciers. Similar deposits are found as far south as Missouri.

Fine glacial sediment is picked up by the wind and deposited over large areas as a sediment called **loess.** There are thick deposits of loess in central North America. The loess layers reveal several intervals of glaciation during the Pleistocene Epoch.

Glaciers also leave evidence in the ocean. When glaciers break off into the ocean, icebergs float out to sea. As the icebergs melt, glacial sediment in the icebergs rains down to the ocean bottom. The glacial sediment is easily recognized because it is much coarser than other ocean-bottom sediment.

Tree Rings

Paleoclimate is also recorded in the annual growth rings in trees. Trees grow more during warm years than during cold years. A drawback to tree rings is that few tree species live long enough to provide a look very far in the past. Bristlecone pines, which can live as long as 5000 years, and giant sequoias, which are also very long-lived, are most often used.

Figure 4 A glacier carries ground-up pieces of rock and sediment into lakes and oceans.

Check Your Understanding

1. How do preserved tree rings indicate changes in climate?
2. List three ways that sediments in the ocean help scientists understand ancient climates.
3. Imagine an ice core taken from the Antarctic ice sheet. A layer of ice called "A" is 100 m below the surface. A layer of ice called "B" is 50 m below the surface. Explain why layer "A" represents the atmospheric conditions of an older climate than layer "B."

Understanding and Applying What You Have Learned

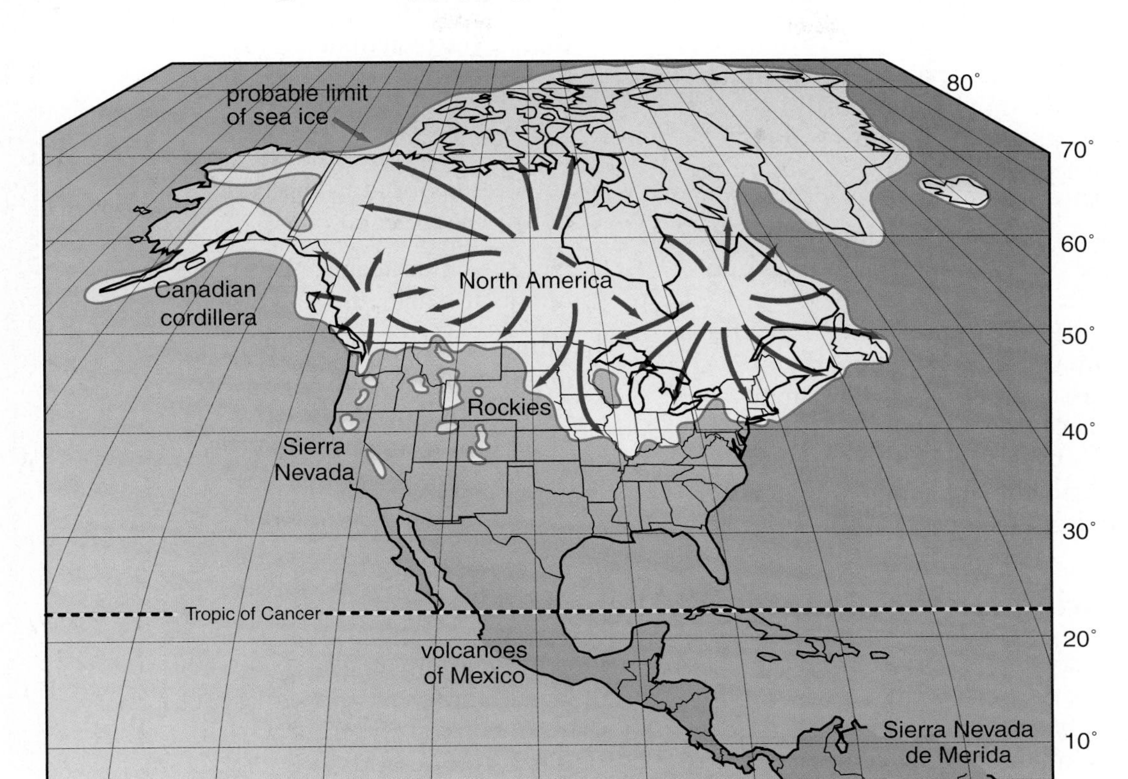

The approximate extent of the ice sheets in North America during the last Pleistocene glaciation.

1. Using evidence from glacial landforms and sediments, geologists have pieced together the maximum advance of glaciers about 18,000 years ago, which is shown in the figure above.
   a) Was your community located under ice during this time period? If not, describe how far your part of the country was from the ice sheet.
   b) What do you think the climate was like in your part of the country during the glacial maximum?

2. Use colored pencils and a ruler to draw a hypothetical series of lake-bottom sediments representing the sequence of climates given on the following page. Use the same colors as you used in the **Investigate** section to represent layers containing different kinds of pollen. Again, assume that it takes 1000 years to deposit 1 cm of sediment.

- 7000–5000 years ago: Warm climate supporting grasses and oaks.
- 5000–3000 years ago: Moderate climate supporting mostly grasses and oaks, with some spruce and alder.
- 3000–2000 years ago: Colder climate supporting mostly spruce and alder, with some grasses and oaks.
- 2000–1500 years ago: Moderate climate supporting mostly grasses and oaks, with some spruce and alder.
- 1500 years ago to the present: Warm climate supporting grasses and oaks.

a) Describe what the climate around the lake was like when each layer of sediment was deposited.

b) From this data, what time period marks the coldest climate recorded in the lake bottom sediments?

c) Did the climate cool at the same rate as it warmed?

d) Think of a hypothesis that might explain your answer to **Part (c)**. What additional observations might help you test this hypothesis?

Preparing for the Chapter Challenge

Using a style appropriate for a newspaper article, write a paragraph or two about each of the following topics, explaining how geologists use them to find out about paleoclimates:

- deep-sea sediments;
- glacial landforms and sediments;
- ice cores from Antarctica;
- pollen studies, and
- tree rings.

Inquiring Further

1. **GISP2 (Greenland Ice Sheet Project)**

   Research GISP2 (Greenland Ice Sheet Project), a project that is collecting and analyzing ice cores. What are the most recent discoveries? How many ice cores have been collected? How many have been analyzed? How far into the past does the data currently reach? Visit the *EarthComm* web site to help you start your research.

2. **Dating deep-sea sediments**

   Investigate some of the techniques geologists use to date deep-sea sediments: carbon-14 dating, isotope dating of uranium, fission-track dating of ash layers, and geomagnetic-stratigraphy dating.

3. **Paleoclimate research in your community**

   Investigate whether any of the paleoclimate techniques discussed in this activity have been used near your community or in your state to research paleoclimates.

Activity 3 How Do Earth's Orbital Variations Affect Climate?

Goals

In this activity you will:

- Understand that Earth has an axial tilt of about 23 1/2°.
- Use a globe to model the seasons on Earth.
- Investigate and understand the cause of the seasons in relation to the axial tilt of the Earth.
- Understand that the shape of the Earth's orbit around the Sun is an ellipse and that this shape influences climate.
- Understand that insolation to the Earth varies as the inverse square of the distance to the Sun.

Think about It

When it is winter in New York, it is summer in Australia.

- Why are the seasons reversed in the Northern and Southern Hemispheres?

What do you think? Write your thoughts in your *EarthComm* notebook. Be prepared to discuss your responses with your small group and the class.

Investigate

Part A: What Causes the Seasons? An Experiment on Paper

1. In your notebook, draw a circle about 10 cm in diameter in the center of your page. This circle represents the Earth.

   Add the Earth's axis of rotation, the Equator, and lines of latitude, as shown in the diagram and described below. Label the Northern and Southern Hemispheres.

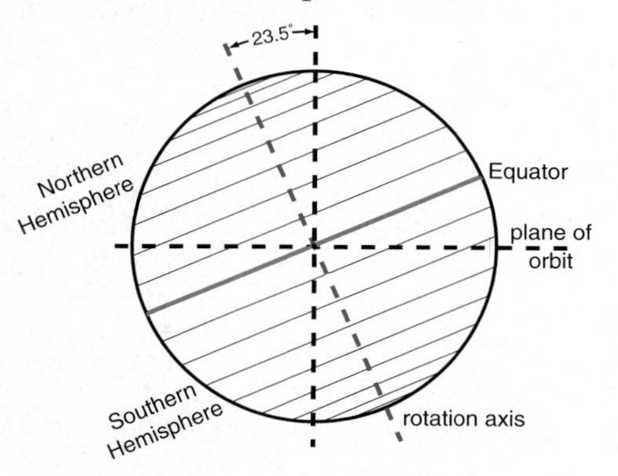

   Put a dot in the center of the circle. Draw a dashed line that goes directly up and down from the center dot to the edge of the circle. Use a protractor to measure 23 1/2° from this vertical dashed line. Use a blue pen or pencil to draw a line through the center of the "Earth" at 23 1/2° to your dashed line. This blue line represents the Earth's axis of rotation. Use your protractor to draw a red line that is perpendicular to the axis and passes through the center dot. This red line represents the Equator of the Earth. Label the Northern and Southern Hemispheres. Next you need to add lines of latitude. To do this, line your protractor up with the dot in the center of your circle so that it is parallel with the Equator. Now, mark off 10° increments starting from the Equator and going to the poles. You should have eight marks between the Equator and pole for each quadrant of the Earth. Use a straight edge to draw black lines that connect the marks opposite one another on the circle, making lines that are parallel to the Equator. This will give you lines of latitude in 10° increments so you can locate your latitude fairly accurately. Note that the lines won't be evenly spaced from one another because latitude is measured as an angle from the center of the Earth, not a linear distance.

2. Imagine that the Sun is directly on the left in your drawing. Draw horizontal arrows to represent incoming Sun rays from the left side of the paper.

3. Assume that it is noon in your community. Draw a dot where your community's latitude line intersects the perimeter of the circle on the left. This dot represents your community.

   a) Explain why this represents noon.

   b) At any given latitude, both north and south, are the Sun's rays striking the Northern Hemisphere or the Southern Hemisphere at a larger angle relative to the plane of orbit?

   c) Which do you think would be warmer in this drawing—the Northern Hemisphere or the Southern Hemisphere? Write down your hypothesis. Be sure to give a reason for your prediction.

   d) What season do you think this is in the Northern Hemisphere?

4. Now consider what happens six months later. The Earth is on the opposite side of its orbit, and the sunlight is now coming from the right side of the paper. Draw horizontal arrows to represent incoming Sun's rays from the right.
5. Again, assume that it is noon. Draw a dot where your community's latitude line intersects the perimeter of the circle on the right.
   a) Explain why the dot represents your community at noon.
   b) Are the Sun's rays striking the Northern Hemisphere or the Southern Hemisphere more directly?
   c) Which do you think would be warmer in this drawing—the Northern Hemisphere or the Southern Hemisphere? Why?
   d) What season do you think this is in the Northern Hemisphere?

Part B: What Causes the Seasons? An Experiment with a Globe

1. Test the hypothesis you made in **Part A** about which hemisphere would be warmer in which configuration. Find your city on a globe. Using duct tape, tape a small thermometer on it. The duct tape should cover the thermometer bulb, and the thermometer should be over the city. With a permanent black marking pen, color the duct tape black all over its surface.
2. Set up a light and a globe as shown.

Use only alcohol thermometers. Place a soft cloth on the table under the thermometer in case it falls off. Be careful not to touch the hot lamp.

3. Position the globe so that its axis is tilted 23 1/2° from the vertical, and the North Pole is pointed in the direction of the light source.
4. Turn on the light source.
   a) Record the initial temperature. Then record the temperature on the thermometer every minute until the temperature stops changing.
5. Now position the globe so that the axis is again tilted 23 1/2° from the vertical but the North Pole is pointing away from the light source. Make sure the light source is the same distance from the globe as it was in **Step 4**. Turn on the light source again.
   a) Record the temperature every minute until the temperature stops changing.
6. Use your observations to answer the following questions in your notebook:
   a) What is the difference in the average temperature when the North Pole was pointing toward your "community" and when it was pointing away?
   b) What caused the difference in temperature?

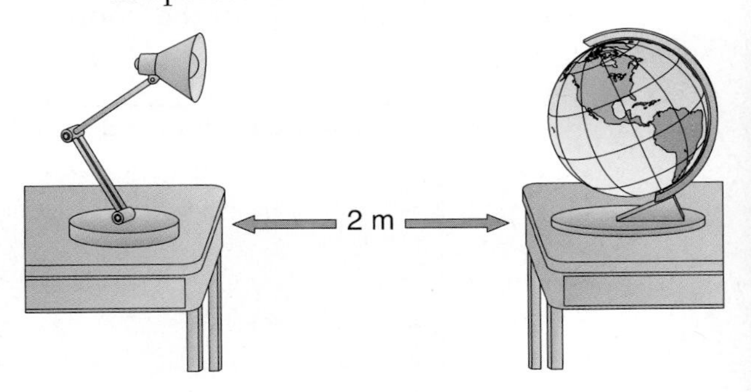

Part C: What Would Happen if the Earth's Axial Tilt Changed? An Experiment on Paper

1. Repeat the experiment you did in **Part A,** except this time use an axial tilt of 10°.

   a) Compared to an axial tilt of 23 1/2°, would your hemisphere experience a warmer or colder winter?

   b) Compared to an axial tilt of 23 1/2°, would your hemisphere experience a hotter or cooler summer?

2. Repeat the experiment you did in **Part A**, except this time use an axial tilt of 35°.

   a) Compared to an axial tilt of 23 1/2°, would your hemisphere experience a warmer or colder winter?

   b) Compared to an axial tilt of 23 1/2°, would your hemisphere experience a hotter or cooler summer?

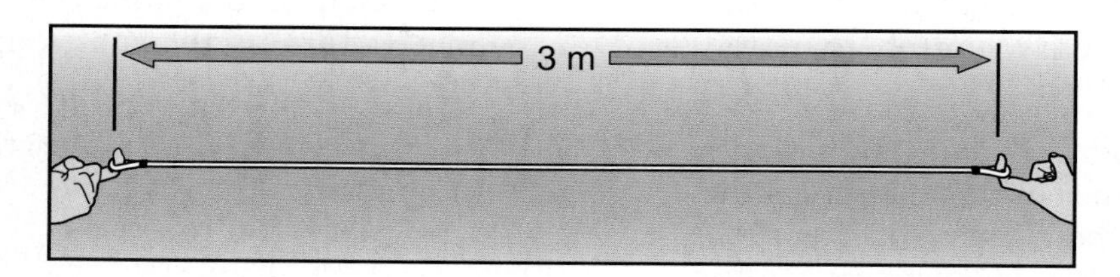

Part D: The Earth's Elliptical Orbit around the Sun

1. Tie small loops in each end of a rope, as shown in the diagram above.

2. Pick a point in about the middle of the floor, and put the two loops together over the point. Put a dowel vertically through the loops, and press the dowel tightly to the floor.

3. Stretch out the rope from the dowel until it is tight, and hold a piece of chalk at the bend in the rope, as shown in the diagram below. While holding the chalk tight against the rope, move the chalk around the dowel.

   a) What type of figure have you constructed?

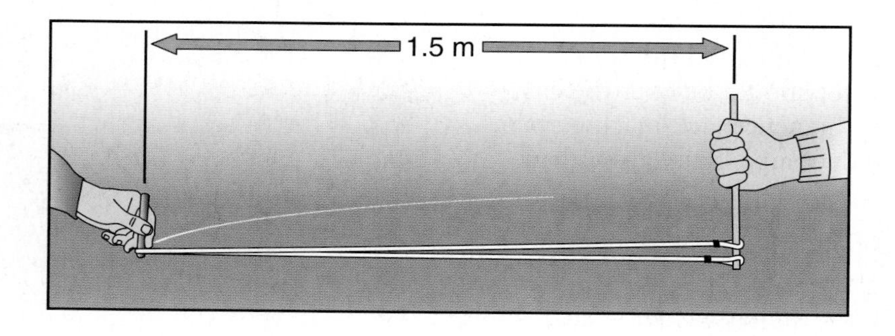

4. Draw a straight line from one edge of the circle that you just made to the opposite edge, through the center of the circle. This line represents the diameter of the circle. Mark two points along the diameter, each a distance of 20 cm from the center of the circle. Put the loops of the rope over the two points, hold them in place with two dowels, and use the chalk to draw a curve on one side of the straight line. Move the chalk to the other side of the line and draw another curve.

   a) What type of figure have you constructed?

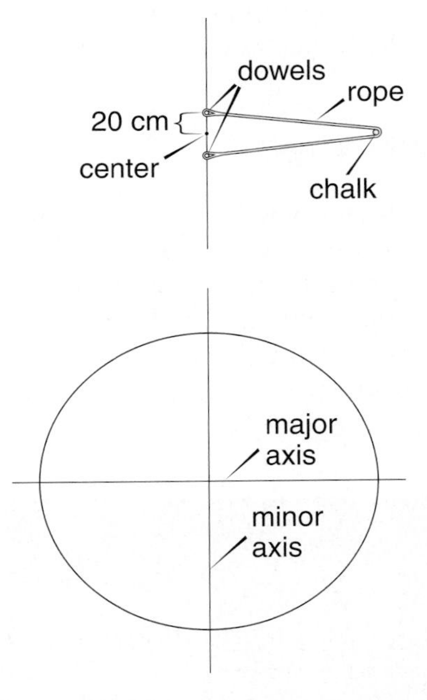

5. Using different colors of chalk, make a few more curved figures in the same way. Choose sets of dowel points that are farther and farther away from the center.

   a) Describe the shapes of the figures you constructed. Make sketches in your *EarthComm* notebook.

   b) What would be the shape of the curve when the dowel points are spaced a distance apart that is just equal to the length of the rope between the two loops?

Part E: How Energy from the Sun Varies with Distance from the Sun

1. Using scissors and a ruler, cut out a square 10 cm on a side in the middle of a poster board.
2. Hold the poster board vertically, parallel to the wall and exactly two meters from it.
3. Position a light bulb along the imaginary horizontal line that passes from the wall through the center of the hole in the poster board. See the diagram.

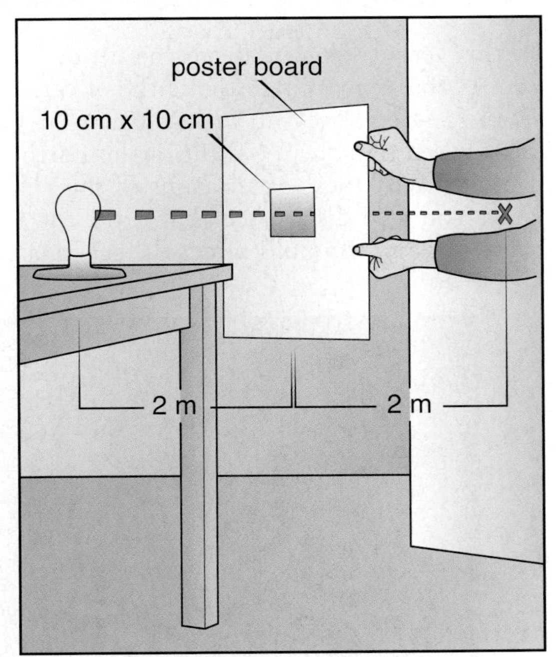

Avoid contact with the hot light bulb. Do not look directly into the light.

4. Turn on the light bulb, and turn off the lights in the room. If the room is not dark enough to see the image on the wall, close any curtains or shades, or cover the windows with dark sheets or blankets.
5. With the chalk, trace the edge of the image the hole makes on the wall.
   a) Measure and record the length of the sides of the image you marked with the chalk.
   b) Divide the length of the image on the wall by the length of the sides of the square in the poster board. Now divide the distance of the light bulb from the wall by the distance of the poster board from the wall. What is the relationship between the two numbers you obtain?
   c) Compute the area of the image on the wall, and compute the area of the square hole in the poster board. Divide the area of the image by the area of the hole. Again, divide the distance of the light bulb from the wall by the distance of the poster board from the wall. What is the relationship between the two numbers you obtain?
6. Repeat **Part E** for other distances.
   a) What do you notice about the relationship between the area of the image and the area of the hole?

Reflecting on the Activity and the Challenge

In this activity you modeled the tilt of the Earth's axis to investigate the effect of the angle of the Sun's rays. You discovered that the axial tilt of the Earth explains why there are seasons of the year. You also discovered that if the tilt were to vary, it would affect the seasons. You also modeled the Earth's elliptical orbit around the Sun. This will help you to understand one of the main theories for explaining why the Earth's climate varies over time. You will need to explain this in your newspaper articles.

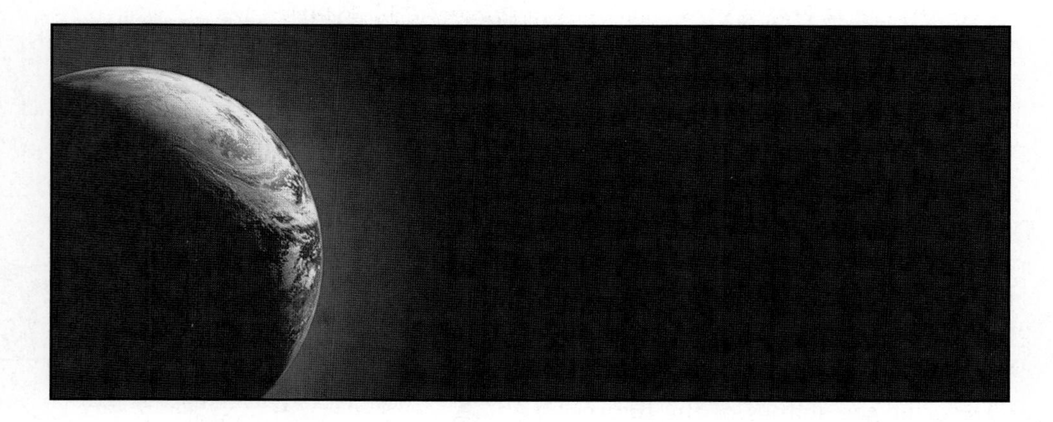

Digging Deeper

THE EARTH'S ORBIT AND THE CLIMATE

The Earth's Axial Tilt and the Seasons

The Earth's axis of rotation is tilted at about 23 1/2° away from a line that is perpendicular to the plane of the Earth's orbit around the Sun (*Figure 1*). This tilt explains the seasons on Earth. During the Northern Hemisphere summer, the North Pole is tilted toward the Sun, so the Sun shines at a high angle overhead. That is when the days are warmest, and days are longer than nights. On the summer solstice (on or about June 22) the Northern Hemisphere experiences its longest day and shortest night of the year. During the Northern Hemisphere winter the Earth is on the other side of its orbit. Then the North Pole is tilted away from the Sun, and the Sun shines at a lower angle. Temperatures are lower, and the days are shorter than the nights. On the winter solstice (on or about December 22) the Northern Hemisphere experiences its shortest day and longest night of the year.

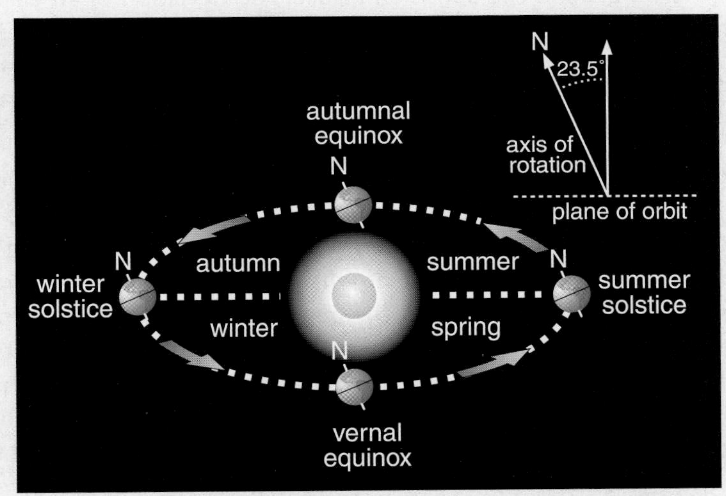

Figure 1 The tilt of the Earth's axis explains the seasons. Note that the Earth and Sun are not shown to scale.

How Do Earth's Orbital Variations Affect Climate?

In **Activity 2** you saw that paleoclimatologists have developed a good picture of the Earth's climatic history. The advances and retreats of the continental ice sheets are well documented. Nevertheless, questions remain.

Why does Earth's climate sometimes become cold enough for ice sheets to advance? Why does the climate later warm up and cause ice sheets to retreat? The answers to these questions are not yet entirely clear. Most climatologists believe that variations in the geometry of the Earth's orbit around the Sun are the major cause of the large variations in climate. These variations have caused the advance and retreat of ice sheets in the past couple of million years.

If the Earth and the Sun were the only bodies in the solar system, the geometry of the Earth's orbit around the Sun and the tilt of the Earth's axis would stay exactly the same through time. But there are eight other known planets in the solar system. Each of those planets exerts forces on the Earth and the Sun. Those forces cause the Earth's orbit to vary with time. The Moon also plays a role. The changes are slight but very important. There are three kinds of changes: **eccentricity**, **obliquity**, and **precession**. These three things are called the Earth's **orbital parameters**.

Eccentricity

The Earth's orbit around the Sun is an ellipse. The deviation of an ellipse from being circular is called its eccentricity. A circle is an ellipse with zero eccentricity. As the ellipse becomes more and more elongated (with a larger major diameter and a smaller minor diameter), the eccentricity increases. The Earth's orbit has only a slight eccentricity.

Even though the eccentricity of the Earth's orbit is very small, the distance from the Earth to the Sun varies by about 3.3% through the year. The difference in **insolation** is even greater. The word insolation (nothing to do with insUlation!) is used for the rate at which the Sun's energy reaches the Earth, per unit area facing directly at the Sun. The seasonal variation in insolation is because of what is called the **inverse-square law**. What you found in **Part E** of the investigation demonstrates this. The area of the image on the wall was four times the area of the hole, even though the distance of the wall from the bulb was only twice the distance of the hole from the bulb. Because of the inverse-square law, the insolation received by the Earth varies by almost 7° between positions on its orbit farthest from the Sun and positions closest to the Sun.

Because of the pull of other planets on the Earth–Sun system, the eccentricity of the Earth's orbit changes with time. The largest part of the change in eccentricity has a period of about 100,000 years. That means that one full cycle of increase and then decrease in eccentricity takes 100,000 years. During that time, the difference in insolation between the date of

Geo Words

eccentricity: the ratio of the distance between the foci and the length of the major axis of an ellipse.

obliquity: the tilt of the Earth's rotation axis as measured from the perpendicular to the plane of the Earth's orbit around the Sun. The angle of this tilt varies from 22.5° to 24.5° over a 41,000-year period. Current obliquity is 23.5°.

precession: slow motion of the axis of the Earth around a cone, one cycle in about 26,000 years, due to gravitational tugs by the Sun, Moon, and major planets.

orbital parameters: any one of a number of factors that describe the orientation and/or movement of an orbiting body or the shape of its orbital path.

insolation: the direct or diffused shortwave solar radiation that is received in the Earth's atmosphere or at its surface.

inverse-square law: a scientific law that states that the amount of radiation passing through a specific area is inversely proportional to the square of the distance of that area from the energy source.

the shortest distance to the Sun and the date of the farthest distance to the Sun ranges from about 2° (less than now) to almost 20° (much greater than now!).

The two points you used to make ellipses in **Part D** of the investigation are called the foci of the ellipse. (Pronounced "FOH-sigh". The singular is focus.) The Sun is located at one of the foci of the Earth's elliptical orbit. The Earth is closest to the Sun when it is on the side of that focus, and it is farthest from the Sun when it is on the opposite side of the orbit. (See *Figure* 2.) Does it surprise you to learn that nowadays the Earth is closest to the Sun on January 5 (called **perihelion**) and farthest from the Sun on July 5 (called **aphelion**)? That tends to make winters less cold and summers less hot, in the Northern Hemisphere.

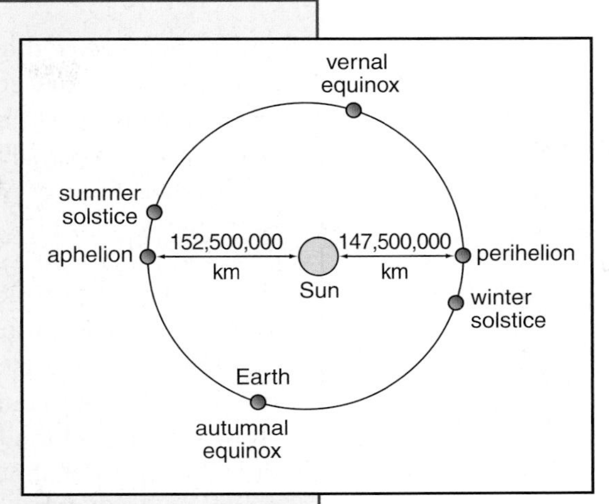

Figure 2 Schematic diagram showing occurrence of the aphelion and perihelion.

Obliquity

The tilt of the Earth's axis relative to the plane of the Earth's orbit is called the obliquity. The axis is oblique to the plane rather than perpendicular to it. In the investigation you discovered that a change in the obliquity would cause a change in the nature of the seasons. For example, a smaller obliquity would mean warmer winters and cooler summers in the Northern Hemisphere. This might result in more moisture in the winter air, which would mean more snow. In cooler summers, less of the snow would melt. You can see how this might lead to the buildup of glaciers.

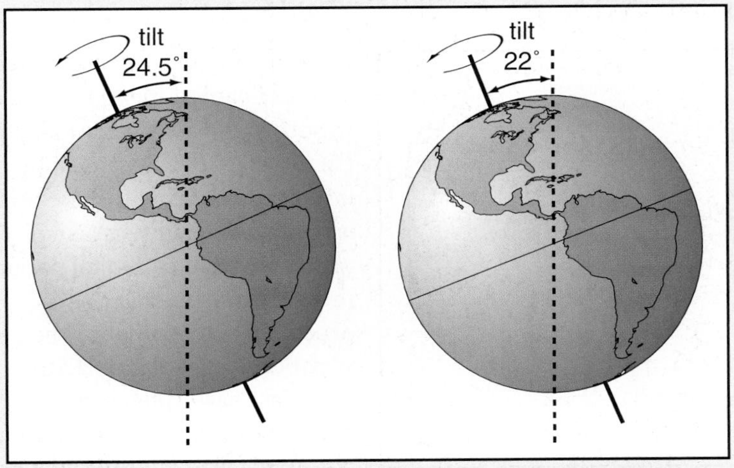

Figure 3 The angle of tilt of the Earth's axis varies between about 24.5° and 22°, causing climatic variations over time.

Geo Words

perihelion: the point in the Earth's orbit that is closest to the Sun. Currently, the Earth reaches perihelion in early January.

aphelion: the point in the Earth's orbit that is farthest from the Sun. Currently, the Earth reaches aphelion in early July.

Geo Words

Milankovitch cycles: the cyclical changes in the geometric relationship between the Earth and the Sun that cause variations in solar radiation received at the Earth's surface.

axial precession: the wobble in the Earth's polar axis.

orbital precession: rotation about the Sun of the major axis of the Earth's elliptical orbit.

Again, because of the varying pull of the other planets on the Earth–Sun system, the Earth's obliquity changes over a period of about 40,000 years. The maximum angle of tilt is about 24 1/2°, and the minimum angle is about 22°. At times of maximum tilt, seasonal differences in temperature are slightly greater. At times of minimum tilt angle, seasonal differences are slightly less.

Precession

Have you ever noticed how the axis of a spinning top sometimes wobbles slowly as it is spinning? It happens when the axis of the top is not straight up and down, so that gravity exerts a sideways force on the top. The same thing happens with the Earth. The gravitational pull of the Sun, Moon, and other planets causes a slow wobbling of the Earth's axis. This is called the Earth's **axial precession**, and it has a period of about 26,000 years. That's the time it takes the Earth's axis to make one complete revolution of its wobble.

There is also another important kind of precession related to the Earth. It is the precession of the Earth's orbit, called **orbital precession.** As the Earth moves around the Sun in its elliptical orbit, the major axis of the Earth's orbital ellipse is rotating about the Sun. In other words, the orbit itself rotates around the Sun! The importance of the two precession cycles for the Earth's climate lies in how they interact with the eccentricity of the Earth's orbit. This interaction controls how far the Earth is from the Sun during the different seasons. Nowadays, the Northern Hemisphere winter solstice is at almost the same time as perihelion. In about 11,000 years, however, the winter solstice will be at about the same time as aphelion. That will make Northern Hemisphere winters even colder, and summers even hotter, than today.

Milankovitch Cycles

Early in the 20th century a Serbian scientist named Milutin Milankovitch hypothesized that variations in the Earth's climate are caused by how insolation varies with time and with latitude. He used what is known about the Earth's orbital parameters (eccentricity, obliquity, and precession) to compute the variations in insolation. Later scientists have refined the computations. These insolation cycles are now called **Milankovitch cycles**. (See *Figure 4.*)

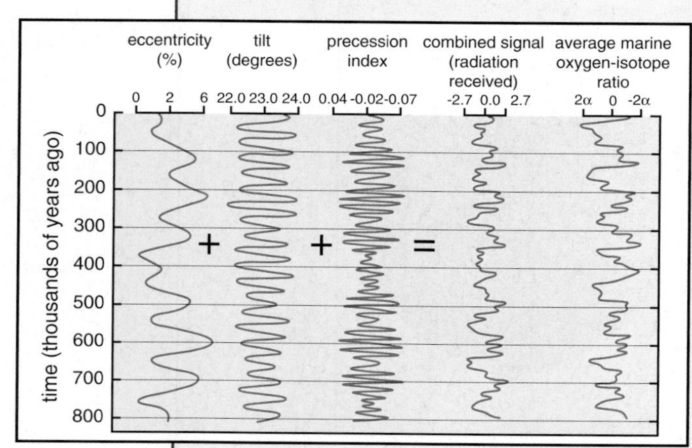

Figure 4 Interpretation of Milankovitch cycles over the last 800,000 years.

Climatologists now generally agree that Milankovitch cycles are closely related to the glacial–interglacial cycles the Earth has experienced in recent geologic time. *Figure 3* in **Activity 2**, based on a long Antarctic ice core, shows how temperature has varied over the past 420,000 years. There is a clear 100,000-year periodicity, which almost exactly matches the eccentricity cycle. Temperature variations seem to have been controlled by the 100,000-year eccentricity cycle during some time intervals but by the 41,000-year obliquity cycle during other time intervals.

Climatologists are trying to figure out how Milankovitch cycles of insolation trigger major changes in climate. The Milankovitch cycles (the "driver" of climate) are just the beginning of the climate story. Many important climate mechanisms must be taken into account. They involve evaporation, precipitation, snowfall, snowmelt, cloud cover, greenhouse gases, vegetation, and sea level. What makes paleoclimatology difficult (and interesting!) is that these factors interact with one another in many complicated ways to produce climate.

Check Your Understanding

1. Explain why the days are longer than the nights during the summer months. Include a diagram to help you explain.
2. What are the three factors in Milankovitch cycles?
3. Explain how Milankovitch cycles might cause changes in global climate.

Understanding and Applying What You Have Learned

1. You have made a drawing of winter and summer in the Northern Hemisphere showing the tilt of Earth.
   a) Make a drawing showing Earth on or about March 21 (the vernal equinox). Indicate from which direction the Sun's rays are hitting the Earth.
   b) Explain why the daytime and nighttime last the same length of time everywhere on the Earth on the vernal equinox and on the autumnal equinox.
2. In **Part E** of the **Investigate** section you explored the relationship between energy from the Sun and distance to the Sun. How would you expect the area of light shining on the wall to change if the light source was moved farther away from the wall (but the cardboard was left in the same place)?
3. The tilt of the Earth varies from about 22° to about 24.5° over a period of 41,000 years. Think about how the solar radiation would change if the tilt was 24.5°.
   a) What effect would this have on people living at the Equator?
   b) What effect would this have on people living at 30° latitude?
   c) What effect would this have on people living at 45° latitude?
   d) What modifications in lifestyle would people have to make at each latitude?

Preparing for the Chapter Challenge

Use a style of writing appropriate for a newspaper to discuss the following topics:

- How does the tilt of the Earth's axis produce seasons?
- How does a variation in this tilt affect the nature of the seasons?
- How might a variation in the tilt affect global climate?
- How does the shape of the Earth's orbit influence climate?

Inquiring Further

1. **Sunspots and global climate**

   Do sunspots affect global climate? There is disagreement over this in the scientific community. Do some research to find out what sunspots are and how the activity of sunspots has correlated with global climate over time. Why do some scientists think sunspot activity affects global climate? Why do some scientists think that sunspot activity does NOT affect global climate?

2. **Milutin Milankovitch**

   Write a paper on Milutin Milankovitch, the Serbian scientist who suggested that variations in the Earth's orbit cause glacial periods to begin and end. How were his ideas received when he first published them?

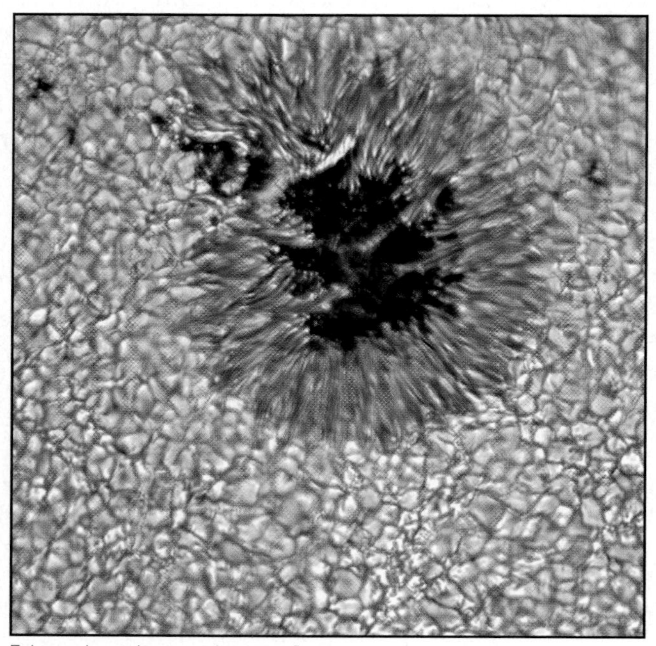

False color telescope image of a sunspot.

Activity 4 How Do Plate Tectonics and Ocean Currents Affect Global Climate?

Goals

In this activity you will:

- Model present and ancient land masses and oceans to determine current flow.
- Explain how ocean currents affect regional and global climate.
- Understand how ocean currents are affected by Earth's moving plates.
- Understand the relationship between climate and Earth processes like moving plates, mountain building, and weathering.

Think about It

Ocean currents help to regulate global climate by transferring heat and moisture around the globe.

- How would a change in the position of a land mass influence global climate?

What do you think? Record your ideas about this question in your *EarthComm* notebook. Be prepared to discuss your responses with your small group and the class.

Investigate

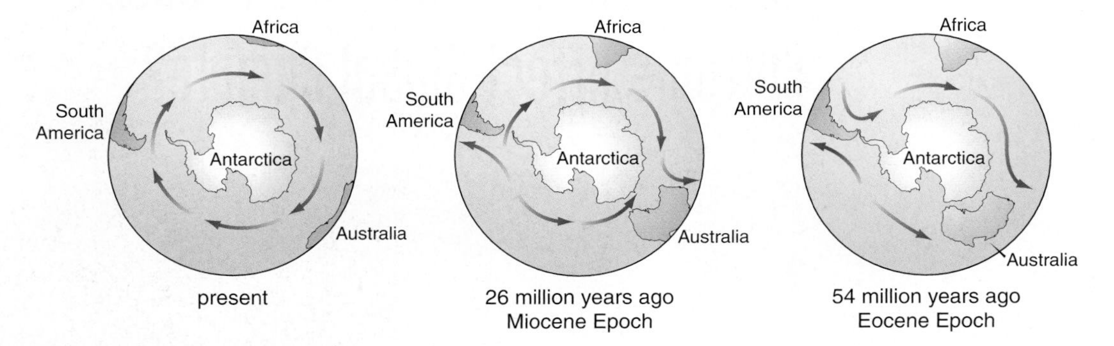

1. Divide your class into three groups. Each group will investigate the flow of water at one of these three periods of time:

   Group 1: The present.

   Group 2: During the Miocene Epoch, 26 million years ago.

   Group 3: During the Eocene Epoch, 54 million years ago.

2. Obtain a copy of the map for your assigned time period. Put this map under a clear plastic container.
3. Using clay, construct the correct land masses inside the container using the map as a template. Make the land masses at least 3 cm high.
4. Remove the map and add water to the container up to the level of the land masses.
5. Obtain a blue-colored ice cube. Place this as close to the South Pole as possible.
   a) As the ice cube melts and the cold water flows into the clear water, draw arrows on the map to record the direction of flow.

   Use only food coloring to dye the ice. Clean up spills immediately. Dispose of the water promptly.

   b) Write a paragraph in which you describe your observations.
6. Present your group's data to the class.
   a) Compare the direction of current flow on your map to those presented by the other groups. How does the current change as the land masses change from the Eocene Epoch to the Miocene Epoch to the present?
7. Use the results of your investigation to answer the following:
   a) Based on the map that shows the position of Australia in the Eocene Epoch and your investigation, what tectonic factors do you think are most important to consider when contemplating Australia's climate in the Eocene Epoch? How might Australia's Eocene Epoch climate have been different from its climate today?
   b) What tectonic factor(s) (for example, the position of the continents, occurrence of major ocean currents, mountain ranges, etc.) is/are most important in affecting the climate in your community today?

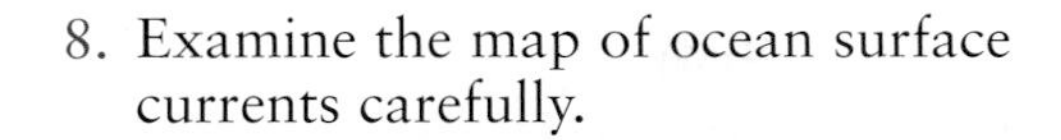

8. Examine the map of ocean surface currents carefully.

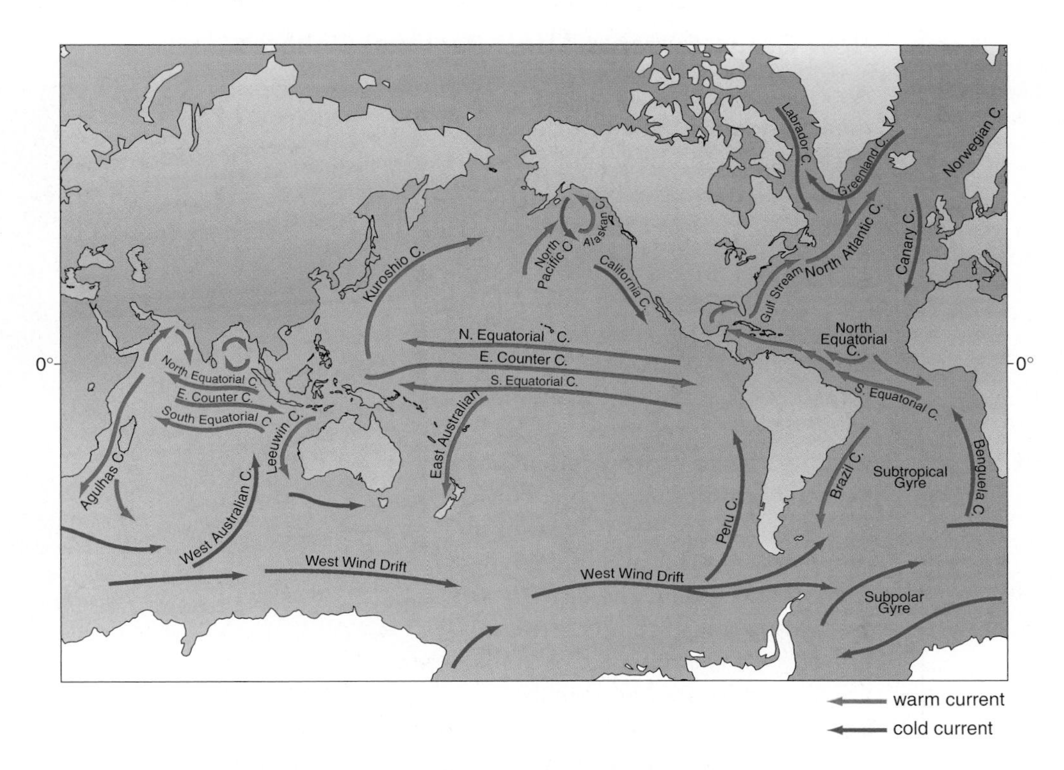

a) How do you think surface ocean currents modify the climatic patterns in the United States?

b) What changes in the surface-current patterns might arise if North America and South America split apart from one another, leaving an open passageway from the Atlantic Ocean to the Pacific Ocean?

c) How might the climate of the East Coast of the United States change if tectonic forces changed the positions of the continents so that the Gulf Stream no longer flowed north?

d) How might some other parts of the world (including your community) be affected if the Gulf Stream stopped flowing north?

Reflecting on the Activity and the Challenge

Ocean currents play a large role in global climate. This activity helped you see how ocean currents change in response to movements of the Earth's lithospheric plates.

Geo Words

thermohaline circulation: the vertical movement of seawater, generated by density differences that are caused by variations in temperature and salinity.

Digging Deeper

CHANGING CONTINENTS, OCEAN CURRENTS, AND CLIMATE

How Ocean Currents Affect Regional Climates

A community near an ocean has a more moderate climate than one at the same latitude inland because water has a much higher heat capacity than rocks and soil. Oceans warm up more slowly and cool down more slowly than the land. Currents are also an important factor in coastal climate. A coastal community near a cold ocean current has cooler weather than a coastal community near a warm ocean current. For example, Los Angeles is located on the Pacific coast near the cold California Current. The city has an average daily high temperature in July of 75°F. Charleston, South Carolina, is located at a similar latitude, but on the Atlantic coast near the warm Gulf Stream. (See *Figure 1*.) Charleston's average daily high temperature in July is 90°F.

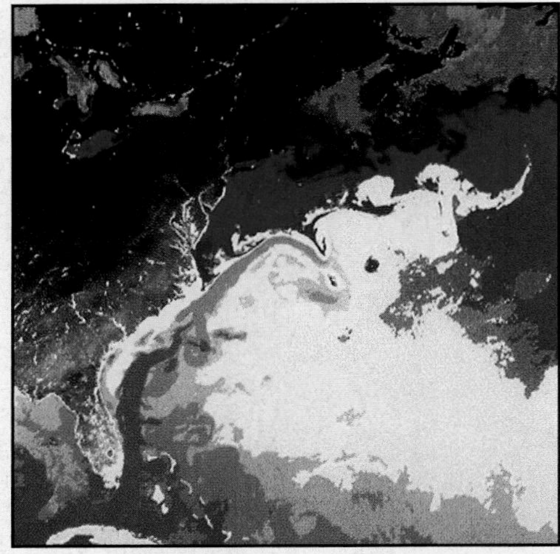

Figure 1 This thermal infrared image of the northwest Atlantic Ocean was taken from an NOAA satellite. The warm temperatures (25°C) are represented by red tones, and the cold temperatures (2°C) by blue and purple tones.

How Ocean Currents Affect Global Climate

Patterns of ocean circulation have a strong effect on global climate, too. The Equator receives more solar radiation than the poles. However, the Equator is not getting warmer, and the poles are not getting colder. That is because oceans and winds transfer heat from low latitudes to high latitudes. One of the main ways that the ocean transfers this heat is by the flow of North Atlantic Deep Water (abbreviated NADW). It works like this: In the northern North Atlantic, the ocean water is cold and salty, and it sinks because of its greater density. It flows southward at a deep level in the ocean. Then at low latitudes it rises up toward the surface as it is forced above the even denser Antarctic Bottom Water. Water from low latitudes flows north, at the ocean surface, to replace the sinking water. As it moves north, it loses heat. This slow circulation is like a "conveyor belt" for transferring heat. This kind of circulation is usually called **thermohaline circulation**. (*thermo* stands for temperature, and *haline* stands for saltiness.)

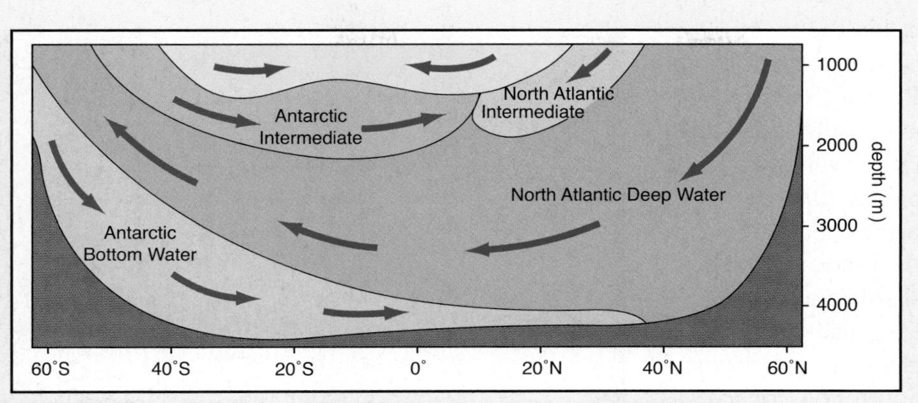

Figure 2 Circulation cell formed by the movement of deep-water masses in the ocean.

When this conveyor belt is disturbed, the entire global climate is affected. For example, about 12,000 years ago, glaciers were melting rapidly, because the Earth was coming out of a glacial age. The melting glaciers discharged a lot of fresh water into the North Atlantic in a short time. The fresh water decreased the salinity of the ocean water thus reducing its density. This decrease was so much that the production of NADW was decreased. This seems to have plunged the world back into a short cold period, which lasted about 1000 years.

How Plate Tectonics Affects Global Climate

The positions of the continents on the Earth change as the Earth's **lithospheric plates** move. (**Plate tectonics** is the study of the movement and interaction of the Earth's lithospheric plates.) The arrangement of the continents has a strong effect on the Earth's climate. Think about the requirements for the development of large continental ice sheets. Glaciers form only on land, not on the ocean. For an ice sheet to develop there has to be large land areas at high latitudes, where snow can accumulate to form thick masses of ice. Where oceans occupy polar areas, accumulation of snow is limited by melting in the salty ocean waters. Polar oceans, like the Arctic Sea, around the North Pole, are mostly covered by pack ice. This ice is no more than several meters thick.

Today, the two continental landmasses with permanent ice sheets are Antarctica, in the Southern Hemisphere, and Greenland, in the Northern Hemisphere. The continent of Antarctica has not always been centered on the South Pole. About two hundred million years ago, all of the Earth's continents were welded together. They formed a single continent,

Geo Words

lithospheric plate: a rigid, thin segment of the outermost layer of the Earth, consisting of the Earth's crust and part of the upper mantle. The plate can be assumed to move horizontally and adjoins other plates.

plate tectonics: the study of the movement and interaction of the Earth's lithospheric plates.

Geo Words

Pangea: Earth's most recent supercontinent, which was rifted apart about 200 million years ago.

called the supercontinent of **Pangea**. Pangea was eventually rifted apart into several large pieces. One of the pieces, the present Antarctica, moved slowly southward. Eventually it moved close enough to the South Pole for ice sheets to form. In recent geologic time, Antarctica has been directly over the South Pole, so the Antarctic ice sheet has remained in existence even during interglacial periods. We know that because otherwise global sea level would have been much higher during the interglacial periods of the past million years.

Figure 3 An ice-core station in Antarctica.

At present, most of the Earth's continental land area is in the Northern Hemisphere. Much of North America and Eurasia is at a high latitude. Ice sheets can form during the parts of Milankovitch cycles that are favorable for decreased global temperatures. During times of increased global temperatures, the North American and Eurasian ice sheets have melted away completely. The picture is very different in the Southern Hemisphere. Except for Antarctica, there is not enough continental land area at high latitudes for large continental glaciers to form.

Figure 4 Mt. St. Helens is an example of a volcano associated with a plate boundary.

Plate tectonics affects climate in other ways besides changing the positions of the continents. Volcanoes like the one shown in *Figure 4* from along active plate margins. Increased activity at these margins causes increased volcanic activity. Volcanoes release carbon dioxide, which is a gas that traps heat in the atmosphere. (You'll learn more about carbon dioxide in the following activity.) In this way, plate tectonics might cause global climate to warm. However, volcanic eruptions also add dust to the atmosphere, which blocks out some solar radiation. This tends to decrease global temperatures.

Continent–continent collisions create huge mountain ranges. The Himalayas and the Alps are modern examples. Many scientists believe that the weathering of such mountain ranges uses up carbon dioxide from the Earth's atmosphere because some of the chemical reactions that break down the rock use carbon dioxide from the atmosphere. This causes global climate to cool. For example, the collisions between continents that produced the supercontinent Pangea resulted in high mountain ranges like the one at the present site of the Appalachian Mountains. The Appalachians were much taller and more rugged when they first formed—perhaps as tall as the Himalayas shown in *Figure 5*. Three hundred million years of erosion have given them their lower and well-rounded appearance. On a global scale, all that weathering (which uses up carbon dioxide) may have contributed to the period of glaciation that began about 300 million years ago and ended about 280 million years ago.

Figure 5 The geologically young Himalayan mountains formed when India collided with Asia.

Check Your Understanding

1. Explain how North Atlantic Deep Water circulates.
2. Why do glaciers form only on continents and not in oceans?
3. Explain how plate tectonics can affect global climate.

Understanding and Applying What You Have Learned

1. In the **Investigate** section, you made models designed to demonstrate how ocean currents were different during the Eocene, Miocene, and today.

   a) Assuming that the maps were accurate, in what ways was your model helpful in exploring possible differences in oceanic currents?

   b) What are some of the drawbacks or problems with your model (how is a model different than the "real world"?)

   c) What improvements could you make to the model so that it would behave in a more accurate way?

2. Increased weathering of rocks uses up carbon dioxide. Decreasing carbon dioxide in the atmosphere contributes to global cooling. Reconstructions of the collision between India and Asia suggest that India first collided with Asia during the Late Eocene but that most of the mountain building took place during the Miocene and later.
   a) When would you expect to observe the greatest changes in weathering rate?
   b) Why?
3. Melting glaciers discharged a lot of fresh water into the North Atlantic in a short period of time about 12,000 years ago. Adding fresh water "turned off" the North Atlantic Deep Water current for about 1000 years.
   a) What other changes could disturb the NADW?
   b) In the event that this happened, what effect would it have had on global climate?
   c) How would this change affect your community? Even if you don't live near the Atlantic Ocean, your physical environment might still be greatly affected.

Preparing for the Chapter Challenge

Using a style of writing appropriate for a newspaper, write several paragraphs containing the following material:

- Explain how the locations of the continents on the Earth affect global climate.
- Explain how ocean currents affect global climate.
- Explain how moving continents change ocean currents.

Inquiring Further

1. **Modeling North Atlantic Deep Water flow**

   Make a physical model of the flow of North Atlantic Deep Water. Experiment with several ideas. Think of how you might do it using actual water, and how you might do it using other materials. If your model idea is large and expensive, draw a diagram to show how it would work. If your model idea is small and simple, construct the model and see if it works.

Activity 5 How Do Carbon Dioxide Concentrations in the Atmosphere Affect Global Climate?

Goals

In this activity you will:

- Compare data to understand the relationship of carbon dioxide to global temperature.
- Evaluate given data to draw a conclusion.
- Recognize a pattern of information graphed in order to predict future temperature.
- Understand some of the causes of global warming.

Think about It

"What really has happened to winter?" You may have heard this type of comment.

- What causes "global warming?"

What do you think? Write down your ideas to this question in your *EarthComm* notebook. Be prepared to discuss your responses with your small group and the class.

Investigate

Part A: Atmospheric Carbon Dioxide Concentrations over the Last Century

Data on 10-year Average Global Temperature and Atmospheric Carbon Dioxide Concentration		
time interval	average global temperature (°F)	atomospheric carbon dioxide (ppm)
1901–1910	56.69	297.9
1911–1920	56.81	301.6
1921–1930	57.03	305.19
1931–1940	57.25	309.42
1941–1950	57.24	310.08
1951–1960	57.20	313.5
1961–1970	57.14	320.51
1971–1980	57.26	331.22
1981–1990	57.71	345.87
1991–2000*	57.87	358.85

*carbon dioxide data only through 1998.

1. Graph the concentration of carbon dioxide in the atmosphere from 1900 to 2000. Put the year on the *x* axis and the CO_2 levels (in parts per million) on the *y* axis.

2. On the same graph, plot the global average temperature for the same period. Put another *y* axis on the right-hand side of the graph and use it for global average temperature.

   a) Is there a relationship between carbon dioxide concentration and global average temperature? If so, describe it.

   b) What do you think is the reason for the relationship you see?

Part B: Atmospheric Carbon Dioxide Concentrations over the Last 160,000 Years

1. Look at the figure showing data from an ice core in Antarctica. The graph shows changes in concentrations of carbon dioxide and methane contained in trapped bubbles of atmosphere within the ice, and also temperature change over the same

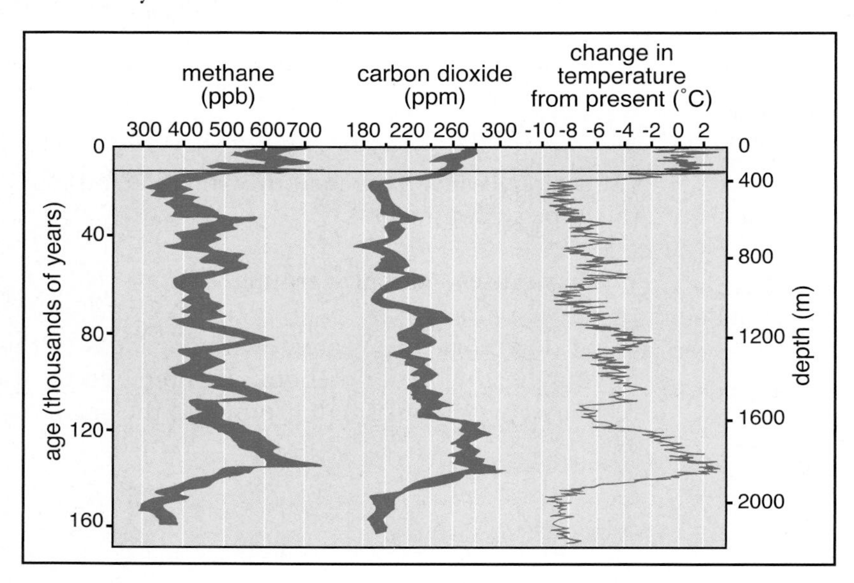

time interval. Data was obtained from the study of an ice core from the Antarctic Ice Sheet. The core was approximately 2200 m long. It was analyzed for methane concentrations (in parts per billion—left graph), carbon dioxide concentrations (in parts per million—middle graph), and inferred change in temperature from the present (in °C—right graph), over the last 160,000 years.

2. Obtain a copy of the graph. Use a straightedge to draw horizontal lines across the three maximum temperatures and the three minimum temperatures.

   a) Describe the correlation between these temperature events and changes in levels of carbon dioxide.

   b) Label likely glacial intervals (low temperatures) and interglacial intervals (higher temperatures).

   c) When did the most recent glacial interval end, according to these graphs?

Part C: The Greenhouse Effect

The phrase "greenhouse effect" is used to describe a situation in which the temperature of an environment (it could be any environment like a room, a car, a jar or the Earth) increases because incoming solar energy gets trapped because heat energy cannot easily escape. The incoming energy easily enters into the environment, but then, once it has been absorbed and is being re-radiated, it is harder for the energy to escape back out of the environment.

1. Work as a group to design an experiment to demonstrate the greenhouse warming in the atmosphere. The experiment should be simple in design, include a control element, and be performed in a short period of time (for example, a class period). The experiment will be presented to the community as a way to show the greenhouse effect.

   a) Record your design in your *EarthComm* notebook. Remember to include a hypothesis. Be sure to also include any safety concerns.

2. Decide on the materials you will use. The materials should be inexpensive and easy to get. The following is a possible list:

   - two identical 2-L plastic bottles with labels removed and tops cut off or two identical beakers
   - water
   - a clear plastic bag
   - a thermometer
   - ice cubes
   - a sunny windowsill or two similar lamps

   a) Record your list in your *EarthComm* notebook.

3. Decide on the measurements that you will make.

   a) Prepare a data table to record your observations.

4. With the approval of your teacher, conduct your experiment.

Have the design of your experiment checked carefully by your teacher for any safety concerns.

5. Use the results of your experiment to answer the following questions:
   a) How did this experiment demonstrate (or fail to demonstrate) the greenhouse effect?
   b) How can this experiment serve as an analogy for atmospheric greenhouse effects?
   c) Was there any difference observed between the greenhouse experiment and the control?
   d) If there was a difference (or differences) describe it (them) in both qualitative and quantitative terms.
   e) How did the data in each case change through time during the experiment?
   f) Did the experiment reach a point of equilibrium where continuing changes were no longer observed? (Note: To answer this question, it may take longer than the class period, or, alternatively, you could hypothesize an answer to this question based on the trends of the data that you were able to gather.)

Reflecting on the Activity and the Challenge

In this activity you designed an experiment to demonstrate the greenhouse effect. You also examined the concentration of atmospheric carbon dioxide to see if it is correlated with changes in global average temperature. You discovered that an increase in carbon dioxide seems to be correlated with an increase in global average temperature. You will need this information to begin writing your article on "What is Global Warming?"

Geo Words

correlation: a mutual relationship or connection.

Digging Deeper

CARBON DIOXIDE AND GLOBAL CLIMATE

Correlation Studies

The relationship between carbon dioxide and global climate was mentioned in previous activities. When there is more carbon dioxide in the atmosphere, global temperatures are higher. When there is less carbon dioxide in the atmosphere, temperatures are lower. A scientist would say that there is a **correlation** between carbon dioxide concentration and global temperature. You might think, "Oh, that's because carbon dioxide concentration affects global temperature." And you might be right—but you might be wrong.

It is important to keep in mind always that a correlation does not, by itself, prove cause and effect. There are three possibilities: (1) carbon dioxide affects temperature; (2) temperature affects carbon dioxide; and (3) both are affected by a third factor, and are independent of one another! Any one of these three possibilities is consistent with the observations. It is the scientists' job to try to figure out which is the right answer. There are good reasons to think that the first possibility is the right one. That is because carbon dioxide is a "greenhouse gas."

Geo Words

greenhouse gases: gases responsible for the greenhouse effect. These gases include: water vapor (H_2O), carbon dioxide (CO_2), methane (CH_4), nitrous oxide (N_2O), chlorofluorocarbons (CF_xCl_x), and tropospheric ozone (O_3).

What Are Greenhouse Gases?

The reason that the Earth is warm enough to support life is that the atmosphere contains gases that let sunlight pass through. Some of these gases absorb some of the energy that is radiated back to space from the Earth's surface. These gases are called **greenhouse gases**, because the effect is in some ways like that of a greenhouse. Without greenhouse gases, the Earth would be a frozen wasteland. Global temperatures would be much lower. Water vapor is the most important contributor to the greenhouse effect. Other greenhouse gases include carbon dioxide, methane, and nitrogen oxides.

How do greenhouse gases work? Most solar radiation passes through the clear atmosphere without being absorbed and is absorbed by the Earth's surface (unless it's reflected back to space by clouds first). There is a law in physics that states that all objects radiate electromagnetic radiation. The wavelength of the radiation depends on the objects' surface temperature. The hotter the temperature, the shorter the wavelength. The extremely hot surface of the Sun radiates much of its energy as visible light and other shorter-wavelength radiation. The much cooler surface of the Earth radiates energy too, but at much longer wavelengths. Heat energy is in the infrared range (*infra-* means "below," and the color red is associated with the longest wavelength in the color spectrum). See *Figure 1*.

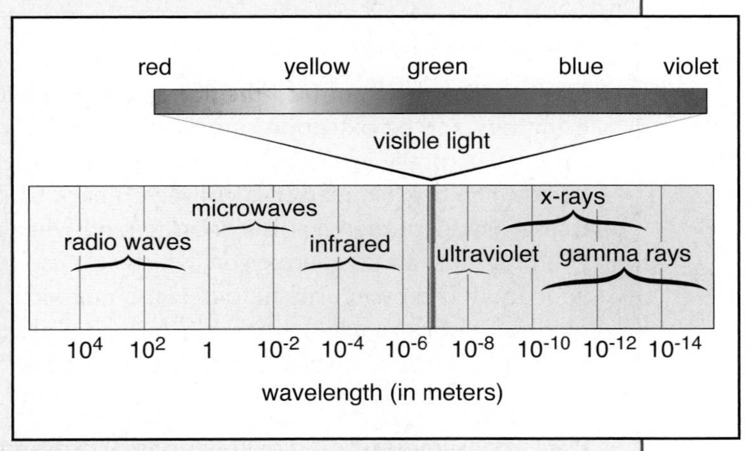

Figure 1 The spectrum of electromagnetic radiation.

Greenhouse gases are those that absorb some of the outgoing infrared radiation. None of them absorb all of it, but in combination they absorb much of it. They then re-radiate some of the absorbed energy back to the Earth, as shown in *Figure 2*. That is what keeps the Earth warmer than if there were no greenhouse gases.

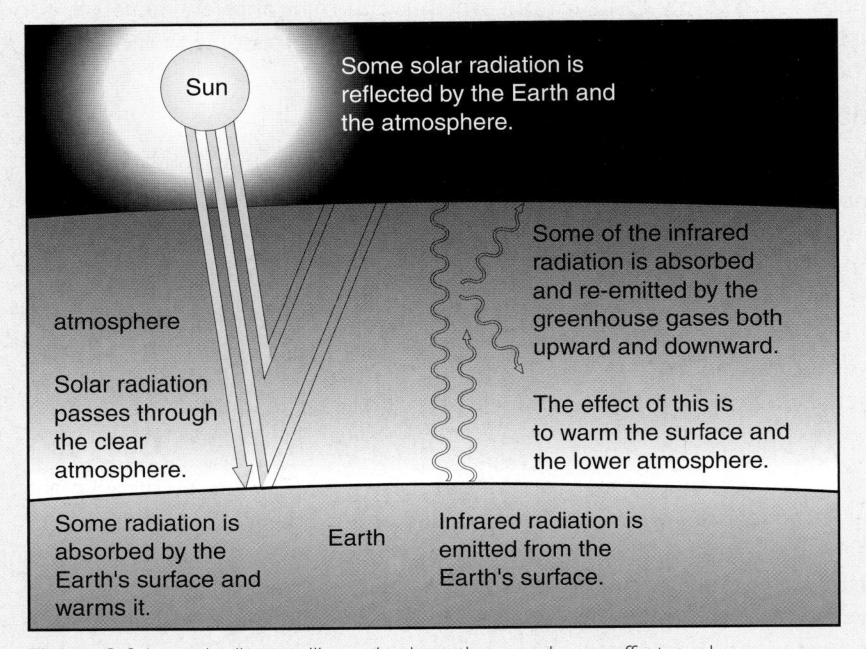

Figure 2 Schematic diagram illustrating how the greenhouse effect works.

The Carbon Cycle

Carbon dioxide is put into the atmosphere in two main ways: during volcanic eruptions, and by oxidation of organic matter. Oxidation of organic matter happens naturally in the biosphere. It occurs when plant and animal tissue decays. The organic matter is converted back to carbon dioxide and water. It also happens when animals breathe (and when plants respire too!). When you breathe, you take in oxygen, which you use to oxidize organic matter—your food. Then you breathe out carbon dioxide. Organic matter is also oxidized (more rapidly!) when it is burned. Carbon dioxide is released into the atmosphere whenever people burn wood or fossil fuels like gasoline, natural gas, or coal.

Plants consume carbon dioxide during photosynthesis. It is also consumed during the weathering of some rocks. Both land plants and algae in the ocean use the carbon dioxide to make organic matter, which acts as a storehouse for carbon dioxide. Carbon dioxide is constantly on the move from place to place. It is constantly being transformed from one form to another. The only way that it is removed from the "active pool" of carbon dioxide at or near the Earth's surface is to be buried deeply with sediments. Even then, it's likely to reenter the Earth–surface system later in geologic time. This may be a result of the uplift of continents and weathering of certain carbon-rich rocks. This transfer of carbon from one reservoir to another is illustrated in the carbon cycle shown in *Figure 3*.

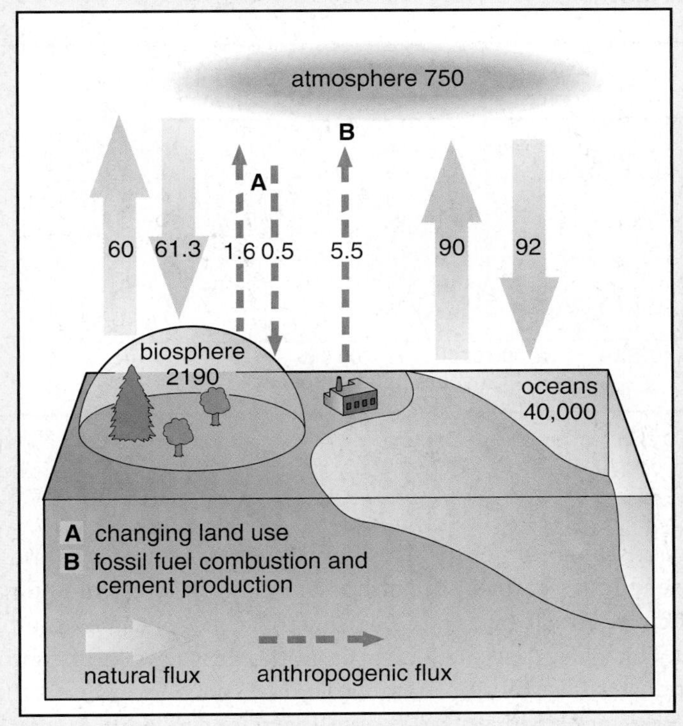

Figure 3 Global carbon cycle. Fluxes are given in billion metric tons per year and reservoirs in billion metric tons.

Carbon Dioxide and Climate

It appears that the more carbon dioxide there is in the environment, the warmer global temperatures are. Scientists have determined this from geologic data like the kind you worked with in the investigation. To what extent is this because carbon dioxide in the atmosphere acts as a greenhouse gas?

It is valuable to look at this question on two different time scales. On a scale of hundreds of thousands of years, carbon dioxide and global temperature track each other very closely. This correlation occurs through several glacial–interglacial cycles (*Figure 4*). It is not easy to develop a model in which carbon dioxide is the cause and global temperature is the effect. It's much more likely that variations are due to Milankovitch cycles. They may well explain the variation in both global temperature and carbon dioxide. On a scale of centuries, however, the picture is different. It seems very likely that the increase in carbon dioxide has been the cause of at least part of the recent global warming.

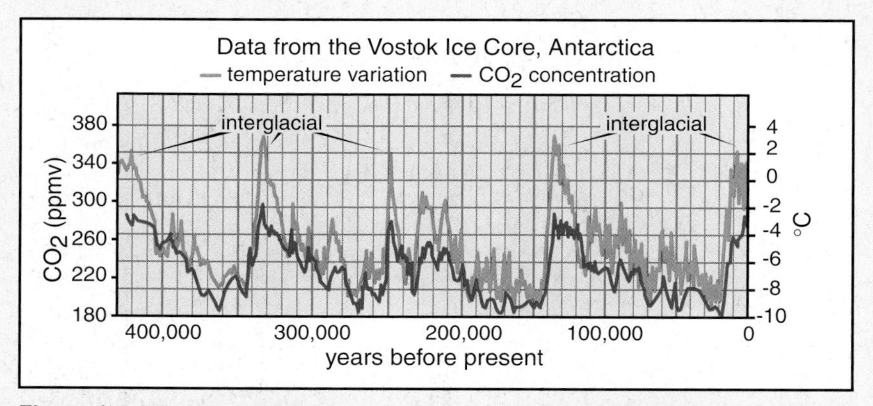

Figure 4 Variations in temperature and carbon dioxide (CO_2) concentration in parts per million by volume (ppmv) over the past 420,000 years interpreted from Antarctic ice cores. Temperature change is relative to the modern surface temperature at Vostok (–55.5°C).

Human emissions of greenhouse gases contribute significantly to the total amount of greenhouse gases in the atmosphere. For a long time humans have been adding a lot of carbon dioxide to the atmosphere by the burning of fossil fuels. This has especially increased in the past couple of centuries. Before the Industrial Revolution, carbon dioxide concentrations in the atmosphere were approximately 300 ppm (parts per million). As of 1995, carbon dioxide concentrations were almost 360 ppm. Scientists are concerned that the temperature of the Earth may be increasing because of this increasing concentration of carbon dioxide in the atmosphere.

Many nations have a commitment to reduce the total amount of greenhouse gases produced. It is their effort to reduce the risk of rapid global temperature increase. The trouble is that the size of the effect is still uncertain. Some people take the position that the increase in carbon dioxide should be reversed. They believe this is necessary even though the size of the contribution to global warming is not certain. It is their belief that the consequences would be very difficult to handle. Other people take a different position. They consider that it would be unwise to disrupt the world's present economy. They consider the future danger to be questionable. The big problem is that no one is certain that rapid global warming will take place. If it does, it may be too late to do anything about it!

Not all of the carbon dioxide released by burning of fossil fuels stays in the atmosphere. Carbon dioxide is also dissolved in ocean water. As carbon is put into the atmosphere, some of it is absorbed by the oceans. That lessens the impact of burning of fossil fuels on climate. Some people have even suggested that enormous quantities of carbon dioxide should be pumped into the oceans. That would tend, however, to just postpone the problem until later generations. Carbon dioxide is also stored by **reforestation**. Reforestation is the growth of forests on previously cleared farmland. Did you know that there is a lot more forested land in the eastern United States now than at the time of the Civil War? The Civil War took place almost 150 years ago. By some estimates, the United States is a sink, rather than a source, for carbon dioxide. Extensive reforestation is occurring east of the Mississippi, despite the continuing expansion of suburbs and shopping malls!

Figure 5 Clear-cut forest area in Olympic National Forest, Washington.

Geo Words

reforestation: the replanting of trees on land where existing forest was previously cut for other uses, such as agriculture or pasture.

Check Your Understanding

1. List four greenhouse gases. Which gas contributes most to the greenhouse effect?
2. Explain how greenhouse gases make it possible for humans to live on Earth.
3. What are two ways in which carbon dioxide is put into the Earth's atmosphere?

Understanding and Applying What You Have Learned

1. Which of the following activities produce carbon dioxide? Which consume carbon dioxide? Explain how each can influence global climate.
   a) cutting down tropical rainforests
   b) driving a car
   c) growing shrubs and trees
   d) breathing
   e) weathering of rocks
   f) volcanic eruptions
   g) burning coal to generate electricity
   h) heating a house using an oil-burning furnace.
2. Describe the carbon cycle in your community. List the ways that carbon dioxide is produced and used up and the organisms responsible for cycling.
3. What are some difficulties involved with predicting concentrations of atmospheric carbon dioxide into the future?
4. Examine the graph your group prepared. You have gathered data through the year 2000. You have seen that this data has changed over time. Using additional graph paper, try to continue this pattern for the next 10 years.
5. The United States has a population of about 280 million people (according to the 2000 census) and uses about 70 billion gigajoules of energy a year. India has a population of about 835 million people (1990) and uses about 7 billion gigajoules of energy a year.
   a) Divide the United States' total yearly energy use by its population to find out the yearly energy use per person.
   b) Calculate the yearly energy use per person for India.
   c) Give as many reasons as you can to explain the difference.
   d) Do you think you use more or less energy than the typical American? Explain.
   e) If you wanted to use less energy, what would you do?
   f) Why is how much energy you use important when considering how much carbon dioxide is in the air?
6. Determine one source of greenhouse gas emission in your community.
   a) What gas is being produced?
   b) How is it produced?
   c) Can you think of a way to determine the level of the gas that is being produced by your community?
   d) Propose a means for limiting emissions of this gas.

Preparing for the Chapter Challenge

1. Using a newspaper style of writing, write several paragraphs in which you:
   - explain how humans have increased the concentration of carbon dioxide in the atmosphere;
   - explain why scientists think that increased carbon dioxide levels might lead to global climate change.
2. Clip and read several newspaper articles containing quotations.
3. Interview a member of your community about global warming. Is this person concerned about global warming? What does he or she think people should do about it? Look over your notes from your interview. Pick out several quotations from the community member that might work well in a newspaper article.

Inquiring Further

1. **Intergovernmental Panel on Climate Change (IPCC)**

   The Intergovernmental Panel on Climate Change (IPCC) is a group of more than 100 scientists and economists from many countries that is investigating the possibility of global warming and proposing ways that the nations of the world should respond. Do some research on the IPCC and what they have reported.

2. **Earth Summit**

   Investigate the 1997 United Nations Earth Summit in New York. What did the world's nations agree to at the Summit? Have the nations stuck to their promises?

Activity 6 How Might Global Warming Affect Your Community?

Goals

In this activity you will:

- Brainstorm the ways that global warming might influence the Earth.
- List ways that global warming might affect your community.
- Design an experiment on paper to test your ideas.
- Explain some of the effects of global warming that computer models of global climate have predicted.
- Understand positive and negative feedback loops and their relationship to climate change.
- Evaluate and understand the limitations of models in studying climate change through time.

Think about It

Some scientists think that the average global temperature may increase by several degrees Fahrenheit by the end of the 21st century.

- How do you think global warming could affect your community?

What do you think? List several ideas about this question in your *EarthComm* notebook. Be prepared to discuss your ideas with your small group and the class.

Investigate

1. In small groups, brainstorm as many effects of higher global temperatures (a few degrees Fahrenheit) as you can. Each time you come up with a possible result of global warming, ask yourselves what effect that result might have. For example, if you think glaciers will recede, ask yourself what the implications of that would be. At this point, do not edit yourself or criticize the contributions of others. Try to generate as many ideas as possible. Here are a few ideas to get your discussion going.

   How might higher temperatures affect the following processes?

   - evaporation
   - precipitation
   - glacial activity
   - ocean circulation
   - plant life
   - animal life.

   a) List all the ideas generated.

2. As a group, review your list and cross off those that everyone in the group agrees are probably incorrect or too far-fetched. The ideas that remain are those that the group agrees are possible (not necessarily proven or even likely, but possible). It's okay if some of the ideas are contradictory. Example: More cloud cover might block out more solar radiation (a cooling effect) vs. More cloud cover might increase the greenhouse effect (a warming effect).

   a) Make a poster listing the ideas that remain. Organize your ideas on the poster using the following headings:

   - geosphere
   - hydrosphere
   - atmosphere
   - cryosphere
   - biosphere.

   b) On a separate piece of paper, write down how each of the possible results might affect your community.

3. Imagine that your group is a group of scientists who are going to write a proposal asking for grant money to do an experiment. Pick one of the ideas on your poster that you would like to investigate.

   a) On paper, design an experiment or project to test the idea. Choose ONE of the following:

   - Design an experiment that you could do in a laboratory that would model the process. Draw a diagram illustrating the model. Tell what materials you would need and how the model would work. Describe what the results would mean. Tell which parts of the experiment would be difficult to design or run, and explain why.

If you plan to perform your experiments do so only under careful supervision by a knowledgeable adult.

- Design a project in which you would gather data from the real world. Include a diagram or sketches illustrating how you would gather data. Tell what kind of data you would gather, how you would get it, how frequently and how long you would collect it, and how you would analyze it. Tell which parts of the project would be difficult to design or carry out, and explain why.

4. Present your poster and your proposal for an experiment or project to the rest of the class.

Reflecting on the Activity and the Challenge

In this activity, you brainstormed ways in which an increase in global temperatures might affect the geosphere, the hydrosphere, the atmosphere, the cryosphere, the biosphere, and your community. Then you designed an experiment or a project for how you might test one of your ideas. This process modeled the way in which scientists begin to think about how to investigate an idea. This will help you to explain which possible effects of global warming would have the greatest impact on your community, and also why it is difficult for scientists to accurately predict climate change.

Digging Deeper

EFFECTS OF GLOBAL WARMING IN YOUR COMMUNITY

Problems with Making Predictions

Many scientists believe that the world's climate is becoming warmer as a result of the greenhouse gases (carbon dioxide, methane, and nitrogen-oxide compounds) that humans are adding to the atmosphere. Because the world's climate naturally experiences warmer years and colder years, it is hard to say for sure whether global average temperature has been increasing. Nowadays, remote sensing of the land and ocean surface by satellites makes it easy to obtain a good estimate of global temperature. The problem is that such techniques didn't exist in the past. Therefore, climatologists have to rely on conventional weather records from weather stations. That involves several problems. Thermometers change. The locations of the weather stations themselves often have to be changed. The move is usually away from city centers. As urban areas have been developed, they become warmer because of the addition of pavement and the removal of cooling vegetation. That is called the **urban heat-island effect**. Climatologists try to make

Geo Words

urban heat-island effect: the observed condition that urban areas tend to be warmer than surrounding rural areas.

corrections for these effects. The consensus is that global average temperature really is increasing. The questions then become: how much of the warming is caused by humankind, and how much is natural? It is known that there have been large variations in global temperature on scales of decades, centuries, and millennia long before humankind was releasing large quantities of carbon dioxide into the atmosphere (*Figure 1*).

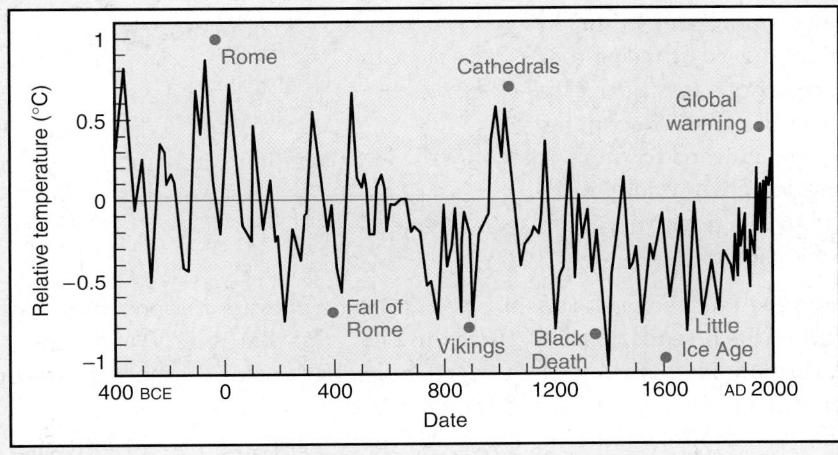

Figure 1 Relative global temperature from 400 BCE to the present.

Scientists who study global warming use very complicated computer models to try to predict what might happen. These computer models do not simply look at carbon dioxide and global temperature. They also do calculations based on many other factors that might be involved in climate change—everything from how much moisture is held in the Earth's soils to the rate at which plants transpire (give off water vapor). The physics of clouds is an especially important but also especially uncertain factor. The workings of the Earth's atmosphere are still much more complex than any computer model. Climatologists are hard at work trying to improve their models.

Drawbacks to the Computer Models

As you have seen, many factors influence the climate on Earth: carbon dioxide and other greenhouse gases, Milankovitch cycles, ocean currents, the positions of continents, weathering of rocks, and volcanic activity. Many of these factors interact with each other in ways that scientists do not fully understand. This makes it hard to make accurate predictions about how the atmosphere will respond to any particular change. The ways that different factors interact in global climate change are called **feedback loops**.

Geo Words

feedback loops: the processes in which the output of a system causes positive or negative changes to some measured component of the system.

Feedback may be positive or negative. Positive feedback occurs when two factors operate together and their effects add up. For example, as the climate cools, ice sheets grow larger. Ice reflects a greater proportion of the Sun's radiation, thereby causing the Earth to absorb less heat. This results in the Earth becoming cooler, which leads to more ice forming. Ice cover and global cooling have a positive feedback relationship.

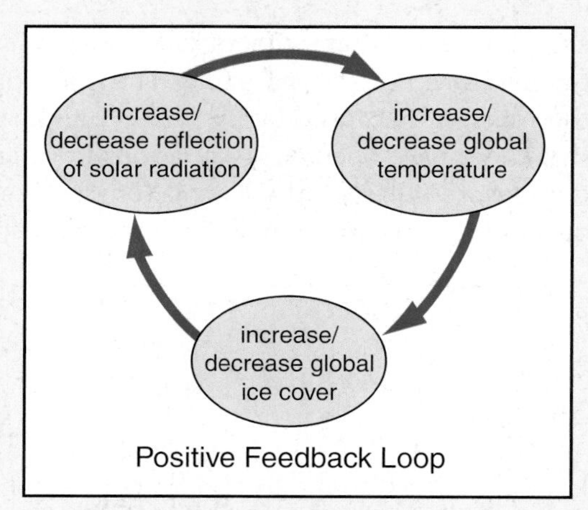

Figure 2 An example of a positive feedback loop.

In a negative feedback relationship, two variables operate in opposition with each other. Each tends to counteract the effects of the other. Weathering and carbon dioxide are one such negative feedback pair. Weathering uses up carbon dioxide, which causes the temperature to drop. When the temperature drops, weathering rates slow down, using up less carbon dioxide, and slowing the rate of temperature decrease. In this sense, weathering acts as a negative feedback for global cooling. The Earth's global climate involves many feedback loops like these. Understanding of such feedback loops is extremely important in understanding how the Earth's physical environment changes through time.

How many feedback loops are there? How do they work? These are questions that scientists are working on every day. The uncertainty about feedback loops is one thing that makes it hard to predict how the Earth's climate might respond to an increase in carbon dioxide. Another major unknown in global warming is clouds. With warmer temperatures, there will be more evaporation and therefore more clouds. Clouds reflect incoming solar radiation into space—a cooling effect. But clouds also act like a blanket to hold in heat—a warming effect. Which effect predominates? Scientists aren't yet sure.

Figure 3 How can clouds influence climate?

What Do the Computer Models Say?

Scientists continue working on their computer models of global climate. They learn as they go and make improvements all the time. Because some of the models have been used for years, scientists can test some of the predictions of past years against weather data collected recently. This helps them make changes to the computer models to make them work better.

Computer climate models have come up with some possible scenarios that may result from the increased concentration of greenhouse gases in the atmosphere. Remember, however, that these scenarios are theoretical outcomes, not certainties.

Changes in Precipitation

Warmer temperatures lead to more clouds. More clouds lead to more rain. Some models predict more rain with global warming. Others predict a change in rainfall patterns—more precipitation in the winter and less in the summer, for example. Some areas of the world would receive more rain, others less. In that respect, some countries would be winners, and others, losers. With the increase in evaporation brought on by warmer temperatures, an increase in extreme events (stronger hurricanes and winter snowstorms) might be likely.

Figure 4 Global warming may cause an increase in the number of extreme winter snowstorms.

Changes in Sea Level

Glaciers around the world have been shrinking in recent years. If the Earth's climate continues to warm, more and more glacier ice sheets will melt. Meltwater is returned to the ocean. This would result in a worldwide rise in sea level. Some models predict a sea-level rise of as much as a meter by 2100.

Changes in Agriculture

In the Northern Hemisphere, where most of the world's cropland is located, warmer temperatures would cause a northward shift of the regions where certain crops are grown. Agriculture would also be affected by changes in rainfall patterns. Some regions might become too dry to support present crops. Other places might become too wet to support present crops. Many areas might continue growing traditional crops but experience declines in productivity. In other words, farmers might still grow corn in Iowa but produce fewer bushels per year. An example of a change that might decrease crop productivity is an increase in nighttime temperatures. Corn and some other grain crops do best when the temperature drops below 70°F at night. Another change that could reduce productivity (or increase costs) is a switch to wetter winters and drier summers.

Changes in Ocean Circulation

The addition of fresh meltwater from glaciers into the North Atlantic could disturb the production of North Atlantic Deep Water. The same thing happened 12,000 years ago (see **Activity 4**). The circulation of North Atlantic Deep Water helps distribute heat from solar radiation evenly around the globe. If this flow is disturbed, there might be far-reaching effects on global climate.

Check Your Understanding

1. Explain how ice cover and global cooling work as a positive feedback loop.
2. How might global warming lead to increased precipitation?
3. Why is it hard to predict how the global climate might react to an increase of carbon dioxide in the atmosphere?

Understanding and Applying What You Have Learned

1. Using the information in **Digging Deeper,** add to the poster you made in the **Investigate** section.
2. For each new item you added to your poster, hypothesize how your community would be affected by that outcome.
3. Using what you learned in **Digging Deeper,** modify the experiment or model you proposed in the **Investigate** section OR design another experiment or model.

Preparing for the Chapter Challenge

1. Using a style of writing appropriate for a newspaper, write a paragraph on each of the following possible effects of global warming. Make sure you make it clear that these are only possible scenarios, not certainties:
   - changes in rainfall patterns
   - increase in extreme events
   - changes in sea level
   - changes in ocean currents
   - changes in agriculture.
2. Write an editorial about how you think your community should respond to global warming. Should your community wait for further research? Should your community take action? What kind of action should be taken? Would these actions benefit your community in other ways, in addition to slowing global warming?

Inquiring Further

1. **Community energy use**

   Make a plan for calculating how much energy your school uses for heating, air conditioning, lights, and other electrical uses. Make a plan for calculating the energy used in gasoline for students, teachers, staff, and administrators to travel to and from school each day. How could you test your estimates to see how accurate they are? What are some ways your school could reduce its energy use? How can a reduction in energy use influence climate?

2. **"CO_2-free" energy sources**

   Investigate some sources of energy that do not produce carbon dioxide, like solar and wind power.

3. **Climate change and crops**

   Call your state's cooperative extension service and find out what are the top three crops grown in your state. Visit the *EarthComm* web site to determine if your state's cooperative extension service has a web site. What are the optimal climatic conditions for maximizing productivity of these crops? How might climate changes due to global warming affect farmers who grow these crops in your state?

Earth Science at Work

ATMOSPHERE: *Plant Manager*
Some companies are providing their workers with retraining to use new equipment that has been designed to reduce the emission of greenhouse gases and control global warming.

BIOSPHERE: *Farmer*
Agriculture is an area of the economy that is very vulnerable to climate change. Climate change that disturbs agriculture can affect all countries in the world. However, there are also steps that farmers can take to reduce the amount of carbon dioxide and other greenhouse gases.

CRYOSPHERE: *Mountaineering Guide*
Regions where water is found in solid form are among the most sensitive to temperature change. Ice and snow exist relatively close to their melting point and frequently change phase from solid to liquid and back again. This can cause snow in mountainous areas to become unstable and dangerous.

GEOSPHERE: *Volcanologist*
Volcanoes can emit huge amounts of carbon dioxide gas as well as sulfur dioxide gas into the atmosphere with each eruption. Both of these substances can have adverse effects on the atmosphere, including global warming as a possible result.

HYDROSPHERE: *Shipping Lines*
Even under "normal" climate conditions, ocean circulation can vary. A changing climate could result in major changes in ocean currents. Changes in the patterns of ocean currents and storm patterns will have important consequences to shipping routes.

How is each person's work related to the Earth system, and to Climate Change?

3

Changing Life ...and Your Community

3

Changing Life ...and Your Community

Getting Started

When you travel across the continental United States by land, you can see obvious changes in the plants and animals. For example, you would expect to find cactus in the Southwest but not in Oregon, alligators in Florida but not in Maine, and sea otters in California but not in Michigan. Organisms live in particular areas because they are adapted to a range of climate conditions. If the range is exceeded, that organism can no longer live in that area.

- Do the plants and animals that live in your community also live in other areas of your state or region? If so, over what geographic area?
- What would happen to these plants and animals in your community, state, or region if climate or geological setting changed?

What do you think? In your *EarthComm* notebook write down the factors that affect what plants and animals live in your community. Consider both the climate and the geography. Be prepared to discuss your ideas with your working group and the class.

Scenario

Scientists have recognized that there are short-term and long-term changes in plants and animals during Earth's history. The United Nations is interested in changes in the number of different organisms and where they live. A special task force has been organized to make recommendations on what should be done over different time scales (decades, centuries, millennia) to maintain the ecosystem in your community. Your class has been asked to help the task force by explaining how and why plants and animals in your community have changed over time. This information will help the task force to make predictions about how life might change in the future. Can the *EarthComm* team meet this challenge?

Chapter Challenge

You have been asked to create a display that illustrates the biological changes your community has experienced over several scales of geologic time. You will need to address:

- Evidence that there has been change in the life forms (trees, shrubs, herbivores, carnivores, etc.) and biodiversity (numbers of different organisms) of your community over time.
- Short-term and long-term factors that have influenced biological changes in your community.
- The natural processes that have been responsible for the appearance and disappearance of life forms throughout geologic time.
- Suggestions as to what could be done to reduce biodiversity loss caused by natural changes in Earth systems.

Assessment Criteria

Think about what you have been asked to do. Scan ahead through the chapter activities to see how they might help you to meet the challenge. Work with your classmates and your teachers to define the criteria for assessing your work. Record all of this information. Make sure that you understand the criteria as well as you can before you begin. Your teacher may provide you with a sample rubric to help you get started.

Activity 1 The Fossil Record and Your Community

Goals

In this activity you will:

- Understand the process of fossilization.
- Determine which plant and animal parts have the highest and lowest potential for becoming fossilized and understand why this is the case.
- Determine which organisms in your community are most likely to become preserved in the fossil record.
- Determine where fossils may be forming within your community.
- Understand the hierarchy of a food chain and how this affects the likelihood that an organism will be preserved in the fossil record.

Think about It

Imagine that it is hundreds of thousands of years into the future. A geologist has just discovered your community and is planning an excavation.

- What evidence would the geologist find to know that life had existed in your community?

What do you think? Record your ideas about this question in your *EarthComm* notebook. Be prepared to discuss your response with your small group and the class.

Investigate

Part A: How Fossils Form

1. With a paper towel, smear petroleum jelly all over the inside of a container.
2. Following the directions on the package, mix plaster in a mixing bowl. Complete **Steps 3** through **6** immediately after this, so that the plaster does not set while you are still preparing the "rock."
3. Fill the container half full of the plaster.
4. With a paper towel, smear a thin coating of petroleum jelly on both surfaces of a clamshell. Place the clamshell in the middle of the container and press it gently into the plaster.
5. Sprinkle some confetti onto the rest of the surface of the plaster, enough to cover about 50% of the surface.

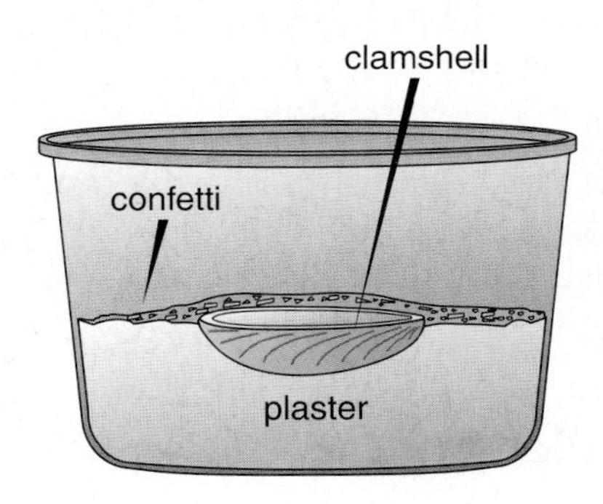

6. Fill the rest of the container with more of the plaster.
7. Let the plaster harden overnight. In the next class, remove the plaster "rock" from the container by turning the container upside down and banging it down against the floor or tabletop.

Wear goggles throughout Part A.

8. Set the plaster on its edge on a hard surface. Hit it gently with the hammer, at about the level where the confetti was sprinkled. It should split along the plane where the clamshell was placed.

Cover the plaster cast with a towel before hitting it with the hammer.

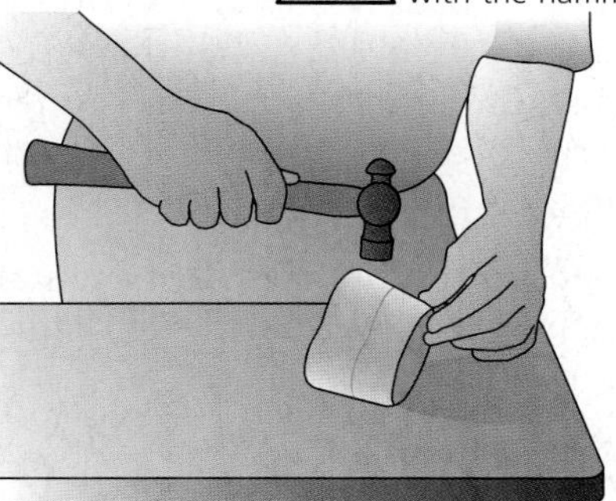

a) Why did the "rock" break along the plane where the fossil is located rather than somewhere else?

b) Do you think that a rock would usually break through the locations of fossils? Why or why not?

c) Clams have two parts to their shell. The parts are called valves. Each valve has an inner surface and an outer surface. How many different kinds of imprints might be seen when the valves of a clam are buried and fossilized as in this investigation?

d) If you had not seen the original shell that went into the plaster and did not see it when the "rock" was split open, how might you reconstruct what the shell looked like, just from studying the fossil evidence?

Part B: Fossils in Your Community

1. Natural ecosystems have different energy levels called trophic levels. Plants belong to the first level because the chemical energy they store comes directly from solar energy. They are primary producers. Organisms that eat only plants are at a higher level. They are called primary consumers or herbivores. Organisms that eat only other animals are higher-order consumers or carnivores. (Omnivores eat both plants and animals.)
   a) Identify the organisms shown at each of these levels in the diagram.

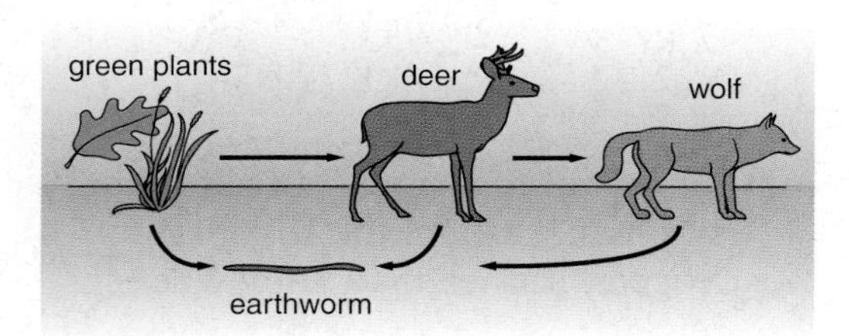

   b) Identify the most common organisms at each of these levels within your community. Try to list at least two or three different organisms for each level.
   c) Draw a diagram that illustrates the connections between each of these organisms. You will have essentially constructed a food web for your community.
2. Think about the organisms and their parts in your community.
   a) Which parts of each organism are the least resistant to decay and decomposition?
   b) Which parts of each organism are the most resistant to decay and decomposition?
3. Compare your list of resistant parts that you have identified for each plant and animal group with others in your group.
   a) Which type of organism has the highest probability of leaving some sort of fossil record?
   b) Why do some plants or animals have a low probability of leaving some sort of fossil record?
   c) How do you think that an organism's position on the food chain affects its likelihood of being fossilized?

Reflecting on the Activity and the Challenge

In this activity you learned how some types of fossils are formed. You also came to understand that not every organism has the same potential for becoming a fossil. Only certain types of parts have a high potential for becoming part of the fossil record. You also realized that the greater the number of parts an organism has, the higher the probability that one individual part may become fossilized at some time if the environmental conditions are right. You are now in a better position to evaluate the kinds of fossil evidence that may be found in your community.

Digging Deeper

FOSSILS

Food Chains and Food Webs

Plants use energy from the Sun to make food, through the process of photosynthesis. Organisms that make their own food are called producers. Other organisms, like animals, are not able to make their own food. They must eat plants, or other animals that eat plants, to obtain energy. Such organisms that rely on plants for food are called consumers. Scientists use a kind of flowchart, called a food chain, to show how organisms are connected to each other by the food they eat. It is a convenient way to show how energy and matter are transferred from producers to the next levels of consumers. In most ecosystems, consumers rely on more than one source of food. Therefore, it is more realistic to show the relationships in the form of a food web. *Figure 1* shows a sample food web.

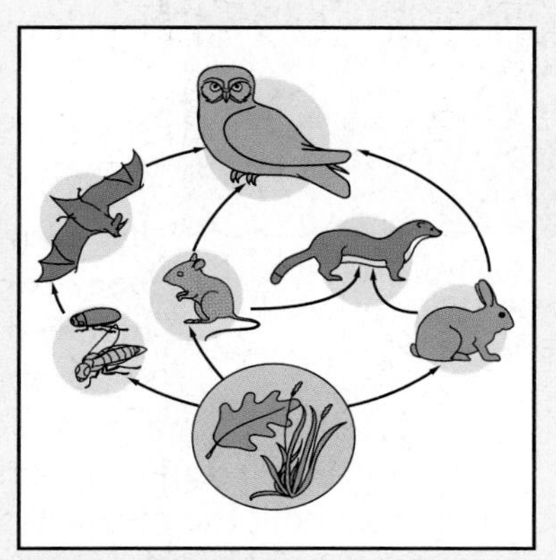

Figure 1 Can you identify the different trophic levels of this food web?

All living things die. Also, consumers generate waste materials from the food they eat. A special group of consumers, called decomposers, obtain the matter and energy they need from wastes and dead plants and animals. Decomposers play an essential role in food webs.

What is a Fossil?

A **fossil** is any evidence of a past plant or animal contained in a sediment or rock. There are two kinds of fossils: **body fossils** and **trace fossils**. Body fossils, like those shown in *Figure 2*, are the actual organisms or some part of them. They may also be the imprint of the organism or some part of it. Bones, teeth, shells, and other hard body parts are relatively easily preserved as fossils. However, even they may become broken, worn, or even dissolved before they might be buried by sediment. The soft bodies of organisms are relatively difficult to preserve. Special conditions of burial are needed to preserve delicate organisms like jellyfish. Sometimes, such organisms fall into a muddy sea bottom in quiet water. There they might be buried rapidly by more mud. Only in circumstances like these can such organisms be fossilized. For that reason, the fossil record of soft-bodied organisms is far less well known. There is a strong **bias** in the fossil record. Some organisms rarely have the chance of becoming fossilized. Under very specific circumstances, however, even these can become part of the fossil record.

Geo Words

fossil: any remains, trace, or imprint of a plant or animal that has been preserved in the Earth's crust since some past geologic or prehistoric time.

body fossil: any remains or imprint of actual organic material from a creature or plant that has been preserved in the geologic record (like a bone).

trace fossil: a fossilized track, trail, burrow, tube, boring, tunnel or other remnant resulting from the life activities of an animal.

bias: a purposeful or accidental distortion of observations, data, or calculations in a systematic or nonrandom manner.

Figure 2 Dinosaur bones collected in Dinosaur National Monument in Utah.

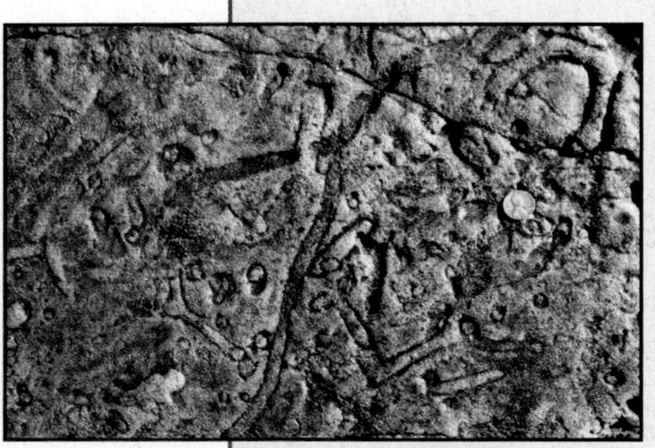

Figure 3 Feeding trails and burrows are kinds of trace fossils. Note the penny for scale.

Trace fossils, like those shown in *Figure 3*, are the record of life activities of organisms rather than the organisms themselves. Tracks, trails, burrows, feeding marks, and resting marks are all trace fossils. Trace fossils are useful for geologists and **paleontologists** because certain kinds of organisms, which live in specific environmental conditions, make distinctive traces.

In relatively young **sediments** and rocks, the actual body parts of an organism are often preserved. You modeled this in **Part A** of the investigation. In older rocks, the body parts are usually dissolved away. They may also be recrystallized or replaced by another kind of mineral. Even so, the imprints of the organisms are still preserved. They can be studied if the rock splits apart in the right place and the right orientation to reveal the imprint. Paleontologists usually collect large numbers of rock pieces. They then split the rock in the laboratory with special mechanical splitting devices to try to find at least a few fossils.

Geo Words

paleontologist: a scientist who studies the fossilized remains of animals and/or plants.

sediments: solid fragmental material that originates from weathering of rocks and is transported or deposited by air, water, or ice, or that accumulates by other natural agents, such as chemical precipitation from solution or secretion by organisms.

Fossilization

Figure 4 Fungi attack a fallen log in the woods.

As you saw when you looked at food chains and food webs, only a very small part of what once lived is spared being a meal for some other organism. There is a very high probability that any organism on Earth will be either consumed by another organism or decomposed by microorganisms following death. Decay affects not only soft body parts but also some of the harder, more resistant body parts. Think of a forest floor like the one in *Figure 4*. Each plant and animal that lives in the forest eventually ends up on the forest floor in some form. Soft tissues of animals, leaves, and flowers are used by decomposers. They decay within several weeks or are used by some other organism as a food source. The most resistant body parts include insect exoskeletons, vertebrate bones, wood, leaf cuticle, seeds, pollen, and spores. They may remain on the forest floor for many years or even centuries. This depends upon the physical and chemical conditions of the soil. A similar situation exists on the ocean floor.

For an organism or body part to become a fossil, it must either live within or be moved to a place where it can be buried. Burial alone does not guarantee that fossilization will occur. Under normal burial conditions all organisms undergo a scientifically predictable decay trend. Hence, there must be other factors operating during or immediately following burial to slow or stop decay. The conditions necessary for fossilization do not exist everywhere all of the time. In fact, they exist in only a few places and for only a tiny fraction of the time.

Despite this, enormous numbers of organisms have become fossilized. The reason lies in the extent of geologic time. It is difficult to imagine how long a million years is. Yet physical, chemical, and biological processes have been operating on Earth not just millions of years but billions of years. *Figure 5* shows a simple geologic time scale that indicates when various kinds of organisms are first seen in the fossil record.

Important Evolutionary Events of Geologic Time
(boundaries in millions of years before present)

Era	Period	Event	Boundary (base)
Cenozoic	Quaternary	modern humans	1.8
	Tertiary	abundant mammals	65
Mesozoic	Cretaceous	flowering plants; dinosaur and ammonoid extinctions	145
	Jurassic	first birds and mammals; abundant dinosaurs	213
	Triassic	abundant coniferous trees	248
Paleozoic	Permian	extinction of trilobites and other marine animals	286
	Pennsylvanian	fern forests; abundant insects; first reptiles	325
	Mississippian	sharks; large primitive trees	360
	Devonian	amphibians and ammonoids	410
	Silurian	early plants and animals on land	440
	Ordovician	first fish	505
	Cambrian	abundant marine invertebrates; trilobites dominant	544
Proterozoic		primitive aquatic plants	2500
Archean		oldest fossils; bacteria and algae	

Figure 5 Important evolutionary events of geologic time. Note that the divisions of geologic time are not drawn to scale.

Geo Words

fossiliferous rock: a rock containing fossils.

sedimentary rock: a rock resulting from the consolidation of accumulated sediments.

Fossiliferous Rocks

A rock, like the one shown in *Figure 6*, that contains fossils is said to be **fossiliferous**. Not all **sedimentary rocks** contain fossils. If you parachuted out of an airplane and landed on sedimentary rock, the chance of your finding a fossil would be rather small.

Figure 6 A fossiliferous limestone composed mostly of brachiopod fossils.

Some kinds of sedimentary rocks contain more fossils than others. Limestones are the most fossiliferous sedimentary rocks. That should not be surprising, because most limestones consist in part, or even entirely, of the body parts of shelly marine organisms. Some shales are fossiliferous as well, because certain organisms like to live on muddy sea floors. Sandstones are usually much less fossiliferous than limestones. Fewer kinds of organisms can tolerate the strong currents and shifting sand beds that are typical of areas where sand is being deposited. For the same reason, conglomerates are the least fossiliferous of sedimentary rocks.

Check Your Understanding

1. What is a food chain, and what role does it play in fossilization?
2. How does the rate of burial relate to the likelihood of fossilization?
3. What proportion of living organisms has the likelihood of becoming fossils?
4. What is the difference between a body fossil and a trace fossil?
5. Which kinds of sedimentary rocks tend to be the most fossiliferous, and which tend to be the least fossiliferous?

Understanding and Applying What You Have Learned

1. A geologic map shows the distribution of bedrock at the Earth's surface. Every geologic map has a legend that shows the kinds of bedrock that are present in the map area. The legend also shows the rock bodies or rock units that these rocks belong to, and their geologic age. Look at a geologic map of your community.
   a) What kinds of rocks are found in your community?
   b) Are fossils likely to be found in these rocks? Why or why not?
2. Revisit the **Think about It** question at the start of the activity.
   a) What evidence do you think the paleontologist would find in your community as proof of past life?
   b) Where would the paleontologist look to find this evidence?
   c) Do the organisms living in your community provide a biological signal that is unique to your area? Explain.
3. Must an organism die as a requirement to be represented in the fossil record? Explain your answer.
4. How are the physical and chemical processes responsible for preservation of plants and soft-bodied animals different from those for organisms that have hard skeletal parts?
5. Why would you expect that organisms living in ponds, lakes, or oceans have a greater chance of becoming part of the fossil record than organisms that live on land?

Preparing for the Chapter Challenge

Write a background summary that introduces fossils and fossilization. Be sure to identify those plant and animal parts in your community that you think are likely to leave some kind of fossil record, and discuss the reasons why you would expect one organism part to be more common than another. Also indicate where you would expect these fossils to be found within your community.

Inquiring Further

1. **Taphonomy**

   Taphonomy is the subdiscipline of paleontology that is concerned with the study of fossilization. The field of forensic science applies these principles to police and detective work. Investigate these subjects further.

2. **Common geological settings for preservation**

   Where on land or in the ocean might an organism be buried "dead or alive?" What are the conditions necessary to preserve soft tissues in these settings? What examples can you find in the fossil record of soft-tissue preservation?

Activity 2 North American Biomes

Goals

In this activity you will:

- Define the major biomes of North America and identify your community's biome.
- Understand that organisms on land and in the ocean have physical and chemical limits to where they live.
- Recognize the most common plants and animals in your community.
- Explore how a change in physical and chemical conditions within your community could alter your community's biome.
- Understand that there are predictable relationships between where the different biomes occur in North America.

Think about It

Humans have adapted to a wide variety of climates. Most plants and animals, too, have a range of climatic tolerances in which they live.

- Why don't the same plants and animals live all over the United States?

What do you think? Record your ideas about this question in your *EarthComm* notebook. Be prepared to discuss your response with your small group and the class.

Investigate

1. A biome is a major biologic community. It is classified according to the main type of vegetation present. The organisms that live in each biome are characteristic of that area. Examine the photographs of the major biomes of North America that have been identified by ecologists.

tundra

taiga

chaparral

desert

grassland

mountain zones

tropical rainforest

temperate deciduous forest

temperate evergreen forest

polar ice

a) Use the photographs above and on the previous page and the *EarthComm* web site to write a short description for each biome, identifying the biome, identifying the predominant plants and animals found in the biome.

b) Which biome most closely resembles your community?

c) How were you able to recognize your biome? What characteristic landscape did you use to pick the most representative photograph?

2. Look at the map of North American biomes. Match the photographs of the biomes with their locations shown on the map.

   a) How do you think that the climate of your community's biome differs from the climates of other biomes in North America? Consider the temperature and precipitation of the area as well as the temperature and precipitation changes from season to season.

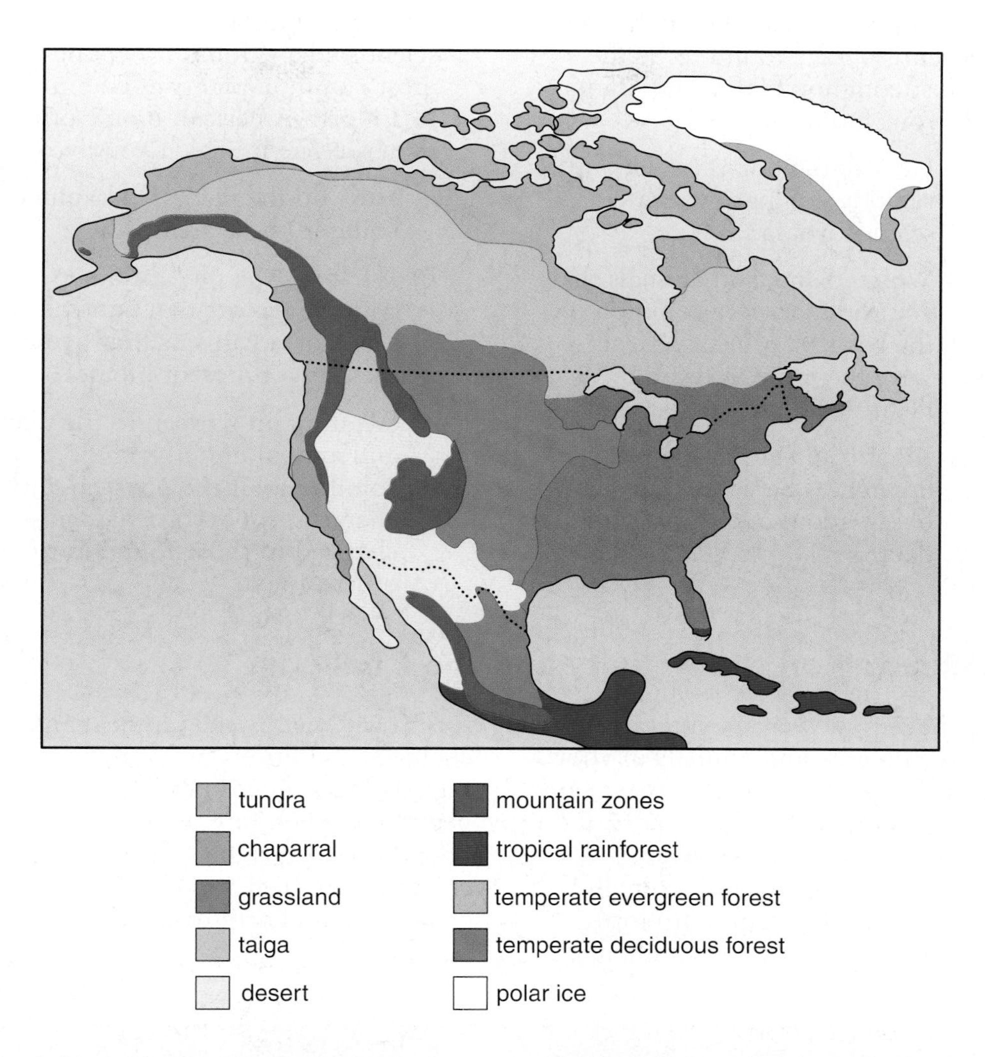

3. As a group, think of the common plants and animals that occur naturally in your community. Use any available resources.
   a) Make a list of the most common naturally occurring plants. Identify at least five plants.
   b) Make a list of the most common naturally occurring animals. Identify at least five animals.

4. Share your lists of common plants and animals with the rest of your class using a class discussion format. To make comparisons between groups easier, each group may want to write their lists on the chalkboard.
   a) From the compiled data, develop a single class list of the 10 most common plants and animals in your community. Rank the plants and animals in order from most common to least common.

5. Compare the class list with what scientists have found to be the most common plants and animals in your biome.
   a) How does your class list and ranking compare with the scientific data?
   b) Which plants and animals that you thought were common on the basis of your experiences were the same as those identified by scientists?
   c) What could account for any differences between your observations and the scientific data?
6. Compare the class list with what scientists have found to be the most common plants and animals in a North American biome other than the one in which you live.
   a) How do the plants and animals living in these areas differ?
   b) What are the physical and climatic factors that control the kinds of organisms that live in these two different biomes?
   c) What might happen to the plant and animal life in your community if the physical and chemical conditions suddenly changed to those found in the other biome?

Reflecting on the Activity and the Challenge

You have now seen the interactions between plants and animals in your community and compared these same relationships with another part of North America. You have also seen that different plants and animals characterize other parts of North America because their requirements for life are met by the physical and chemical conditions presently in those regions. You should now be able to begin to explain how biome boundaries can change in response to changing physical and chemical conditions on Earth over time.

Digging Deeper

CLIMATE AND BIOMES

An **ecosystem** is a community of plants and animals together with the physical and chemical environment in which that community exists. In the broadest sense, there are two major kinds of ecosystems on Earth today: aquatic and terrestrial ecosystems. Both of these kinds of ecosystems can be traced far back in geologic time. Each can be split into smaller subdivisions, using different criteria. For example, aquatic ecosystems can be subdivided into two categories using water chemistry: freshwater ecosystems and saltwater ecosystems. Each of these can be further subdivided. The aquatic ecosystem shown in *Figure 1* is a freshwater ecosystem. The terrestrial ecosystem can be divided into many categories, using the tallest plants in the area.

Figure 1 An example of a freshwater ecosystem.

Biological subdivisions on land are called **biomes**. Many present-day biomes parallel the lines of **latitude**. Climate conditions across the globe at these latitudes tend to be similar. In other words, there is a relationship between the physical and chemical processes that operate within latitude belts and the plants and animals that are adapted to these conditions. There is a slight difference in the case of mountains, because temperature decreases with elevation, as shown in *Figure 2*. Often, lowland biomes at high latitudes

Figure 2 How do you think the organisms found at the base of a mountain differ from those found near the top of the mountain?

Geo Words

ecosystem: a unit in ecology consisting of the environment with its living elements, plus the nonliving factors that exist in it and affect it.

biome: a recognizable assemblage of plants and animals that characterizes a large geographic area on the Earth; a number of different biomes have been recognized, and the distribution of the biomes is controlled mainly by climate.

latitude: a north-south measurement of position on the Earth. It is defined by the angle measured from the Earth's equatorial plane.

Geo Words

climate: the characteristic weather of a region, particularly as regards temperature and precipitation, averaged over some significant interval of time.

weather: the condition of the Earth's atmosphere, specifically, its temperature, barometric pressure, wind velocity, humidity, clouds, and precipitation.

(closer to the poles) are very similar to highland biomes at lower latitudes (closer to the Equator). For example, plant communities on the rocky, windswept peaks in the Appalachians are similar to those in many places in subarctic Canada.

Climate involves the long-term characteristics of the **weather** in a given region of the Earth. The most important factors in climate are temperature and precipitation. Both the average values and the deviations from the average are important. Everybody knows that some places are hot and some are cold. Some places are wet and others dry. What is also important is how much temperature and precipitation change from season to season. The average rainfall in an area may be high, but there may also be a long dry season. Plants need adaptations for surviving those dry periods before renewing their growth during the wet season. In areas with rainy seasons and dry seasons, it is also important for plant communities whether the wet season coincides with summer or with winter. You can easily see why climate plays the most important role in the distribution of biomes.

What links common animals with common plants is the dietary requirements of the plant eaters (herbivores). Similarly, the dietary requirements and taste preferences of meat eaters (carnivores) are linked to the specific herbivores that live in a biome. The most common plants, like the ferns, mosses, and trees shown in *Figure 3*, provide food as well as

Figure 3 Temperate evergreen forest, Hoh Rainforest, Washington.

Figure 4 How do you think an organism like an owl responds to climate change?

shelter for many animals in the biome. When plants and animals die, decomposers break down the dead and discarded organic matter, recycling it to the soils to be used again by plants. The rate of decay and the release of nutrients into the soil are also affected by the climate.

Global climate change is in the news nowadays. There is abundant evidence, however, that climate has changed continually in the past, long before the development of modern human society. These changes have taken place on time scales that range from as short as decades to as long as hundreds of millions of years. **Climate Change and Your Community**, investigates climate change in greater depth.

How do biomes respond to climate change? Generally, animals in a biome can migrate quite rapidly in response to climate change. Plants migrate much more slowly than animals. Plants are not able to respond immediately when the climate conditions change and the plant's tolerances are exceeded! Plants can shift their range only by dispersal of seeds and spores by the wind or by animals. Keep in mind that all animals depend, directly or indirectly, on plants as their food source. If climate change is slow, biomes can respond without much disruption as they shift in position. If climate change is more rapid, there can be great changes in the makeup of the plant and animal communities in a given area.

Geo Words

global climate: mean climatic conditions over the surface of the Earth as determined by the averaging of a large number of observations spatially distributed throughout the entire region of the globe.

Check Your Understanding

1. What are the two major kinds of ecosystems found on Earth?
2. Define, in your own words, the term biome.
3. What factor plays the greatest role in determining the distribution of biomes on Earth?
4. How do biomes respond to changes in climate?

Understanding and Applying What You Have Learned

1. Why might different teams in your class have developed different rankings of the common plants and animals in your community? What role might the location of your home have in what you determined to be the most commonly occurring organisms?
2. Determine the date when your town or city was incorporated.
   a) What plants and animals were common before that date?
   b) How have the common organisms in your community changed since that date?
   c) How do you explain these changes?
3. What are the main climate factors that restrict the distribution of plants and animals within your biome?
4. Refer back to the biome map of North America.
   a) What proportion of the United States lives in the same biome as your community?
   b) What is the relationship between biome distribution and latitude?
   c) What is the relationship between biome distribution and elevation?
5. What changes in the animals and plants might you expect if the climate of your community became colder? Became warmer?

Preparing for the Chapter Challenge

Write a short paper in which you describe the present biome in your community. Indicate which physical and chemical factors determine the limits of your biome, and indicate how variations in these factors could cause changes in the distribution of organisms in your community. For example, suppose that your community is now located in a mild climate. How might a decrease in the yearly average temperature affect the organisms in your community?

Inquiring Further

1. **Herbivores and carnivores in your community**

   What relationships do herbivores and carnivores have with plants in your biome? Do all consumers have the same preference for food, or do different groups of animals use different food sources?

2. **Animal adaptations to climate**

   Animals in Alaska must survive the extreme cold of the arctic and subarctic climate. What adaptations have Alaskan animals developed to survive? How do you think these adaptations might affect how these organisms would be preserved in the fossil record?

Activity 3 Your Community and the Last Glacial Maximum

Goals

In this activity you will:

- Investigate how changes in climate are linked to shifts in the distribution of oak and spruce trees in North America, 20 ka (20,000 years ago) to the present.
- Understand how scientists can use data collected from pollen and spores to reconstruct past environments.
- Understand how the climate of your community has changed over the past 18,000 years.
- Understand how this climate change has affected the plants and animals of your community.

Think about It

The Earth's climate has varied significantly over geologic time. As recently as 18,000 years ago much of North America was covered by thick glacial ice.

- What plants and animals (if any) lived in your community when large parts of North America were covered by ice?
- As the continental glaciers retreated northward, what changes in the plants and animals happened over time?

What do you think? Record your ideas about these questions in your *EarthComm* notebook. Be prepared to discuss your response with your small group and the class.

Investigate

Part A: Can You Recognize a Forest without the Trees?

1. Examine the maps. Pollen and spore records were recovered from lakes, ponds, and bogs across North America. Coupled with carbon-14 radio-isotopic dating, scientists have reconstructed the changes in plant types over space and time. The maps show sample sites across North America for approximately the last 20,000 years. The percentages of spruce and oak pollen are plotted on each map set.

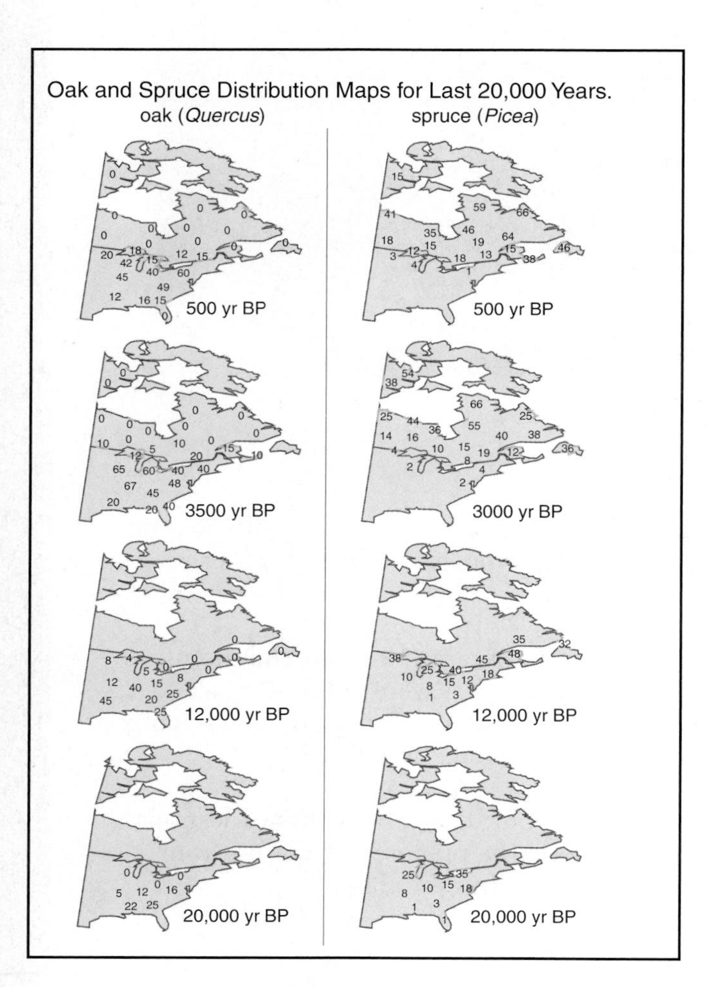

   a) Obtain copies of the maps. Using a different color for each plant, draw the 5% and 20% contour lines neatly across each map.

   b) Where is the maximum percentage of spruce found 500, 3000, 12,000 and 20,000 years ago?

   c) Where is the maximum percentage of oak found 500, 3500, 12,000 and 20,000 years ago?

2. Oak trees are characteristic of deciduous forest, and spruce trees are representative of the taiga.

   a) Do the general trends in oak parallel the changes in spruce? Explain.

   b) What other plants and animals might you expect to find in the fossil records in the collection sites 3000 and 20,000 years ago?

Part B: Your Community Biome: Pleistocene to Present

1. Develop a hypothesis concerning the plants and animals that may have lived in your community when North America experienced the transition from a glacial period to the present day (nonglacial period).

   a) Record your prediction and the reason for your prediction in your *EarthComm* notebook.

2. Visit the *EarthComm* web site to find a list of databases and animations regarding pollen distribution from the Late Pleistocene to the present.

   a) Make a list of plants that have been found in your community from approximately 18,000 years ago to the present.

b) How does this list compare to the list of plants commonly found in your community, which you developed in **Activity 2**?

c) How do you account for the differences between the plants found in your community today and the plants that were found in your community in the past?

3. Databases also exist that document the fossil animals found across North America through time. Visit the *EarthComm* web site to find a list of databases that show animal distribution in North America.

a) Why are fossil animal records not as comprehensive as pollen records?

b) Make a list of animals that have been found in your community from approximately 18,000 years ago to the present.

c) How does this list compare to the list of animals commonly found in your community, which you developed in **Activity 2**?

d) How do you account for the differences between the animals found in your community today and the animals that were found in your community in the past?

4. Evaluate the hypothesis you generated at the beginning of **Part B** of the investigation.

a) Do the data you have collected verify or refute your hypothesis?

b) What modification(s) to your hypothesis must be made, in light of the data you found? Revise your hypothesis according to your limited data set.

Reflecting on the Activity and the Challenge

You have learned that the plants and animals that once lived in your community were probably very different than today. As changes in climate accompanied the retreat of the glaciers, there were changes in the composition of the ecosystem. Compiling information on past communities helped you to think about the potential change in plants and animals in your state, region, and community under future changes in climate.

Geo Words

spore: a typically unicellular reproductive structure capable of developing independently into an adult organism either directly if asexual or after union with another spore if sexual.

pollen: a collective term for pollen grains, which are microspores containing the several-celled microgametophyte (male gametophyte) of seed plants.

Digging Deeper

ORGANISM RESPONSE TO CLIMATE CHANGE

Pollen and Spores

The reproduction in plants is different from that in animals. As part of the plant life cycle, **spores** (from mosses, ferns, and horsetails) or **pollen** (from most other kinds of plants) are produced in the hundreds of thousands per individual plant each year. Spores and pollen can be dispersed in a number of ways. The most common way is by wind. This is made easier by the very small size of individual pollen or spore grains, generally less than 50 μm (0.05 mm). *Figure 1* shows a 15 μm pollen grain magnified 1600 times its actual size.

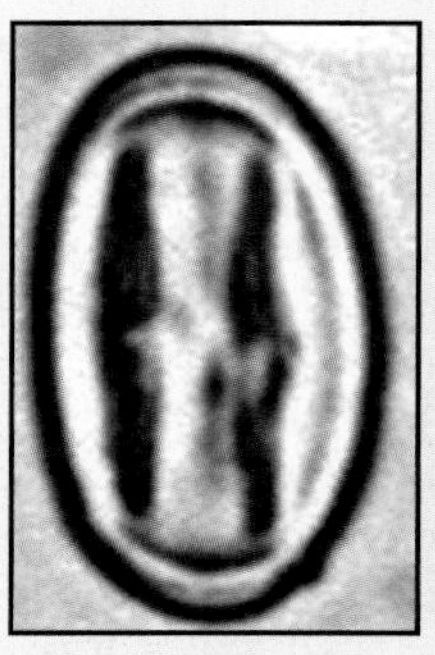

Figure 1A Transmitted-light image of a pollen grain (of the genus *Cupuliferoipollenites*), 15 μm in diameter.

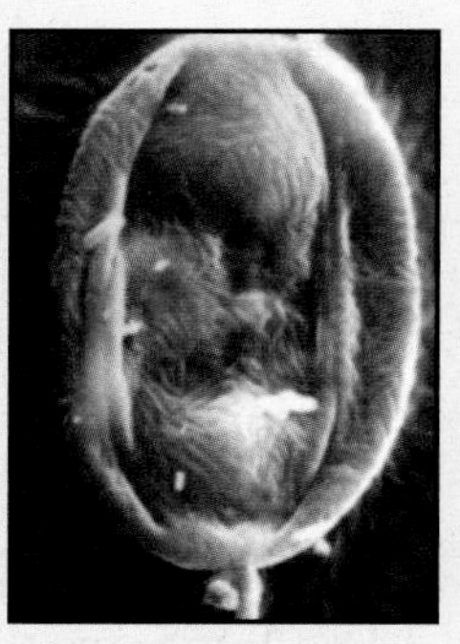

Figure 1B Scanning electron microscope image of the same pollen grain.

All species are genetically unique. The genes of an organism control its specific growth and development. When plants reproduce, each plant species produces a unique spore or pollen grain. Spores and pollen grains differ greatly in their shape, size, and surface features. Some plants produce spores that are triangular, and others produce spores that are football shaped, as shown in *Figure 2*. Some spores are covered with short spines or club-shaped structures, and

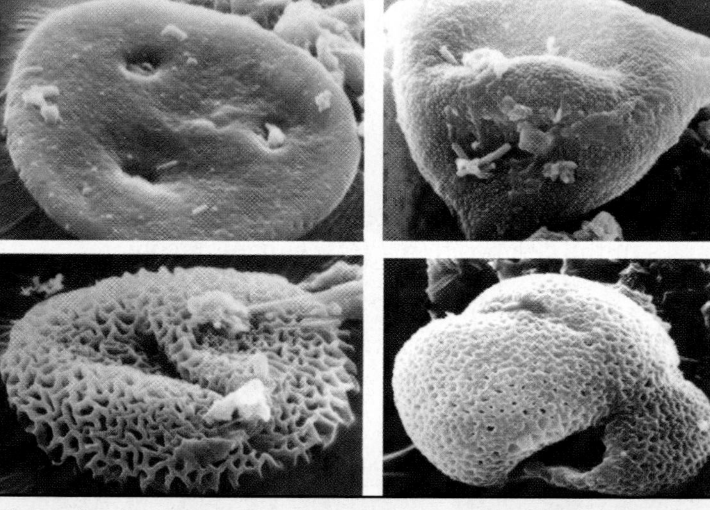

Figure 2 Scanning electron microscope image of four different pollen grains, ranging in diameter from 18 to 38 μm.

others are perfectly smooth. All plants of the same species produce the same type of pollen each year throughout each generation. The pollen grains are transported to ponds, lakes, and rivers where they settle with the sediment and are buried.

Pollen and spores are very resistant to decay. Pollen and spores can be preserved for hundreds of millions of years under the right burial conditions. This resistance to decay is the main reason why spores and pollen can be extracted not only from recently deposited sediments but also from ancient sedimentary rocks from the geologic past.

Paleoclimate

Paleoclimatology is the study of the Earth's past climates. Climatologists know from several kinds of evidence that the Earth's climate has changed continually, and substantially, through geologic time. Over the past million years (a short time, geologically!) there have been several colder periods when vast ice sheets, as much as a few kilometers thick, expanded to cover much of northern and central North America and Eurasia. In between these **glacial periods**, there were shorter periods, called **interglacials**, when global temperature was much warmer and ice sheets disappeared from North America and Eurasia. *Figure 3* shows temperature changes over the past 420,000 years as interpreted from data recovered from the Vostok ice core (Antarctica).

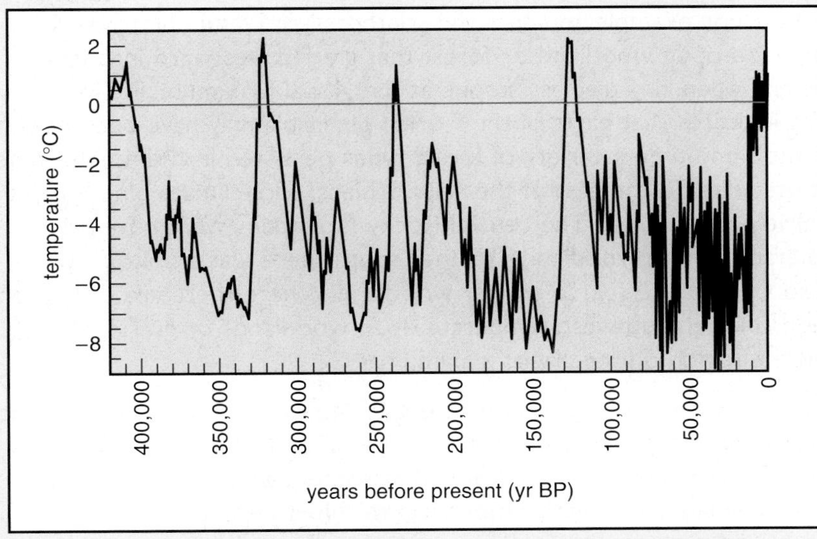

Figure 3 Temperature variation over the past 420,000 years relative to the modern surface temperature at Vostok (−55.5°C).

Geo Words

paleoclimatology: the scientific study of the Earth's climate during the past.

glacial period: an interval in time that is marked by one or more major advances of glacier ice. Note that the time interval is not necessarily of the same magnitude as the "Period" rank of the geologic time scale.

interglacial period: the period of time during an ice age when glaciers retreated because of milder temperatures.

Geo Words

steppe: an extensive, treeless grassland found in semiarid mid-latitude regions. Steppes are typically considered to be drier than the prairie.

Glacial periods last longer than interglacials. Ice sheets build up slowly over several tens of thousands of years, and then retreat rapidly from their maximum advance over a brief period of several thousand years. Interglacials last only 20,000 to 25,000 years. The last glacial maximum was about 20,000 years ago. By about 7000 years ago, almost all of the Northern Hemisphere continental ice sheets, except for the Greenland ice sheet, had melted away. Since then, global climate has been relatively warm, although with significant fluctuations over decades, centuries, and millennia.

The Response of Biomes to Climate Change

Each glacial advance modified the climate across North America. In response to these climatic changes, there was a change in biome boundaries. Each plant species has evolved to be adapted to a particular range of climatic conditions. Outside of that range of conditions, the climate is too cold or warm, or too wet or too dry, for that species to survive over the long term. As climatic belts shift during times of global climate change, plant species shift accordingly in their range of distribution. If the change is too rapid, they become extinct. The animals that rely upon specific kinds of plants for food or shelter are similarly affected. Some of the organisms that lived during the most recent glacial and interglacial periods continue to exist today in North America. Some of the organisms that lived in the past are now extinct.

The distribution of biomes during the last glacial period, as determined from the fossil record across North America, was very different from the present distribution. For example, much of the southeastern United States was covered in an open woodland or forest that graded westward into forest **steppe** and open, dry treeless steppe at the glacial maximum. Fossil evidence indicates that parts of the Florida peninsula may have been desert-like. At the same time a variety of forest types persisted in the northernmost part of the state. To the east of the Appalachians, cool-climate pine forests and prairie herbs existed. The central Rocky Mountains were covered in cold-tolerant conifer woodlands. Farther south there was a mixture of semidesert scrub and sparse conifer woodland. Where there are dry desert conditions in the southwest, temperate open woodlands or grasslands covered the region during glacial maximum.

Not only was biome distribution different in response to the shift in climate belts across North America, but many of the organisms within them were also different. Fossil evidence indicates that some associations between plants and animals, particularly among ice-age North American mammals that exist today did not exist in the Pleistocene. For example, the snowshoe hare still found in Minnesota and Wisconsin is known from fossil evidence in

Missouri, Kentucky, and Virginia when these states were covered in taiga. It's known that extinction of large mammals in North America occurred at the time when the last continental ice sheets melted. Two hypotheses have been proposed to account for this. One hypothesis proposes that ecological changes at the end of the Pleistocene disrupted the biological balance. This hypothesis suggests that at the time of biome readjustment, a large number of species were unable to adapt. The other hypothesis, the "overkill model," proposes that the expansion of human migrants from Asia into North America resulted in overhunting and depletion of the mammal populations. Some authors suggest that humans killed everything in their path, whereas others hypothesize that humans reduced mammal populations to low numbers. Such low populations, in turn, were unable to maintain birthrates high enough for the species to survive. Whatever the reason (or reasons), North America today is very different from the North America of just 10,000 years ago.

Check Your Understanding

1. What are the two hypotheses to account for the extinction of some animals at the end of the Pleistocene?
2. Why are pollen and spores conducive to long-term preservation?
3. How was the climate of North America different during the last full glacial episode? What impact did this have on organisms in the area?

Understanding and Applying What You Have Learned

1. A scientist is like a detective. Rarely do detectives have all of the information to solve the case. However, they usually get enough evidence to show beyond a reasonable doubt what happened and when it happened. Use evidence that you have gathered to answer the following questions:

   a) How can you demonstrate that the biome in which you now live was different during and after the last glacial maximum?
   b) How many different kinds of fossil data have been found in your community or region?
   c) What do these fossil data tell you about the plants and animals that once lived where you do?
   d) What happened to the Late Pleistocene plants and animals that today are not found in your community?

2. It has been stated that the fossil record is biased. What biases might have affected the Late Pleistocene fossil record of your community's recent past?

3. How good is the Pleistocene fossil record? How much data have been collected and used to characterize the North American Pleistocene?

4. a) For the Pleistocene plants and animals that no longer live in your community and survived the Pleistocene extinction, what part(s) of the food web do they occupy? Why might they no longer be found in your community naturally?

   b) For the Pleistocene plants and animals that lived in your community but didn't survive the Pleistocene extinction, what part(s) of the food web did they occupy? Why might they have become extinct?

Preparing for the Chapter Challenge

Think about what you have learned about the plants and animals in the present and past biome of your area. Prepare an illustration that shows the relationships between the organisms of the past and present. Can you document a progression from the organisms of the past to the organisms of the present? Include a narrative in which you explain why the organisms changed through time.

Inquiring Further

1. **Carbon-14 dating techniques**

   Search your library and the web for information about how it is possible to determine the carbon-14 age of organic matter. Prepare a report for the class on the physical and chemical principles upon which the technique is based, and how advances in analytical equipment have reduced the uncertainty in the age estimates.

2. **Late Pleistocene extinctions**

   Approximately 11,000 years ago, a variety of animals across North America became extinct. These were mostly large mammals (over 50 kg), like saber-toothed cats, mammoths, mastodons, and giant beavers. What mechanisms have been proposed to have caused the extinctions in North America? Did Late Pleistocene extinctions occur in places other than North America? If so, where and what types of organisms became extinct?

Activity 4 The Mesozoic—Cenozoic Boundary Event

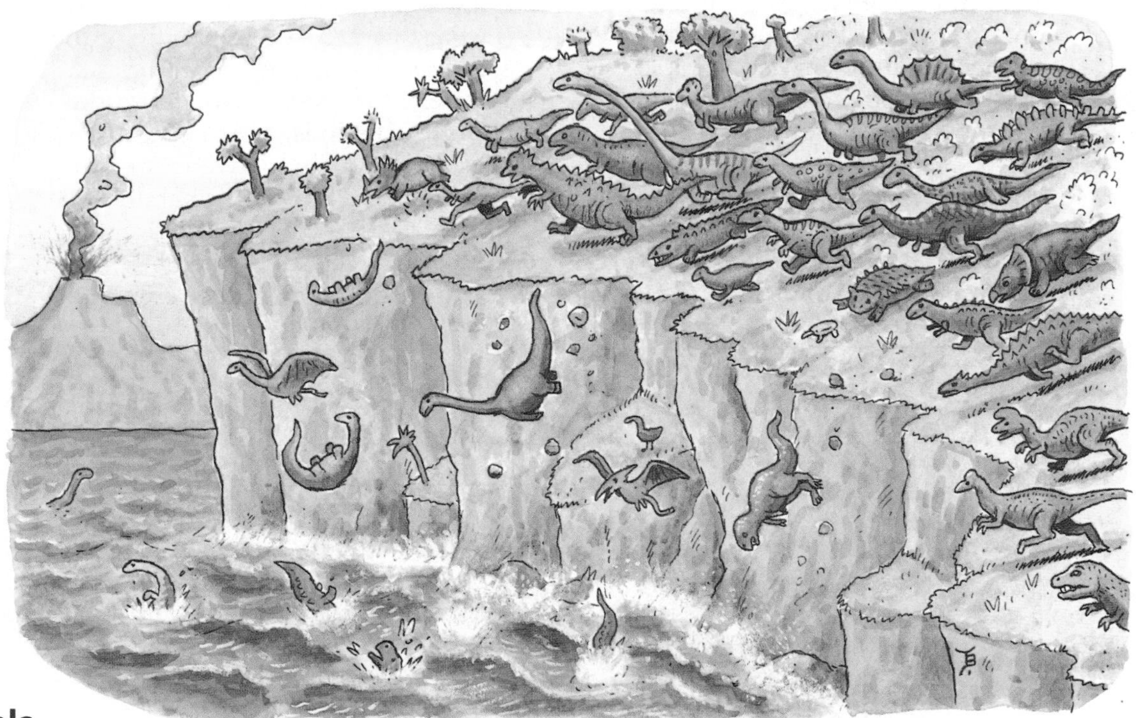

Goals

In this activity you will:

- Understand how changes in the Earth's climate have affected organisms throughout geologic time.
- Understand that the organisms that dominate the continents today differ from the organisms that dominated the Earth in the deep geologic past.
- Understand that severe ecological disruptions alter the history of life, resulting in extinction followed by the evolution and appearance of new organisms.
- Understand that newly evolved organisms develop similar body features that allow them to use the same resources as those organisms that became extinct.

Think about It

Perhaps no other interval of geologic time has captured the attention of popular culture as much as the Mesozoic Era. This was the time when dinosaurs roamed the Earth.

- If the dinosaurs were such successful creatures, what happened to them, and why?

What do you think? Record your ideas about this question in your *EarthComm* notebook. Be prepared to discuss your ideas with your working group and the class.

Investigate

Part A: Changes in Climate and Life at the End of the Mesozoic Era

1. Look at the geologic time scale. Familiarize yourself with the terms used to name different geologic time intervals. The boundary between the Mesozoic Era and the Cenozoic Era, about 65 million years ago, represents one of the most catastrophic extinction events in Earth history. In this part of the investigation you will focus on the changes in climate and organisms between the end of the Cretaceous Period (the end of the Mesozoic), and the beginning of the Tertiary Period (the beginning of the Cenozoic). You may sometimes see different names used to refer to the Cenozoic Periods. That is because there are two alternative ways that the Cenozoic Era can be subdivided into periods. In the older terminology, the Cenozoic is subdivided into the Tertiary and Quaternary Periods.

Major Divisions of Geologic Time
(boundaries in millions of years before present)

Boundary	Epoch	Period		Era	Eon
	Holocene	Neogene	Quat.	Cenozoic	Phanerozoic
.01	Pleistocene				
1.8	Pliocene		Tertiary		
5.3	Miocene				
24	Oligocene	Paleogene			
34	Eocene				
56	Paleocene				
65		Cretaceous		Mesozoic	
145		Jurassic			
213		Triassic			
248		Permian		Paleozoic	
286		Pennsylvanian			
325		Mississippian			
360		Devonian			
410		Silurian			
440		Ordovician			
505		Cambrian			
544				Proterozoic	Cryptozoic (Precambrian)
2500				Archean	

Note that the divisions of geologic time are not drawn to scale. Quat. = Quaternary

The newer terminology, however, subdivides the Cenozoic Era into the Paleogene and Neogene Periods. It is important to note, as is shown in the "Major Divisions of Geologic Time" diagram, that the Period boundaries are not in the same place for the two different terminologies.

2. In your group, visit the *EarthComm* web site to collect data on paleoclimate from the pre-boundary time interval (the Late Cretaceous) and the post-boundary time interval (the earliest Tertiary, also known as the Paleogene Period).
   a) How did the climate change at the end of the Mesozoic?
   b) What evidence is there on how fast the climate changed?
   c) From your findings, what do you think are the dominant plants and animals in North America before the boundary?
   d) From your findings, what do you think are the dominant plants and animals in North America after the boundary?
3. In your group, visit the *EarthComm* web site to collect data on the fossil record of organisms from the pre-boundary time interval (the Late Cretaceous) and the post-boundary time interval (the Paleogene).
   a) How accurate were your predictions on the dominant plants and animals in North America before and after the boundary? Explain any differences.
   b) How do you think that the interactions between organisms differed before and after the boundary? Why?
   c) Which organisms from the fossil record are still living (extant) and which ones are no longer known to be alive on Earth today (extinct)?
   d) Why do you think that some organisms survived beyond the boundary, but others did not?
   e) How do you think that the dominant plants and animals in your community differed before and after the boundary? Why?

Part B: Consumers of the Mesozoic and Cenozoic

The adaptions of organisms that have evolved to function within a given ecosystem appear to have several common features that allow success. You have now seen that plants before and after the extinction between the Cretaceous and Tertiary Periods (the K/T extinction) were very similar but that the dominant animals were very different. Or were they?

1. Examine the photographs of the skull and jaw structure of both a hare (rabbit) and a wolf (shown on the next page).

Hare skull.

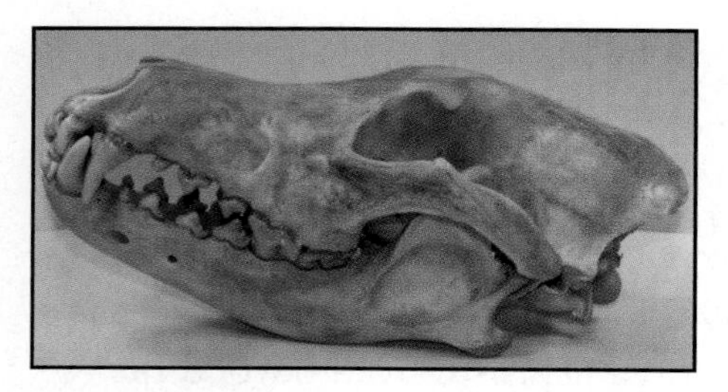

Wolf skull.

a) What are the differences in the organization of teeth in these animals?

b) Which animal is a carnivore and which is a herbivore?

c) How does the tooth arrangement reflect their levels within the ecosystem?

2. Examine the photographs of extinct fossil mammal skulls collected from Paleogene rocks of North America.

   a) How are these skulls similar to modern rabbits and wolves?

   b) In which level(s) of the ecosystem would you predict that each of these animals existed? That is, which ones were herbivores and which ones were carnivores? How can you make these interpretations?

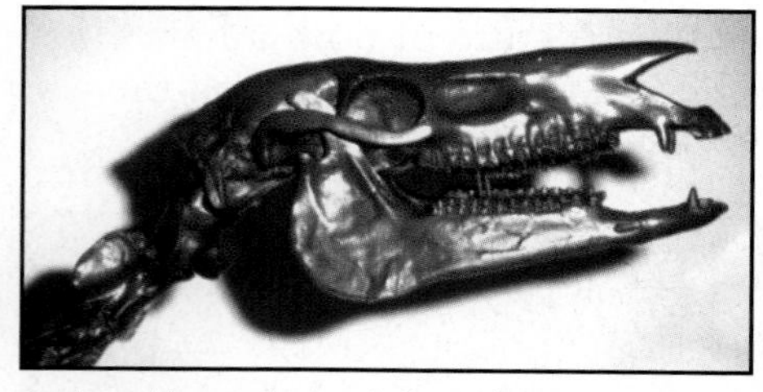

Archaeohippus (an early horse).

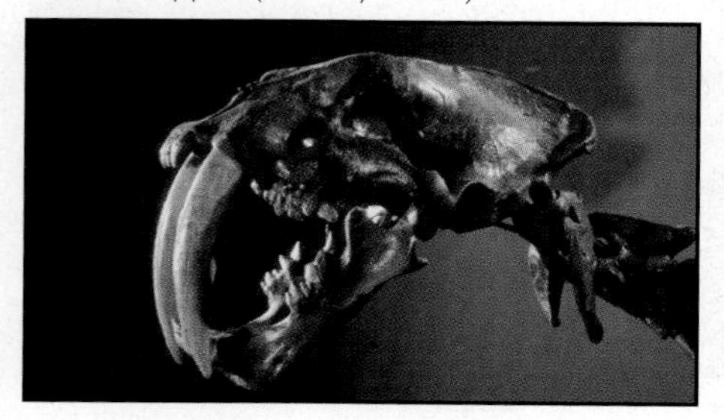

Smilodon *californicus* (saber-toothed tiger).

3. Examine the photographs of dinosaur skulls collected from Mesozoic rocks of North America.

   a) What features do these skulls have that are similar to the modern rabbits and wolves, and to the extinct mammals from the Tertiary Period?

   b) Which dinosaurs exhibit skull, jaw, and teeth similar to the herbivores?

   c) Which dinosaurs exhibit skull, jaw, and teeth similar to the carnivores?

   d) In which level(s) of their ecosystem would you predict that each of these animals existed? How can you make these interpretations?

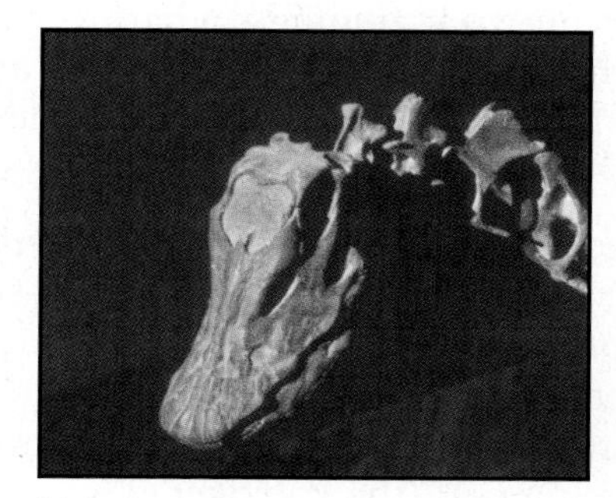

Diplodocus.

Tyrannosaurus *rex*.

Reflecting on the Activity and the Challenge

This activity gave you the chance to explore the fact that life forms that dominated Earth in the deep past differ in several ways from what exists at the present. However, these extinct life forms occupied all the parts of the same ecological levels that exist today. It also showed you that at certain times in the geologic past, climate and life changed over very short geological times. New organisms evolved to take the ecological place of those that became extinct. How did this activity change your ideas about life on Earth? You will need to explain the appearance of new animals and plants in Paleogenic (post-boundary) rocks that have no fossil record in Cretaceous (pre-boundary) rocks as part of your **Chapter Challenge.**

Digging Deeper

THE EXTINCTION OF SPECIES

The success or failure of a particular species through time is impossible to predict. Many physical and biological factors interact with one another in complicated ways over a long time to determine success or failure. Additionally, the data from the fossil record seem to indicate that different kinds of organisms have different rates of overall success. Paleontologists have documented that some species have persisted for tens of millions of years. Others span only a few million years, or even only a few hundred thousand years.

The Extinction Event at the End of the Mesozoic

You have learned that there was a marked difference in the landscapes of the Mesozoic Era and the Cenozoic Era. (The terms come from the Greek *meso-*, meaning middle, and *kainos-*, meaning new.) The groups of animals that dominated Earth for nearly 130 million years during the Mesozoic in the pre-boundary biosphere mysteriously disappeared. The disappearance was sudden; almost overnight, in terms of geologic time. This extinction affected some plants and many groups of animals that lived on land. It similarly affected much of the food web in the oceans. Groups from lowly **phytoplankton** to top carnivores disappeared from the fossil record, never to be seen again except as fossil material. The post-boundary biosphere, which became established early in the Paleogene was very different in its aspect. It took several million years for the plant and animal groups

Geo Words

phytoplankton: small photosynthetic organisms, mostly algae and bacteria, found inhabiting aquatic ecosystems.

known to exist at present to evolve and fill all of the ecological spaces opened by the end-of-the-Mesozoic extinction event.

Geo Words

background extinction: normal extinction of species that occurs as a result of changes in local environmental conditions.

biodiversity: the diversity of different biologic species and/or the genetic variability among individuals within each species.

mass extinction: a catastrophic, widespread perturbation in which major groups of species become extinct in a relatively short time compared to normal background extinction.

Figure 1 Dinosaur tracks provide evidence for these prehistoric life forms.

If you were to examine the post-boundary fossil record in more detail, you would find that changes in the kinds of animals now extinct are related to the appearance and disappearance of their food source(s). The fossil record indicates that when evolution changes the basic composition of plants in a community, those dependent organisms must find a new food source, change how they process food for nutrition, or face extinction. For example, there is no physical evidence in the extensive fossil record, to date, for many grazing animals until the mid-Cenozoic. The first evidence for the evolution and appearance of grasslands appears at about this time. Following this, many new groups of animals whose diets include these plants are found for the first time in the fossil record, not before. In North America such animals include camels, rhinoceroses, horses, and a wealth of other mammals that are now known to be extinct. The extinction of a few species now and then appears to be a normal phenomenon. Scientists refer to the appearance and disappearance of a few species at any time as **background extinction**. When the fossil record documents a significant change in Earth's **biodiversity**, these events are termed **mass extinctions**. There have been five major mass extinctions during the history of life. During these mass extinctions up to 90% of the known biodiversity was lost. The extinction event at the end of the Paleozoic Era, between the Permian and the Triassic Periods, was even more devastating to life on Earth than the extinction at the end of the Mesozoic. The general patterns of evolution and recovery in each of the post-extinction worlds are similar, including the appearance of new organisms with similar adaptions for living. Understanding these patterns may provide insight into how the Earth's biosphere responds to severe trauma.

Figure 2 The woolly mammoth became extinct near the end of the Pleistocene Epoch.

Incidentally, are you aware that the subdivision of geologic time into the Paleozoic, Mesozoic, and Cenozoic Eras at the particular places in the span of geologic time was adopted just because of the magnitude and abruptness of the extinctions? To a great extent, the accepted scheme of subdivision of geologic time isn't just arbitrary; it has a natural basis.

The Cause of the Extinction

Geologists and paleontologists have known for more than 150 years about the mass extinction at the end of the Mesozoic. The cause, or causes, of the extinction, however, remained obscure until very recently. In the 1970s a scientist name Luis Alvarez and his coworkers proposed that the extinction was caused by collision of a gigantic asteroid with the Earth. They based that hypothesis on studies of cores and outcrop sections through sedimentary rocks whose deposition spanned the Mesozoic–Cenozoic boundary. In several such sections they found geochemical evidence, in the form of unusually high concentrations of the chemical element iridium, that pointed toward a catastrophic collision. Iridium is known to be introduced into the Earth system during certain meteorite and asteroid impacts. Later, what most geoscientists consider to be the "smoking gun" was found in the Yucatán Peninsula of Mexico: the remnants of a colossal impact structure, called

the Chicxulub crater. The idea is that the collision placed such quantities of dust and ash into the atmosphere that the climate became dramatically cooler and it is believed that light was unable to reach Earth's surface. The Earth's ecosystems were stressed for long times after the collision, leading to widespread extinction of species of many kinds of organisms. In the Gulf of Mexico, evidence of sediment movement and deposition by a gigantic sea wave, presumably caused by the impact, has strengthened the hypothesis.

Figure 3 The Chicxulub Crater, on the Yucatan Peninsula of Mexico, provides evidence for a large impact event at the end of the Cretaceous. Here the crater is magnified using detailed measurements of the Earth's gravitational field and is revealed as the roughly circular feature in the center of this image.

Many aspects of the extinction remain unclear. Extinction of species continued for geologically long times after the collision. One needs to appeal to long-term ecological effects to explain this circumstance. In addition, there is also good evidence for increased volcanic activity around the time of the Mesozoic–Cenozoic boundary. A minority of geoscientists favor an alternative hypothesis for the great extinction, having to do with climate change induced by the volcanism. Evidence is still being gathered to resolve the details of the extinction and its causes.

Check Your Understanding

1. How long do species last?
2. What happens to animal species when the food sources of the animals change?
3. How can you document that as different groups of organisms evolved they developed similar structures, allowing them to occupy a position in the ecological hierarchy?
4. What is the difference between background extinction and mass extinction?
5. What kinds of evidence can be used to document a major meteorite or asteroid impact in the geologic past?

Understanding and Applying What You Have Learned

1. Think about the Mesozoic and Cenozoic fossil records you have uncovered. How do the organisms of the deep geologic past differ from those in your present community? Why?
2. What is the probability of one or more of the organisms that lived just before or just after the time of the Mesozoic–Cenozoic boundary still existing today somewhere in the world? Explain your answer.
3. If you were to examine the rock record of another planet, how would you identify that a mass extinction occurred when you were unfamiliar with that planet's life forms?
4. How are new data discovered, where are these data reported, and what is the purpose of independent review of a scientist's data and ideas?

Preparing for the Chapter Challenge

With your group, develop a poster presentation that illustrates both the Cretaceous (pre-boundary) and Paleogene (post-boundary) organisms within their ecological context. Indicate on the poster all of the data sources you consulted, the approximate duration of time that is represented by your fossil communities (how many millions of years), and the major differences between pre-boundary and post-boundary biodiversity. Compare these data to the data you collected for the last glacial maximum to the present. Over how many years is it thought that both of these extinction events occurred? Consider which organisms or types of life strategies appear to have been affected most by the extinction events.

Inquiring Further

1. **Hypotheses to explain the Mesozoic–Cenozoic boundary event**

   Using available resources, investigate further the hypotheses that have been proposed to explain the Mesozoic–Cenozoic boundary event. (It would be useful to know that the event has often been called the "Cretaceous–Tertiary boundary event," because in earlier usage, which still exists today, the Cenozoic was divided into the Tertiary and Quaternary periods.) How do the hypotheses differ? What scientific evidence and data exist to support each hypothesis?

2. **Other mass-extinction events**

   Research other mass extinctions that paleontologists have identified in the rock record. When in time did these occur? What groups of organisms were affected?

Activity 5 How Different Is Your Community Today from that of the Very Deep Past?

Goals

In this activity you will:

- Understand that many different plants and animals evolved and became extinct during the Phanerozoic.
- Recognize that plant and animal fossil assemblages of the very deep geologic past (for example, in the Paleozoic) are unlike organisms alive today.
- Describe the variety of plants and animals that once lived in your community.
- Relate the increases and decreases in the numbers of different organisms during the Phanerozoic to geologic causes.
- Describe several lines of fossil evidence to support the theory of evolution based on the fossil record.

Think about It

You have been learning about how life has been changing throughout geologic time.

- If you could be transported back in time, would you recognize your community?

What do you think? Record your ideas in your *EarthComm* notebook. Be prepared to discuss your ideas with your working group and the class.

Investigate

1. Examine a geologic map of your region.
   a) Which geologic time periods are represented by the rocks in your community?
   b) Choose one of these periods as your "target" in the geologic past. Make sure that all of the time periods represented by the rock record in your community are selected by one or more of your classmates.
2. Using available resources, including textbooks, encyclopedias, the *EarthComm* web site, and museum displays (if possible), collect data on the kinds of plants and animals that have been found as fossils in your target geologic period.
3. Construct either a natural habitat (for land-dwelling animals) and aquarium (for marine-dwelling organisms), or a botanical garden (for land plants) diorama depicting your community during your selected interval of geologic time. Your display should present the biodiversity of the single period of time you chose to investigate.
4. Once you and your classmates have constructed your displays, assemble them in your classroom in order according to the geologic time scale.
   a) How has your community changed over time? Take notes on the similarities and differences between adjacent "exhibits."

Have your teacher approve the materials you plan to use to construct your diorama.

Reflecting on the Activity and the Challenge

You may have seen that some major groups of organisms are restricted to one or two geologic periods. Other groups may have lived during several geologic periods. You may also find that some geological periods do not show evidence for one or more particular groups of organisms. This may be a result of the kinds of sedimentary rocks preserved in your region. Check the geologic map of your area, again, to see if all the geologic periods are represented by rock. Once you have finished comparing and contrasting the data, decide for yourself if you could recognize your community at any given point in geologic time on the basis of the plants and animals known from its fossil record.

Geo Words

species: a group of organisms, either plant or animal, that may interbreed and produce fertile offspring.

taxonomy: the theory and practice of classifying plants and animals.

morphology: the (study of the) features that comprise or describe the shape, form, and structure of an object or organism.

anatomy: the (study of the) internal structure of organisms.

biodiversity curve: a graph that shows changes in the diversity of organisms as a function of geologic time.

Digging Deeper

BIODIVERSITY AND CLIMATE CHANGE

The classification of plants and animals into **species**, and the grouping of species into related groups, is called **taxonomy**. The species is the basic taxonomic unit. A species can be defined as a population of organisms that can interbreed to produce fertile offspring. Species that seem to be closely related, in terms of how they evolved, are grouped into a higher taxonomic unit called a genus (plural: genera). In turn, genera that are thought to be closely related are grouped into families, and so on; see *Figure 1*. In today's world, biologists can (in theory, at least!) determine what a species really is. Paleontologists can't do that, because individuals of the species are no longer alive. Paleontologists try to recognize features using both external appearances (**morphology**) and internal structures (**anatomy**). Keep in mind also that genera and higher taxonomic units cannot be directly verified, either for fossil organisms or for living organisms.

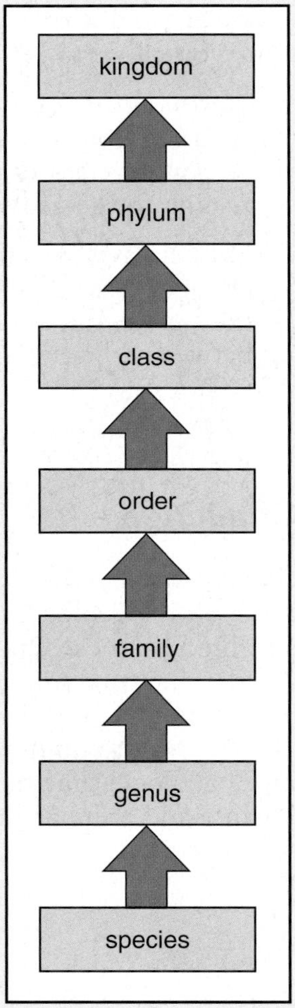

Figure 1 The hierarchy of taxonomic units.

All evolutionarily related species and genera (for example, *Homo sapiens*) are placed into the next highest larger category called a family (*H. sapiens* belongs to the family Hominidae). Related families are grouped into a higher category called an order (Hominidae belong to the order Primates). Hence, there can be many families in any order and many genera within any family. Paleontologists have tried to construct graphs, called biodiversity curves that show biodiversity as a function of geologic time. These **biodiversity curves** are based on the fossil record of higher-order taxonomic units like families or orders. That is, rather than using the number of species or genera known from an interval in geologic time, paleontologists have used the number of families or orders that have been identified.

Your community's biodiversity as reflected in the fossil record is related both to the geologic period you have examined and to the physical and chemical environment that existed there. There is incontrovertible evidence that the number, diversity, and complexity of both marine and terrestrial organisms found in the fossil record have changed through the hundreds of millions of years of Earth's history. Many of the "big groups" that are found today in your community have fossil records that extend deep into the past. You can certainly identify vertebrates in your community, and vertebrates certainly have a fossil record that is first found in the Paleozoic. You have also discovered in this activity, however, that there are many strange and interesting creatures that once lived but are no longer found on Earth today.

Figure 2 Reconstruction of what life may have been like 540 million years ago in British Columbia, Canada, based on the fossils found within the Burgess Shale.

New fossil data are published weekly all over the globe. These data do not originate only from scientists working in North America, nor are the scientific journals in which they are found published only here. Rather, there are thousands of paleontologists (in China alone, for example) studying Earth's fossil record on all continents, including Antarctica, and there are nearly 100 different scientific journals that publish data on fossils. The results of these scientists' investigations continue to support and reinforce the general trends you have analyzed in this activity.

One principle of scientific inquiry is that a hypothesis may be falsified sometime in the future if some new, scientifically reviewed and published data are found. These new investigations may challenge certain specifics concerning an individual fossil or the group to which it belongs.

One recent example of this is the controversy about whether or not feathers are only found in birds or if non-avian reptiles also evolved feathers. Nevertheless, the overall evolutionary pattern of life on Earth has not been dramatically altered or changed because of new scientific data.

Figure 3 Fossil of *Archaeopteryx*, which is considered to be the first ancestor to modern birds. *Archaeopteryx* is about 150 million years old.

Many different organisms have had the same or similar life strategies throughout Earth history. Some of these groups that dominated the landscape for several million years have gone extinct or have been placed into a minor role, following some change in global conditions. For example, "club moss" trees, which grew to heights greater than about 40 m (about 120 ft.), are the plants that formed the Mississippian and Pennsylvanian Period coals of North America, Europe, and China. (These two periods are sometimes jointly referred to as the Carboniferous Period because of the extensive hydrocarbon desposits of this time.)These fossil trees are known to have lasted in peat swamps for nearly 30 million years without change. Without these trees in the very deep past, there would have been no coal. Imagine how many trees must have lived over an interval of 30 million years to result in the vast resources of coal that the United States relies on now for electricity. At one time heating, cooking, and steam engines were all powered by coal. So, why are "club moss" trees, like the ones reconstructed in the coal swamp shown in *Figure 4*, not found outside of your school today? If they were successful for 30 million years, why aren't these trees living in your community or somewhere, anywhere, on Earth?

Figure 4 Reconstruction of middle to late Carboniferous tropical coal swamp showing different plant communities.

The loss of biodiversity in the recent past and the deep geologic past appears to be related to changes in global climate. In the case of the club moss trees, their extinction is associated with climate change associated with Late Paleozoic ice ages. These trees were replaced across the Northern Hemisphere by the gymnosperm trees that reproduced by seeds. Seed plants are better adapted to dry environments, similar to the environments that are interpreted to have accompanied global deglaciation and the assembly of Pangea. Some of these same gymnosperm families are still living today, like the cycads and ginkgos, but most are extinct. Distant relatives of the club moss trees are found today as groundcover in the north temperate woodlands and elsewhere. (Change in global climate is not the only factor that contributes to the loss of biodiversity. Many species are endangered or extinct through the actions of humans.)

Figure 5 Photograph of a ginkgo tree in autumn.

Check Your Understanding

1. What factors influence biodiversity as reflected in the fossil record?
2. Why did the "club moss" tree become extinct?
3. How does the work of scientists contribute to the understanding of how life has changed through time?

Understanding and Applying What You Have Learned

1. Large-scale changes in biodiversity due to extinctions and evolution of new organisms thereafter have altered the aspects of your community through geologic time.
   a) How different was your community biome in the record of the deep geologic past you investigated?
   b) Are the common plants and animals you identified in your present community the same as in the fossil record? Explain.
   c) How many times during Earth's history has there been a major change in the biodiversity of your community?
2. When a group of organisms is no longer found in the fossil record, how do new organisms adapt to take its place in the ecosystem?
3. If all soft-bodied organisms and those with hard skeletons that lived in your community in the deep geologic past had been fossilized, how would this change your perspective about life on Earth?
4. What would your present community be like if one or more of the extinction events in the deep geologic past had not occurred?

Preparing for the Chapter Challenge

As a group, synthesize what you have learned about the changes in biodiversity over geologic time and the causes proposed for the mass extinction events documented in the fossil record. Review your notes on how your community has changed over time. Write an essay in which you explain whether organisms with similar life strategies have similar fossil records (i.e., do organisms with the same life strategies respond the same). Explain any differences.

Inquiring Further

1. **Extinctions**

   Is extinction common throughout Earth history, or are extinctions limited to only the "Big Ones?" One paleontologist has asked the question whether extinction is "bad timing, or bad luck?" What do you think? How do mass extinctions change the balance of Earth's ecosystems?

2. **Originations**

   What natural mechanisms are responsible for the evolution and origination of new organisms following extinction events? Is the rate of origination of new organisms slow or rapid following an extinction event? What kinds of organisms appear first in the fossil record following a mass extinction event?

Earth Science at Work

ATMOSPHERE: *Birdwatcher*
Climate change might change the distribution of birds in North America. Birdwatchers can help scientists determine if the range of a given species of bird has changed, or if there has been a change in the time of bird migration.

BIOSPHERE: *Paleontologist*
When detectives investigate a murder, they do not just look at the dead body. They look at a great deal of other evidence. Similarly, when paleontologists find a dinosaur's bones, they can pick up extra clues from trace fossils. These clues to lifestyles help scientists conjure a much richer picture of prehistoric life.

CRYOSPHERE: *Park Ranger*
The Arctic ecosystem is in a very delicate balance and is particularly vulnerable to climate change. Climate change is likely to threaten both marine and terrestrial wildlife. From plankton to polar bears, many species could suffer or disappear entirely.

GEOSPHERE: *Urban Planner*
Urban expansion can result in deforestation of large areas. Also, pollution can make existing trees more susceptible to freeze injury and in turn to drought and insects. Urban planning needs to take into account not only such social and economic factors, but environmental factors as well.

HYDROSPHERE: *Fishing Industry*
Fisheries already face the problems of overfishing, diminishing wetlands, pollution, and competition among fleets for uses of aquatic ecosystems. The industry is constantly monitoring seafood supplies for any further changes.

How is each person's work related to the Earth system, and to Changing Life?

Glossary
Charts/Graphs/Tables
Index

Glossary

absolute zero: the temperature at which all vibrations of the atoms and molecules of matter cease; the lowest possible temperature.

acid: a compound or solution with a concentration of hydrogen ions greater than the neutral value (corresponding to a pH value of less than 7).

air pressure: the cumulative force of a multitude of air molecules colliding with a unit surface area of any object in contact with air.

angle of repose: the maximum slope or angle at which loose material remains stable, commonly ranging between 33° and 37° on natural slopes.

anthropogenic: generated or produced by human activities.

anticline: part of the fold that is convex upward.

aqueduct: a system of large surface pipes and channels used to transport water.

aquiclude: a body of rock that will absorb water slowly, but will not transmit it fast enough to supply a well.

aquifer: a body of porous rock or sediment that is sufficiently permeable to conduct ground water and to provide an adequate supply of water from wells.

ash: fine pyroclastic material (less than 2 mm in diameter).

asthenosphere: the part of the mantle beneath the lithosphere. The asthenosphere undergoes slow flow, rather than behaving as a rigid block, like the overlying lithosphere.

avulsion: a major change in the course of a river when the river breaks out of its levees during a flood.

backflood deposits: sediments deposited when flooding causes water to pour over dams and into lakes.

basal slip: the part of movement of a glacier that is caused by sliding of the glacier over the material beneath it, aided by a thin lubricating layer of water.

basalt: a kind of volcanic igneous rock, usually dark colored, with a high content of iron.

base level: the elevation of a river's mouth.

base: a compound or solution with a concentration of hydrogen ions less than the neutral value (corresponding to a pH value of greater than 7).

batholiths: large masses of intrusive igneous rock with irregular shapes.

bed load: sediment particles that travel along the stream bed, by sliding, rolling, and bouncing.

bedrock: solid rock that is connected continuously down into the Earth's crust, rather than existing as separate pieces or masses surrounded by loose materials.

biomass: the total mass of living matter in the form of one or more kinds of organisms present in a particular habitat.

brook: a term used for a small stream.

caldera: a large basin-shaped volcanic depression, more or less circular, the diameter of which is many times greater than that of the included vent or vents.

calving: the breaking away of a mass of ice from a glacier.

carbon cycle: the global cycle of movement of carbon, in all of its forms, from one reservoir to another.

catchment (also drainage basin, watershed): the land area from which rainfall collects to reach a given point along some particular river.

cement: new mineral material precipitated around the particles of sediment when it is buried below the Earth's surface.

cementation: the process by which sediments are converted into rock by the precipitation of a mineral "cement" among the grains of sediment.

chemical energy: energy stored in a chemical compound, which can be released during chemical reactions, including combustion.

chemical sedimentary rock: a sedimentary rock formed by direct chemical precipitation of minerals from a solution.

clast: an individual grain or fragment of sediment produced by the physical disintegration of a larger rock mass.

clastic sedimentary rock: a sedimentary rock made up mostly of fragments derived from pre-existing rocks and transported mechanically to their places of deposition.

cleavage: the breaking of a mineral along regularly oriented planes of weakness, thus reflecting crystal structure.

climate: the long-term average of weather in a particular region of the Earth, over years, decades, or centuries.

climatologists: scientists who study the Earth's climate.

closed system: a system in which material moves from place to place but is not gained or lost from the system.

coal: a combustible rock that had its origin in the deposition and burial of plant material.

compaction: the reduction in bulk volume or thickness of fine-grained sediments owing to increasing weight of overlying material that is continually being deposited.

composite cone (stratovolcano): a volcano that is constructed of alternating layers of lava and pyroclastic deposits.

compression force: a force that tends to push material together.

conchoidal fracture: a type of mineral fracture that gives a smoothly curved surface.

condensation nuclei: tiny solid and liquid particles that water vapor can condense on.

condensation: the change of state of matter (in this case water from a gas (water vapor) to a liquid. Heat is released.

conduction: a process of heat transfer by which the more vigorous vibrations of relatively hot matter are transferred to adjacent relatively cold matter, thus tending to even out the difference in temperature between the two regions of matter.

cone of depression: the conical shape of the water table near a well, caused by the drawing down of the water table as water is pumped from the well.

consumptive water: water that is returned to the atmosphere as water vapor.

continental accretion: the growth of a continent along its edges.

contour interval: the vertical distance between the elevations represented by two successive contour lines on a topographic map.

contour line: a line on a map that connects points of equal elevation of the land surface.

convection cell: a pattern of motion in a fluid in which the fluid moves in a pattern of a closed circulation.

convection: motion of a fluid in which the fluid moves in a pattern of closed circulation as a result of density differences caused by heating from below.

convergent plate boundary: a plate boundary where two plates move toward one another.

core: the solid, innermost part of the Earth, consisting mainly of iron.

Coriolis force: the apparent force caused by the Earth's rotation which serves to deflect a moving body on the surface of the Earth to the right in the Northern Hemisphere and to the left in the Southern Hemisphere.

creek: a term used for a small stream.

crude oil: (see petroleum)

crust: the thin outermost layer of the Earth. Continental crust is relatively thick and mostly very old. Oceanic crust is relatively thin, and is always geologically very young.

cyclone: a large low-pressure weather system in which surface winds blow counterclockwise and inward viewed from above in the Northern Hemisphere.

decomposition: the chemical process of separation of matter into simpler chemical compounds.

deform: to change shape.

density: the mass per unit volume of a material or substance.

desalination: the process of removing dissolved salts from sea water in order to make it potable.

dikes: sheets of igneous rock that intruded along fractures that cut through any existing rock.

discharge: the volume of water that flows past a point on a river per unit of time.

distributary system: an outflowing branch of a river, such as what occurs characteristically on a delta.

divergent plate boundary: a plate boundary where two plates move away from one another.

downcutting: erosion of a valley by a stream.

downstream fining: the decrease in sediment size downstream in a stream or river.

downwelling: the downward movement of ocean water from the surface.

drainage basin (also catchment, watershed): the land area from which rainfall collects to reach a given point along some particular river.

drainage divide: the boundary between adjacent drainage basins.

drainage network: the collection of all the streams in a watershed.

drip irrigation: a form of irrigation in which water is supplied slowly to the soil through special piping and tubing that is punctured with many small holes that allow the water to drip or ooze out.

earthquake: a sudden motion or shaking in the Earth caused by the abrupt release of slowly accumulated strain.

eddies: swirling masses of turbulent fluid.

efficiency: the ratio of the useful energy obtained from a machine or device to the energy supplied to it during the same time period.

Ekman drift: horizontal movement of ocean water that results from the balance between wind stress and the Coriolis force.

elastic rebound: The return of a bent elastic solid to its original shape after the deforming force is removed.

electric energy: energy associated with the generation and transmission of electricity.

electric power: power associated with the generation and transmission of electricity.

electromagnetic radiation: the movement of energy, at the speed of light, in the form of electromagnetic waves.

epicenter: the point on the Earth's surface directly above the focus of an earthquake.

erosion: the wearing away of soil or rock by weathering, mass wasting (downhill movement of material under the influence of gravity), and the action of streams, glaciers, waves, wind, and underground water.

evaporation: the change of state of matter from a liquid to a gas. Heat is absorbed.

evapotranspiration: loss of water from a land area through transpiration of plants and evaporation from the soil and surface water.

extrusive igneous rock: an igneous rock that has formed by eruption of lava onto the surface of the Earth.

fault: a fracture or fracture zone in rock, along which rock masses have moved relative to one another and parallel to the fracture.

feedstocks: raw materials (for example, petroleum) that are supplied to a machine or processing plant that produces manufactured material (for example, plastics).

First Law of Thermodynamics: the law that energy can be converted from one form to another but be neither created nor destroyed.

flash flood: a sudden rise in the water level of a stream, a river, or a man-made drainage channel in response to extremely heavy rains.

flood stage: the river stage (water level) at which a river rises above its banks and begins to cause a flood.

floodplain: the area of a river valley next to the channel, which is built of deposited sediments and is covered with water when the river overflows its banks at flood stage.

flux: the movement of water from one reservoir to another.

focus: the point of an earthquake within the Earth where rupture first occurs to cause an earthquake.

fold: a bend in a planar feature in rocks. A fold is usually a result of deformation.

foliation: the tendency for a metamorphic rock to split along parallel planes.

force: a push or pull exerted on a body of matter.

fossil fuel: fuel derived from materials (mainly coal, petroleum, and natural gas) that were generated from fossil organic matter and stored deep in the Earth for geologically long times.

friction: the force exerted by a body of matter when it slides past another body of matter; the force resists the motion of one surface against another surface.

front: a narrow zone of transition between air masses that contrast in temperature and/or humidity.

geochemist: one who studies the distribution and amounts of chemical elements in minerals, ores, rocks, soils, water, and the atmosphere, and their circulation in nature.

geologic map: a map on which the distribution, nature, and age relationships of rock units are recorded.

geophone: a seismic detector placed on or in the ground that responds to ground motion at its point of location.

geophysicist: one who studies the physical properties of the Earth, or applies physical measurement to geological problems.

geostrophic currents: in the ocean, a current in which the horizontal pressure force is balanced by the equal but opposite Coriolis force.

geothermal energy: energy derived from hot rocks and/or fluids beneath the Earth's surface.

glacial till: poorly sorted, unlayered sediment carried or deposited by a glacier, usually consisting of a mixture of clay, silt, sand, gravel, and boulders ranging widely in size and shape.

glacier: a large mass of ice on the Earth's surface that flows by deforming under its own weight.

Global Positioning System (GPS): a satellite-based system for accurate location of points on the Earth.

gradient: the slope of a stream or river expressed as a loss in elevation of the stream or river with distance downstream.

gravimeter: an instrument for measuring variations in Earth's gravitational field.

greenhouse gases: gases in the Earth's atmosphere that absorb certain wavelengths of the long-wavelength radiation emitted to outer space by the Earth's surface.

gust front: the leading edge of the rain-cooled gusty air preceding a thunderstorm.

gyre: a circular motion of water in each of the major ocean basins.

hazard: a natural event, like an earthquake, that has the potential to do damage or harm.

headwaters: the areas of the river system that are farthest away from the mouth of the river, where the river empties into the ocean.

headworks: an engineering structure built to control the flow of river water out of a river channel during a flood.

heat capacity: the amount of heat needed to raise the temperature of a substance.

hot spot: a fixed source of abundant rising magma that forms a volcanic center that has persisted for tens of millions of years.

hydrogen bond: a weak chemical bond between a hydrogen atom in one polar molecule and an electronegative atom in a second polar molecule.

ice age: a time of extensive glacial activity.

ice sheet: a large glacier that forms on a broad land area at high latitudes where summers are cool enough that not all of the previous winter's snow is melted.

igneous rock: rock or mineral that solidified from molten or partly molten material, i.e., from magma.

inflow: the amount of water entering a reservoir.

interglacial: the time between ice ages.

internal deformation: the part of movement of a glacier that is caused by change of shape within the glacier.

intrusive igneous rock (plutonic igneous rock): igneous rock formed at considerable depth by the crystallization of magma.

isoseismal map: a map showing the lines connecting points on the Earth's surface at which earthquake intensity is the same.

lahar: a wet mixture of water, mud, and volcanic rock fragments, with the consistency of wet concrete, that flows down the slopes of a volcano and its river valleys.

lapilli: pyroclastics in the general size range of 2 to 64 mm.

lava flow: an outpouring of molten lava from a vent or fissure; also, the solidified body of rock so formed.

lava: magma that reaches the Earth's surface from a volcano or fissure.

levee: a natural or man-made embankment built along the bank of a river to confine the river to its channel and/or to protect land from flooding.

lithification: the conversion of unconsolidated sediment into a coherent, solid rock.

lithosphere: the outermost layer of the Earth, consisting of the Earth's crust and part of the upper mantle. The lithosphere behaves as a rigid layer, in contrast to the underlying asthenosphere.

lithospheric plate: a rigid, thin segment of the outermost layer of the Earth, consisting of the Earth's crust and part of the upper mantle. The plate can be assumed to move horizontally and adjoins other plates.

load (of a glacier): loose rock and mineral material that is carried by the glacier.

loam: in general, a fertile, permeable soil composed of roughly equal portions of clay, silt, and sand, and usually containing organic matter.

lowstand: the interval of time during a cycle of sea-level change when sea level is about at its lowest position.

magma: naturally occurring molten rock material generated within the Earth. Magma also contains dissolved gases, and sometimes solid crystals.

mantle: the zone of the Earth below the crust and above the core. It is divided into the upper mantle and lower mantle with a transition zone between.

map projections: the process of systematically transforming positions on the Earth's spherical surface to a flat map while maintaining spatial relationships.

mass movement: downslope movement of soil, sediment, or rock at the Earth's surface by the pull of gravity.

meander bend: one of a series of curves or loops in the course of a mature river.

meander scar: low ridges on the part of the floodplain inside the meander bend caused by deposition of sediment on the point bar during a flood.

meandering stream: a stream with a channel that curves or loops back and forth on a wide floodplain.

Mercator projection: a map projection in which the Equator is represented by a straight line true to scale, the meridians by parallel straight lines perpendicular to the Equator and equally spaced according to their distance apart at the Equator, and the parallels by straight lines perpendicular to the meridians and the same length as the Equator. There is a great distortion of distances, areas, and shapes at the polar regions.

mesocyclone: a counterclockwise (viewed from above) circulation that develops in a supercell thunderstorm; may evolve into a tornado.

mesoscale convective complex (MCC): a nearly circular cluster of thunderstorm cells covering an area that may be a thousand times that of an individual cell.

metamorphic rock: rock that has been changed (metamorphosed) into a different rock type, without actually melting, by an increase in temperature and/or pressure, and/or the action of chemical fluids.

meteorologists: scientists who study the weather.

microburst: an intense downdraft impacting a relatively small area (4 km or less across).

mid-ocean ridge: a chain of undersea ridges extending throughout all of the Earth's ocean basins, and formed by sea-floor spreading.

mineral: a naturally occurring inorganic, solid material that consists of atoms and/or molecules that are arranged in a regular pattern and have characteristic chemical composition, crystal form, and physical properties.

moraine: a mound or ridge of chiefly glacial till deposited by the direct action of glacial ice.

nickpoint (also knickpoint): an interruption or break of slope, especially an abrupt change in the longitudinal profile of a stream or its valley.

normal fault: a fault formed by tension forces that cause the body of rock above the fault plane to slide down relative to the body of rock below the fault plane.

ocean current: a predominantly horizontal movement of ocean water.

oceanographers: scientists who study the Earth's oceans.

organic sedimentary rock: a sedimentary rock consisting mainly of the remains of organisms.

orographic: relating to mountains.

outflow: the amount of water leaving a reservoir.

overland flow: the flow of water on the Earth's surface in the form of thin, slow-moving sheets, rather than in localized channels.

oxbow lake: a crescent-shaped body of standing water situated in the abandoned channel (oxbow) of a meander after the stream formed a neck cutoff and the ends of the original bend were plugged up by sediment.

paleomagnetism: the record of the past orientation and polarity of the Earth's magnetic field recorded in rocks containing the mineral magnetite.

Pangea: Earth's most recent supercontinent which was rifted apart about 200 million years ago.

percent grade: the ratio of the vertical and horizontal distance covered by a given slope multiplied by 100.

physical weathering: the processes of weathering by which rock is broken down by physical forces or processes, including gravity, water, ice, wind, or human actions at or near the Earth's surface.

phytoplankton: tiny plants that float in the near-surface waters of the ocean.

plate tectonics: the study of the movement and interactions of the Earth's lithospheric plates.

polar molecule: a molecule with a negative charge on one side and a positive charge on the other.

postglacial rebound: the gradual rebound (uplift) of the Earth's crust after the melting of glaciers.

precipitation: the process of forming solid mineral constituents from a solution by evaporation.

primary wave (P wave): a seismic wave that involves particle motion (compression and expansion) in the direction in which the wave is traveling. It is the fastest of the seismic waves.

protolith: the rock from which a metamorphic rock was formed.

pycnocline: a layer of water in the ocean characterized by a rapid change of density with depth.

pyroclastic flow: a high-density mixture of hot ash and rock fragments with hot gases formed by a volcanic explosion or aerial expulsion from a volcanic vent.

regelation: a two-fold process involving the melting of ice under excess pressure and the refreezing of the derived meltwater upon the release of pressure.

regional metamorphism: a general term for metamorphism affecting an extensive region.

relief: the physical configuration of a part of the Earth's surface, with reference to variations of height and slope or to irregularities of the land surface.

reservoir: a natural or artificial storage place of water.

resonance: a condition in which a vibration affecting an object has about the same period as the natural vibration period of the object.

reverse fault: a fault formed by compression forces that cause the body of the rock above the fault plane to slide upward relative to the body of rock below the fault plane.

rift valley: a large, long valley on a continent, formed where the continent is pulled apart by forces produced when mantle material rises up beneath the continent.

risk: the potential impact of a natural hazard on people or property.

river: a relatively large flow of water in natural channels.

rock unit: a body of rock that consists predominantly of a certain rock type, or a combination of types.

salinity: the concentration of dissolved salts in seawater, expressed as parts per thousand.

seamount: a peaked or flat-topped underwater mountain rising from the ocean floor.

secondary wave (S wave): a seismic wave produced by a shearing motion that involves vibration perpendicular to the direction in which the wave is traveling. It does not travel through liquids, like the outer core of the Earth. It arrives later than the P wave.

sediment: solid fragments or particles that are transported and deposited by wind, water, or ice.

sedimentary rock: a rock, usually layered, that results from the consolidation or lithification of sediment, for example, a clastic rock like sandstone, or a chemical rock like rock salt, or an organic rock like coal.

seismic (earthquake) waves: a general term for all elastic waves in the Earth, produced by earthquakes or generated artificially by explosions.

seismogram: the record made by a seismometer.

seismology: the study of earthquakes and of the structure of the Earth.

seismometer: an instrument that measures seismic waves. It receives seismic impulses and converts them into a signal like an electronic voltage.

shear force: a force that tends to make two masses of material slide past each other.

shear strength: the shear force needed to break a solid material.

shield volcano: a broad, gently sloping volcanic cone of flat-dome shape, usually several tens or hundreds of square miles in extent.

silica: material with composition SiO_2.

silicate: a compound whose basic structure consists of very tightly bonded units consisting of silicon and oxygen (called silica) that are bonded less strongly to various other atoms.

sills: sheets of igneous rock that intruded along layers of sedimentary rocks.

snow line: the boundary between the zone of accumulation and the zone of ablation.

soil horizon: a layer of soil that is distinguishable from adjacent layers by characteristic physical properties such as structure, color, texture, or chemical composition.

solar radiation: energy from the Sun available to be absorbed by all of the Earth systems.

squall line: an elongated band of thunderstorm cells.

stage: the height of the water surface in a river channel, relative to sea level, at a given place along the river.

stillstand (in a glacier): the balance in accumulation and ablation when the volume of a glacier is constant and the terminus stays in the same place.

stratosphere: the outer layer of the atmosphere overlying the troposphere. The air temperature is at first constant with altitude and then increases with altitude.

stream discharge: the volume of water passing a point along the river in a unit of time.

stream: a small or large flow of water in natural channels.

strike-slip fault: a fault formed by horizontal shear forces that cause the bodies of rock on either side of the fault plane to slide past each other horizontally.

subduction: the movement of one plate downward into the mantle beneath the edge of the other plate at a convergent plate boundary. The downgoing plate always is oceanic lithosphere. The plate that stays at the surface can have either oceanic lithosphere or continental lithosphere.

subduction zone: a long, narrow belt in which one plate descends beneath another.

subsidence: the process by which local areas of the Earth's crust can be slowly lowered by large-scale forces acting within the Earth or the cooling of rocks.

supercell: a single thunderstorm cell that is much larger and longer-lasting than an ordinary cell.

supercontinent: a large continent consisting of all of the Earth's continental lithosphere. Supercontinents are assembled by plate-tectonic processes of subduction and continent–continent collision.

surface runoff: the part of the water that travels over the ground surface without passing beneath the surface.

surface wave: a seismic wave that travels along the surface of the Earth.

suspended load: material that travels down a stream suspended in the water.

suture zone: the zone on the Earth's surface where two continents have collided and have been welded together to form a single continent

syncline: part of the fold that is concave upward.

teleconnections: the effects that climate change in one region of the Earth have on the climate in distant regions (tele- means far or distant, as in television, "distant seeing").

tension force: a force that tends to pull material apart.

tephra: a collective term for all the particles ejected from a volcano and transported through the air. It includes volcanic dust, ash, cinders, lapilli, scoria, pumice, bombs, and blocks.

terminus: the downstream end of a glacier.

terrain analysis: the process of interpreting a geographic area to determine the effects of the natural and man-made features on a planned activity.

thermal convection: a pattern of movement in a fluid caused by heating from below and cooling from above. Thermal convection transfers heat energy from the bottom of the convection cell to the top.

thermocline: the zone of rapid change from warm water to cold water with increasing depth in the ocean.

threshold velocity: the velocity of flow that is needed to move certain particles along the bed of a stream.

thrust fault: a reverse fault in which the fault plane is nearly horizontal.

tools: rock and mineral particles that are carried at the base of the glacier and abrade the bedrock.

topographic map: a map showing the topographic features of the land surface.

transform fault: a vertical surface of slippage between two lithospheric plates along an offset between two segments of a spreading ridge.

transform plate boundary: a plate boundary where two plates slide parallel to one another.

transposition: the process by which lines or planes within a material become more nearly parallel when it is sheared.

tributary: a small stream that drains into a larger stream.

tributary system: a group of streams that contribute water to another stream.

tropopause: the top boundary of the troposphere.

troposphere: the portion of the atmosphere next to the Earth's surface, in which temperature generally decreases rapidly with altitude.

trunk stream: a major river, fed by a number of fairly large tributaries; the main stream in a river system.

trunk stream: the main stream in a watershed.

tsunami: a great sea wave produced by a submarine earthquake (or volcanic eruption or landslide).

turbulence: the irregular motion of water.

unconformity: the contact between an earlier rock and younger sedimentary and/or volcanic layers.

unconsolidated material: sediment that is loosely arranged, or whose particles are not cemented together, either at the surface or at depth.

uplift: the process by which local areas of the Earth's crust can be slowly raised by large-scale forces acting within the Earth or the heating of rocks.

upwelling: the upward movement of ocean water from deep in the ocean to the surface.

urban: relating to a city.

urban-heat-island effect: the fact that many cities have higher average temperature than the surrounding countryside.

urbanization: the process by which humans convert natural lands into areas developed for specific human uses. This can include building, paving, or other construction. It also includes partitioning, grading, and clearing areas of land that were previously in a natural state.

valley glacier: a smaller glacier that forms in mountainous areas and flows down valleys to lower elevations.

viscosity: the property of a fluid to offer internal resistance to flow.

volcanic bomb: a blob of lava that was ejected while viscous and received a rounded shape (larger than 64 mm in diameter) while in flight.

Volcanic Explosivity Index: the percentage of pyroclastics among the total products of a volcanic eruption.

Walker Circulation: circulation cells within the equatorial atmosphere caused by differences in climate.

water mass: a large region of the ocean with about the same temperature and salinity of water.

watershed (also, catchment, drainage basin): the land area from which rainfall collects to reach a given point along some particular river.

weather: the atmospheric conditions at a particular time, day to day.

weathering: the destructive process by which rocks are changed on exposure to atmospheric agents at or near the Earth's surface.

wind shear: a sudden change in wind speed or direction with distance.

wind stress: the friction force exerted on the ocean surface by the wind.

zone of ablation: the area in the lower part of a glacier where there is net removal of glacier ice year after year.

zone of accumulation: the area in the upper part of a glacier where there is net addition of new glacier ice year after year.

zooplankton: tiny animals that live in the near-surface waters of the ocean. They consume phytoplankton.

Charts/Graphs/Tables

Abbreviations: G= Dynamic Geosphere, E= Earth System Evolution, F= Fluid Spheres, R= Natural Resources, U= Understanding Your Environment

Index

Abbreviations: **G= Dynamic Geosphere, E= Earth System Evolution, F= Fluid Spheres, R= Natural Resources, U= Understanding Your Environment**

D

E

F

G

Unit 1

Photos and Illustrations

G2 Krakatoa: Source National Geophysical Data Center (NGDC) (P. Hedervari)
G5, G8, G10, G15, G63, G68, G70, G75, G76, G79, G82, G83 Fig. 5, G93, G107, G108, G113, G114, G115, G116, G123, G124, G127, G132, G133, G135, G139, G148, G149, G157, G158, G168 illustrations by Stuart Armstrong
G6, G16, G18, G22, G25, G32, G35, G39, G44, G47, G69, G90 source United States Geological Survey (U.S. Geological Survey)
G7 photo by R.T. Holcomb U.S. Geological Survey Hawaii Volcanoes Observatory (U.S. Geological Survey HVO) 1974.
G9, G11, G29, G32, G33, G34, G42, G49, G55, G66, G81, G83 Fig. 4, G91, G92, G96, G99, G151 illustrations by Stuart Armstrong, source U.S. Geological Survey
G10 Fig 4 photo by Lina Galtieri.
G13 photo by D.W. Peterson U.S. Geological Survey HVO 1972.
G16, G45, G57, G134 photos by Bruce Molnia
G18 adapted from Earth Science, 7th Edition, Tarbuck and Lutgens, 1994.
G19 Fig 2 Photo by Jim. D Griggs U.S. Geological Survey HVO, Fig 3 photo by Gary Smith
G20 Fig 4 University of Colorado Fig 5.
G21 Fig 6 photo by Gary Smith
G27 Photos by Jim D. Griggs U.S. Geological Survey HVO. 1989, 1990.
G28 Fig 4. Photo by Department of Natural Resources, State of Washington.
G35 Fig 2 (left) photo by T.J. Casadevall 1991 (right) photo by R.L. Rieger, U.S. Navy
G44 photos by Rich Busch, map source U.S. Geological Survey
G46, G47, G87 photos by Rich Busch
G50 photo courtesy Dia Met Minerals Ltd.
G52 Montserrat Observatory
G63, G64 illustrations by Stuart Armstrong, source MIT Global Time Series webpage
G60 photo by R.W. Griggs U.S. Geological Survey HVO 1989
G61 photo by Jim D. Griggs U.S. Geological Survey HVO 1971
G69 Fig 2 photo by Laszlo Kesthelyi U.S. Geological Survey 1997.
G71, G72 National Ocean and Atmospheric Administration
G84 photo by P. Rona, NOAA, OAR/National Undersea Research Program (NURP)
G97 illustration by Stuart Armstrong, source *This Dynamic Planet* Map – USGS, Smithsonian,
US Naval Research Laboratory
G102 photo by Jennifer Wang 1998.
G111, G120 Photo Disk
G121 source University of Colorado
G123 photo by Eric Shih, American Geological Institute
G128 illustration by Stuart Armstrong adapted from B.A. Bolt.
G130 photo courtesy Veritas DGC, Inc.
G136 illustration by Stuart Armstrong, source American Geological Institute
G137 photo by Peter Bormann
G140 Reinsurance Company, Munich, Germany
G144 photo by J. Dewey U.S. Geological Survey Northridge Earthquake 1994
G154 EROS Data Center, U.S. Geological Survey.
G159 IRIS
G163 United State Department of the Interior
G166 photo by Roger Hutchison
G167 (top) Reinsurance Company, Munich, Germany; (middle) University of California, Berkeley; (bottom) NOAA NGDC

Unit 2

Photos and Illustrations

U77 Adobe Image Library, Water Everywhere
U22 (middle-top), U44, U54 Fig. 1-2, U93, U105 Fig. 2 from AGI Photo CD
U7, U34 by American Association of Petroleum Geologists and the U.S. Geological Survey
U18 Fig. 1, U24, U27, U40, U42, U45 Fig. 2, U50, U51, U52, U61 (top), U62, U78, U85, U87, U91, U92, U96, U97, U104, U106, U127, U132, U147, U157, U159 (bottom) illustrations by Stuart Armstrong
U4, U14, U23, U33, U39, U49, U57, U70, U81, U90, U100, U113, U124, U131, U138, U146, U155, U163 illustrations by Tomas Bunk
U6, U13 (top) Rock Samples, U15 Figs. 1-6, U20, U21 Figs. 1-6, U25 Figs. 1-4, U30 Figs. 5a-b, U31 Figs. 1-2 photos by Richard Busch
U41, U63 Fig. 2, U71, U82, U94 by Caitlin Callahan
U56 by Emily Crum
U74 (top) Culbertson Museum, Culbertson, MT
U63 Fig. 1 Geological Survey of Canada, Natural Resources Canada
U160 Historic National Weather Service Collection, NOAA and US Public Health Service
U25 (bottom), U29, U47 (bottom), U55 Figure 4a John Karachewski
U58 Map and Key Source Philip B. King and Helen M. Beikman, 1974
U47 Top, U55 Figure 3 Michelle Markley
U46 Fig. 4a by Maryland Geological Survey
U10 by Robert McMahon
U142 (bottom) by Marcus Milling
U8, U9, U22 (bottom), U28, U37 Fig. 1, U55 Fig. 4b photos by Bruce Molnia
U105 Fig. 3, U135 (bottom) by NASA
U170 by National Archives and Records Administration
U114 by Louie Palu, Echo Bay Mines
U2, U3, U11, U13 (bottom), U16, U32, U48, U61 (bottom), U64, U65, U68, U69, U84, U86, U89, U98, U101, U103, U107, U108, U109 Top-Bottom, U110, U115, U116, U119, U122, U123, U128, U134, U135 Top, U141, U142 Top, U144, U150, U153, U154, U158, U159, U161, U166, U167 by PhotoDisc
U18 Figs. 2-3 by Royalty Free Images, Dynamic Graphics-Rocks, Marble, Granite
U151 by R.L. Schuster, US Geological Survey
U45 Fig. 3 by Michael Smith
U46 Fig. 4b by Hannah Thomas
U37 Fig. 2 by Basil Tikoff
U139 by TOPO! National Geographic Maps
U74 (bottom) by Upper Midwest Environmental Sciences Center, U.S. Geological Survey
U72, U79, U102, U111, U125, U149 U.S. Geological Survey
U143 by Vern Whitten Photography, Fargo, North Dakota

Unit 3

Photos and Illustrations

F2, F3, F16, F22, F26-F27, F51, F55, F59 Fig. 2, F61, F62, F64, F65, F66, F99 Fig. 2, F110, F118 Fig. 4, F124, F126, F153, F154 Fig. 3, F161, F189 photos from PhotoDisc
F4, F13, F23, F37, F48, F56, F66, F77, F88, F95, F103, F113, F128, F138, F149, F158, F167, F174, F180 illustrations by Tomas Bunk
F5, F8, F9, F15, F19, F21, F25, F26, F29, F31, F33, F34, F40, F43, F45, F46 Fig. 4, F49, F53, F60 Fig. 4, F68 (top), F79, F106, F116, F120 Figs. 6-7, F129, F130, F131, F134 Fig. 3, F139, F140, F144, F145 Fig. 5, F152, F154 Fig. 2, F155, F159, F162, F168, F169, F172, F175, F181 illustrations by Stuart Armstrong
F6 illustration by Burmar
F10 photo by NOAA, Dr. Robert Embley
F12 photo by OAR/National Undersea Research Program
F18 photo by NOAA/ Rear Admiral Harley D. Nygren
F24 photo by Jim White and Curt Ebbesmeyer, BeachCombers Alert!
F27 (top), F28, F92 Fig. 3 photos by NOAA
F30 photo by Caitlin Callahan
F35 image from Philip Richardson, 1980. Benjamin Franklin and Timothy Folger's First Printed Chart of the Gulf Stream. Science (207) 643-645.
F42 figure from "Rise and Fall of the '97–'98 El Niño as Tracked by TOPEX/POSEIDON." NASA/JPL, CNES.
F44 photo by Jorge Tapia
F46 (bottom) images by NOAA
F54, F60 Fig. 3, photos by United States Geological Survey
F58 figure by NASA/Goddard Space Flight Center
F59 Fig. 1, photo by NOAA/ Dr. James P. McVey
F65 (top), F163 images by NASA
F68 (bottom), F184 images from U.S. Geological Survey
F69, F71, F72, F73 (bottom), F81, F84, F85, F92 Fig. 2, F92 Fig. 4, F93 Fig. 5 illustrations by Stuart Armstrong, source: Moran, J.M. and Morgan, M.D. "Meteorology, the Atmosphere and the Science of Weather." Prentice Hall, NJ. 1997
F73 (top), F74, F76, F97, F108, F109, F111, F118 Fig. 5, F78 (right), F82 (top left), F83 (bottom), F87 photos from Digital Stock Royalty Free Images CD
F78 (left three), F135 photos by Doug Sherman
F82 (middle left, top right, bottom right), F83 (middle left) photos by NOAA, Ralph Kresge
F82 (bottom left), F83 (top left, top right), F127, F132, F143 Fig. 2, F145 Fig. 4, F146, F148, F157, F171, F186 photos by Bruce Molnia
F89 illustration by Stuart Armstrong, source: Joe Moran
F91 photo by Joe Moran
F94, F122 image by National Weather Service
F96, F165 map by U.S. Geological Survey
F99 Fig. 3, photo by Robert Jarrett
F100 photo by John Weaver
F107 image by Los Alamos National Laboratory
F134 Fig. 4 images from Kenneth Libbrecht, Caltech
F137 photo by Kathy Hurlburt
F143 Fig. 1 photo by NOAA, Ardo X. Meyer
F150, F151 images by Paul Morin, Richard Peltier
F170 photo by NASA
F176 photo by Martin Miller
F177 photo by Paul Herr
F185 photo by Marge Potter
F187 photo by Larry Parson, Citrus Research and Education Center

Unit 4

Photos and Illustrations

R5, R6, R9, R30, R46, R48, R64, R66, R73, R74 (anemometer), R78 Fig. 1–2, R98, R99, R105, R119, R120, R125, R128, R137, R138 (left, right), R153 (bottom), R157, R159, R160, R164 (top, bottom), R167 (top, bottom), R178, R181 Figs. 3, 4a–4b, R182 Figs. 5a–5b, R185, R186, R197, illustrations by Stuart Armstrong
R40, illustration by Stuart Armstrong, source: Energy Information Administration
R23, R58, R59, illustrations by Stuart Armstrong, source: Energy Information Agency
R31, R32, R41 Fig. 2, illustrations by Stuart Armstrong, source: Kentucky Geological Survey
R74, map illustrations by Stuart Armstrong, source: National Renewable Energy Laboratory
R47, illustration by Stuart Armstrong, source: Oak Ridge National Laboratories
R44, R200, illustrations by Stuart Armstrong, source: United States Environmental Protection Agency
R39, R112, R148, R173, illustrations by Stuart Armstrong, source: United States Geological Survey (U.S. Geological Survey)
R126, photo by Atlantic-Richfield Company (ARCO)
R115, R132, R140, photos by ASARCO
R60, photo by The Bakersfield Californian
R10 Fig. 2, photo by Charles Bickford, The Taunton Press
R163, photo by John Buchanan
R4, R16, R25, R35, R43, R53, R62, R72, R88, R96, R111, R118, R127, R136, R146, R156, R169, R177, R184, R196, illustrations by Tomas Bunk
R20, R21, R22, R27, maps by Bureau of Economic Geology, source: Energy Information Agency
R56, R81, maps by Bureau of Economic Geology at the University of Texas at Austin
R123 Fig. 2, illustration by Burmar Technical Corporation
R97 Figs. A, C–E, G–L, R103, R110 Fig. 2, photos by Richard Busch
R166, R175 Fig. 5, R189, R190, R191, R193 Figs. 5–6, photos by California Department of Water Resources
R95, photo by Caitlin Callahan
R102, photo by Corbis
R201, photo by Cornerstone Water Systems
R12 Fig. 5, R70 Fig. 3, R79 Fig. 3, photos by Department of Energy
R19, R187, photos by Digital Stock Royalty Free Images
R29, R59 Fig. 2, R69, R70 Fig, 2, photos by Digital Vision Royalty Free Images
R133, photo by Gold Institute
R139 Fig. 2, photo by George James
R192, photo by Maureen Keller, Bigelow Laboratory for Ocean Sciences
R91, photo by Nathan Lamkey
R124, photo by Michelle Markley
R162, photo by Gene McSweeney
R151, R199, photos by Bruce Molnia
R174 Fig. 4, photo by Martin Muller
R130, photo by Alma Paty
R86, (left, bottom), R87 (bottom), R90, R93, R113, R139 Fig. 1, R144 (left, center), R145 (right), R152, R2, R3, R10 Fig. 3, R13 Fig. 6, R76, R80, R80, R153 (top), R174 Fig. 3, R180 Figs. 1–2, photos by PhotoDisc
R26 A–D, R49, R87 (top), R92, R97 (B, F), R103 Fig. 3, R104, R106 Figs. 6A–D, R107 Figs. 7A–D, R108 Figs. 8–9, R26 A–D, R49, R114 Fig. 2, R164, R174 Fig. 2, photos by Doug Sherman, Geo File Photography
R129, photo by Mark Snyder Photography
R175 Fig. 6, photo by South Florida Water Management District
R15, photo by Tom Sponheim

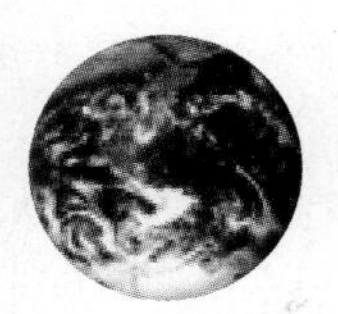

R123 Fig. 3, photo by George Springston, Vermont Geological Survey
R154, photo by Leslie Taylor
R114 Fig. 3, R122, R165, photos by US Geological Survey
R41 Fig. 3, R131, photos by Peter Warwick, U.S. Geological Survey
R11 Fig. 4, photo by D. R. "Doc" Young Photography

Unit 5

Photos and Illustrations

E175, photo by Cody Arenz, Nebraska Wesleyan University
E10, E11, E20, E22, E24, E29, E30, E31, E33, E34 Fig. 4, E50, E52, E59, E64, E71, E75 Fig. 2, E100, E103, E106, E107, E108, E109, E111, E113, E114, E118, E119, E121, E129, E130, E131, E140 Fig. 2, E149, E150, E151 Fig. 1, E184, illustrations by Stuart Armstrong
E85, illustration by Stuart Armstrong, source: Köppen Climate Map
E97 (top and bottom right), illustration by Stuart Armstrong; source: R. S. Bradley and J. A. Eddy based on J. T. Houghton et al., Climate Change: The IPCC Assessment, Cambridge University Press, Cambridge, 1990 and published in EarthQuest, vol. 5, no. 1, 1991.
E100 Fig. 1, by Stuart Armstrong, source: redrafted from Muller and MacDonald, Ice Ages and Astronomical Causes, Springer-Praxis, 2000
E101 Fig. 3, E132, E169, illustrations by Stuart Armstrong, source: Jouzel, J.C., et al, 1987, Nature 329:403-8; Jouzel, J.C., et al, 1993, Nature 364:407-12; Jouzel, J.C., et al, 1996, Climate Dynamics 12:513-521; and Jouzel, J.C., et al, 1999, Nature 399: 429-436
E114, illustration by Stuart Armstrong, source: Skinner and Porter, The Dynamic Planet, John Wiley & Sons, 2nd Edition, February 2000
E126 Part A Table, illustration by Stuart Armstrong, source: Global Historical Climatology Network 1701-07/2000 (meteorological stations only) and Hansen, J., Sato, M., Lacis, A., Ruedy, R., Tegen, I. & Mathews, E. (1998) Climate forcings in the Industrial era, Proc. Natl. Acad. Sci. USA, 95, 12753-12758.
E126 Part B graph, illustration by Stuart Armstrong, source: Methane data: Chappellaz et al, Nature, v. 345, p. 127-131, 1990; CO2 data: Barnola et al., Nature, v. 329, p. 408-414, 1987; Change in temperature data: Jouzel et al, Climate Dynamics, 12:8, p. 513-521, 1996.
E139, redrafted by Stuart Armstrong from a diagram by John Woolsey of Woolsey Studio, Boston, source: Stanley Chernicoff
E159, illustration by Stuart Armstrong, source: Tom Crowley and Consequences Magazine E161
E166, illustration by Stuart Armstrong, source: Gastaldo, Savrda, Lewis. Deciphering Earth History. Page 13-11.
E66, Figs. 4, 5, Dana Berry, Space Telescope Science Institute
E4, E14, E28, E37, E47, E58, E69, E105, E117, E125, E136, E148, E156, E165, E173, E182, illustrations by Tomas Bunk
E176 Tyrannosaurus Rex, University of California Museum of Paleontology
E51, photo by Digital Stock Corporation Royalty Free Images: Backgrounds from Nature
E18, E43, E54, photos by Digital Stock Corporation Royalty Free Images: Space Exploration
E9, E19, photos by Digital Vision Royalty Free Images: Astronomy and Space
E157 desert, photo by Digital Vision Royalty Free Images: North American Landscapes
E161 Fig. 2, photo by Digital Vision Royalty Free Images: American Highlights
E75 Fig. 4, E76 Figs. 5A, 5C, photo by European Southern Observatory
E176, Archaeohippus, Florida Museum of Natural History
E176 Wolf skull, Bill Forbes, Biology Department, Indiana University of Pennsylvania
E180, image courtesy of Geological Survey of Canada
E76 Fig. 5B, photo by Pat Harrington at University of Maryland
E77, photo by Jeff Hester and Paul Scowen, Arizona State University, NASA, and National Space Science Data Center
E34 Fig. 5, International Astronomical Union
E97 (top left), photo courtesy of Laboratory of Tree Ring Research, University of Tennessee
E92, E102 Fig. 4, E140 Fig. 3, E152 Fig. 3, E157 grassland, taiga, photos by Bruce Molnia
E187 Fig, 5, photo courtesy of the Morton Arboretum, Lisle, Illinois.
E7, photo by NASA and European Space Agency, A. Dupree and Ronald Gilliand
E8, E13, photos by NASA and the Hubble Heritage Team
E41, Photo by NASA and National Space Science Data Center
E42 Table 1, E44 Table 2, data from NASA
E45, E65 Fig. 3, photos by NASA
E53, photo by NASA Goddard Space Flight Center
E83 (top right), image by NASA
E2, E3, E25-27, E32, E40, E55, E68, E84, E96, E82 (left, bottom), E110, E122 Fig. 4, E135, E146, E147, E157 tundra, chaparral, mountain zones, E160, E161 Fig. 1, E163, E171, by PhotoDisc
E187 Fig. 4, from a painting by Alice Prickett and published in black and white in Phillips and Cross (1991, pl. 4) Phillips, T.L., and Cross, A.T., 1991, Paleobotany and paleoecology of coal, in Gluskoter, H.J., et al., eds., Economic Geology: U.S.: Boulder, Colorado, Geological Society of America, Geology of North America, v. P-2, p. 483-502.
E168, photos courtesy of F.W. Potter and D.L. Dilcher, Florida Museum of Natural History
E116, photo courtesy of T. Rimmele, M. Hanna/NOAO/AURA/NSF
E93, E133, E151 Fig. 2, E152 Fig. 4, E154, E158 tropical rainforest, temperate evergreen forest, temperate deciduous forest, E162, E178, photos by Doug Sherman, Geo File Photography
E141, photo by Mike Smith
E120, photo courtesy of Walter Smith, NASA and David Sandwell, Scripps Institution of Oceanography
E179, E185, image courtesy of Smithsonian Institution
E186, photo courtesy of Smithsonian Institution
E176 Smilodon Californicus and Diplodocus, photos courtesy of Smithsonian Institution
E73, photo by Susan Tereby, Extrasolar Research Group, NASA, and National Space Science Data Center
E100 Fig. 2, E122 Fig. 3, E158 polar ice, photos by Mark Twickler, Institute for the Study of Earth, Oceans and Space, University of New Hampshire
E65 Fig. 2, photo by Dave Westpfahl, New Mexico Tech University, and Dave Finely, National Radio Astronomy Observatory
E23, photo by Jerome Wycoff
E75 Fig. 3, photo by Hui Yang, University of Illinois and NASA
E60, table modified with permission from "Scaling the Spectrum," courtesy of Donna Young, Tufts University
E123, photo by Barbara Zahm